Grand Expectations

The Oxford History of the United States

C. Vann Woodward, *General Editor*

GRAND EXPECTATIONS

The United States, 1945–1974

JAMES T. PATTERSON

New York   Oxford
OXFORD UNIVERSITY PRESS
1996

Oxford University Press

Oxford   New York
Athens   Auckland   Bangkok   Bombay
Calcutta   Cape Town   Dar es Salaam   Delhi
Florence   Hong Kong   Istanbul   Karachi
Kuala Lumpur   Madras   Madrid   Melbourne
Mexico City   Nairobi   Paris   Singapore
Taipei   Tokyo   Toronto

and associated companies in
Berlin   Ibadan

Published by Oxford University Press, Inc.,
198 Madison Avenue, New York, New York 10016

Oxford is a registered trademark of Oxford University Press, Inc.

Library of Congress Cataloging-in-Publication Data
Patterson, James T.
Grand expectations : the United States, 1945–1974 /
James T. Patterson.
p. cm.—(Oxford history of the United States ; v. 10)
Includes bibliographical references and index.
ISBN 0-19-507680-X
1. United States—History—1945–   I. Title.   II. Series.
E173.094   vol. 10   [E741]   973.92—dc20   95-13878

1 2 3 4 5 6 7 8 9

Printed in the United States of America
on acid-free paper

To Cynthia, with love

Preface

My title, *Grand Expectations*, tries to capture the main theme of this book, that the majority of the American people during the twenty-five or so years following the end of World War II developed ever-greater expectations about the capacity of the United States to create a better world abroad and a happier society at home. This optimism was not altogether new: most Americans, living in a land of opportunity, have always had great hopes for the future. But high expectations, rooted in vibrant economic growth, ascended as never before in the 1950s and peaked in the 1960s, an extraordinarily turbulent decade during which faith in the wealth of the United States—and in the capacity of the federal government to promote progress—aroused unprecedented rights-consciousness on the home front. America's political leaders, meanwhile, managed to stimulate enormous expectations about the nation's ability to direct world affairs. More than ever before—or since—Americans came to believe that they could shape the international scene in their own image as well as fashion a more classless, equal opportunity society.

I call this grand quest for opportunity at home a rights revolution. It affected all manner of Americans, including people who were disadvantaged—minorities, the poor, women, and many others—and who demanded greater access to the ever-richer society that was glittering around them. The quest resulted in significant and lasting improvement in the economic and legal standing of millions of people. No comparable period of United States history witnessed so much economic and civic progress. In this golden age it often seemed that there were no limits to what the United States could do both at home and abroad.[1]

Throughout these years, however, the revolution in expectations confronted stubborn forces that blocked the most grandiose of personal dreams. There were limits after all. In the postwar era, as before, social cleavages beset the United States, one of the world's vastest and most heterogeneous nations. Racial conflicts in particular polarized American life. Other long-standing divisions—of gender, region, religion, ethnicity, and class—grew increasingly glaring, especially in the 1960s. And frightening international tensions, anchored in a Cold War, lasted throughout the postwar years. These tensions inspired some creative statecraft, but they also nourished extremes, such as McCarthyism, and they provoked terrible blunders, notably vast escalation of war in Vietnam. Both the internal divisions and the blunders aroused dissension and enlarged the gap between what people expected and what they managed to accomplish.

Many of the grand expectations survived the turbulent 1960s; activists for environmental protection and women's rights, for instance, achieved considerable visibility in the early 1970s. Also surviving, however, were strongly held traditional ideas: faith in the virtue of hard work, belief in self-help and individualism, conservative religious values. Popular doubts intensified about the postwar rise of large, centralized government. The rights revolution, moreover, helped to stir backlash from people who resented what they considered to be the demands of groups for special privileges. And the Vietnam War widened a "credibility gap" between what America's leaders said they were doing and what in fact they were doing. This gap, already profound by 1968, grew enormous when President Richard Nixon tried to cover up the involvement of his aides in the scandal of Watergate. These events deepened a popular distrust of government—and of elites in general—that in varying forms has lasted to our own times.

The economy, a driving force behind the rise of expectations from 1945 to the late 1960s, also developed worrisome problems by the early 1970s. These problems—which stymied economic growth in the mid- and late 1970s—did not destroy either the grand expectations or the rights-consciousness that had mushroomed since 1945. Demands for rights,

1. Eric Hobsbawm, in *Age of Extremes: The Short Twentieth Century, 1914-1991* (London, 1994), applies the term "Golden Age" to much of Western history between 1947 and 1991.

sharply whetted during the previous decades, remained as enduring legacies of the postwar era. But popular uneasiness about the economy did
more than any other development of the 1970s to dull the extraordinary
optimism that had peaked in the mid-1960s. Therein lay a central feature
of the more somber culture that emerged after 1974: rising tension between still grand expectations on the one hand and unyielding social
divisions, traditional beliefs, and economic uncertainty on the other.
From the early 1970s to our own times Americans have displayed an often
rancorous disillusion. Much of the older optimism has abated. We live in
a more troubled and often more contentious society.

Providence, R.I.   Jim Patterson
October 1995

Acknowledgments

Many people have helped to make this book possible. I am grateful first to the expert staff of the Brown University History Department: Camille Dickson, Cherrie Guerzon, Karen Mota, and Fran Wheaton, who handled my many requests—especially concerning printing, copying, and mailing—with efficiency and good humor. Several graduate students in the History Department served expertly as research assistants and critics of earlier drafts. They are Lucy Barber, James Sparrow, David Witwer, and Bernard Yamron, who also expertly compiled the index. Larry Small, a Brown undergraduate at the time, proved to be an outstanding research aide as a summer intern. India Cooper was a first-rate copy editor and Joellyn Ausanka provided excellent additional editorial assistance. Andrew Albanese, assistant editor of the Oxford University Press, ably took charge of many important matters, including photographs and maps, and shepherded the manuscript through its many stages of production.

The Woodrow Wilson International Center for Scholars awarded me a fellowship which enabled me to take time off from teaching and work full time at research. My thanks to the center, and especially to Michael Lacey, its Director of United States Studies, who gave me enthusiastic support and many good ideas. Brown University also provided important research assistance.

Several scholars who are authorities about aspects of postwar United States history offered discerning comments on earlier drafts of the book. They include James Giglio, George Herring, Townsend Ludington,

Charles Neu, David Patterson, John Rowett, Luther Spoehr, and William Stueck. Alan Brinkley and Alonzo Hamby read large portions of the manuscript, greatly improving it in the process. I am especially grateful to my friends and fellow historians John Dittmer, Steven Gillon, and David Kennedy, each of whom read the entire manuscript, covering it with acute observations and criticisms. C. Vann Woodward, general editor of the Oxford History of the United States, and Sheldon Meyer, senior vice-president of the Press, also read every word and saved me from more blunders than I care to remember. I thank finally my wife, Cynthia, whose constant encouragement enabled me to carry the book through to completion.

J. T. P.

Contents

Illustrations appear following pages 270 and 558.

Maps

Editor's Introduction

The writing of recent history surely needs no defense. A few historians may shy away from the present as venturing too close to the brink of the future, but James T. Patterson, author of this volume in *The Oxford History of the United States*, is clearly not one of them. He might, had he felt the need for it, have cited the precedent set by Thucydides, father of the profession, who wrote the history of his own times. A special need is served by the historian who addresses the recent past, since it is one of the favorite breeding places of mythology. That is particularly true of the period treated in this volume.

The three decades following the Second World War were prolific breeders of myth. The two great military victories on opposite sides of the globe, followed by unparalleled prosperity at home and world leadership abroad, bred a national euphoria, even hubris in some, capable of the boast that America could do anything: "The impossible takes a little longer." Older myths enjoyed new life—national invincibility and national innocence, for example. Americans won their wars—all of them, so they believed—and fought them all for righteous purposes. New crusades were inspired against old domestic ills and injustices. A War Against Poverty was officially declared, and campaigns were waged to assure equal rights and justice for all.

The last decade treated in *Grand Expectations*, however, proved to be crowded with shattered expectations and hopes. The country's longest and most unpopular war, one difficult to call righteous, ended not in victory but in defeat. Fear of nuclear attack in the Cold War drove citizens into bomb shelters at times. The civil rights movement broke apart, and violent mobs set cities aflame, including the capital. The plight of a black underclass became worse. A President was assassinated and so was his brother while seeking the same office. Assassination also proved the fate of the two foremost black leaders of the period, each gunned down in his prime at the age of thirty-nine. Threatened with impeachment for misconduct in the White House, a President resigned in disgrace.

A period so crowded with contradictions and complexities, so befogged with myths to glorify successes and expectations, as well as myths to justify failures and disgraces, demands talents of a high order on the part of the historian. James Patterson meets those demands with remarkable qualities of skill and courage. No myth is too sacred, no reputation so exalted as to escape his unsparing analysis and plain speaking. At the same time he is always ready to acknowledge good intentions and achievements of high order. His readers will finish this book with a new and deeper understanding of this country and its history.

A few changes in the plans originally announced for the series of volumes in *The Oxford History* should be noted. Instead of nine volumes there will be ten to cover that many periods, and there will be one volume on economic history. There is no change in the plan to publish each book as it is completed and to leave each author free of any expectation of conformity in interpretation or point of view.

C. Vann Woodward

Grand Expectations

Prologue: August 1945

At 7:00 P.M. EWT (Eastern War Time) on August 14, 1945, President Harry Truman announced to a packed press conference that World War II had ended. It was Victory over Japan (V-J) Day. Hearing the news, crowds that had stood all day in front of the White House set up a chant, "We want Harry." Truman, with his wife Bess at his side, shortly appeared on the lawn. "This is a great day," he said. "The day we've all been looking for. . . . We face the greatest task . . . and it is going to take the help of all of you to do it." [1]

Joyous celebrations followed, enlivening a two-day holiday proclaimed by the President. In Harlem, the *New York Times* reported, "couples jived in the streets and the crowd was so large that traffic was halted and sprinkler trucks . . . were used to disperse pedestrians." In Italian-American sections of Brooklyn "tables were brought to the streets and food, wine, and liquor were offered to passersby." In other cities the story was much the same. Office workers in St. Louis dumped waste paper and bags filled with water from their windows, and cars dragged tin cans over the pavements. San Franciscans lit bonfires, pulled trolleys from the tracks, and spun the city's cable cars around on their turntables. In Seattle a navy petty officer walked hand-in-hand with his wife down a main

1. *Newsweek*, Aug. 20, 1945, pp. 32–33.

3

street. Someone asked about his plans for the future. "Raise babies and keep house!" he shouted happily as he stopped to kiss his wife.[2]

Not everyone, of course, was so joyful. In Memphis a woman sat dejectedly on a park bench, a Navy Department telegram clutched in her hand. She was the latest of millions of Americans who lamented the loss of loved ones: 405,399 United States military personnel died as a result of war-related fighting, and 670,846 suffered non-fatal wounds. These were small numbers in the wider context of history's bloodiest war, which cost the lives of an estimated 60 million people throughout the world, including some 6 million European Jews murdered by the Nazis.[3] Still, American casualties were heavy in contrast to other twentieth-century wars: World War I, for instance, had killed 116,516 Americans and wounded 204,002.[4]

Many people in the United States had other cause for concern in August 1945: uncertainty about the future. Some worried about the ability of Truman, new to the presidency, to cope with the postwar world— and especially with the Soviet Union. Other Americans were scared about the economy. Government defense spending, by far the largest public works project in the nation's history, had brought great prosperity to a nation that had suffered through the Depression in the 1930s. But officials at the War and Navy departments, frightened that surpluses would pile up, now began to cancel war orders. Some economists feared that the cutbacks, combined with the return to civilian life of 12.1 million military personnel, would lead to unemployment of 8 million people by early

2. Jon Teaford, *The Twentieth-Century American City: Problem, Promise, and Reality* (Baltimore, 1986), 97.
3. Recent accounts of World War II tend to devote considerable attention to its savagery. See Gerhard Weinberg, *A World at Arms: A Global History of World War II* (Cambridge, Eng., 1994); Paul Fussell, *Wartime: Understanding and Behavior in the Second World War* (New York, 1989); and Michael Adams, *The Best War Ever: America and World War II* (Baltimore, 1994). Adams's title is deliberately ironic. Weinberg, 894, estimates deaths at 60 million, including more than 25 million people from among the many ethnic groups in the Soviet Union, 15 million Chinese, and 6 million Poles. The number of Germans killed is estimated at 4 million, the number of Japanese at 2 million, the number from Yugoslavia at 1.5 to 2 million, the number from the United Kingdom at 300,000.
4. American casualties in later wars were still more modest, especially given the larger populations: 54,246 killed and 103,284 wounded in Korea, and 58,151 killed and 153,303 wounded in Vietnam. All figures include war-related non-battlefield deaths. The population of the United States was 139.9 million in 1945, 151.7 million in 1950, 194.3 million in 1965, and 204.9 million in 1970.

1946.[5] That would have been around 13 percent of the labor force. To people who vividly remembered the Great Depression, this prospect was unsettling indeed. The writer Bernard De Voto recognized this and other concerns as sources of a "fear which seems altogether new. It is not often acknowledged," but "it exists and it may well be the most truly terrifying phenomenon of the war. It is a fear of the coming of peace."[6]

Racial tensions aroused further nervousness in 1945. During the war masses of blacks had fled poverty-stricken areas of the South to work in northern and western defense plants, where conflicts over jobs and housing occasionally broke into violence. Race riots had erupted in Harlem and Detroit in 1943. Many other blacks had joined the armed services, where they protested against systematic segregation and discrimination. Secretary of War Henry Stimson cried in alarm, "What these foolish leaders of the colored race are seeking is at bottom social equality."[7] One black man exclaimed bitterly, "Just carve on my tombstone, here lies a black man killed fighting a yellow man for the protection of a white man."[8] Another wrote a "Draftee's Prayer":

> Dear Lord, today
> I go to war:
> To fight, to die,
> Tell me what for?
> Dear Lord, I'll fight,
> I do not fear,
> Germans or Japs;
> My fears are here.
> America![9]

Some Americans in August 1945 worried especially about the legacy of the most momentous event of that time: the near obliteration by atomic bombs of Hiroshima on August 6 and of Nagasaki on August 9. Could the world survive with atomic weapons? Truman, sailing back to the United States from a deadlocked meeting with the Soviet Union at Potsdam, seemed unconcerned. "This [the bombing of Hiroshima] is the greatest

5. *Newsweek*, Aug. 20, 1945, p. 33; Aug. 27, 1945, p. 29.
6. Lawrence Wittner, *Rebels Against War: The American Peace Movement, 1941–1960* (New York, 1969), 114. See also Alan Winkler, *Life Under a Cloud: American Anxiety About the Atom* (New York, 1993).
7. Richard Polenberg, *War and Society* (Philadelphia, 1972), 124.
8. William Leuchtenburg et al., *The Unfinished Century: America Since 1900* (Boston, 1973), 454.
9. Wittner, *Rebels Against War*, 47.

thing in history," he told crew members on the ship. The sailors, foreseeing the end of war, cheered. But Truman was probably more uneasy than he let on. After learning of the first successful test of the A-bomb, at Alamagordo, New Mexico, on July 16, he had written in his diary, "I hope for some sort of peace—but I fear that machines are ahead of mortals. . . . We are only termites on a planet and maybe when we bore too deeply into the planet there'll [be] a reckoning—who knows?" A week later he brooded apocalyptically on "the most terrible thing ever discovered. . . . It may be the fire destruction prophesied in the Euphrates Valley era, after Noah and his fabulous ark."[10]

Truman, of course, was not alone in reflecting on destruction and doom. J. Robert Oppenheimer, "father of the Bomb" at Alamagordo, was moved to quote from the *Bhagavad Gita*, "If the radiance of a thousand suns were to burst into the sky, that would be like the splendor of the Mighty One. . . . I am become Death, destroyer of worlds."[11] Following the slaughter—mostly of civilians—at Hiroshima, *Newsweek* editorialized, "There was a special horror in the split second that returned so many thousand humans to the primeval dust from which they sprang. For a race which still did not entirely understand steam and electricity it was natural to say: 'who next?'" *Time*'s cover of August 20 was bleak: a harsh black X painted across the center of the sun.[12]

Though few Americans said so at the time, it was clear that the decision to drop the bombs reflected the broader hatreds that been unleashed during the savagery of fighting. As early as February 1942, war-inspired fears had prompted the forcible removal of 112,000 Japanese-Americans, the majority of them American citizens, to "relocation centers," mainly in dismally arid regions of the West. This was the most systematic abuse of constitutional rights in twentieth-century United States history. Later in 1942 General Leslie McNair, director of training for American ground

10. Paul Boyer, "'Some Sort of Peace': President Truman, the American People, and the Atomic Bomb," in Michael Lacey, ed., *The Truman Presidency* (Washington, 1989), 192; Robert Ferrell, *Harry Truman and the Modern American Presidency* (Boston, 1983), 54–56. See also Boyer's much lengthier account of American attitudes toward atomic energy between 1945 and 1950, *By the Bomb's Early Light: American Thought and Culture at the Dawn of the Atomic Age* (New York, 1985).
11. David Halberstam, *The Fifties* (New York, 1993), 34.
12. *New York Times*, Aug. 20, 1945, p. 19; *Time*, Aug. 20, 1945. See chapter 5 for a fuller account of American decisions to drop the Bomb on Hiroshima and Nagasaki.

forces, told servicemen, "We must lust for battle; our object in life must be to kill; we must scheme and plan night and day to kill." Admiral William "Bull" Halsey, a commander in the Pacific, was even more blunt. He told his men, "Kill Japs, kill Japs, and then kill more Japs." After the ceremony of Japanese surrender on the battleship *Missouri*, Halsey told reporters that he would "like to have kicked each Jap delegate in the face."[13]

Truman, too, experienced the toughening that came with the war. When an official of the Federal Council of Churches, upset by news of Hiroshima, urged him not to bomb again, the President (knowing that Nagasaki or some other Japanese city was about to be destroyed) replied, "Nobody is more disturbed over the use of the atomic bomb than I am, but I was greatly disturbed over the unwarranted attack by the Japanese on Pearl Harbor, and then murder of our prisoners of war. The only language they seem to understand is the one we have been using to bombard them. When you have to deal with a beast, you have to treat him as a beast."[14]

Most Americans, agreeing with Truman, hailed news of Hiroshima and Nagasaki. A poll immediately thereafter discovered that 75 percent were glad that the bombs had been dropped. Like Truman, they thought the Japanese deserved what they got and that use of the Bomb hastened the end of the war, saving innumerable lives in the process.[15] People further delighted in the fact that the United States, indisputably the number one military and economic power in the world, was sole possessor of the Bomb and could use it to enforce peace in the years ahead. Many Americans expected that the United States would preside over what *Time* magazine publisher Henry Luce had called in 1941 the "American Century"—the spread of democracy and capitalism throughout the world. Walter Lippmann, a widely read columnist, predicted in 1945,

13. Wittner, *Rebels Against War*, 105.
14. Boyer, "'Some Sort of Peace,'" 176.
15. Many historians believe that Japan was on the verge of surrender before Hiroshima and that America's use of the bomb was unnecessary. For more on this angry debate see chapter 5. Critical accounts of American policy include Barton Bernstein, ed., *The Atomic Bomb: The Critical Issues* (Boston, 1976); Gar Alperovitz, *Atomic Diplomacy and the Decision to Use the Bomb* (New York, 1995); Robert Messer, *The End of an Alliance: James F. Byrnes, Roosevelt, Truman, and the Origins of the Cold War* (Chapel Hill, 1982); Martin Sherwin, *A World Destroyed: The Atomic Bomb and the Grand Alliance* (New York, 1975); and Kai Bird, *John J. McCloy: The Making of the American Establishment* (New York, 1992).

"What Rome was to the ancient world, what Great Britain has been to the modern world, America is to be to the world of tomorrow."[16]

Most Americans in 1945 also believed firmly that the fighting had been worth it—it had been a Good War. Domestic tensions notwithstanding, World War II had promoted not only scientific-technological marvels such as the Bomb (and penicillin) but also unparalleled prosperity. Some people, downplaying the persistence of class divisions, thought that the collective effort had inspired greater social solidarity. "We are all in this together" was a common phrase during the war. Together the American people had produced magnificently, fought valiantly, and destroyed their evil enemies. They would join harmoniously to make things better and better in the years ahead.

There seemed ample reason in August 1945 for these high expectations. Although the government was cutting back on orders, it was also lifting irritating wartime regulations. The day after V-J Day the War Production Board revoked many of its controls on industry. Gasoline rationing came abruptly to an end. So did the thirty-five-mile-an-hour speed limit, restrictions on travel to sports events, even a ban on singing telegrams. Magazines jubilantly reported accelerating production of all sorts of consumer goods that had been hard to buy in the war: washing machines, electric ranges, cotton goods, girdles and nylons, cameras and film, shoes, sporting goods, toys (such as electric trains), and a fantastic array of home appliances. Automobile production, sharply limited before the Japanese surrender, was expected to boom to 3 million or more by 1947.

It also seemed in 1945 that Americans had succeeded in forming an uneasy consensus behind a degree of governmental stimulation of the economy. Franklin D. Roosevelt's New Deal, while stymied by conservatives since 1937, appeared safe from repeal. In 1944 Congress had already approved legislation—the so-called GI Bill of Rights—that promised millions of veterans generous government aid for higher education and home-buying. Builders anticipated a boom in construction that would stimulate the entire economy. Truman, meanwhile, promised to fight for a "full employment" bill during the congressional session that was set to convene on September 5.[17]

Grand expectations indeed lifted the mood of August 1945. Americans,

16. Alan Brinkley, "For America, It Truly Was a Great War," *New York Times Magazine*, May 7, 1995, pp. 54–57.
17. *Time*, Aug. 20, 1945, p. 21; *Newsweek*, Aug. 27, 1945,29, pp. 34–35.

having fought to win the war, expected to dominate the world order to come. Although worried about a return of economic depression, they had reason to hope that wartime prosperity would continue. The enemies had been defeated; the soldiers and sailors would soon return; families would reunite; the future promised a great deal more than the past. In this optimistic mood millions of Americans plunged hopefully into the new postwar world.

1

Veterans, Ethnics, Blacks, Women

Many things that middle-class Americans took for granted by the 1960s scarcely existed for the 139.9 million people who inhabited the forty-eight states in 1945 or for the 151.7 million in 1950. Consider a few of these things: supermarkets, malls, fast-food chains, residential air-conditioning, ranch-style homes, freezers, dishwashers, and detergents. Also ballpoint pens, hi-fis, tape recorders, long-playing records, Polaroid cameras, computers, and transistors. And four-lane highways, automatic transmissions and direction signals, tubeless tires, and power steering. In 1945 only 46 percent of households had a telephone; to get long distance, people paid a good deal and asked for an operator. In 1950, 10 percent of families had television sets and 38 percent had never seen a TV program. Although 33 million of America's roughly 38 million households in 1945 had radios, these were for the most part bulky things cased in wooden cabinets, and they took time to warm up. Some 52 percent of farm dwellings, inhabited by more than 25 million people, had no electricity in 1945.[1]

The United States in 1945 had become a more urban nation than

1. J. Ronald Oakley, *God's Country: America in the Fifties* (New York, 1986), 20–22; William O'Neill, *American High: The Years of Confidence, 1945–1960* (New York, 1986), 1–4. Except where otherwise indicated statistics here and elsewhere come from *Statistical History of the United States, from Colonial Times to the Present* (New York, 1976).

earlier in the century. The Census Bureau reported that 96.5 million people, or nearly two-thirds of the population, lived in "urban" areas in 1950. But this definition counted as "urban" all places having 2,500 or more residents. The number living in places with 10,000 or more was 73.9 million, less than half the total population. And the number in places with 50,000 or more totaled only 53.3 million, a little more than one-third of the population. In many of the towns and villages the elm trees still stood in stately power, not yet destroyed by blight. Most American cities presented architecturally stolid fronts featuring a good deal of masonry and little aluminum or glass. Only a few, such as New York and Chicago, had much of a skyscraper center. Suburbs had long surrounded major cities, but there had been relatively little residential building in the 1930s and early 1940s, and the fantastic sprawl of suburbia was only beginning by the mid-1940s. Culturally as well as demographically the United States remained in many ways a world of farms, small towns, and modest-sized cities—places where neighbors knew each other and in which people took local pride. Mail came twice a day to homes.

Many aspects of daily life for most Americans had changed little between the early 1930s and the mid-1940s, years of depression and war. There were 25.8 million cars registered in 1945, nearly one for every three adults. But this was only 2.7 million more cars than in 1929, when there had been 18 million fewer people. Not many Americans in 1945, as in 1929, dared to travel by air; if they lacked a car, they took a bus or a train, or they stayed close to home. Most still consumed "American" cuisine: roasts, fried chicken, burgers, fries, corn, tomatoes, pie, and ice cream.[2] People did not eat out much, and the TV dinner did not arrive until 1954. Americans dressed in clothes made from natural fibers, which needed ironing and wrinkled badly in the heat. Business and professional men always wore coats and ties in public and never (save when playing tennis) appeared in shorts. Almost everyone, men and women alike, wore hats outdoors. People still thought in small sums: annual per capita disposable income in current dollars was $1,074 in 1945. At that time it cost three cents to mail a letter and a nickel to buy a candy bar or a Coke. Relatively few Americans had hospital insurance or company pensions, though Social Security was beginning to become of some use to the elderly who had been employed. In 1945 urban families spent an average of $150 a year on medical care. All Americans did without such later developments

2. Harvey Levenstein, *Paradox of Plenty: A Social History of Eating in Modern America* (New York, 1993), 119–22.

as polio vaccines, birth control or hormone pills, and legal abortions, and they expected as a matter of course that their children would get measles, chicken pox, and mumps.

Young people listened avidly to popular new singers like Frank Sinatra, but so, too, did older Americans: as yet there was no sharply defined "teenage" music. Irving Berlin's "White Christmas," introduced in 1942, remained one of the best-selling songs ever, and Bing Crosby, Perry Como, Rosemary Clooney, and the Andrews Sisters sang hit after hit in a thriving pop music business that turned out 189 million records in 1950, some 80 million more than five years before. "Country and western" music (no longer called "hillbilly") was also booming, with Hank Williams producing a series of million-record favorites before dying of drugs and alcohol in the back seat of a car on New Year's Day 1953. Gene Autry, the singing cowboy, led the charts in late 1950 with "Rudolph the Red-Nosed Reindeer."[3]

Until the late 1940s, movies continued to be a favored form of popular entertainment, attracting a weekly attendance of 85 to 90 million people a year between 1945 and 1949. Entertainment remained rather tame, at least by contrast to later standards: it was virtually impossible in the late 1940s to find nudity in films or magazines. No one at that time could have foreseen a popular culture featuring rock 'n' roll, let alone a world of big-selling magazines such as *Playboy* (which arrived on the newstands in 1953 with its famous centerfold of Marilyn Monroe). One historian has concluded: "The United States in 1950 still bore a resemblance—albeit a rapidly fading one—to the small-town America idealized in the Norman Rockwell paintings that graced the covers of the highly popular *Saturday Evening Post*."[4]

"A CULTURE," THE CRITIC Lionel Trilling wrote in 1951, "is not a flow, nor even a confluence; the form of its existence is struggle, or at least debate—it is nothing if not a dialectic." The sociologist Daniel Bell later elaborated on this theme of culture-as-contest in maintaining that the United States remained a "bourgeois" *society* in the postwar years, even as it was developing an adversarial "modernist" *culture*.[5] Their observations are relevant to American society and culture in the late 1940s,

3. Oakley, *God's Country*, 11–13.
4. Ibid., 21.
5. Lionel Trilling, *The Liberal Imagination* (New York, 1951), 9; Daniel Bell, "The Culture Wars: American Intellectual Life, 1965–1992," *Wilson Quarterly* (Summer 1992), 74–117.

which were complex, diverse, and rent with anomalies and contradictions. The United States during these years—and later—was a bewilderingly pluralist society that rendered any static vision, such as Norman Rockwell's, largely irrelevant.

Begin with an especially numerous and visible group: servicemen and their families. In all, 16.4 million Americans, the vast majority of them young men, joined the armed services during World War II. More than 12.1 million of them were still in uniform in early August 1945. This was nearly two-thirds of *all* American men aged 18 to 34 at the time. Young, numerous, male in a male-dominated culture, and eager to make up for time "lost" during the war (and, for many, during the Depression), the returning veterans placed a firm stamp on American culture and society during the 1940s and thereafter. Their experiences, while varying according to regional, racial, class, and personal circumstances, offer revealing angles of vision into cultural ambiguities in the postwar era.

Most of these young men had volunteered or been taken without a fuss by the draft. Like most Americans, they were deeply patriotic, and they had served because it was their duty. Many had fought bravely. But most of them, polls suggested, had not cherished idealistic notions about destroying fascism or building a brave new world. One poll in September 1945 found that 51 percent of American soldiers still in Germany thought that Hitler, while wrong in starting the war, had nonetheless done Germany "a lot of good." More than 60 percent of these men had a "very favorable" or "fairly favorable" view of Germans—about the same percentage that viewed the French in this way.[6] Many American soldiers also resented the special privileges enjoyed by officers.[7] *Stars and Stripes* said, "A caste system inherited from Frederick the Great of Prussia and the 18th century British navy is hardly appropriate to the United States . . . the aristocracy-peasantry relationship characteristic of our armed forces has a counterpart nowhere else in American life."[8]

In late 1945 the soldiers and sailors wanted above all to come home, get out of the service, and rejoin their families. Many deluged hometown newspapers and members of Congress with demands for transport home and release from military duty. "No boats, no votes." Their wives and girlfriends were equally anxious to get on with "normal" life. Many wives

6. Joseph Goulden, *The Best Years, 1945–1950* (New York, 1976), 19–49; Frederick Siegel, *Troubled Journey: From Pearl Harbor to Ronald Reagan* (New York, 1984), 36.
7. I. F. Stone, *The Haunted Fifties, 1953–1963* (Boston, 1963), 185.
8. Goulden, *Best Years*, 31.

sent angry pleas, along with baby booties, through the mail to Capitol Hill. An anonymous GI poet added:

> Please Mr. Truman, won't you send us home?
> We have captured Napoli and liberated Rome;
> We have licked the master race,
> Now there's lots of shipping space,
> So, won't you send us home?
> Let the boys at home see Rome.[9]

The clamor of GIs largely succeeded. Demobilization proceeded at a very rapid pace. By June 1946 the number in service had dropped to 3 million, and Congress had agreed to authorize an army of only 1 million by July of 1947. For a while the returning troops were treated as heroes. But like veterans throughout history, they found that life had gone on without them. Many, yanked from home for years, deeply resented civilians who had stayed out of the service and prospered. Seizing chances to move ahead, more than 8 million "vets" took advantage of the "52–20" provision of the GI Bill of Rights, which provided $20 per week for up to fifty-two weeks of unemployment (or earnings of less than $100 a month). A form of affirmative action (a phrase of later years), the GI Bill cost $3.7 billion between 1945 and 1949.[10] Other veterans, including thousands who had married hastily while on wartime leave, could not adjust to married life. The divorce rate in 1945 shot up to double that of the prewar years, to 31 divorces for every 100 marriages—or 502,000 in all. Although the divorce rate dropped in 1946 and returned to prewar levels by the early 1950s, its jump in 1945 exposed the rise of domestic tensions in the immediate aftermath of war.

Many of these tensions were captured in a revealing Hollywood film, *The Best Years of Our Lives* (1946). Based on a novel by MacKinlay Kantor, it won nine Academy Awards. As befitting a product of Hollywood, it ended on an upbeat note by affirming the quest for security of three veterans returning to Boone City, an archtypical American community. But the title is also ironic and the story disquieting—so much so, in fact, that the right-wing House Committee on Un-American Activities later considered questioning the writer Robert Sherwood about the script.

In the course of readjusting to civilian life the movie's three veterans encounter, sometimes bitterly, what they perceive as the runaway materialism and lack of patriotism of postwar American society. One veteran

9. Ibid.
10. Ibid., 46–49.

(Fredric March) gets a job as a loan officer at a bank, only to be chastised by higher-ups for softness to struggling veterans seeking assistance. "Last year," he complains, "it was kill Japs. This year it's make money!" He ultimately copes, with the help of his understanding wife (Myrna Loy) and his grown children. The second vet (Dana Andrews) at first cannot find his wife (Virginia Mayo), whom he had married after a brief courtship during the war. When he locates her—she is a nightclub performer—he realizes that she is tough and self-centered. Soon she leaves him. He finally lands menial "women's work" in a heartless chain store, but there he encounters a grouchy male customer who criticizes the war and all who fought in it. Furious, the veteran slams him in the jaw and is fired. In the end he finds a job helping a company use discarded war planes for the building of prefabricated houses. The third veteran lost both hands in the war and manages with hooks instead. But he feels useless in an acquisitive society, faces terrible problems of readjustment, and survives only because of the love of his loyal girlfriend next door.[11] Though the ending is schmaltzy, there was bite enough in the film to distinguish it from a Norman Rockwell vision of the nation. *The Best Years of Our Lives* captured rather well the stresses encountered by many veterans and their families in the immediate aftermath of war.

THE EXPERIENCES of America's diverse ethnic and racial groups, while defying easy categorization, also revealed some of the tensions of postwar American society. The nation's population of 139.9 million in 1945 included nearly 11 million foreign-born and 23.5 million people of foreign-born or mixed parentage. Most of these 34.5 million people, 25 percent of the population, were of European descent, including some 5 million whose roots were in Germany, 4.5 million from Italy, 3.1 million from Canada, 2.9 million from Poland, 2.8 million from Great Britain, 2.6 million from the USSR, and 2.3 million from Ireland (Eire). Substantial numbers also hailed originally from Austria, Hungary, Czechoslovakia, Sweden, and Norway. Many more Americans, of course, had European roots dating from the third generation or farther back. Negroes, as most people then called African-Americans, numbered nearly 14 million, or 10 percent of the population. The census identified a much smaller number, 1.2 million, as people of Mexican background, though there were many others (no one knew how many) who made themselves

11. A useful book on postwar film is Nora Sayre, *Running Time: Films of the Cold War* (New York, 1982). See 50–52.

scarce at enumeration time. The Mexicans and Mexican-Americans were concentrated in a few places, mostly in Texas, the Southwest, and southern California. In Los Angeles they were already numerous enough during the war to frighten white residents, who launched gang attacks on them in the streets. By contrast Asians, most of whom had long been excluded from the United States by racist immigration laws, were numerically tiny in 1945: Chinese-Americans numbered around 100,000, Japanese-Americans around 130,000. There were approximately 350,000 people who told the census-takers that they were Indian (Native American).[12]

Some of these people, such as Japanese-Americans, suffered greatly during the war. Others, such as the majority of Indians, continued to live in an especially dismal poverty. But many other ethnic groups were better off in the late 1940s—or at least felt a little more at home—than in the prewar era. The war, in so many ways a powerful force in the domestic history of twentieth-century America, was an engine that accelerated acculturation. Millions of Negroes and first- and second-generation Americans served in the armed forces or pulled up stakes to work in defense plants, thereby leaving their enclaves and mixing for the first time with "old-stock" white people. Having joined in the war effort, they also came to identify more emotionally with the United States. As Cold War tensions mounted over the next two decades, many European-Americans, especially those who had roots behind the iron curtain, emerged as among the most patriotic—and super-patriotic—of United States citizens.[13]

Still, it was wrong to assume, as many hopeful observers did at the time, that the war and acculturation were working some kind of amal-

12. Here, as in later chapters, I often use terms widely used at the time, such as "Negro" or "Indian." Through 1950 the classification of population by "race" was usually obtained by the enumerator's observation. Persons of mixed white and "other" parentage were normally not counted as white. The category "Indian" included unmixed American Indians together with persons of mixed white and Indian ancestry if they were enrolled on an Indian reservation or Indian agency roll. Persons who were part Indian were considered Indian if they were one-fourth or more Indian, or if they were regarded as Indians in the community in which they resided. Starting in 1960 (and completely in 1970) the census relied more on people's self-classification. With the rise of ethnic and racial self-pride and assertiveness, especially after 1965, the number of people who called themselves "Indian" or "Native American" rose dramatically. The census missed many people, especially the poor. African-Americans, Mexicans, and Mexican-Americans, and some other groups were therefore undercounted.

13. Gary Gerstle, *Working-Class Americanism: The Politics of Labor in a Textile City, 1914–1960* (New York, 1989).

gamating magic. Regional tensions and differences, especially North versus South, remained profound. So did ethnic feelings. Laws from the 1920s had drastically reduced legal immigration, thereby cutting the percentage of foreign-born people in the United States in 1945 to around 8 percent. This was the lowest percentage—to that time—in twentieth-century American history. But the nation was still far from having become a melting pot in which ethnic and religious differences had fused into a common "American" nationality.[14]

Religious differences, indeed, remained very strong in the 1940s. Some 71.7 million Americans, more than half the population, said they belonged to religious groups in 1945, roughly 43 million of them in Protestant denominations, 23 million Catholic, and nearly 5 million identifying themselves as Jewish.[15] These people inhabited an increasingly secular world in which theological dictates carried less weight than in earlier generations but in which church membership was nonetheless increasing, from 49 percent of the population in 1940 to 55 percent in 1950 (and to an all-time high of 69 percent by 1959).[16] Whether church-going much affected personal behavior of course sparked many debates, but the upward trend in attendance was noteworthy and impressive. More and more Americans obviously considered it important to their self-identities to be members of an organized religion. Few Western populations, including the Catholic countries of Europe, came close to matching America's record of church-going in the postwar years.

It was difficult, moreover, to find much of an ecumenical spirit among these religious Americans. Protestant denominations still evoked strong loyalties in the 1940s and early 1950s. Conservative evangelical groups became more active, forming in 1947 the Fuller Theological Seminary in Pasadena and benefiting from the formidable recruiting talents of spellbinders like the youthful Billy Graham, then in the conservative wing of American Protestantism.[17] Anti-Catholic feelings remained strong. Paul

14. The record high census year of foreign-born was 1910, with 14.7 percent. The record low census year was 1970, with a percentage of 4.7. See Rubén Rumbaut, "Passages to America: Perspectives on the New Immigration," in Alan Wolfe, ed., *America at Century's End* (Berkeley, 1991), 212ff, for a useful survey.
15. Figures on religious affiliation depended heavily on reports submitted by the churches themselves; these varied in completeness and need to be read with caution. The gross aggregates here, as included in *Statistical History*, 391–92, are probably reasonably accurate.
16. Stephen Whitfield, *The Culture of the Cold War* (Baltimore, 1991), 83–84.
17. James Hunter and John Rice, "Unlikely Alliances: The Changing Contours of American Religious Faith," in Wolfe, ed., *America at Century's End*, 318–39;

Blanshard's polemically anti-Catholic *American Freedom and Catholic Power* (1949) was a best-seller for six months. It attacked the Catholic Church for what Blanshard considered its support of reactionary governments, its repressive attitude toward questions of personal morality, and its hierarchical organization, all of which Blanshard said were intrinsically un-American. Blanshard focused especially on the inflammable contemporary issue of state aid to parochial schools, which the Supreme Court upheld by a 5-to-4 decision in 1947.[18] Jews, too, felt the sting of criticism and exclusion. They confronted systematic discrimination in gaining entrance to prestigious colleges, universities, and professional schools, and in securing teaching tenure. It was hardly surprising that most Jews and Catholics—many of them among the first- and second-generation immigrant population—clung, often resentfully, to their churches, synagogues, clubs, and neighborhoods.

Many of these "new-stock" Americans, though relatively poor at the close of World War II, had acquired property, which they cherished as a sign of their social mobility and which further deepened their commitment to their neighborhoods. (In Chicago, foreign-born residents actually had higher rates of home-ownership than others in the city).[19] These and other first- and second-generation Americans embraced often quite separate subcultures featuring neighborhood festivals, schools, churches, and above all their extended families.[20] They cherished their own cuisine and modes of dress and supported a flourishing foreign-language press. In the early 1940s there were 237 foreign-language periodicals in New York City, 96 in Chicago, 38 in Pittsburgh, and 1,000 in the nation as a whole, with a circulation of 7 million. Roughly 22 million people, one-seventh of the population, told census enumerators in 1940 that English was not their native tongue.[21]

George Marsden, "Evangelicals and the Scientific Culture: An Overview," in Michael Lacey, ed., *Religion and Twentieth-Century Intellectual Life* (Washington, 1989), 23–48.

18. *Everson v. Board of Education*, 330 U.S. 1 (1947); Diane Ravitch, *The Troubled Crusade: American Education, 1945–1980* (New York, 1983), 29–39.

19. Arnold Hirsch, *Making the Second Ghetto: Race and Housing in Chicago, 1940–1960* (New York, 1983), 185–200.

20. Herbert Gans, *The Urban Villagers: Group and Class in the Life of Italian-Americans* (New York, 1962), remains a classic depiction of a postwar ethnic community of this sort, in Boston.

21. Richard Polenberg, *One Nation Divisible: Class, Race, and Ethnicity in the United States Since 1938* (New York, 1980), 34–36.

THE LIVES of black Americans in the late 1940s, like those of America's more recent immigrants, also improved on the average. Thanks in part to the rapid mechanization of cotton production in the early 1940s, which ultimately threw millions of farm laborers out of work, and in part to the opening up of industrial employment in the North during the wartime boom, roughly a million blacks (along with even more whites) moved from the South during the 1940s. Another 1.5 million Negroes left the South in the 1950s. This was a massive migration in so short a time—one of the most significant demographic shifts in American history—and it was often agonizingly stressful.[22] The black novelist Ralph Ellison wrote in 1952 of the hordes of blacks who "shot up from the South into the busy city like wild jacks-in-the-box broken loose from our springs—so sudden that our gait becomes like that of deep-sea divers suffering from the bends."[23]

Still, many of the migrants gradually reaped unprecedented benefits. The number of Negroes employed in manufacturing jumped from 500,000 to 1.2 million during the war. The percentage of employed black women who worked as domestic servants—before the war one of the few jobs they could get—declined from 72 to 48 during the same period. Blacks also advanced on other fronts, which seem token in retrospect but represented notable achievements at the time. In 1944 for the first time a black reporter was admitted to a presidential press conference; in 1947 blacks gained access at last to the Senate press gallery.[24] Thanks in part to legal pressure from the National Association for the Advancement of Colored People (NAACP), the Supreme Court in 1944 outlawed the "white primary," a ploy that had enabled states in the South to exclude blacks from all-important Democratic primary races.[25] In 1946 the Court

22. Nicholas Lemann, *The Promised Land: The Great Black Migration and How It Changed America* (New York, 1991).

23. Ralph Ellison, *Invisible Man* (New York, 1952), 332.

24. Harvard Sitkoff, *The Struggle for Black Equality, 1954–1992* (New York, 1993), 3–19; Manning Marable, *Race, Reform, and Rebellion: The Second Reconstruction in Black America, 1945–1990* (Jackson, 1991), 13–39; David Goldfield, *Black, White, and Southern: Race Relations and Southern Culture, 1940 to the Present* (Baton Rouge, 1990), 45–62; and William Harris, *The Harder We Run: Black Workers Since the Civil War* (New York, 1982), 123–89, are four of many books that deal in part with postwar race relations. See also James Jones, *Bad Blood: The Tuskegee Experiment, a Tragedy of Race and Medicine* (New York, 1981), for a particularly egregious story of racist science.

25. *Smith v. Allwright*, 321 U.S. 649 (1944).

ruled against segregation on conveyances engaged in interstate travel. [26] In 1945 Branch Rickey of the Brooklyn Dodgers signed the black baseball star Jackie Robinson to a minor league contract. It was understood that if he was good enough he would become the first Negro player in modern times to play in the Big Leagues. In 1947 he did, beginning a career of stardom with the Brooklyn team. [27]

Many of these changes came because blacks themselves demanded them. As early as 1941 A. Philip Randolph, head of the all-Negro sleeping-car porters' union, had threatened a "march on Washington" if the federal government did not act against rampant discrimination in the armed services and publicly contracted employment. To prevent the march, President Roosevelt gave in and issued an executive order against such treatment. He also set up a Fair Employment Practices Commission (FEPC) to oversee things. The order was widely evaded, but Randolph's boldness nonetheless encouraged blacks to pursue further protest. The *Pittsburgh Courier*, a leading black newspaper, demanded a "Double V" during the war, victory over fascism and imperialism abroad and over racism at home. Civil rights leaders recognized that ordinary blacks were growing more and more restless and angry. Roy Wilkins, a leader of the NAACP, wrote a fellow activist in 1942, "It is a plain fact that no Negro leader with a constituency can face his members today and ask full support for the war in light of the atmosphere the government has created."[28] The groundswell of protest was indeed growing: membership in the NAACP, by far the most important civil rights organization, increased from 50,000 to 450,000 during the war.

Students of the "Negro problem" in the early 1940s had grand expectations about the potential for this groundswell. This feeling especially gripped the scholars who collaborated with Gunnar Myrdal, the Swedish social scientist who published in 1944 *An American Dilemma*. This was a

26. *Morgan v. Virginia*, 328 U.S. 373 (1946).
27. Jules Tygiel, *Baseball's Great Experiment: Jackie Robinson and His Legacy* (New York, 1983). No blacks had been permitted to play in the National Football League between 1934 and 1945. They were readmitted (in very small numbers) thereafter mainly because of competition from the newly formed All-American Football Conference, which signed two black players in 1946 and five more in 1947. The National Basketball League (Association after 1950) had one black player, beginning in 1946, but did not otherwise open up its gates until 1950. See Arthur Ashe, Jr., *A Hard Road to Glory: A History of the African-American Athlete Since 1946* (New York, 1988); and Richard Davies, *America's Obsession: Sports and Society Since 1945* (Ft. Worth, 1994), 35–62.
28. Walter Jackson, *Gunnar Myrdal and America's Conscience: Social Engineering and Racial Liberalism, 1938–1987* (Chapel Hill, 1990), 235.

much-acclaimed study of race relations in the United States. The "dilemma," Myrdal thought, stemmed from the historic conflict between the "American Creed" of democracy and equality and the reality of racial injustice. Myrdal amply documented the power of such injustice, identifying the "vicious circle" of prejudice and discrimination that had victimized black people in the United States. But he had faith in American ideals, and he was optimistic about the future. Negroes, he argued, could no longer be regarded as a "patient, submissive majority. They will continually become less well 'accommodated.' They will organize for defense and offense. They will become more and more vociferous." Whites, he added, would surely resist change. "The white man can humiliate the Negro; he can thwart his ambitions; he can starve him." But whites did "not have the moral stamina to make the Negro's subjugation legal and approved by society. Against that stands not only the Constitution and the laws which could be changed, but also the American Creed which is firmly rooted in the Americans' hearts." Not since Reconstruction, Myrdal wrote with emphasis, *"has there been more reason to anticipate fundamental changes in American race relations, changes which will involve a development toward American ideals."*[29]

With the advantage of hindsight it is clear that *An American Dilemma* had its limitations as analysis. Myrdal and his collaborators were first of all too affirmative, too optimistic about the potential of the "American Creed." White racial prejudice and structural discrimination proved to have great staying power. Second, Myrdal assumed that whites would lead the way to change: like most people in the 1940s, he underestimated the rage and determination of blacks, who were stirring to take matters into their own hands. As Ellison's black protagonist exclaimed in *Invisible Man,* "You ache with the need to convince yourself that you do exist in the real world, that you're part of all the sound and anguish, and you strike out with your fists, you curse and you swear to make them [whites] recognize you."[30] This did not happen much in the 1940s, but it did in the 1960s, when advocates of "black power" pushed whites out of the civil rights movement.

Myrdal also accepted conventionally unflattering views of African-American culture. *"In practically all its divergences,"* he wrote, *"American Negro culture is not something independent of general American cul-*

---

29. Gunnar Myrdal, *An American Dilemma: The Negro Problem and American Democracy* (New York, 1944), lxi; Jackson, *Gunnar Myrdal and America's Conscience,* 161–73, 197–201.
30. Ellison, *Invisible Man,* 7–8.

ture. It is a distorted development, or a pathological condition, of the general American culture." An American Dilemma deplored the "high Negro crime rate," as well as the "superstitition, personality difficulties, and other characteristic traits [that] are mainly forms of social pathology." Myrdal concluded, "*It is to the advantage of American Negroes as individuals and as a group to become assimilated into American culture, to acquire the traits held in esteem by the dominant white Americans.*"[31]

During the racial confrontations of the 1960s, An American Dilemma encountered rising criticism from activists and scholars who disputed Myrdal's optimism about white liberalism, as well as his negative statements about certain aspects of African-American culture. In the mid- and late 1940s, however, the study received virtually unsparing praise. W.E.B. Du Bois, the nation's most distinguished black historian and intellectual, hailed the book as a "monumental and unrivaled study." So did other black leaders, ranging from the sociologist E. Franklin Frazier, whose criticisms of lower-class black culture influenced Myrdal's arguments, to the novelist Richard Wright, whose bitter autobiography, *Black Boy*, appeared in 1945. Prominent white intellectuals—the theologian Reinhold Niebuhr, the sociologist Robert Lynd, the historian Henry Steele Commager—concurred in this approbation. The near unanimity in support of Myrdal's message reflected the rising expectations among liberals for racial and ethnic progress at the close of the war.

Amid this atmosphere of hope, activists for racial justice pressed for change on a number of fronts in the mid- and late 1940s. One front was to desegregate the military. Some blacks, like Bayard Rustin, refused to submit to the draft—partly on pacifist grounds, partly in protest against Jim Crow in the armed services. He went to prison for his temerity. Most American Negroes, however, were ready and willing to fight: they composed 16 percent of Americans enlisted in the armed services during the war, though they were only 10 percent of the population. Approximately a million Negroes served between 1942 and 1945. But they confronted discrimination at every turn. The navy accepted Negroes only for menial tasks, often to serve as mess attendants. The army took in blacks but set up segregated training camps and units and refused to train blacks as officers. It also assumed that Negroes were poor fighters and hesitated to send them into combat. Secretary of War Stimson explained that blacks should serve under white officers, because "leadership is not imbedded in the Negro

31. Myrdal, *American Dilemma*, 928–29; Jackson, *Gunner Myrdal and America's Conscience*, 170, 225–26.

race yet and today to make commissioned officers lead men into battle— colored men—is only to work a disaster to both."[32]

By 1944 the protests of blacks—for Randolph and other leaders military desegregation was a top priority—had a modest effect on the armed services. The navy slowly moved toward integrated units. The army, at a loss for manpower during the Battle of the Bulge in December 1944, pressed blacks into combat, with positive results. But segregation persisted in the army, and racial tensions became intense. "My God! My God!" army chief of staff General George Marshall exclaimed, "I don't know what to do about this race question in the Army." He added, "I tell you frankly, it is the worst thing we have to deal with. . . . We are getting a situation on our hands that may explode right in our faces."[33] Though Marshall did nothing about the situation, he correctly assessed the more militant mood. A black Alabama corporal explained in 1945, "I spent four years in the Army to free a bunch of Dutchmen and Frenchmen, and I'm hanged if I'm going to let the Alabama version of the Germans kick me around when I get home. No sirreee-bob! I went into the Army a nigger; I'm comin' out a *man*."[34]

Expectations such as these unavoidably sharpened racial conflict in the postwar South, where more than two-thirds of American Negroes still lived—mass migrations notwithstanding—in the late 1940s. There, little had changed since the late nineteenth century. Most southern blacks—at least 70 percent—lived in poverty in 1945.[35] Virtually everything remained segregated: schools, churches, parks, beaches, buses, trains, waiting rooms, restaurants, hotels, rest rooms, drinking fountains, and other

32. Jackson, *Gunnar Myrdal and America's Conscience*, 234–36; Polenberg, *One Nation Divisible*, 76; John Diggins, *The Proud Decades: America in War and Peace, 1941–1960* (New York, 1988), 28; Charles Moskos, "From Citizens' Army to Social Laboratory," *Wilson Quarterly* (Winter 1993), 10–21.

33. Polenberg, *One Nation Divisible*, 77.

34. Goulden, *Best Years*, 353.

35. There were no official government definitions of poverty in the United States until 1963. A statistic such as this reflects later government estimates of what it had meant to be "poor"—living below (arbitrary) government notions of contemporary standards of minimum subsistence—in the 1940s. Contemporary definitions of poverty in the 1940s were harsh compared with those of the much more affluent 1960s, when many people defined as "poor" owned automobiles, televisions, and other household conveniences. If 1960s definitions of poverty had been applied to the less affluent 1940s, the percentage of African-Americans defined as "poor" at that time would have been higher than 70 percent. The point, of course, is that most blacks in the 1940s lived, as always, at the bottom of the economic pyramid.

public accommodations. All but a few white southerners believed theirs was the superior race, with a natural right to supremacy.[36] Mississippi senator James Eastland, later to become an influential national spokesman for white racism, expressed this view without embarrassment in a wartime speech against the FEPC: "What the people of this country must realize is that the white race is a superior race, and the Negro race is an inferior race."[37] Myrdal conceded that whites in the South "do not see the handwriting on the wall. They do not study the impending changes; they live again in the pathetic illusion that the matter is settled. They do not care to have any constructive policies to meet the trends."[38] Racist feelings promoted institutional discrimination and a virtual totality of white power. Deep South states in the early 1940s admitted almost no black lawyers, judges, or policemen. Notwithstanding the Supreme Court decision against white primaries, Negroes in the lower South faced a range of ruses and outrages—poll taxes, impossibly designed "literacy" tests, violent intimidation—that deprived them of any voice in politics. The emblem of the Democratic party in Alabama (Republicans did not matter) was a lusty gamecock under a scroll that read WHITE SU-PREMACY.[39]

Resting very close to the surface of these white concerns, especially in the South, were complicated feelings about sex between the races. There was irony here, of course, for white men continued, as they had throughout American history, to demand sexual favors from economically and legally defenseless black women. Miscegenation was the great open secret of sexual life in the South.[40] But state laws criminalized interracial sex as well as racially mixed marriages. (Until 1956 Hollywood's Motion Picture Code forbade interracial marriage to be shown; no black man embraced a white woman on screen until 1957.)[41] And woe to black men in the South who seemed too friendly with white women. By 1945 whites less often retaliated against such behavior by lynching—there were nineteen reported lynchings of Negroes between 1940 and 1944, compared to

36. Polenberg, *One Nation Divisible*, 26, 74.
37. Ibid., 119.
38. Myrdal, *American Dilemma*, 660–62; Ravitch, *Troubled Crusade*, 117.
39. Polenberg, *One Nation Divisible*, 27–28.
40. So widespread was this sexual mixing over the years that it calls into question the very notion of "race." But that is another story: throughout the period described in this book, most Americans took for granted the assumption that "whites" and "blacks" were of different "races."
41. Sayre, *Running Time*, 43; Thomas Cripps, *Making Movies Black: The Hollywood Message Movie from World War II to the Civil Rights Era* (New York, 1993).

seventy-seven between 1930 and 1934 and forty-two between 1935 and 1939—but all black American men knew that white violence was an ever-present possibility following any kind of "uppity" behavior, no matter how exaggerated by whites, especially if it was thought to threaten the supposed purity of southern white womanhood. Southern blacks who escaped violence, only to be brought to trial for such alleged offenses, faced all-white judges and juries and had virtually no possibility of justice.

Some young black men, often led by veterans, dared to challenge these patterns of racial discrimination in the South in the immediate postwar years. Rustin and others in the Fellowship of Reconciliation, a pacifist group dedicated to racial justice, embarked on highly dangerous "freedom rides" in the Upper South in 1947 to test the Supreme Court's decision against segregation in interstate travel. Other blacks came home, still in uniform, and tried to register to vote. Medgar Evers sought to cast his ballot in Mississippi, even though local whites said they would shoot him if he tried. In parts of the Upper South these efforts paid modest dividends. In most places, however, whites countered with threats or with violence. Rustin was arrested, jailed, and sent to work in a chain gang. Evers and four others were driven away by whites wielding pistols. Whites in 1946 killed three blacks—and two of their wives—who sought to vote in Georgia. Eugene Talmadge won his race for governor in Georgia at that time by bragging that "no Negro will vote in Georgia for the next four years."[42]

As later events were to demonstrate, this kind of crackdown on Negro protest in the South by no means dampened the determination of blacks to struggle against institutionalized discrimination. On the contrary, many continued to resist: clamoring to register, fighting discrimination in employment, seeking to join unions, challenging segregation. Moreover, southern blacks continued in these postwar years to build up their own institutions—schools, churches, community organizations—that served as bases for black self-pride and solidarity. All-black schools, for instance, had already succeeded in cutting illiteracy among Negroes, from 70 percent in 1880 to 31 percent in 1910 to around 11 percent in 1945.[43] But

42. William Chafe, *The Unfinished Journey: America Since World War II* (New York, 1991), 86–87.

43. These are estimations of "literacy" used by the census, in *Statistical History*, 364–65, 382. One may quibble a bit with the percentages—definitions differed some over time—but the trend was unmistakable and remarkable. For historical background, see James Anderson, *The Education of Blacks in the South, 1860–1935* (Chapel Hill, 1988), esp. 148–85.

these efforts did not begin to temper the intransigence of most southern whites during and after World War II. Refusing to bend, the whites drove black protests below the surface in the South: in 1950, as in 1940, white supremacy seemed secure in Dixie.

Black protest in the North was by contrast much more open in the late 1940s. The Congress of Racial Equality (CORE), founded in 1942, led "sit-ins" at Chicago restaurants as early as 1943. Activists were especially militant at the local level, primarily in areas where large numbers of blacks congregated during the mass migrations of the era. Between 1945 and 1951, eleven states and twenty-eight cities enacted laws establishing Fair Employment Practices Commissions, and eighteen states approved legislation calling for the end of racial discrimination in public accommodations.[44] The NAACP, CORE, and Urban League succeeded in 1947 in forcing the *Chicago Tribune* to cease its practice of negative "race labeling" in stories about black activities, including crime. At the same time, Chicago adopted an ordinance outlawing the publication of "hate" literature.[45] A year later the Supreme Court, in a decision hailed by civil rights leaders, ruled that "restrictive covenants," private pacts used by whites to keep blacks (or other "undesirables") out of residential neighborhoods, were not legally enforceable in the courts.[46]

The battle against restrictive covenants revealed a key fact about post–World War II racial conflicts in the North: many such fights centered on the efforts of blacks, crowding into northern cities in record numbers, to find decent housing. Negroes faced not only the covenants but also the systematic adoption by lending institutions of "red-lining," a practice that blocked off large areas of the cities from blacks seeking home mortgages. Blacks also confronted the racist policies of developers, many of whom refused to sell to blacks, and of city officials, who tightened zoning restrictions so as to limit the building of low-cost housing. Those few builders who sought to construct such housing ordinarily demanded public subsidies—construction, they said, would otherwise not pay—only to be rebuffed by local officials.

The federal government played a key role in these conflicts over housing. Some federal officials, notably Interior Secretary Harold Ickes, who exercised control over the housing division of the Public Works Adminis-

44. Kenneth Jackson, *Crabgrass Frontier: The Suburbanization of the United States* (New York, 1985), 279.
45. Hirsch, *Making the Second Ghetto*, 175–79.
46. Ibid., 30–31; Peter Muller, *Contemporary Sub/Urban America* (Englewood Cliffs, N.J., 1981), 89–92. The case was *Shelley v. Kraemer*, 334 U.S. 1 (1948).

tration until 1946, tried to promote relatively liberal policies concerning race relations. But even Ickes, confronting widespread hostility against the movement of blacks to white neighborhoods, dared not support the building of public housing projects open to blacks in white areas. Instead, he followed a so-called neighborhood composition rule, which approved public housing for Negroes only in areas that were already predominantly black.[47] When such projects were constructed, they led to further crowding in these areas. Meanwhile, the Federal Housing Adminstration, which distributed billions of dollars in low-cost mortgage loans in the late 1940s, thereby underwriting much of the suburban expansion of the era, openly screened out applicants according to its assessment of people who were "risks." These were mainly blacks, Jews, or other "unharmonious racial or nationality groups." In so doing it enshrined residential segregation as a public policy of the United States government.

All these policies helped to hasten the growth of large, institutional ghettos—cities within the central city—in some of the bigger urban areas of the North after 1945. Few such ghettos had existed before then. These areas became increasingly crowded, especially by contrast to white areas of these cities. In Chicago the number of white people declined slightly, by 0.1 percent between 1940 and 1950, yet the number of dwelling units occupied by whites increased by 9.4 percent. During the same years the number of black people in Chicago, a mecca for southern migrants, increased by 80.5 percent, but they occupied only 72.3 percent more dwelling units than in 1940. The percentage of non-whites in "overcrowded" accommodations (defined as more than 1.5 people per room) grew during that time from 19 percent to 24 percent. The number of dwelling units without private bath facilities increased by 36,248. Black residents complained of huge invasions of rats. Fires in Negro areas of Chicago killed 180 slum dwellers, including sixty-three children, between 1947 and 1953. For the dubious privilege of living in such crowded areas, blacks in Chicago, lacking market options, faced rents ranging between 10 and 25 percent higher than those paid by whites for comparable shelter.[48]

The Negro writer Ann Petry wrote a novel, *The Street* (1946), that described this sort of living. It focused on West 116th Street in Harlem, a grim place that blighted the experiences of Lutie Johnson, a single, black, working mother, and of Bub, her eight-year-old son. Children, keys

47. Hirsch, *Making the Second Ghetto*, 13–15, 214–15.
48. Ibid., 24–28.

strung around their necks, walked home to empty apartments and waited until their parents—too poor to afford a sitter—got home after work. Men with bottles of liquor in brown paper bags loitered about the stoops, waiting to prey on the unwary. "The men stood around and the women worked," Petry wrote.

> The men left the women and the women went on working and the kids were left alone. The kids burned lights all night because they were alone in small, dark rooms, and they were afraid. Alone. Always alone. . . . They should have been playing in wide stretches of green park and instead they were in the street. And the street reached out and sucked them up.[49]

While blacks crowded into ghettos, whites found ample space in the mushrooming suburbs. In Chicago, 77 percent of home-building between 1945 and 1960 took place in suburban areas. As of 1960 only 2.9 percent of people in these suburbs were black, roughly the same percentage as had lived in Chicago suburbs in 1940. "White flight," indeed, was rendering restrictive covenants unnecessary even before the Supreme Court decision in 1948. Many white urban residents, anxious to escape the influx of blacks, sold their houses to blacks and—racial covenant or no—took off for the suburbs. This process led to creation of a few "salt-and-pepper" areas of racial mixing, but neighborhood desegregation rarely lasted for long. In some places realtors engaged in "block-busting," warning whites in adjacent neighborhoods of the coming black "invasion." Fearful whites then sold in droves at market-bottom prices to the realtors, who cut the houses into smaller and smaller units and charged high rents for what soon became dilapidated slum housing. The neighborhood transformation was almost always rapid.

Some working-class whites, of course, could not afford to move. Many of these people lived in closely knit, ethnically homogeneous neighborhoods where ownership of property was both a treasured value and a primary asset.[50] They could not—would not—leave. They banded together to preserve the state of their neighborhood, relying less on covenants—a middle-class ploy—than on direct action. The result was what one careful study has called an "era of hidden violence" and of "chronic urban guerilla warfare."[51] In Detroit between 1945 and 1965, racially transient neighborhoods witnessed some 120 incidents of violence, featuring rock-throwing, cross-burnings, arson, and other attacks

49. Ann Petry, *The Street* (Boston, 1946), 388.
50. Use of "ethnic" here includes English-American.
51. Hirsch, *Making the Second Ghetto*, 40–67.

on property.[52] In Chicago, which had the largest Negro population of any American city, racially motivated bombing or arson disturbed the peace every twenty days during the late 1940s. Whites also staged large-scale "housing riots" to push blacks away from their neighborhoods. One of these, in Cicero near Chicago, drew a looting and burning mob of between 2,000 and 5,000—to drive one black family out of an apartment. Only the police and 450 National Guardsmen brought an end to the violence. Another Chicago riot, in 1947, aimed to stop blacks from getting space in heretofore white public housing. It attracted 1,500 to 5,000 whites who assaulted blacks, injuring thirty-five. This riot required the intervention of 1,000 policemen, who stopped the trouble after three days of mayhem. In both cases blacks got the message; it was impossible not to.

This grim picture tends to focus only on discrimination and therefore to underplay the sense of possibility that many blacks nonetheless cherished at the time. More black people than ever before, after all, were escaping the specially vicious world of Jim Crow in the South. The North *was* different! Millions of blacks, at last, had jobs in the industrial sector, which seemed prosperous in the late 1940s. For such people a stable family life, with a future for the children, seemed within reach: most Negro families at that time were headed by two parents. Blacks who wanted no part of whites—thousands felt this way—could find a variegated world of black institutions in the growing Negro neighborhoods of the major metropolises. To many blacks who remember those times, in fact, places like Harlem or Chicago's South Side were neither slums nor ghettos; they were black *communities* that nearly glittered with promise, especially by contrast to the rural South. Northern blacks, moreover, could vote. Chicago's South Side had sent a black man, Oscar DePriest, to Congress as early as 1928. Charles Dawson, also black, represented Chicago on Capitol Hill from 1942 to 1970.[53] In 1945 the erratic, militant Reverend Adam Clayton Powell, Jr., a Negro, began his long and tempestuous career as a congressman from Harlem.[54] There and elsewhere in urban areas of the North the rising number of black voters forced

52. Thomas Sugrue, "The Structures of Urban Poverty: The Reorganization of Space and Work in Three Periods of American History," in Michael Katz, ed., *The "Underclass" Debate: Views from History* (Princeton, 1993), 85–117.
53. Lemann, *Promised Land*, 74–76.
54. Charles Hamilton, *Adam Clayton Powell, Jr.: The Political Biography of an American Dilemma* (New York, 1991).

white politicians to take notice, thereby transforming the patterns of city politics. [55]

A wholly negative picture of race relations, by focusing on oppression, also tends to slight the vibrancy of certain aspects of black cultural life during the 1940s. These were years of notable vitality and creativity among black artists, writers, intellectuals, and—most audibly—musicians. Charlie "Bird" Parker, Thelonious Monk, Louis Armstrong, Ella Fitzgerald, Nat "King" Cole, Mahalia Jackson, Dizzy Gillespie, and many others experimented with a range of musical forms—jazz, blues, gospel, bebop—that had distinctly African-American roots. They drew whites as well as blacks to the cabarets and nightclubs in black sections of the city. Contemporaries identified much of this as "race music." Some of it had a hard and driving beat, lyrics that were sexually suggestive, and talk about "rocking and rolling," a phrase (like "jazz") that blacks understood to signify sexual relations.

Given these many variations of race and ethnic relations, it is risky to offer sweeping generalizations about their essence in America during the late 1940s. Still, two developments seem irrefutable. First, many people—from liberals like Myrdal to ethnics and blacks themselves—anticipated the possibilities of progress: World War II seemed a turning point in the nation's quest for greater ethnic acculturation and racial equality. Like the veterans who considered 1945 a chance—at last—for the Good Life, many Negroes and "new-stock" Americans of the era were decidedly hopeful, especially in contrast to the discouraging years of the immediate past. They were developing unusually high expectations.

Still, and second, it is foolish to wax romantic about the rate of ethnic acculturation, or especially about the status of black people in the 1940s. Most blacks, northern as well as southern, remained very poor; the vast majority never went to a nightclub or stayed in a hotel; many lacked radios—or even electricity; they encountered discrimination and rejection almost every day of their lives. Even their music was segregated: only in 1949, when the popularity of "race music" was becoming increasingly obvious, did *Billboard* magazine drop this category, refer to it instead as "rhythm and blues," and print charts of its best-sellers. [56] By the early 1950s, Antoine "Fats" Domino, Chuck Berry, and others were slowly leading "race music" into a larger mainstream that swept popular music

55. Robert Garson, *The Democratic Party and the Politics of Sectionalism, 1941–1948* (Baton Rouge, 1974).
56. Oakley, *God's Country*, 12.

from its moorings and placed rock 'n' roll on the high seas of American popular culture.

Americans who turned on the radio—the major purveyor of popular culture in the 1940s—also encountered a widespread marginalization of black people. Many listeners regularly tuned in "Amos 'n' Andy," one of the top programs of the era, or Jack Benny, a leading radio comedian. Amos and Andy were black characters portrayed in "colored" dialect by white announcers. They had redeeming qualities, and many blacks as well as whites greatly enjoyed their antics from the 1930s into the 1950s. The program, however, mainly presented blacks as unreliable and hapless. It remained on the radio until November 1960 and even found a life (this time with black actors) on national TV in 1951–52. Rochester, Benny's Negro valet, was similarly a complex character. He was shrewd and manipulative in his own way, and he became more assertive as the Benny show (which enjoyed a long run on TV) tried to keep pace with rising black militancy in the late 1950s and 1960s. But especially in the 1940s Rochester often came across as servile, more a butt than a source of humor.[57] The consistently enthusiastic response by whites to shows such as these, featuring mostly denigrating stereotypes of African-Americans, was yet another indication that Myrdal's optimism about the potential for liberalization of white attitudes was exaggerated.[58] Not until the 1960s, when blacks succeeded in shaking off some of the yoke of inequality, did the popular media begin to take them a little more seriously.

READING ABOUT American women in the immediate postwar years, one quickly encounters polar interpretations. Conservative writers tend to celebrate the late 1940s (and 1950s) as a wonderful era of domesticity, in which women happily accepted the rewarding duties of child-rearing and home management. Most feminists disagree, perceiving instead a world of widespread gender discrimination. Millions of women, they say, chafed as wives and mothers in homes that, in Betty Friedan's memorable phrase, were little more than "comfortable concentration camps."[59]

---

57. Melvin Ely, *The Adventures of Amos 'n' Andy: A Social History of an American Phenomenon* (New York, 1991), 245–47; Joseph Boskin, *Sambo: The Rise and Demise of an American Jester* (New York, 1986), 176–97.
58. The radio popularity of the Lone Ranger, who was aided by a mostly deferential Tonto, an Indian, suggests other contemporary uses of racial and ethnic stereotypes.
59. For excellent interpretations, see Arlene Skolnick, *Embattled Paradise: The American Family in an Age of Uncertainty* (New York, 1991); William Chafe,

One trouble with both these polar views, as most writers readily ac-knowledge, is that generalizations fail to catch the varied histories of individual women. In 1945 (the last year in which men outnumbered women in the United States), there were 69.9 million women in the forty-eight states. They differed greatly in age, class standing, race, regional background, and family situation. Their attitudes were naturally complex and often ambivalent, and their experiences obviously changed over time. It will not do to erect some transhistorical "model types" by which most American women can be categorized.

These caveats in mind, four generalizations about women's experi-ences in the 1940s seem valid. First, World War II—in so many ways a driving social force—changed the lives of millions of women, bringing them into the marketplace in record numbers and into new and some-times better-paying kinds of jobs. Second, demobilization drove many of these women from such jobs, but it only briefly slowed what was already a powerful long-range trend toward greater female participation in the mar-ket. Third, neither during the war nor afterward did most women think they were making an either-or choice between family and work. The majority gradually came to need both, at least for stretches of their lives, and encountered all the satisfactions and tribulations of juggling the two. Fourth, American women did such juggling amid a dominant cultural milieu that continued to place traditional notions of femininity above feminist quests for equal rights. As in the past, it was ordinarily expected that women be nurturant, deferential, and maternal. Some things change, some remain pretty much the same.

The war indeed accelerated the paid employment of women. The percentage of women (14 and older) who were part of the work force increased from 26 percent in 1940 to 36 percent in 1945. This was a jump in numbers from 13 million to 19.3 million. Prior to the war, working women had been predominantly young, single, and working-class. They had mainly found jobs in sex-segregated "women's work," ranging from poor-paying employment as laundresses, cleaners, farm laborers, and waitresses to jobs as office workers, nurses, and grade school teachers. Only a small percentage cracked gender barriers to work in business management or the professions. The war changed some of these patterns, accelerating the numbers of middle-class married women who for the first time entered the work force. While some of these tendencies had arisen

The Paradox of Change: American Women in the 20th Century (New York, 1991); and Carl Degler, *At Odds: Women and the Family in America from the Revolution to the Present* (New York, 1980).

prior to the 1940s, the war nonetheless represented a turning point of some magnitude in the modern history of American women.[60]

Some wartime work available to women was unprecedented. For the first time women in large numbers found better-paying factory jobs. Pioneers like the much-fabled "Rosie the Riveter" worked alongside men at welding, ship-building, and a few other blue-collar trades formerly reserved for men. Probably the majority of these women were from the working classes; they had been elsewhere in the market before the war. Now they left their poorly paid jobs in large numbers; as early as 1942 more than 600 laundries had to close for lack of help. The number of black women employed as farm workers declined by 30 percent between 1940 and 1945 while the number employed in the metals, chemicals, and rubber industries jumped from 3,000 to 150,000, an increase of 5,000 percent. Black female employment in government rose in the same years from 60,000 to 200,000.[61] Women doing this kind of work often felt proud and empowered. No wonder that Irving Berlin, composing for *Annie Get Your Gun* in 1946, has Annie Oakley sing out, "Anything you can do, I can do better."

The sudden end to war quickly changed these patterns. By early 1946 approximately 2.25 million women workers had quit their jobs, either because they wanted to or because they saw the handwriting on the wall. Another million were laid off. The biggest losers were the women who had found industrial work during the war: these jobs either disappeared amid the rush of demobilization or were given to veterans who returned to civilian life. Sex-segregated employment, still widespread during the war, became ever more the norm: by the late 1950s, 75 percent of women worked at female-only jobs, especially in the rapidly growing service sector. As one historian put it, "Rosie the Riveter had become a file clerk."[62] Gender segregation at work was by then greater than in 1900 and sharper than segregation by race.[63]

All of these trends confirm the influential lament of Friedan, whose *The Feminine Mystique* (1963) exposed the gender discrimination of

60. Chafe, *Paradox*, 172. See also Karen Anderson, *Wartime Women: Sex Roles, Family Relations, and the Status of Women During World War II* (Westport, Conn., 1981); and D'Ann Campbell, *Women at War in America: Private Lives in a Patriotic Era* (Cambridge, Mass., 1984).
61. Degler, *At Odds*, 420–21.
62. Cynthia Harrison, *On Account of Sex: The Politics of Women's Issues, 1945–1968* (Berkeley, 1988), 4–5.
63. James Davidson and Mark Lytle, *After the Fact: The Art of Historical Detection* (New York, 1986), 310.

postwar American life. But this was only part of the story, for the trends were complex. While demobilization adversely affected many women workers, it did not stop the steadily increasing desire of women, especially middle-aged and married women, to test the marketplace. By 1950 there were 18 million women working for pay, only a million or so fewer than in 1945. More than half of them, for the first time in American history, were married. By then the percentage of women who worked had risen to 29, three points higher than in 1940. The percentage kept going up, to 35 percent in 1960 and 42 percent in 1970. The rise in female employment was one of the most powerful demographic trends of the postwar era.

Few of these women, polls suggested, went to work because the war had begun a process of consciousness-raising that turned housewives into career women. On the contrary, even during the war few homemakers showed much enthusiasm for entering the job market. Most did so because they needed the money, because they were patriotic, or, with husbands or boyfriends in the service, because they were bored or lonely. The government, concerned over impending labor shortages, had to launch a propaganda campaign about patriotic duty in an effort to get many women out of their homes.[64] These women appear to have been numerous among the 2.25 million who gave up their jobs once the war was over.

Similarly varied motives impelled the millions of women who entered the job market in the late 1940s (and later). For many, work was a means to greater self-esteem. But for most, especially the large number of married women, the decision to work for pay rested mainly on economic needs. These were baby boom years. Women in the 1940s were marrying younger, having more children, buying an ever-greater variety of consumer goods, and looking for ways to supplement household income. Once their youngest child was in school, millions went to work to help with the bills. This is not to argue that women considered "feminist" reasons and rejected them; motives are not so easily disentangled. It is rather to say that most women continued, as they normally had done throughout American history, to place family concerns first.

Women who felt this way echoed cultural values widely held in American life at the time. A much-noted poll in 1945 concluded that 63 percent of Americans did not approve of married women working if their husbands could support them. (A similar poll in 1973 showed that 65 percent of people *did* approve).[65] Americans in the 1940s showed a similar lack of

64. Diggins, *Proud Decades*, 25–27.
65. Degler, *At Odds*, 420–21.

enthusiasm about women in politics: in 1949 fewer than 3 percent of state legislators were women. There were eight representatives (of 435) on Capitol Hill and one woman senator, Margaret Chase Smith of Maine. She had replaced her husband, who died in office.[66] President Truman expressed a common view in 1945. Women's rights, he said, were "a lot of hooey." When asked whether women might become President, he had a stock one-liner in reply: "I've said for a long time that women have everything else, they might as well have the presidency."[67]

Attitudes such as these abetted continuing discrimination. Early in the war the army refused to commission women doctors, until an act of Congress forced its hand in 1943. Women were frozen out of important governmental decision-making during and after the war. Private institutions discriminated openly. Medical schools closed their doors to women or maintained very low quotas. Bank training programs, law schools, and many businesses also established quotas. The perpetuation of sex-segregated employment contributed to low wages for women, who earned only slightly over half as much as the average for men.

Women seeking greater sexual freedom encountered equally resistant reactions in the 1940s. By then middle-class married women in many places could go to their doctors and get fitted for birth control devices. But single women normally could not, nor could the millions of poorer women without access to private doctors. Birth control clinics were few and far between. In some states it was illegal to sell—or even to use—birth control devices. Though these laws were widely ignored, it was not until 1965 that the Supreme Court ruled against such a law in Connecticut, thereby setting that issue to rest.[68] Women wishing to terminate unwanted pregnancies in the 1940s (and 1950s and 1960s) had to consider illegal abortion, the only kind there was at the time. This was a secret and frequently a dangerous business. Millions nonetheless undertook the risk. Contemporary data from the studies of Alfred Kinsey suggested that roughly 22 percent of married women had induced abortions, most of them early in the marriage or late in the child bearing years, and that the majority of single women in his survey who became pregnant resorted to abortion.[69]

Amid such a cultural milieu people who called themselves feminists

66. Harrison, *On Account of Sex*, 23–35.
67. Oakley, *God's Country*, 31.
68. *Griswold v. Connecticut*, 381 U.S. 479 (1965).
69. John D'Emilio and Estelle Freedman, *Intimate Matters: A History of Sexuality in America* (New York, 1988), 253.

found little popular support. A sign of the times was the respectful attention given to *Modern Women: The Lost Sex,* written in 1947 by the
sociologist Ferdinand Lundberg and the psychoanalyst Marynia Farnham. It denounced career women. "The independent woman is a contra-
diction in terms," the authors argued. Women should strive instead for
"receptivity and passiveness, a willingness to accept dependence without
fear or resentment, with a deep inwardness and readiness for the final goal
of sexual life—impregnation." Women who rejected this kind of femininity were "sick, unhappy, neurotic, wholly or partly incapable of dealing with life." Lundberg and Farnham went on to say that bachelors in
their thirties should get psychotherapy and that spinsters should be legally
forbidden to teach, on grounds that they were emotionally incompe-
tent.[70]

Few women in the mid-1940s could accept such extreme views, but
most well-known female political leaders readily asserted that women
were in some ways a "weaker" sex that needed protection under the law.
These leaders, including Eleanor Roosevelt and Labor Secretary Frances
Perkins, had fought hard in earlier years for "protective legislation," much
of it on the state level, that opposed exploitation of women on the job.
Laws set maximum working hours, outlawed night work, and disallowed
the employment of women at jobs involving strenuous tasks thought to be
appropriate for men. Perkins maintained that "legal equality . . . between the sexes is not possible because men and women are not identical
in physical structure or social function, and therefore their needs cannot
be identical."[71]

Perkins, Roosevelt, and leaders of the government's Women's Bureau
therefore opposed approval of the gender-blind Equal Rights Amendment, first proposed in 1923, which affirmed that "equality of rights
under the law shall not be denied or abridged by the United States, or by
any State, on account of sex." Such an amendment, they assumed, would
gut the protective laws. Frieda Miller, head of the Women's Bureau,
denounced the ERA as "radical, dangerous, and irresponsible." Another
Women's Bureau official memo at the time identified backers of the ERA
as a "small but militant group of leisure class women [venting] their
resentment of not having been born men."[72]

70. New York, 1947. Many scholars have stressed the attention given to this book,
    including Chafe, *Paradox,* 177–79, and Skolnick, *Embattled Paradise,* 71.
71. Chafe, *Unfinished Journey,* 16.
72. Harrison, *On Account of Sex,* 18; Women's Bureau memo cited in Chafe, *Unfin-
    ished Journey,* 85.

Alice Paul, head of the militantly pro-ERA National Women's Party, fought back by arguing that women were equal in every way and needed no "protection." She insisted that protective laws encouraged employers to deny work to women and instead to hire men, who faced fewer restrictions. The debates between the pro- and anti-ERA camps were bitter and sometimes nasty. Paul occasionally resorted to playing on contemporary Cold War fears. Among the opponents of the ERA, she pointed out, were members of the American Communist party. Alerting the rabidly anti-Communist House Un-American Activities Committee, she disparaged the loyalty of her rivals.[73]

The rancor that divided these women leaders was but one of many forces that helped assure defeat of the ERA in the 1940s and early 1950s. The amendment came closest to passage in 1946, when Paul and other leading women—Georgia O'Keeffe, Margaret Mead, Margaret Sanger, Pearl Buck, and Katharine Hepburn among them—demanded its adoption. In that year a small majority of the Senate, 38–35, voted for it. But the initiation of constitutional amendments by this means requires approval by two-thirds of both houses of Congress. Some of the senators who voted in favor in 1946 did so knowing it would not receive two-thirds. In the House, meanwhile, the ERA got nowhere. When it failed to win approval in the Senate, the *New York Times* breathed a sigh of relief. "Motherhood," it declared, "cannot be amended, and we are glad the Senate didn't try."[74]

ERA supporters kept up the fight in the late 1940s and early 1950s, but the struggle commanded little public attention. ERA was not an issue in the 1948 campaign. Truman made no mention of it. In 1950 the Senate again approved it, this time by a substantial margin of 63 to 11. But they were voting on a new version that included an amendment exempting protective legislation. In this form the National Women's Party opposed it, and the House ignored the issue. The same series of events took place again in 1953. After that the ERA was all but dead until the resurgence of feminism in the 1960s.

The fate of the ERA, which never had a realistic chance in the late 1940s, hardly stood as a main event in the lives of American women at the time. Much more important for people was the quest for satisfactory personal lives—a quest that was driving a boom in consumer goods and sending ever-larger numbers of women into the work force. In time, the

73. Richard Fried, *Nightmare in Red: The McCarthy Era in Perspective* (New York, 1990), 165.
74. Harrison, *On Account of Sex*, 16–23.

experience of working outside the home broadened the expectations of many American women. Their children, moreover, grew up in homes where mothers were different role models: wage-earners as well as house-keepers. But though these social changes were powerful, they were gradual and slow to affect the culture in the 1940s. Values about matters like gender roles—or religion and race—normally evolve only gradually, and the postwar era was no exception. In this sense, defeat of the ERA in the 1940s and 1950s was more a symbol of continuing cultural patterns than it was a major political battle. Expectations aside, some things indeed remain much the same.

2

Unions, Liberals, and the State: Stalemate

In 1942 the left-wing Almanac Singers performed a new song, "UAW-CIO," about the United Automobile Workers, a militant union of the Congress of Industrial Organizations:

> I was there when the Union came to town.
> I was there when old Henry Ford went down;
> I was standing at Gate 4
> When I heard the people roar:
> "Ain't nobody keeps us Autoworkers down!"
>
> It's that UAW-CIO
> Makes the Army roll and go—
> Turning out the jeeps and tanks and airplanes every day
> It's that UAW-CIO
> Makes that Army roll and go—
> Puts wheels on the USA
>
> I was there on that cold December day
> When we heard about Pearl Harbor far away
> I was down on Cadillac Square
> When the Union rallied there
> To put them plans for pleasure cars away
>
> *Chorus*
>
> There'll be a union-label in Berlin
> When the union boys in uniform march in;

And rolling in the ranks
There'll be UAW tanks—
Roll Hitler out and roll the union in!

The song captured well the genuine pride and patriotism of many wartime blue-collar workers in the United States. Just as the war helped to acculturate many first- and second-generation immigrants, it brought millions of Americans into a much-expanded work force, gave them meaningful jobs to do, and engaged them in a common effort against the enemy. The war virtually ended unemployment and led to considerable improvements in the average real wages of workers. Many of these workers gladly proclaimed their pride in "being an American."[1]

The song revealed a second key fact about many American blue-collar workers during the war: they took great pride in their unions. These had grown greatly during the 1930s, when the American labor union movement at last surged ahead. Thanks in considerable part to the militancy of the newly formed CIO, which recruited millions of unskilled and semi-skilled workers, total union membership increased between 1930 and 1940 from 3.4 million to 8.7 million, or from 11.6 to 26.9 percent of non-agricultural employment. Growth during the war was even greater, indeed unmatched before or since. By 1945 there were 14.8 million union members. In that year they composed 35.5 percent of total non-agricultural employment and 21.9 percent of employment overall.[2]

Unions had not only grown by 1945; they also constituted a vanguard of American liberalism. This is not to say that they were egalitarian in all that they said and did: unions, like other American institutions, had a generally poor record when it came to admitting women, blacks, and other minorities.[3] Nor were American workers ordinarily class-conscious in a Marxian sense. But many union members—as well as other low-income Americans—felt the sting of inequality, and they proudly identified themselves as "working class." Given a sense of their rights as citizens during the New Deal, they backed a range of liberal social policies, voted for liberal office-seekers, and exercised considerable political power

1. Gary Gerstle, *Working-Class Americanism: The Politics of Labor in a Textile City, 1914–1960* (New York, 1989), 278–309 (on Woonsocket, R.I.).
2. Nelson Lichtenstein, *Labor's War at Home: The CIO in World War II* (New York, 1982).
3. Nancy Gabin, *Feminism and the Labor Movement: Women and the United Automobile Workers, 1935–1975* (Ithaca, 1990); Dorothy Cobble, *Dishing It Out: Waitresses and Their Unions in the 20th Century* (Urbana, 1991).

within the Democratic party. More than at any other time in American history, the union movement in 1945 defined the left liberal of what was possible in politics.[4]

Indeed, unions were considerably more powerful at the time than left-wing organizations such as the Socialist and Communist parties. Other activist, progressive organizations, such as the Congress of Racial Equality, the NAACP, and the Southern Conference for Social Welfare, which battled to help poor and black people in the South, also confronted widespread hostility: none of these had much influence in politics. The postwar years frustrated the American Left, which faced increasingly harsh Red-baiting after 1945.

Yet even unions, poised as they were for further gains in 1945, contended with large obstacles in the late 1940s, a time of strong corporate and conservative resistance to social reform. The travails of unions in those years reveal much about the stalemate that came to characterize American labor relations as well as politics in the late 1940s.[5]

THE DREAMS of the United Automobile Workers, as expressed by one of the union's most articulate leaders, Walter Reuther, were especially compelling to liberals at that time. Reuther, thirty-eight years old in 1945, differed from stereotypes that portrayed labor union "bosses" as crude, cigar-chomping, and semi-literate. A one time tool-and-die maker who had worked in auto plants in Detroit (and, briefly, in the Soviet Union during his youth), he was fastidious in his personal habits and ill-at-ease in what one historian has described as the "blowzy comradeliness of smoke-filled union halls and blue-collar taverns."[6] He commanded the respect, but not always the personal affection, of the rank-and-file. Some close associates worried about his great ambitions. The story circulated that Reuther, Truman, and CIO head Philip Murray met about this time. When Reuther briefly left the room, Truman took it upon himself to give Murray a warning. "Phil," he told Murray, "that young man is after your

4. Nelson Lichtenstein, "From Corporatism to Collective Bargaining: Organized Labor and the Eclipse of Social Democracy in the Postwar Era," in Steve Fraser and Gary Gerstle, eds., *The Rise and Fall of the New Deal Order, 1930–1980* (Princeton, 1989), 30; Jack Bloom, *Class, Race, and the Civil Rights Movement* (Bloomington, Ind., 1987).

5. Melvyn Dubofsky, *The State and Labor in Modern America* (Chapel Hill, 1994), 188–95.

6. Robert Zieger, *American Workers, American Unions, 1920–1985* (Baltimore, 1986), 160.

job." Murray replied quietly, "No, Mr. President, he really is after *your* job."[7]

But friends as well as foes recognized that Reuther was an articulate, principled, and even visionary man. A former Socialist, he had been beaten by Ford Motor Company thugs during the strikes of the 1930s. By the 1940s he had rejected socialism as impractical in the United States, but he maintained good relations with democratic socialists like Norman Thomas, the longtime head of the Socialist party. Reuther believed that organized labor must take the lead in promoting a more progressive society that would improve the lives of blacks and other poor and powerless people.[8] His vision incorporated much that the Left was seeking at the time: a governmentally guaranteed "annual wage," a much-expanded welfare state, civil rights protection for abused minorities, federal legislation to promote better education and health care, and workers' control of key decisions about production and technological development.[9]

Though relatively few American workers had such large visions as Reuther, many responded enthusiastically to the bread-and-butter demands that he and other labor leaders agitated for in the 1940s. Militant movements for better wages and working conditions blazed like wildfires rising from the grass roots. In 1944 these caused a record number of 4,956 work stoppages involving 2.12 million workers, or 4.8 percent of all those who were employed. Not only conservatives, who retaliated with bills to draft strikers, but also Roosevelt and many other liberals expressed mounting alarm about disruption of wartime production.

The end of the war brought a host of new problems for American labor. Layoffs reaffirmed the specter of depression. Workers still on the job lost overtime pay, which had been vital to their earnings during the war. Take-home pay for many such workers declined in late 1945 by 30 percent. Rumors of corporate greed fanned the flames of egalitarian outrage. Murray, a soft-spoken but determined advocate, estimated that the after-tax profits of steel corporations were 113 percent higher between 1940 and 1944 than they had been in the previous four years. Charles E. Wilson, head of General Motors, was known to have received a salary in 1943 of

7. Ibid., 107.
8. Ibid., 104–5, 159–60; Joseph Goulden, *The Best Years, 1945–1950* (New York, 1976), 114; Godfrey Hodgson, *America in Our Time* (Garden City, N.Y., 1976), 92.
9. Lichtenstein, "Corporatism"; David Halle and Frank Romo, "The Blue-Collar Working Class: Continuity and Change," in Alan Wolfe, ed., *America at Century's End* (Berkeley, 1991), 152–78; Zieger, *American Workers*, 100–101.

$459,014. In late 1945 around one-fifth of working families in American cities received less than $1,500 in total cash income for the year—at a time when the average for full-time employees was $2,190.[10]

Anxieties such as these provoked a rash of angry disputes and walkouts in late 1945. Joseph Goulden's sprightly book on the era captures the turbulence well: "At the Stock Exchange, the tickers fell silent when 400 clerical employees walked out. Barbers, butchers, bakers struck. Stoppages interrupted the production of copper wire, Campbell's Soup, castor oil, Christmas toys. The Pittsburgh Pirates baseball team, near the cellar in the National League, took a strike vote in midsummer, but decided to continue playing." A strike of 3,500 electric company workers in Pittsburgh threw an additional 100,000 out of work. The city's trolleys came to a halt, the streetlights went out, and office buildings shut down for fear of elevator failure. "This is a disaster," Mayor David Lawrence said in imploring workers to end the strike. In Manhattan, simultaneous strikes by elevator operators, truck drivers, and maritime workers stopped shipments in and out of the city. In all, there were 4,750 work stoppages in 1945, only 200 or so fewer than in 1944. The disruptions upset Truman. "People are somewhat befuddled and want to take time out to get a nerve rest," he wrote his mother. "Some want a guarantee of rest at government expense and some I'm sorry to say just want to raise hell and hamper the return to peacetime production to obtain some political advantage."[11]

The wave of strikes crested in early 1946, involving 1.8 million workers in such major industries as meat-packing, oil-refining, electrical appliances, steel, and automobile manufacture. The year 1946 ultimately became the most contentious in the history of labor-management relations in the United States, with 4,985 stoppages by 4.6 million workers, or about one of every fourteen Americans in the labor force. The number of worker-days lost was 116 million, three times the previous high in 1945. More than in 1945, these were strikes called by union leaders after lengthy but futile negotiations with management. Contemporaries were impressed by the order and discipline of the masses of union members who

10. Goulden, *Best Years*, 110–12; Zieger, *American Workers*, 102. Other estimates put the average at $2,374. See *Statistical History of the United States, from Colonial Times to the Present* (New York, 1976), 164.
11. Alan Wolfe, *America's Impasse: The Rise and Fall of the Politics of Growth* (New York, 1981), 13–79; Alonzo Hamby, *Beyond the New Deal: Harry S. Truman and American Liberalism* (New York, 1973), 66–68; Robert Donovan, *Conflict and Crisis: The Presidency of Harry S. Truman, 1945–1948* (New York, 1977), 119–22, 166–68; Goulden, *Best Years*, 113.

attended meetings, walked the picket lines, and undertook the sacrifices of refusing to work.[12]

The disputes mainly involved wages. Union leaders cited wartime and postwar inflation as justification for seeking wage hikes of 30 percent. For auto workers—a little better paid than most manufacturing workers at the time—this would have meant average pay increases of approximately thirty-three cents an hour, from $1.12 to $1.45, or from $44.80 to $58 for a forty-hour week. That would have been $3,016 for a full year. General Motors countered with an unsatisfactory offer of ten cents an hour, and the strike began in December 1945, followed in January by a wave of walkouts in other industries. Government fact-finders then estimated that inflation since the start of the war might justify raises of around nineteen cents. Steel industry leaders, authorizing a price hike, settled there, as did bargainers in most other industries. The auto workers hung on but were isolated. After 113 days Reuther agreed to wage increases of eighteen and a half cents an hour.

Labor leaders at the time complained that the settlements, which resulted in average wages in manufacturing in 1946 of around $50 a week ($2,600 for a fifty-two-week working year) were far too stingy. Actually, these workers did well by comparison with many other Americans at the time—well enough, in fact, to think in time of buying a small home. By contrast, regularly employed farm, forestry, and fishery workers at the same time averaged $1,200, domestic workers $1,411, medical workers (mainly nurses) $1,605. Public school teachers and principals averaged only $1,995. Still, it was true that a wage of $50 a week left little margin: temporary unemployment, injury, and illness were among the many setbacks that could quickly cause impoverishment. Aging workers, fearing replacement by more productive younger people, were especially vulnerable. At that time few American corporations had pension plans. Longtime workers (in covered jobs) who retired could expect Social Security pensions of around $65 to $70 per month, or a maximum of $840 per year.[13]

Reuther and other American liberals especially lamented their inability to prevail in the larger battle over control of practices on the shop floor: managers refused to budge on this important issue. Reuther also insisted that corporations had reaped great profits in the war and could afford to pay better wages without increasing their prices. Management, he as-

12. Zieger, *American Workers*, 104.
13. Goulden, *Best Years*, 118–20.

serted, should "open its books" to expose its costs. This claim especially outraged business leaders, who recognized that it threatened their control of basic decisions. Almost to a man they resisted Reuther's "socialistic" approach to labor-management relations.[14]

It is difficult to say whether Reuther was correct in claiming that corporations could afford to pay better wages without increasing prices. Business leaders adamantly refused to open their books. Instead, they tended to yield to demands for modest wage increases, whereupon they raised the prices of their goods and passed on wage costs to consumers. Then and later, labor leaders reluctantly acquiesced in this kind of compromise. Starting in 1948 unions compromised further, agreeing to COLA (cost-of-living agreements) clauses in annual contracts. These used measures of inflation in order automatically to readjust wages. In so doing labor surrendered effective efforts to have a say in pricing. This pattern of annual wage and price settlements—almost always upward—reflected an important reality that dominated the economics and politics of the postwar years: well-established interest groups ultimately agreed to accommodate each other while giving lip service at best to the needs of the unorganized. Subsequent labor negotiations, especially in manufacturing industries, led to similar patterns in the 1950s.

The labor contracts of the war years, the late 1940s, and the 1950s also gave slowly increasing numbers of union members better benefits: health insurance, life insurance, paid vacations, and old-age pensions. Employer contributions to these plans were passed on to consumers in the form of higher prices. These were important gains for those American workers, mainly men in the larger manufacturing industries such as steel, rubber, and automobiles, who received them. But better benefits could be a mixed blessing: some covered employees felt "locked in" and therefore afraid to look elsewhere for jobs. Moreover, smaller corporations usually lacked the resources to extend such benefits to their workers, and non-union workers lacked the clout to demand them. Gulfs that historically had divided American workers—union and non-union, skilled and unskilled, full-time and part-time, male and female, manufacturing and non-manufacturing, white and black—if anything widened over time.

It also became increasingly clear that these gains, like COLA clauses, were achieved at the cost of larger visions. In 1945 many American liberals hoped that the end of war might bring expansion of New Deal

14. John Barnard, *Walter Reuther and the Rise of the Auto Workers* (Boston, 1983), 101–7.

social programs, including extension of Social Security to millions of uncovered workers such as waitresses, domestic workers, and farm laborers. Liberals also supported federal aid to education, higher minimum wages, and even some form of government-provided health insurance. Some were showing interest in civil rights action for minorities.[15] Many labor leaders remained officially committed to such liberal programs, and blue-collar workers continued to vote heavily for liberal Democratic candidates; it was not accurate to claim (as some did) that "labor" had dropped its political agenda. But the vision was becoming dimmer. More and more, union leaders concentrated on securing better *private* benefits. The 1940s, a time of significant expansion of governmental social welfare in many western European countries, in fact solidified the privatization of social welfare in the United States. In 1948 almost half of America's workers still had no coverage from federal retirement plans. Those who did worried greatly that inflation might cut the purchasing power of their pensions.[16]

Unions, moreover, grew only slowly after World War II. Leaders like Reuther and American Federation of Labor (AFL) head George Meany, an uninspiring but shrewd and hard-working labor bureaucrat, maneuvered as effectively as they could both at bargaining tables and as lobbyists for better labor legislation. But the union movement did not keep pace with growth in the labor force. In 1950 union membership was 15 million, only barely higher than in 1945. Unions then included only 31.5 percent of non-agricultural employees, 4 percent less than in 1945. By 1960 the unions had 17 million members but only 31.4 percent of a larger non-agricultural force.

The fundamental sources of this stall, and of rapid union decline thereafter, were structural and political. World War II, promoting huge expansion in manufacturing, boosted employment—and therefore growth in union membership—in heavy industries that had already begun to unionize in the 1930s. In the late 1940s and 1950s, however, these sectors of the economy expanded only slowly. Growth areas after 1945 were increasingly in white-collar work and service employment. Workers in these jobs were scattered, often in relatively small companies. Many were part-time employees. Thousands of difficult-to-organize middle-aged married

15. Alan Brinkley, *The End of Reform: New Deal Liberalism in Recession and War* (New York, 1995); Brinkley, "The New Deal and the Idea of the State," in Fraser and Gerstle, eds., *Rise and Fall*, 85–121.
16. Nelson Lichtenstein, "Labor in the Truman Era: Origins of the 'Private Welfare State,'" in Michael Lacey, ed., *The Truman Presidency* (Washington, 1989), 151.

women moved in and out of the work force. Higher percentages of Americans labored in rapidly growing southern and western areas, many of which were hostile to union organizers. Add to all these structural realities the traditional problems facing union activists—racial and sexual discrimination in unions, interethnic hostilities, the resistance of many white-collar Americans to labor organization, the determination of anti-union business groups and politicians, the divisive influence of Cold War anti-communism at home—and one begins to understand the stall of union organization as well as of liberal politics in the United States after 1945.

Four conflicts during the late 1940s highlighted the problems faced by organized labor at the time. The first pitted a few union leaders against a sometimes angry President Truman. The President, a liberal New Dealer, ordinarily supported blue-collar demands for change. But he had been distressed by the strikes in January 1946, which he believed threatened his efforts at reconversion, and he grew alarmed when faced with strikes by coal miners two months later and by railroad workers in May.[17] Truman could not prevent the miners from striking, but he was determined to stop the railroad workers. Trains, after all, were vital to the American economy, for both passenger and freight travel, in the days before widespread superhighway transportation.

After weeks of bickering with the carriers and twenty rail unions, Truman thought he had an agreement in hand. But leaders of the two largest unions, A. F. Whitney of the Brotherhood of Railroad Trainmen and Alvanley Johnston of the Brotherhood of Locomotive Engineers, refused to toe the line. Both were old friends of the President, who brought them to the Oval Office three days before the strike deadline and lectured them, "If you think I'm going to sit here and let you tie up this whole country, you're crazy as hell." Whitney said he was sorry: "We've got to go through with this, Mr. President. Our men are demanding it." Truman responded by giving them forty-eight hours to reach a settlement. "If you don't, I'm going to take over the railroads in the name of the government."[18]

When the workers nonetheless went out on strike, Truman became perhaps as angry as any American President of recent times. On Friday, May 24, he marched into a Cabinet meeting and announced he would go to Capitol Hill and seek an extraordinarily draconian law. It would have

17. Barton Bernstein, "The Truman Administration and Its Reconversion Wage Policy," *Labor History*, 4 (1965), 216–25.
18. Donovan, *Conflict and Crisis*, 208–16; Robert Ferrell, *Harry S. Truman and the Modern Presidency* (Boston, 1983), 91–92; David McCullough, *Truman* (New York, 1992), 493–506; Goulden, *Best Years*, 121–22.

authorized him to draft strikers into the military, without regard to age or number of dependents, whenever a walkout threatened to create a national emergency. Truman even had aides prepare a speech to be given over the radio that evening. It was virtually irrrational, assailing the patriotism of "effete union leaders" like Whitney, Johnston, Murray and others, whom he also associated with Communism. "Every single one of the strikers and their demigog [sic] leaders have been living in luxury, working when they pleased and drawing from four to forty times the pay of a fighting soldier." Truman's draft closed, "Let's give the country back to the people. Let's put transportation and production back to work, *hang a few traitors*, and make our country safe for democracy." Charles Ross, Truman's old friend and press secretary, was "horrified" by the diatribe. Clark Clifford, Truman's top policy adviser, recalled that it was "surely one of the most intemperate documents ever written by a President."[19]

Ross and others managed to calm Truman down, and his radio address that evening was forceful but temperate. Still, he insisted on going to Capitol Hill the next day, by which time everyone knew his intent. As he entered the House, he received a standing ovation, whereupon he outlined his tough proposals. When he was almost finished, his chief labor adviser, John Steelman, informed Clifford, who was in an anteroom off the House floor, that the unions had settled on the terms proposed by the President. Clifford wrote a note about the settlement and had it delivered to Truman, who paused and then read it to thunderous applause. Clifford and others considered it a victory for Truman, who dropped his demands for legislative action. But many liberals and labor leaders were deeply frightened. The *New Republic*, a leading liberal journal, called Truman's congressional message the "most vicious piece of anti-union legislation ever introduced by an American President."[20]

If railroads were important, coal was vital to the economy in 1946. It still drove 95 percent of locomotives and furnished 62 percent of electric power. Labor-management strife had long affected the coal industry, which by the 1940s correctly perceived the very real threat of oil. John L. Lewis, head of the mine workers, was a tenacious, relentless, and successful advocate of safety, better wages, and benefits for the dwindling number of miners who managed to survive the decline of the industry. Unlike Reuther and many other labor leaders at the time, he did not trust the government, liberals included, to be reliable in support of the workers'

19. Clark Clifford, "Serving the President: The Truman Years (1)," *New Yorker*, March 25,1991, pp. 54–55.
20. Richard Pells, *The Liberal Mind in a Conservative Age: American Intellectuals in the 1940s and 1950s* (New York, 1985), 59.

interests, and he had had no hesitation even during the war in urging his following to strike.[21] This they often did, for the miners loved and trusted him: many miners' homes displayed only two pictures—of John L. Lewis and the Virgin Mary. The historian Joseph Goulden has remarked, "To Americans of the 1930s and 1940s Lewis's coal strikes were as much an annual ritual as the first sighting of the ground hog, or a President throwing out the first baseball of the season."[22]

Lewis was a fierce-looking, hulking, and majestic man whose ice-blue eyes glared at people under huge and bushy eyebrows. He spoke in rolling, theatrical cadences rich in biblical phrases and did not care who he insulted. In 1937 he had called no less a figure than Vice-President John Garner a "poker-playing, whiskey-drinking, labor-baiting, evil old man." On another occasion, he said that the "AFL has no head; its neck just growed up and haired over." Lewis relished power and had used it readily over the years. "When we control the production of coal," he said, "we hold the vitals of our society right in our hands. . . . I can squeeze, twist, and pull until we get the inevitable victory." By 1946 he was a widely feared figure. Truman loathed Lewis, regarding him as a "racketeer," and aimed his May 25 diatribe to Congress against him as well as the trainmen.[23]

When the coal miners settled their strike in May, it seemed possible that further conflict would be avoided, at least in 1946. But in November Lewis made it clear that he wanted to change the settlement and would call out the miners, 400,000 strong, if he did not get his way. Truman was almost desperately determined to resist. Winter was fast approaching. Equally important, Truman saw Lewis's action as a deliberate test of wills. He intervened by securing a court order stopping the proposed strike. Lewis, ignoring the order, sent the miners on strike and was cited for contempt of court. A federal judge fined the union an astounding $3.5 million and Lewis personally $10,000. The strike, the judge added, was "an evil, demoniac, monstrous thing that means hunger and cold and unemployment and disorganization of the social fabric. . . . if actions of this kind can be successfully persisted in, the government will be overthrown."[24] Lewis then sought to negotiate, but the President refused

21. Melvyn Dubofsky and Warren Van Tine, *John L. Lewis: A Biography* (New York, 1978); Lichtenstein, "Labor," 138.
22. Goulden, *Best Years*, 123.
23. Ibid., 123–27.
24. Ibid.; Donovan, *Conflict and Crisis*, 239–42; McCullough, *Truman*, 528–29; Truman, *Memoirs of Harry S. Truman*, Vol. 1, *Year of Decisions* (Garden City, N.Y., 1955), 552–56.

to talk to him or to any of his representatives. The struggle held front-page attention for more than two weeks.

On Pearl Harbor Day 1946 Lewis capitulated by calling off the strike. Truman was grimly delighted. "The White House," he told aides at an impromptu celebration, "is open to anybody with legitimate business, but not to that son of a bitch." Editorial writers also celebrated. Arthur Krock of the *New York Times* wrote that the President "has greatly regained stature as a national leader." Joseph and Stewart Alsop, widely syndicated columnists, added that Lewis's surrender was the "first break that he [Truman] has had in considerably more than a year."[25]

Truman and his aides indeed had grounds for self-congratulation. In May he had threatened the trainmen and had faced them down. Six months later he had brought the imperious John L. Lewis to heel. These actions by no means revolutionized labor-management-government relations. On the contrary, Reuther, Lewis, and other leaders of big industrial unions continued to fight against corporate powers. Over these struggles, which set larger patterns for wage-price increases in the United States, the government had relatively little control, whether in the 1940s or in later years. In the longer run, moreover, the ongoing structural changes in the economy proved especially decisive in determining the fate of unionization. Still, Truman's stiffened posture in 1946—especially his humbling of Lewis—struck a political blow against Big Labor, as opponents called it, in the United States.

Hostility to Big Labor proved especially powerful on Capitol Hill, where the second revealing conflict over labor rights erupted in 1947. Influential congressmen since the late 1930s had chafed openly at what they considered the arrogance of organized labor. During the war and again in 1946, Democratic Congresses had passed legislation that restricted union power, only to encounter presidential vetoes. Voters in 1946 then elected Republican majorities to both houses for the first time since 1930. Supporting these Republicans were many conservative Democrats, especially from the South and other areas of the country where labor unions were politically unpopular. The coalition seeking to curb labor benefited from intense lobbying by employer groups and was led in the House by Fred Hartley, a strongly anti-union Republican from New Jersey. In the Senate the key leader was Robert Taft of Ohio, son of the former President. Acting quickly, they drafted the so-called Taft-Hartley bill in early 1947.

25. Clifford, "Serving (1)," 57–59.

Taft-Hartley was a bold effort to weaken the pro-labor Wagner Act of 1935. A highly publicized clause authorized the President to call for eighty-day "cooling-off" periods before strikes could be called that might affect the national interest. Another title banned closed shops, workplaces that required workers to be members of unions at the time of hiring. The bill also prohibited secondary boycotts, which had enabled workers to boycott the goods of allegedly anti-labor companies. An especially contro- versial provision, Section 14b, authorized states to prohibit union shops, which were workplaces requiring employees to join unions within a short period of time following initial hiring. States instead could pass what became known as "right-to-work" laws that were expected to impose major obstacles to union organizing. Reflecting widespread fears that radi- cals dominated the labor movement, the bill required union leaders to sign non-Communist affidavits if they wanted their workers to receive access to the activities of the National Labor Relations Board (NLRB). This access had proved important and usually beneficial to unions, many of which since 1935 had come to rely on the NLRB to police labor- management relations, including fair elections seeking to certify union locals.[26]

Taft-Hartley aroused storms of outrage from labor leaders, who damned it as "fascistic" and as a "slave labor act." CIO counsel Lee Pressman explained, "When you think of it merely as a combination of individual provisions, you are losing entirely the full impact of the program, the sinister conspiracy that has been hatched."[27] Murray cried that the law was "conceived in sin." Like other labor leaders, he feared that the act would end a brief era in American history, during which the state had served as a neutral or pro-labor broker in relations with management. The rules of the game would be changed.

The angry reaction of unions prompted strenuous, often frantic lobby- ing. Truman, too, opposed the bill as an unreasonable assertion of corpo- rate interests. Having damaged his standing with unions in 1946, he was also anxious for political reasons to mend fences. When the bill passed in mid-1947, he vetoed it. But both houses quickly and decisively overrode the veto, and the bill became law. Some union leaders then considered calling strikes in protest, only to back off from such a defiant step. This was probably a prudent decision, for rank-and-file attitudes were uncer-

26. Christopher Tomlins, *The State and the Unions: Labor Relations, Law, and the Organized Labor Movement in America, 1880–1960* (New York, 1985).
27. Lichtenstein, "Labor," 421; Lichtenstein, "Corporatism," 134.

tain.[28] Union leaders decided instead to fight the law in the political arena, by urging subsequent Congresses to amend or repeal it.

In fact, Taft-Hartley was far from a "slave labor act." The ban on closed shops, to be sure, lessened the control that a few strong unions had had over hiring. But most unions managed to live with the law. Swallowing civil libertarian scruples, they signed the non-Communist affidavits and continued to avail themselves of the procedures and protections of the NLRB. The major industrial unions bargained as before with employers. Aggressive, expanding unions such as the Teamsters—ultimately to become by far the nation's largest union—flourished even in right-to-work states. By the 1950s most observers agreed that Taft-Hartley was no more disastrous for workers than the Wagner Act had been for employers. What ordinarily mattered most in labor relations was not government laws such as Taft-Hartley but the relative power of unions and management in the economic marketplace. Where unions were strong, they usually managed all right; when they were weak, new laws did them little additional harm. [29]

Still, the success of Taft-Hartley revealed that the political power of organized labor was waning. Although the law did not repeal the Wagner Act—the conservatives who wrote it reaffirmed labor's basic right to bargain collectively—its passage exposed broad popular suspicion of Big Labor that remained strong thereafter. Unions, despite sustained lobbying, failed to secure revision of the law in the Democratic Congress of 1949— or even in the heady years of liberal triumph in the mid-1960s.[30]

A third conflict of the late 1940s, so-called Operation Dixie, further revealed the limits of labor power at the time. This was the quest of the CIO and the AFL, competing with one another, to break the anti-union hold of employers in much of the South. The CIO spent $1 million and took on 200 organizers at the peak of its efforts in 1946 and 1947. From the beginning, however, Operation Dixie ran into determined opposition from state governments and from the textile industry. Agricultural employers also fought forcefully against the effort. Anti-union leaders appealed shamelessly to racist feelings by linking the CIO with efforts for desegregation, thereby driving wedges between white and black workers.

28. Zieger, *American Workers*, 114.
29. James Patterson, *Mr. Republican: A Biography of Robert A. Taft* (Boston, 1972), 352–66; R. Alton Lee, *Truman and Taft-Hartley: A Question of Mandate* (Lexington, Ky., 1966).
30. Benjamin Aaron, "Amending the Taft-Hartley Act: A Decade of Frustration," *Industrial and Labor Relations Review*, 11 (April 1958), 352–60.

In Birmingham, Alabama, a major steel city, white steel workers rejected an insurgent, largely black local of workers who organized under the Communist-dominated Mine, Mill, and Smelter Workers Union. By the end of 1947 the hopes for interracial steel union locals were all but vanishing in the South.[31]

By 1948 it was clear that Operation Dixie had failed completely. Indeed, unions enrolled slightly smaller percentages of non-agricultural labor in the South in 1955—around 18 percent—than they had in 1945. Equally worrisome to liberals, the failure exposed the enduring divisiveness of race, which had managed—as so often in United States history— to damage the potential for working-class solidarity. The lack of such solidarity, in turn, weakened liberal southern politicians, who after 1947 faced an anti-union coalition of conservatives roused to potitical action by Operation Dixie. Liberal office-holders who dared to present themselves as pro-union racial moderates often went down to defeat—as Florida's Claude Pepper and North Carolina's Frank Graham did in races for the United States Senate in 1950. Other political figures concluded that they had to soft-pedal liberal ideas about unions and about blacks if they hoped to stay in office.[32]

The struggle in Birmingham revealed the fourth, and sometimes nastiest, conflict that damaged labor solidarity in the immediate postwar years. This was the battle between Communist and anti-Communist union leaders.[33] Corresponding with rising Cold War fears, it simmered as early as the war years. By 1945 a number of unions, including the United Electrical Workers, the International Longshoremen's Union, the National Maritime Union, and the Mine, Mill, and Smelter Workers Union either had Communist leaders or followed the Soviet party line on political and economic issues. Most of these were in the CIO. In all, twelve of thirty-five CIO affiliates in 1946 had a Communist or strongly pro-Soviet leadership. Nearly a million workers (most of them non-Communist) belonged to these unions.[34]

31. Robert Norrell, "Caste in Steel: Jim Crow Careers in Birmingham, Alabama," *Journal of American History*, 73 (Dec. 1966), 669–701.
32. F. Ray Marshall, *Labor in the South* (Cambridge, Mass., 1967); Philip Taft, *Organizing Dixie: Alabama Workers in the Industrial Era* (Westport, Conn., 1981); Lichtenstein, "Corporatism"; Ira Katznelson, "Was the Great Society a Lost Opportunity?," in Fraser and Gerstle, eds., *Rise and Fall*, 185–211.
33. David Oshinsky, "Labor's Civil War: The CIO and the Communists," in Robert Griffith and Athan Theoharis, eds., *The Specter: Original Essays on the Cold War and the Origins of McCarthyism* (New York, 1974), 116–51.
34. Zieger, *American Workers*, 123–27.

By this time most leading labor figures, including Reuther, Meany, and Murray, had fought many battles with Communist unionists. So had others on the non-Communist Left, including Socialists. All these leaders believed that Communists, parroting the Soviet party line, sought to seize control for their own sectarian purposes. In 1947 and 1948 these internal struggles became bitter and occasionally violent, leading Murray to tell the CIO executive board, "If communism is an issue in your unions, throw it to hell out, and throw its advocates out along with it." In 1949 he called Communists "sulking cowards," "apostles of hate," and "dirty, filthy traitors." With the support of anti-Communist allies he expelled eleven unions, with 900,000 members, from the CIO.[35]

Murray and his allies were right about one thing: Communist or pro-Communist leaders of these unions generally followed the Soviet party line on a range of domestic and international issues. In so doing they did not reflect the political views of most of the workers they were representing, and they exposed the union movement as a whole to loud charges that it was "soft" on Communism. For these reasons Murray and other top unionists decided to fight against Communist influence. In expelling Communist officers, however, Murray and his allies acted without much attention to democratic procedures. Some leaders of the purged unions, after all, had been elected to office by democratic votes of their members. The purges also threw a number of effective union organizers out of the movement and stifled internal discussion. As Paul Jacobs, a union activist who reluctantly supported the expulsions, said later, the purges brought "all serious debate within the CIO to a standstill. In some unions it became a habit to brand as a Communist anyone who opposed the leaders."[36]

By mid-1950 the Red Scare in unions had triumphed, leaving anti-Communists in firm control of the labor movement in the United States. For this and other reasons, many unions operated thereafter more as special interest groups than as supporters of broad-based liberal ideas such as those that Reuther had emphasized in 1945. Strong within the Democratic party in industrial areas, unions were also something of a prisoner of the party, unable to do much in politics without it. In non-industrial areas—and among the masses of blacks, working women, and other non-unionized Americans—they had little influence at all. To committed social reformers in the United States, the stagnation of organized labor by

35. Ibid., 131; Richard Polenberg, *One Nation Divisible: Class, Race, and Ethnicity in the United States Since 1938* (New York, 1980), 106.
36. William Chafe, *The Unfinished Journey: America Since World War II* (New York, 1991), 107.

1950 was both cause and consequence of a broader political stalemate that blocked liberal goals at the time.[37]

IN 1945 IT WAS NOT altogether easy to predict the development of such a stalemate. On the contrary, some signs suggested that liberals might advance. The Democrats after all, continued to have a clear majority of the electorate, having grown mightily under the leadership of Roosevelt and the New Deal. They could count on the allegiance, some of it strong in those days of intense partisan loyalties, of large majorities of politically numerous people: blacks, poor farmers, blue-collar workers, unionists, Jews, and Catholics. Of course, there were reasons for Democrats to be concerned. Voter turnouts favoring Democrats, after rising in the 1930s, had fallen again during the war. Truman, although a loyal New Dealer, was an uncertain political quantity as a leader. And Republicans enjoyed the backing of most businessmen; with the aid of conservative Democrats the GOP had blocked New Deal measures in Congress since the late 1930s. Still, conservatives seemed to be in a minority among the electorate; the Democratic-dominated political universe looked stable in 1945.

Popular willingness to call on the State, a source of liberal social policies since the 1930s, also seemed to grow during the war. As passage of the GI Bill of Rights suggested, a sense of rights-consciousness had grown during the previous decade.[38] By 1945, the majority of people accepted the once heretical ideas that government had a responsibility to fight unemployment, that economists and other "experts" had the know-how to manage economic life, even that short-term governmental deficits were tolerable in times of distress.[39] Total federal spending rose from around $9 billion in 1939 to $95 billion in 1945; expenditures during the war years were twice the total spent during the previous 150 years of United States history. This did not mean that Americans accepted or even understood the often arcane ideas of liberal economists like John Maynard Keynes, who had argued for compensatory public spending to relieve recessions. It

37. Steven Gillon, *Politics and Vision: The ADA and American Liberalism, 1947–1985* (New York, 1987), 3–32; Harvey Levenstein, *Communism, Anti-Communism, and the CIO* (Westport, Conn., 1981); and Mary Sperling McAuliffe, *Crisis on the Left: Cold War Politics and American Liberalism, 1947–1954* (Amherst, Mass., 1978). See chapter 7 for fuller discussion of the postwar Red Scare.
38. Sidney Milkis, *The President and the Parties: The Transformation of the American Party System Since the New Deal* (New York, 1993).
39. Herbert Stein, *The Fiscal Revolution in America* (Chicago, 1969), 194–96.

did mean, however, that people had developed a matter-of-fact appreciation of the larger role of the federal government.[40] Deficits, indeed, had
been a fiscal reality since 1930, had mushroomed during the war to a
scarcely conceivable peak of $54 billion in 1943, and were widely be-
lieved to have been an engine of wartime prosperity. After all, the gross
national product (GNP) boomed at the time, from $100 billion in current
prices in 1939 to $212 billion in 1945.

The Supreme Court by 1945 had unambiguously accepted these ideas
about the economic obligations of the State. One case in point was
Wickard v. Filburn in 1942, in which the Court unanimously laid to rest
any lingering doubts about the constitutionality of intrusive federal management of the economy. It ruled that farmers who had agreed to federally
determined production quotas could be penalized if they grew wheat in
excess of such quotas, even if they used the excess only for home consumption.[41] Comparably broad interpretations of the commerce clause
had sparked explosive disagreements as recently as the 1930s. After 1942,
however, they evoked virtually none at all. Then and later the Court,
dominated by Roosevelt appointees, endorsed expansive notions about
governmental activism in economic matters.[42]

Americans seemed to support such governmental activity even when it
came to taxation, which grew dramatically during the war. In 1939 only
3.9 million Americans had paid individual federal income taxes. At no
time in the 1930s had this tax revenue amounted to more than 1.4 percent
of GNP. Federal corporate taxes had never exceeded 1.6 percent. By
1943, 40.2 million Americans paid federal income taxes, which had risen
by then to 8.3 percent of GNP. Starting in 1944 Americans went along
with the withholding of taxes—also a wartime innovation—as a normal
fact of life. After 1945 federal taxes dipped a little but never sank to prewar
levels: income taxes reached a postwar low of 5.9 percent of GNP in 1949,
by which time 35.6 million people were filing tax returns. Scarcely looking back, the nation went from a system of class taxation to one of mass
taxation during the war.[43]

40. Robert Collins, *The Business Response to Keynes, 1929–1964* (New York, 1981);
    Byrd Jones, "The Role of Keynesians in Wartime Policy and Postwar Planning,
    1940–1946," *American Economic Review*, 62 (1972), 125–33.
41. 317 U.S. 111 (1942).
42. Richard Adelstein, "The Nation as an Economic Unit: Keynes, Roosevelt, and the
    Management Ideal," *Journal of American History*, 78 (June 1991), 160–87.
43. C. Eugene Steuerle, *The Tax Decade: How Taxes Came to Dominate the Public
    Agenda* (Washington, 1992), 13–15; John Witte, *The Politics and Management of
    the Federal Income Tax* (Madison, 1985), 252.

With the advantages of hindsight, however, it is now clear that war-related growth in government also had malignant consequences. As early as the mid-1930s Roosevelt had begun making wider use of the Federal Bureau of Investigation (FBI) headed by J. Edgar Hoover. Shrewd, deeply suspicious, a consummate bureaucratic infighter, Hoover had already developed a number of carefully guarded special files, including an "Obscene File," a "Sex Deviate File," and other "Personal and Confidential" files that contained all manner of rumor, gossip, and information about people. When FDR asked Hoover to investigate the activities of isolationists and political opponents in 1940, the FBI went so far as to tap the phone of Harry Hopkins, FDR's chief adviser. There is no evidence that all this skulduggery contributed to the war effort. What it mainly did was vastly to increase Hoover's power, as well as his command of information that he could use to silence potential opponents. The FBI was to become an extraordinarily dangerous force in American politics after 1945.[44]

Some of the vengefulness that characterized the FBI pervaded postwar politics in general. Optimists who thought that partisan rivalry was just politics as usual underestimated how bitterly many Republicans (and conservatives generally) resented Roosevelt and the New Deal. Although relatively few of these conservatives were prepared to turn back the clock by abolishing such landmark legislation as Social Security, they hungered to regain political power after so many years in the minority. Theirs was a politics of revenge that led easily to excesses of Red-baiting, even during the war itself. FDR, in turn, hardly tried to conceal his contempt for conservative rivals during much of the war. When he died in April 1945, many of his opponents quietly exulted and regrouped to recapture the White House in 1948. Truman and most liberal Democrats fought back with equal fervor. Partisanship flared as intensely in the late 1940s and early 1950s as at any other time in modern United States history.

As passage of Taft-Hartley in 1947 made clear, conservatives possessed important assets in these battles. One of these was the support of resurgent business leaders. This development, too, owed much to the war, which pulled corporate leaders out of the self-doubt that possessed them following the Crash and Depression. Large corporations benefited especially, garnering the lion's share of lucrative, cost-plus defense contracts, renegotiating contracts to their advantage when the contracts went badly, gaining ownership of patents achieved with the aid of government subsidy, and buying government-owned factories for a song at the close of war. Many

44. Barton Bernstein, "The Road to Watergate and Beyond: The Growth and Abuse of Executive Authority Since 1940," *Law and Contemporary Problems*, 40 (Spring 1976), 58–76.

liberals worried about the rise in corporate power during the war, but top
government officials acquiesced in it as a way of speeding up the war
effort. "If you are going to . . . go to war . . . in a capitalist country,"
Secretary of War Henry Stimson explained, "you had better let business
make money out of the process or business won't work."[45]

Stimson and others might well have pondered more fully a contemporary adage: "The hand that signs the war contract is the hand that shapes
the future." For the fact of the matter was that the war accelerated development of what many later critics, President Dwight Eisenhower among
them, called a military-industrial complex. After the war many corporate
leaders lost defense contracts. But they had amassed considerable power
and prestige in the war years, and they reasserted themselves thereafter
with uncommon relish, spending large sums on lobbying, campaign finance, litigation, public relations, advertising, philanthropy, and spon-
sorship of research in efforts to broaden their influence.

This was no corporate monolith. As always, American corporations
competed vigorously with one another. Many small businessmen resented bitterly the gains of big business during the war. The National
Association of Manufacturers, a leading conservative group, frequently
clashed with slightly more liberal business lobbies such as the Committee
for Economic Development and the United States Chamber of Commerce. Nor were business elites consistently successful in the political
arena. Often divided over particular issues, they were frequently clumsy at
politics, and they lost any number of battles over the years. But in most
situations the leading corporate figures of the postwar era were agreed in
their opposition to expansion of the New Deal.[46] America, one business
spokesman said in 1944, must "rid the economy of injurious or unnecessary regulation, as well as administration that is hostile or harmful," and
pursue "constructive fiscal, monetary, and other policies that provide a
climate in which a private enterprise system can flourish." In effect these
leaders were pressing for a government that would largely do the bidding
of big business. In so doing they put together what one historian has called
"the largest and most systematic deployment of corporate power in the
history of the United States."[47]

45. John Blum, V Was for Victory: Politics and American Culture During World
    War II (New York, 1976), 122.
46. Howell Harris, The Right to Manage: Industrial Relations Policies of American
    Business in the 1940s (Madison, 1982).
47. Robert Griffith, "Forging America's Postwar Order: Domestic Politics and Politi-
    cal Economy in the Age of Truman," in Lacey, ed., Truman Presidency, 64–66.
    Businessman cited is William Barton.

ONE MAY STILL ASK how these conservatively inclined groups succeeded in holding Truman, union leaders, and liberal allies to stalemate in the late 1940s. A final answer is that the economy was flourishing, thereby encouraging people to rely more on private effort than on governmentally sponsored social changes.[48]

The relationship of economic growth to social policy is hardly simple or predictable. Take the 1960s for an example. The vitality of the market, which flourished throughout most of the 1940s and 1950s, seemed so fantastic by the early 1960s that reformers, playing on ever-greater popular expectations about the future of the economy, generated wide political support for government, social, and educational programs. After all, the economic pie seemed huge, and the country seemed well able to afford a rise in social spending. Many liberals in the heady climate of the early 1960s even imagined that governmental programs, spurring the market, would virtually eliminate poverty and discrimination.

A rather different scenario dominated the late 1940s, however. At that time economic growth was only beginning to develop, and popular expectations, while rising, were considerably less utopian. As the compromises of labor unions revealed, the emergent prosperity of the immediate postwar years slowly softened class conflicts and resentments. It also helped to moderate popular pressure for State-sponsored liberal programs. Cautiously optimistic about the future, most Americans did not think that government could—or should—intervene very far in economic matters. Heeding the rhetoric of conservative interest groups, they believed instead that the market (with only modest help from the State) would ordinarily manage on its own to promote further economic growth. Many liberals, too, lowered their sights, hoping that fiscal tinkering—adjusting spending, taxes, and interest rates—would be sufficient to fine-tune the economy and assure expansion. They assumed that tougher measures such as more progressive taxation, heightened anti-trust activity, long-range social planning, and stringent governmental regulation, would not be necessary and might even be counter-productive.[49]

For all these reasons, the majority of politically influential Americans in the late 1940s (and 1950s) championed a social order that they believed rewarded individual effort and an economic order that continued to be the most thoroughly commercialized in the world.[50] They largely accepted a political arena that was controlled by well-established interest groups:

48. Frank Levy, *Dollars and Dreams: The Changing American Income Distribution* (New York, 1987), 23–44.
49. Brinkley, *End of Reform*, 227–71.
50. On these points see David Potter, *People of Plenty: Economic Abundance and the*

businessmen, large commercial farmers, physicians, veterans, some unionized workers.[51] The rise of "interest group liberalism," as it has been called, dominated American politics after 1945.[52] Much of the time this led to turf wars between the groups and to political stalemate.

There were of course other reasons for the political stalemate of the late 1940s, high among them the rise of Cold War issues to the top of the national agenda. By 1949 these issues had done much to divert popular attention from social and economic reforms. But as early as 1946 the stalemate was already fairly clear. Then and later the most socially progressive Americans recognized ruefully that World War II had strengthened many of the conservative values and interest groups of the prewar culture. With a buoyant economy beginning to erode memories of the Great Depression, it was not an opportune time for dramatic political change.

American Character (Chicago, 1954); and Gordon Wood, The Radicalism of the American Revolution (New York, 1992), 313.

51. Brian Balogh, "Reorganizing the Organizational Synthesis: Federal-Professional Relations in Modern America," Studies in American Political Development, 5 (Spring 1991), 119–72; Katznelson, "Was the Great Society a Lost Opportunity?"

52. Theodore Lowi, The End of Liberalism (New York, 1979); Samuel Lubell, The Future of American Politics (New York, 1952).

3

Booms

Economic growth was indeed the most decisive force in the shaping of attitudes and expectations in the postwar era. The prosperity of the period broadened gradually in the late 1940s, accelerated in the 1950s, and soared to unimaginable heights in the 1960s. By then it was a boom that astonished observers. One economist, writing about the twenty-five years following World War II, put it simply by saying that this was a "quarter century of sustained growth at the highest rates in recorded history." Former Prime Minister Edward Heath of Great Britain agreed, observing that the United States at the time was enjoying "the greatest prosperity the world has ever known."[1]

By almost any standards of measurement the postwar economic power and affluence of the United States were indeed amazing. With 7 percent of the world's population in the late 1940s, America possessed 42 percent of the world's income and accounted for half of the world's manufacturing output. American workers produced 57 percent of the planet's steel, 43 percent of electricity, 62 percent of oil, 80 percent of automobiles. Dominating the international economy like a colossus, it had three-quarters of the world's gold supplies. Per capita income in the United States in mid-1949, at $1,450, was much higher than in the next most prosperous group of nations (Canada, Great Britain, New Zealand, Switzerland, and Sweden), at between $700 and $900. Unemployment was estimated at

1. Daniel Yankelovich, *The New Morality* (New York, 1974), 166.

61

1.9 percent of the civilian labor force in 1945 and slightly under 4 percent from 1946 through 1948. Urban Americans at that time consumed more than 3,000 calories in food per day, including about as many fruits and vegetables per capita as forty years later. This caloric intake was around 50 percent higher than that of people in much of western Europe. [2]

Social stability also seemed fairly well assured for Americans in the postwar era. Thanks to strict immigration laws dating to the 1920s, few people in the 1940s worried that masses of strangers would take away jobs or disrupt the social order. [3] Although youth gangs troubled some cities, most streets remained safe. Since the early 1930s, violent crime had fallen greatly in the United States. The murder rate had halved by 1945. As became clear later, this situation was abnormal, stemming in part from the fact that the United States had a relatively small cohort of young men—those most prone to crime—at the time. This in turn was the result of a trend to smaller families, which had developed earlier in the century, and of the war, which put the boys in uniform and sent millions overseas. Later, when crime rates exploded in the 1960s, people looked nostalgically to the 1940s, failing to realize the special demographic reasons that had helped to make the low rates possible. Still, it was a fact that most neighborhoods were able to control crime in the late 1940s; public disorder was only here and there a major worry. [4]

From the perspective of later years, it is clear that nostalgia for the late 1940s can be misplaced in other ways, for millions of people—especially blacks and Mexican-Americans—did not share in the blessings of prosperity. If there had been a "poverty rate" at that time, it would have identified at least 40 million people, 30 percent of the population, as "poor" by the standards of the era. Those standards, moreover, were harsher than in later years of higher expectations about life. In 1947 one-third of American homes had no running water, two-fifths had no flush toilets, three-fifths lacked central heat, and four-fifths were heated by coal or wood. Most people lived in rented housing. They ate considerably less

2. Alan Wolfe, *America's Impasse: The Rise and Fall of the Politics of Growth* (New York, 1981), 155; Morris Janowitz, *The Last Half-Century: Societal Change and Politics in America* (Chicago, 1978), 411–12; Richard Easterlin, *Birth and Fortune: The Impact of Numbers on Personal Welfare* (New York, 1980), 32–34.
3. Immigration between 1932 and 1945 had ranged from a low of 23,068 in 1933 to a high of 82,998 in 1939. It rose in the late 1940s, but only to a high of 188,317 in 1949.
4. Charles Silberman, *Criminal Violence, Criminal Justice* (New York, 1978), 30–33; Landon Jones, *Great Expectations: America and the Baby Boom Generation* (New York, 1980), 144.

beef and chicken than would later generations. Almost half worked at demanding physical tasks on farms, in factories, in mines, or in construction.[5] Working-class wives typically arose at 5:00 or 6:00 A.M. to begin a day of labor that scarcely abated until they turned in at night.

Life on the farm continued to be especially hard. Mechanization, isolation, and poverty had already driven millions of farmers and farm workers off the land in earlier decades of the twentieth century. A total of 24.4 million Americans nonetheless depended on their livelihoods from the soil in 1945, or 17.5 percent of the population. Beginning in the late 1940s, the current of people fleeing from farm to town and city swelled into a flood—one of the most dramatic demographic shifts of modern American history. By 1970 only 9.7 million people, or 4.8 percent of the overall population, worked on the land. The number of farms fell from 5.9 million at the close of World War II to 3 million twenty-five years later.[6] Those farmers who remained included a small minority who enjoyed the bountiful benefits of governmentally supported large-scale commercial agriculture. Highly organized politically, these magnates of agribusiness amassed formidable power in Congress, where rural states of the South and West had ample representation, especially in the Senate. Most smaller farmers, however, struggled to make a living. And farm workers (many of them black or Hispanic) suffered from widespread exploitation: their poverty rates remained much higher than the national average. There were other languishing sectors of the American economy in these years—mining, for one—but none that suffered more in the postwar years than small-scale agriculture.

Even American families with annual earnings around the median (slightly over $3,000 in money income in 1947) lived carefully at that time. Many had vivid memories of the Depression years, when it had been easy to fall over the cliff into ruin. These people saved what money they could. Their children wore Keds, simple and inexpensive footwear. Playthings tended to be uncomplicated: yo-yos, Mr. Potato Head kits, cheap board games. Fancy toys were anything that required a battery. If

5. Frank Levy, *Dollars and Dreams: The Changing American Income Distribution* (New York, 1987), 25; Robert Collins, "Growth Liberalism in the Sixties," in David Farber, ed., *The Sixties: From Memory to History* (Chapel Hill, 1994), 11–15; James Patterson, *America's Struggle Against Poverty, 1900–1994* (Cambridge, Mass., 1995), 78–96.
6. Kirkpatrick Sale, *Power Shift: The Rise of the Southern Rim and Its Challenge to the Eastern Establishment* (New York, 1975), 20–21; Marty Jezer, *The Dark Ages: Life in the United States, 1945–1960* (Boston, 1982), 55.

the children had bikes, they were models with one speed. Few homes had more than one radio or telephone. If the family took a vacation, it was likely to be nearby, not to the Caribbean (and surely not to Europe). To want not in the early postwar years was to waste not.

Still, the dominant, increasingly celebrated trend of these years was economic progress that ultimately—in the 1950s and 1960s—shot millions of people into the ranks of the home-owning, high-consuming, ever-better-educated middle classes. Poverty, while remaining more serious than contemporaries recognized, declined steadily, falling to around 22 percent in 1959 (and to a postwar low of 11 percent in 1973). Economic expansion, greatly accelerated during the war, boosted many industries—notably aircraft, electrical and electronics firms, and chemicals. Tobacco and food-processing companies made enormous gains. Developments during the war unleashed a fantastically growing pharmaceuticals industry and hastened research that culminated in the arrival of the first digital computer in 1946 and the transistor in 1947.[7]

Government spending greatly assisted this expansion. Although federal outlays declined from $95.2 billion in fiscal 1945 to a postwar low of $36.5 million in 1948, they remained much higher than prewar levels—only $9.4 billion in 1939—and they then rose again, to $43.1 billion by the outset of the Korean War in 1950.[8] State and local spending, meanwhile, more than doubled between 1945 and 1948, to $21.3 billion in 1948. Much of this went to support schools and road-building. And private investment boomed, especially in science and technology. The number of scientists and engineers engaged in industrial research leapt from fewer than 50,000 in 1946 to around 300,000 fifteen years later.[9] Finally, the United States had a hard-working labor force. All these assets accounted for perhaps the most revealing statistic of all: gains in productivity. Output per employee increased by a remarkable 3.3 percent per year between 1947 and 1965, compared to rates of between 2 and 2.5 percent between 1900 and 1940 (and 1.4 percent between 1973 and 1977).[10]

7. Alfred Chandler, "The Competitive Performance of U.S. Industrial Enterprises Since the Second World War," *Business History Review*, 68 (Spring 1994), 1–72.
8. Figures in current dollars, from *Statistical History of the United States, from Colonial Times to the Present* (New York, 1976), 1105.
9. Richard Nelson and Gavin Wright, "The Rise and Fall of American Technological Leadership: The Postwar Era in Historical Perspective," *Journal of Economic Literature*, 30 (Dec. 1992), 1931–64.
10. Levy, *Dollars and Dreams*, 48; Thomas Edsall, *The New Politics of Inequality* (New York, 1984), 213–14.

Impressive as such economic statistics are, they cannot convey the broader, though admittedly hard to quantify, sense of well-being that the majority of Americans were beginning to feel by the late 1940s.[11] These feelings reflected real improvements: the average American of the late 1940s earned more in real dollars, ate better, lived more comfortably, and stayed alive longer than his or her parents. Increases in earnings during the war, along with savings (totaling an enormous $140 billion in 1945) set this stage. Having money meant having the chance to buy more things, conferring on people a wonderful sense of entitlement that many had been deprived of in the 1930s and even in the early 1940s. Some younger people became so optimistic that they boldly went into debt, sure that they would do well enough to pay it off in the future. Most of their parents, especially in the working classes, found this attitude almost incomprehensible. As the boom psychology spread it may even have made people work harder, thereby benefiting the economy as a whole. In any event, the mood among the ever-larger middle classes was positive. Americans felt increasingly flush as the years passed.

Millions of people in the late 1940s also thought that the American dream was alive and well. This was not a dream of rags to riches; few sensible citizens had ever imagined that. Nor did it imagine the abolition of privileges and special distinctions: Americans at that time, as earlier, tolerated open and unapologetic rankings in schools, in the military, in job descriptions. Rather, it was defined by the belief that hard work would enable a person to rise in society and that children would do better in life than parents. The United States was indeed the land of opportunity and high expectations.[12]

The dream rested also on a widespread sense among white Americans that the United States did not suffer from a rigid, impermeable class structure. Some working-class people, to be sure, were keenly aware of continuing class distinctions and held an "us" versus "them" attitude characteristic of European peasant societies. Many others clung to their ethnic, working-class subcultures. Perceiving schools as dominated by "outsiders," these people were cool to the very American faith that education could or should lead to individual advancement. Work was something one did, not a way to personal "careers." Folks like these resented young people who tried to break away from the neighborhood and strike out on their own. They especially believed in sticking close to the ex-

11. Janowitz, *Last Half-Century*, 155.
12. Gordon Wood, *The Radicalism of the American Revolution* (New York, 1992), 234.

tended family.[13] But even these people were likely to believe that American society offered opportunities—acquiring property, for instance—to the person with spunk and drive.[14]

Many forces sustained this faith in economic opportunity, the core of the American dream: the fantastic material abundance of the country, the egalitarian ideals of the American Revolution, the high aspirations of energetic and entrepreneurial immigrants, the existence (for whites) of political rights, the real possibility of getting ahead. All these forces had historically made Americans a restless, enterprising, and geographically mobile people. In the postwar years this mobility may have declined a little, mainly owing to increased home-ownership. But the United States continued to be—along with other "new immigrant" societies such as Canada and Australia—the most geographically mobile country in the world. From the 1940s to the 1970s roughly 20 percent of Americans changed residence every year.[15]

The dream depended finally on the sense that social mobility, too, was possible. Rags to riches made no sense, but rags to respectability did. Whether this belief represented reality depends on one's standards of measurement. Studies of twentieth-century income distribution, for instance, tend to show high levels of inequality, as measured by the percentage of income controlled by various percentiles of the income pyramid. The money income of the wealthiest 5 percent of American families (and unrelated individuals) in 1947 was 19 percent of the national total; the richest 20 percent had 46 percent of income ; the lowest 20 percent had 3.5 percent. Studies of the mobility of *individuals*, however, reveal a vast amount of movement up (and down) the income scale.[16] Millions of optimistic Americans, especially younger people, *thought* that they could get ahead.

13. Herbert Gans, *The Urban Villagers: Group and Class in the Life of Italian-Americans* (New York, 1962), x, 122–25, 181–86, 250–55.

14. See David Brody, "The Old Labor History and the New: In Search of an American Working Class," *Labor History*, 20 (1979), 111–26.

15. Claude Fischer, "Ambivalent Communities: How Americans Understand Their Localities," in Alan Wolfe, ed., *America at Century's End* (Berkeley, 1991), 79–90.

16. Carole Shammas, "A New Look at Long-Term Trends in Wealth Inequality in the United States," *American Historical Review*, 98 (April 1993), 412–31; Isabel Sawhill and Mark Condon, "Is U.S. Inequality Really Growing?," *Policy Bites* (Urban Institute, Washington, 1992); James Patterson, "Poverty and the Distribution of Income and Wealth in Twentieth-Century America," in Stanley Kutler, ed., *Encyclopedia of the United States in the Twentieth Century* (New York, 1995); Christopher Jencks, *Rethinking Social Policy: Race, Poverty, and the Underclass* (Cambridge, Mass., 1992), 6–7.

These were years, too, of grander expectations about the blessings of science, technology, and expertise in general. Many scientists were certain that the harnessing of atomic energy had all sorts of beneficial peacetime possibilities, ranging from use as a cheap source of power to medical "breakthroughs." Faith in medicine and in the goodness of physicians rose dramatically among the middle classes who could afford them. The war, after all, had seen the introduction of "wonder drugs" such as streptomycin and penicillin; now it was time to conquer other dread diseases, such as polio and cancer. A few days after the bombing of Nagasaki, executives of General Motors announced a $4 million grant to Memorial Hospital in New York, to establish the Sloan-Kettering Institute for Cancer Research there.[17] "The automobile industry," they said, "starting from nothing has developed into one of our greatest industries. We should like to place at the disposal of the medical profession, which has done and is doing such a magnificent job, any of our particular type of research techniques which it feels can be used advantageously to help it to conquer this so-called 'incurable' disease."[18]

Nowhere was this faith in progress more clear than in the area of education. Public schools, of course, had long been celebrated as central to the American dream. But there had always been a good deal of idle lip service to this ideal, and it is striking to recall the more prosaic reality of American education in the early 1940s. According to the 1940 census, only one-third of the 74.8 million Americans who were 25 years of age or older at that time had gone beyond the eighth grade. Only one-fourth were high school graduates; one-twentieth had graduated from four-year colleges or universities.[19] Teenagers by that time were staying in school much longer than their elders had, but still only 49 percent of 17-year-olds graduated from high school. Class and racial distinctions remained sharp. Most black children in the South struggled to learn in segregated and poorly financed institutions, taking instruction from teachers who commonly earned less than $600 per year.

The war did not improve things much, if at all. Thanks to wartime interruptions, a slightly smaller percentage of 17-year-olds (47.4 percent) graduated from high school in 1946 than had done so before the war. Some 350,000 teachers had left their jobs for the service or better jobs. Advocates of better schools complained that teachers earned less on the

17. James Patterson, *The Dread Disease: Cancer and Modern American Culture* (Cambridge, Mass., 1987), 141–44.
18. *Newsweek*, Aug. 13, 1945.
19. Richard Polenberg, *One Nation Divisible: Class, Race, and Ethnicity in the United States Since 1938* (New York, 1980), 20–21.

average than truck drivers, garbage collectors, or bartenders.[20] The United States spent a smaller percentage of its national income on schools than did either Great Britain or the Soviet Union. By 1945 teacher morale was reported to have hit an all-time low, and by 1946 teachers' strikes, heretofore scarcely imaginable, were beginning to break out.[21]

Some of this improved—and suddenly—after the war. A key to the change was passage of the GI Bill of Rights in 1944.[22] This was a remarkably broad piece of legislation, which not only offered veterans aid in purchasing housing, and loans to start businesses, but also provided monthly stipends for veterans who wanted help with educational costs. These stipends were not huge: $65 per month at first for single veterans, $90 for those with dependents, and a maximum of $500 per year for tuition and books. But they were a real inducement, especially for veterans with savings, and millions jumped at the opportunity. By 1956, when the programs ended, 7.8 million veterans, approximately 50 percent of all who had served, had taken part. A total of 2.2 million (97.1 percent of them men) went to colleges, 3.5 million to technical schools below the college level, and 700,000 to agricultural instruction on farms. The GI Bill spent $14.5 billion, a huge sum in those years, for educational benefits between 1944 and 1956.[23]

The GI Bill indeed promoted an educational boom. Colleges and universities were nearly swamped by the change; almost 497,000 Americans (329,000 of them men) received university degrees in the academic year 1949–50, compared to 216,500 in 1940. The influx jolted faculty and administrators, who had to reach out beyond the predominantly upper-middle-class young people whom they previously had served, to deal with older students, to offer married housing, to accelerate instruction, and to provide a range of more practical, career-oriented courses. The GI Bill was almost certainly worth it economically, helping millions of Americans to acquire skills and technical training, to move ahead in life, and therefore to return in income taxes the money advanced to them

20. These were complaints by advocates for teachers. Opponents rejoined that teachers did not work a full year and that they were often poorly trained themselves.
21. Diane Ravitch, *The Troubled Crusade: American Education, 1945–1980* (New York, 1983), 3–7, 324.
22. GI stood for Government Issue—clothing and equipment for military personnel in the war.
23. Ravitch, *Troubled Crusade*, 14; Joseph Goulden, *The Best Years, 1945–1950* (New York, 1976), 55–60.

by the government. It was the most significant development in the modern history of American education.

Predictably there were voices that dissented from the chorus of hosannas about educational progress. A few universities swelled to previously unimaginable size: as early as 1948, ten of them enrolled 20,000 students or more. These were no longer "villages with priests" but impersonal and bureaucratic "multi-versities."[24] Some of the science faculties at leading universities such as MIT and Stanford relied so heavily on military funding that it was fair to speak of them as part of a "military-industrial-academic complex."[25] *Time* magazine asked in 1946, "Is the military about to take over U.S. science lock, stock, and barrel, calling the tune for U.S. universities and signing up the best scientists for work fundamentally aimed at military results?"[26]

Critics also decried changes at the secondary level, especially what they thought was the anti-intellectual focus on non-academic "life adjustment" courses pushed by "progressive" educators. In Denver, Colorado, high school students took part in a unit on "What is expected of a boy on a date?" It dealt with matters such as "Do girls want to pet?" In Des Moines, Iowa, teachers presented "correct social usage" as part of a course in "Developing an Effective Personality." Advocates of such courses maintained that they taught socially acceptable behavior. Critics, who grew louder in the 1950s, retorted that schools were abandoning rigorous academic standards and "dumbing down" the curricula.[27]

It is hard to determine whether criticisms such as these applied to the majority of public schools in the 1940s. There has always been an extraordinary amount of uninformed opinion about what actually goes on in the widely varying classrooms of American public schools. Critics, however, were on target in complaining about educational inequality: educational spending, which depended mainly on local property taxes, varied enormously. Middle-class and upper-middle-class school districts spent much more money per student than did working-class or lower-class districts. Black communities, especially in the South, received far less money than white areas. Liberal interest groups such as the National Education Association stepped up pressure for federal aid that would reduce such in-

24. Quotes of Clark Kerr, head of University of California, Berkeley, in Ravitch, *Troubled Crusade*, 183.
25. Stuart Leslie, *The Military-Industrial-Academic Complex at M.I.T. and Stanford* (New York, 1993).
26. Goulden, *Best Years*, 265.
27. Ravitch, *Troubled Crusade*, 68.

equalities, but they ran into a number of obstacles—ideological, fiscal, racial, and otherwise—in Congress. No such bill passed until the 1960s.

These were valid complaints that rang true decades later. But more optimistic voices dominated public discussion about education in the early postwar years. Contemporary news stories rejoiced over the extraordinary growth in higher education and over the apparently broader popular support for schools. Per pupil spending for public education more than doubled between 1944 and 1950. Optimists were also happy about the steady rise in the number and percentage of teenagers who completed high school (up to 57.4 percent in 1950). They predicted—correctly, as it turned out—that these trends would continue: by 1970, 75.6 percent of 17-year-olds were graduating from high school, and nearly 48 percent of 18-year-olds were enrolled in some kind of college or university. If these optimists confused quantitative growth with qualitative improvement, that was understandable. Americans tend to do that. But their optimism was catching. The astonishing growth of education in the late 1940s (and thereafter) seemed yet another sign that the American dream was alive and well.

THESE WERE ABOVE ALL years of nearly unimaginable consumption of goods. As the historian Fred Siegel pointed out, "A good deal of what had once seemed science fiction became everyday life." In the five years after the war Americans were presented with such new items as the automatic car transmission, the electric clothes dryer, the long-playing record, the Polaroid camera, and the automatic garbage disposal unit. Rising numbers of people were buying vacuum cleaners, refrigerators, electric ranges, and freezers. Millions bought frozen foods, which were marketed widely for the first time. It was a booming new age of "wonder" fibers and plastics: nylon for clothing, cheap food wraps, light new Styrofoam containers, inexpensive vinyl floor coverings, and a wide array of plastic toys.[28] Between 1939 and 1948 clothing sales jumped three-fold, furniture four-fold, jewelry four-fold, liquor five-fold, household appliances, including TV, five-fold.

The postwar years were an automobile age on an unprecedented scale. New car sales in 1945 totaled 69,500. In 1946 they leaped to 2.1 million, in 1949 to 5.1 million, a figure that broke the record of 4.5 million set in 1929. Sales kept going up, to 6.7 million in 1950 and to 7.9 million in

28. Fred Siegel, *Troubled Journey: From Pearl Harbor to Ronald Reagan* (New York, 1984), 93.

1955. Very few of these cars (only 16,336 in 1950) were foreign-made, just 300 of them Volkswagens. Americans instead liked big, roomy automobiles with powerful eight-cylinder, 100-horsepower engines, sealed-beam headlights, radios, and heaters. Most of these came from the Big Three: General Motors, Ford, and Chrysler. Their cars were not cheap, considering family incomes. New Chevrolets and Fords started at around $1,300, around two-fifths of median family income at the time.[29] The buying nonetheless went on, with most Americans purchasing new cars at full list prices. Many buyers slipped extra cash to dealers in hopes of assuring prompt delivery. By 1950 there were 40.3 million cars registered to 39.9 million families.[30]

By then the boom in automobiles was threatening the financial health of downtown retail districts and hotels, reducing ridership of buses and urban mass transit, and badly damaging the already fragile railroad industry. It was doing wonders, however, for the oil business, gasoline stations, roadside hotels and restaurants, the trucking industry, and highway departments, which received massive federal aid by the late 1950s. The boom also started the runaway spread of suburban shopping centers; there were eight of these in 1946, and more than 4,000 by the late 1950s.[31] As in the 1920s, but on a much larger scale, automobiles not only hastened the adoption of new patterns of living; they also did much to stimulate the remarkable economic growth of the era.

The amazing growth of the automobile industry helped drive another powerful engine of economic growth: construction. Some of this building was in cities, especially on the rapidly growing West Coast, where millions of servicemen and their families had discovered the blessings of warm-weather living. Some was construction of apartment or urban public housing for the poor, which received a modest boost from congressional subsidization in 1949. Much of the building, however, was of single-family homes in suburbs. The number of new single-family homes started in 1944 had been 114,000. This rose to 937,000 in 1946 and to nearly 1.7 million in 1950.[32] Some 15 million housing units were constructed in the United States between 1945 and 1955, leading to historic

29. J. Ronald Oakley, *God's Country: America in the Fifties* (New York, 1986), 9.
30. Polenberg, *One Nation Divisible*, 130. There were also 8.6 million trucks registered in 1950.
31. Russell Jacoby, *The Last Intellectuals: American Culture in the Age of Academe* (New York, 1987), 40–42.
32. Jon Teaford, *The Twentieth-Century American City: Problem, Promise, and Reality* (Baltimore, 1986), 100–101.

highs in home-owning. By 1960, 60 percent of American families owned
their own homes, compared to slightly fewer than 50 percent in 1945.[33]

The government had much to do with this boom. Congress authorized
increased spending for home loans by the Federal Housing Administration (FHA) and the Veterans Administration (VA). The terms that these
agencies offered were extraordinarily generous, especially by contrast to
the policies of private bankers before the war. Then, prospective home-
buyers had often had to put down substantial down payments, of 50
percent or more, and to pay off mortgages in a short time, often ten years.
The FHA and VA revolutionized the old system, offering mortgages of up
to 90 percent of the value of the home and allowing up to thirty years to
pay them off. Interest rates on the mortgages were normally around 4 to
4.5 percent. The VA permitted many veterans to purchase homes with
virtually no down payment. By 1950 these two agencies were insuring 36
percent of all new non-farm mortgages; by 1955 they were handling 41
percent.[34]

Private builders tried eagerly to meet the apparently insatiable demand
for new housing. Some had vast ambitions, as did the developers of Park
Forest outside of Chicago: it was a planned city for 30,000 people. But no
builders were more renowned than William Levitt, an enterprising salesman, and his brother Alfred, an architect. They had been small-scale
builders until World War II, when they witnessed the fantastic gains in
productivity associated with assembly line techniques. After the war they
applied their expertise to the mass production of suburban housing. Their
lumber came from forests they had purchased; they also made their own
nails and cement. They hired non-union workers so as avoid strikes and
paid them well above prevailing wage scales. The Levitts formed these
workers into as many as twenty-seven teams, each with a carefully defined
task. At peak the teams, using pre-assembled materials, including plumbing systems, could put up a house in sixteen minutes. Their first "Levittown" sprang up on onetime potato fields thirty miles from New York City
near Hempstead, Long Island. This was then—and remained fifty years
later—the largest single housing development erected at one time by an
American builder. It had 17,000 homes, which accommodated more
than 80,000 residents. The Long Island Levittown also featured seven
village greens and shopping centers, fourteen playgrounds, nine swimming pools, two bowling alleys, and a town hall. The Levitts sold land at
cost for schools and donated space for churches and fire stations. Other

33. William O'Neill, *American High: The Years of Confidence, 1945–1960* (New
   York, 1986), 15–18.
34. Polenberg, *One Nation Divisible*, 131; O'Neill, *American High*, 17.

Levittowns, the biggest of which were near Bucks County, Pennsylvania, and Willingboro, New Jersey, were almost as vast.[35]

Levittowners got the buy of the century. The Levitts' early basic model (slightly larger and more expensive by the late 1940s) cost $7,990. This was a little more than two and one-half times the median family income at the time and $1,500 or so less than buyers would have had to pay elsewhere for comparable housing. Purchasers using the FHA or the VA had little problem making down payments, which tended to be around $90, or managing payments, which were in the range of $58 per month for twenty-five years. When houses were placed on sale, people lined up in advance the night before, as if waiting for World Series tickets. What they acquired for their patience was first of all a 60' by 100' lot on which were planted small fruit trees or evergreens. Lot sizes were nearly twice as big as those in most turn-of-the-century "streetcar suburbs."[36] Buyers also got a four-and-one-half-room, two-story house 25' by 30' in dimension and Cape Cod in style: kitchen, living room, two bedrooms, bathroom, and expandable attic. Later models had a carport. The houses were well constructed and generous for that time in their amenities. The Levitts laid the houses on concrete slabs, from which copper coils provided radiant central heating, and included built-in bookcases, closets, and eight-inch Bendix television sets, as well as refrigerators, stoves, fireplaces, and washing machines.

The Levitts understandably attracted reams of public notice, but they were only the most well-known of builders in a virtual paradise for suburban developers at the time. Suburbs had existed long before the 1940s. But they had spread relatively slowly until then, housing 17 percent of the population in 1920 and still only 20 percent in 1940. By 1960, 33 percent of Americans lived in what the Census Bureau defined as suburban areas. During the early postwar years some central cities in the West and Southwest grew substantially. But many other American cities expanded only a little, and some, including New York and Chicago, did not grow at all. Instead, Americans were streaming into the suburbs, some of whose populations increased at rates of 50 to 100 percent in these years.[37]

35. Polenberg, *One Nation Divisible*, 132–34; Teaford, *Twentieth-Century American City*, 102; Alexander Boulton, "The Buy of the Century," *American Heritage*, July/Aug. 1993, pp. 62–69.

36. Kenneth Jackson, *Crabgrass Frontier: The Suburbanization of the United States* (New York, 1985), 58–64.

37. Peter Muller, *Contemporary Sub/Urban America* (Englewood Cliffs, N.J., 1981), 51; Polenberg, *One Nation Divisible*, 128.

Such a spectacular development as suburbanization inevitably invited widespread attention, some of it hostile. Critics deplored the early suburban landscapes without tall trees and the absence of life on the streets. Suburbs, they said, were built for cars, not for the interaction of people: the front of many suburban houses was a garage. There was little place in most of the new suburbs for older folk, and intergenerational extended families and neighborhoods were thought to suffer.[38] Critics especially yearned for a greater mix in the new suburbs of residential and non-residential areas—little shops, corner drugstores, cafés and restaurants—that existed in more varied or colorful urban neighborhoods. They objected to the essential sameness of the houses: "ticky-tacky, all in a row," the folksinger Malvina Reynolds later said.

Opponents of suburban development disliked above all the forced conformity that they claimed to find in some of these places. Levittown rules at first required home-owners to mow their lawns every week, forbade the building of fences, and outlawed the hanging of wash outside on weekends. Lewis Mumford, a sharp-tongued critic, was especially appalled by the passing of "community" that he found in more pluralistic urban neighborhoods. Places like Levittown, he thought, were bland, conformist, inhuman nightmares. When he first saw Levittown in Long Island, he was said to have pronounced that it would become an "instant slum." In 1961 he denounced the

> multitude of uniform, unidentifiable houses lined up inflexibly, at uniform distances, on uniform roads, in a treeless communal waste, inhabited by people of the same class, the same income, the same age group, witnessing the same television performances, eating the same tasteless pre-fabricated foods, from the same freezers, conforming in every outward and inward respect to the same common mold.[39]

A more serious criticism deplored the racial exclusiveness of many suburban developments. This exclusiveness tightened a "white noose" around minorities in many American cities.[40] Blacks were barred from the Levittowns, and people who rented out their homes there were told to specify that the premises were not to be "used or occupied by any person other than members of the Caucasian race."[41] Other suburbs relied on

---

38. Jackson, *Crabgrass Frontier*, 62–64.
39. Lewis Mumford, *The City in History: Its Origins, Its Transformation, and Its Prospects* (New York, 1961), 486.
40. Muller, *Contemporary Sub/Urban America*, 69.
41. Boulton, "Buy of the Century."

zoning or on restrictive covenants, even after the Supreme Court had stripped such covenants of legal standing in the courts. Racist patterns set at that time persisted long after civil rights activity changed much else in America: the 1990 census showed that there were only 127 African-Americans in the Levittown, Long Island, population of more than 400,000. The United States, blacks then complained, had become a nation of "chocolate cities and vanilla suburbs."[42]

Critics of racial policies in Levittowns thoroughly angered the builder. "The Negroes in America," William Levitt explained,

> are trying to do in 400 years what the Jews in the world have not wholly accomplished in 600 years. As a Jew I have no room in my mind or heart for racial prejudice. But . . . I have come to know that if we sell one house to a Negro family, then 90 or 95 percent of our white customers will not buy into the community. That is their attitude, not ours. . . . As a company our position is simply this: we can solve a housing problem, or we can try to solve a racial problem, but we cannot combine the two.[43]

Levitt's defense was of course cold comfort to black people. But he was surely correct about white attitudes. Other developers and realtors, fearing to challenge similar attitudes, did as Levitt did. Racial segregation of American neighborhoods was virtually ubiquitous, especially in the new suburbs, and harder to change than any other aspect of race relations. It reflected a culturally powerful desire of people to have neighbors like themselves—similar in class as well as race. Most of the postwar suburban developments were indeed homogeneous economically, whether stable working-class like the Levittowns, middle-class like much of Park Forest, or upper-middle-class.

Critics were unfair to single out suburban developers as the main cause of subsequent urban decay, especially in those central city areas that became a dumping ground for poverty-stricken minority groups. True, suburbs drained many cities, especially in the East and Midwest, of middle-class people and of urban tax bases. Downtown merchants complained bitterly about the new suburban developments and malls. So did central city movie operators, who lost business as early as 1947: suburbs (and the contemporaneous baby boom, which kept people at home), not

42. Phrase attributed to the musical artist George Clinton, by Cornel West in *New York Times Book Review*, Aug. 2, 1992.

43. David Halberstam, *The Fifties* (New York, 1993), 141.

TV, killed the downtown movie palaces.[44] But residential movement has had a long history, and suburban growth would have occurred in the more prosperous postwar era with or without the FHA, the VA, or entrepreneurs like the Levitts.[45] This is because people who can afford to—and many more could in the postwar era—seem naturally to desire lots of space around them. Some moved to the Levittowns of the world to get more house for the money, others to escape urban problems, others to find new and better schools. Most sought greater privacy and autonomy; they were not "conformists."[46]

Levittowners, and others who moved in the millions to suburbs in the postwar years, pulled up stakes, finally, because they were seeking a more satisfying family life. Most were very glad that they did. As one perceptive scholar concluded, Levittown "permits most of its residents to be what they want to be—to center their lives around the home and the family, to be among neighbors whom they can trust, to find friends to share leisure hours, to participate in organizations that provide sociability and the opportunity to be of service to others."[47] A Long Island Levittowner remembered forty-seven years later the thrill of moving from a furnished apartment in Brooklyn. "We were proud," he recalled. "It was a wonderful community—and still is."[48]

The quest for a satisfying family life indeed proved strong during the postwar years. As early as 1940, marriage rates, which had been low throughout the hard times of the 1930s, moved upward.[49] They peaked in 1946, when 2.2 million couples said their vows. This was a record that stood for thirty-three years. Though the rate then tailed off a bit, it remained high into the early 1950s. Divorce rates, which had increased fairly steadily since 1900, exploded to record highs in 1945–46, dropped sharply thereafter, and remained low until the mid-1960s.[50] This was a

44. Douglas Gomery, "Who Killed Hollywood?," Wilson Quarterly (Summer 1991), 106–11.
45. Herbert Gans, The Levittowners: Ways of Life and Politics in a New Suburban Community, rev. ed. (New York, 1982), 408–13.
46. Ibid., 408–9.
47. Ibid., 413.
48. New York Times, Jan. 28, 1994.
49. Marriage rates are defined here, as in the census, as number of marriages per year per 1,000 unmarried females aged 15 or older. The rate was 73.0 in 1939, 82.8 in 1940, 93.0 in 1942, and 118.1 in 1946. Between 1947 and 1951 marriage rates ranged from 106.2 to 86.6, slowly declining thereafter to late 1930s levels by 1958.
50. Census definition of divorce rates, used here, is rate of divorces per year per 1,000 married females. It was 8.5 in 1939, 10.1 in 1942, 17.9 in 1946, 13.6 in 1947, 9.9 in 1951, and between 8.9 and 9.6 from 1953 through 1963.

striking, unanticipated reversal of what had seemed to be a long-range trend.

The baby boom that ensued was perhaps the most amazing social trend of the postwar era. Demographers, aware of long-term declines in fertility in the United States (and in other urban-industrial nations), thought that the relatively small cohort born in the 1920s and the very small cohort born in the Depression years would not produce a boom in the 1940s and 1950s. They therefore expected at most a short-term blip in child-bearing following the war.[51] But birth rates ("furlough babies") rose in 1942 and 1943. And then: the boom. In May 1946, nine months after V-J Day, births increased from a low in February of 206,387 to 233,452. In June they rose again, to 242,302. By October they numbered 339,499 and were occurring at a record rate. Landon Jones, historian of the boom, notes that "the cry of the baby was heard across the land."[52] By the end of the year an all-time high number of 3.4 million babies had been born, 20 percent more than in 1945. They came in time to abet the sales of a new book, one of the nation's great publishing success stories: *The Common Sense Book of Baby and Child Care* by Benjamin Spock, M.D.

The babies kept on arriving: 3.8 million in 1947, 3.9 million by 1952, and more than 4 million every year from 1954 through 1964, when the boom finally subsided. Birth rates—estimated total live births per 1,000 population—had ranged between 18.4 and 19.4 per year between 1932 and 1940. They rose to a postwar peak in 1947 of 26.6, the highest since 1921, stayed at 24.0 or higher from then until 1959, and were still at 21.0 in 1964, after which the rates dropped to 1930s levels. The total number of babies born between 1946 and 1964 was 76.4 million, or almost two-fifths of the population in 1964 of 192 million.

Historians and demographers have offered a range of explanations for this astonishing interruption of long-range trends. One hypothesis sees it as the result of a quest for "normalcy" immediately after the war. This explanation fails to account for the increase of the early 1940s or for the duration of it after 1946 into the early 1960s. A second theory points to wartime propaganda by the Office of War Information and other government agencies that called for Americans to build up the population. This explanation, however, tends to see people as tools and understates the depth of yearning by young adults for marriage and children. A third hypothesis accounts for the postwar boom by seeing it as part of a quest by Americans for psychological security (and by men to keep women at

51. Easterlin, *Birth and Fortune*, 48.
52. Jones, *Great Expectations*, 11.

home) amid the fears and tensions of the atomic, Cold War age. This view, too, is both facile and conspiratorial.[53]

The boom in fact happened mainly because of decisions made by two different groups of Americans in these years. The first, who played a major role in the immediate postwar boom, were older Americans who had been forced by the Depression and World War II to delay marriage and child-rearing. Fertility for women in this group increased greatly after 1945. The second group, which accounted for the remarkable duration of the boom, was composed of younger folk, in their late teens or early twenties in the late 1940s. These people were more likely to marry than young people had been in the 1930s; by 1960, 93 percent of women over 30 were or had been married, as opposed to 85 percent in 1940. They tended to marry younger; the average age of marriage for women dropped from 21.5 years in 1940 to a low of 20.1 in 1956. They were more likely to have children; 15 percent of married women had remained childless in the 1930s, compared to only 8 percent in the 1950s. And they gave birth earlier in their marriages; first children were born an average of thirteen months following matrimony by the late 1950s, as opposed to an average of two years in the 1930s.[54] The boom was not the result of parents having huge families, but rather of so many people deciding to marry young, to start a larger family quickly, and to have two, three, or four children in rapid succession.

Why did these groups act as they did? The behavior of the older parents was fairly easy to understand: deprived of "conventional" family life by the Depression, they sought to "catch up" in the 1940s. The younger cohorts, especially those who reached child-bearing age in the 1950s, often recalled no such deprivations, yet they seemed to yearn for marriage and children. This yearning appears to have been especially deep among men; women, after all, were the ones who shouldered most of the burden of child-rearing. But most women, too, seemed content enough with their decisions at that time. Many of these young adults may have have sought familial intimacy in order to cope with the pressures of an increasingly complex and bureaucratic world.

53. Arlene Skolnick, *Embattled Paradise: The American Family in an Age of Uncertainty* (New York, 1991), 64–68; Randall Collins and Scott Cottrane, *Sociology of Marriage and the Family*, 3d ed. (Chicago, 1991), 164–66; Elaine Tyler May, *Homeward Bound: American Families in the Cold War Era* (New York, 1988); and May, "Cold War—Warm Hearth: Politics and the Family in Postwar America," in Steve Fraser and Gary Gerstle, eds., *The Rise and Fall of the New Deal Order, 1930–1980* (Princeton, 1989), 153–81.
54. Jones, *Great Expectations*, 23–35.

The most satisfactory explanation of the boom is offered by Jones, who highlights the "great expectations" of the younger generation at the time. Not only the veterans but also their younger brothers and sisters maturing in the next few years developed rising aspirations amid the increasingly prosperous economic climate of the 1940s and 1950s. Most of them knew that they were better off, relatively speaking, than their parents had been at their age. They sensed that they could afford to marry, buy a house, start a family, and educate their children.[55] In this way as in so many others, the health of economy—as well as optimistic perceptions of continuing prosperity—drove social change in postwar America.

No economic interpretation, of course, can entirely explain why large numbers of people decide to marry and have families. Black women, most of them very poor, continued to have considerably higher birth rates (averaging around 34 from the late 1940s to 1960) than white women. But the increasingly exuberant spirit of the age clearly was important. The biggest jumps in fertility occurred among well-educated white women with medium to high incomes. Demographers point out also that the only societies experiencing baby booms at that time, aside from the United States, were other dynamic, fairly prosperous nations with comparable moods of "high expectations": Canada, New Zealand, and Australia.[56]

The baby boom was in some ways cause as well as consequence of prosperity in the postwar years. The rise in births unleashed a dynamic "juvenile market," especially for producers of toys, candy, gum, records, children's clothes, washing machines, lawn and porch furniture, televisions, and all manner of household "labor-saving" devices. It helped drive the surge in suburban home-building and automobile-buying, as well as a gradual boom in school construction.[57] Diapers alone were a $50 million business by 1957. The juvenile market peaked around that time at approximately $33 billion a year.[58] *Life* magazine in 1958 ran a cover story under a banner headline, KIDS: BUILT-IN RECESSION CURE—HOW 4,000,000 A YEAR MAKE MILLIONS IN BUSINESS. The story estimated that an infant was a prodigious customer, "a potential market for $800 worth of products." While in the hospital, the baby had "already rung up $450 in medical expenses." Four-year-olds represented a "backlog of business orders that will take two decades to fulfill."[59]

By that time the phrase "baby boom generation" was common. De-

55. Easterlin, *Birth and Fortune*, 39–53; Skolnick, *Embattled Paradise*, 64–67.
56. Jones, *Great Expectations*, 21.
57. Jackson, *Crabgrass Frontier*, 231–45.
58. Jones, *Great Expectations*, 38.
59. Ibid., 37.

mographers and journalists have subsequently been fascinated by it, seeing the boom as a "cutting edge" of social change, a "Goliath generation
stumbling awkwardly into the future," a "pig in a python" that bulged
through many decades thereafter.[60] The phrase is in some ways fairly
clumsy. To generalize about a "generation," as if sharp differences of
class, race, gender, and region did not exist within an age cohort, is
foolish. Were people born in 1946 so sharply distinguishable (except in
numbers) from people born in 1945? Moreover, the span of years from
1946 to 1964 obviously covers a good deal of ground. Early "baby
boomers" had very different life experiences from later ones. Americans
born in 1946 confronted the turmoil of the 1960s and of the Vietnam
War. Most of them entered a job market that was attractive in part because
of prosperity, which persisted until the early 1970s, and in part because
there was a relatively small cohort of older people ahead of them.
Boomers born in 1956, by contrast, entered the world of work in the mid-
to late 1970s. These were recession years, made all the more traumatic for
job-seekers because millions of older boomers ahead of them clogged up
the employment market.

It is also clear, mainly in retrospect, that the baby boom years were
hardly the untroubled years of domesticity that television shows such as
"The Adventures of Ozzie and Harriet" made them out to be. As the
increase in female employment indicated, millions of married women
could not afford the luxury—or endure the routine—of staying at home
full-time. Millions of children, with both father and mother off at work,
saw less of their parents than did the happy young boys of "Leave It to
Beaver," another TV celebration of postwar domestic life. Families,
moreover, continued to encounter not only the unavoidable, day-to-day
stresses that have characterized all households throughout history but also
the strains that accompanied the scrambling quest for security, advancement, and consumption in the postwar age. Meanwhile, the sexual revolution was quietly but steadily advancing, inciting innumerable conflicts
between young people and between them and their parents. By the 1950s
newspapers and magazines were carrying headlines about rising juvenile
delinquency, "peer group pressures," and even a "youth culture" (though
consisting mostly at that time of children born *before* 1946) that threatened to reject the older generation. Many of these stories rested on dubious and often exaggerated generalizations; rates of juvenile delinquency, for instance, were not on the rise. But some of the stories, such as

60. Ibid., 2.

many of those concerning sexual behavior, did not. The articles suggested trends that became obvious when the baby boomers came of age in the 1960s: family life in the 1940s and 1950s was considerably more complicated and less idyllic than nostalgia-mongers have cared to admit.[61]

Still, these problems became widely appreciated only later. In the late 1940s the baby boom excited contemporaries. Resting in part on postwar affluence, it promoted still greater prosperity and fed the huge rise of suburbia. And the divorce rate did decline, suggesting to many optimists that the durable two-parent family with children was a norm that would become stronger than ever. Although it later became clear that these developments were an anomalous interlude amid more lasting historical social trends that resumed in the 1960s (later marriages, lower birth rates, smaller families, ever-higher divorce rates), that was far from obvious before 1950, or even 1960. On the contrary, the baby boom symbolized a broader "boom" mentality of many younger Americans, especially whites and the ever-larger numbers of people moving upward into the middle classes. They were developing expectations that grew grander and grander over time.

61. Skolnick, *Embattled Paradise*, 51–52, 169–71; Judith Stacy, "Backward Toward the Postmodern Family," in Wolfe, ed., *America at Century's End*, 17–34.

4

Grand Expectations
About the World

World War II did more than usher in unparalleled prosperity for the United States. It transformed America's foreign relations. The war devastated the Axis nations, which took years to recover. It also savaged America's allies, including the Soviet Union, which lost an estimated 25 million people during six years of fighting. Alone of the world's great powers the United States emerged immeasurably stronger, both absolutely and relatively, from the carnage. In a new balance of power it was a colossus on the international stage.

Few Americans at the end of the war fully understood how vast a role the United States would play on this stage in the future. Top policymakers at first did not talk much about a *Pax Americana*, about "worldwide Communist expansion," or even about the "American Century" that Henry Luce had anticipated in 1941. But it was obvious that advances in air power, rocketry, and atomic weapons ended America's history of relatively free security, and that the United States might have to fill at least part of the postwar power vacuum. Most political leaders recognized that they lived on an interconnected planet, in which a spark in one corner of the world could ignite explosions in many other corners. America, Secretary of War Stimson observed, could "never again be an island to itself. No private program and no public policy, in any sector of our national life, can now escape from the compelling fact that if it is not framed with reference to the world, it is framed with perfect futility."[1]

1. William Leuchtenburg, "Franklin D. Roosevelt," in Fred Greenstein, ed., *Leadership in the Modern Presidency* (Cambridge, Mass., 1988), 10.

82

After 1945 this recognition dominated official American approaches to the world. But it was a recognition that arrived somewhat rudely forced on a reluctant nation by the dramatic events of the era. The change was indeed rapid. As late as 1938 Romania had supported a larger army than did the United States. Until the war America had only a few cryptographers and no national intelligence service.[2] By 1945 leaders in the realm of foreign policy knew, like Stimson, that major changes were in store and that the national interests of the United States might expand almost immeasurably. But in 1945 they were unsure what these interests were or how to defend them.

Then and in later years these leaders sometimes felt insecure. That may seem odd, given America's awesomely preponderant power after the war. Indeed, policy-makers did not fear military attack in the late 1940s. After all, no nation then had the airplanes to rain bombs on the continental United States, and none yet had the Bomb. But in an age of formidable military and technological prowess these things could change quickly. And *political* pressures, especially from the Soviet Union, increasingly unsettled the nerves of high officials in the West.

In 1945 American leaders worried especially about the will of the citizenry to support major involvement of the country in overseas controversies. As it turned out, public opinion shifted decisively toward acceptance of substantial American engagement with the rest of the world: the people, following their leaders, developed large expectations about the role of American foreign policy. But this shift in opinion was hard to predict before 1947, and the unease of top officials pervaded the conduct of American foreign relations at the time.[3]

Worries about the Soviet Union prompted by far the greatest unease, not only in the United States but also among America's Western allies. Even before the end of the war, the alliance between the United States and the Soviet Union, the world's number two power, had grown tense. By early 1946 these tensions had badly strained Soviet-American relations, and by 1947 the Cold War, as it was then called, had arrived—to dominate international politics for more than forty years. Diplomats of both nations struggled to steer a safe course through the storms unleashed by the war, but they often lacked the charts—or even the compass—to

2. Ernest May, "Cold War and Defense," in Keith Nelson and Robert Haycock, eds., *The Cold War and Defense* (New York, 1990), 9.
3. John Gaddis, "The Insecurities of Victory: The United States and the Perception of the Soviet Threat After World War II," in Michael Lacey, ed., *The Truman Presidency* (Washington, 1989), 235–72.

guide them reliably. Insecure, often confused, they frequently misperceived the course of the other side. Several times they almost collided. Given the megatons of weaponry that they ultimately overloaded themselves with, it is amazing that they did not destroy each other, in the process raining catastrophe on the rest of the world.[4]

Notwithstanding these feelings of insecurity, which were especially obvious in the immediate aftermath of the war, the leaders of America's postwar foreign policy—a group that came to be known as the Establishment—developed a self-confidence that occasionally bordered on self-righteousness. Their rising certitude rested on the belief that the Soviet Union was a dangerous foe, that the United States had large interests in the world, and that it must assert these interests strongly; "appeasement" led inevitably to disgrace and defeat. Leaders of the Establishment did not always define these interests clearly: where, indeed, must the United States risk war? But they were confident that America possessed the economic and military resources to outlast and ultimately to overcome a host of potential enemies. In their approach to international relations they developed very grand expectations that they managed to fashion into official American policy.

To say that the Cold War stemmed in part from international insecurities is to make the obvious point that both sides followed nervous, sometimes wrong-headed courses in the postwar years. Most American political leaders in the late 1940s, however, would have hotly rejected such a non-judgmental view of the Cold War. So, too, have many among the host of scholars who have pored over the history of the early Cold War years. Until the early 1960s most American writers tended to blame the Soviets. Then, under the influence of frightening events such as the missile crisis over Cuba in 1962, and the Vietnam War, revisionists sharply challenged this patriotic view, either assigning blame to both sides or finding the United States the more "guilty" of the two.[5] "Post-revisionists" broadened the scope of inquiry and tried to strike a

4. Daniel Yergin, *Shattered Peace: The Origins of the Cold War and the National Security State* (Boston, 1977); John Gaddis, *The United States and the Origins of the Cold War, 1941–1947* (New York, 1972).
5. Including Thomas Paterson, ed., *Cold War Critics: Alternatives to American Foreign Policy in the Truman Years* (Chicago, 1971); Walter La Feber, *America, Russia, and the Cold War, 1945–1966* (New York, 1967); and Richard Freeland, *Foreign Policy, Domestic Politics, and Internal Security, 1946–1948* (New York, 1970).

balance between polar interpretations. Although the debates seemed relatively tame by the 1990s—when the Cold War at last subsided—they were
by no means dead or irrelevant, and they may be briefly summarized
here.[6]

Those who blame the Soviet Union for the Cold War make several
assertions. Josef Stalin, they note, did much to bring on World War II
when he signed a non-aggression pact with Germany in 1939. The two
nations followed by cynically carving up Poland. The Soviet Union also
seized the Baltic states of Estonia, Lithuania, and Latvia, and parts of
Finland and Romania as well. Critics of Stalin emphasize correctly that
he was not only a ruthless dictator but also in many ways a barbarian.
Deeply suspicious of rivals for power, he purged associates, executing
some of them in the late 1930s after widely publicized show trials. He put
in place the infamous Gulag Archipelago, a system of slave labor camps
that imprisoned all manner of people thought to be dangerous to his
regime.[7]

Stalin's behavior during the war also appalled his critics. When German soldiers invaded the Soviet Union in 1941, they found in the Katyn
Forest the bodies of thousands of Polish officers. They had been murdered
under orders from Stalin. Four years later, as rapidly driving Russian
troops stormed to the edge of Warsaw, Polish underground forces arose in
open rebellion against the Nazis. Stalin, however, ordered his tanks to
remain on the outskirts. The Germans then ruthlessly crushed the uprising. The full extent of Stalin's crimes was not widely recognized in 1945,
but enough was known for informed people to stamp Stalin as a cruel and
often merciless tyrant.

The war over, Stalin's critics say, he remained true to form. Although
apparently agreeing at the Yalta conference in early 1945 to the holding of
free elections in Soviet-occupied eastern Europe—he signed a Declaration on Liberated Europe—he clamped down on democratic elements in
much of the region. His repression of Poland, over which World War II
had begun in 1939, outraged many people, including the millions of
Polish-Americans with friends and relatives in the old country. Seeking to
stamp out ethnic differences in the Soviet Union, Stalin forced from their
homes hundreds of thousands of non-Russian people. He tried to close off

6. John Gaddis, "The Emerging Post-Revisionist Synthesis on the Origins of the Cold
   War," *Diplomatic History*, 7 (Summer 1983), 171–90.
7. Stephen Whitfield, *The Culture of the Cold War* (Baltimore, 1991), 2; Michael
   Beschloss, *The Crisis Years: Kennedy and Khrushchev, 1960–1963* (New York,
   1991), 322.

Soviet borders, as if fearing that any contact with the outside world would undermine his regime. Stalin's frightening behavior, some diplomats speculated, was consistent with a diagnosis of clinical paranoia.[8]

Western diplomats worried especially about Stalin's foreign policies. He seemed determined not only to fasten iron control on eastern Europe, including the Soviet zone of eastern Germany, but also to widen Soviet influence in Manchuria, Iran, Turkey, and the Dardanelles. Apparently distrusting everyone, he displayed little interest in diplomacy—that is, in the give-and-take of negotiation. How could any other nation work with such a man? Averell Harriman, America's ambassador to the Soviet Union, sounded an alarm in 1944. "Unless we take issue with the present policy," he wrote in September, "there is every indication that the Soviet Union will become a world bully wherever their interests are involved." When Stalin failed to permit free elections in Poland and Romania, he tested the considerable patience of Roosevelt, who had tried hard to accommodate his wartime ally. "Averell is right," FDR complained three weeks before his death in April 1945. "We can't do business with Stalin. He has broken every one of the promises he made at Yalta."[9] At the same time, however, FDR refused to break with Stalin: it remains impossible to know what he would have done about the Soviet Union had he lived.

Why Stalin acted as he did remains a source of debate among historians and other Kremlinologists. Some think his hostility to the West stemmed primarily from the Marxist-Leninist ideology that he and fellow Soviet leaders shared. This held that history moved inexorably toward the revolutionary overthrow of capitalism and the establishment of Communism throughout the world. Such a view permitted the possibility of transitory peaceful coexistence with capitalistic powers; after all, the crash of capitalism was part of the design of history. Stalin, therefore, did not welcome the idea of war with a nation such as the United States. But Communists did not believe in a passive policy that would enshrine the status quo. History, they thought, could—must—be moved along, and the Soviet Union, as the leading force in the making of this history, must compete and stay ahead of rivals. Until the late 1970s, long after Stalin's death in 1953, Soviet leaders refused to accept the notion of military parity with the United States.[10]

8. William Chafe, *The Unfinished Journey: America Since World War II* (New York, 1991), 61.
9. Gaddis, "Insecurities," 249.
10. Raymond Garthoff, *Detente and Confrontation: American-Soviet Relations from Nixon to Reagan* (Washington, 1985), 20, 38–41.

Western critics in the mid-1940s pointed to Stalin's military policies as evidence of his driven need to compete. Having built the world's biggest army in his efforts to stop Hitler, Stalin maintained much of it after the war was over. Estimates placed its size at 3 million men. The Soviets, indeed, had an enormous manpower advantage over Western occupation forces in Europe during the postwar years. Whenever they wanted to, alarmists said, Communists could overrun the Continent. At the same time Stalin made sure that the West knew of his highest military priority: to build up Soviet offensive capabilities, especially nuclear weapons, submarines, and long-range (3,000-mile) bombers. These were modeled on the American B-29, three of which had crashed in the Soviet Union following wartime raids on Japan.[11] Nothing that Stalin did, fearful Americans came to think, more clearly demonstrated the unrelentingly aggressive thrust of Communism, including the use of using military power to promote worldwide revolution.

Other critics of Stalin doubt, probably correctly, that ideological considerations alone do much to explain Stalin's behavior in the late 1940s. They think the Soviets mainly pursued imperial policies similar to those of the tsars.[12] Stalin, they say, did not seek worldwide revolution so much as control of adjacent areas that posed a threat to Russian national security. Chief among these, of course, were eastern Europe, Iran, and Turkey, all of which had long feared Russia. Revisionists who seek to understand Stalin's own concerns also emphasize that he was highly insecure, that Soviet pressure on adjoining nations was largely defensive in nature. All states, after all, yearn to protect themselves against potentially hostile neighbors and to fill power vacuums into which enemies might tread. Stalin's critics reply that he nonetheless went about his business in an especially uncompromising and provocative way. George Kennan, a leading diplomatic expert on the Soviets, explained in July 1946, "Security is probably their [the Soviets'] basic motive, but they are so anxious and suspicious about it that the objective results are much the same as if the motive were aggression, indefinite expansion. They evidently seek to weaken all centers of power they cannot dominate, in order to reduce the danger from any possible rival."[13]

11. May, "Cold War and Defense," 54–61.
12. Vojtech Mastny, *Russia's Road to the Cold War: Diplomacy, War, and the Politics of Communism, 1941–1945* (New York, 1979); Gaddis, "Insecurities," 268–70; "Comments" by Gaddis and by Bruce Kuniholm, in *American Historical Review*, 89 (April 1984), 385–90.
13. Gaddis, "Insecurities," 261.

Stalin's critics, then and later, rest their case finally on the fact that he was a powerful, dour, often brutal dictator. The very fact of this dictatorship deeply offended Americans, who cherished their freedoms, who sympathized with the oppressed masses of eastern Europe, and who earnestly hoped to promote the spread of democracy. "Totalitarian" states, they believed, habitually relied on force to get their way in world affairs. "It is not Communism but Totalitarianism which is the potential threat," *New York Times* publisher Arthur Hays Sulzberger said. "Only people who have a Bill of Rights are not the potential enemies of other people." President Truman agreed, noting privately in November 1946, "Really there is no difference between the government which Mr. Molotov represents and the one the Czar represented—or the one Hitler spoke for."[14]

Truman's analogy with Hitler would have made eminent sense to many fellow citizens at the time. Why shed the blood of Americans to rid the world of one dictator, only to have another tyrant take over? The analogy went still further, for many people blamed "appeasement" in the 1930s for the rise in Nazi power that led to the war. This must not happen again. "No more Munichs" was virtually a battle cry to alarmed and anxious Americans throughout the postwar era of conflict with the Soviet Union.

American anger at Soviet dictatorship went well beyond fears of appeasement, great though these were in 1945 and thereafter. It was also righteous and passionate. A specially religious people, many Americans approached foreign policies in a highly moralistic way. This was not only because Communism embraced atheism, though that mattered, especially to Catholics and other religiously devout citizens. It was also because many Americans believed so fervently in the rightness of their political institutions and the meaning of their history. America, as the Puritans had said, was a City on a Hill, a special place that God had set aside for the redemption of people. It followed that the United States had a God-given duty—a Manifest Destiny, it had been called in the nineteenth century—to spread the blessings of democracy to the oppressed throughout the world. The power of this messianic feeling lent a special urgency—indeed an apocalyptic tone—to American Cold War diplomacy as well as to repression of Communists at home.[15]

Despite these sources of tension, the United States did not dare take too hard a line against the Soviet Union in the immediate aftermath of the war. The Truman administration (and its successors) felt obliged to acquiesce in the Soviet oppression of eastern Europe. Millions of people there

14. Ibid., 257.
15. Eric Hobsbawm, *Age of Extremes: The Short Twentieth Century, 1914–1991* (London, 1994), 236.

remained in thrall for more than forty years. Still, virtually all American foreign policy leaders—from the Truman administration through the 1980s—expressed their anger and outrage at what they considered to be grossly excessive Soviet behavior. All believed that the Soviets must not be allowed to go farther. The alternative, appeasement, would encourage aggression and World War III.

Revisionists make several points in rejoinder to defenders of American policies.[16] They highlight first the entirely understandable fear and hatred that Russians felt concerning Germany. In 1914 and again in 1941 Germany had swept across northern Europe to invade the Motherland. World War II had ended with the destruction of 1,700 Russian towns, 31,000 factories, and 100,000 collective farms. This was staggering devastation, especially in contrast to the relatively benign experience of the United States, which had no fighting on its soil. No wonder Stalin stripped eastern Germany of its industrial potential in 1945 and insisted on dominating East Germany in later years. No wonder, too, that he insisted on controlling his east European neighbors, especially Poland, through whose flat and accommodating terrain the Nazi armies had torn just four years earlier.[17]

Many revisionists stress three other arguments. First, there was little that the West could do about Soviet domination of eastern Europe: Soviet armies, having driven to the heart of Germany during the war, were in control of the area, just as Western armies were in control of western Europe, and could not be dislodged. The division of Europe was yet another powerful legacy of the war, one which statesmen might deplore but which "realists" should have had the sense to live with. Britain's Winston Churchill had done so in 1944, signing an agreement with Stalin that ceded the preeminence of Soviet interests in Bulgaria and Romania. Just as Americans had their own "sphere of interest," including all of the Western Hemisphere, so, too, the oft-invaded Russians desired theirs.[18]

Many revisionists stress a second point: that Stalin's foreign policies

16. Bruce Cumings, "Revising Postrevisionism, or, The Poverty of Theory in Diplomatic History," *Diplomatic History*, 17 (Fall 1993), 539–69; Gabriel Kolko, *The Limits of Power: The World and United States Foreign Policy, 1945–1954* (New York, 1972); Lloyd Gardner, *Architects of Illusion: Men and Ideas in American Foreign Policy, 1941–1949* (Chicago, 1970).
17. A well-argued summation of many revisionist points is Melvyn Leffler, "Reply," *American Historical Review*, 89 (April 1984), 391–400.
18. The American sphere, of course, was much more consensual; the Soviets imposed theirs.

were more flexible than anti-Communist Americans could admit, either at the time or later. There was truth—some—to this argument. Brutal to opponents at home, Stalin was more cautious, conservative, and defensive abroad. This was in part because he had to focus on serious economic and ethnic problems at home. In 1945 Stalin did demobilize some of his armed forces. He acquiesced until 1948 in a coalition government in Czechoslovakia. Rebellious Finland managed to secure some autonomy. Stalin gave little aid to Communist rebels in Greece, who were ultimately defeated. His pressure on Iran and Turkey, while frightening to government leaders there, was intermittent; when the United States protested strongly in 1946, he backed off. Stalin lent little support, either moral or military, to Communist rebels under Mao Tse-tung in China. Instead, he formally recognized Mao's bitter enemy, Chiang Kai-shek. The sum total of these policies suggests that Stalin did not hold strongly, if at all, to Leninist doctrines of worldwide Communist revolution.

Critics of American hard-line reactions stress finally that the policies of the United States magnified Stalin's already heightened sense of insecurity. During the war Roosevelt had delayed opening up a second front in western Europe until 1944, thereby forcing Russian soldiers to bear the brunt of the fighting. This was probably a sensible military decision; to have moved earlier on Normandy might have been disastrous to Allied forces. But the delay fed Stalin's already deep suspicions. The United States and Britain, moreover, had refused to share their scientific work on atomic weaponry with the Soviet Union—or even to tell the Soviet government about it. When the European war ended, the Truman administration abruptly cut off lend-lease shipments to the Soviet Union and refused to extend a loan that Stalin urgently needed. To the Soviets, highly suspicious of capitalist behavior, the United States seemed to be ganging up with nations like Britain and France to develop an empire in the West.

In trying to make sense of these often angry debates about the origins of the Cold War, one must understand the situation as it existed in 1945. This means returning to the starting point: World War II left a new and highly unsettled world that bred insecurity on all sides. The United States and the Soviet Union, by far the strongest powers in the world, suddenly found themselves face-to-face. Dissimilar ideologically and politically, the two nations had been especially cold to each other since the Bolshevik Revolution in 1917, and they had different geopolitical concerns in 1945. Conflict between the two sides—a Cold War—was therefore unavoidable.

That this conflict could have been managed a little less dangerously is true. America's leaders frequently whipped up Cold War fears that were grossly exaggerated, thereby frightening its allies on occasion and deepening divisions at home. Soviet officials, too, often acted provocatively. Whether the Cold War could have been managed *much* less dangerously, however, is doubtful given the often crude diplomacy of Stalin and his successors and given the refusal of American policy-makers to retreat from their grand expectations about the nature of the postwar world.[19]

MANY OF FRANKLIN D. ROOSEVELT'S opponents always thought he was something of a lightweight: charming, buoyantly optimistic, politically deft, but intellectually soft. Critics of his diplomacy see the same traits. The British politician and diplomatist Anthony Eden wrote later that FDR knew a good deal of history and geography, but his conclusions therefrom "were alarming in their cheerful fecklessness. He seemed to see himself disposing of the fate of many lands, Allied no less than the enemy. He did this with so much grace that it was not easy to dissent. Yet it was too like a conjurer, skillfully juggling with balls of dynamite, whose nature he failed to understand."[20]

Roosevelt's detractors especially lament what they think was his credulous attitude toward Soviet behavior during the war. After meeting Stalin for the first time at the Teheran conference in late 1943, he told the American people, "I got along fine with Marshal Stalin . . . I believe he is truly representative of the heart and soul of Russia; and I believe that we are going to get along very well with him and the Russian people—very well indeed."[21] In March 1944 Roosevelt dismissed the thought that the Soviets would be aggressive after the war: "I personally don't think there's anything in it. They have got a large enough 'hunk of bread' right in Russia to keep them busy for a great many years to come without taking on any more headaches."[22]

Optimistic statements like this, of course, could be expected from a leader who needed reliable allies during the war. What else could he have said? Moreover, many Americans did admire the courage of the Russian people. *Life* magazine, a Luce publication, declared that the Russians

19. Jacob Heilbrun, "Who Is to Blame for the Cold War?," *New Republic*, Aug. 15, 1994, pp. 31–38.
20. Cited in Frank Freidel, *Frankin D. Roosevelt: A Rendezvous with Destiny* (Boston, 1990), 466.
21. Chafe, *Unfinished Journey*, 41.
22. Gaddis, "Insecurities," 243.

were "one hell of a people . . . [who] to a remarkable degree . . . look like Americans, dress like Americans, and think like Americans."[23] Other Americans saw no good alternative to working with Stalin. Max Lerner, a liberal journalist, said in 1943, "The war cannot be won unless America and Russia win it together. The peace cannot be organized unless America and Russia organize it together."[24]

This was essentially FDR's attitude. He was indeed a bit feckless, especially in expecting that his own personal charm could forge a strong personal bond between Stalin and himself. He worked at that project with little success both at Teheran and at Yalta. But Roosevelt was hardly a naive idealist. In seeking Soviet-American cooperation after the war he tended to think that Stalin was moved less by ideological passions than by considerations of national interest. He refused, therefore, to worry much about Communism, which he thought had little appeal in the West. If Stalin caused difficulties, the United States had carrots and sticks. Overoptimistically, FDR hoped that the threat to withhold economic aid could keep the Soviet Union in line.

In other ways, too, Roosevelt eschewed flights of idealism in his approach to the postwar order. Although he supported creation of the United Nations, established after his death in 1945, he did not expect it to resolve major controversies. He hoped instead that the United States and the Soviet Union, along with China and Great Britain, would act as "Four Policemen" to secure a postwar peace.[25] Here, too, Roosevelt was overoptimistic, especially about China, which for many years was far too divided to do any effective policing. But he was surely correct about a major geopolitical reality: without cooperation between the world's strongest powers, much of the heroism of World War II might be wasted.

This is not to argue, as some have, that FDR's death in April 1945 prevented the United States from working out a better relationship with the Soviet Union. Given the array of formidable problems emanating from the war, it is doubtful that one person, however charming or wise, could have made a great difference. Moreover, in his foreign policies (as in his New Deal at home) FDR often acted deviously, confusing not only the American people but also advisers who tried to apprehend his think-

23. Fred Siegel, *Troubled Journey: From Pearl Harbor to Ronald Reagan* (New York, 1984), 13.
24. Alonzo Hamby, *Beyond the New Deal: Harry S. Truman and American Liberalism* (New York, 1973), 17.
25. John Gaddis, *Strategies of Containment: A Critical Appraisal of Postwar American National Security Policy* (New York, 1982), 5–13.

ing. His neglect of Truman in early 1945, one of his greatest failings, compounded the difficulties that arose later in the year. He told him virtually nothing about his thinking and kept him totally in the dark about the Bomb.

Roosevelt also misled the American people, largely hiding from them the growing strains in Soviet-American relations that alarmed Harriman and others in early 1945. Nowhere is this more clear than in his glowing public report in February 1945 on the Yalta conference. The Allied leaders there could not agree on many matters, including postwar arrangements for Germany. They postponed decisions and awaited further developments. The Declaration on Liberated Europe, they recognized, was hardly a ringing endorsement of democracy. It committed the powers only to *consult* on ways to help democracy develop among "liberated" people. Admiral William Leahy, a key military adviser, complained to FDR that the Declaration was "so elastic that the Russians can stretch it all the way from Yalta to Washington without ever technically breaking it."[26] FDR did not disagree; he knew Leahy was right. But he never let that be known to Truman, the Congress, or to the American people, most of whom assumed that the Declaration was a Soviet commitment. When the Russians "broke" it in the next few months, many Americans were stunned and angry. Among them was Truman, who berated the Soviets in the probably mistaken assumption that Roosevelt would have done the same.

ON THE AFTERNOON of April 12, 1945, when Roosevelt died suddenly in Georgia, Harry S. Truman left the Senate chamber, where he had presided as Vice-President, and headed for the office of his old friend Sam Rayburn, Speaker of the House. He intended to have a drink with a few other Democratic leaders, who frequently met to "strike a blow for liberty." When he arrived, Rayburn told him that White House aide Stephen Early had phoned and wanted him to call back. Truman did so and was asked to come to the White House. There he was taken to see Eleanor Roosevelt. "Harry," she said, "the President is dead." Truman fought to collect his emotions, then asked, "Is there anything I can do for you?" Mrs. Roosevelt responded, "Is there anything *we* can do for *you*? For you are the one in trouble now."[27]

Truman was indeed in trouble, for contemporaries struggled to cope

26. Chafe, *Unfinished Journey*, 47.
27. Harry S. Truman, *Memoirs*, Vol. 1, *Year of Decisions* (Garden City, N.Y., 1955), 5.

with the shock of change in presidential leadership. Millions of Americans had come to think that Roosevelt, inaugurated in January for an unprecedented fourth term as President, was the only leader for the country. Many had little knowledge of Truman, a sixty-year-old party regular who had secured the vice-presidency as a compromise choice following unusually convoluted, last-minute back-room politicking at the 1944 Democratic convention. Those who did know him included prominent Democratic liberals who associated Truman with the malodorous political machine of Thomas Pendergast of Kansas City. David Lilienthal, director of the Tennessee Valley Authority, confessed privately that he felt "consternation at the thought of that Throttlebottom, Truman." Max Lerner added, "Can a man who has been associated with the Pendergast machine be able to keep the panting politicians and bosses out of the gravy?"[28]

Lerner and others exaggerated Truman's association with the corruption in Kansas City: the new President seems personally to have steered clear of it. Furthermore, Truman was hardly an unknown quantity to people who followed national politics. From 1935 through 1944 he had been a senator from Missouri. During World War II he impressed observers with his fair-minded leadership of a special Senate Committee to Investigate the National Defense Program. Close associates found him direct, plain-spoken, and refreshingly modest. The day after Roosevelt's death he was honest to reporters about his feelings, confessing that he felt as if "the moon, the stars, and all the planets had fallen on me." Years later he recalled, "I was plenty scared, but, of course, I didn't let anybody see it."[29]

In coping with his fears Truman had models from a lifetime of reading military history and biography. History for him was largely the accomplishments of strong and honorable leaders: Cincinnatus, the Roman warrior who supposedly preserved the Roman state and then laid down his arms; Cato, a model of virtue; and George Washington, a patriot who led the country without being caught up in the quest for personal power.[30] People who came in contact with Truman over the years were impressed with his interest in history and geopolitics and his reverence for the institution of the presidency. His idea of decisive presidential lead-

28. Hamby, *Beyond the New Deal*, 54–55.
29. Robert Griffith, "Forging America's Postwar Order: Domestic Politics and Political Economy in the Age of Truman," in Lacey, ed., *Truman Presidency*, 60.
30. John Diggins, *The Proud Decades: America in War and Peace, 1941–1960* (New York, 1988), 96; Alonzo Hamby, "The Mind and Character of Harry S. Truman," in Lacey, ed., *Truman Presidency*, 20ff.

ership, "The buck stops here," which was displayed in a sign on his desk in the Oval Office, reflected his reading about the past.

Truman had other qualities that associates came to appreciate. Unlike Roosevelt, who was often devious, playing subordinates off against one another, Truman was accessible and forthright. In this sense he was an orderly administrator. He disliked pomp and pretense, especially among the "brass hats" of the military and the "striped-pants boys" in the State Department. He was informal, plain, and unassuming, liking simple food and pleasures.[31] While President he especially enjoyed climbing aboard the presidential yacht, the *Williamsburg*, and cruising up and down the Potomac playing poker with old friends like Fred Vinson, whom Truman appointed as Treasury Secretary in 1945 and as Supreme Court Chief Justice in 1946. Sometimes Truman stayed aboard from Friday afternoon through Sunday. When reporters asked him what he had been doing, he was open: "Some of the boys and I were playing a little poker." Asked what they had been drinking, he replied, "Kentucky bourbon."[32]

Later, when many Americans were sick of presidential excesses—lying about Vietnam, Watergate—Truman was often lionized. People especially admired his directness and decisiveness. President Jimmy Carter retrieved Truman's THE BUCK STOPS HERE sign from the archives and put it on his desk in the Oval Office. This adulation, however, would have surprised many contemporaries, not only in 1945 but at most later stages of his seven-year presidency. Truman then seemed a poor contrast to FDR, with whom he was incessantly and disadvantageously compared. Bespectacled, apparently short, he looked more like a scholar than a dynamic leader of men.[33] Though he could be an effective extemporaneous speaker when his partisan instincts were aroused, he more often spoke much too fast and stumbled, in part because his poor eyesight made it hard for him to read the text. Clark Clifford, a key White House aide, recalled that "he generally read poorly from prepared texts, his head down and his words coming forth in what the press liked to call a 'drone.' He waved his hand up and down as if he were chopping wood."[34]

31. Robert Ferrell, *Harry S. Truman and the Modern Presidency* (Boston, 1983), 179–81; Clark Clifford, "Serving the President: The Truman Years (1)," *New Yorker*, March 25, 1991, pp. 49–52.
32. Clifford, "Serving (1)," 49.
33. Truman was actually five feet, ten inches tall—hardly short. But photos made him seem smaller than that.
34. Clark Clifford, "Serving the President: The Truman Years (2)," *New Yorker*, April 1, 1991, p. 60; J. Ronald Oakley, *God's Country: America in the Fifties* (New York, 1986), 25.

Some people, then and later, also perceived Truman as truly the provincial and mostly average man that enemies often made him out to be.
His background was hardly the sort to inspire large confidence. He had
been raised on farms in western Missouri before his family moved to
Independence, a town near Kansas City, when Harry was six. His father
suffered severe financial reverses in 1900, when Harry was a senior in high
school, and later returned to manage a farm. Harry could not afford
college and was to became the only modern American President without
higher education. Instead, he spent eight hard years—until his father's
death in 1914—helping on the land.[35] Efforts to make more money,
including investments in an Oklahoma zinc mine and oil-drilling, failed.
A member of the National Guard, Truman proved to be a remarkably
successful leader of men as an artillery officer during brief combat in
France in World War I. He returned and opened up a haberdashery shop
in Kansas City. This, too, failed, victim of postwar recession in 1922.
Truman was then thirty-eight years old, three years married, and without
many prospects in life.

At this point, Truman was rescued by the Pendergast political organization in Kansas City, with which he had been associated since before
World War I. In the mid- and late 1920s he rose rapidly in the machine,
well aware of its corruption but apparently taking no part in it. He became
a presiding judge—really an administrator—of Jackson County outside of
the city, where he proved an honest and competent public official. The
machine rewarded him in 1934 by making him Democratic nominee for
the Senate and by getting out the vote in a dirty and hotly contested
primary. (The machine found him an estimated 40,000 "ghost votes" in
Kansas City.)[36] When Truman entered the Senate in 1935, some of his
colleagues considered him damaged goods and shunned him. Although
he established himself as a hard-working and loyal New Dealer, he got no
help from the White House when he faced a tough primary battle before
reelection in 1940. In all, this was not an especially impressive résumé for
a President.[37]

Critics have wondered even about Truman's most widely praised trait:
decisiveness. Some have speculated that he celebrated his capacity for

35. Alonzo Hamby, "An American Democrat: A Reevaluation of the Personality of
    Harry S. Truman," *Political Science Quarterly*, 106 (1991), 33–55.
36. Alonzo Hamby, "Harry S. Truman: Insecurity and Responsibility," in Green-
    stein, ed., *Leadership*, 47–48.
37. David McCullough, *Truman* (New York, 1992), 15–34; Hamby, "Mind and
    Character," 20ff.

decision as compensation for a deeper insecurity, rooted perhaps in his childhood. As a youth he had often called himself a sissy, and even at age twenty-nine he told his fiancée, Bess Wallace, that he was a "guy with spectacles and a girl mouth."[38] His parents' marginal economic situation created other insecurities: the socially more prominent family of Wallace, whom he courted for years before marrying, seems always to have looked down on him. His own stumbles in finding success in life, and his ever-embarrassing association with the Pendergasts, further placed him on the defensive. Truman was honest, ambitious, and very determined. But especially on little things he could fly into rages that scared his associates. He was sometimes short-tempered, combative, resentful, and extraordinarily touchy.[39]

Most serious scholars do not wish to sail too far out on this sort of psychological tack. Truman was probably more insecure than some of his best-known political contemporaries—FDR and Eisenhower come to mind—but it is not clear that slights in his early years much affected his presidential actions. Indeed, Truman's decisiveness as President has been exaggerated. Having received little help from Roosevelt, he felt his way carefully for nearly two years after 1945.[40] During this difficult time he depended heavily on the advice of others, and even later he frequently took his time before reaching big decisions, such as committing American ground troops to Korea in 1950 or firing General Douglas MacArthur from his Pacific command in 1951.[41] In his foreign policies Truman is best described not as a heroic man-of-decision-the-likes-of-which-we-may-never-see-again-in-the-White-House, but as a patriotic, conscientious, and largely colorless man whose fate it was to cope, sometimes imaginatively and sometimes imprudently, with some of the most difficult foreign policy problems in American history.

TRUMAN DID NOT KNOW much about world affairs in 1945. Aside from his European service in 1918—and a trip to Central America in 1939—he had never left the United States. As a senator he had concentrated on domestic issues in the Depression years and on defense policies during the war. Perhaps his best-known venture into foreign policy ques-

38. Hamby, "Harry S. Truman," 47.
39. Ronald Steel, New Republic, Aug. 10, 1992, pp. 34–39.
40. Arthur McClure and Donna Costigan, "The Truman Vice Presidency: Constructive Apprenticeship or Brief Interlude?" Missouri Historical Review, 65 (1970), 318–41.
41. Hamby, "Mind and Character," 41; Hamby, "Harry S. Truman," 47–48.

tions had come in June 1941, when he offered his reaction to Germany's invasion of the Soviet Union: "If we see that Germany is winning we ought to help Russia, and if Russia is winning we ought to help Germany and that way let them kill as many as possible, although I don't want to see Hitler victorious under any circumstances."[42]

This statement has received a fair amount of attention from historians, some of whom argue that it revealed Truman to be among other things a shrewd (or cynical) practitioner of *Realpolitik*. That makes too much of an off-hand remark—one that many Americans in 1941 found attractive. Other evidence in fact points to traces of Wilsonian idealism in his thinking. For many years he carried in his wallet a copy of part of Tennyson's "Locksley Hall," a poem that foresaw a "Parliament of Man, the Federation of the world." Truman explained, "We're going to have that some day. . . . I guess that's what I've really been working for ever since I first put that poetry in my pocket."[43]

Although Truman meeded experience in foreign policy concerns in 1945, he did have strong feelings. Like most people, he hated repressive dictatorships and aggressive behavior by other nation-states. That much was clear from his comment in 1941, and it remained clear throughout his presidency. That the Soviet Union was Communist bothered him; that it was "totalitarian" bothered him more. Well before entering the Oval Office he mistrusted the Soviets because they crushed dissent and freedom. His mistrust was controlled: Truman, like most Americans in 1945, did not want to fight the Soviets. But it strongly affected his thinking. Moral concerns about freedom abroad pervaded his presidency.

Truman felt especially deeply about another thing in 1945: it was his duty to carry out the foreign (and domestic) policies of his predecessor. This made sense; Vice-Presidents generally do this, or think they are doing it. But following FDR's ideas in foreign policy was much easier said than done, for Truman had little idea of what those ideas were. For this reason, and because he lacked experience, he turned to top advisers for guidance. The influence of these advisers, often called the Establishment in later years, became powerful by 1946 and had extraordinary staying power that lasted well beyond the Truman administration.

Like any so-called Establishment, the elite had a varied cast of characters.[44] One of its leading lights in 1945, Secretary of War Stimson, was

42. Chafe, *Unfinished Journey*, 57.
43. Gaddis, *Strategies*, 56.
44. Walter Isaacson and Evan Thomas, *The Wise Men: Six Friends and the World They Made* (New York, 1986), which focuses on Dean Acheson, Charles Bohlen,

aging but still a force to be reckoned with in government. Stimson had a very long pedigree. He had been War Secretary under President William Howard Taft and Secretary of State under Herbert Hoover. Stimson was a Republican, a New York corporate lawyer, and a conservative, brought into the highest councils of state by Roosevelt to lend an air of bipartisanship to foreign policy in 1940. It was Stimson who stayed around after a Cabinet meeting in April 1945 and told Truman—eleven days after he became President—about the Bomb.

Stimson's influence extended well beyond his own office. Thanks to his good reputation, he attracted to the War Department a number of businessmen, bankers, and lawyers who dominated foreign and defense policy-making then and later. One was Robert Lovett, a decorated aviator in World War I, a Yale graduate, and a Wall Street investment banker with Brown Brothers, Harriman. Stimson made him Assistant Secretary for Air during the war. Lovett was shy, self-effacing, and deadpan but confirmed in his belief that the United States must maintain a strong defense posture in the postwar years. Widely respected by men in his circle, he served as Undersecretary of State in 1947–48, as Deputy Defense Secretary during the Korean War in 1950–51, and as Defense Secretary from then until early 1953. He remained a major power-broker then and for many years thereafter. One of his protégés during World War II, Robert McNamara, was too junior to sit in on high-level meetings at that time. But his experiences during the war, supervising the logistics behind bombing raids, deeply affected his thinking about the potential for air power. McNamara was soon to become a "whiz kid" at the Ford Motor Company, moving up to be its president in 1960, and later to be Secretary of Defense under Presidents John F. Kennedy and Lyndon Johnson.

Other members of the Establishment had equally strong ties with the military. Chief among these was General George Marshall, army chief of staff during the war. Marshall was an aloof, grave, and unusually formal professional soldier who rarely addressed even close associates by their first names. Though possessing a fierce temper, he struggled successfully to keep it under control, and he almost never raised his voice, preferring instead to listen carefully and to seek consensus. As a major architect of military victory during the war, Marshall elicited admiration bordering on adulation among most of his contemporaries. Many younger officers,

Averell Harriman, George Kennan, Robert Lovett, and John McCloy; Kai Bird, *John J. McCloy: The Making of the American Establishment* (New York, 1992).

including Dwight D. Eisenhower, owed their rapid advancement and preferment to him. Like many others, Eisenhower found Marshall uniquely temperate and coolly impersonal in his judgments. Stimson told Marshall, "I have seen a great many soldiers in my day, and you, sir, are the finest . . . I have ever known." Winston Churchill considered him "the greatest Roman of them all." Truman was awed by Marshall, calling him "the greatest living American." He picked Marshall as a special envoy to China in 1946, as Secretary of State in 1947, and as Secretary of Defense after the Korean War broke out in 1950.[45]

James Forrestal was the closest the navy had to a Stimson or a Marshall. Like many leading Establishmentarians, he had an Ivy League education—in his case Princeton—and a prewar career on Wall Street as an extraordinarily successful bond salesman. Ambitious and driven, Forrestal was a workaholic and a loner whose intensity worried people around him. His marriage was a shambles, and he had no time for his children. He, too, entered government service in 1940, as Assistant Secretary of the Navy. In 1944 he became Secretary. By 1946 he had become an influential voice in the ever-louder chorus that demanded tough policies against the Russians. In 1947 Truman named him as America's first Secretary of Defense, whereupon the tensions involved in trying to curb interservice rivalries worsened his erratic behavior. Truman finally removed him in March 1949. Two months later, as a patient at the naval hospital at Bethesda, Forrestal jumped from a high window to his death.

All these men identified easily with tough-minded approaches to many foreign policy questions. So, too, did many other leading American diplomatists. Few took a harder line in 1945 than Harriman, whom Truman kept on as Ambassador to the Soviet Union until March 1946. Harriman then served Truman as Ambassador to Great Britain and as Secretary of Commerce. Long after that Harriman remained influential in Democratic politics, both as governor of New York in the mid-1950s and as a dark horse for the Democratic presidential nomination in 1956. He surfaced again as Lyndon Johnson's negotiator in the forlorn quest for peace in Vietnam during the late 1960s.

Harriman was the son of railroad tycoon Edward Henry Harriman and was educated at Groton School (FDR's alma mater) and Yale. He was tall, rich, imperious, bored with the railroad business, and eager—detractors said sycophantically and desperately eager—to make a name for himself

45. Forrest Pogue, *George C. Marshall: Statesman, 1945–1959* (New York, 1987); Mark Stoler, *George C. Marshall: Soldier-Statesman of the American Century* (Boston, 1989); Gardner, *Architects of Illusion,* 139–70.

in government. He entered public service by working for the New Deal during the 1930s and was posted in 1941 to Great Britain as FDR's "defense expeditor." Named Ambassador to the Soviet Union in 1943, Harriman never acquired much expertise about the Soviet system, and until mid-1944 he agreed with what he took to be Roosevelt's quest for accommodation and cooperation with Stalin. Later that year, however, Harriman convinced himself that the Soviets were not to be trusted, and he fired off cables calling for get-tough policies. These did not much affect Truman in 1945, but they echoed the strong anti-Soviet feelings that many other Establishment advisers were developing at the same time.[46]

Dean Acheson, in many ways the most powerful Establishmentarian of them all, was also a product of Groton and Yale, followed by the Harvard Law School and later by two years as secretary to Supreme Court Justice Louis Brandeis. Acheson, too, had entered public service during the 1930s, but he found New Deal economic policies far too liberal for his conservative tastes and resigned as Undersecretary of the Treasury in 1933. He returned to government service as Assistant Secretary of State between 1941 and 1945 and as Undersecretary from August 1945 to mid-1947. Acheson was quick-witted, widely read, arrogant, and condescending. He dressed in well-tailored clothes and spoke with a clipped accent that betrayed his unapologetic Anglophilia. Many Republicans (and some Democrats) literally loathed him. But Truman found him a forceful adviser and a faithful admirer. When the President returned from Missouri to the Washington train station in November 1946 after Republicans had swept the mid-term elections, Acheson stood virtually alone on the platform to greet him. Truman never forgot this show of faith and relied ever more heavily on his counsel thereafter. Acheson served as Secretary of State during Truman's second term.[47]

Acheson held conservative views about most domestic policies and had little but contempt for Marxist or Communist ideas. He was certain that there was little if anything to be gained by trying to negotiate seriously with the Soviets. More than most diplomatists of his generation he had formed a self-assured and broad view of world history that he thought explained Russian behavior. The Soviets, he believed, practiced old-

46. Rudy Abramson, *Spanning the Century: The Life of W. Averell Harriman, 1891–1986* (New York, 1992).
47. Melvyn Leffler, "Negotiating from Strength: Acheson, the Russians, and American Power," in Douglas Brinkley, ed., *Dean Acheson and the Making of U.S. Foreign Policy* (New York, 1993), 178–86; Gaddis Smith, *Dean Acheson* (New York, 1972); Gardner, *Architects of Illusion*, 202–31.

fashioned Russian geopolitics aimed at securing warm-water ports and greater influence in Iran, Turkey, and eastern Europe. They had to be countered with equally determined tenacity by the West.[48] Acheson's greatest influence on the Truman administration came later, in 1947 and again after 1949. But by 1946 he, too, was offering a highly self-confident voice for firmness against the Soviet Union.

In 1945 there were other, lesser-known Establishment figures who stood in the wings of policy-making and who became prominent later. Among them were John Foster Dulles, a Princeton-educated New York attorney who served as a special adviser in the State Department in the Truman years and Secretary of State under President Eisenhower; Allen Dulles, his younger brother, another Princeton-educated lawyer, who became a dashing wartime intelligence agent, headed the influential Council on Foreign Relations between 1946 and 1949, and later ran the Central Intelligence Agency; Charles Bohlen, a career diplomat who served as an interpreter of Russian at Yalta and became Ambassador to the Soviet Union in 1953; George Kennan, a cerebral Princeton graduate who shared Bohlen's special competence in Russian affairs; Dean Rusk, a Georgian who became a Rhodes Scholar and served as an army officer in the Burma-India-China theater during the war, then rose steadily in the State and War departments, becoming Assistant Secretary of State for Far Eastern Affairs in March 1950; and many other well-educated young men whose self-assurance was whetted, though usually not in combat, during the dire days of World War II.[49]

Most of these people, too, had gone to private schools and elite universities. (Roughly 75 percent of State Department recruits between 1914 and 1922—senior officials after 1945—had prep school backgrounds.) They tended to be Europe-oriented and Anglophilic and were often well connected to major eastern law firms, banks, and investment houses. Some of them, including Acheson, Rusk, and the Dulles brothers, were sons of Protestant ministers: like Woodrow Wilson before them, they brought a high-toned moralism to their duties.[50] All moved ahead rapidly in the expanding foreign policy and War Department bureaucracies. Then and later their network of contacts helped them to shift easily

48. Dean Acheson, *Present at the Creation: My Years in the State Department* (New York, 1969), 3–5, 274–75.
49. Wilson Miscamble, *George F. Kennan and the Making of American Foreign Policy, 1947–1950* (Princeton, 1992); Charles Bohlen, *Witness to History, 1929–1969* (New York, 1973).
50. Godfrey Hodgson, *America in Our Time* (Garden City, N.Y., 1976), 144.

from one such bureaucracy to another, or at least to communicate informally by congregating at one of the exclusive clubs to which they belonged.

By the mid-1940s many of these officials—Acheson and Kennan above all—tended to become critical, even contemptuous, of the less well educated, democratically elected "politicians" who had traditionally played major roles in American foreign policy-making during the more relaxed and amateur days before World War II.[51] In this way the war, and the Cold War years that followed, did much to change the way that foreign policy was conducted. Henceforth it was to depend more on non-elected officials who circulated in and out of private life (especially the law and high finance in New York and Washington), and less on members of Congress and other elected politicians, who necessarily (Establishmentarians thought) catered to ill-informed constituencies and pressure groups at home. A foreign policy run by knowledgeable elites who knew and trusted each other, the officials thought, could be protected from the dangerous explosions of popular opinion.

The Establishment, needless to say, was hardly monolithic. Truman had a very different background from those of most of his advisers. Some influential officials, such as Kennan and Rusk, did not come from the Northeast. Marshall was a Virginian and a professional soldier. Forrestal, though a Princetonian, had a middle-class Irish-American background (which he resolutely tried to erase from his consciousness). They differed also in their views on domestic policies—Harriman was a liberal Democrat, Acheson a conservative, the Dulleses Republicans—and on many controversies involving foreign and defense policies.

Still, there is merit in using the term "Establishment." Save some older men like Stimson, most of these key people formed their opinions during World War II, a hot forge of patriotism. They emerged from the conflict believing that the United States had fought nobly in a good and necessary war and that America—democratic and well-meaning—stood for what was morally correct in the world. They especially deplored appeasement, which they thought had encouraged the bullies of the 1930s. The message of history was therefore clear: dangerous foreign leaders like Stalin must be met with power and firmness, so that war did not break out again. However these officials may have differed, in age, background, and political persuasion, they also shared a central faith in the capacity of well-educated, sophisticated "experts" like themselves to band together and

51. May, "Cold War and Defense," 10–14.

conduct an "enlightened" foreign policy based on the essential goodness of American principles.[52]

Underlying their thinking were two other assumptions. The first was that the United States must maintain a strong economic and military posture. Without this, policy would not be *credible*. This quest for "credibility"—a consistent concern of virtually all American leaders after 1945—lay at the center of United States diplomacy throughout the Cold War years.[53] The second was that the United States had the means—economic, industrial, and military—to control the behavior of other nations. This belief, which became widely shared by the American people during the postwar years, helped the Establishmentarians to shape what the historian John Gaddis calls the "inner-directed" nature of postwar American foreign policy: it often depended less on what other nations did or did not do than on what the experts thought the United States had the capacity to do.[54] This capacity, the Establishmentarians thought, was vast, and they developed correspondingly grand expectations concerning the ability of the nation to prevail in the international arena.[55]

It helped the generally high reputations of these officials that they seemed not only self-assured but also fairly selfless. Mostly rich and well-connected, they could move in and out of government without worrying much about salary. As appointive officials they did not need to cater to voters. They could appear to be—and truly thought they were—disinterested, far-seeing, and patriotic public servants. Indeed, they thought of themselves as missionaries of a gospel that could save the world; Acheson later entitled his memoirs *Present at the Creation*. For all these reasons the Establishment exercised a special influence in postwar America, most especially in the confused and turbulent years that confronted the brand-new Truman administration after 1945.

52. Melvyn Leffler, A *Preponderance of Power: National Security, the Truman Administration, and the Cold War* (Stanford, 1992), 179.
53. Frank Ninkovich, "The End of Diplomatic History?," *Diplomatic History*, 15 (Summer 1991), 439–48; John Gaddis, "The Tragedy of Cold War History," ibid., 17 (Winter 1993), 1–16.
54. Gaddis, *Strategies*, 355; Leffler, *Preponderance*, 51–54; John Thompson, "America's Cold War in Retrospect," *Historical Journal*, 37 (1994), 745–55.
55. Ninkovich, "End of Diplomatic History?"

5

Hardening of the Cold War, 1945–1948

In coping with the Soviet Union between 1945 and 1948, Truman's foreign policies went through three interrelated phases. The first, lasting until early 1946, exposed a good deal of floundering and inconsistency as Truman sought to find himself. The second, dominant through the end of 1946, revealed a little more floundering and uncertainty but also a stiffening of purpose. Although Truman and his advisers still hoped to ameliorate gathering tensions, they made only half-hearted efforts to accommodate the Soviets, or even to negotiate seriously with them. In the third phase, clear by February 1947, the administration hit on a more consistent, clearly articulated policy: containment. The essential stance of the United States for the next forty years, the quest for containment entailed high expectations. It was the most important legacy of the Truman administration.

WHEN TRUMAN TURNED his attention in 1945 to what was happening in eastern Europe, his first instincts were to act firmly. Stalin, he thought, was violating the Yalta accords, especially in Poland. Roosevelt, he was certain, would have resisted. The President was determined to be decisive and called in Vyacheslav Molotov, the Soviet Foreign Minister, to tell him how he felt.

The meeting took place within two weeks of Roosevelt's death and was one of the most fabled of Cold War contacts. Truman wasted no time on small talk and told Molotov that the USSR was breaking the Yalta agreements. Molotov was shaken by Truman's tone and replied, "I have never been talked to like that in my life." Truman retorted, "Carry out your agreements and you won't get talked to like that." Molotov, a Truman aide recalled, turned "a little ashy." But Truman did not relent. "That will be all, Mr. Molotov. I would appreciate it if you would transmit my views to Marshal Stalin."[1]

Truman, who liked to let people know he was tough, was pleased at this encounter. "I gave him the one-two, right to the jaw," he told a friend.[2] He followed with two other moves that signified his faith in the sticks of economic diplomacy. The first postponed (as Roosevelt had) an American response to urgent Soviet requests for an advance of $6 billion in credits. The second, taken quickly after Germany's surrender in early May, called off lend-lease shipments to the USSR. Truman maintained that American law tied lend-lease to the existence of war in Europe and that he had had no choice but to terminate the aid. But in fact he possessed more flexibility than that, and he moved more decisively than he had to: ships already bound for the Soviet Union were told to reverse and come home. Stalin later said that the American action was "brutal," done in a "scornful and abrupt manner."[3]

These actions demonstrated Truman's capacity, on occasion, for quick decision-making. They also exposed his lack of interest in the subtleties and ambiguities of diplomacy, which he never made much effort to practice. The irony was that he thought he was doing as Roosevelt would have done. But Roosevelt had been more indirect and tactful. To the already intensely suspicious Stalin and his fellow leaders in the Kremlin it appeared that Truman, far from carrying out the policy of FDR, was dramatically reversing it.

Truman's toughness, moreover, had little effect on Soviet behavior. On the contrary, Stalin soon imprisoned sixteen of the twenty Polish

1. The meeting is often described. See Stephen Ambrose, *Rise to Globalism: American Foreign Policy Since 1938*, 4th rev. ed. (New York, 1985), 61–63; William Chafe, *The Unfinished Journey: America Since World War II* (New York, 1991), 57; and John Gaddis, "The Insecurities of Victory: The United States and the Perception of the Soviet Threat After World War II," in Michael Lacey, ed., *The Truman Presidency* (Washington, 1989), 235–72.
2. Harry Truman, *Memoirs of Harry S. Truman*, Vol. 1, *Year of Decisions* (Garden City, N.Y., 1955), 77–82.
3. Chafe, *Unfinished Journey*, 60.

leaders of the anti-Communist government who had returned to their homeland from wartime exile in London. He claimed they had been inciting resistance to Soviet occupation forces in Poland. Stalin then permitted the other four to take part in a Soviet-dominated puppet government. By imposing Soviet control on Poland—and similarly pro-Communist regimes on Romania and Bulgaria—Stalin underscored two Cold War realities: first, that he was determined to ensure his domination over bordering nations in Europe; and second, that he had the military power to do so, no matter what the Allies said in protest. For Truman, as well as for many later American Presidents, this was a sobering and greatly frustrating reality.

At this point Truman backed and filled a little. Recognizing that he had been brusque with Molotov, he sent Harry Hopkins, Roosevelt's closest adviser, to Moscow, in the hope that the gesture would reassure the suspicious Soviet leadership. Following Hopkins's advice, he then accepted the new arrangements in Poland. He further conceded that Stalin could dismantle factories in eastern Germany and other parts of eastern Europe and that the Allies were to be frozen out of any significant role in all these areas.

Truman's acquiescence reflected several key realities that continued to complicate a consistently get-tough policy in the next few months. The first was that the United States, which had not yet tested the Bomb, was anxious to secure the Soviets as allies in the Pacific. At Yalta Stalin had promised to enter the war against Japan within three months of the defeat of Germany. That would presumably be in early August. Second, military advisers reminded Truman of the obvious: the huge Soviet army already occupied much of central and eastern Europe and would not be dislodged. Third, with the war over in Europe, pressures in the United States for demobilization and reconversion were mounting rapidly. Then and in 1946 popular opinion in America showed little stomach for new military commitments, especially in areas to the east of Germany that seemed remote.[4]

With these thoughts in mind Truman went in July to Potsdam, in Germany, for his first (and only) face-to-face meeting with Stalin. Later Truman maintained that he returned from this conference deeply disillusioned with the Russians. At the time, however, he seemed mostly pleased. "I can deal with Stalin," he noted in his diary at Potsdam. The

4. Alonzo Hamby, *Beyond the New Deal: Harry S. Truman and American Liberalism* (New York, 1973), 113–15; and John Gaddis, *Strategies of Containment: A Critical Appraisal of Postwar American National Security Policy* (New York, 1982), 16.

President remarked to an associate that Stalin was "as near like Tom Pendergast as any man I know." This was in many ways intended as a compliment, for Pendergast kept his word. Then and later Truman seemed to think that Stalin meant well but was a prisoner of a recalcitrant Politburo.[5]

Some of Truman's optimism at that time derived from news of the long-awaited atomic test. This took place at Alamagordo, New Mexico, on July 16, the day before the Potsdam conference opened, and was powerful beyond all expectations. It imbued Truman with great self-confidence. Churchill later observed, "When [Truman] got to the meeting after having read the report he was a changed man. He told the Russians just where they got off and generally bossed the whole meeting."[6] Other top Americans at the conference, notably Secretary of State James Byrnes, were also emboldened by having the Bomb. They imagined that monopoly of such a weapon, along with the sticks of economic diplomacy, would bring Stalin to his senses.

Possession of the Bomb nonetheless confronted Truman with difficult decisions. Should he tell Stalin about it? What sort of warning, if any, should he give to the Japanese? Should such an awesome weapon be used at all in war? If so, when and under what circumstances? In coping with the first question Truman resolved to tell Stalin no more than he had to. At Potsdam he mentioned to him casually that the United States had a big new bomb but made no effort to engage him in discussion about it. Stalin, who had spies working for him, expected the news and showed little interest. In deciding what to do with the Bomb Truman apparently pondered the moral questions a little more than he later cared to admit. But not very deeply, and he did not let moral scruples stop him from his determination to drop the Bomb on Japan. At Potsdam he warned the Japanese to surrender or face "prompt and utter destruction." This was scarcely a clear warning, especially to leaders of a nation whose cities had already been devastated by fire-bombs of napalm and jellied gasoline, the most horrific of which in March had killed 100,000 Tokyo civilians. When Japan did not respond to the warning, Truman made no effort to postpone the bombing of Hiroshima, which was destroyed on August 6. When Japan, confused by what had happened, still did not surrender, the bombing of Nagasaki went off according to standing orders, on August 9.

5. Hamby, *Beyond the New Deal*, 115; Gaddis, "Insecurities," 251; Robert Ferrell, *Harry S. Truman and the Modern Presidency* (Boston, 1983), 52.
6. Cited in Barton Bernstein and Allen Matusow, eds., *The Truman Administration: A Documentary History* (New York, 1966), 25–26.

The attacks killed an estimated 135,000 people, most of them civilians. Another 130,000 or so died within the next five years of radiation sickness and other bomb-related causes.

These decisions have naturally stimulated a vast amount of controversy and second-guessing, much of it many years after the fact. Some of Truman's critics deplore as immoral the use of such weapons at all. Others argue that he could have issued more explicit warnings, waited longer for a Japanese response, or arranged for top Japanese leaders to see a demonstration in an uninhabited area. Even writers who are relatively friendly to Truman can find little excuse for his refusal to reconsider standing orders to bomb a second city (Nagasaki). Still other critics insist that use of the bombs was unnecessary, because moderate factions among Japanese leaders had already sent out diplomatic signals indicating a desire to surrender, providing that they could retain their Emperor (which Japan ultimately was allowed to do). Truman's defenders retort that a demonstration might not have worked, which at the least would have been embarrassing. At that time, moreover, the United States had only two A-bombs ready for use. Above all, Truman insisted later that he decided to drop the bombs in order to stop the carnage of fighting as soon as possible. His defenders cite official American estimates—based in part on the suicidal resistance between April and June of Japanese forces in Okinawa—that a bloody land invasion of Japan (not scheduled to start until November) would otherwise have been necessary to end the war. The A-bombs, they claim, ended the war immediately and saved hundreds of thousands or even millions of American and Japanese lives.[7]

Use of the atomic bomb has also prompted angry and unending debates concerning Soviet-American relations. Some revisionist writers dispute Truman's claim that he authorized the bombings solely to defeat Japan as soon as possible. They argue that he wanted to force a Japanese surrender before the Soviets could make good on their promise to enter the war in the Pacific. (As it happened, the Soviets declared war on Japan on August 8.) Ending the war quickly would help to prevent the Soviet Union from claiming important concessions in Asia. Revisionist critics add that Truman used the Bomb in order to show the Russians that the United States

7. Virtually the entire Japanese garrison at Okinawa—some 107,500 men—fought to the death, causing United States casualties of nearly 12,000 dead and 37,000 wounded. Another 150,000 civilians perished. See Michael Adams, *The Best War Ever: America and World War II* (Baltimore, 1994), 112; and William O'Neill, *A Democracy at War: America's Fight at Home and Abroad in World War II* (New York, 1993), 413–15.

would indeed drop it against an enemy in war. That would make them respect American resolve in the future. Truman, they say, should not have used the Bomb until he saw what the Japanese would do, especially after the much-feared Russians joined the Allied forces against them. That he did not wait, revisionists conclude, indicates that he was playing Atomic Diplomacy.[8]

Several conclusions seem fair concerning these heated debates. First, Truman had no serious qualms about using the Bomb against Japan. During the months before the successful test at Alamagordo he gave only cursory attention to the counter-arguments of those scientists who raised moral concerns. He took this position in part because the Bomb had been developed with wartime use in mind (though not, at first, against the Japanese), and in part because the weapon was ready as of August in 1945. The majority of Truman's aides, moreover, favored dropping it on Japan; expressed doubts about the wisdom or morality of doing so mostly came later, after the fact. In effect, neither Truman nor his advisers resisted the powerful bureaucratic momentum that had accumulated by mid-1945. Truman also decided as he did because he thought that the Japanese— whose most influential military leaders seemed determined to fight on— were "savages, ruthless, merciless, and fanatic." Like many people in 1945, the President was swept up in the passionate emotions of a long and catastrophic war. Finally, Truman felt a considerable responsibility as commander-in-chief. He believed it his duty to put an end to the fighting—especially to American casualties—as soon as possible.

The decision to use the Bomb at Hiroshima and Nagasaki was politically popular in the United States—no doubt of that. And it quickly ended the war. Amid the awful passions of the time, it is hardly surprising that Truman acted as he did. Still, revisionism persists. In retrospect, it seems clear (though this is debated) that he could have waited longer—to give the shaken and bewildered Japanese time to figure out what had happened at Hiroshima—before approving use of the Bomb that leveled Nagasaki. It also seems clear that he would have risked little by postponing the bombings in order to ascertain whether Japanese moderates in Tokyo might

8. For a cogent defense of Truman's atomic policies, see Robert Maddox, "The Biggest Decision: Why We Had to Drop the Bomb," *American Heritage*, May/June 1995, 71–76. A key revisionist account is Gar Alperovitz, *Atomic Diplomacy and the Decision to Use the Atomic Bomb* (New York, 1995). See also Paul Boyer, "'Some Sort of Peace': President Truman, the American People, and the Atomic Bomb," in Lacey, ed., *Truman Presidency*, 174–204; and Martin Sherwin, *A World Destroyed* (New York, 1975), 193–219.

succeed in their efforts to reach a peace. Postponement would also have given Truman time to assess the impact on the Japanese of Russian engagement in Asia. The land invasions, after all, were not scheduled to take place for another three months, during which time America was unlikely to suffer much in the way of casualities. That Truman pushed ahead, however, does not prove that he was playing Atomic Diplomacy with the Soviets; the best evidence suggests instead that he wanted to stop the fighting as soon as possible. Moreover, subsequent studies of official Japanese decision-making suggest that most top leaders in Tokyo adamantly opposed peace in August 1945: only the A-bombs, bringing on the intercession of Emperor Hirohito, finally forced the Japanese to surrender. For these reasons, the revisionist arguments, while understandable given the horror of nuclear weapons, command only partial acceptance among scholars of the subject.

Still more ironies. Still more Western frustration. Dropping the bombs did not in fact change or soften Soviet behavior. Apparently unimpressed by America's atomic might, Stalin clamped down harder on Romania and Bulgaria. Having absorbed a large slice of eastern Poland, the Soviets compensated the Poles by giving them a chunk of eastern Germany. They made a satellite of northern Korea, resisting Western efforts to reunify the nation. The Soviet Union refused to take part in the World Bank or the International Monetary Fund, Western-dominated institutions that the United States deemed central to economic recovery. By late 1945 Stalin was intensifying pressure on Turkey for greater control of the Dardanelles and on Iran for a sphere of interest.[9]

These Soviet moves placed the United States in a reactive position. Truman and his advisers were not only frustrated but irresolute. Part of the problem, Truman gradually came to believe, was Secretary of State Byrnes. Truman had known him well since the 1930s when Byrnes, a South Carolinian, had served with him in the Senate. Byrnes had then assumed other high-ranking positions, including a place on the Supreme Court in 1941 and head of the Office of War Mobilization after 1943. Many people, Truman included, had expected Byrnes to be Roosevelt's choice for vice-presidential nominee in 1944. In July 1945 he appointed Byrnes as his Secretary of State.[10]

Given Byrnes's experience, it seemed a logical appointment. But Byrnes had no good strategy for dealing with the Soviets, other than

9. Gaddis, "Insecurities," 251.
10. David Robertson, *Sly and Able: A Political Biography of James F. Byrnes* (New York, 1994).

hoping that economic pressures would lead them to make concessions. By the end of 1945 it was becoming ever more clear that this was not happening. Leading senators, including Tom Connally of Texas, head of the Foreign Relations Committee, and Arthur Vandenberg of Michigan, top Republican on the committee, came to Truman to complain that Byrnes was too willing to horse-trade with the Soviets. Acheson, who was Undersecretary of State, meanwhile fretted at Byrnes's frequent absences from Washington and at his inattention to orderly administration. "The State Department fiddles while Byrnes roams," wags were saying.[11] Acheson and others in the State Department also tired of Byrnes's improvisational style of diplomacy. That might have worked well in the Senate but, Acheson believed, was wholly inappropriate in dealing with the Russians. Byrnes, who had expected to be Vice-President, compounded his problems by patronizing Truman and by failing to keep him informed of his discussions with allies and enemies. Truman, always sensitive to slights, grew increasingly irritated, referring privately to his "able and conniving secretary of state."[12]

By early 1946 Truman was rapidly losing patience with the Soviet Union. Ordering Byrnes to keep him better informed, he also made it clear that he intended to hold firm against Soviet pressure in Iran, the Mediterranean, and Manchuria. He exclaimed, "Unless Russia is faced with an iron fist and strong language another war is in the making. Only one language do they understand—'how many divisions have you?' . . . I'm tired of babying the Soviets."[13]

It was characteristic of Truman's persistent uncertainty at this time, however, that he did not get rid of Byrnes, who stayed on as Secretary for all of 1946. Truman also kept on other high officials, including Commerce Secretary Henry Wallace—FDR's Vice-President between 1941 and 1945—who were calling for a much more accommodating policy toward the Russians. The continuing presence of such diversity of opinion in the Cabinet testified to the President's difficult quest for clear direction in policy-making. Here, as in many other matters in 1946, he was hardly the super-decisive, "buck stops here" President of legend. As late as January 1946 few observers foresaw an end to the irresolution that marked

11. John Gaddis, *The United States and the Origins of the Cold War, 1941–1947* (New York, 1972), 347.
12. Ernest May, "Cold War and Defense," in Keith Nelson and Robert Haycock, eds., *The Cold War and Defense* (New York, 1990), 20.
13. Hamby, *Beyond the New Deal*, 117–19; Gaddis, "Insecurities," 251–52.

American foreign relations during the new President's first nine months on the job.

"THE ULTIMATE AIM of Soviet foreign policy," Navy Secretary Forrestal wrote a friend in April 1946, "is Russian domination of a communist world."[14]

Forrestal's opinion by no means determined American policy; he was just one high official among many. But it reflected a growing consensus among top-ranking American officials that developed fairly rapidly in February and March of 1946. During that critical period a rapid-fire sequence of events convinced all but a few American leaders that Soviet behavior was offensive, not defensive, and that the United States had to act decisively if it hoped to avert the sad spectacle of appeasement of the 1930s.

The first two events came a week apart, on February 9 and 16. On the ninth Stalin gave a major speech in which he blamed "monopoly capitalism" for the onset of World War II and implied that it must be replaced by Communism if future wars were to be avoided. Many Americans, now perceiving an unrelenting ideological thrust behind Stalin's behavior, reacted in alarm. Supreme Court Justice William Douglas, a liberal, proclaimed Stalin's speech the "Declaration of World War III." A week later Canada announced that it had arrested twenty-two people on charges of trying to steal atomic secrets for the Soviet Union during and after World War II. The announcement intensified the investigatory zeal of the anti-Communist House Committee on Un-American Activities, which made headlines for the next several years as it probed allegations of Soviet influence in American life.[15]

At this critical juncture there arrived in Washington one of the key documents of the early Cold War: the so-called Long Telegram of George F. Kennan, minister-counselor of the American embassy in Moscow. Kennan was one of a handful of well-trained experts on Russian history and language, having studied them since his graduation from Princeton in 1925 and entry into the foreign service a year later. Much of his

14. Melvyn Leffler, "The American Conception of National Security and the Beginnings of the Cold War, 1945–1948," *American Historical Review*, 89 (April 1984), 366–38.

15. Fred Siegel, *Troubled Journey: From Pearl Harbor to Ronald Reagan* (New York, 1984), 39–40; William O'Neill, *American High: The Years of Confidence, 1945–1960* (New York, 1986), 66; Gaddis, *Cold War*, 300.

subsequent diplomatic career had concentrated on the study of Soviet behavior, which he observed from posts in eastern Europe and the Soviet Union itself. Learned and eloquent, Kennan went on to have a distinguished career as a diplomat and historian. He was conservative in the sense that he doubted the capacity of democratic governments, driven by the dangerous winds of popular opinion, to chart a steady and well-informed course in the world. He preferred instead a foreign policy guided by experts such as himself. He was also revulsed by the Soviet system, which he considered brutal and uncivilized. His Long Telegram offered an anguished statement of his views, which fell on especially receptive ears at that time.

The Soviet Union, Kennan wrote, was an "Oriental despotism" in which "extremism was the normal form of rule and foreigners were expected to be mortal enemies." The Kremlin used Marxism as "the fig-leaf of their moral and intellectual respectability" to justify military growth, oppression at home, and expansion abroad. The USSR was "a political force committed fanatically to the belief that with the United States there can be no permanent *modus vivendi,* that it is desirable and necessary that the internal harmony of our society be destroyed, the international authority of our state be broken, . . . if Soviet power is to be secure."[16]

It would exaggerate the influence of the Long Telegram to say that it formulated American foreign policy for the future. But thanks in large part to Forrestal, who circulated it aggressively among American leaders, it received widespread attention. It gave them a congenial theoretical explanation for what they already considered to be Stalin's anti-Western behavior: that it stemmed from a combination of ideological and totalitarian imperatives deeply rooted in Russian as well as more recent Soviet history. This explanation was simple, clear, and therefore psychologically satisfying to American policy-makers, already irate at Soviet actions.

The explanation also offered Americans a way of coping with the Soviet Union. This was what later became known as the policy of "containment," which Kennan himself, identified only as "Mr. X," elaborated on in a famous article in *Foreign Affairs* in July 1947.[17] The containment approach assumed that the Soviets, not the Americans, were responsible for the breakdown of wartime cooperation and that the USSR was an implacable totalitarian regime. In dealing with it the United States must be firm, thereby confronting the Soviets "with unalterable counter-force

16. Robert Pollard, "The National Security State Reconsidered: Truman and Economic Containment, 1945–1950," in Lacey, ed., *Truman Presidency,* 210.
17. X, "The Sources of Soviet Conduct," *Foreign Affairs,* 25 (July 1947), 566–82.

at every point where they show signs of encroaching upon the interests of a peaceful and stable world."[18]

Later in his career Kennan was to complain that American policy-makers—mainly after 1950—overemphasized the military thrust of containment, thereby erecting an enormous edifice of military alliances that pitted the so-called Free World against the threat of worldwide Communist revolution. This military emphasis was not his in 1946 or 1947. The West, he said, should be watchful, and it should respond quickly to aggressive moves. Kennan, indeed, favored covert actions by United States intelligence agents in the Communist bloc. But the West should not overreact by building up huge stores of atomic weapons or making military moves that would provoke a highly suspicious Soviet state into dangerous counter-actions. The United States should above all be prudent and patient, vigilantly containing Soviet expansion and awaiting the day—which Kennan thought would come—when the Communist world would fall apart because of its own internal contradictions and brutalities.

Two weeks after Kennan's Long Telegram, Winston Churchill, speaking at Westminster College in Missouri, offered yet another voice for firmness against the Soviets. Churchill had been voted out of office in 1945, but he remained not only leader of the Conservative party but a towering symbol of wartime Allied unity and to Americans a much-admired foreign leader. As head of Great Britain during the war he had often been shrewdly realistic—insofar as British interests were concerned—in dealing with Stalin. But long before 1946 he was also known to be highly suspicious of Soviet intentions. At Westminster he voiced these suspicions in a memorable address that featured one of the most enduring metaphors of the Cold War, the "iron curtain":

> From Stettin in the Baltic to Trieste in the Adriatic, an iron curtain has descended across the continent. From what I have seen of our Russian friends and allies during the war, I am convinced that there is nothing they admire so much as strength, and there is nothing for which they have less respect than weakness, especially military weakness.[19]

Churchill's clarion call seemed especially significant because it had been delivered at the invitation of Truman himself, who arranged for Churchill's appearance at the college (in the President's home state) and who rode out on the train, playing poker the while with him, from

18. Ibid.; Gaddis, "Insecurities," 262.
19. Widely cited. See Joseph Goulden, *The Best Years, 1945–1950* (New York, 1976), 257–58.

Washington. Truman introduced Churchill, sat behind him on the dais while he was speaking, and applauded at several times during the presentation. Without explicitly endorsing what Churchill said, Truman seemed to signal that he agreed on the need to take a strong stand against the Soviets.[20]

These developments in February and March of 1946 led to mostly firmer, containment-like policies in the following months. During this time the United States moved the Sixth Fleet to the eastern Mediterranean and stepped up its protests against Soviet pressure in the area. This response seemed to bring results. By late 1946 the Soviet Union withdrew its troops from northern Iran and seemed less insistent on its demands from Turkey. The Truman administration also refused to let the Soviets play a significant role in the postwar occupation of Japan and resisted Communist pressures to reunite Korea under North Korean domination. Though cool to the Chinese Nationalist regime of Chiang Kai-shek, which was widely known to be corrupt, Truman encouraged Marshall, his emissary in China, to try to resolve the civil war there and agreed to significant congressional appropriations—$3 billion between 1945 and 1949—in aid to Chiang. In Germany the United States stopped shipment of reparations out of its zone of occupation and began moving to an anti-Soviet consolidation of American, British, and French zones.[21]

In these months the Truman administration also took enduring steps to harden its atomic shield. Some of Truman's advisers, notably Stimson, had argued in late 1945 that the United States consider sharing control of atomic weapons with the Soviet Union, which was certain to develop an A-bomb within a few years. "The chief lesson I have learned in a long life," Stimson said, "is that the only way you can make a man trustworthy is to trust him and the surest way to make him untrustworthy is to distrust him and show your distrust." He warned: "If we fail to approach them now and continue to negotiate with . . . this weapon rather ostentatiously on our hip, their suspicions and their distrust of our purposes will increase."[22] Other advisers, including some leading scientists, backed Stimson, pointing out that United States, by building and storing atomic weapons, was guaranteeing the escalation of a dangerous new arms race. Still others pointed out a flaw in relying heavily on such weapons: they

20. Clark Clifford, "Serving the President: The Truman Years (2)," *New Yorker*, April 1, 1991, pp. 37–38.
21. Gaddis, *Strategies of Containment*, 22.
22. Henry Stimson and McGeorge Bundy, *On Active Service in Peace and War* (New York, 1948), 644; Chafe, *Unfinished Journey*, 63.

were clumsy at best as deterrents in the vast majority of diplomatic disputes.

Truman at first had seemed open to such arguments, and he appointed Acheson and David Lilienthal, a liberal, to devise a plan to be presented to the United Nations. They recommended setting up an international Atomic Development Authority that would be able to control all raw materials used in making such weapons, including those in the Soviet Union, and would ban all subsequent A-bomb-making. It is very likely that the Soviet Union would have rejected the plan, because it permitted the United States to keep its own small stockpile while preventing the Soviets from developing theirs. Truman, however, ensured Soviet rejection when he authorized Bernard Baruch, a strongly anti-Communist financier, to present a revision of the Acheson-Lilienthal plan to the UN. The new plan authorized sanctions against violators and stipulated that no nation could use the Security Council's veto power to escape punishment for such violations.[23] When the Soviets insisted on the veto power in such matters, it seemed to Americans that Stalin was bound and determined to develop atomic weaponry on his own and that no agreement was possible. By the end of 1946 passed the last hopes, small though they had been, for amelioration of the atomic arms race that thereafter frightened the world.

ALTHOUGH THE EARLY MONTHS of 1946 marked a turning point in official American attitudes toward the Soviet Union, they did not commit the United States to a completely coherent, overtly anti-Soviet policy. Truman made this privately clear, writing his mother and sister after the iron curtain oration, "I am not yet ready to endorse Mr. Churchill's speech."[24] Until early 1947 he was firmer than he had been earlier, but he remained a little tentative, in part because he was still learning on the job, and he resisted making any dramatic changes in policy that would greatly increase tensions with the Soviet Union.

Some of this hesitation came from considerations of public opinion at home. This was in fact difficult to judge throughout most of 1946, but the majority of Americans were probably less concerned about Soviet behavior at that time than were top Truman officials. There were signs, to be sure, that some people yearned for a get-tough policy if that would bring order to international relations. Leading radio commentators, including H. V. Kaltenborn and Edward R. Murrow, seemed sympathetic to such

23. John Diggins, *The Proud Decades: America in War and Peace, 1941–1960* (New York, 1988), 62.
24. Cited in Goulden, *Best Years*, 257.

ICELAND
Reykjavik

North Sea

NORWAY
Oslo

Received U.S. loan of $3.5 billion, 1946
Exploded first atomic bomb, 1952
Joined Common Market, 1973

DENMARK
Copenhagen

Joined Common Market, 1973

IRELAND
Dublin
Joined Common
Market, 1973

GREAT
BRITAIN

NETHERLANDS
Amsterdam C

WEST
GERMANY
C

Bonn

EAST
GERMANY
West
Berlin

Berlin blockade
1948-1949

London

BELGIUM
Brussels C

LUX.
Luxembourg
C

Joined NATO,
1955

Paris

Atlantic

Ocean

FRANCE
C

Exploded first atomic bomb, 1960
Withdrew from NATO, 1968

Bern
SWITZERLAND

Zones of occupation
ended, 1955

ITALY
C

PORTUGAL
Joined Common
Market, 1986

Madrid

Rome

SPAIN
C

Lisbon

Joined NATO, 1982
Joined Common Market, 1986

Mediterranean Sea

0   200   400
         miles

SWEDEN

FINLAND

Helsinki

Stockholm

Baltic Sea

UNION OF SOVIET
SOCIALIST REPUBLICS
Exploded first atomic bomb, 1949

o Moscow

POLAND

o Warsaw

o Prague
CZECHOSLOVAKIA
Communist coup, 1948
U.S.S.R. invasion, 1968

Vienna o

AUSTRIA

o Budapest

HUNGARY
Revolution, 1956

RUMANIA

Belgrade o

YUGOSLAVIA
Tito-Stalin schism, 1948

Bucharest o

Black Sea

BULGARIA
o Sofia

Withdrew from WP, 1968

Tirane o

ALBANIA

o Ankara

GREECE

TURKEY

Truman Doctrine, 1947
Joined NATO, 1952
Joined Common Market, 1981

Athens

Truman Doctrine, 1947
Joined NATO, 1952

Europe
1946–1989

Member of NATO (1949) *

Member of the Warsaw Pact (1955) *

C   Member of The European
Common Market (1958) *

* Years are dates of formation.
NATO is the North Atlantic Treaty Organization.

an approach, especially with regard to western Europe—which virtually all shades of American opinion considered the most strategically vital area of the world.[25] The Luce publications, *Life* and *Time*, offered highly slanted accounts of perfidious Communist activities, especially in Asia. Luce refused to print dispatches that were unflattering to Chiang Kai-shek, leading a top reporter, Theodore White, to resign in disgust.[26]

Other, less partisan writers also produced work in 1946 that may have hardened opinion among well-read Americans. Brooks Atkinson, a leading cultural critic, and John Fischer, a well-known magazine editor, traveled separately to the Soviet Union in 1946 and wrote articles and books describing their findings. Fischer had long been a student of Russia; neither he nor Atkinson was known to be anti-Soviet. Yet both found the Soviet Union to be a completely closed society. Fischer's book, *Why They Behave Like Russians* (1946), was a Book-of-the-Month Club selection and sold hundreds of thousands of copies.[27]

Many liberals, however, resisted a tougher policy. Three liberal senators, Claude Pepper of Florida, Glen Taylor of Idaho, and Harley Kilgore of West Virginia, issued a joint statement following the iron curtain speech: "Mr. Churchill's proposal would cut the throat of the Big Three, without which the war could not have been won and without which the peace cannot be saved." Some liberals truly feared an anti-Soviet policy would bring war. "If somebody doesn't call a halt," the reporter Thomas Stokes noted privately, "the interests in this country who seem hell-bent on a war with Russia—and soon—will get their way. Lots of people seem to have gone completely mad."[28]

Many of these liberals had little faith in the good intentions of the Soviet leadership. They were strongly anti-Communist, and they opposed any spread of Soviet influence in western Europe. But they considered it foolish to throw money to Chiang, and they seemed prepared to accept as irreversible the Soviet sphere of influence in eastern Europe. One such writer was the theologian and intellectual Reinhold Niebuhr, who was by far the most highly regarded of anti-Communist liberals during the late 1940s. In his many writings at the time Niebuhr roused a generation of younger liberals—the historian Arthur Schlesinger, Jr., was the best known of these—to greater awareness of the Soviet threat. But Niebuhr

25. James Baughman, *The Republic of Mass Culture: Journalism, Filmmaking, and Broadcasting in America Since 1941* (Baltimore, 1992), 33–34.
26. Stephen Whitfield, *The Culture of the Cold War* (Baltimore, 1991), 159–60.
27. Goulden, *Best Years*, 258.
28. Hamby, *Beyond the New Deal*, 102.

was uncomfortable with what he considered to be excessively moralistic responses from Washington. In September 1946 he wrote in the *Nation* that the United States should end its "futile efforts to change what cannot be changed in Eastern Europe, regarded by Russia as its strategic security belt."

> Western efforts to change conditions in Poland, or in Bulgaria, for instance, will prove futile in any event, partly because the Russians are there and we aren't, and partly because such slogans as "free elections" and "free enterprise" are irrelevant in that part of the world. Our copybook versions of democracy are frequently as obtuse as Russian dogmatism. If we left Russia alone in the part of the world it has staked out, we might actually help, rather than hinder, the indigenous forces which resist its heavy hand.[29]

Intellectuals like Niebuhr did not much affect policy-makers in 1946; the caution of Congress, however, could not be ignored. This caution reflected the continuing aversion of constituents to jump from the fires of World War II into yet another conflagration, as well as the determination of Congress to reduce defense spending. Congress made sharp cuts in military expenditures of all kinds in 1945 and 1946. The navy had to sell 4,000 ships, mothball 2,000 more, and shut down eighty-four shipyards. Near mutinies in the army in early 1946—some veterans even took out paid ads demanding release—hastened the demobilization of soldiers. In April 1946 Congress extended the draft through March 1947 but called for voluntary recruitment between April 1947 and August 1948. Then and later it rejected efforts by Truman to introduce universal military training. Some opponents thought such a system "un-American." For all these reasons defense spending declined from $81.6 billion in fiscal 1945 (ending June 30 of that year) to $44.7 billion in 1946 to $13.1 billion in 1947, remaining at that low level through the fiscal year ending June 1950. Aided by such reductions, the federal government actually ran small surpluses between 1947 and 1949.

All these actions drained the military establishment, setting off round after round of fierce and unedifying interservice fighting for scarce resources. By mid-1947 the armed forces of the United States totaled only 1.5 million, most of whom were needed to man bases at home or to stand occupation duty in Europe and Japan. Although America retained the world's largest navy and air force, it lacked the ground forces, as one historian has observed, "to intervene in anything greater than a minor

29. Ibid., 111.

conflict, such as the territorial dispute between Italy and Yugoslavia over Venezia Giulia."[30]

Even the nation's atomic monopoly was of questionable military value in these years. Until mid-1950 the United States relied heavily on World War II–vintage B-29s, all of which were based in Louisiana, California, or Texas, too far to fly safely to the Soviet Union. Military experts privately estimated that in a war it could take two weeks to drop an atomic bomb on the USSR, by which time the large Russian armies could have swept to Paris. Not until after the start of the Korean War in June 1950 did the United States have newer, longer-range B-36 bombers completely equipped for action over the USSR.[31]

America's atomic shield was indeed thin in those years. By mid-1946 the United States had around seven atomic bombs of the Nagasaki type; by mid-1947 it had around thirteen. These were not easy to use. They had to be transported in parts; a team of seventy-seven specialists had to work for a week on the final assembly of an A-bomb before it was ready for use. Only specially designed planes could carry the bombs, which were hardly precise: a test A-bomb at Bikini in the Pacific in 1946 missed its target by two miles. Uranium, necessary for fission bombs of the day, was known to be scarce, and future production was expected to be slow. Top-ranking advocates of strategic bombing anticipated that the weapons of World War II, mainly TNT and incendiaries, would have to be relied on heavily in a forthcoming war.[32]

Supporters of a get-tough policy against the USSR also found mixed support at best from major interest groups. The armed services, of course, battled for higher appropriations. And some top officials, such as Forrestal, held very broad views of what was necessary for long-range national security, including control of the Western Hemisphere, the Atlantic and Pacific oceans, a system of outlying bases, and access to the resources and markets of Eurasia.[33] Still, the Pentagon turned out to be relatively weak on Capitol Hill throughout the late 1940s. This was in part because the services fought so fiercely among themselves. It was also because the military-industrial complex, a villain of much revisionist history, lacked cohesion following World War II. Many leading businessmen had been intent on reconversion to lucrative civilian production as early as 1943,

30. Pollard, "National Security State," 208; May, "Cold War and Defense," 29.
31. May, "Cold War and Defense," 8.
32. Ibid., 49; Pollard, "National Security State."
33. Leffler, "American Conception of National Security," 379.

and others competed vigorously for the rapidly growing consumer home market after the war. American exports in these years actually fell below the norm (as a percentage of GNP) of the pre-Depression years, never exceeding 6.5 percent between 1945 and 1950. With some exceptions business leaders at the time imagined the country to have a vast, growing, and largely self-sufficient domestic market. Confident of profits at home, they did not lobby very hard to push American economic influence abroad in the postwar years.[34]

It was in this context of domestic irresolution and military retrenchment that Truman faced his last major foreign policy decision of 1946: what to do about Henry Wallace, his dovish Secretary of Commerce. Wallace was one of the most remarkable figures in the history of twentieth-century American politics. The son of Harding's and Coolidge's Secretary of Agriculture, he grew up an Iowa Republican and, as a young man in the 1920s, became a well-known farm editor. In 1928, however, he supported Democratic presidential candidate Al Smith, and in 1932 he backed Roosevelt against Hoover. A progressive as well as a renowned scientist in the field of plant genetics, he became FDR's Secretary of Agriculture from 1933 to 1940, and then Vice-President in Roosevelt's third term. There he remained a visible New Deal spokesmen and administrator. But many main-line Democratic politicians found him increasingly insufferable. He was shy, dreamy, tousle-haired, sloppy in dress, and largely incapable of small talk. Sometimes he fell asleep in conferences. He was above all an idealist and a deeply religious man who was drawn to the ritual of high Episcopalianism, the mysticism of a White Russian guru, and the moral concerns of the social gospel.[35] If he had a model, it was the prophet Isaiah.

By 1944 Wallace had many followers among Democratic liberals, who admired his concern for the downtrodden of the world. In 1942 he had proclaimed, "The century on which we are entering . . . is the century of the common man." He added, "The people's revolution is on the march, and the devil and all his angels can not prevail against it. They can not prevail, for on the side of the people is the Lord." But moderates and conservatives had had enough of idealistic perorations such as this, and they opposed his renomination as Vice-President in 1944. When Roosevelt reluctantly gave way, accepting Truman instead, he compensated

34. Gaddis, "Insecurities," 265–66.
35. Hamby, *Beyond the New Deal*, 22–24.

Wallace by naming him Commerce Secretary early in 1945. There Wallace remained, working for the man who had replaced him as Vice-President, into the late summer of 1946.

Well before then Wallace had grown deeply interested in foreign affairs, and he brooded over the collision course that the wartime allies were on, especially the accelerating arms race. In July 1946 he wrote a long letter to Truman in which he urged more conciliatory policies toward the Soviet Union. His appeal was passionate:

> How do American actions since V-J Day appear to other nations? I mean by actions concrete things like $13 billion for the War and Navy Departments, the Bikini tests of the atomic bomb and continued production of bombs, the plan to arm Latin America with our weapons, production of B-29s and planned production of B-36s, and the effort to secure air bases spread over half the globe from which the other half of the globe can be bombed. I cannot but feel that these actions must make it look to the rest of the world as if we were paying only lip service to peace at the conference table.

Wallace went on to stress the understandable desire of the Soviet Union, like Russia before 1917, to seek warm-water ports and security on its borders. The United States should offer "reasonable . . . guarantees of security" to the Soviets and "allay any reasonable Russian grounds for fear, suspicion, and distrust. We must recognize that the world has changed and that today there can be no 'One World' unless the United States and Russia can find some way of living together."[36]

Truman might have listened carefully, bringing Wallace into the loop of foreign policy-making. Or he might have told him to mind his own business. Neither option, however, was palatable to him. Instead, he kept Wallace in the Cabinet and ignored his unwelcome advice. Wallace then acted again, warning the President that he was about to make a major speech before a Soviet-American friendship rally at Madison Square Garden in New York City on September 12. In the speech he offered some criticisms of the Soviet Union and insisted that the United States should not concede political control of western Europe to the Communists. But he otherwise elaborated on his letter in July, adopting a spheres-of-interest approach that accepted Soviet political (not economic) domination of eastern Europe. "We should recognize that we have no more business in the political affairs of Eastern Europe than Russia has in the

---

36. John Blum, ed., *The Price of Vision: The Diary of Henry A. Wallace* (Boston, 1973), 589–603; Robert Donovan, *Conflict and Crisis: The Presidency of Harry S. Truman, 1945–1948* (New York, 1977), 219–28.

political affairs of Latin America, Western Europe, and the United States."[37]

At Madison Square Garden Wallace rather maliciously mentioned that the President had read his speech in advance and had said it represented the policy of his administration.[38] This revelation launched a barrage of editorials that blasted Truman for encouraging such dovish ideas. Truman had earlier told a press conference that he "approved the whole speech," but he now backed away, and he grew increasingly unsettled. By September 19 he was furious, noting privately that Wallace was "fuzzy," a "pacifist 100%," and a "dreamer." "All the 'Artists' with a Capital A, the parlor pinks and the soprano voiced men are banded together. . . . I am afraid they are a sabotage front for Uncle Joe Stalin." At this point Byrnes, trying to stand firm in negotiations with the Soviets in Paris, insisted angrily that Truman decide between Wallace and himself. European allies, Byrnes pointed out, had parliamentary systems in which governments were supposed to speak with one voice.[39]

Truman then demanded Wallace's resignation and replaced him with Harriman. But the episode damaged him and his administration. To Byrnes and others, including America's allies, he had seemed vacillating. To Wallace and many in the press he seemed dishonest in trying to claim (inaccurately) that he had not looked at the speech in advance. One characteristically snide judgment on his performance purported to answer a question as to why Truman had been late to a press conference that day: "He got up this morning a little stiff in the joints and he is having difficulty putting his foot in his mouth."[40]

It may be argued that the experience was cleansing for the Truman administration, at least in the long run. After all, Truman was now rid of Wallace, whose passionately held views were implacable, and could try to formulate more unified policies. When he finally relieved Byrnes in December, replacing him with Marshall, this unity indeed developed. But that was still several months away, and Truman, having jettisoned a dissenter, had henceforth to cope with a martyr of sorts who led the forces against containment over the next two years.

37. Hamby, *Beyond the New Deal*, 126–30.
38. As Wallace recognized (but did not say), Presidents do not have time to review lengthy speeches while Cabinet members sit across from the desk. Wallace knew that his speech defied administration policy.
39. Robert Griffith, "Harry S. Truman and the Burden of Modernity," *Reviews in American History*, 9 (Sept. 1991), 298.
40. Goulden, *Best Years*, 217–23.

The episode, moreover, did not result in a reformulation of policy. Four days after Wallace's departure Truman received a lengthy digest of foreign policy options from advisers, including the man who was by then his top White House aide, Clark Clifford. Much of the report was temperate, holding out the hope that the United States might some day join the Soviet Union in a "system of world cooperation." But it concluded, as Kennan had in his Long Telegram, that the Soviets were strengthening themselves "in preparation for the 'inevitable' conflict, and . . . were trying to weaken and subvert their potential opponents by every means at their disposal. So long as these men adhere to these beliefs, it is highly dangerous to conclude that hope of international peace lies only in 'accord,' 'mutual understanding,' or 'solidarity' with the Soviet Union." It urged the United States to develop a strong military presence and to "confine" the Soviet Union. Armed with such a report, Truman might have used it to push for greater military appropriations. But he did not, for he still remained unsure what to do. He ordered all copies be put in the White House safe, where they remained for the rest of his administration.[41]

IT HAPPENS OFTEN in history that nations take bold moves only when external events force their hand. That happened to the Truman administration in early 1947, when Great Britain sent word that it no longer had the resources to maintain political stability in Greece and Turkey, areas that the British had until then considered parts of their sphere of interest. The specter of rising Communist influence, if not control, of the eastern Mediterranean suddenly loomed before American officials. Worst-case scenarios included the downfall of the pro-Western monarchy of Greece, then embroiled in civil war against Communist insurgents; renewed Soviet pressure on Turkey, a key buffer for Greece and the gateway to the Middle East; and even, perhaps, Soviet domination of Iran and the oil-rich nations surrounding it. At that time western Europe, struggling to recover from World War II, obtained 75 percent of its oil from the area.[42]

Truman's now more harmonious team of foreign policy advisers, led by Secretary Marshall and Undersecretary Acheson, quickly determined

---

41. Gaddis, "Insecurities," 254.
42. Bruce Kuniholm, "U.S. Policy in the Near East: The Triumphs and Tribulations of the Truman Age," in Lacey, ed., *Truman Presidency*, 299–338; Donovan, *Conflict and Crisis*, 279–91; Truman, *Memoirs*, Vol. 2, *Years of Trial and Hope* (Garden City, N.Y., 1956), 128–36; Daniel Yergin, *Shattered Peace: The Origins of the Cold War and the National Security State* (Boston, 1977), 282–86.

that the United States must step into Britain's shoes and provide military aid to Greece and Turkey. But the administration worried about Congress. In the Senate, where the major battle was expected, the key Democrat was Connally, ranking Democrat on the Foreign Relations Committee. Although Connally generally backed administration policies, he was not an altogether reassuring entity. He affected the manners and habits of the old school, including wide black hat, string tie, and oversized black jacket. His white hair curled over his collar. Though he had some sense of humor, he reveled in flattery, and he was easily caricatured. Some thought that the cartoonist Al Capp's Senator Throttlebottom was modeled after him.[43]

A much larger problem was the attitude of Republicans, who had swept to majorities in both houses in the 1946 elections. That made Arthur Vandenberg of Michigan head of the Foreign Relations Committee. Vandenberg, a senator since 1928, had been a prominent anti-interventionist prior to World War II. The war changed his mind, and he supported most administration foreign policies after 1945. But Vandenberg, too, was hardly easy to deal with. Vain and pompous, he needed even more flattery than Connally. Nor was it at all clear that Vandenberg could carry with him other Republicans, including the dominant GOP personality in the Senate, Robert Taft of Ohio. Many of these Republicans, especially those from the Midwest, opposed significant enlargement of American commitments in Europe.

Truman, recognizing trouble ahead, called key congressional leaders to a meeting at which they heard the administration's case. Marshall, grave and dignified, led off with an earnest but apparently unconvincing review of the situation. Acheson then jumped into action with a dramatic and deliberately florid statement of what was later to be known as the "domino theory" of foreign interconnection. "We are met at Armageddon," he began:

> Like apples in a barrel infected by one rotten one, the corruption of Greece would infect Iran and all to the East. It would also carry infection to spread through Asia Minor and Egypt, and to Europe through Italy and France. . . . The Soviet Union was playing one of the greatest gambles in history at minimal cost. . . . We and we alone were in a position to break up this play.[44]

43. May, "Cold War and Defense," 19.
44. Dean Acheson, *Present at the Creation: My Years in the State Department* (New York, 1969), 219; Wilson Miscamble, "Dean Acheson's *Present at the Creation: My Years in the State Department*," *Reviews in American History*, 22 (Sept. 1994), 544–60.

Acheson's peroration apparently stunned those present, Vandenberg included. He turned to Truman and told him there was "only one way to get" what he wanted: "That is to make a personal appearance before Congress and scare the hell out of the American people."[45]

This was good political advice, and Truman followed it. On March 12 he appeared before Congress and called for $400 million in military aid to Greece and Turkey. He justified this request in sweeping language:

> I believe that it must be the policy of the United States to support free peoples who are resisting attempted subjugation by armed minorities or by outside pressures.
>
> I believe that we must assist free people to work out their own destinies in their own way.
>
> I believe that our help should be primarily through economic and financial aid which is essential to economic stability and orderly political processes.[46]

Privately Truman was jubilant to have acted boldly at last. The next day he wrote his daughter Margaret, "The attempt of Lenin, Trotsky, Stalin, et al., to fool the world and the American Crackpots Association, represented by . . . Henry Wallace, Claude Pepper and the actors and actresses in immoral Greenwich Village, is just like Hitler's and Mussolini's so-called socialist states. Your pop had to tell the world just that in polite language."[47]

Truman's speech set off impassioned debate. Critics of the new policy included many Republicans as well as people on the left who followed Wallace. They blanched at the cost, which was in fact high—1 percent of the total federal budget of nearly $40 billion—and which was expected to escalate as subsequent requests came in. Opponents complained that the United States was assuming the imperialist interests of Great Britain, that Truman was exaggerating and distorting the internal problems of Greece and Turkey, that the program bypassed the United Nations, and especially that it seemed so vast and unlimited. Did Truman have in mind *all* "free people?" Critics on the left demanded to know why the United States sought to provide military (as opposed to economic) aid, and why it should go to a monarchy in Greece and a dictatorship in Turkey. Fiorello La Guardia, the liberal former mayor of New York City, declared that it was

45. Siegel, *Troubled Journey*, 56.
46. Bernstein and Matusow, eds., *Truman Administration*, 251–56; Truman, *Memoirs*, 2:129–30.
47. Gaddis, *Strategies of Containment*, 66.

not worth a single soldier to keep the Greek king on the throne. Wallace pronounced that the aid would lead to a "Century of Fear."[48]

Truman, however, had succeeded in seizing a political middle ground between anti-interventionists, most of whom were on the right, and Wallaceites on the left. Anti-Communists in both parties swallowed their reservations and generally supported the President. Many were liberals who were then in the process of joining the newly formed Americans for Democratic Action, which became an important pressure group for liberal programs at home and anti-Communist policies abroad.[49] Vandenberg and most East Coast Republicans also backed the President. All these supporters were convinced, given the history of Soviet-American tensions over the previous eighteen months, that Communism posed a real threat in the eastern Mediterranean and that it was time to act. Congress gave the plan resounding majorities in May, 67 to 23 in the Senate and 284 to 107 in the House.

The Truman Doctrine was not by itself a turning point in American foreign policy. That had already occurred, in early 1946, when Truman had embarked, however uncertainly, on a policy of containment. Still, the Truman Doctrine was a highly publicized commitment of a sort the administration had not previously undertaken. Its sweeping rhetoric, promising that the United States should aid all "free people" being subjugated, set the stage for innumerable later ventures that led to globalistic commitments. It was in these ways a major step.

Equally important was the other half of American foreign policy ventures in 1947, the so-called Marshall Plan of economic aid to western Europe. In announcing the plan at Harvard University's commencement in June, the Secretary spoke as usual in a soft, almost inaudible fashion. He scarcely looked up at his audience. But his call was bold, offering American assistance to all of Europe, including the Soviet Union. Marshall emphasized that the goal was humanitarian:

> Our policy is directed not against any country or doctrine, but against hunger, poverty, desperation, and chaos. . . . At this critical point in history, we of the United States are deeply conscious of our responsibilities to the world. We know that in this trying period, between a war that is over and a peace that is not yet secure, the destitute and oppressed of the earth

48. Goulden, *Best Years*, 269–73; Hamby, *Beyond the New Deal*, 175, 198.
49. Steven Gillon, *Politics and Vision: The ADA and American Liberalism*, 1947–1985 (New York, 1987), 3–32.

look chiefly to us for sustenance and support until they can again face
life with self-confidence and self-reliance.[50]

Humanitarian motives indeed formed part of the motivation behind the
Marshall Plan. The winter of 1946–47 had been the worst in memory for
western Europeans. Blizzards and cold in Great Britain, France, and
Germany had brought commerce and transportation virtually to a standstill, creating frightening shortages of winter wheat, coal, and electricity.
The gears of Big Ben froze, and England at one point was but a week away
from running out of coal. People were cold, hungry, and desperate. In
May 1947 Churchill described Europe as "a rubble-heap, a charnel
house, a breeding ground of pestilence and hate."[51]

The Marshall Plan, however, had bigger ambitions than the relief of
destitution, important though that was. Indeed, the United States had
already spent large sums of money on postwar European relief and recovery, funneling much of it through such international agencies as the
World Bank and the United Nations Relief and Rehabilitation Administration. Marshall and others thought that aid on a much vaster scale was
necessary to rebuild the economies of European nations and in the long
run to promote the economic and political integration of western Europe.
Supporters of the plan were frank at the time in stressing that the money
would not go "down a rathole." On the contrary, the aid would give the
Europeans the means not only to rebuild themselves but also to buy
American goods. The Marshall Plan, in short, would abet American
prosperity as well as European recovery.[52]

Strategic motives also drove Truman and his advisers. They worried
especially that Europeans might turn to Communism, which seemed to
thrive on economic discontent. France already had Communists in the
Cabinet, including the Ministry of Defense. Italy seemed even more
unstable. Economic aid to Europe was therefore a political complement
to the military aid of the Truman Doctrine. Provided directly by the
United States, not by an international agency, it could be directed in ways

50. Bernstein and Matusow, eds., Truman Administration, 257–60; Donovan, Con-
    flict and Crisis, 133–48; Michael Hogan, The Marshall Plan, America, Britain,
    and the Reconstruction of Western Europe, 1947–1952 (New York, 1987).
51. Ferrell, Harry S. Truman, 71.
52. Michael Hogan, "The Search for a 'Creative Peace': The United States, European
    Unity, and the Origins of the Marshall Plan," Diplomatic History, 6 (Summer
    1982), 284; Charles Maier, "Alliance and Autonomy: European Identity and
    U.S. Foreign Policy Objectives in the Truman Years," in Lacey, ed., Truman
    Presidency, 278.

suitable to American political interests. It would hold off the Communist threat and create an integrated trading bloc, including a revived western Germany, that would serve as a beacon of the blessings of capitalism and free markets.

Marshall made it clear that the United States required joint proposals of needs from the European countries, as part of a European Recovery Plan (ERP). At first the Soviets seemed ready to take part, and Foreign Minister Molotov and aides appeared at a conference in Paris to make known their desires. At the last moment, however, Molotov received a telegram from home and marched his delegation out. Although it is not wholly clear why Stalin refused to take part, two concerns may have moved him. First, he knew that the Soviet Union would not get very much in the way of needed short-term credits through ERP. Second, the Soviets would have been required to share information about their resources and to relinquish some of their control over economic management. Skeptics who doubt the humanitarian rationale behind the Marshall Plan think that American planners deliberately included these requirements with the expectation that the troublesome East bloc would stay out of the planning. Perhaps, perhaps not—it is still hard to say—but the offer had been made and could hardly have been withdrawn if the Soviets had accepted it. Their refusal made the ERP a much more attractive proposition to the American people and to the Congress. In this sense Stalin was short-sighted to reject it. [53]

Western European nations, by contrast, eagerly embraced Marshall's offer. In late August they came in with a proposal for $29 billion spread over four years. This was an enormous, politically unacceptable sum, and American officials pared it down to $17.8 billion. Even this was an extraordinary amount of money, especially to budget-conscious Republicans (and many Democrats) who were then trying to cut federal spending and taxes. Some conservatives and isolationists derided the plan, calling it a "sob sister proposal" and a "European TVA." Wallace was ambivalent—after all, this was economic, not military aid, and it could do much to reduce suffering. But he cooled to the idea, proclaiming that ERP meant Erase the Russian Peril and that the Marshall Plan ought to be known as the Martial Plan. [54]

Critics of the plan subjected it to extensive debate that lasted through the winter of 1947–48. Doubters continued to wonder if such massive aid

53. Hogan, "Search."
54. Siegel, *Troubled Journey*, 62–64.

was necessary. But congressmen who visited Europe that winter returned with reports of widespread suffering, including starvation. Moreover, the American economy was healthy, and the federal budget under control; it seemed the United States could afford the help. And after more than two years of Cold War Americans were increasingly ready to believe the worst of Stalin. Everett Dirksen, an influential Republican congressman from Illinois, backed the plan in late 1947 in the course of crying in alarm of "this red tide" that was "like some vile creeping thing which is spreading its web westward."[55] Stalin indeed seemed more dangerous than ever, rigging elections to assure a pro-Soviet regime in Hungary in August 1947 and promoting a Communist coup in Czechoslovakia in February and March 1948. This provocative act guaranteed passage of the plan. In April 1948 Congress approved a fifteen-month appropriation for $6.8 billion.

It is arguable, especially in retrospect, that the Marshall Plan had some unfortunate, though unintended, consequences. Together with the Truman Doctrine, it greatly alarmed Stalin, who more than ever suspected that these American efforts were part of a concerted conspiracy to encircle him.[56] Stalin, expecting that the ERP would bolster European prosperity, may have called for the coup in Czechoslovakia to prevent the Czechs from joining with the West. It is also true, as revisionists have emphasized, that the Marshall Plan was "selfish" in the sense that it did much for well-placed American business interests. ERP, along with American military aid, which escalated after 1949, greatly revived the capacity of Europeans to buy American goods.[57]

It is also possible to exaggerate the impact of ERP on the European economies. Americans, certain of their rectitude, power, and wealth, tend to do this without recognizing the important role that the industrious and efficient west Europeans played in their own recovery. Indeed, the Europeans deserve much of the credit for their economic revitalization after 1948. The plan gave them considerable autonomy and initiative, and they took it, reviving their historical possibilities in rapid time.[58] In later years, when the United States directed aid at other less developed parts of the world, the results were by no means so felicitous.

55. Marty Jezer, *The Dark Ages: Life in the United States, 1945–1960* (Boston, 1982), 47.
56. Scott Parrish, "The Turn Toward Confrontation: The Soviet Reaction to the Marshall Plan, 1947," Cold War International History Working Paper no. 9, March 1994, Woodrow Wilson Center, Washington, D.C.
57. Alan Wolfe, *America's Impasse: The Rise and Fall of the Politics of Growth* (New York, 1981), 152.
58. Maier, "Alliance," 298.

But most of these qualifications do not detract from the remarkable success of the Marshall Plan, which funneled $13.34 billion in aid to western Europe between 1948 and 1952. Welcomed eagerly by suffering European nations, the assistance hastened a very impressive recovery. It probably promoted greater political stability, if for no other reason than that it demonstrated the commitment of the United States to that part of the world. Compared especially with the selfish reaction of the United States to the plight of Europe after World War I, the Marshall Plan represented a remarkably enlightened effort. Few other postwar foreign policies of the United States can claim as much.

DRAMATIC THOUGH THE Truman Doctrine and Marshall Plan were, they still had little effect on the overall defense policy of the United States. Leading American officials for the most part did not expect the Russians to stage a military attack. Rather, like Kennan, they worried mainly about the psychological appeal of Communism to frightened citizens of unstable countries. Hence the need for "patient" containment, mainly via the means of economic aid.[59]

The struggle for a National Security Act, finally approved in 1947, reveals how little these defense policies changed. In seeking such legislation Truman and others hoped that Congress would strike a blow against interservice battling by creating the office of Secretary of Defense to design and coordinate military policy. Instead the services, notably the Navy Department under Forrestal, fought tenaciously against centralization, and the final bill left the services with considerable autonomy. Forrestal became the nation's first Defense Secretary in September 1947, at which point he battled for the powers that he had opposed as Navy Secretary. He got nowhere before he was replaced eighteen months later.

The National Security Act created two other agencies that were later to become important parts of America's defense bureaucracy. One, the National Security Council (NSC), was to be controlled by the White House, not—as Forrestal had wanted—by the Pentagon. The other, the Central Intelligence Agency (CIA), promised to give the United States—at last—a permanent intelligence-gathering bureaucracy. The statute left fuzzy the degree to which the CIA would be subject to meaningful congressional oversight, and it implicitly authorized covert activities. A key clause said the CIA could "perform other such functions and duties related to intelligence affecting the national security as the National Security Council may from time to time direct."[60]

59. Gaddis, *Strategies of Containment*, 33–40.
60. Clifford, "Serving (2)," 52–55.

It was revealing, however, that Truman made little use of these agencies in the late 1940s. He attended only twelve of the NSC's first fifty-seven meetings.[61] He had disbanded the CIA's wartime predecessor, the Office of Strategic Services, not long after the fighting ended in 1945 and had dispersed its activities thereafter to the contentious and non-cooperating armed services. While he accepted creation of the CIA, which quickly engaged in covert activities in Italy, he paid it little attention for most of his presidency. It was only later, in the 1950s and 1960s, that the provisions of the National Security Act turned out to be significant additions to the centralized power of the State.

Truman also continued to fight against large defense expenditures and to rely primarily on economic aid to conduct the Cold War. This was true even when the Russians staged their coup in Czechoslovakia in February 1948. Truman wrote his daughter, "We are faced with exactly the same situation in which Britain and France were faced in 1938–39 with Hitler. Things look black. A decision will have to be made. I am going to make it."[62] But he recognized that there was little, given the conventional military power of the Soviets, that he could do about the coup. It was much the same on Capitol Hill. Members of Congress angrily denounced the Soviets and reauthorized the draft, but they took no major steps to increase American preparedness.

Shortly thereafter, in June 1948, the Russians clamped a blockade on West Berlin. They did so for many reasons, mainly because they were frightened by Western plans then underway to establish West Germany as an independent nation. The Soviet blockade touched off even larger fears among policy-makers in the United States, some of whom recommended that American troops force their way into the beleaguered city. The Truman administration resisted this course of action, responding instead with a heroic and ultimately sucessful airlift, which involved sending hundreds of planes daily to relieve the West Berliners.[63] As before, Truman had refused to take steps that would have greatly militarized the Cold War. Defense spending continued to be modest, and the Soviets maintained a large edge in military manpower in Europe.

61. Alonzo Hamby, "Harry S. Truman: Insecurity and Responsibility," in Fred Greenstein, ed., *Leadership in the Modern Presidency* (Cambridge, Mass., 1988), 61.
62. Chafe, *Unfinished Journey*, 70; David Halberstam, *The Fifties* (New York, 1993), 19.
63. The Soviets finally ended the blockade in September 1949.

EVENTS SUCH AS THESE showed how dangerous the Cold War had become by mid-1948. It was not only raising the risk of armed conflicts but also souring the political atmosphere at home. Right-wing opponents accused the administration of being "soft" on Communism. The House Committee on Un-American Activities eagerly investigated a wide range of alleged subversive activities. Unions, universities, and other large institutions were gearing up to purge themselves of leftists.[64] Soviet athletes in 1948 stayed away from the Olympic games in London and St. Moritz.

The Truman administration had not always dealt very adeptly with the events that led to this sour state of affairs. In 1945 Truman, inexperienced and insecure, had tended to vacillate. In early 1946 he and top advisers devised a policy of containment, showing firmness concerning Soviet pressures on Iran and Turkey, but still seemed unsure of how to apply it. In 1947 and 1948 the United States acted more forcefully, especially with the Truman Doctrine and the Marshall Plan. These policies, countering Soviet coldness, symbolized the polarization of East and West.

In its defense policies the Truman administration was largely consistent, preferring for the most part to combat the Soviets by monopolizing the Bomb and by relying otherwise on foreign aid. This approach had potential flaws, for it left the United States without a flexible military deterrent in many parts of the world. Still, the alternative, building up a large military apparatus, was politically impossible before 1948. Controls over defense spending appealed to a population—and its congressional representatives—that was anxious and increasingly angry at the Soviets but that also feared yet another war and its attendant sacrifices. Advocates of a get-tough policy found these popular attitudes frustrating, for they did set limits on American responses.

To focus on mistakes of the Truman administration or on the role of American popular opinion, however, is to miss the most significant source of the Cold War in the 1940s. That was the uniquely difficult and bipolar world that suddenly arose after World War II: two very different societies and cultures found themselves face-to-face in a world of awesome weaponry. In part because they worried about their security, the Soviets proceeded to oppress their eastern European neighbors and to threaten Western interests in the Mediterranean and Middle East. The Americans, believers in democracy, had high expectations about their capacity and duty to contain such threats. They also came to fear that the

64. See chapter 7.

USSR was bent on even wider territorial expansion that would endanger the economic and political supremacy of the United States. Leaders on each side, frequently believing the worst of the other, proved unable to curb the escalation of tensions.

To stop there, however—assigning responsibility for these tensions to both sides—is to ignore the apocalyptic tone that came to surround Soviet-American relations in the late 1940s and thereafter. This stemmed in part from the tendency of the Truman administration, anxious to stand firm, to whip up domestic fears in order to secure political support for containment. Truman and his advisers knew that war with the Soviet Union was highly unlikely. Yet they had to have domestic political backing for firmness; the United States, after all, had a democratic system. To secure this support they resorted to a fair amount of harping—and some exaggeration—about the dangers that the Soviet Union and international Communism posed to the "Free World."

The apocalyptic character of the Cold War owed even more, however, to the peculiarly suspicious, dictatorial, and often hostile stance of Stalin. This truly alarmed policy-makers and over time aroused popular attitudes. In these years it was the Soviet Union, more than the United States, whose behavior—especially in eastern Europe—seemed alarming in the world. Not just the United States but also other Western nations concluded that "appeasement" would be disastrous. "Credibility" required that they resist. A defter American administration might have coped more sure-handedly with these problems than did Truman's, thereby muting to some degree the extremes of Cold War hostility. Given the understandable determination of the United States and its allies to contain the Soviets, however, there is no way that serious friction could have been avoided. Some sort of "Cold War"—even a quasi-apocalyptic one—seems as close to being inevitable as anything can be in history.

6

Domestic Politics: Truman's First Term

The shadow of Roosevelt fell heavily on Truman early in his first term. Jonathan Daniels, one of FDR's White House aides, remembers seeing Truman in the Oval Office shortly after Roosevelt died. "He swung around in the President's chair as if he were testing it, more uncertain than even I was about its size."[1]

Truman's uncertainty was entirely understandable, for to millions of Americans, especially the poor, Roosevelt had been an almost saintly father figure. Liberals had regarded him as the very model of strong presidential leadership in battles for social change. Although FDR had lost most of these battles after 1937, progressive Americans blamed Congress and "the interests," not the President. As recently as 1944 Roosevelt had rallied reformers—and raised popular expectations—by calling for an "Economic Bill of Rights" after the war. It is not too much to say that liberals were shattered when FDR died in April 1945. For them and many other Americans, no man could have filled Roosevelt's chair.[2]

---

1. William Leuchtenburg, *In the Shadow of Roosevelt: From Harry Truman to Ronald Reagan* (Ithaca, 1983), 1–40; Robert Griffith, "Harry Truman and the Burden of Modernity," *Reviews in American History*, 9 (Sept. 1981), 295–306.
2. Richard Pells, *The Liberal Mind in a Conservative Age: American Intellectuals in the 1940s and 1950s* (New York, 1985), 45–47; Alonzo Hamby, *Beyond the New Deal: Harry S. Truman and American Liberalism* (New York, 1973), xix, 40–41.

Any liberal successor to Roosevelt faced especially large structural obstacles to change. Unlike many west European countries, the United States lacked a strong political Left. The Socialist party headed by Norman Thomas had been undercut by the New Deal in the 1930s and by factional splits during the war, which Thomas had opposed. It was barely alive.[3] The Communist party, though stronger in 1945 than ever before, remained tiny; most Americans feared to associate with it.[4] The labor unions had record membership but by 1945 were beginning to lose momentum as a progressive political force. Roosevelt, indeed, had enjoyed advantages denied to Truman. During the Depression FDR could generate enthusiasm from "have-not" groups, such as restless workers and poor farmers, and during the war he could appeal to their patriotism. By 1945, however, many of these people, such as upwardly mobile blue-collar workers, were becoming "haves"—interest groups with much to be gained from supporting the status quo. In this way, as in many others, the recovery of the American economy reshaped American politics—for the most part toward the center and the right.[5]

A special obstacle to liberals in 1945, as throughout much of postwar American history, was Congress. On the surface this did not necessarily seem to be the case. Truman had comfortable Democratic majorities in both houses, 242 to 190 in the House and 56 to 38 in the Senate. In the House he could count on Speaker Sam Rayburn of Texas, a bald-headed bachelor who devoted his life to the chamber he had first entered in 1913. Popular with his colleagues, moderately liberal, a partisan Democrat, Rayburn was a forceful leader. In the Senate Truman could rely on Democratic majority leader Alben Barkley of Kentucky. Barkley was more easy-going and less effective than Rayburn. He was also aging, turning sixty-eight in 1945. But Barkley, too, had put in long service on the Hill, dating to 1913 when he, like Rayburn, had first entered the House. He had moved to the Senate in 1927. A moderate, he was liked by most of his colleagues and by Democratic party leaders throughout the country. De-

3. Murray Seidler, *Norman Thomas: Respectable Rebel* (Syracuse, 1961); Bernard Johnpoll, *Pacifist's Progress: Norman Thomas and the Decline of American Socialism* (Chicago, 1970).
4. David Shannon, *The Decline of American Communism: A History of the Communist Party in the United States Since 1945* (New York, 1959); Maurice Isserman, *Which Side Were You On? The American Communist Party During the Second World War* (Westport, Conn., 1982).
5. William Leuchtenburg, *A Troubled Feast: American Society Since 1945* (Boston, 1973), 14; Alan Wolfe, *America's Impasse: The Rise and Fall of the Politics of Growth* (New York, 1981), 23; Hamby, *Beyond the New Deal*, 510–11.

spite Barkley's age, Truman asked him to be his running mate in 1948.

But since 1937 power on Capitol Hill had usually belonged to a coalition of Republicans and conservative Democrats, many of them from the South. They returned to Capitol Hill in September 1945 in an uncooperative mood. They were especially tired of aggressive presidential leadership.[6] Most of the Republicans could hardly wait until 1948, when they expected—at last—to recapture the White House. The late 1940s were among the most partisan, rancorous years in the history of modern American politics.

Some observers of Truman thought that he was essentially a partisan who accepted this state of affairs. One was the journalist Samuel Lubell, who described Truman as "the man who bought time." "Far from seeking decision, he sought to put off any possible showdown, to perpetuate rather than to break the prevailing political stalemate."[7] This is accurate insofar as it captures Truman's sometimes zigzag approach, as he alternately tried to satisfy and to fend off the claims of interest groups. It is a little unfair, however, as a description of Truman's motivation. The new President had been a loyal New Dealer in the 1930s, and he believed in strong presidential leadership. He sincerely supported most of the liberal programs that he introduced during his years in office.[8]

For a variety of reasons, however, the President failed to persuade many liberals that he was one of them, at least not in 1945–46. While Truman wanted to protect the New Deal, he was uneasy around some of the liberals—a "lunatic fringe," he called them—who had risen to high office under Roosevelt. One was Wallace, another the curmudgeonly Interior Secretary, Harold Ickes. Truman was uncomfortable even with words like "liberal" or "progressive." He preferred "forward-looking." Correctly sensing the temper of the times, he also doubted that major reforms had much of a chance right after the war. "I don't want any experiments," he told his adviser Clark Clifford. "The American people have been through a lot of experiments, and they want a rest from experiments."[9]

Some of Truman's views also put him at loggerheads with liberals. One was his fiscal conservatism. As an administrator in Jackson County, Mis-

6. Allen Drury, A Senate Journal, 1943–1945 (New York, 1963).
7. Samuel Lubell, The Future of American Politics (New York, 1952), 8–28.
8. Robert Griffith, "Forging America's Postwar Order: Domestic Politics and Political Economy in the Age of Truman," in Michael Lacey, ed., The Truman Presidency (Washington, 1989), 86–87; Alonzo Hamby, "The Mind and Character of Harry S. Truman," in ibid., 44–52.
9. Hamby, Beyond the New Deal, 82–84.

souri, he had prided himself on his attempts to balance the budget. He was a man of modest means—perhaps the poorest member of the United States Senate while he was in it—and he had always had to be careful with money. Truman's fiscal conservatism was good politics: most Americans at that time believed that the government, like a household, should normally spend no more than it took in. Moreover, few politicians in Truman's lifetime (Roosevelt included) favored deficit spending in times of prosperity. But Truman's conservative feelings on the subject were powerful and genuine, rooted in all of his experience. He remained cautious about advancing liberal social programs that would cost a good deal of money.

Truman also believed strongly that he was President of *all* the people. This did not mean that he claimed, as President Eisenhower later did, to be above politics. On the contrary, Truman was never happier than when in the company of fellow politicians, and he was intensely partisan. But he considered it his duty as President to rise above what he thought were the more local, provincial concerns of members of Congress and to resist interest groups that acted against what he considered the *national* well-being. This feeling provoked him to oppose union wage demands in 1946—opposition that did him temporary harm with liberal supporters of the labor movement.

As much as anything, Truman's personal style discouraged liberal Democrats in 1945–46. Roosevelt had been Harvard-educated, eloquent, and charming. People warmed to the glow of his buoyant personality. By contrast, Truman had risen from machine politics and had reached the White House by accident. Harry Dexter White, Undersecretary of the Treasury, expressed this feeling well in 1946. When FDR had been alive, White said, "we'd go over to the White House for a conference on some particular policy, lose the argument, and yet walk out of the door somehow thrilled and inspired to go on and do the job the way the Big Boss had ordered." Now, White lamented, "you go in to see Mr. Truman. He's very nice to you. He lets you do what you want to do, and yet you leave feeling somehow dispirited and flat."[10]

No one was unhappier with Truman than the cantankerous journalist I. F. Stone, who wrote columns for liberal journals such as *PM* and *The Nation*. Under Truman, he wrote, the New Dealers

> began to be replaced by the kind of men one was accustomed to meeting in county courthouses. The composite impression was of big-bellied, good-natured guys who knew a lot of dirty jokes, spent as little time in their

10. I. F. Stone, *The Truman Era* (New York, 1972 ed.), xx.

offices as possible, saw Washington as a chance to make useful "contacts," and were anxious to get what they could for themselves out of the experience. They were not unusually corrupt or especially wicked—that would have made the capital a dramatic instead of a depressing experience for a reporter. They were just trying to get along. The Truman era was the era of the moocher. The place was full of Wimpys who could be had for a hamburger.[11]

This was an unfair comment. Truman in fact made many distinguished appointments, especially in the area of foreign affairs, where he relied heavily on experienced advisers. In stating it, however, Stone reflected a characteristically liberal view of presidential leadership: that White House dynamism was in and of itself a key to progress. Liberals also assumed wrongly that there was great reform sentiment among the people, just waiting to be aroused by an inspiring leader. They forgot that Roosevelt, their idol, had struggled unsuccessfully since 1937, and they ignored signs in 1945 and 1946 that many Americans wanted a respite from the excitement and intrusiveness of governmental activism.

Still, liberals like Stone were correct that Truman in 1945–46 seemed indecisive and uncertain in domestic matters, just as he did in foreign affairs at the time. Again, Roosevelt loomed as their standard. FDR, Max Lerner said, had given the country a "confident sense of direction." Truman lacked this ability. *Progressive* magazine added, "A curious uneasiness seems to pervade all levels of the Government. There is a feeling at times that there is no one at the wheel."[12]

ALTHOUGH TRUMAN WAS ABSORBED in foreign policy matters for much of 1945 and 1946, he wasted little time in advancing an ambitious domestic agenda. On September 6, 1945, he stamped himself in the mold of Roosevelt by praising FDR's Economic Bill of Rights and calling on Congress to approve a host of reforms. These included laws that would expand federal control of public power, increase minimum wages, provide funds for public housing, broaden coverage of Social Security, and establish a national health program. Truman also made it clear that he expected Congress to confer permanent status on the wartime Fair Employment Practices Commission and to approve the so-called full employment bill, which would commit the government to promote policies against joblessness.[13]

11. Joseph Goulden, *The Best Years, 1945–1950* (New York, 1976), 257.
12. Hamby, *Beyond the New Deal*, 63, 83–85.
13. Ibid., 59–62.

Conservatives erupted in dismay at Truman's proposals. House Republican Leader Joseph Martin of Massachusetts cried, "Now, nobody should have any more doubt. Not even President Roosevelt ever asked so much at one sitting. It is just a case of out–New Dealing the New Deal."[14] Martin was a conservative, partisan legislator who had marched in torchlight parades for William McKinley in the late 1890s and had befriended Calvin Coolidge while serving with him in the Massachusetts legislature. He would have opposed most of these programs no matter how cautiously they had been introduced. But others, including loyal Democrats, were also taken aback by Truman's sweeping requests. Truman, they grumbled, was asking too much too fast, expecting Congress to do his bidding and setting the stage to blame it if it did not. This was hardly the way to develop a harmonious working relationship along the length of Pennsylvania Avenue.

Complaints such as these punctuated all seven years of Truman's presidency, which witnessed unusually antagonistic relations between the White House and Congress. Truman vetoed 250 bills in his seven years, third only to FDR, who vetoed 631 in twelve, and Grover Cleveland, who vetoed 374 in eight.[15] Twelve of his vetoes were overridden, the most since the days when Andrew Johnson defied the Radical Republicans over Reconstruction. Truman, however, acted as if these complaints did not bother him. "What the country needed in every field," he said, "was up to me to say . . . and if Congress wouldn't respond, well, I'd have done all I could in a straight-forward way."[16] He was a little unwise to sound so cavalier, both because members of Congress resented his attitude and because they had trouble distinguishing between what he really wanted and what he demanded. Truman, like many who followed him to the Oval Office, did not always delineate his priorities.

It is unlikely, however, that the deftest of presidential leadership would have made much of an impression on the conservative coalition or on the well-established interest groups that dominated Capitol Hill. Southern senators launched a filibuster against the FEPC bill, ultimately preventing its consideration. Business interests, which had grown powerful during the war, had special influence. Oil companies and state political leaders pressed for a "tidelands" bill that would have turned over to states

14. Alfred Steinberg, *Man from Missouri: The Life and Times of Harry S. Truman* (New York, 1962), 262.
15. Cleveland had vetoed a host of special pension bills.
16. Richard Neustadt, *Presidential Power: The Politics of Leadership* (New York, 1960), 177.

oil-rich "submerged lands" off their coasts; the bill passed twice during Truman's presidency, was vetoed twice, and finally was enacted when Eisenhower signed it in 1953. The electric utility lobby took the lead in successful efforts against proposed new federal power authorities in the Missouri and Columbia valleys. Railroad interests urged passage of a bill that would have exempted many of their practices from anti-trust prosecution. This, too, later passed over a Truman veto.[17]

The fate of Truman's quest for a national system of health insurance clearly revealed the power of special interests. His proposal was fairly conservative, calling for care to be financed by a tax of 4 percent on the first $3,600 of personal income. General government revenue would assist many among the poor. A powerful medical lobby headed by the American Medical Association (AMA) attacked the plan as socialistic, and conservatives in Congress agreed. The plan never came close to passage.[18] Instead, the AMA backed the so-called Hill-Burton bill, which Congress approved in 1946. It provided federal aid for hospital construction, thereby pleasing building company interests as well as medical leaders. Between 1946 and 1975 some $4 billion in federal funds supported this program, which ultimately helped to produce large excess in hospital beds. The Hill-Burton Act mainly benefited doctors, hospital administrators, and the rising network of medical insurers such as Blue Cross–Blue Shield.[19]

Congress also gutted liberal versions of the employment bill. The final act, passed in 1946, deliberately omitted mention of a government commitment to "full" employment, as well as provisions mandating public spending to supplement private expenditures. Instead, it called for creation of a three-person Council of Economic Advisers, a Joint Economic Committee of Congress, and an annual presidential report on the state of the economy. The Employment Act represented a step in the direction of governmental responsibility for economic welfare, a principle that would have seemed almost revolutionary as recently as the 1920s. But it was a far more cautious and noncommittal step than many reformers had hoped for.[20]

17. Griffith, "Forging," 76–82.
18. Monte Poen, *Harry S. Truman Versus the Medical Lobby* (Columbia, Mo., 1979).
19. Wolfe, *America's Impasse*, 88–90.
20. Crawford Goodwin, "Attitudes Toward Industry in the Truman Administration: The Macroeconomic Origins of Microeconomic Policy," in Lacey, ed., *Truman Presidency*, 101–2; Hamby, *Beyond the New Deal*, 69.

Liberals were disappointed with Truman's handling of many of these issues. They were especially upset that he did not denounce the fili-busterers against FEPC and that he acquiesced in the revisions to the full employment bill. Truman, indeed, focused on foreign affairs and did not fight very effectively for domestic programs on Capitol Hill. He also showed little interest in listening to "experts" on the economy. It was six months before he got around to appointing people to the Council of Economic Advisers, and he paid them relatively little attention thereafter.

No domestic issue of these years did Truman more damage than the highly contentious question of what to do about wartime restraints on prices, which had been supervised by the Office of Price Administration (OPA). Businessmen generally wanted controls lifted so that they could take full advantage of the huge surge in demand that was expected follow-ing the war. Free-market conservatives agreed, arguing that a less regu-lated world of supply and demand ought to be restored. Many liberals hotly dissented. Vast pent-up wartime savings, they said, would intensify demand beyond the capacity of businesses to supply, with runaway price increases and large corporate profits the result.

Truman mostly agreed with the liberals. Fearing inflation, he seemed to support the OPA. But John Snyder, a conservative aide who was supervising reconversion policies, temporarily lifted controls on supplies of building materials, stimulating large demand among builders seeking to engage in profitable commercial construction instead of residential building. Meanwhile, Reuther and other union leaders were demanding large wage increases. The insistence of these and other interests at the time would have pressured almost any chief executive, especially one so inexperienced. It surely confounded Truman, who was buffeted by events. OPA head Chester Bowles, an ardent liberal, complained to Tru-man in January, "The Government's stabilization policy is not what you have stated it to be, but is instead one of improvising on a day-to-day, case-by-case method, as one crisis leads to another—in short, . . . there is really no policy at all."[21]

More backing and filling on the issue followed until June 1946, when conservatives in Congress passed a bill that extended OPA beyond its statutory lifetime on June 30 but that stripped the agency of real authority. Barkley told Truman to approve it: "Harry, you've got to sign this bill. Whether you like it or not, it's the best bill we can get out of this Congress,

21. Bowles to Truman, Jan. 24, 1946, in Barton Bernstein and Allen Matusow, eds., *The Truman Administration: A Documentary History* (New York, 1966), 65–66.

and it's the only one you're going to get." Truman refused and vetoed the bill. Prices and rents, no longer controlled, shot up immediately. Steak went from fifty-five cents a pound to a dollar, butter from eighty cents a pound to a dollar. The *New York Daily News* carried a headline, PRICES SOAR, BUYERS SORE, STEERS JUMP OVER THE MOON.[22]

Three weeks later Congress passed another bill that restored the OPA, again with emasculated powers. This time Truman signed it, but everyone knew that it was too little and too late, for prices had soared, and the new controls were ineffective. Some wags branded the OPA as OCRAP, the Office for Cessation of Rationing and Priorities.[23] When OPA tried to curb hikes in meat prices, farmers and ranchers refused to deliver their products to market. Consumers erupted in outrage, much of it aimed at the government. The *Washington Times-Herald* captured these feelings with a headline, ONLY 87 MEATLESS DAYS UNTIL CHRISTMAS. Truman was privately furious with the "reckless group of selfish men" who resisted controls. He prepared an angry speech, blasting the American people for sacrificing "the greatest government that was ever conceived for a piece of beef, for a slice of bacon." Businessmen and labor leaders, he added, greedily sought profit "from the blood and sacrifice of the brave men who bared their breasts to the bullets."[24] The President, however, then thought better of making such incendiary statements. He surrendered to pressure in mid-October and removed the controls on meat. By the end of the year the OPA was dead.[25]

The issue of controls was but one of many domestic concerns that hurt Truman's relations with liberals in 1946. But shortages and controls affected people in especially personal ways, and Truman suffered badly from criticism during the acerbic election campaigns of 1946. Jokes reflected the popular mood. "Would you like a Truman beer? You know, the one with no head." "To err is Truman." Remembering FDR, people asked, "I wonder what Truman would do if he were alive." Republicans summed up their message in a widely used slogan, "Had Enough? Vote Republican." They swept to victory, taking control of both houses of Congress for the first time since 1930: 245 to 188 in the House and 51 to 45 in the Senate. So decisive was the repudiation of Truman's leadership that Arkansas senator J. William Fulbright, a Democrat, suggested that

22. Goulden, *Best Years*, 102–6.
23. Robert Ferrell, *Harry S. Truman and the Modern American Presidency* (Boston, 1983), 85.
24. Ferrell, *Harry S. Truman*, 89.
25. Goulden, *Best Years*, 106–7.

Truman consult with the Republicans to name a new Secretary of State, resign, and let the new Secretary take over. (That was then the prescribed line of presidential succession.) Truman naturally ignored the un- welcome advice from "Halfbright," as he called him. But he could not hide the obvious: voters had repudiated his administration.

LITTLE THAT HAPPENED in the next few months promised to improve the President's political prospects. In late December sympa- thizers of Wallace announced the formation of Progressive Citizens of America (PCA), which outlined an ambitious progressive domestic agenda and called for worldwide disarmament and immediate destruction of all nuclear weapons. This was the Soviet position. It was obvious that PCA hoped to run Wallace against Truman in 1948, an effort that seemed certain to divide the Democratic vote and likely to throw the election to the Republicans.

Liberals countered a week later by creating the Americans for Demo- cratic Action (ADA). Like the PCA, the ADA favored progressive domes- tic legislation, but it distanced itself far from the Soviet Union: "We reject any association with Communists or sympathizers with communism in the United States as completely as we reject any association with Fascists or their sympathizers." Among the founders of the ADA were Franklin D. Roosevelt, Jr., CIO unionist David Dubinsky, the liberal economist John Kenneth Galbraith, the historian Arthur Schlesinger, Jr., and the young liberal mayor of Minneapolis, Hubert H. Humphrey. The ADA became a vociferous and well-organized pressure group for liberal causes in the 1940s and 1950s. But many of its members openly disdained Truman. The President had little reason to rejoice in its creation at the time. [26]

The new 80th Congress gave Truman even less cause to feel confident. Its members included a few Democratic newcomers who later achieved fame, among them young Congressman John F. Kennedy of Massa- chusetts. But the feisty "Class of 1946" featured a cast of conservative Republicans. Among the new GOP senators were John Bricker of Ohio, who had run as the Republican vice-presidential nominee in 1944, and the then little-known Joseph McCarthy of Wisconsin. Among the new Republican congressmen was Richard Nixon of California. Like others in his party, he demanded that the administration root out leftists in the United States and get tough against the Soviet Union overseas.

26. Steven Gillon, *Politics and Vision: The ADA and American Liberalism, 1947–1985* (New York, 1987), 3–32; Hamby, *Beyond the New Deal*, 159–62.

More influential in the 80th Congress was an older and generally conservative generation of Republican leaders. In the House they included Martin of Massachusetts and a band of midwesterners led by Charles Halleck of Indiana and Everett McKinley Dirksen of Illinois. These Republicans sided with business interests on most domestic issues; many also resisted foreign policy initiatives such as the Marshall Plan. In the Senate the towering Republican figure was Robert Taft of Ohio. Taft was the son of former President and Supreme Court Chief Justice William Howard Taft. He had finished first in his class at Yale and at Harvard Law School, apprenticed as a regular Republican politician in his native city of Cincinnati, and entered the Senate in 1939. Partisan and hard-working, he had risen quickly and challenged boldly but unsuccessfully for the Republican presidential nomination in 1940.

Then and later Taft opposed significant American commitments in Europe. This attitude had hurt him in his race for the GOP presidential nomination in 1940 and placed him at odds with the Truman administration. But he mainly deferred to Vandenberg, his Republican colleague, on such issues. He focused instead on domestic matters, where his towering self-assurance helped to give him extraordinary influence in the Senate. No conservative of his generation evoked more admiration. Critics, Truman among them, responded by painting him as a reactionary. Taft, they charged, "had the best eighteenth-century mind in the Senate."[27]

Taft was slightly less conservative than his liberal critics thought: by 1949 he supported liberal bills to increase funds for public housing and federal aid to education. But in 1947 he stood solidly on the right concerning the major issues of the day, notably labor legislation and tax policy. He was also an adamant partisan, so much so that he became known as "Mr. Republican." In 1946 he had led GOP forces in the fight against OPA, and in 1947 he drove to passage the Taft-Hartley Act, a hotly debated tax cut that especially benefited the wealthy, and other measures that he championed as ways of limiting the influence of Big Government. Taft acted because he detested the liberalism of Roosevelt and Truman. He also hoped to get the GOP presidential nomination. Under his highly partisan leadership the Republican 80th Congress expected to discredit the President.

In doing so the Republicans underestimated Truman, who counterattacked vigorously after his party's decisive defeat in the 1946 elections. It

27. James Patterson, *Mr. Republican: A Biography of Robert A. Taft* (Boston, 1972), 301–68.

was then, in mid-November and early December, that he faced down John L. Lewis and the United Mine Workers. This triumph greatly enlivened him and intensified his zest for political combat. A few months later he acted decisively in foreign policy as well, announcing the Truman Doctrine. Throughout the angry, partisan struggles of early 1947 he showed much more zest for battle than he had in 1945 and 1946.

Nothing did more for Truman's standing among liberals than his ringing veto message condemning the Taft-Hartley bill in June. "T.R.B.," columnist for the *New Republic*, was delighted: "Let's come right out and say it, we thought Truman's labor veto thrilling." James Wechsler, a leading liberal journalist, added, "Mr. Truman has reached the crucial fork in the road and turned unmistakably to the left."[28] Within the month Truman twice vetoed the Republican tax bills. Although Congress overruled his veto of Taft-Hartley, Truman had shown his fighting spirit. It carried him much more confidently into his battle for re-election in 1948.

CLARK CLIFFORD, A NATIVE of St. Louis, had worked as a lawyer before serving in the navy during the war. He entered the Truman administration in 1945 as a junior naval aide. He was handsome, polished, and politically shrewd. His views on the issues—resist the Soviets, promote liberal domestic legislation—accorded with Truman's. By 1947 he was serving officially as Truman's special counsel and unofficially as his most powerful and trusted adviser on matters relating to politics and domestic policy. He stayed in this important role until returning to law practice in February 1950.

In November 1947 Clifford, James Rowe, and other White House aides turned over a forty-three-page memorandum to Truman. It laid out carefully what the President must do if he hoped to win the election of 1948. It was in many ways the most revealing source of Democratic electoral strategy for the forthcoming presidential campaign.

The memorandum was hardly infallible. It blithely took for granted the loyalty of the so-called Solid South: "As always the South can be considered safely Democratic. And in formulating national policy, it can be safely ignored." But the memo was otherwise sound in emphasizing the central fact of American political life since the rise of the New Deal: the potential electoral strength of the Democratic coalition. If Truman could attract the major interest groups in that coalition—blue-collar workers, black people, Jews, other ethnic groups, farmers, and poor people

28. For Truman, labor, and Taft-Hartley, see chapter 2. Also Robert Donovan, *Conflict and Crisis: The Presidency of Harry S. Truman, 1945–1948* (New York, 1977), 299–304; Hamby, *Beyond the New Deal*, 180–87.

generally—he could triumph in 1948, just as FDR had done in the four presidential elections since 1932. This meant that the President must continue to resist the Russians. It especially meant that he must confront the Republican Congress, with the expectation not of getting laws but of winning the election:

> The Administration should select the issues upon which there will be conflict with the majority in Congress. It can assume it will get no major part of its own program approved. Its tactics must, therefore, be entirely different than [sic] if there were any real point to bargaining and compromise. Its recommendations—in the State of the Union message and elsewhere—must be tailored for the voter, not the Congressman; they must display a label which reads "no compromises."[29]

Satisfying the elements of the Democratic coalition called for careful maneuvering, as Truman quickly found out with two such groups in early 1948: blacks and Jews. The volatile issue of race, while far less central to America's national politics at that time than it was to become later, was already creating strains in partisan alignments. After Truman failed in his efforts to secure a permanent FEPC, he named a liberal committee in December 1946 to advise him on civil rights policies. The committee's report, "To Secure These Rights," was released in October 1947 and demanded a range of measures against racism in America. These included legislation eliminating discrimination and segregation in employment, housing, health facilities, interstate transportation, and public accommodations; a law making lynching a federal crime; abolition of the poll tax; federal protection of voting rights; creation of a permanent FEPC; and issuance of executive orders against racial discrimination in the federal civil service and the armed forces.

The report sparked great excitement in liberal circles. The New Republic wrote, "For those who cherish liberty, freedom and forebearance; for those sickened by the sight of reaction ruling the land; for those who feel alone and for those who are afraid, here is a noble reaffirmation of the principles that made America." Truman, too, seemed pleased and ready to act. "Every man," he proclaimed, "should have the right to a decent home, . . . the right to a worthwhile job, the right to an equal share in making public decisions through the ballot."[30]

Truman's support for civil rights did not include social mixing of the races. "The Negro himself knows better than that," he once had ex-

29. Clark Clifford, "Serving the President: The Truman Years (2)," New Yorker, April 1, 1991, p. 60.
30. Hamby, Beyond the New Deal, 188; William Chafe, "Postwar American Society: Dissent and Social Reform," in Lacey, ed., Truman Presidency, 167.

plained, "and the highest types of Negro leaders say quite frankly that they prefer the society of their own people."[31] In private conversation he used "nigger" and other racial slurs from time to time. His Justice Department did little to investigate or prosecute the many violations of civil rights in the country. Still, Truman's appointment of such a liberal committee, and his endorsement of the report, stamped him as a friend of civil rights. No American President before him, FDR included, had taken such a strong stand.

Speaking for civil rights, however, was not the same as taking decisive action. When it came to that, Truman moved slowly. In February 1948 he sent a message to the Hill calling for enactment of some of the committee's recommendations, including passage of an anti-lynching law, a permanent FEPC, and laws against poll taxes and discrimination in interstate transportation. He said he would issue executive orders against discrimination in the armed forces and the civil service. But he did not follow up by introducing legislation, which would have provoked a filibuster, and throughout the spring and early summer of 1948 he failed to issue the executive orders. At the Democratic National Convention in July he gave his support to a civil rights plank so vague that liberals like Hubert Humphrey erupted in protest. Only then, faced with open rebellion, did Truman turn about and support a more liberal plank.[32]

Only then, too, did he finally issue his executive orders, in a move whose political motives were obvious: to prevent loss of black votes in the North. But here, too, Truman moved cautiously, for the issues remained volatile. The order affecting the civil service called for an end to discrimination, not immediately to segregation. More important was his order against segregation in the armed services, into which millions of impressionable young men were later to be drafted. Advocates of civil rights, relatively optimistic in those days, hoped that greater interracial contact among young men would gradually diminish prejudice. They hailed Truman's move.

But this order, too, was implemented only slowly, in part because of resistance to it among top military leaders, who were frightened that

31. Hamby, *Beyond the New Deal*, 46.
32. Barton Bernstein, "The Ambiguous Legacy: The Truman Administration and Civil Rights," in Bernstein, ed., *Politics and Policies of the Truman Administration* (Chicago, 1970), 269–314; Donald McCoy and Richard Ruetten, *Quest and Response: Minority Rights and the Truman Administration* (Lawrence, 1973); and William Berman, *The Politics of Civil Rights in the Truman Administration* (Columbus, Ohio, 1970).

desegregation would damage military discipline and provoke fighting among the troops. The day after Truman issued his order, army chief of staff Omar Bradley warned, "The Army is not out to make any social reform. The Army will not put men of different races in the same companies. It will change that policy when the nation as a whole changes it."[33] Resistance such as this delayed widespread implementation of Truman's order until after the North Korean invasion of South Korea in the summer of 1950, when the American army had to scramble to put together units with any available troops. Even then, black draftees piled up in Japan, barred by the army from joining white units even as battlefield commanders begged for help. Not until 1954 was the process of desegregation in the Army complete in the sense that no unit was more than one-half black. Thereafter blacks continued to form only a very small percentage of the army's officer corps.

Truman's caution on the issue of civil rights greatly bothered many liberals. But his backtracking rested on a political reality that had paralyzed Roosevelt, too: the Democratic party had always been badly divided on the issue of race. In a few northern cities, such as Chicago, Detroit, and New York, the massive south-to-north migrations of black people heightened their potential at the polls. In some of these areas, Clifford had pointed out, black voters could make the difference between victory and defeat. Most whites in the North, however, had not yet been drawn to the cause of racial justice; that began to happen only later. And in the South, where a majority of African-Americans still lived, almost all whites adamantly opposed moves to liberalize race relations. When Truman belatedly accepted the liberal civil rights plank at the convention in July, thirty-five delegates from Alabama and Mississippi marched out of the hall waving battle flags of the Confederacy. They spearheaded a move that culminated in the nomination for President of Governor J. Strom Thurmond of South Carolina on the States' Rights Democratic ticket. "Dixiecrats," as opponents branded them, carried four Deep South states (Alabama, Louisiana, Mississippi, and South Carolina) for Thurmond in November. So much for Clifford's predictions about the loyalty of what was obviously the no-longer-so-solid South.

By contrast to blacks, Jews were a small group. In 1948 they numbered fewer than 5 million people, or around 3.5 percent of the total population. (Blacks then numbered nearly 15 million, or 11 percent.) Jews differed among themselves in the depth and nature of their religious

33. *New York Times*, Feb. 1, 1993.

commitments. But most Jews had greatly admired FDR and the New Deal; by 1948 they were overwhelmingly Democratic. More than blacks, they were concentrated in a few northern urban areas, and they were politically active. Rowe and Clifford had mentioned Jews as potentially vital to Democratic prospects in 1948, especially in the electorally important state of New York.

By this time most politically engaged American Jews had become supporters of Zionism, a movement that called for creation of an independent Jewish state in Palestine—the Holy Land—which had been under mandate to the British. Many Zionists thought such a state had been promised to the Jews in the Balfour Declaration issued by the British Foreign Secretary in 1917. Many others, with the Holocaust fresh in their memories, greatly stepped up their appeals after World War II. The American Zionist Emergency Council (AZEC) embarked on a well-financed publicity campaign that helped by late 1947 to generate large majorities—some polls said over 80 percent—of the American people in favor of such a homeland. AZEC's efforts helped induce thirty-three state legislatures to pass resolutions favoring a Jewish state in Palestine. In addition, forty governors, fifty-four senators, and 250 members of congress signed petitions to Truman on the issue. [34]

All this activity took place amid growing violence between Arabs and Jews in the region, which prompted the British in late 1947 to turn to the United Nations for help. In November the UN supported partition of the region. It was immediately clear, however, that partition—which involved creation of a Jewish state—would drive the Arabs to war. UN leaders then sought to craft a plan that would have placed the area under UN trusteeship, thereby deferring creation of Jewish independence. Virtually all top foreign policy officials in the United States—Secretary of State Marshall, Undersecretary Lovett, Defense Secretary Forrestal, Kennan—also resisted creation of a sovereign Jewish state out of part of Palestine. Helping to set up an independent nation for the Jews, they thought, would endanger American relations with the Moslem world, thereby undermining Truman Doctrine efforts to promote stability in Turkey, Iran, and Arab countries. Behind these concerns was the unthinkable: the cutting off by angry Moslems of shipments of oil to western Europe and the United States.

Forrestal was especially adamant, both because he worried about oil

34. Bruce Kuniholm, "U.S. Policy in the Near East: The Triumphs and Tribulations of the Truman Age," in Lacey, ed., *Truman Presidency*, 324.

supplies and because he (and Marshall, too) was certain that creation of a Jewish state would mean war, which he thought the Jews would lose. "You fellows over at the White House," he exclaimed to Clifford over breakfast one morning, "are just not facing up to the realities in the Middle East. There are thirty million Arabs on one side and about six hundred thousand Jews on the other. It is clear to me that in any contest the Arabs are going to overwhelm the Jews. Why don't you face up to the realities? Just look at the numbers!"[35] Clifford had indeed looked at the numbers. But he was listening more carefully to two ardent Zionists in the White House, presidential assistant David Niles and Max Lowenthal, an old friend of Truman. Clifford sympathized with the plight of the Jews, and he was very conscious of the importance of the Jewish vote. Niles, Lowenthal, and Clifford fed Truman memoranda on the subject and drafted some of his statements.[36]

How much Truman knew of this in-house activity is unclear, but throughout the winter and early spring of 1948 he did not give the matter great attention. It gradually became clear, however, that though he was irritated at times by Zionist pressure, he sympathized with the Jewish position. This predisposition had many sources. He appreciated the suffering of the Jews, their apparent commitment to democracy, and their desire to establish a new world for themselves. His selective reading of history inclined him to believe that the Jews had the best claim to a homeland in the area. And he was well aware of domestic political considerations, including the importance of campaign contributions from the Jews. He once told State Department people, "I have to answer to hundreds of thousands who are anxious for the success of Zionism. I do not have hundreds of thousands of Arabs among my constituents."[37]

For all these reasons Truman made little effort to see things from the perspective of Arabs, who hotly demanded to know why the President was ignoring their deeply held feelings—Palestine was a Holy Land for Moslems (and Christians) as well as for Jews—at the same time that he was resisting large-scale immigration of Jewish refugees to America. Deaf to such complaints, Truman also downplayed the ferocity of regional hatreds in the Middle East. He shared with many other American liberals the naive hope that Jews and Arabs could learn to cooperate and that the United States could manage to get along amicably with both sides.

35. Clark Clifford, "Serving the President: The Truman Years (1)," *New Yorker*, March 25, 1991, p. 59.
36. Kuniholm, "U.S. Policy," 324.
37. Ibid., 324.

The issue came to a head on May 12, when Truman assembled his top advisers for a key meeting. By this time the imminent departure of the British (on May 14) made it impossible to put off decisions any longer. It was one of the most explosive confrontations of his presidency. When Marshall found Clifford at the meeting, he grew angry and accused him—and Truman—of favoring a Jewish state for political reasons. "Unless politics were involved," he said in what Clifford later called a "righteous goddam Baptist tone," "Mr. Clifford would not even be at this meeting."[38] Astounding those present, Marshall then said to his commander-in-chief, "If you follow Clifford's advice and if I were to vote in this election, I would vote against you." According to Clifford's recollection, this outburst—all the more shocking because it came from the normally grave and judicious Marshall—so stunned those present that the meeting ended then and there.[39]

The confrontation deeply upset the President, who revered Marshall. Truman especially feared that Marshall might make a public statement on the issue, thereby exposing the divisions within his inner circles. Worse, Marshall might resign, causing political damage to his administration. Calling in Clifford, the President asked him to work out a solution. Clifford turned to Lovett, a friend, as a go-between with Marshall. Two very tense days followed, after which Marshall finally signaled that he would not rock the boat. On May 14 Israel proclaimed itself a state, which the Truman administration instantaneously recognized de facto. Arabs then attacked Israel, which surprised a lot of people by resisting effectively and ultimately winning the war.

Many American Jews wanted Truman to go further in the summer of 1948—by recognizing the new state de jure and giving arms to Israel. But they appreciated Truman for his immediate de facto regonition, and they voted in large majorities for him in November. (The President, however, lost New York State.) Whether the President deserved great credit for his policies nonetheless remains debatable. He was not well informed about Middle Eastern history and politics; he did not take charge of policy-making on the matter; and he let political considerations affect important questions of national security. Here as at other times in his presidency, Truman vacillated, exhibiting little of the "buck stops here" decisiveness with which he has been credited.

38. David McCullough, *Truman* (New York, 1992), 614.
39. Clifford, "Serving (1)," 60–64; Ernest May, "Cold War and Defense," in Keith Nelson and Robert Haycock, eds., *The Cold War and Defense* (New York, 1990), 11.

Truman's policies had controversial long-range results. Hoping for the best, Truman aligned the United States with the Jews and therefore against the Arabs.[40] In so doing he associated the United States with the continued survival of Israel. This alignment greatly bolstered the power—henceforth substantial—of pro-Israeli groups in American politics. It also inflamed Moslem wrath against the United States, which confronted crisis after crisis in the Middle East thereafter. Still, there were no easy solutions. Jewish and Moslem nationalism were colliding inexorably in the Holy Land, and there was no policy for America that would not offend one side or the other. In choosing Israel Truman acted out of humanitarian and political concerns and in the hope that the presence of a democratic, pro-American nation in the Middle East would promote long-range Western security in the Cold War. The humanitarian and political motives were hard to resist at the time; whether Truman's hopes concerning security were correct continued to be debated decades later.

WHILE TRUMAN WAS STRUGGLING with civil rights, Palestine, and the 80th Congress in early 1948, few people gave him much of a chance of winning the election ahead.[41] Indeed, he faced open rebellion from a number of onetime Democrats, who by then were gearing up to support Henry Wallace for President on a Progressive party ticket. Wallace had announced his candidacy in December 1947, exhorting his "Gideon's Army, small in number, powerful in convictions," to turn Truman out of office.[42]

Some liberal Democrats who were cool to Wallace also searched for ways to dump Truman in early 1948. In March Elliott Roosevelt and Franklin D. Roosevelt, Jr., sons of FDR, publicly endorsed General Eisenhower for the Democratic presidential nomination. A month later the New Republic ran a front-page editorial titled AS A CANDIDATE FOR PRESIDENT, HARRY TRUMAN SHOULD QUIT. At the same time, the board of Americans for Democratic Action called for an open convention. A few of these political activists were ready to back Eisenhower; a larger number openly favored William Douglas, a liberal justice of the Supreme Court. They liked most of Truman's policies but deplored his "leadership" and

40. Kuniholm, "U.S. Policy," 336.
41. Irwin Ross, The Loneliest Campaign: The Truman Victory of 1948 (New York, 1968).
42. Curtis MacDougall, Gideon's Army (New York, 1965); Norman Markowitz, The Rise and Fall of the People's Century: Henry A. Wallace and American Liberalism, 1941–1948 (New York, 1973); Hamby, Beyond the New Deal, 201–2.

were sure he would lose in November. James Wechsler, a liberal journalist, explained that "Mr. Truman's place in history may be written in Mike Gonzales' ageless remark about a rookie ballplayer: good field, no hit."[43]

Before the opening of the Democratic National Convention in Philadelphia in July, other leading Democrats joined in a "Stop Truman" effort. They included liberals such as Chester Bowles, who was running for governor of Connecticut; Hubert Humphrey, who was seeking a Senate seat from Minnesota; Florida senator Claude Pepper; and the UAW's Walter Reuther. Some favored Eisenhower, some Douglas.[44] Rarely in the modern history of American politics had so many leading party figures rebelled openly against an incumbent President seeking renomination.

The craze for Eisenhower was ironic, for Truman himself had once been so enamored of him—and so unsure of himself—that he had offered to back him for the presidency. "General," he told "Ike" at Potsdam, "there is nothing that you may want that I won't try to help you get. That definitely and specifically includes the presidency in 1948."[45] By 1948 Truman had no such notions in mind, and Eisenhower, who had recently been selected as president of Columbia University, resisted all entreaties. Privately Ike noted that the Democrats had been "desperately searching around for someone to save their skins" but that his friends "would be shocked and chagrined at the very idea of my running on a Democratic ticket for anything."[46] When Douglas, too, spurned admirers, rebellious liberal Democrats at the convention in July were without a candidate. Determined to make a difference, they overturned the moderate civil rights plank, in so doing driving the Alabamans and Mississippians out of the hall. But they had no choice except to join in the nomination of Truman and Barkley.

From this unpromising point the campaign turned around for Truman and the Democrats—for four reasons: the parochialism of the Dixiecrats, the political ineptitude of Wallace, the even greater ineptitude of the Republicans, and Truman's own spirited counter-attacks. The result was his remarkable victory in November, a triumph scarcely imagined by most people in July.

First, the Dixiecrats. Thurmond was a young, vigorous, and energetic campaigner. For the most part he tried to focus on states' rights, not just on race. Truman's civil rights programs, he said, would threaten the

43. Hamby, Beyond the New Deal, 225–26.
44. Ibid., 242; Goulden, Best Years, 381–91.
45. Dwight Eisenhower, Crusade in Europe (Garden City, N.Y., 1948), 444.
46. Stephen Ambrose, Eisenhower: The President (New York, 1984), 278.

prerogatives of states. Thurmond also appealed to anti-Communist feelings, which had become strong among Americans by that time. Radicals, subversives, and Reds, he insisted, had captured the Democratic party. Truman's civil rights program had "its origin in communist ideology" and sought "to excite race and class hatred" and thereby "create the chaos and confusion which leads to communism."[47] Thurmond's linkage of civil rights and Communism was to become a staple of right-wing thinking over the next several decades. Beyond the Deep South, however, he had little credibility as a presidential candidate. Even there, representatives and senators were reluctant to bolt the Democratic party, lest they be denied seniority and other trappings of power in the congressional session of 1949. Truman, having accepted the more liberal civil rights plank at the convention, shortly thereafter issued his executive orders against discrimination and hoped for the best in the South. As Rowe and Clifford had counseled, he concentrated instead on attracting a good-sized vote in the North, where the election was likely to be close.

Wallace proved to be almost as non-threatening to Truman as Thurmond. Many liberal Democrats were originally attracted to him, for he took bold stands in favor of civil rights and other progressive issues. As the campaign developed, however, Wallace's views on foreign policy alarmed many of these supporters. Some were already upset by his opposition to what he called the "Martial Plan." Others thought him a virtual tool of the Communists. Reuther explained, "Henry is a lost soul. . . . Communists perform the most complete valet service in the world. They write your speeches, they do your thinking for you, they provide you with applause, and they inflate your ego."[48] Wallace seemed unworried in 1948 by the Communist coup in Czechoslovakia and by the fate of West Berlin. Throughout the campaign he seemed blind to the political impression that his association with Communists was making. "If they [Communists] want to support me," he said, "I can't stop them."[49]

Truman tried to ignore Wallace, but it was hard for him, and he succumbed to the temptation to Red-bait. The President said, "I do not want and I will not accept the political support of Henry Wallace and his Communists." "A vote for Wallace," he added later, " . . . is a vote for all the things for which Stalin [and] Molotov . . . stand." On another

47. Goulden, Best Years, 402–3; Richard Polenberg, One Nation Divisible: Class, Race, and Ethnicity in the United States Since 1938 (New York, 1980), 110.
48. John Diggins, The Proud Decades: America in War and Peace, 1941–1960 (New York, 1988), 105.
49. Hamby, Beyond the New Deal, 230–32, 245.

occasion he departed from a prepared text to urge Wallace "to go to the country he loves so well and help them out against his own country if that's the way he feels."[50]

Wallace's close association with Communist ideas indeed cost him dearly in the Cold War climate of 1948. Well before November, many Progressive party candidates for Congress withdrew in favor of liberal Democrats. Only a handful of well-known figures, including the black singer and Communist Paul Robeson, stuck with Wallace. The socialist Irving Howe dismissed Wallace as a "completely contrived creature of Stalin." John Dewey, America's most eminent philosopher, added, "There can be no compromise, no matter how temporary, with totalitarianism. Compromise with totalitarianism means stamping an imprimatur on the drive for a *pax Sovietica*."[51]

The political ineptitude of Wallace, who was known for his idiosyncrasies, might have been predicted. That of the Republican presidential nominee, Thomas E. Dewey, was not. He was after all a seasoned campaigner and office-holder, having twice been elected governor of New York, where he was generally popular. Indeed, he went on to win a third term in 1950. In 1944 Dewey had undertaken the formidable task of challenging FDR for the presidency and had come closer to winning than any of Roosevelt's other opponents for that office. He captured the nomination again in 1948, defeating Taft and Harold Stassen, the moderately liberal former governor of Minnesota, in primaries and in the GOP convention. Dewey was liberal, especially by contrast to Taft and other leading congressional Republicans, and he endorsed a GOP platform in 1948 that was very progressive on civil rights. Although Dewey implied that the Democrats were not tough enough on Communism, he refrained from Red-baiting. In a key debate with Stassen during the primary campaign Dewey had refused to endorse the outlawing of the American Communist party.[52]

But Dewey exhibited two fatal flaws. First, he was personally cold, pompous, and virtually without charisma. While machine-like in his efficiency, he appeared uninterested in people around him. Alice Roosevelt Longworth, Theodore Roosevelt's daughter, memorably called him "the little man on the wedding cake." Even smiling seemed to come with

---

50. Fred Siegel, *Troubled Journey: From Pearl Harbor to Ronald Reagan* (New York, 1984), 69; Hamby, *Beyond the New Deal*, 223.

51. Pells, *Liberal Mind*, 107–12; Hamby, *Beyond the New Deal*, 263.

52. Richard Smith, *Thomas E. Dewey and His Times* (New York, 1982).

difficulty. A photographer once said, "Smile, governor." "I thought I was," he responded.[53]

Dewey's other liability was overconfidence. Virtually all the pundits gave Truman no chance to win, and Dewey believed them. He did not get his campaign underway until mid-September and did not work hard at it thereafter. His speeches were bland in the extreme and offered no reason for voters to choose him over Truman. Neither he nor his running mate, Governor Earl Warren of California, paid much attention to farm-state voters, who were restive in 1948. One reporter quipped that Dewey didn't run; he walked. Another called him "Mr. Hush of politics."[54] The *Louisville Courier-Journal* later summed up his campaign: "No presidential candidate in the future will be so inept that four of his major speeches can be boiled down to these four historic sentences: Agriculture is important. Our rivers are full of fish. You cannot have freedom without liberty. The future lies ahead. (We might add a fifth: the TVA is a fine thing, and we must make certain that nothing like it happens again.)"[55]

Truman's campaign offered a dramatic contrast. He launched it right away by insisting on addressing party delegates following his renomination. It was by then 2:00 A.M., and he had sat patiently in the wings awaiting his opportunity. Then he electrified the faithful with a spirited partisan attack on the conservative Congress, which he proceeded to call into special session. "It was a great speech for a great occasion," Max Lerner said, "and as I listened I found myself applauding." T.R.B. of the *New Republic* added, "It was fun to see the scrappy little cuss come out of his corner fighting . . . not trying to use big words any longer, but being himself and saying a lot of honest things."[56]

After the special session—which deadlocked—Truman waged an extraordinarily energetic campaign. Between September and election day he traveled a record 31,700 miles, much of it on trains that "whistle-stopped" across the country. Standing on the back of the train, Truman assailed Congress, after which he asked the crowds if they would like to meet his family. Then he introduced Bess, his wife, as "the Boss," and Margaret, his daughter, "who bosses the boss." As the train pulled out Margaret would toss a red rose into the crowd.[57]

53. David Halberstam, *The Fifties* (New York, 1993), 6.
54. Diggins, *Proud Decades*, 107.
55. Siegel, *Troubled Journey*, 69.
56. Ferrell, *Harry S. Truman*, 87–104; Donovan, *Conflict and Crisis*, 395–439; Hamby, *Beyond the New Deal*, 244.
57. Clifford, "Serving (2)," 62–63.

Truman's speeches constantly reminded listeners of all the programs that he supported and that conservative Republicans opposed: expansion of Social Security, additional public housing, higher minimum wages, controls on inflation, more progressive taxation. He regularly emphasized his toughness—including the Berlin Airlift, which was front-page news during the campaign—against the Soviets. And he relished partisan attacks on his opponents. The morning he left on the first of his long train trips Barkley came down to the station to wish him well. "Go out there and mow 'em down," Barkley advised. "I'll mow 'em down, Alben," Truman replied, "and I'll give 'em hell." Reporters heard the conversation and put it in their stories, and by the time the train reached the West Coast people were yelling, "Give 'em hell, Harry."[58]

Sometimes this hell could be hot. Seeking to rally the working classes at the core of the Democratic coalition, Truman charged that Republicans were "Wall Street reactionaries," "gluttons of privilege," "bloodsuckers," and "plunderers." GOP legislators in the 80th Congress, he said, were "tools of the most reactionary elements" who would "skim the cream from our natural resources to satisfy their own greed." Dismissing Dewey, "whose name rhymes with hooey," Truman said, "If you send another Republican Congress to Washington, you're a bigger bunch of suckers than I think you are." "Give 'em hell, Harry!" the people shouted back. "Pour it on!"[59]

Pollsters and pundits, however, paid little attention to the popular enthusiasm that Truman's travels were generating. They continued to believe that Truman would lose. The Elmo Roper polling service predicted on September 9 that Dewey would get 44.3 percent of the then decided vote to Truman's 31.4 percent, Wallace's 3.6 percent, and Thurmond's 4.4 percent. Roper added that he would continue polling but would not report the results "unless something really interesting happens. My silence on this point can be construed as an indication that Mr. Dewey is still so clearly ahead that we might just as well listen to his inaugural."[60] A month later *Newsweek* polled fifty top political journalists, all of whom picked Dewey as the winner. The final polls showed Dewey winning by 52.2 to 37.1 percent (Roper), 49.5 to 44.5 percent (Gallup), and 49.9 to 44.8 percent (Crosby). Reinhold Niebuhr spoke for

58. Goulden, *Best Years*, 393.
59. William O'Neill, *American High: The Years of Confidence, 1945–1960* (New York, 1986), 100; Ferrell, *Harry S. Truman*, 100–1; Hamby, *Beyond the New Deal*, 248–52.
60. Goulden, *Best Years*, 398.

many lukewarm Truman supporters in early November: "We wish Mr. Dewey well without too much enthusiasm and look to Mr. Truman's defeat without too much regret."[61]

On election eve the experts were still sure that Dewey would win, no one more so than Colonel Robert McCormick, the ultra-conservative publisher of the *Chicago Tribune*, perhaps the most influential newspaper in the Midwest. On the first of its eleven election-night editions the *Trib* bannered, DEWEY WINS ON BASIS OF FIRST TALLY. By 10:00 P.M. it decided to pull out all the stops: DEWEY DEFEATS TRUMAN. An early election-eve edition of the *San Francisco Call-Bulletin* carried a cartoon depicting a jubilant elephant and a doleful donkey. Only in the next edition was there a fast retouch: the elephant was now startled, the donkey joyous.[62]

When the results became clear on the morning after the election, most Democrats were indeed joyous. Truman received 24,179,345 votes to Dewey's 21,991,291. This was 49.6 percent of the popular vote to Dewey's 45.1 percent. Thurmond got 1,176,125 popular votes and Wallace 1,157,326. Thereafter the Progressive party disappeared, and Wallace ceased to be a force in American politics. Norman Thomas, running for the last time as the Socialist candidate, received 139,572 votes. His weakness, too, exposed the pitiful state of the political Left in America. Truman, carrying twenty-eight states to Dewey's sixteen, won 303 electoral votes to Dewey's 189. Thurmond's four states brought him 39 electoral college votes. No other candidate broke into the electoral college in 1948.

It was a highly satisfying triumph for Truman and for the Democratic party, which regained control of Congress. Among the liberal newcomers was Humphrey of Minnesota. The comedian Fred Allen chortled, "Truman is the first President to lose in a Gallup and win in a walk." Entraining to Washington from Missouri, the President gleefully posed for photographers while he held up the *Chicago Tribune* headline. When he arrived at Union Station, a huge crowd cheered him, and the *Washington Post* displayed a big sign: MR. PRESIDENT, WE ARE READY TO EAT CROW WHENEVER YOU WANT TO SERVE IT. Dewey had already gone to New York's Grand Central Station to board a train for Albany. He smiled wanly but declined to wave for photographers.[63]

61. Ibid., 415.
62. *Newsweek*, Nov. 15, 1948, p. 56.
63. Goulden, *Best Years*, 421; Dewey, in J. Ronald Oakley, *God's Country: America in the Fifties* (New York, 1986), 36.

The Democrats had reason to chortle, but it was also clear that Truman had not won by much. The election was the closest since 1916. Truman ran behind his ticket in many key states and failed to draw a surge to the polls: voter turnout, at 53 percent of the registered electorate, was the lowest since 1924. With fewer than 50 percent of the votes he was a minority President without a strong mandate for his second term. Except for Massachusetts and Rhode Island, Truman failed to carry the normally Democratic strongholds of the industrial Northeast. Wallace, strongest in New York City, may have cost Truman New York State, and he probably hurt him in Maryland, New Jersey, and Michigan. Truman won three other key states, Ohio, California, and Illinois, by margins of 7,107, 17,865, and 33,612 respectively. If Dewey had taken these very closely contested states, he would have carried the electoral college.

The fact remained, however, that Truman did win. Why remains debatable, for all sorts of reasons can be adduced in explanation. Among these, all agree, were Truman's courageous campaign and Dewey's lethargic one. The candidacies of Wallace and Thurmond may actually have helped Truman, by reminding voters that the President, no extremist of Left or Right, stood steadfast against the Russians and solidly in the center of moderate-to-liberal opinion in the United States. As election day approached and Wallace and Thurmond voters—normally Democratic—realized their candidates had no chance of winning, many swallowed their doubts and chose Truman over Dewey and the GOP.

Most analyses of the 1948 election stress above all, as Rowe and Clifford had, the importance of the Democratic coalition. Here, as in so many other ways, the shadow of Roosevelt loomed over the political landscape. Truman carried the thirteen largest cities, doing especially well in the poorest and working-class wards. Like Roosevelt, Truman also had success in the border states and in the West, carrying every state west of the Plains save Oregon. Again like Roosevelt, Truman was particularly popular among Catholics, Jews, and European-American voters. Blacks were stronger for Truman in some cities than they had been for FDR in 1944. They greatly helped Truman in the key states of Illinois, Ohio, and California.

Two groups in the Democratic coalition were probably of special importance to the Democrats in 1948. One was organized labor, which except for Lewis's United Mine Workers was pro-Truman. Labor, to be sure, was hardly all-powerful: Truman even lost Michigan, stronghold of the UAW. But labor organizers worked hard for Truman and against

Wallace, often Red-baiting him. The Political Action Committee of the CIO effectively registered union members and got them to the polls. Although the AFL issued no formal endorsement, it created Labor's League for Political Action, which printed and distributed reams of literature critical of the GOP and of the Taft-Hartley "slave labor" Act. The AFL had never before been so active in American politics.[64]

The other group were commercial farmers. While Dewey and Warren were largely ignoring the rural areas, Truman made some eighty speeches in farm states during the campaign. Again and again he rapped the Republican Congress for its apparent indifference to farm prices, which thanks to bumper crops sagged in the late summer of 1948. Truman excoriated Congress for cutting off the authority of the Commodity Credit Corporation (CCC) to acquire additional storage space for crop surpluses. This action prevented many farmers from storing their excess until such time as market prices might improve and from receiving additional CCC loans to tide them over. Truman's assault on the Congress was demagogic, for the storage issue had been non-controversial when it was decided early in the summer. But he correctly read the fear and frustration of many farmers in the fall. In November he carried three states that Dewey had won in 1944, Ohio, Wisconsin, and Iowa.[65]

Truman succeeded, finally, because the majority of Americans were better off in 1948 than they had been in earlier years. The postwar boom—in cars, household appliances, suburbanization, education, real wages—was gathering steam. Memories of the Great Depression were slowly fading, and millions of people were moving hopefully up the ladder to new positions in life. This is not to say that bread alone moves votes—it does not. But political incumbents are blamed for recessions and praised for progress. It was Truman's good fortune, for he had little to do with what was happening to the economy, to be President in relatively good times. As the incumbent in a society of rising expectations he was carried back into office.

ADMIRERS OF TRUMAN contend that he was not only a bold campaigner but also a savior of liberalism and the New Deal. The events of his first term lend a little credibility to this view. He fought back against the conservative coalition in Congress, especially after 1946, and headed

64. Goulden, *Best Years*, 409–12.
65. *Newsweek*, Nov. 15, 1948, p. 25.

a triumphant Democratic victory in 1948. But Truman's role in these events should not be exaggerated, for he was slow to find his way before 1947, often insecure, and far less decisive than legend has it. Polls demonstrate that he was never very popular personally.

Much more important in preserving the New Deal were political forces established before Truman took office. By 1945 most Americans had accepted Rooseveltian programs such as Social Security. Only a small minority of reactionaries thought of tearing down the rudimentary welfare state that FDR had set up in the 1930s. The Roosevelt years had also revolutionized political allegiances by creating the Democratic electoral coalition. This, too, was solid enough to survive as a central fact of American politics after the 1940s. Truman would have had to bungle badly to lose its support in 1948.

All this is to offer the heresy that the role of presidential leadership, yet another shadow cast by the Roosevelt years, is often exaggerated. Presidents of course can take executive actions, especially in foreign affairs, that have dramatic effects. But only sometimes, for many snags—bureaucratic inertia, the capriciousness of public opinion, partisan opposition, interest group pressures, Congress—hem in presidential designs. In domestic policies the hemming-in is ordinarily tight indeed, as it was during Truman's first term. A decent, moderately liberal man, Truman labored to prevent the unraveling of the political design that Roosevelt and the New Deal had stitched together. In this modest holding action he largely succeeded.

7

Red Scares Abroad
and at Home

The radical social activist Michael Harrington once commented that "1948 was the last year of the thirties." He meant specifically that labor unrest and class-consciousness abated amid rising prosperity after 1948. So did chances for substantial extension of the New Deal. The political Left, already weak, reeled under sustained assaults. In place of reform activity, Cold War fears rose to the center of American society, politics, and foreign policy in 1949 and early 1950, generating a Red Scare that soured a little the otherwise optimistic, "can-do" mood of American life until 1954.

TRUMAN DID NOT KNOW how strong the Red Scare was to be when he returned seriously to the political fray in January 1949. Buoyed by the election, he hoped that the new Democratic Congress—54 to 42 in the Senate and 263 to 171 in the House—would support a wide range of domestic programs that he christened as the Fair Deal in his State of the Union address. Two weeks later he was inaugurated on a brilliantly clear day that seemed dazzling with promise. The first full-scale inaugural since the war, it was also the first to be seen on television. An estimated 10 million people as far west as St. Louis—where the television coaxial cable then terminated—watched the ceremonies. Millions more heard them on radio. At age sixty-four Truman seemed to brim with vitality and optimism.[1]

1. David McCullough, *Truman* (New York, 1992), 723–24.

From the beginning, however, the President had troubles with Congress. The Fair Deal was a long and liberal laundry list: repeal of the Taft-Hartley Act, a more progressive tax system, a seventy-five-cent minimum wage (it was then forty cents), agricultural reform, resource development and public power, broadening of Social Security, national medical insurance, federal aid to education, civil rights, and expansion of federal housing programs.[2] By the end of the 1949–50 congressional sessions Truman partially achieved three of these goals: public housing, a hike in the minimum wage, and expansion of Social Security.[3] Otherwise the coalition of Republicans and conservative Democrats continued to dominate. At the start of the session, the Senate stymied liberal efforts to make cloture (cutting off of debate) possible by a simple majority instead of two-thirds. Congress's action killed chances for civil rights legislation, which barely simmered on congressional back burners until the late 1950s. Special interest groups helped defeat the other programs. Congress refused to repeal Taft-Hartley, to pass a farm reform program, or to approve federal aid to education. National health insurance continued to be strongly opposed by the American Medical Association and failed to pass; that, too, faded as a visible legislative issue until the late 1950s.

The fate of efforts for agricultural reform illustrated the constellation of forces, especially well-organized interest groups, that stymied much of the President's Fair Deal. The reform took the name of the Brannan Plan, named after Truman's liberal Agriculture Secretary, Charles Brannan. He sought to scrap the costly system of production controls, government price supports, and benefit payments that had been enacted during the 1930s. In its stead, Brannan proposed, farmers raising perishable crops would be encouraged to produce as much as the market would bear—an effort that was expected to increase supply and drive down prices for consumers. In return, the government would compensate these farmers with direct income payments, up to maximums per producer. With these maximums he expected to limit the amount of benefits that would go to big producers and to attract the support of smaller "family farmers." Brannan's larger goal was political: to cement the Democratic alliance between small farmers, urban workers, and consumers that had appeared to be developing in the 1948 election.

A determined coalition of interests, however, opposed the plan. It included the majority of Republicans, who resisted Brannan's ill-con-

2. Alonzo Hamby, *Beyond the New Deal: Harry S. Truman and American Liberalism* (New York, 1973), 293.
3. Alan Wolfe, *America's Impasse: The Rise and Fall of the Politics of Growth* (New York, 1981), 84–87. See chapter 11 for a discussion of the housing act.

cealed political objectives; the Farm Bureau Federation, which represented the large growers; and many farmers, food processors, and middlemen, who feared the imposition of new and possibly complicated controls and who predicted that the costs of the plan would bankrupt the government. Some urban Democrats, too, were cool to a program that proposed to direct federal money at rural areas. A number of southern Democrats, worried that the plan would end up reducing government subsidies for cotton, also joined the anti-Brannan coalition. All these opponents defeated the plan in the House in 1949. While it appeared to have some chance in the Senate, the outbreak of the Korean War shunted it aside in 1950. It then died, leaving the old system in place. Thereafter, as in the past, powerful interests remained firmly in control of America's agricultural system.[4]

Truman fared a little better in his quest for anti-Communist foreign policies. In July 1949 the Senate overwhelmingly (82 to 13) ratified American participation in the North Atlantic Treaty Organization (NATO). The pact committed the twelve signatories to treat an attack on one as an attack upon all.[5] This was a historic commitment for the United States, which since 1778 had refused to join military alliances in time of peace. When Cold War tensions increased in 1950, Truman sought to develop the military potential of the pact. After a "great debate" in early 1951, American troops were assigned in April to NATO forces in Europe, where they remained for decades.[6]

Truman's foreign and military policies otherwise ran up against formidable pressures. One such policy was Point Four, so named because it was the fourth point in his 1949 inaugural address. It called on Congress to appropriate funds for American technical assistance to so-called underdeveloped nations. Truman occasionally entertained vast, idealistic notions of transforming the Euphrates, Yangtze, and Danube valleys into models of the American TVA. But he added Point Four at the last minute and had done little to explain his goals to the State Department. Dean

4. Alan Matusow, *Farm Policies and Politics in the Truman Years* (New York, 1970), 191–221; Robert Griffith, "Forging America's Postwar Order: Domestic Politics and Political Economy in the Age of Truman," in Michael Lacey, ed., *The Truman Presidency* (Washington, 1989), 75; Hamby, *Beyond the New Deal*, 305–10.
5. Aside from the United States, these were Belgium, Canada, Denmark, France, Great Britain, Iceland, Italy, Luxembourg, Netherlands, Norway, and Portugal. Greece and Turkey joined in 1952, West Germany in 1955.
6. John Gaddis, *Strategies of Containment: A Critical Appraisal of Postwar National Security Policy* (New York, 1982), 72–73; Robert Pollard, "The National Security State Reconsidered: Truman and Economic Containment, 1945–1950," in Lacey, ed., *Truman Presidency*, 223–25.

Rusk, charged with helping to coordinate the program, later complained that "we in the State Department had to scurry around and find out what the dickens he was talking about and then put arms and legs on his ideas."[7] This was hard to do, in part because many conservatives and business leaders were cool to Point Four. Such a program would spend taxpayers' money; technical aid might assist potential competitors. Finally approved in May 1950, Point Four was poorly funded and sputtered along as a small and insignificant addition to overseas lending oriented mainly to Cold War concerns.[8]

Truman's military programs in 1949–50 provoked further controversy, most of it within his administration. When Forrestal was forced to step down as head of Defense in early 1949, Truman replaced him with Louis Johnson, a loyal fund-raiser during the whistle-stop campaigning in 1948. Johnson was blunt, blustery, and highly ambitious, and he aroused a storm when he canceled the "supercarrier" that the navy was counting on as its key weapon of the future. Top naval officers dared insubordination by openly resisting Johnson and by opposing air force development of the B-36 bomber. The interservice brawling became ugly. General Omar Bradley, Chairman of the Joint Chiefs, sided with the air force and called navy leaders "fancy dans" who refused to play on a team "unless they can call the signals." A compromise was finally reached in 1950, but the fighting exposed the divisions that still plagued the military establishment and revealed the inability of the Defense Secretary to discipline the Pentagon.[9]

The fire and smoke emanating from these battles partially obscured a continuing reality: American military defense remained unbalanced. Under General Curtis LeMay, a tough, hard-driving Cold Warrior who took over the Strategic Air Command in late 1948, America's long-range bombing potential gradually acquired some efficiency. Nuclear tests in 1948–49 also encouraged planners: for the first time they could look ahead to quantity production of nuclear bombs that could be handled safely. But even that was a few years ahead.[10] And fiscal considerations helped keep overall military expenditures in check. The defense budget in 1949–50 was around $13 billion, less than half the amount requested by the services. Low appropriations especially demoralized the army, whose

7. Dean Rusk, as told to Richard Rusk, *As I Saw It* (New York, 1990), 141.
8. Hamby, *Beyond the New Deal*, 371; Wolfe, *America's Impasse*, 174–76.
9. Ernest May, "Cold War and Defense," in Keith Nelson and Robert Haycock, eds., *The Cold War and Defense* (New York, 1990), 44–51.
10. Ibid, 44–54.

strength had sunk to a low of 591,000 men by the time the Korean War broke out in June 1950. Given America's grand expectations of leading the so-called Free World, the modest size of its military establishment was ironic. Acheson had earlier put his finger on these contradictions when he said that postwar American foreign policy could be summed up in three sentences: "1. Bring the boys home; 2. Don't be Santa Claus; 3. Don't be pushed around."[11]

AS IN 1947, WHEN THE BRITISH decided they could no longer assure the security of Greece or Turkey, two events abroad in late summer and early fall of 1949 had momentous effects in the United States. These were the intelligence in late August that the Soviets had successfully exploded an atomic bomb, and the collapse of the Nationalist regime of Chiang Kai-shek that culminated on October 1 in the creation of the Communistic People's Republic of China. These developments drove many people in the administration to reconsider their reliance on economic aid and to think about substantially greater militarization of the Cold War. The events also unleashed a gathering spate of criticism from anti-Communist groups in the United States who blamed Truman for having done too little too late. Some saw spies under the tables of state. A Red Scare, already an undercurrent in American life, rose ominously in 1949–50, ultimately diverting national politics—and much else— throughout the next four years.

If top administration leaders had spoken more frankly about what they knew about Soviet science, the explosion in the USSR would not have come as a big surprise. They recognized that the Russians understood the basic science involved, and military leaders were aware that Stalin had given nuclear development very high priority. Moreover, the Soviet achievement did not change very much in the short run. The Pentagon recognized that the USSR still lacked the long-range bombers necessary to wage an air attack on the United States and that Soviet air defenses—to say nothing of the Soviet economy—were weak. Still, when Truman informed the American people in September, many were deeply alarmed. The cover of the *Bulletin of Atomic Scientists*, which until then had displayed a clock with the minute hand pointing to eight minutes before twelve, the hour of doom, now moved the hand to 11:57.[12] Many other

11. Evan Thomas and Walter Isaacson, *The Wise Men: Six Friends and the World They Made* (New York, 1986), 338.
12. Robert Jungk, *Brighter Than a Thousand Suns: The Story of the Men Who Made the Bomb* (New York, 1958), 265.

Americans simply refused to believe that the Communists—whose system was surely technologically inferior—could have managed the feat by themselves: spies must have done it for them.

Within the administration the news reinforced the hand of advocates who were demanding the strengthening of American armed forces. After all, Stalin still seemed tyrannical and unyielding. He had promoted the coup in Czechoslovakia and threatened West Berlin. Who could tell what he would do when he had the planes to deliver the Bomb? Kennan, who then headed the State Department's Policy Planning Staff, had been opposing major militarization of containment and had been calling on the administration to step up negotiations with the Soviets. He also recommended thinking about the reunification and demilitarization of Germany, as a way of reducing a major source of Cold War tensions in central Europe. News of the Soviet bomb destroyed his hopes, and he resigned, discouraged and defeated, at the end of the year.[13] Henceforth America's European policy moved rapidly toward the militarization of NATO, the rearming of the Federal Republic of Germany (West Germany), which joined NATO in 1955, and American acceptance of the apparently permanent division of Germany and Europe.

The victory of Mao Tse-tung in China should have been even less surprising. Since the end of World War II his Communist forces had steadily beaten back the Nationalists under Chiang Kai-shek, who ultimately fled to the island of Taiwan, where he imposed a harsh rule on the natives. Well before 1949 many Americans close to the scene had been disgusted by Chiang, a corrupt and increasingly unpopular leader with his own people. General Joseph "Vinegar Joe" Stilwell, America's chief military adviser in China during World War II, had complained at the time that the Nationalists were more interested in battling the Communists than the Japanese. In coded messages he had referred contemptuously to Chiang as "the Peanut."[14] In 1945–46 Truman had hoped that America could help end the civil war and sent Marshall to China as an emissary. There was no stopping the fighting, however, and Truman lost all faith in Chiang. They were "all thieves, every last one of them," he said privately of the Nationalists in 1948.[15]

13. Gaddis, *Strategies of Containment*, 82–83; Pollard, "National Security State."
14. Barbara Tuchman, *Stilwell and the American Experience in China, 1911–1945* (New York, 1970), 3–5, 283, 494.
15. Tang Tsou, *America's Failure in China, 1941–1950* (Chicago, 1963); Dorothy Borg and Waldo Heinrichs, *Uncertain Years: Chinese and American Relations, 1947–1950* (New York, 1980); Robert McMahon, "The Cold War in Asia: Towards a New Synthesis," *Diplomatic History*, 12 (Summer 1988), 307–27.

By then Truman recognized that the hatreds dividing Chiang and Mao were implacable and that the United States could not save the venal Nationalist regime.[16] Acheson, who replaced Marshall as Secretary of State in Truman's second term, issued a government White Paper in August 1949 that asserted this pessimistic perspective in no uncertain terms. "The unfortunate but inescapable fact," the paper said, "is that the ominous result of the civil war in China was beyond the control of the government of the United States. Nothing that this country did or could have done within reasonable limits of its capabilities could have changed the result. . . . It was the product of internal Chinese forces, forces which this government tried to influence but could not."[17]

This assessment was in some ways disingenuous. Most of Truman's top people were committed Anglophiles and Europe-firsters. They consistently focused on aiding western Europe, where United States interests were paramount, not on helping Chiang. Still, Acheson's paper was accurate in most respects. The Truman administration had tried to help Chiang's regime, to the tune of nearly $3 billion in aid since the war, only to watch the aid wasted by the corrupt and uninspiring Nationalist leadership. The President and Acheson were correct in saying that Chiang was his own worst enemy and that the United States did not have the economic or military capacity to save him.

Unfortunately for Acheson and Truman, Americans were in no mood to accept the White Paper's version of history. Alarmed by the rise of Communism, they had also been developing high expectations about the capacity of the country to have its way in the world. Henry Luce of *Life* and *Time*, raised in China as the son of Presbyterian missionaries, had long demanded greater American commitment to Chiang, and with others in a loosely organized but well-financed "China lobby" he led rising criticism of the administration's Asia policies after Chiang's defeat. Conservative Republicans, including Congressman Walter Judd of Minnesota, a former medical missionary in China, joined him. Many of these Republicans had been Asia-oriented since the days of President McKinley.[18] Democratic Congressman John F. Kennedy, a Catholic anti-Communist, also assailed the President. He explained to an audience in Boston that "pinks" had betrayed American policy in China. "This is the

16. William Stueck, *The Wedemeyer Mission: American Politics and Foreign Policy During the Cold War* (Athens, Ga., 1984).
17. Barton Bernstein and Allen Matusow, eds., *The Truman Administration: A Documentary History* (New York, 1966), 300–309.
18. Richard Fried, *Nightmare in Red: The McCarthy Era in Perspective* (New York, 1990), 87–89.

tragic story of China, whose freedom we once fought to preserve. What our young men had saved, our diplomats and our Presidents have frittered away."[19]

These critics had varied motivations. Some were highly partisan Republicans. Shocked and embittered by Truman's unexpected victory in 1948, they were eager to tar the administration however they could. More generally, Americans were frustrated. Why couldn't the United States, the most powerful and wealthy nation in the world, prevent bad things from happening? As one observer put it, people had an "illusion of American omnipotence." When setbacks occurred—the Bomb in the USSR, "losing" China—the United States must have done something wrong. From this simplistic starting point it was an easy next step to lash out at scapegoats, including spies, "pinks," and "Commie sympathizers" in the government.

In dealing with frustrations such as these, high-level administration officials tried to muddle through. Acheson, while an ardent foe of the Soviet Union, not only defended the White Paper but also considered recommending that the United States eventually recognize, as many Western allies did, Mao's regime. Such a move, he hoped, might encourage Mao to act as a sort of "Asian Tito" and drive a wedge into international Communism.[20] In January 1950 Acheson made a widely noted speech in which he excluded Taiwan (and South Korea) from the "defense perimeter" that he said the United States ought to protect.

The United States, however, did not recognize Red China. Mao, a revolutionary, acted hostilely toward the United States. Most Americans, moreover, believed in the existence of a worldwide Communist conspiracy, in which Mao and Stalin were twin demons. "Credibility" demanded that the United States stand firm against such a threat. For all these reasons the People's Republic continued to be treated as a major enemy. The United States turned a blind eye to Chiang's despotism in Taiwan and refused to support the admission of the People's Republic to the United Nations, whereupon the Soviet Union stalked out of the Security Council in January 1950. Fear of China also caused the Truman administration to stiffen its posture against Communist activity in neighboring Indochina, then under the uneasy rule of the French. In May 1950 the United States began sending military aid to Bao Dai, the puppet

19. Fred Siegel, *Troubled Journey: From Pearl Harbor to Ronald Reagan* (New York, 1984), 72.
20. Gaddis, *Strategies of Containment*, 68–70; McMahon, "Cold War in Asia."

anti-Communist head of Vietnam.[21] Although the aid at first was small—
it was hardly noted at the time—it marked a further militarization and
globalization of American foreign policy, and it quietly set in motion an
ever-greater American commitment against Communist influence in
Southeast Asia.

THESE COMMITMENTS paled before the two most important and
long-range policy consequences of the events of 1949: the Truman administration's decision to go ahead with development of the hydrogen
bomb, or "Super," in January 1950, and the consensus of top military and
foreign policy planners behind one of the key documents of the Cold
War, National Security Council Document 68, in April.

Unlike the A-bomb, which almost everyone in the know had favored
developing in the early 1940s, the idea of producing a hydrogen bomb
evoked passionate arguments in late 1949 and early 1950. Scientists expected that the Super, a fission or "thermonuclear" weapon, would be an
awesomely destructive horror that could unleash the equivalent of several
million tons of TNT. This was hundreds of times more powerful than
atomic bombs. A few well-placed hydrogen bombs could kill millions of
people.

Among the foes of development were famous scientists who had supported atomic development during World War II. One was Albert Ein-
stein, who took to the radio to say that "general annihilation beckons."[22]
Another was James Conant, the president of Harvard, who served on a
general advisory committee of the Atomic Energy Commission (AEC).
He opposed developing the Super on moral grounds, arguing that "there
are grades of morality." He also believed that the H-bomb was unnecessary because the United States already had enough atomic power to deter
all aggressors. Also influential in the fight against the super was J. Robert
Oppenheimer, who was widely known for his scientific expertise, his
literary talents (he had learned seven languages, including Sanskrit, as a
prodigy at Harvard), and his managerial skills as the director of atomic
bomb manufacture at Los Alamos during the war. "Oppie," as he was
known to his friends, had many left-wing associates. His brother and his
wife had been Communists in the 1930s. But his opposition to develop-

21. Robert McMahon, "Toward a Post-Colonial Order: Truman Administration Poli-
    cies Toward South and Southeast Asia," in Lacey, ed., *Truman Presidency*, 352–
    55. Vietnam was a part of French Indochina; it included Annam, Tonkin, and
    Cochin China.
22. McCullough, *Truman*, 761.

ment of the Super was not politically inspired. It rested, like Conant's, on a combination of moral revulsion and practical policy considerations. Their arguments carried the day in the advisory committee, which recommended against development. [23]

Leading government officials, too, had doubts about the hydrogen bomb. Among them was Kennan, who wrote a seventy-nine-page memorandum opposing the Super before he left the government in January 1950. Kennan believed in what was later called "minimum deterrence," which he thought possible with a decent arsenal of atomic bombs. He urged the United States to say that it stood for "no first use" of nuclear weapons. David Lilienthal, who headed the AEC, agreed with Kennan. He favored negotiating with the Soviet Union in the hope that both countries would agree not to develop the new weapons. [24]

Other government officials, however, strongly urged development. Eleanor Roosevelt, whom Truman had named to America's UN delegation, came out for it in January. Lewis Strauss, a dissenter from the AEC report, considered it "unwise to renounce unilaterally any weapon which an enemy can reasonably be expected to possess." The Joint Chiefs maintained that the bomb would be a deterrent as well as "an offensive weapon of the greatest known power possibilities." Senator Brien McMahon, chairman of the Joint Committee on Atomic Energy, expressed a common viewpoint on Capitol Hill when he wrote Truman, "Any idea that American renunciation of the super would inspire hope in the world or that 'disarmament by example' would earn us respect is so suggesstive of an appeasement psychology and so at variance with the bitter lessons learned before, during and after two recent world wars that I will comment no further." No statement more clearly revealed the fear of "appeasement," rooted in the "lessons of history," that lay behind a host of Cold War decisions by American officials in the postwar era. [25]

On January 31, 1950, Truman decided in favor of development. He was influenced in part by the position of the Joint Chiefs, particularly by General Bradley, whom Truman admired greatly. He was also keenly aware, as was Dean Acheson, of the criticism he would get from conservatives and other anti-Communists if he opposed the H-bomb. Most important, no one could be sure that the Soviets would not go ahead on their own. "Can the Russians do it?" he asked his final advisory committee of

23. David Halberstam, *The Fifties* (New York, 1993), 29–33.
24. Gaddis, *Strategies of Containment*, 79–82.
25. Ibid., 81.

Acheson, Lilienthal, and Defense Secretary Johnson. All nodded yes. "In that case," Truman replied, "we have no choice. We'll go ahead." Truman later explained to his staff, "[We] had to do it—make the bomb—though no one wants to use it. But . . . we have got to have it if only for bargaining purposes with the Russians."[26]

When Truman announced his decision, many liberals were appalled. Max Lerner wrote, "One of the great moral battles of our time has been lost. To move toward the ultimate weapon could mean only an ever-escalating arms race, the possible decay of democracy in a garrison atmosphere . . . and the possibilities of unimaginable horror." Other liberals, however, backed the President. Arthur Schlesinger, Jr., replied to critics like Lerner by asking, "Does morality ever require a society to expose itself to the threat of absolute destruction?"[27] Schlesinger's answer, of course, was no, as was Truman's. Given the frigid Cold War atmosphere of early 1950, the decision to go ahead with the hydrogen bomb seems to have been virtually unavoidable.

Development, as it turned out, proved complicated, in part because of formidable mathematical problems involved. But scientists and mathematicians, including the strongly anti-Communist Hungarian refugees Edward Teller and John von Neumann, persisted. With the help of more powerful computers, which were becoming vitally important in the high-tech world of American weaponry, they moved rapidly ahead. The world's first thermonuclear explosion took place on November 1, 1952, at Eniwetok Atoll in the Marshall Islands of the Pacific.

The explosion exceeded all expectations, throwing off a fireball five miles high and four miles wide and a mushroom cloud twenty-five miles high and 1,200 miles wide. Eniwetok disappeared, replaced by a hole in the Pacific floor that was a mile long and 175 feet deep. Scientists figured that if the blast had been detonated over land, it would have vaporized cities the size of Washington and leveled all of New York City from Central Park to Washington Square.

Eight months later, on August 12, 1953, the Soviets followed suit, setting off a blast in Siberia. Premier Georgi Malenkov announced, "the United States no longer has a monopoly on the hydrogen bomb." His boast was somewhat misleading, for the Soviets (like the Americans) did not yet have the capacity to make a "bomb" light enough to be delivered on a target. Still, development raced ahead in the next few years, not only

26. Halberstam, *Fifties*, 46; Gaddis, *Strategies of Containment*, 82.
27. Hamby, *Beyond the New Deal*, 374.

in the United States and the Soviet Union but also in other nations. The age of nuclear proliferation and of maximum possible destruction was near at hand. [28]

The Super represented one half of the plans in 1950 for America's future military posture. National Security Council Document 68 (NSC-68), which called for vast increases in defense spending, was the other. It, too, had its roots in late January. Truman then authorized a study of defense policy and named Paul Nitze, who had succeeded Kennan as head of the State Department's Policy Planning Staff, to head the effort. Nitze, a close associate of Acheson, was yet another Establishmentarian—private school and Harvard graduate, Wall Street investment banker, official since 1940 in the Navy and the State departments, and vice-chairman of the postwar Strategic Bombing Survey that had explored the impact of air raids during World War II. Another key adviser in the process that led to NSC-68 in April was Robert Lovett, who later that year left his own investment banking business to return to government as the Deputy Secretary of Defense.

Nitze, Lovett, and the others who worked on NSC-68 in early 1950 were virtually fixated on the Soviet atomic explosion, and they adopted a worst-case scenario for the world. Asserting that the USSR would have the capacity to deliver 100 atomic weapons on the United States by 1954, they rejected arguments that a moderate mix of economic, military, political, and psychological measures would be sufficient to contain the Soviet Union and keep major areas of industrial-military value—mostly in western Europe—out of hostile hands. [29] They insisted instead that the Soviet Union was an aggressive, implacable, and dangerous foe that either directly or indirectly (by infiltration and intimidation) sought domination of the world. As Lovett put it in an apocalyptic memo:

> We must realize that we are now in a mortal conflict; that we are now in a war worse than any we have experienced. Just because there is not much shooting as yet does not mean that we are in a cold war. It is not a cold war; it is a hot war. The only difference between this and previous wars is that death comes more slowly and in a different fashion. [30]

28. J. Ronald Oakley, *God's Country: America in the Fifties* (New York, 1986), 45. The British conducted a successful atomic test off the coast of Australia in October 1952; the French followed in February 1960, with tests in the Sahara. China become the fifth nuclear power in 1964. See chapter 10 for discussion of testing of H-"bombs" in the mid-1950s.
29. Gaddis, *Strategies of Containment*, 91–99.
30. Samuel Wells, "Sounding the Tocsin: NSC 68 and the Soviet Threat," *International Security*, 4 (Fall 1979), 129–30.

The obvious conclusion was that the United States and its allies must build up not only their nuclear power but also their more conventional forces "to a point at which the combined strength will be superior . . . to the forces that can be brought to bear by the Soviet Union and its satellites." This amounted to what was later called a policy of "flexible response." Although the committee did not include cost estimates for this policy, advocates understood that military spending would have to quadruple to around $50 billion a year, which would "provide an adequate defense against air attack on the United States and Canada and an adequate defense against air and surface attack on the United Kingdom and Western Europe, Alaska, the Western Pacific, Africa, and the Near and Middle East, and on long lines of communication to those areas."[31]

This was a breathtaking and revolutionary document, full of emotional language contrasting the "slave society" of Communists to the blessings of the "Free World." The USSR, "unlike previous aspirants to hegemony, is animated by a new fanatic faith, antithetical to our own, and seeks to impose its absolute authority over the rest of the world." Soviet fanaticism necessitated globalistic responses: "The assault on free institutions is world-wide now, and in the context of the present polarization of power a defeat of free institutions anywhere is a defeat everywhere."

The conclusions of NSC-68 rested on one key assumption, which reflected the grand expectations that pervaded America in the postwar era: economic growth in the United States made such a huge expansion of defense spending easy to manage, and without major sacrifices at home. One of Lovett's memos strongly made this case: *"There was practically nothing the country could not do if it wanted to do it."*[32] While drafting the document, Nitze communicated regularly with Leon Keyserling, chairman of Truman's Council of Economic Advisers. Keyserling had great faith in the ability of government spending to stimulate the economy. Then, as throughout in the postwar era, grand expectations about American economic and industrial growth promoted globalistic foreign and military policies.

NSC-68 was seriously flawed in many respects. As Kennan complained at the time, it assumed the worst of Soviet foreign policy, which for the most part remained cautious, concentrating on tightening control of eastern Europe and other sensitive regions close to Soviet boundaries. NSC-68 also defined United States defense policies in terms of hypotheti-

31. Gaddis, *Strategies of Containment*, 99.
32. Ibid., 93–94. The emphasis is mine.

cal Soviet moves rather than in terms of carefully defined American interests. This approach required the United States to be prepared to put out fires all over the globe.[33]

The report's assumptions about the relationships between Soviet and American power were especially questionable. In 1949 the American GNP was roughly four times as great as that of the Soviet Union, which remained an inefficient and relatively unproductive society. Although the Soviets were devoting perhaps twice as much of their GNP to military spending, this was being done at terrific costs at home and could not make them serious economic rivals of the United States in the foreseeable future. The Soviets maintained a much bigger army, but they had used it to stamp out dissent in their spheres of interest, not to invade new territories. There was no clear indication in 1950 that this largely defensive posture would change. America had much the greater arsenal of nuclear weapons, by far the superior navy, much stronger allies, and incomparably greater economic health. As it turned out, moreover, the Soviets did not make a big effort to improve their long-range bombing forces until the mid-1950s; NSC's worries about nuclear attack as early as 1954 were way off the mark.[34]

When Truman received the report in early April, he neither endorsed nor rejected it. Instead, he passed it along for economic analysis. If the Korean War had not broken out two months later, it might not have been acted on; Truman still hoped to curb defense spending. Still, NSC-68 commanded the support of virtually all high-ranking American officials (Defense Secretary Johnson excepted) at the time it was delivered. It was music to the ears of the armed services. The Korean War then cinched the case for defense spending along the lines urged by the report. By fiscal 1952 the United States was paying $44 billion for national defense; by 1953, $50.4 billion, roughly the amount privately anticipated by advocates of NSC-68. Spending declined a little when the Korean War ended but still ranged between $40 and $53.5 billion every year between 1954 and 1964. Along with the decision for the Super, the logic of NSC-68 reflected the rapid militarization in American foreign policies following the Soviet atomic explosion and the "fall" of China.

THE TOUGHENING of American attitudes toward the Soviets in early 1950 did not exist in a cultural or political vacuum. On the contrary,

33. Ibid., 100–101; Wells, "Sounding the Tocsin."
34. Wells, "Sounding the Tocsin."

events heated up already flammable anti-Communist emotions and ignited a Red Scare of considerable fire and fury. On January 21, ten days
before Truman decided for the Super, a federal jury brought thirteen
months of hotly contested litigation to a close by finding Alger Hiss,
accused of having been a spy for the Soviets in the 1930s, guilty of
perjury. Hiss, a middle-rank Establishment figure in foreign policy councils during the mid-1940s, was sentenced to five years in prison. On
January 27 Klaus Fuchs, a German-born English atomic scientist who
had worked on the A-bomb, was arrested for turning over secrets to the
Soviets during and after the war. He was later tried in England, convicted,
and imprisoned. On February 9 Senator Joseph McCarthy of Wisconsin
alleged that Communists infested the American State Department. His
accusations, offered to the Ohio County Women's Republican Club of
Wheeling, West Virginia, increased pressure on the Truman administration to get tough with the Soviets. The Red Scare of "McCarthyism"
helped to besmirch American politics and culture for much of the next
five years.

These dramatic events, while of great significance in fanning the flames
of anti-Communism in the United States, have to be seen in a longer
historical context. McCarthy was in fact a Joe-Come-Lately to the Red
Scare, whose roots require a quasi-archaeological probe into the American past. Americans have periodically lashed out at radicals, alleged subversives, aliens, immigrants, blacks, Catholics, Jews, and other vulnerable groups who could be blamed for complex problems. The Red
Scare in America following the Bolshevik Revolution was only the most
flagrant of many outbursts, driven both by the government and by popular
vigilantism, against left-wing activists. These outbursts revealed the volatility of popular opinion, the growing capacity of the State to repress
dissent, and the frailty of civil libertarian thought and action in the United
States.

The turbulent years of the 1930s and especially of World War II did
much to lay the foundation for the Red Scare of the 1940s and 1950s.
From the mid-1930s on, right-wing politicians and intellectuals readily
associated the New Deal with socialism and Communism. The House
Committee on Un-American Activities investigated left-wingers following
its establishment in 1938.[35] In 1940 Congress approved the Smith Act,
which made it a criminal offense for anyone to "teach, advocate, or

35. Walter Goodman, *The Committee: The Extraordinary Career of the House Committee on Un-American Activities* (New York, 1964).

encourage the overthrow or destruction of . . . government by force or violence." People accused under the act did not have to be shown to have *acted* in any way, only to have *advocated* action. The Smith Act was used by the Roosevelt administration against alleged Nazis as well as against American Trotskyists—prosecutions that Communists applauded.

At the same time, Roosevelt unleashed FBI director J. Edgar Hoover to check into potentially subversive people and groups. In 1941 Congress authorized the army and navy summarily to dismiss any federal employee considered to be acting contrary to the national interest. This was the start of governmental "security risk" programs, which cost 359 employees their jobs in 1942. The Justice Department began developing in 1942 the "Attorney General's list" of groups considered disloyal. By the middle of the year the FBI had helped the AG to name 47 such groups.[36] Even the American Civil Liberties Union (ACLU), which had been formed following World War I to protect dissenters, joined in the patriotic efforts of the war years. As early as 1941 it had excluded Communists from membership. From 1942 on Morris Ernst, its head, corresponded with Hoover on a "Dear Edgar" basis, in which he passed on information about alleged Communists in the ACLU.

Wartime patriotism spurred other, much more flagrant violations of civil liberties, notably the incarceration of Japanese-Americans in "relocation" camps during most of the war. Less obvious but of long-run significance was the hyper-patriotism that developed among many American people. For some this patriotism arose during military service. For others it followed years of work in defense plants. Either way, large numbers of people, including many European-Americans, came to feel a larger sense of belonging to the United States. The patriotic wartime injunction "Be American" competed with earlier ethnic or class identifications.

When the Cold War arose after 1945, Americans were often quick to join the "get-tough" chorus. The atheistic dogmas of orthodox Marxism repelled Catholics and other religious believers. The subjection of the "old countries" offended many others. More generally, Americans who were trying to get ahead—going to college, raising families, moving to suburbia, acquiring consumer goods—were all the more ready to believe fervently that the United States was a free and mobile society and that Communism, which took away private property, was not only totalitarian but also a threat to their social and economic futures. In this way the

36. Fried, *Nightmare in Red*, 50–56.

hopes for social mobility that pervaded the postwar years stimulated both grand expectations and nervous feelings about Reds. The quests for personal security and domestic security became inextricably interrelated.[37]

World War II had lasting effects in one other, less definable way: like most armed conflicts it toughened popular feelings. The fighting, people concluded, had been necessary. Sacrifice was noble. "Appeasers" were "soft." Long after the war many Americans tended to glorify the the "manly" virtues of toughness. Those who were "soft" ran the risk of being defined as deviant. Arthur Schlesinger, Jr's. widely admired liberal manifesto The Vital Center (1949) made this clear. Liberals, he said, showed "virility," leftists and rightists "political sterility." Communism was "something secret, sweaty, and furtive like nothing so much, in the phrase of one wise observer of modern Russia, as homosexuals in a boys' school."[38] The homophobia that pervaded American culture had many sources, but some of it rested on the view that homosexuals were not only perverted but also subject to blackmail. In the early 1950s they were specifically included among the categories of people who could be fired from sensitive positions as "security risks."[39]

Many postwar forces abetted these wartime developments. One was disturbing evidence of Communist espionage. In June 1945 the FBI arrested several employees of Amerasia, a left-wing magazine close to the American Communist party, as well as John Stewart Service, a China expert in the State Department. The magazine's offices contained 600 secret and top-secret documents, some of which contained information concerning American plans for bombing in Japan. When it became known that federal agents had illegally entered the magazine's offices, the case against the editors fell apart. Evidence concerning Service was too skimpy, and he was released. The result was small fines for three Amerasia staff members for illegal possession of government documents.[40]

In part because the Justice Department was embarrassed about its own illegal activities in the case, the Amerasia matter did not get widespread

37. Gary Gerstle, Working-Class Americanism: The Politics of Labor in an Industrial City, 1914–1960 (Cambridge, Eng., 1989), 278–309.

38. Arthur Schlesinger, Jr., The Vital Center: The Politics of Freedom (Boston, 1949), 151; Stephen Whitfield, The Culture of the Cold War (Baltimore, 1991), 43.

39. Alan Berube, Coming Out Under Fire: The History of Gay Men and Women in World War II (New York, 1990).

40. Joseph Goulden, The Best Years, 1945–1950 (New York, 1976), 278–88; Robert Ferrell, Harry S. Truman and the Modern American Presidency (Boston, 1983), 134–35.

publicity at the time. But it was a worrisome affair for government officials. When Igor Gouzenko, a file clerk in the Soviet Embassy in
Toronto, defected in early 1946, their worries intensified. Gouzenko
produced evidence that the Soviets had spied on atomic energy research
in Canada and elsewhere during the war. Neither the *Amerasia* nor the
Gouzenko affair proved that any Americans—let alone government
officials—were guilty of espionage. Indeed, no American public officials
were convicted of spying at any time during the postwar Red Scare. But
Gouzenko's revelations did show that the Soviets had spied on America
during and after World War II. Evidence such as this later played into the
hands of Red Scare activists. [41]

The heat of partisan politics further intensified the postwar Red Scare.
Running for the presidency in 1944, Dewey had linked Communism,
FDR, and the New Deal. The Democrats had fired back by associating
the GOP with fascism and "fifth column" activities. Red-baiting in the
1946 campaign smeared many candidates, including Congressman Jerry
Voorhis, Richard Nixon's opponent in southern California. Voorhis denied Nixon's unfounded accusations, but to no avail. His defeat, like that
of others attacked by anti-Communists in 1946, provided an obvious
object lesson: Red-baiting could pay off at the polls. [42]

Ardent foes of Communism often enjoyed substantial support from
conservative interest groups, many of which worked closely with Hoover.
The American Legion was one, the United States Chamber of Commerce
another. Right-wing publishers such as Colonel Robert McCormick of
the *Chicago Tribune* and aging, melancholic William Randolph Hearst,
who ran a nationwide chain of papers, regularly (and sometimes hys-
terically) raised the alarm against subversives at home and Communists
abroad. Patriotic organizations such as the Daughters of the American
Revolution chimed in. Leading prelates of the Catholic Church as well as
the Knights of Columbus were especially outraged by the atheism of
Communism. Francis Cardinal Spellman of New York was a sort of
chaplain of the Cold War and actively assisted the FBI to root out Reds
from American institutions. [43]

By mid-1946 a number of anti-communist liberals and leftists joined in
this chorus against Communism at home and abroad. They included
union leaders, intellectuals, and others who joined the ADA and who

41. David Caute, *The Great Fear: The Anti-Communist Purge Under Truman and
    Eisenhower* (New York, 1978), 55–56.
42. Fried, *Nightmare in Red*, 57–67.
43. Ibid., 85, 97; Whitfield, *Culture of the Cold War*, 92–94.

worried about Communist influence in the labor movement and other high places.[44] These liberals opposed the extreme and sometimes irrational fulminations about Communism that emanated from the Far Right. They detested Red-baiters like Nixon. Unlike many conservatives, they did not worry much that Communism threatened private property in the United States. But, having tried to work with Communists in progressive causes, they were certain that American Communists got their marching orders from Moscow.[45] Irving Howe, a democratic socialist, explained, "Those who supported Stalinism and its pitiful enterprises either here or abroad, helped befoul the cultural atmosphere, helped bring totalitarian methods into trade unions, helped perpetrate one of the great lies of the twentieth century, helped destroy whatever possibilities there might have been for a resurgence of a serious radicalism in America."[46]

Whether the American Communist party was as alien an organization as Howe and others maintained still divides historians. Some point out accurately that the party, after reaching a peak in membership in 1945, declined in the late 1940s. Although influential among certain unions until 1949, it was hardly a potent force in American politics, and its leaders did not receive a great deal of aid from Moscow. Many American Communists in union locals, moreover, were honest, effective, and popular with the rank-and-file. Still, Howe and others were on target in lamenting the lockstep consistency with which leading party officials—as opposed to many lesser party members—followed the Moscow line on all major questions, including the coup in Czechoslovakia and the Berlin Airlift.[47] Some of the leaders had indeed surrendered their intellectual independence—and their patriotism—to the Kremlin. And Howe was surely correct in bewailing the baneful effect of such a rigid and uncompromising party on the chances for revival of an independent Left in the United States.

Another issue dividing historians concerns the degree to which the "public at large" promoted a Red Scare. Should one emphasize the role of elites—partisan Republicans, interest group leaders, anti-Communist

44. David Oshinsky, "Labor's Cold War: The CIO and the Communists," in Robert Griffith and Athan Theoharis, eds., *The Specter: Original Essays on the Cold War and the Origins of McCarthyism* (New York, 1974), 116–51. See also chapter 2.
45. As an example, Schlesinger, *Vital Center*, 102–30.
46. Whitfield, *Culture of the Cold War*, 114.
47. David Shannon, *The Decline of American Communism: A History of the Communist Party in the United States Since 1945* (New York, 1959).

liberals—or see the elites as mainly reflecting the "voice of the people"? There is no sure answer to this question, in part because polls of public opinion on the subject offer conflicting evidence. Analysts who focus on elites, however, probably have the better case, for the majority of Americans in the late 1940s and early 1950s only slowly grew worried about subversion. As William Levitt remarked of his suburbanites, people were too busy getting ahead to fret a great deal about Communists, let alone to lead crusades against Reds. A survey at a height of the Red Scare in 1954 found that only 3 percent of Americans had ever known a Communist; only 10 percent said they had known people they even suspected of being a Communist. The same survey concluded, "The internal communist threat, perhaps like the threat of crime, is not directly felt as personal. It is something one reads about and talks about and even sometimes gets angry about. But a picture of the average American as a person with the jitters, trembling lest he find a Red under the bed, is clearly nonsense."[48]

Still, anti-Communism cut fairly deeply in some ways. As early as 1946 polls showed that 67 percent of Americans opposed letting Communists hold government jobs. A poll in 1947 revealed that 61 percent of respondents favored outlawing the Communist party.[49] Political leaders and anti-Communist pressure groups had helped to rouse such popular feelings; they did not spring up on their own. But there was a good deal of evidence to suggest that ardently patriotic and anti-Communist emotions were not hard to whip up once Cold War fears mounted. When the Truman administration sent the so-called Freedom Train around the country in 1947–48, it was greeted by enthusiastic crowds, brass bands, and patriotic speeches. An estimated 4 million people turned out to look at the train's exhibits, which included the Declaration of Independence, the Constitution, and the Truman Doctrine.[50]

It is especially clear that most American people continued to care little for civil liberties in the immediate postwar years. Americans in the 1940s and early 1950s may not have worried deeply that there were a lot of Communists hiding under the bed, but they were often ready to believe that party members and sympathizers were dangerous to the Republic. Acting on such beliefs, liberal organizations moved summarily to purge Communists in these years. By 1949 labor unions, the NAACP, the

48. Samuel Stouffer, *Communism, Conformity, and Civil Liberties: A Cross-Section of the Nation Speaks Its Mind* (Garden City, N.Y., 1955), 59–87.
49. Fried, *Nightmare in Red*, 59–60
50. Ibid., 97–99; Marty Jezer, *The Dark Ages: Life in the United States, 1945–1960* (Boston, 1982), 85.

Urban League, and the Congress of Racial Equality had largely succeeded in doing so, and in 1950 the NAACP decided to expel Communist-dominated chapters.[51]

Civil liberties also came under siege in the world of education. Those who knew much about the history of education in the United States were not surprised by this development, for taxpayers long had demanded that schools and colleges promote national values. Patriotism, as taught in the schools from the salute to the flag and the Pledge of Allegiance, was where young people learned the American way. In 1940 twenty-one states required loyalty oaths of teachers. There was therefore little reason to suppose that school boards, principals, or college administrators would behave much differently from other American officials caught up in the Red Scare.

What began happening in the late 1940s was nonetheless unsettling to beleaguered civil libertarians in the academic world. In 1948 the University of Washington fired three teachers—two of them with tenure—when they refused to answer questions by state legislators about whether they belonged to the Communist party. The teachers never found another academic job. Later in the same year the American Federation of Teachers, a leading union, voted against allowing Communists to teach, and the Board of Regents of the University of California system required its faculty to take a non-Communist oath. Faculty members who refused to sign got caught up in a long and internecine controversy. A total of thirty-one, including some with tenure, were ultimately fired.[52]

Although legislators and other outsiders led these anti-Communist crusades, they found prominent educators ready to go along with much of their agenda. Charles Seymour, president of Yale, announced, "There will be no witch hunt at Yale, because there will be no witches. We do not intend to hire Communists." Presidents Conant of Harvard and Eisenhower of Columbia headed a blue-ribbon panel which concluded in 1949 that Communists were "unfit" to teach. The American Association of University Professors (AAUP) opposed loyalty oaths and the firing of teachers for taking the Fifth Amendment, but it granted the right of university administrators to expect its professors to answer questions about

51. Mary McAuliffe, "The Politics of Civil Liberties: The American Civil Liberties Union During the McCarthy Years," in Griffith and Theoharis, eds., *Specter*, 152–71; Fried, *Nightmare in Red*, 164–66.
52. Diane Ravitch, *The Troubled Crusade: American Education, 1945–1980* (New York, 1983), 94ff; Ellen Schrecker, *No Ivory Tower: McCarthyism and the Universities* (New York, 1986), 105–25, 308–37; Fried, *Nightmare in Red*, 101–4.

their politics. Moving with excruciating slowness, the AAUP did not censure universities that violated civil liberties until 1956.[53]

All who took stern positions essentially endorsed the view that Communists, as minions of Moscow, had surrendered their independence of thought. Sidney Hook, a well-known philosopher, defended the autonomy of educational institutions, but he said that university administrators could and should police their campuses against such influences. Hook reiterated that Communists were not free to think for themselves: there was a party line "for every area of thought from art to zoology."[54] The Socialist leader Norman Thomas, a veteran of bitter fights with Communists over the years, agreed: "The right of the Communist to teach should be denied because he has given away his freedom in the quest for truth. . . . He who today persists in Communist allegiance is either too foolish or too disloyal to democratic ideals to be allowed to teach in our schools."[55]

It is an exaggeration to conclude that the Red Scare terrorized American academe in general.[56] Most universities—and many individual faculty members—defended academic freedoms.[57] Still, the Red Scare in education was a demoralizing episode, especially at the not-so-ivory towers of universities where academic freedom had been supposed to be safe. Schools and colleges feared to keep on their faculties anyone who had refused to deny that he or she was a Communist. When confronted with such issues, a number of administrators and non-Communist faculty members were quick to assume that all Communists were the same, without asking if Communist teachers made any effort to indoctrinate students and without distinguishing between those faculty who were good scholars and teachers and those who were not. Teachers grew cautious, and some suffered greatly.[58] Though estimates vary, it is thought that 600 or so public school teachers and professors in these years lost their jobs because they were smeared by accusations that they were Communists or

53. Richard Pells, *The Liberal Mind in a Conservative Age: American Intellectuals in the 1940s and 1950s* (New York, 1985), 288.
54. Ravitch, *Troubled Crusade*, 96.
55. Ibid., 97–98.
56. Schrecker, *No Ivory Tower*, 339–41.
57. Willam O'Neill, *American High: The Years of Confidence, 1945–1960* (New York, 1986), 165–68.
58. Russell Jacoby, *The Last Intellectuals: American Culture in the Age of Academe* (New York, 1987), 125–26.

Communist sympathizers. Blacklists often ensured that they would not be hired elsewhere.[59]

MUCH OF THE IMPETUS for McCarthyism in the early 1950s developed between 1947 and 1949 and arose from the activities of government. Some of these activities, such as the efforts of the House Committee on Un-American Activities (HUAC), came from right-wing zealots and political opportunists in Congress.[60] Others came from the Justice Department (including the FBI) of the Truman administration. Well before McCarthy took his place on the national stage, ever more energetic governmental scene-setters had rung up the curtain on a Red Scare drama that appeared to play popularly at the polls.[61]

By all odds the most durable villain of the drama was Hoover, who had begun his hunt for subversives when Woodrow Wilson's Red-hunting Attorney General, A. Mitchell Palmer, placed him in charge of the Justice Department's newly created General Intelligence Division in 1919. Hoover was then 24.[62] He moved quickly to set up special files on virtually all radicals known to the country and did the legwork that much facilitated the Red Scare of 1919. By 1924 he was head of the FBI, a post he retained for forty-eight years until his death in office in 1972. Hoover was vain, surrounded by sycophants, obsessed with order and routine. People who met him in his later days at the FBI were led through his many "trophy rooms" to his office, which glowed with a purplish insect-repelling light that Hoover, a hypochondriac, had installed to "electrocute" bad germs.[63] Hoover sat regally behind a desk on a six-inch-high dais and looked down on his visitors. Throughout his career he employed very few blacks or other minorities. Those who did get hired spent much of their time driving his limousine, handing him towels, or swatting flies.

Hoover worked harder at rooting out subversion than at any other activity. In doing so he used a vast and intricate network of informers,

59. Pells, *Liberal Mind*, 288; John Diggins, *The Proud Decades: America in War and Peace, 1941–1960* (New York, 1988), 166.
60. This was of course an inaccurate acronym.
61. Richard Freeland, *The Truman Doctrine and the Origins of McCarthyism: Foreign Policy, Domestic Politics, and Internal Security, 1946–1948* (New York, 1970), 117–34.
62. Richard Powers, "Anti-Communist Lives," *American Quarterly*, 41 (Dec. 1989), 714–23.
63. David Oshinsky, *New York Times Book Review*, Sept. 15, 1991.

some of them undercover agents, others—like Cardinal Spellman—well-known public figures anxious to cooperate in the anti-Communist crusade. In return he gave them information about subversives in their midst. No rumor, it seemed, was too trivial for Hoover to follow up, especially if it involved sexual activities. Much of the information—fact, hearsay, trivia—went into his secret files.

These many flaws were well known to critics in the Truman years. Bernard De Voto lashed out in 1949 against Hoover's use of "gossip, rumor, slander, backbiting, malice and drunken invention, which, when it makes the headlines, shatters the reputations of innocent and harmless people. . . . We are shocked. We are scared. Sometimes we are sickened. We know that the thing stinks to heaven, that it is an avalanching danger to our society."[64] Truman complained privately, "We want no Gestapo or Secret Police. FBI is tending in that direction. They are dabbling in sex life scandals and plain blackmail. . . . *This must stop.*"[65] But Truman made no effort to fire him. He refrained from criticizing Hoover openly, even when he realized that the director was feeding information about alleged subversives to enemies of his administration. Truman even depended on the FBI to check the loyalty of federal employees and to help prosecute Communist leaders.[66] When Truman left office at the peak of the Red Scare, Hoover and the FBI were stronger than they had been in 1945.

The reasons for Hoover's success were not hard to discover. One was his carefully cultivated reputation as a crime-fighter. Another was his apparently large pile of dirt about people high in public places. Hoover was also a consummate bureaucrat. More than most high officials in the 1920s and 1930s, he had then mastered the art of public relations. When FBI agents killed John Dillinger, "Public Enemy Number One," Hoover took personal credit. Equally important, Hoover was hardly a rogue elephant. He ordinarily made sure that authority for aggressive activities such as bugging and wiretapping came from above. Again and again he got such assurances from Presidents and Attorneys General who recognized that Hoover commanded information they needed—or thought they needed—to have.

A co-villain in the Red Scare drama was HUAC, which had attracted

64. Jezer, *Dark Ages*, 84.
65. Robert Griffith, "Harry S. Truman and the Burden of Modernity," *Reviews in American History*, 9 (Sept. 1981), 299.
66. Fried, *Nightmare in Red*, 83–84; Halberstam, *Fifties*, 335–42.

some of the most reactionary and bigoted men in public life.[67] A senior Democrat, John Rankin of Mississippi, was an especially rabid anti-Semite and racist. Denouncing civil rights activity in 1950, Rankin exclaimed, "This is a part of the communistic program, laid down by Stalin approximately thirty years ago. Remember communism is Yiddish. I understand that every member of the Politburo around Stalin is either Yiddish or married to one, and that includes Stalin himself."[68] Another HUAC power, Republican J. Parnell Thomas of New Jersey, was later convicted of illegally padding his payroll. A third member thereafter was Nixon, who did more than anyone else after 1946 to make HUAC an aggressive agent of anti-Communism. His persistent labors gave him a national reputation.

In 1947 HUAC concentrated on probing into left-wing activity in Hollywood. Announcement of the committee's intention inspired protests among American entertainment figures. "Before every free conscience in America is subpoenaed," Judy Garland cried, "please speak up! Say your piece. Write your Congressman a letter! Airmail special." Frank Sinatra asked, "Once they get the movies throttled, how long will it be before we're told what we can say and cannot say into a radio microphone? If you make a pitch on a nationwide radio network for a square deal for the underdog, will they call you a Commie? . . . Are they going to scare us into silence?" Fredric March demanded, "Who's next? Is it your minister who will be told what he can say in his pulpit? Is it your children's school teacher who will be told what he can say in his classroom? . . . Who are they after? They're after more than Hollywood. This reaches into every American city and town."[69]

The hearings that opened in October began relatively quietly with the testimony of "friendly witnesses" who cooperated with the committee. The actor Gary Cooper, terse as ever, said that he opposed Communism "because it isn't on the level." Walt Disney contended that the Screen Cartoonists Guild was Communist-dominated and had earlier tried to take over his studio and make Mickey Mouse toe the party line. Ronald Reagan, head of the Screen Actors Guild, tried to straddle the fence. He criticized the deviousness of Communists but added that he hoped never to see Americans "by either fear or resentment . . . compromise with

67. Halberstam, *Fifties*, 12.
68. *Time*, March 16, 1950, p. 17.
69. Les Adler, "The Politics of Culture: Hollywood and the Cold War," in Griffith and Theoharis, eds., *Specter*, 240–61; Goulden, *Best Years*, 297.

any of our democratic principles."[70] In the next few years Reagan turned more decisively to the right, enthusiastically enforcing a blacklist on actors accused of being Communists and identifying to the FBI actors and actresses who "follow the communist party line."[71] His cooperation with HUAC and Hoover was a milestone on his road from New Deal liberalism to the Republican Right.

Donnybrooks broke out when unfriendly witnesses confronted HUAC. Some cited the Fifth Amendment, which protected them against possible self-incrimination. Ten others took the much riskier path of claiming the right of freedom of speech under the First Amendment. They refused to give straight answers to a range of committee questions about whether they were Communists. The "Hollywood Ten," as they became known, included talents such as the screenwriters Alvah Bessie, Dalton Trumbo, and Ring Lardner, Jr. Some of them rudely insulted HUAC members. The screenwriter Albert Maltz likened Rankin and Thomas to Goebbels and Hitler.[72]

The stand of the Ten evoked admiration among left-liberal colleagues in the entertainment industry, including such stars as Humphrey Bogart, Lauren Bacall, Katharine Hepburn, and Danny Kaye, who formed the Committee for the First Amendment. But public opinion seemed hostile, and the studio heads, who feared for the image of the industry, closed ranks against them. The Ten, along with 240 or so others, were blacklisted by the industry, many of them for years. HUAC cited the Ten for contempt. When they lost their appeals in 1950, they went to prison for terms ranging between six months and a year.

If Hoover and HUAC were the villains of the anti-Communist drama, Truman and his advisers clumsily—and sometimes recklessly—acted as spear-carriers. Even as the Red-hunters looked for subversives in 1947 the administration was scouting out the "loyalty" of federal employees. Their efforts rested on Executive Order No. 9835 issued on March 22, 1947, nine days after announcement of the Truman Doctrine. The order set up "loyalty boards" in government agencies that employed some 2.5 million people. It seemed reasonable on the surface. Employees (and potential new hires) who were called before the boards had the right to a hearing and to counsel. They were to be informed of the specific charges against them and had the right to appeal to a Loyalty Review Board under the auspices of the Civil Service Commission. They could be fired if the boards found "reasonable grounds . . . for belief that the person in-

70. Fried, *Nightmare in Red*, 74–77; Caute, *Great Fear*, 487–516.
71. Whitfield, *Culture of the Cold War*, 142.
72. Fried, *Nightmare in Red*, 74–77.

volved is disloyal to the Government of the United States." This involved such activities as sabotage, espionage, treason, advocacy of violent revolution, performance of duties "so as to serve the interests of another government," and affiliation with any group "designated by the Attorney General as totalitarian, fascistic, communistic, or subversive."[73]

In practice Truman's loyalty program was careless of civil liberties. The very word "loyalty" was problematic, encouraging zealots to bring charges on vague and imprecise grounds. While employees had the right to hear of charges against them, accusers could withhold anything they designated as secret. Government workers did not have the right to know the identity of their accusers—often agents of the FBI—or to confront them in the hearings. "Evidence" used against them often amounted to no more than a dossier available only to members of the loyalty boards.[74] Many governmental employees investigated by the loyalty boards were guilty of nothing more than having belonged to liberal organizations on an AG list that grew rapidly after 1947.

Critics of Truman have argued that his loyalty program was intended as a tough domestic counterpart to the Truman Doctrine. This was not the case. The order had been considered for some time and stemmed from recommendations of a Temporary Commission on Employee Loyalty that Truman had established in November 1946. Moreover, the program was hardly new; for the most part it broadened and codified orders that Roosevelt had issued during the war. The AG's list dated to 1942. In tightening these wartime procedures the President hoped to fend off right-wingers who said that he was "soft" on Communism and to keep management of the program out of the hands of the FBI. That is why the Civil Service Commission, not Hoover, controlled the program. Hoover felt slighted and angry about the order.

Still, the President and his aides should have taken greater care to protect people. Truman knew that government employees deserved fair procedures and asserted that his loyalty program would provide them. But he nonetheless expanded an already flawed set of procedures and did nothing to stop other agencies of government from establishing even more arbitrary loyalty programs: the armed services were allowed to investigate civilian employees of defense contractors and to order firings without giving any account of the charges against the suspects.[75] By mid-1952 Truman administration loyalty boards had investigated many thousands

73. Ibid., 66–68.
74. Goulden, *Best Years*, 309.
75. Hamby, *Beyond the New Deal*, 388.

of employees, of whom around 1,200 were dismissed and another 6,000 resigned rather than undergo the indignities and potential publicity of the whole process. None of them was proved to be a spy or a saboteur.[76] The program reflected badly on the administration's awareness of civil liberties and encouraged subsequent apocalyptic thinking about subversion. It was ironic indeed that Truman's partisan opponents scored political points by charging him with being "soft" on Communism.

A year later, during the 1948 campaign, the Truman administration went still further to demonstrate its Americanism, by prosecuting top leaders of the American Communist party. This effort led to drawn-out litigation that culminated in review by the Supreme Court in June 1951. At every level of the court system the eleven defendants lost their case. In New York the judge ruled that the leaders had violated the Smith Act by urging overthrow of the government "as speedily as circumstances permit" and therefore represented a "clear and present danger" to American society. Judge Learned Hand of the appeals court agreed and cited events of the Cold War as evidence of the Communist threat. When the Supreme Court upheld the convictions in the case of *Dennis v. U.S.*, the leaders were fined $10,000 each and sentenced to prison terms of five years.[77]

Almost no one in those troubled times was eager to stand up and defend these leaders of the American Communist party. That was understandable, not only because of Cold War tensions but also because the party itself had never cared for civil liberties. Americans thought the Communists had it coming. But the prosecutions were disturbing to advocates of free speech, for they depended on the ill-constructed Smith Act, which took aim at organizations deemed to be engaged in the teaching or advocacy of violent revolutionary activity. The government, unable to show that the defendants had committed any overt acts of violence or crime, fell back on the argument that belonging to the Communist party made them part of a conspiracy to commit such acts in the future. In so doing the government was stifling speech, and Supreme Court Justice Hugo Black said so in a dissent that he registered with William Douglas. He hoped that "in calmer times, when present pressures, passions, and fears subside, this or some later court will restore the First Amendment liberties to the high preferred place where they belong in a free society."[78]

76. Oakley, *God's Country*, 67.
77. 341 U.S. 494.
78. Fried, *Nightmare in Red*, 114; Whitfield, *Culture of the Cold War*, 45–51; Schrecker, *No Ivory Tower*, 6.

Having put the top leadership in jail, the Truman administration then went after other well-known Communists. By the end of 1952 it had secured thirty-three more convictions. Ultimately 126 were indicted and ninety-three convicted before Cold War fears abated and a more liberal Supreme Court discouraged such litigation in the mid- and late 1950s.[79] In this sense the prosecutions were successful; they not only sentenced Communists to prison but forced the party to throw vast amounts of its time and money toward their defense. Meanwhile the party committed suicide by supporting Soviet foreign policies, including the crushing of the Hungarian revolution in 1956. Thereafter it was estimated that party membership plummeted to 5,000, of whom so many were FBI agents that Hoover considered taking over the party by massing his men at its next convention.[80]

But the prosecutions were otherwise unfortunate, for two reasons. First, they engaged the government in further attacks on civil liberties. Second, they drove the remaining leaders underground, where it proved harder to keep track of their activities. Indeed, the prosecutions represented a remarkable overreaction. They revealed, as did the broader McCarthyism that complemented them after 1950, the growing force of the Red Scare in America, a force that owed some of its strength to activities of the administration of Harry S. Truman.

All these actors in the drama against subversion—Hoover and the FBI, HUAC, administration loyalty boards and prosecutors—gained the attention of an increasingly alarmed American audience between 1947 and early 1950. The most compelling actors, however, were the protagonists in a prolonged and bitter legal fight that episodically grabbed center stage between the summer of 1948 and January 1950: the tribulations of Alger Hiss. Many decades later this fight stands out as among the most dramatic in the history of the Red Scare.

HUAC opened the action in August 1948 when it brought a number of confessed ex-Communists before it to testify. One of these was Whittaker Chambers, who said he had spied for the Soviets in the 1930s. Rejecting the party in 1938, Chambers embraced Christianity and emerged as a knight-errant against atheistic, brutal Communism. A facile writer, he worked for nine years at *Time* magazine before resigning as a senior editor in 1948. Chambers was pudgy, rumpled, disheveled, sad-faced, and emotionally unstable to the point of frequently considering suicide. In sensa-

79. Whitfield, *Culture of the Cold War*, 45–51.
80. Ibid., 50.

tional testimony he identified for HUAC a number of people as fellow Communists in the 1930s.[81]

One of those named was Alger Hiss, then the much-respected head of the Carnegie Endowment for International Peace. Hiss was the antithesis of Chambers. He had been educated at Johns Hopkins University and the Harvard Law School. A protégé of Harvard professor Felix Frankfurter, he became a clerk for Supreme Court Justice Oliver Wendell Holmes. During the 1930s he worked in a number of New Deal departments, including State after 1936. Although not quite a top-ranking official, he attended a number of international conferences, including Yalta, and was a promising member of the State Department when he left to direct the Carnegie Endowment in 1947.

Hiss was an Establishmentarian. Among his friends were Acheson and others in the elite of foreign policy-makers in the Roosevelt-Truman years. Associates marveled especially at his poise. He had handsome, well-defined features and the facility in speech of a well-trained attorney. Murray Kempton, a respected liberal columnist, said that Hiss "gave you a sense of absolute command and absolute grace." Alistair Cooke, a friendly journalist, observed that Hiss "had one of those bodies that without being at all imposing or foppish seem to illustrate the finesse of the human mechanism."[82] Much of the drama that followed Chambers's testimony stemmed from the apparently impeccable credentials of Alger Hiss. If such a man had been a Communist, then nothing the government did was safe.[83]

When Hiss heard of Chambers's accusations, he insisted on responding. Under oath he denied all before HUAC, whose members he openly disdained, and challenged Chambers to repeat his charges without benefit of congressional immunity. When Chambers did so, Hiss sued him for libel. Hiss's many friends were outraged at Chambers's accusations; Truman himself denounced HUAC's fishing as a "red herring." But Nixon was suspicious of Hiss, regarding him as the epitome of the liberal eastern Establishment, and he pressed the case against him. The FBI worked closely with Nixon, apparently feeding him sensitive material—and denying it to Hiss—whenever it was needed.

Chambers then fought back. In November 1948 he said that Hiss had

81. Alistair Cooke, A Generation on Trial: U.S.A. v. Alger Hiss (New York, 1950); Allen Weinstein, Perjury: The Alger Hiss Case (New York, 1978); Goulden, Best Years, 322–34; Hamby, Beyond the New Deal, 379–81.
82. Goulden, Best Years, 324.
83. Pells, Liberal Mind, 271.

not only been a Communist but had committed espionage by turning over confidential government documents to the Soviets in the late 1930s. In one of the most theatrical moments of the controversy Chambers brought reporters to a field on his Maryland farm and showed them microfilmed documents that he had hidden in a hollowed-out pumpkin. These, he said, were copies of State Department documents that Hiss had turned over to him in 1937 and 1938. The "Pumpkin Papers," as they were called, made for sensational newspaper coverage.

Hiss was now on the defensive. The same grand jury that heard evidence concerning the top Communist leaders weighed Chambers's accusations in December 1948 and decided to press ahead with his case. The statute of limitations had lapsed for charges of espionage, but the jury indicted Hiss on two counts of perjury: for denying that he had ever given Chambers any government documents, and for claiming that he had never seen Chambers after January 1, 1937. The trial in June 1949 ended in a hung jury, but when he was tried again he was convicted on January 21, 1950. Hiss later served three years of his five-year sentence and spent more than forty years thereafter stoutly asserting his innocence.

Whether Hiss was innocent in fact remained a much-disputed matter years later. The political legacy of the case, however, was clear. The prolonged and often sensational struggle refurbished the facade of HUAC, which grew bolder in its anti-Communist probes. It advanced Nixon, whose instincts about Hiss seemed justified by the results. It established that Chambers and others had indeed been engaged in espionage for the USSR in the 1930s. When Klaus Fuchs was arrested on charges of atomic espionage six days after Hiss's conviction in 1950, it was easy for people to imagine the existence of a vast and subterranean conspiracy that had to be exposed.

The Hiss trials had still wider symbolic value for many conservatives and anti-Communists in the United States.[84] To them Hiss's conviction seemed long-overdue validation of all that they had been saying about rich, elitist, well-educated, eastern, Establishmentarian New Dealish government officials. "For eighteen years," HUAC Republican Karl Mundt of South Dakota exploded, the United States "had been run by New Dealers, Fair Dealers, Misdealers and Hiss dealers who have shuttled back and forth between Freedom and Red Fascism like a pendulum on a Kukoo clock."[85]

84. Siegel, *Troubled Journey*, 74.
85. Fried, *Nightmare in Red*, 22.

WITH A STAGE SO WELL SET in February 1950, it was hardly surprising that one of these angry partisans in Congress should have marched to the center and stolen the scene. Senator Joseph R. McCarthy of Wisconsin wasted no time doing so when he spoke to the Republican women in Wheeling on February 9. There he waved around papers that he said documented the existence of widespread subversion in government. His exact words on that occasion are disputed, but he appears to have said, "I have here in my hand a list of 205—a list of names that were made known to the Secretary of State as being members of the Communist Party who nevertheless are still working and shaping policy in the State Department."[86]

These were sensational charges. McCarthy, after all, was a United States senator, and he claimed to have hard evidence. Intrigued, reporters asked for more information. Doubters denounced him and demanded to see the list. McCarthy brushed them off and never produced one. His information, indeed, was at best unreliable, probably based on FBI investigations of State Department employees, most of whom were no longer in the government. In subsequent speeches—he was on a "Lincoln Birthday" tour—McCarthy changed the figure from 205 to fifty-seven. When he spoke on the subject in the Senate on February 20, he rambled for six hours and bragged that he had broken "Truman's iron curtain of secrecy." The numbers changed again—to eighty-one "loyalty risks" in the State Department—but McCarthy remained aggressively confident. "McCarthyism" was on its way.[87]

People who knew Joe, as he liked to be called, were hardly surprised by the brazenness of his behavior. A lawyer and controversial circuit court judge before the war, McCarthy had served in the marines during World War II. In 1946, still only thirty-seven, he beat Robert La Follette, Jr., the incumbent, in a Republican primary that featured lies about La Follette's campaign finances. McCarthy then swept to victory in the anti-Truman backlash of that year. His campaign relied heavily on lies about his war record as a marine officer in the Pacific. Advertising himself as "Tail Gunner Joe," he falsely maintained that he had flown up to thirty combat missions when in fact he had gone on none. Later he often walked with a limp that he said had been caused by "ten pounds of shrapnel" that had earned him a Purple Heart. In fact he had hurt his foot by falling down the

86. Thomas Reeves, The Life and Times of Joe McCarthy: A Biography (New York, 1982), 223–28; Fried, Nightmare in Red, 120–24.
87. Reeves, Life and Times, 235–42; Richard Fried, Men Against McCarthy (New York, 1976), 43–57.

stairs at a party. He had seen very little combat action and had never been wounded. This did not faze him: in the Senate, he used his political influence to get an Air Medal and the Distinguished Flying Cross.[88] McCarthy was in fact a pathological liar throughout his public life.

Colleagues also knew that McCarthy was crude and boorish. Thickset, broad-shouldered, and saturnine, he was often unshaven and rumpled in appearance. He spent more time playing poker and acccepting favors from lobbyists than he did on Senate business. A heavy drinker, he regularly carried a bottle of whiskey in the dirty briefcase that he said was full of "documents." He bragged about putting away a fifth of whiskey a day. McCarthy liked above all to be thought of as a man's man. Many of the subversives, he said, were "homos" and "pretty boys." When attractive women appeared before his committees, he leered at them and jokingly told aides to find out their telephone numbers. Being a man's man, he seemed to believe, meant being rough and profane: he thought nothing about using obscenities or belching in public.

McCarthy was no stranger to Red-baiting, having resorted to it himself in 1946. But he was intellectually lazy and had never bothered to learn much about Stalin or the American Communist party. Above all, he was unscrupulous and ambitious. With re-election facing him in 1952, he cast about for an issue that would fortify his otherwise weak record. For a while he seems to have thought crime would be the issue, but Senator Estes Kefauver of Tennessee preempted that possibility by staging highly publicized hearings on organized crime. Dining with friends in January 1950, McCarthy was advised to go after subversives. "That's it," he said, his face lighting up. "The government is full of Communists. We can hammer away at them."[89]

He kept up the hammering, rarely relenting for long, for more than four years. He was remarkably inventive and imaginative in doing so. As before he did not worry about lying. Again and again he would stand up, pulling a bunch of documents from his briefcase, and improvise while he went along. As his audience grew, he became more and more animated and skillful at spinning stories. When opponents demanded to see documents, he refused on the ground that they were secret. When caught in an outright lie, he attacked his accuser or moved on to other lines of investigation. Few politicians have been more adept in their use of rhetoric that

88. David Oshinsky, A Conspiracy So Immense: The World of Joe McCarthy (New York, 1983), 30–35; Oakley, God's Country, 60.
89. Richard Rovere, Senator Joe McCarthy (New York, 1959), 122–23.

makes good headlines. Repeatedly he blasted "left-wing bleeding hearts," "egg-sucking phony liberals," and "communists and queers."[90]

McCarthy did not much care who he attacked. Once he referred to Ralph Flanders, a liberal Republican colleague from Vermont, as "senile—I think they should get a man with a net and take him to a good quiet place." Robert Hendrickson, a Republican from New Jersey, was "a living miracle in that he is without question the only man who has lived so long with neither brains nor guts."[91] During McCarthy's four-year tear through American institutions he also assailed the army, the Protestant clergy, and the civil service. He reveled in super-masculine imagery of fighting and bloodshed, bragging that he went "for the groin" and that he would "kick the brains out" of opponents.[92]

The senator from Wisconsin saved most of his hard knocks, however, for the Democrats. Acheson, a special target, was the "Red Dean," a "pompous diplomat with his striped pants and phony British accent." Marshall, who had "lost" China, was part of a "conspiracy so immense and an infamy so black as to dwarf any such previous venture in the history of man." The "Democratic label," he said, "is now the property of men and women . . . who have bent to the whispered pleas from the lips of traitors." The Democratic years were "twenty years of treason." When Truman fired General Douglas MacArthur from his Asian command in 1951, McCarthy said of the President, "The son of a bitch ought to be impeached."[93]

If there was a core of consistency to McCarthy, it was an emotional one of class and regional resentments. A Roman Catholic and a midwesterner, he seems genuinely to have detested the well-educated, wealthy, and mainly Protestant eastern Establishment. This is why Acheson and other "striped-pants" Anglophiles who dominated the State Department were such inviting targets. McCarthy underscored his feelings at Wheeling: "It is not the less fortunate members of minority groups who have been selling their nation out, but rather those who have had all the benefits the wealthiest nation on earth has had to offer. . . . This is glaringly true of the State Department. There the bright young men who are born with silver spoons in their mouths are the ones who have been worst."[94]

90. Oakley, God's Country, 61.
91. William Leuchtenburg, A Troubled Feast: American Society Since 1945 (Boston, 1973), 36.
92. Rovere, Senator Joe McCarthy, 49; Siegel, Troubled Journey, 77.
93. Rovere, Senator Joe McCarthy, 11; Oakley, God's Country, 61.
94. Godfrey Hodgson, America in Our Time (Garden City, N.Y., 1976), 42–43.

But McCarthy was not an ideologue. He was above all a demagogue seeking attention, re-election, and—maybe in the future—the presidency. A loose cannon, he had no organization to speak of, and he rarely followed up on any of his charges. Challenged to name a true subversive, he announced in March 1950 that he would "stand or fall" by his accusation that Owen Lattimore was the "top Russian agent" in the United States. This was a bizarre and unfounded accusation. Lattimore was a little-known scholar of Asia who had been uncritical in some of his writings about Stalin and Mao Tse-tung. But McCarthy could produce no documentary evidence that the professor had ever been a Communist.[95] Thereafter McCarthy made no serious effort to name people who were supposedly ruining the United States, and he never identified a single subversive. It was simpler to scatter his shot around the landscape.

The scattering made him the most controversial public figure in the country by the spring of 1950. Both *Time* and *Newsweek*, while critical of him, put him on their covers. A Gallup poll in May reported that 84 percent of respondents had heard of his charges and that 39 percent thought they were good for the country.[96] This was a remarkably high level of public recognition, and it placed the Truman administration on the defensive. Was there a way to counter McCarthy's reckless accusations?

Some people thought so. Given the attention McCarthy was receiving, they said, Truman and others should immediately have recognized the danger and appointed an impartial blue-ribbon investigating committee to weigh his charges.[97] But that would probably have required giving such a committee access to sensitive personnel files. To the President, this was unthinkable. Democrats instead tried to refute McCarthy. In February Truman retorted that there was "not a word of truth" to McCarthy's accusations. In late March he said McCarthyites were the "greatest asset the Kremlin had."[98] Senate Democrats set up a committee headed by Millard Tydings of Maryland to investigate the charges. Testimony before the Tydings Committee exposed many of McCarthy's lies and exaggerations, which the majority report later concluded were a "fraud and a hoax perpetrated on the Senate of the United States and the American people."[99]

95. Stanley Kutler, *The Inquisition: Justice and Injustice in the Cold War* (New York, 1982), 183–214; Fried, *Nightmare in Red*, 125–28.
96. Oakley, *God's Country*, 58.
97. Fried, *Nightmare in Red*, 128–30.
98. Hamby, *Beyond the New Deal*, 396–97.
99. Robert Griffith, *The Politics of Fear: Joseph R. McCarthy and the Senate* (Lexington, Ky., 1970), 100.

McCarthy and his allies, however, brushed aside the Tydings Committee by charging that it was a partisan cover-up. Senator William Jenner of
Indiana, a fervently anti-Communist Republican, accused Tydings of
heading the "most scandalous and brazen whitewash of treasonable conspiracy in our history." McCarthy branded the committee report as a
"green light to the Red fifth column in the United States" and a "sign to
the traitorous Communists and fellow travelers in our Government that
they need have no fear of exposure."[100] Reactions such as these showed
why it was so difficult to discredit McCarthy and his allies. So long as the
President refused to turn over personnel files, McCarthy could savage
whatever committee tried to refute his allegations.

Others who lament McCarthy's climb to fame have blamed the press.
Reporters, they say, should more insistently have demanded that he produce evidence. Some newspeople were indeed appalled by his behavior.
But McCarthy generally did well at manipulating the press. Many publishers were deeply conservative and believed in what McCarthy was
saying. Moreover, reporters were not editorialists, and they felt obliged to
record what a United States senator, who was "news," had to say. Again
and again his charges got headlines that proclaimed the rise of a Red
menace in the United States.[101]

Rigorous investigative journalism of the sort that arose in the 1960s and
1970s would probably have weakened McCarthy. But it is ahistorical to
expect such journalism to have existed in the 1950s. Reporters were ill
paid in those days and lacked the resources in staff or money to dig deeply
into McCarthy's charges. The Washington press corps was small. It was
not until later, amid growing anger about a "cover-up" during the Vietnam War, that significant numbers of reporters became obstreperous in
challenging "official" sources. Only in the 1970s, in the aftermath of
Watergate, did this attitude become widespread among political journalists in the United States.

Others analysts of McCarthyism in retrospect have concluded
pessimistically that it demonstrated the susceptibility of the American
people to demagogic appeals. There is evidence for such gloomy indict-
ments of democracy, but it is limited. McCarthy's attacks on the eastern
Establishment indeed set off responsive echoes, especially among conservative Republicans. Like McCarthy, some of these Republicans literally

100. Ibid., 101.
101. Edwin Bayley, *Joe McCarthy and the Press* (Madison, 1981); James Baughman,
     *The Republic of Mass Culture: Journalism, Filmmaking, and Broadcasting in
     America Since 1945* (Baltimore, 1992), 6.

loathed Acheson. "I look at that fellow," Republican Senator Hugh Butler of Nebraska said. "I watch his smart aleck manner and his British clothes and that New Dealism, everlasting New Dealism in everything he says and does, and I want to shout, Get Out! Get Out! You stand for everything that has been wrong in the United States for years."[102] The managing editor of McCarthy's hometown Appleton newspaper explained, "We don't want a group of New Yorkers and Easterners to tell us whom we are going to send to the Senate. That is our business, and it is none of theirs."[103] The anger that underlay such comments suggested that regional resentments still burned strongly indeed in the United States.

McCarthy's rampage also appealed to people who nursed hostility toward elites, especially in government. This feeling reflected enduring class, ethnic, and religious tensions, which periodically broke into the open amid more superficial manifestations of popular consensus in the United States. Working-class people who had struggled to get ahead after the war resented it when "educated" liberals appeared to look down on their accomplishments and their styles of life. A number of east-European-Americans, moreover, responded warmly to McCarthy's claims that Democrats had "sold out" the masses behind the iron curtain. Many Catholics, hating "Godless" Communism, also seemed to support his crusades. McCarthy, like Alabama's George Wallace in the 1960s, often appealed to all these groups by highlighting the influence of those who were wealthier and more influential. The sociologist Jonathan Rieder observes correctly that McCarthy advanced a "rhetoric of plebian contempt for things effete" and "hurried the movement of the Right toward a conservatism conspicuously more majoritarian than previously."[104]

The phenomenon of McCarthyism, however, should not be seen as a broadly popular movement or as one that was essentially working-class, Catholic, or ethnic in membership. Millions of such people, after all, still tended to vote Democratic and to reject McCarthyite visions of the world. Rather, three things may be said about McCarthyism. First, it derived much of its staying power from the frightened and calculating behavior of political elites and of allied interest groups, not from the

102. Siegel, *Troubled Journey*, 73.
103. Halberstam, *Fifties*, 52–53.
104. Jonathan Rieder, "The Rise of the Silent Majority," in Steve Fraser and Gary Gerstle, eds., *The Rise and Fall of the New Deal Order, 1930–1980* (Princeton, 1989), 247; Pells, *Liberal Mind*, 333.

people at large. Second, many partisan Republicans took the lead in backing their reckless colleague. Third, McCarthyism rode on anti-Communist fears—again, strongest among elites—that were already cresting in early 1950.[105]

The role of political leaders was indeed important. Those who had to run for office were often very cautious. Most of them did not like McCarthy personally and were appalled by his behavior. But the ferocity of his attacks—and his apparent invulnerability to criticism—shook many of them. They were reluctant to stand up and be counted against him, especially in an election year. Some of the most fearful office-holders came from working-class and Catholic districts. Representative John F. Kennedy, whose father was a friend and a patron of McCarthy, was among them. "McCarthy may have something," Jack said. Neither as a representative nor (after 1952) as a senator did Kennedy speak out against McCarthy.

Many Republican senators eagerly supported their colleague. For some this was only natural: they had been making similar charges for some time. On the same day that McCarthy spoke in Wheeling, Homer Capehart of Indiana rose in the Senate to ask, "How much more are we going to have to take? Fuchs and Acheson and Hiss and hydrogen bombs threatening outside and New Dealism eating away at the vitals of the nation! In the name of Heaven, is this the best America can do?"[106] Then and later Capehart, Jenner, and other conservatives gladly reinforced their Wisconsin colleague. When Truman nominated Marshall to be his Secretary of Defense following the outbreak of war in Korea, Jenner denounced the former Secretary of State as a "living lie" and as a "front man for traitors."[107]

Had these Republicans been the only ones to stand up for McCarthy, he might have had a more difficult time of it in the Senate—and with the press and the American people. But McCarthy also got the backing of Robert Taft, "Mr. Republican," the most influential GOP politician on Capitol Hill. Taft was not close to McCarthy—or to zealots like Jenner—and he did not think that subversion threatened the nation. But Taft, like most of his Republican colleagues on the Hill, strongly opposed the drift of American foreign and domestic policies since the New Deal. He very much disliked Acheson, one of McCarthy's favorite targets. Shocked by the unexpected victory of the Democrats in 1948, Taft longed to embar-

105. Griffith, *Politics of Fear*, 101–14.
106. Hodgson, *America in Our Time*, 34.
107. Griffith, *Politics of Fear*, 115.

rass and defeat them. He also hoped to win the GOP presidential nomination in 1952. And he knew that anti-Communism was politically popular.
For all these reasons Taft refused to denounce his colleague. McCarthy,
he said, should "keep talking and if one case doesn't work out, he should
proceed with another one." This was an irresponsible position that reflected the especially harsh partisan atmosphere of the times. [108]

Taft, although influential among his Republican colleagues, could not
silence all senatorial opposition to McCarthy. In June 1950, seven liberal
Republican senators led by Margaret Chase Smith of Maine issued a
"Declaration of Conscience" that complained about the Senate being
used as a "publicity platform for irresponsible sensationalism." Moreover,
it is doubtful that Taft—or anyone else—could have silenced McCarthy,
who reveled in the attention that he aroused. Still, the support of McCarthy by the GOP, especially in the Senate, did much to lend a veneer of
political respectability to McCarthyism from 1950 through 1954.

To highlight the role of elites in the support of McCarthy is to challenge the notion that he aroused great popular support. Polls, indeed,
showed that he did not; only once, in 1954, did more than 50 percent of
Americans say that they backed him. Still, office-holders knew that it paid
off politically to be loudly and insistently against Communism, especially
following the alarms that rang through American society in late 1949 and
early 1950: the Soviets had the Bomb, the Reds had China, Hiss had lied,
Fuchs was a spy. These were widely known, profoundly alarming events
that were already promoting Red Scares—in unions, in schools and universities, in Hollywood, within the Truman administration itself—well
before McCarthy made his headlines. It was in this highly charged Cold
War atmosphere of fear and suspicion that McCarthy and his well-placed
allies were able to run amok.

LOOKING BACK ON MCCARTHYISM and the postwar Red Scare,
George Kennan came close to despair:

> What the phenomenon of McCarthyism did . . . was to implant in my
> consciousness a lasting doubt as to the adequacy of our political sys-
> tem. . . . A political system and a public opinion, it seemed to me, that
> could be so easily disoriented by this sort of challenge in one epoch would
> be no less vulnerable to similar ones in another. I could never recapture,
> after these experiences of the 1940s and 1950s, quite the same faith in the

108. James Patterson, *Mr. Republican: A Biography of Robert A. Taft* (Boston, 1972),
    455–59.

American system of government and in traditional American outlooks
that I had had, despite all the discouragements of official life, before
that time.[109]

Other observers have been a little less pessimistic than Kennan, who
had always doubted the capacity of democracy to cope with crisis. Indeed,
McCarthy ultimately overreached himself and crashed into popular disgrace in 1954.[110] Thereafter the Red Scares that had sullied American
politics and society abated. Still, Kennan had good cause for pessimism,
for McCarthy's fall occurred more than four years after he started his
rampage at Wheeling, more than five years after anti-Communists moved
to cleanse the unions, schools, and colleges, more than six after government started using the Smith Act to put Communist leaders in jail, and
more than seven after Truman tightened loyalty progams and HUAC
assailed Hollywood. During these eight years it is estimated that a few
thousand people lost their jobs, a few hundred were jailed, more than 150
were deported, and two, Julius and Ethel Rosenberg (Communists who
were arrested in 1950 following further revelations about the Fuchs
case) were executed in June 1953 on charges of conspiracy to commit
espionage.[111]

Kennan, moreover, was correct in lamenting two broader results of the
postwar Red Scare. First, it constricted public life and speech. Before the
Red Scare peaked, many public figures had been both liberal and anti-
Communist, without worrying much about being labeled as a "pink" or
being accused of "disloyalty." During the Red Scare, however, liberal
politicians and intellectuals became vulnerable to the charge of being
"soft" on Communism—or worse. Some muted their liberalism, especially in the 1950s. As Diana Trilling said many years later, "McCarthy
not only deformed our political thinking, he . . . polluted our political
rhetoric. [He] had a lasting effect in polarizing the intellectuals of this
country and in entrenching anti-anti-communism as the position of
choice among people of good will."[112]

Second, McCarthyism helped to tie a straitjacket of sorts on America's
foreign and defense policies. How much the coat confined remains disputed. Some of the major policy initiatives of 1949–50—militarization of

109. George Kennan, *Memoirs, 1950–1963* (Boston, 1972), 228.
110. See chapter 9.
111. Caute, *Great Fear*, 62–68; Ronald Radosh and Joyce Milton, *The Rosenberg
    File: A Search for the Truth* (New York, 1983). The Rosenbergs were the first
    American civilians to suffer the death penalty in an espionage trial.
112. *Newsweek*, Jan. 11, 1993, p. 32.

NATO, non-recognition of the People's Republic, development of the Super, support for NSC-68—would have occurred anyway, as alarmed reactions of government officials confronted with a foe like Stalin, especially after the Soviets got the Bomb in 1949. The straitjacket was nonetheless tight. The Red Scare helped to turn understandable concerns about Communist intentions into demands for the toughest kinds of responses. Could the arms race, both nuclear and otherwise, have been less dangerous than it became after 1950? Could the United States have cautiously built bridges to the People's Republic, thereby driving a wedge between the Soviet Union and China? These and other options would have been politically perilous after the Cold War hardened in 1946, but the Red Scare made certain that they were not seriously explored. Especially after 1949, politicians, scholars, and writers who dared suggest initiatives that seemed "naive" or "soft" on Communism were even more than before at risk of losing office or reputation.

The Red Scare, finally, dampened a little the otherwise upbeat, can-do mood of American life at the time. "A little" is the way to put it, for postwar prosperity increased even more rapidly from 1950 through 1954 than it had between 1945 and 1948. The rising personal expectations of millions of Americans—most of them unaffected by the Red Scare—grew ever more grand. From this perspective the Red Scare may be seen as a shameful saga of overreaction and intolerance; it left scars. Still, in the longer run it did not stop the majority of Americans from their expectant pursuit of the Good Life.

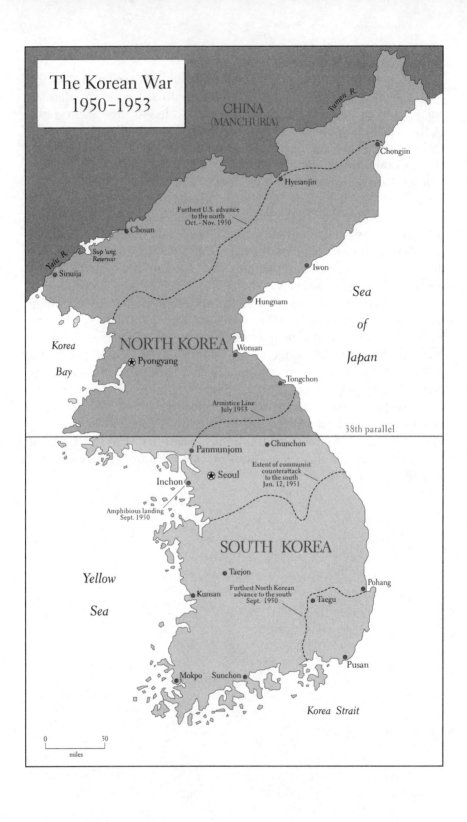

The Korean War
1950–1953

CHINA
(MANCHURIA)

Tumen R.

Chongjin

Hyesanjin

Furthest U.S. advance
to the north
Oct. - Nov. 1950

Chosan

Yalu R.

Sup'ung
Reservoir

Sinuiju

Iwon

Hungnam

Sea

of

Japan

Korea
Bay

NORTH KOREA

Wonsan

Pyongyang

Tongchon

Armistice Line
July 1953

38th parallel

Panmunjom

Chunchon

Seoul

Extent of communist
counterattack
to the south
Jan. 12, 1951

Inchon

Amphibious landing
Sept. 1950

SOUTH KOREA

Yellow

Taejon

Pohang

Sea

Kunsan

Furthest North Korean
advance to the south
Sept. 1950

Taegu

Pusan

Mokpo   Sunchon

Korea Strait

0   50

miles

8

Korea

There are few monuments in the United States commemorating the Korean War. By the 1960s most Americans had tried to put the war out of memory. Many who later saw "M*A*S*H," the popular TV series about an American medical unit in Korea, assumed that the episodes were set in Vietnam. Other Americans recalled the war as a relatively insignificant "police action," as Truman called it on one occasion. One book on the conflict is entitled *The Forgotten War*.[1]

This national amnesia is understandable, for the Korean War, which pitted American and allied troops against North Korea and China between June 1950 and July 1953, seems inconsequential compared to the two world wars and to America's ten-year battle in Vietnam. At the time, however, the Korean conflict loomed large both at home and abroad. On several occasions during the war Truman and his advisers feared that it might escalate into World War III. Determined to stem the tide, as they saw it, of worldwide Communism, they briefly considered using nuclear weapons. Although they prevented the conquest of South Korea, they did not achieve a wider objective—reunifying the peninsula under non-Communist control—that they undertook in the fall of 1950. The war had lasting diplomatic, economic, and domestic consequences. Far from an insigificant little police action, it was a brutal, bloody conflict that devastated Korea and inflicted nearly 4 million casualties (dead,

1. Clay Blair, *The Forgotten War: America in Korea, 1950–1953* (New York, 1987).

wounded, and missing), more than half of whom were civilians. It left 33,629 Americans dead from battle and 103,284 wounded.[2]

THE KOREAN CONFLICT was rooted in World War II. When the Second World War ended in August 1945, the United States and the Soviet Union (having entered the fight against Japan at the last minute) assumed responsibility for the surrender of enemy forces on the Korean peninsula, a largely mountainous, mineral-rich land that Japan had annexed and ruled brutally since 1910. Pentagon officials looked hurriedly at the map and decided on the 38th parallel as a line to divide the country for occupation purposes—the USSR in the North, the United States in the South—until it could be reunited in the future. This was a midway line roughly 300 miles south of the Yalu River, which formed much of Korea's northern border with Manchuria, and 300 miles north of the southernmost areas of the coast, which jutted into the Sea of Japan toward the southwest portions of Japan. About 10 million Koreans lived in each half of the country, with most of the industry in the North and much of the agriculture in the South.[3]

The Cold War quickly dashed hopes for the reunification and independence of Korea. Instead, the 38th parallel became a frontier that separated two hostile regimes. Kim Il Sung, a charismatic young Communist, took over control in the North; Syngman Rhee, an American-educated anti-Communist, gathered the reins of power in the South. In 1948 the regions became separate regimes; the People's Democratic Republic in the North and the Republic of Korea (ROK) in the South. Kim ran a tyrannical regime; opponents were commonly executed without trial. Rhee's was slightly less autocratic but equally bent on reunification by conquest. Fighting between partisans of both sides rent the unhappy peninsula between 1945 and 1950, killing perhaps 100,000 people. When North Korean troops invaded the South on the night of June 25 (June 24 United States time), they greatly widened a conflict that had long taxed the patience of the occupying powers in both areas.[4]

What decided the North Koreans on attack remains debated more than forty years later. At the time American high officials, convinced that Kim

2. David Rees, *Korea: The Limited War* (Baltimore, 1964), 460.
3. David McCullough, *Truman* (New York, 1992), 785–86.
4. Rosemary Foot, "Making Known the Unknown War: Policy Analysis of the Korean Conflict in the Last Decade," *Diplomatic History*, 15 (Summer 1991), 411–31; William Stueck, *The Korean War: An Internatonal History* (Princeton, 1995).

was a pawn of Moscow, thought that Stalin had masterminded the invasion. Although the Soviets had removed their own troops from from the area in 1949, they had continued to give North Korea substantial military assistance in early 1950, including T-34 tanks, which were devastating offensive weapons. America's ambassador to the Soviet Union, Alan Kirk, cabled home on June 25 that the attack represented a "clear-cut Soviet challenge which . . . US should answer firmly and swiftly as it constitutes direct threat [to] our leadership of free world against Soviet Communist imperialism."[5]

Critics of American policy in Korea—then and later—add that the attack took place because the Soviets thought that the United States would not defend the South. Kim had reason for such optimism in 1950. American troops had been pulled out of South Korea in June 1949, and Truman refused to commit the United States to a security pact with Rhee or to support his urgent and angry requests for better arms. Like his top advisers, the President worried about Rhee's own aggressive designs. Truman administration officials also remained staunch Europe-firsters; committing major military resources to Korea, they thought, would weaken defenses in the West. For these reasons, Truman withheld substantial military aid from South Korea.

Stalin and Kim may have paid special attention to Dean Acheson's "defense perimeter" speech in January 1950. In this widely noted address the Secretary of State omitted South Korea from areas that the United States would automatically defend against aggression. Those who carefully read the speech perceived that Acheson was deliberately murky in certain passages, holding out the possibility that unspecified places (such as Korea) might expect American help—if they could not defend themselves—from "the entire civilized world under the Charter of the United Nations."[6] Still, it was unwise of Acheson to lay out publicly what the United States was likely to do in the world; it would have been better to keep people guessing. In doing so he made it clear that defense of South Korea was a low-priority item in the United States.

It now seems that American leaders misread Stalin's role in the invasion. To be sure, Soviet military assistance made such an attack possible. Indeed, the Soviets drew up the plan of attack once it had been decided to move. But the initiative for the invasion came from Kim, who appears to have thought that an invasion would touch off a revolution against Rhee's

5. Barton Bernstein, "The Truman Administration and the Korean War," in Michael Lacey, ed., *The Truman Presidency* (Washington, 1989), 419.
6. Ibid., 417.

autocratic rule in the South. After resisting Kim's appeals, Stalin gave his assent to the attack, apparently believing that the fighting would be brief and that the United States would not intervene.[7]

The war, in short, was as much a dramatic extension of the civil conflict in Korea as it was a deliberate provocation from the Kremlin. Still, it is highly improbable that Kim would have acted without Soviet approval, and American leaders were therefore correct at the time to heap a good deal of the blame on Moscow. Then and throughout the war they were deeply worried that the USSR had fomented the fighting so as to tie up American forces in Korea, thereby opening western Europe to Communist advances.[8]

Whatever the sources of North Korea's decision, it was clear that Kim and Stalin had badly misjudged the situation. Although some South Koreans supported the invaders, most did not. The South Korean army, while badly outgunned, remained loyal to Rhee. And the United States confounded enemy expectations by deciding to help the South Koreans. Kim's misjudgment, based at least in part on the irresolute signals of the Truman administration, was one of the most portentous in the history of the Cold War.

THE NORTH KOREAN ONSLAUGHT in the darkness of June 25 was a well-organized, smashing offensive spearheaded by 150 of the Soviet T-34 tanks, ROK rockets from bazookas bounced harmlessly off the tanks. Estimates were that some 90,000 well-trained, well-camouflaged North Korean troops took part in the assault. Many were battle-hardened, having served as "volunteers" for Mao in the Chinese civil war. They overwhelmed the ill-equipped ROK forces, some of whom had to rush back from furlough. About all that the defenders could do was stand and resist, fall back, resist a little more, and fall back again. Within a few days the North Koreans overran Seoul, the South Korean capital, smashed down the peninsula, and seemed destined to push ROK forces into the sea.

Truman received news of the invasion at 9:30 P.M. June 24 at home in Independence, where he had gone for a weekend visit with his family.

7. Sergei Gocharov, John Lewis, and Xue Litai, *Uncertain Partners: Stalin, Mao, and the Korean War* (Palo Alto, 1994); Jacob Heilbrun, "Who's to Blame for the Korean War? The Revision Thing," *New Republic*, Aug. 15, 1994, pp. 31–38; Bruce Cumings, *The Origins of the Korean War: The Roaring of the Cataract, 1947–1950* (Princeton, 1990).
8. Robert McMahon, "Toward a Post-Colonial Order: Truman Administration Policies Toward South and Southeast Asia," in Lacey, ed., *Truman Presidency*, 364.

Acheson, telephoning the bad news, told him that he had already asked the UN Security Council to call for an end to the fighting and the withdrawal of North Korean forces to the 38th parallel. The next day, a Sunday, Truman flew back to Washington and went straight to Blair House, where he was living while the White House across the street was undergoing major renovations. By then the Security Council had already approved the American-sponsored resolution, 9–0. But the North Koreans paid no attention and smashed ahead to the south. Top military and diplomatic officials joined the President for dinner at Blair House and for the first of many tense meetings that Truman convened there over the next few days. The key players included Acheson, Dean Rusk, who was Assistant Secretary of State for Far Eastern Affairs, Omar Bradley as Chairman of the Joint Chiefs, the service chiefs, Defense Secretary Louis Johnson, and other officials from the executive branch. No member of Congress was invited.[9]

From the start all these men favored taking a strong stand against North Korean aggression. Their motives varied a little. A few worried that Communist control of South Korean air bases would pose a greater threat to the security of Japan, which the United States was building up as a bastion of capitalism in Asia. But most of those at the meeting were not much concerned about Japan, and they cared little for South Korea in itself. They worried instead that the North Korean invasion, like Nazi moves in the 1930s, brazenly challenged the will and the credibility of the "Free World." If the United States pursued appeasement, Armageddon was at hand.[10] Truman told an aide that "Korea is the Greece of the Far East. If we are tough enough now, if we stand up to them like we did in Greece three years ago, they won't take any next steps. . . . There's no telling what they'll do, if we don't put up a fight right now."[11]

Despite such talk, Truman and his advisers at first hoped that the North Koreans could be stopped without having to involve the United States in ground combat. This hope reflected their acute awareness of America's military weakness, especially its army's. The President made no military commitments at the meeting on June 25. By the next night, however, the situation in Korea had deteriorated badly. General Douglas MacArthur,

9. William Stueck, *The Road to Confrontation: American Policy Toward China and Korea, 1947–1950* (Chapel Hill, 1981), 177–220; Glenn Paige, *The Korean Decision: June 24–30, 1950* (New York, 1968); Robert Donovan, *The Tumultuous Years: The Presidency of Harry S. Truman, 1949–1953* (New York, 1982), 187–240.

10. McMahon, "Toward a Post-Colonial Order," 339–65; Foot, "Making Known."

11. Bernstein, "Truman Adminsitration," 422.

the American military commander in Asia, urgently requested American help. Truman responded with more decisive steps, committing American air and naval forces to the South and greatly increasing aid to Indochina and the Philippines. He also ordered the Seventh Fleet to the waters between mainland China and Taiwan, thus guaranteeing Chiang Kai-shek naval protection for the first time since his flight the previous year. Within two days of the invasion the United States was significantly expanding and militarizing its foreign policies in Asia.

The next morning Truman widened his circle of advisers a little by bringing in some congressional leaders for a meeting. But his goal was mainly to inform them of the decisions the previous night, not to seek their advice. They offered no serious objections, and the meeting lasted only half an hour. Later that day Congress seemed enthusiastic in support of the President's actions. That evening the UN passed a resolution supporting the dispatch of air and naval forces to aid the beleaguered South Koreans.

But these moves, too, failed to slow the advance from the North, and MacArthur grew ever more agitated, calling for immediate dispatch of two American army divisions from occupation duty in Japan to Korea. His plea reached the Pentagon at 3:00 A.M. June 30, Washington time. Truman, up and shaved, received it at Blair House at 4:47 and approved it without hesitation or further consultation. His decision produced the development that virtually all American political and military leaders, MacArthur included, had dreaded until that time: United States soldiers were to fight on the land masses of Asia.

Thanks to the desperate military situation in Korea in the five days following the invasion, Truman and his advisers were acting under formidable pressures of time. It is hardly surprising, therefore, that they made some mistakes. One was Truman's failure adequately to consult Congress. Neither then nor later did the President request what some people thought the Constitution demanded in such a situation: a congressional declaration of war. When he asked Senator Tom Connally, head of the Foreign Relations Committee, if he should do so, Connally said no. "If a burglar breaks into your house, you can shoot at him without going down to the police station and getting permission. You might run into a long debate by Congress, which would tie your hands completely. You have the right to do so as Commander in Chief and under the UN Charter."[12]

12. Robert Ferrell, *Harry S. Truman and the Modern American Presidency* (Boston, 1983), 124.

Connally merely told Truman what he wanted to hear. The President did not involve Congress in the decision to intervene. He later explained, "I just had to act as Commander-in-Chief, and I did."[13] This was a plausible argument, for time was indeed of the essence, and he did not know then how bloody a conflict lay ahead. Still, he set a bad precedent. Presidents in the past had maneuvered in ways that committed American troops to battle. But Truman had gone farther than that, asserting that his constitutional role as commander-in-chief justified executive action alone. Many later Presidents, notably Lyndon Johnson, followed in Truman's footsteps by committing the United States to fighting without securing congressional sanction.

Truman's executive decisions angered a few doubters, such as Taft, who had called for congressional consultation before the United States dispatched its ground forces. At the time, however, these doubters did not say much, and Taft did not press his case. Most people in the frightening, urgent days of late June believed that Truman had to act quickly. When the war went badly, however, Taft and others—mostly conservative Republicans—sharpened their attacks on Truman's display of presidential authority. The President's failure to consult Congress in 1950 added to his political difficulties over the next two and a half years, during which time his foes again and again branded the morass as "Truman's War."

Truman also erred in calling the war a "police action." When he held a press conference on June 29, he declared, "We are not at war." A reporter then asked if it would be accurate to call the fighting a police action under the United Nations. "Yes," Truman replied. "That is exactly what it amounts to."[14] His response seemed harmless at the time. But he had already sent in air and naval help, and within twenty-four hours he had committed the first of what became many divisions of American ground forces. Moreover, the brunt of the "UN" effort was borne by South Korean and American men, who were commanded by MacArthur and subsequent American generals. It was therefore misleading to call the war a UN initiative. When the fighting became stalemated, costing many thousands of American lives, it was understandable that people tore into him for calling a "war" a "police action." Words can be big weapons in politics.

In June and July, however, Truman had little to worry about at home, for his decisions evoked widespread approval. Prominent public figures

13. Bernstein, "Truman Administration," 425.
14. McCullough, *Truman*, 782–83.

as otherwise different as Thomas Dewey, George Kennan, and Walter Reuther hailed his moves. Eisenhower, then president of Columbia, said, "We'll have a dozen Koreas soon if we don't take a firm stand." Taft, although upset at the bypassing of Congress, indicated he was for American intervention. Even Henry Wallace emerged from relative obscurity to take a hawkish stand. He declared, "I am on the side of my country and the United Nations." By August Wallace was in favor of using nuclear bombs if necessary, and by November he called for massive American rearmament.[15]

Going to war seemed equally agreeable to the American people. Polls indicated that nearly three-quarters of the public approved of Truman's actions. *Newsweek* questioned individuals, most of whom were delighted that the United States had taken a stand. "After China, the Russians thought they could get away with it," an auto worker in Detroit exclaimed. Another worker angrily replied, "It should have been done two years ago." A businessman agreed: "Truman was in a spot where he couldn't do anything else, but he did all right." A man on a street corner concluded, "I think it's one of the few things the President's done that I approve of, and that seems to be the general feeling among people."[16]

Opinions such as these revealed a profound truth about Americans in the post–World War II era: they were not only patriotic but also eager—in the short run—to back decisive presidential actions in the field of foreign affairs. Later Presidents, indeed, came to understand that "police actions" and "surgical strikes" could greatly (though briefly) revive sagging ratings in the polls. This was emphatically not Truman's motive for going to war, but his firm and "presidential" resolve helped temporarily to raise his popular standing in the summer of 1950.

Americans above all seemed pleased that the United States had taken—finally—a strong stand against Communism. When Truman's decision to send in troops was announced, members of both houses of Congress stood up and cheered, even though they had not been consulted. When he asked them in mid-July for an emergency defense appropriation of $10 billion (nearly as much as the $13 billion budgeted for the entire year), they stood up and cheered again. Both houses approved his request almost unanimously. Congress also authorized him to call up the reserves, extended the draft, and gave him war powers similar to those exercised by Roosevelt during World War II. Members of Congress were above all

15. Ibid., 781–83; Alonzo Hamby, *Beyond the New Deal: Harry S. Truman and American Liberalism* (New York, 1973), 404–8.
16. *Newsweek*, July 10, 1950, p. 24.

pleased that America was standing firm against Communism. Joseph Harsch, an experienced reporter for the *Christian Science Monitor*, summed up the feeling in Washington: "Never before have I felt such a sense of relief and unity pass through the city."[17]

IN KOREA, HOWEVER, the war went badly for the United States and its UN allies in the first few weeks. MacArthur had been optimistic; like many Americans he had a low opinion of Asian soldiers, and he thought the United States could clean things up quickly. But he had done a poor job of preparing his occupation forces in Japan.[18] The troops who were rushed from Japan to Korea—most of them to the port of Pusan on the southeast corner of the peninsula—were poorly equipped and out of shape. Colonel John "Mike" Michaelis, a regimental commander, complained that many of the soldiers did not even know how to care for their weapons. "They'd spent a lot of time listening to lectures on the differences between communism and Americanism and not enough time crawling on their bellies on maneuvers with live ammunition singing over them. They'd been nursed and coddled, told to drive safely, to buy War Bonds, to give to the Red Cross, to avoid VD, to write home to mother— when someone ought to have been telling them how to clear a machine gun when it jams."[19]

If conditions had been better, the troops might have had a little time, once in Pusan, to train more intensely. But they were rushed to the front lines. There they were torn up by the well-planned North Korean advance. Ignorant of the terrain, UN forces also struggled against drenching rains, which turned roads into mud and created near-chaotic snarls for retreating vehicles. Temperatures hovered around 100 degrees during the day. Thirsty American soldiers drank standing water from rice fields that had been fertilized with human waste: many were wracked with dysentary. In the first two weeks of savagery, much of it in nighttime close combat, United Nations forces suffered 30 percent casualties and reeled back toward Pusan.[20]

During late July the North Koreans kept driving south, inflicting devas-

17. Ronald Oakley, *God's Country: America in the Fifties* (New York, 1986), 78; William O'Neill, *American High: The Years of Confidence, 1945–1960* (New York, 1986), 118; on Harsch, Fred Siegel, *Troubled Journey: From Pearl Harbor to Ronald Reagan* (New York, 1984), 80.
18. Callum MacDonald, *Korea: The War Before Vietnam* (New York, 1986), 203.
19. David Halberstam, *The Fifties* (New York, 1993), 74.
20. McCullough, *Truman*, 785–87.

tation on the UN forces, the vast majority of whom were Americans. But the UN was slowly evening the odds. Rapid dispatch of troops from Japan boosted manpower; UN troops outnumbered North Korean in the South by early August. Artillery and anti-tank weapons gradually neutralized the T-34 tanks. And the UN forces had overwhelming superiority in the air. This they used to full advantage, raising havoc with North Korean supply lines. UN air superiority remained vital throughout the war, enabling the dropping of 635,000 tons of bombs (and 32,557 tons of napalm)—more than the 503,000 tons dropped in the Pacific theater during all of World War II.[21] The bombing followed a consciously devised "scorched earth" policy that wiped out thousands of villages and deliberately destroyed irrigation necessary for the all-important rice economy of the peninsula. Thousands of Koreans suffered from starvation and slow death; many survivors cowered in caves. The number of civilian deaths—estimated at around 2 million—approximated 10 percent of the prewar population of the peninsula. The ratio of civilian deaths to total deaths in the Korean conflict was considerably higher than in World War II or Vietnam.[22]

Firepower such as this inflicted especially heavy casualties on the North Korean forces, which (according to later estimates) suffered 58,000 dead or wounded by early August. They lost around 110 of their 150 tanks. Increasingly reliant on green conscripts, they also stretched their supply lines. At this point the UN forces stiffened within a small but defensible perimeter bounded in part by the Naktong River on the north and the Sea of Japan to the east. The perimeter protected Pusan, where supplies and troops were off-loaded at a furious pace. General Walton Walker, commander of the American Eighth Army, had the advantage of mobility within the perimeter. Having broken North Korean codes, his forces often knew where enemy would attack. By mid-August the UN no longer feared a Dunkirk-like evacuation.[23]

Disaster having been averted, MacArthur began pressing his superiors in Washington for approval of a retaliatory operation that he had conceived early in the war: a surprise amphibious landing at the port of Inchon, some thirty miles west of Seoul. Such an attack, he argued,

21. Foot, "Making Known."
22. Robert Divine, "Vietnam Reconsidered," *Diplomatic History*, 12 (Winter 1988), 79–93.
23. Robert Leckie, *Conflict: The History of the Korean War, 1950–1953* (New York, 1962), 50–73; Rees, *Korea*, 36–54; Stueck, *Road to Confrontation*, 223–31; MacDonald, *Korea*, 48–50.

would catch the enemy by surprise, outflank them, trap them between UN forces to the north and to the south, and obviate need for the alternative: a long and bloodier counter-offensive straight up the peninsula.

MacArthur's plan at first seemed too risky for Bradley and other top-ranking military officials in Washington. Admirals were especially nervous about it, for Inchon had no natural beach—only seawalls that protected the city. Worse, the tides at Inchon were enormous, up to thirty-two feet. An amphibious assault would have to be timed to coincide with the highest of high tides, either September 15, September 27, or October 11. If anything went wrong, such as a sunken ship blocking the harbor, the assault could be stalled and the landing ships left grounded on an open expanse of mud. It was rumored, moreover, that the Russians were mining the harbors. "Make up a list of amphibious don'ts," one naval officer grumbled, "and you have an exact description of the Inchon operation."[24]

In considering the Inchon option, Truman, Bradley, and others also had to consider its source: MacArthur himself. MacArthur, then seventy years old, was one of the most distinguished soldiers in American history. Graduating number one in his class from West Point in 1903, he served in the Philippines, East Asia, and Mexico and reached the rank of major by the time of America's entrance into World War I in 1917. As commander of the famed 42d Infantry (Rainbow) Division during the war, he proved a brave and dashing leader. Twice wounded, he was decorated thirteen times and emerged from the war a brigadier general. He then became superintendent of West Point, held various other high-ranking posts in the Philippines and the United States, and was elevated to chief of staff of the army in 1930. At age fifty he was the youngest man ever to hold that post.[25]

When his term as chief of staff ended in 1935, MacArthur was named military adviser to the newly created Philippine Commonwealth. Though he retired from the United States Army in 1937, he stayed on in the Philippines with the rank of field marshal. Famed for his gold-braided hats, aviation-style sunglasses, and corncob pipes, he was a fabled figure even before World War II. When war broke out, he returned to active duty to become commander of United States Army forces in Asia.

24. Trumbull Higgins, *Korea and the Fall of MacArthur: A Précis in Limited War* (New York, 1960), 44.
25. For MacArthur, see D. Clayton James, *The Years of MacArthur*, Vol. 3 (Boston, 1985); and Douglas Schaller, *MacArthur: The Far Eastern General* (New York, 1989).

Though driven out of the Philippines after the bloody battles of Bataan and Corregidor in 1942, he escaped to Australia and commanded American soldiers in the successful island-hopping assaults that battered Japan in the Pacific. By war's end he had returned to the Philippines and become a five-star general. Thereafter he served as commander of American occupation forces in Japan, where he developed a hugely favorable reputation as a firm but benevolent destroyer of Japanese militarism. By 1950, when the Korean War broke out, he had been in Asia, never returning to the United States, for more than thirteen years.

This impressive record of achievement gave MacArthur an almost legendary reputation. Many observers of his work in postwar Japan hailed him as an American Caesar. But Bradley and others who knew him recognized that MacArthur was also vain, arrogant, and domineering. Having spent much of his life in Asia, MacArthur was certain that it was crucial to the long-run security of the United States. He was equally sure that he understood the "mind of the Oriental," as he called it, better than anyone in Washington. He was surrounded by sycophants and reveled in publicity, much of it provided by photographers who took pictures of him that flatteringly emphasized the jut of his jaw and by journalists who turned out copy that highlighted his personal achievements without saying much about the contributions of others. Eisenhower, asked if he knew MacArthur, later said, "Not only have I met him. . . . I studied dramatics under him for five years in Washington and four in the Philippines."[26]

Truman, too, had his doubts about MacArthur. In 1945 he had referred to him in his journal as "Mr. Prima Donna, Brass Hat" and as a "play actor and bunco man."[27] When MacArthur issued an unauthorized statement in August 1950 on the need for American defense of Taiwan—a sensitive foreign policy question—Truman was so livid that he personally dictated a message calling on MacArthur to withdraw his statement. As Truman conceded later, by which time MacArthur had many times overstepped his position as a military commander, he would have done well to have removed the general then and there.

Instead, Truman not only kept MacArthur on but also—following approval by the Joint Chiefs—authorized the Inchon operation. Given the go-ahead, MacArthur moved quickly, and the assault took place on September 15. Protected by American air power, some 13,000 marines

26. Oakley, God's Country, 79; Halberstam, Fifties, 79–82; Stephen Ambrose, Eisenhower: Soldier and President (New York, 1990), 42–52.
27. McCullough, Truman, 792–94.

poured ashore and overwhelmed the small, inexperienced garrison of troops who were defending the area. Only twenty-one Americans lost their lives in the landing. Inchon fell within a day, and additional UN forces stormed east into Seoul. At the same time, American and ROK troops broke through the Pusan perimeter and chased the now retreating North Koreans northward. By September 26 the devastated city of Seoul was again in UN hands; by September 27 half of the North Korean army was trapped; by October 1 UN forces were back at the 38th parallel. Although 40,000 or so North Korean troops escaped into the North—a severe disappointment to the United States—the Inchon assault had turned the war around. MacArthur had organized a remarkably successful military operation.[28]

At this point in late September the Truman administration made one of the most fateful decisions of the postwar era: to unify Korea under Western auspices. This meant crossing the 38th parallel, destroying Kim's army, and driving to the North Korea–China border along the Yalu. MacArthur insisted on such a strategy, and his opinion—now that he had managed such a magnificent operation—carried weight. No one at the Pentagon dared to challenge him. But MacArthur's goals merely echoed those of others. Virtually all American officials, flush with excitement after Inchon, believed that the enemy must be destroyed; aggression must not go unpunished; the credibility of the "Free World" was at stake.[29] Public opinion, too, seemed to demand that the UN finish the job. The *New Republic*, dismissing rumors that the Chinese might intervene, agreed: "War with China would certainly be a disaster for the West. Yet war cannot be averted by conceding to illegal aggression."[30] America's UN allies concurred with the decision to cross the parallel, as did the UN itself, which formally approved it in early October.

As these approvals made clear, the decision to go north was not taken in haste. Still, Truman and his advisers might have proceeded with more care, for the goal of unification changed the original goal of UN intervention. In particular, Truman and his top advisers would have done well to ponder the meaning of ambiguous but repeated Chinese threats and warnings. They might have done more consulting with Congress—which again was left out of the loop. And they should have thought more about long-range problems. How close to the Yalu should UN forces go? If the drive succeeded, how would Korea be unified? Did the United States

28. Donovan, *Tumultuous Years*, 268–80; James, *Years of MacArthur*, 476.
29. Bernstein, "Truman Administration," 429–32.
30. Hamby, *Beyond the New Deal*, 407.

intend to stay in Korea after the fighting in order to protect what it had won? These and other questions received relatively little consideration in the heady excitement that followed success at Inchon.

At first the fighting went well enough. With the X Corps driving up the east and the Eighth Army sweeping up the west of North Korea, UN forces advanced in great strides. Bombing further savaged the enemy armies and devastated civilian life in the North. By the end of October a few ROK units were already near the Yalu. Although the Chinese government was threatening to intervene, military intelligence and the CIA picked up little evidence of major Chinese troop movements toward Korea.

Truman nonetheless decided to talk personally with MacArthur about the situation, and in mid-October he flew with top aides 14,425 miles to Wake Island in the Pacific. MacArthur, Truman later remembered, was arrogant and condescending at Wake. Nonetheless, the two men and their aides managed to reach consensus in only an hour and a half. MacArthur was reassuring, predicting that the Eighth Army could be out of Korea by Christmas.[31] When Truman asked about the possibility of Chinese intervention, the general replied, "Very little." He added, "We are no longer fearful of their intervention. . . . The Chinese have 300,000 men in Manchuria. Of these probably not more than 100,000–125,000 men are distributed along the Yalu River. Only 50,000–60,000 could be gotten across the Yalu River. They have no Air Force. . . . If the Chinese tried to get down to Pyongyang [the North Korean capital] there would be the greatest slaughter."[32] When the conference ended, Truman lauded MacArthur and gave him a Distinguished Service Medal.

Two weeks after the Wake Island conference, however, ROK units began capturing soldiers who were Chinese. Interrogation suggested that the invaders were arriving in force. Why they intervened is yet another debated question about the war. One motivation may have been Chinese gratitude to the North Koreans for sending in more than 100,000 "volunteers" to help against Chiang in the late 1940s. Surely more important were Chinese concerns about security, concerns that mounted greatly when MacArthur, disobeying orders, sent American (as opposed to ROK) troops close to the Yalu in late October.[33] Whatever the Chinese motives,

31. Harry S. Truman, Memoirs, Vol. 2, Years of Trial and Hope (Garden City, N.Y., 1956), 414–20; Stueck, Road to Confrontation, 238–39; Donovan, Tumultuous Years, 284–88; MacDonald, Korea, 57–59.
32. Higgins, Korea and the Fall, 58; McCullough, Truman, 800–807.
33. Foot, "Making Known."

they were compelling enough for Mao to persist even when the Soviets delayed in making good on earlier assurances of air cover to sustain Chinese intervention.[34]

MacArthur was relatively unmoved even when the Chinese attacked in force on November 1. He was simply egotistical. Having assured himself that the Chinese would not dare to intervene, he refused to believe that they might prevail. When he returned to Tokyo (where he maintained his headquarters) after a visit to Korea on November 24, he confidently announced a final UN offensive. "If successful," he proclaimed, "this should for practical purposes end the war."[35] It was the day after Thanksgiving.

MacArthur's offensive at first met little resistance. But two days later Mao's tough, battle-seasoned men poured into combat in full force. They wore warm padded jackets to fend off the bitter cold, with wind-chill temperatures as low as twenty or thirty degrees below zero, and with howling winds that froze the weapons and batteries of defenders. Drawn from poor peasant stock, the Chinese were used to privation. They carried only eight to ten pounds of equipment—as opposed to sixty pounds for many UN soldiers—and moved very quickly. Accustomed to the lack of air cover, they were skilled at holding still when planes were overhead. When the planes went away, they swarmed as close as possible to the enemy, blasted away with automatic fire, and engaged in terrifying hand-to-hand combat. Often fighting at night, they overran UN soldiers huddled on the frozen ground and stabbed them to death through their sleeping bags. The Chinese were unimaginably numerous, and they seemed fearless. UN forces likened the hordes before them to an endless wave of humanity that seemed oblivious to danger or death.

The fighting that followed for the next few weeks was among the bloodiest in the annals of American military history. Some of this carnage stemmed from faulty generalship. In his overconfidence MacArthur had left a gap between his forces on the east and on the west, thereby endangering their flanks. In his haste to get to the Yalu he had stretched his supply lines and thinned his forces. UN soldiers were cold, increasingly ill fed and ill supplied, and in many cases cut off from reinforcements. Overwhelmed by the suddenness and surprise of the Chinese assault, they scrambled desperately for cover. In places retreat became rout, as UN

---

34. This was among the many causes of the Sino-Soviet split that only later became apparent to the West.
35. Halberstam, *Fifties*, 104–7.

forces jammed lowland roads and exposed themselves to withering fire from masses of Chinese on the hillsides.

The bravery of these beleaguered troops has become the stuff of legends. One was an epic "fighting retreat" by members of America's First Marine Division from the Chosin Reservoir, where they had been trapped, to the port of Hungnam forty miles away, where they were evacuated. The marines sustained 4,418 battle casualties, including 718 deaths, and (together with their air support) inflicted an estimated 37,500 casualties on the enemy, two-thirds of them fatal. Still bloodier was the lot of men in the United States Army's Seventh Infantry Division, which retreated for some sixty miles over twisting, mountainous roads. Their worst ordeal was a gauntlet of six miles where the Chinese held high ground on both sides and blasted American soldiers with mortars, machine guns, and small-arms fire. On one day some 3,000 American soldiers were killed, wounded, or lost. Overall the 7th Division suffered 5,000 casualties (one-third of its total force) in the last three days of November.[36]

The sudden turnabout staggered MacArthur, whose overweening self-confidence disappeared overnight. "We face an entirely new war," he lamented. Thereafter he fired off plaintive, self-serving messages that blamed Washington for giving him meager support. He demanded that Chiang's army be "unleashed" to fight alongside the UN in Korea, that China be blockaded, and that Chinese industrial targets be bombed, if necessary with atomic weapons. "This group of Europhiles," he complained of Truman's advisers, "just will not recognize that it is Asia which has been selected for the test of communist power and that if all Asia falls Europe would not have a chance—either with or without American assistance."[37]

Truman did his best to keep his emotions under control during these desperate weeks. But the strain was severe, and domestic provocations had already compounded his troubles. On the unseasonably hot afternoon of November 1, when he was napping at Blair House, two fanatic Puerto Rican nationalists fired at guards outside the house. White House police jumped into action, and a wild gun battle erupted on the steps and the sidewalk. When the shooting was over, one of the would-be assassins was dead, the other wounded. One policeman was shot dead at close range and two wounded. Truman maintained his schedule as if nothing had

36. Leckie, *Conflict*, 178–96; Rees, *Korea*, 161–77.
37. John Gaddis, *Strategies of Containment: A Critical Appraisal of Postwar American National Security Policy* (New York, 1982), 118; MacDonald, *Korea*, 71–90.

happened. But the attempted assassination sharply constricted his freedom. There were no more walks across the street to work at the White House—instead he rode in a bullet-proof car. He confided to his diary, "It's hell to be President." The biographer David McCullough thinks that the months of November and December 1950 were "a dreadful passage for Truman . . . the most difficult period of his Presidency."[38]

A few days after the Blair House shootings the off-year elections confirmed the President's shaky standing with voters. At that time Chinese intervention was only beginning; it was not yet a big concern. But Republicans nonetheless attacked him for his conduct of the war and for being "soft" on Communism. In Illinois the conservative anti-Communist Everett Dirksen defeated Scott Lucas, who was then Senate majority leader. In Ohio Taft won a sweeping victory that set him up as a major presidential contender for 1952. In California Richard Nixon beat Congresswoman Helen Gahagan Douglas, a liberal Democrat, in a nasty fight for a seat in the Senate. She called him "Tricky Dick," a name that stuck. He called her the "Pink Lady" and charged that she was "pink down to her underwear."[39] McCarthy helped to engineer the defeat of Maryland's Millard Tydings, who had chaired the Senate committee that had investigated him, by circulating faked photographs showing Tydings chatting with Earl Browder, a head of the American Communist party. Republicans cut Truman's majority in the Senate from 12 to 2 and his majority in the House from 92 to 35. Pundits interpreted the results as a repudiation of the President.

Already tense in early November, Truman faltered in the aftermath of the post-Thanksgiving Chinese offensive a few weeks later. At a press conference on the morning of November 30 reporters asked him if the United States might fight back by using the atomic bomb. The President responded, "There has always been active consideration of its use. I don't want to see it used. It is a terrible weapon and it should not be used on innocent men, women, and children who have nothing to do with this military aggression." A reporter persisted, "Did we understand you clearly

38. McCullough, *Truman*, 808–13.
39. Stephen Ambrose, *Nixon: The Education of a Politician, 1913–1962* (New York, 1987), 197–223; Ingrid Winther Scobie, *Center Stage: Helen Gahagan Douglas: A Life* (New York, 1992), 221–52; Roger Morris, *Richard Milhous Nixon: The Rise of an American Politician* (New York, 1990), 515–624; Richard Fried, "Electoral Politics and McCarthyism: The 1950 Campaign," in Robert Griffith and Athan Theoharis, eds., *The Specter: Original Essays on the Cold War and the Origins of McCarthyism* (New York, 1974), 190–223.

that use of the bomb is under active consideration?" Truman replied, "Always has been. It is one of our weapons." When asked whether targets would be civilian or military, he said, "I'm not a military authority that passes on those things. . . . The military commander in the field will have charge of the use of weapons, as he always has."[40]

Truman's comments unleashed international consternation. Clement Attlee, Prime Minister of Great Britain, was so alarmed that he flew to the United States for talks. In Washington, Truman's top aides began damage control. Charles Ross, presidential press secretary, issued a release to clarify Truman's comments. The President, he said, knew full-well that only the commander-in-chief, not generals in the field, determined whether nuclear weapons might be used in crisis.

By the time Attlee arrived in Washington on December 4, it was generally understood that Truman did not intend to deploy the Bomb. Shaken by the Chinese onslaught, he had spoken carelessly. But his remarks revealed the edginess that he felt at the time. This became especially clear when Ross, an old and valued childhood friend, suddenly collapsed and died of a coronary occlusion following a press briefing on the afternoon of December 5. Though badly shaken, Truman and his wife pulled themselves together and went off to hear their daughter Margaret, an aspiring singer, give a recital before a gala crowd at nearby Constitution Hall.

Margaret, uninformed about Ross's death, sang to repeated applause. But not all listeners were impressed. Paul Hume, music critic for the *Washington Post*, published a polite but nonetheless devastating review of her performance in the next morning's paper. Truman, up early as usual, read the review at 5:30 A.M. and exploded by writing a 150-word diatribe to Hume that he immediately sealed and stamped (three cents) and gave to a messenger to post. A key passage said, "Some day I hope to meet you. When that happens you'll need a new nose, a lot of beefsteak for black eyes, and perhaps a supporter below."

This was by no means the first time that Truman had lost his temper and confided his fury to paper. But he had never gone so far as to post something like this. Hume and his editors at the *Post* did nothing about it, but copies were made, and the letter soon appeared on the front page of the *Washington News*. Publication of it unleashed a deluge of angry mail

40. Paul Boyer, "'Some Sort of Peace': President Truman, the American People, and the Atomic Bomb," in Lacey, ed., *Truman Presidency*, 192; McCullough, *Truman*, 820–22.

that descended on the White House. Some of it raged about the war in Korea:

HOW CAN YOU PUT YOUR TRIVIAL PERSONAL AFFAIRS BEFORE THOSE OF ONE HUNDRED AND SIXTY MILLION PEOPLE. OUR BOYS DIED WHILE YOUR INFANTILE MIND WAS ON YOUR DAUGHTER'S REVIEW. INADVERTENTLY YOU SHOWED THE WHOLE WORLD WHAT YOU ARE. NOTHING BUT A LITTLE SELFISH PIPSQUEAK.

Perhaps the hardest letter for Truman to bear arrived with a Purple Heart enclosed. It read:

Mr. Truman: As you have been directly responsible for the loss of our son's life, you might just as well keep this emblem on display in your trophy room, as a memory of one of your historic deeds.

One major regret at this time is that your daughter was not there to receive the same treatment as our son received in Korea.

Truman put the letter in his desk drawer, where it stayed for several years.[41]

The letter to Hume, however regrettable, was but a diversion compared to the mayhem that was continuing in Korea. When news of the Chinese assault reached the President, he knew that the whole war had changed. "We've got a terrific situation on our hands," he commented on November 28. "The Chinese have come in with both feet."[42] Although he resisted MacArthur's appeals for a widening of the war, he demanded drastic action at home. On December 15 he went on television to declare a national emergency and to call for all-out mobilization. This, he said, would necessitate a build-up of the army to 3.5 million men and the imposition of economic controls. "Our homes, our nation, all the things we believe in are in great danger." Clinging to the belief that the Soviets were to blame, he added, "This danger has been created by the rulers of the Soviet Union."[43]

Meanwhile the retreat continued. By Christmas UN forces had been pushed below the 38th parallel—a fallback of more than 300 miles in less than a month. Soldiers and millions of cold and panic-stricken refugees jammed the roads. In the first week of January Seoul had to be evacuated by the UN forces for the second time, and Rhee fled with his government to Pusan. For Truman and the United States it was the bleakest time in the long history of the war.

41. McCullough, *Truman*, 826–31.
42. Ibid., 815–16.
43. Oakley, *God's Country*, 84.

ACCIDENTS AND INDIVIDUALS sometimes make a difference amid the larger determinants of history.

The accident happened on an icy road near Seoul on December 23. It killed General Walton Walker, head of the American Eighth Army, who was riding in a jeep. The individual was his replacement, Lieutenant General Matthew Ridgway, who was then vice-chief of staff of the army in Washington. As soon as Ridgway was appointed he flew from Washington to Tokyo, arriving Christmas Day. After talking with MacArthur the next morning, he flew to Korea that afternoon.

Ridgway, fifty-five, already enjoyed a distinguished military record. A protégé of Marshall, he was well read and thoughtful. He was also brave and superbly fit. During World War II he had planned and executed the airborne invasion of Sicily. In June 1944 he jumped himself as leader of his division's D-Day airborne assault on Normandy. A soldier's soldier, he thrived in the field and in the midst of his men. In Korea he stalked the front lines, a hand grenade strapped to his chest, and tried to rally his dispirited forces.

Ridgway was appalled by the low morale, bad food, inadequate clothing, and poor intelligence-gathering of the Eighth Army and moved quickly to improve things. He insisted on more patrolling, so as better to locate and shoot at the enemy. He got some of his men off the roads and up onto the hillsides and brought in the air force and artillery on a much larger scale. He was also fortunate, for enemy supplies, especially of gasoline, ran low, and the advance stalled. Ridgway mounted Operation Killer, which directed heavy artillery fire on enemy soldiers, who by then had become stretched out and more exposed. Thousands were slaughtered. "I'm not interested in real estate—just killing the enemy," he explained. By mid-January Ridgway had helped to restore morale, stopped the retreat, and brightened the UN prospects. He then followed with Operation Ripper, a counter-attack that ground its way back north. By the end of March his troops had regained most of the territory south of the 38th parallel, including Seoul, as well as bits here and there of the North. There the front stabilized, changing little over the remaining twenty-eight months of stalemate on the peninsula.[44]

MacArthur, however, seemed to find relatively little to cheer about amid this recovery. In December, following his public demands for escalation, he had been ordered to say nothing without prior clearance. But

44. Ferrell, *Harry S. Truman*, 124; Leckie, *Conflict*, 204–26; Rees, *Korea*, 176–95; McCullough, *Truman*, 831–33.

MacArthur continued to give interviews to journalists in Tokyo. These repeated his main themes: Asia was the major battleground of the Cold War, and "limited" war was unthinkable. As before, his complaints angered Truman, Bradley, and other top advisers. But they, too, were shocked by the ferocity of the Chinese assault, and they were afraid to remove so legendary a figure as MacArthur, hero of Inchon, or even seriously to reprimand him. One who did complain was Ridgway, at a meeting shortly before his appointment in Korea. As the meeting was breaking up he grabbed General Hoyt Vandenberg, the air force chief, and asked him why the Joint Chiefs didn't *tell* MacArthur what to do. Vandenberg shook his head. "What good would that do? He wouldn't obey the orders. What can we do?" Ridgway replied, "You can relieve any commander who won't obey orders, can't you?" Vandenberg just looked at him, amazed. Acheson later observed, "This was the first time that someone had expressed what everybody thought—that the Emperor had no clothes on."[45]

There matters rested uneasily until March 20, at which time Truman told MacArthur that he planned to seek a negotiated settlement with the Chinese. MacArthur sabotaged this idea four days later by issuing a statement of his own in which he offered to meet with the Chinese and to work out a settlement. If they refused, he said, his troops might invade China.[46] When Truman heard of this, he was enraged. So were Acheson, Lovett, and other advisers, who wanted the general fired. Truman now knew this would have to be done. "I've come to the conclusion that our Big General in the Far East must be recalled," he wrote in his diary.[47] He told Democratic Senator Harley Kilgore of West Virginia, "I'll show that son of a bitch who's boss. Who does he think he is—God?"[48] But Truman took his time. His ratings in the polls were at an all-time low of 26 percent, and he shrank from the firestorm that would erupt if he got rid of the general. Instead, he sent a mild reprimand and awaited a more flagrant act of insubordination that would justify removal.

MacArthur had already written a letter that was just such an act. He sent it on March 20 to Joseph Martin, the House Republican leader, in response to a speech that Martin had given in February and then relayed to Tokyo for MacArthur's reaction. The United States, Martin had proclaimed, must be in Korea to win! If not, "this administration should be

45. Halberstam, *Fifties,* 109.
46. Oakley, *God's Country,* 87.
47. McCullough, *Truman,* 835–37; Halberstam, *Fifties,* 113–14.
48. Oakley, *God's Country,* 87.

indicted for the murder of American boys." MacArthur's reply heartily endorsed Martin's sentiment and nicely summarized his overall views:

> It seems strangely difficult for some to realize that here in Asia is where the Communist conspirators have elected to make their play for global conquest, and that we have joined the issue thus raised on the battlefield; that here we fight Europe's war with arms while the diplomats there still fight it with words; that if we lose the war to Communism in Asia the fall of Europe is inevitable, win it and Europe most probably would avoid war and yet preserve freedom. As you point out, we must win. There is no substitute for victory.[49]

In writing such a letter to a partisan foe of the President—and in placing no restrictions on its publication—MacArthur sealed his doom as commander in Asia. When Martin read the letter on the floor of the House on April 5, Truman knew he had to act.[50] Still, he moved deliberately, first consulting not only his military advisers but also Vice-President Barkley and House Speaker Rayburn. He even sought the opinion of Fred Vinson, who was Chief Justice of the Supreme Court. All said that the President had no choice but to remove MacArthur, political firestorm or no. When the Joint Chiefs finally recommended removal—on military grounds—on April 9, Truman had the papers drawn up, but he still hoped to get word to MacArthur privately before proclaiming the deed to the world. When a leak threatened to foul up this strategy, the firing was announced sooner than Truman had planned—at 1:00 A.M. on the morning of April 11. That was almost six days after Martin had aired the letter.[51]

In removing MacArthur, Truman noted the important policy issues, especially whether or not to limit the war, that separated the two men. These differences were profound, involving the relative strategic importance to the United States of Europe and Asia, the use or non-use of nuclear weapons, and the taking of other provocative actions of war against China. MacArthur may have been correct to think that stepping up the war with China, especially the threat of nuclear weapons, would have induced Mao to ease off or to back down. But MacArthur had been wrong about a lot of things in 1950, including his predictions that the Chinese would not intervene. And his demands for escalation were frightening not only to Truman and his advisers but also to America's allies and

49. Ferrell, *Harry S. Truman*, 127; McCullough, *Truman*, 837–39; James, *Years of MacArthur*, 589–90.
50. Donovan, *Tumultuous Years*, 352.
51. Truman, *Memoirs*, 2:499–510.

the UN. Had Truman followed MacArthur's counsel, he would have damaged relations with his NATO allies and weakened Western defenses in Europe. He would have faced an even more costly war against China, and he might have had to fight the Soviet Union as well. To get bogged down in a major war in Asia would have been senseless, and Truman knew it.

Instead of dwelling on these policy disputes, however, Truman fired MacArthur because he wished to preserve the important constitutional principle of civilian control over the military. MacArthur had repeatedly disobeyed orders. He had been insubordinate, directly challenging the President's constitutional standing as commander-in-chief. Truman derived no special pleasure in taking action, which he should have done much earlier. "I was sorry to have a parting of the ways with the big man in Asia," he wrote Eisenhower on April 12, "but he asked for it, and I had to give it to him." Although the firing required a certain amount of political courage, Truman later explained to a reporter that "courage had nothing to do with it. He was insubordinate, and I fired him."[52]

When MacArthur got news of his removal, he was at a luncheon in Tokyo. He said to his wife, "Jeannie, we're going home at last."[53] A few days later he took off for the United States, receiving a hero's welcome in Tokyo, Hawaii, and San Francisco before arriving in Washington shortly after midnight on April 19. There he was met by the Joint Chiefs, who had unanimously recommended his firing, and a substantial crowd. Around noon that day he went to Capitol Hill to give an address to both houses of Congress. It was a scene of high drama, and MacArthur did not disappoint his admirers. He strode confidently down the aisle, whereupon the Congress gave him a standing ovation. He spoke for thirty-four minutes, during which time he was interrupted by applause thirty times. Those present were struck by his control as he outlined his now familiar differences with American policies. He ended with dramatics. "I am closing my fifty-two years of military service," he said. "The world has turned over many times since I took the oath on the Plains at West Point. . . . But I still remember the refrain of one of the most popular barracks ballads of that day which proclaimed most proudly that—'Old soldiers never die; they just fade away.' And like the old soldier of that ballad, I now close my military career and just fade away—an old soldier who tried to do his duty as God gave him the light to see that duty. Good-bye."

Not everyone, of course, found the speech thrilling. Truman privately

52. Oakley, God's Country, 87.
53. Newsweek, April 30, 1951, p. 18; Oakley, God's Country, 88.

pronounced it "a hundred percent bullshit." But some congressmen, including people who had wanted him fired, wept openly. Dewey Short, a conservative Republican from Missouri, said, "We heard God speak here today, God in the flesh, the voice of God." From New York came the verdict of former President Herbert Hoover, who described Mac-Arthur as "a reincarnation of St. Paul into a great General of the Army who came out of the East."[54] A Gallup poll found that 69 percent of the American people sided with MacArthur in the controversy.[55]

There was more to come. After MacArthur spoke he rode triumphantly down Pennsylvania Avenue, where an estimated 300,000 people cheered him. Jet bombers and fighters flew in formation overhead. In New York the next day, he received a ticker-tape parade the likes of which the city had never seen before. Some estimates placed the crowds at 7.5 million people. Office workers and residents clustered on balconies, rooftops, and fire escapes and threw down blizzards of torn paper. Men shouted, "God bless you, Mac!" On the river, tugs and ocean-going boats tooted, adding to the din of the occasion. The general, his cap white with paper, climbed onto the folded top of the open car and acknowledged the adoration. At City Hall he accepted a gold medal and exclaimed, "We shall never forget" the tremendous reception.[56]

While the homecoming orgy was taking place, Americans throughout the country were letting Truman and Congress know what they thought about the issue. Within twelve days of the firing the White House received more than 27,000 letters and telegrams, which ran twenty to one against the President. Many of these were so hostile and abusive that they were shown to the Secret Service. Members of Congress got another 100,000 messages during the first week, many of which demanded Truman's impeachment:

IMPEACH THE JUDAS IN THE WHITE HOUSE WHO SOLD US DOWN THE RIVER TO LEFT WINGERS AND THE U.N.

SUGGEST YOU LOOK FOR ANOTHER HISS IN BLAIR HOUSE.[57]

Rancor also colored debate in Congress. Senator Robert Kerr, a freshman Democrat, dared to defend the firing. "General MacArthur," he

54. Oakley, God's Country, 90; Newsweek, April 30, 1951, p. 20; Halberstam, Fifties, 115; McCullough, Truman, 848–51.
55. McCullough, Truman, 847–48.
56. Newsweek, April 30, 1951, p. 18.
57. Oakley, God's Country, 88.

said, "claimed that if we started a general war on Red China, Russia would not come to her rescue. . . . I do not know how many thousand American GIs are sleeping in unmarked graves in North Korea. . . . But most of them are silent but immutable evidence of the tragic mistake of 'The Magnificent MacArthur' who told them that the Chinese Communists just across the Yalu would not intervene."[58] Jenner, however, shouted, "Our only choice is to impeach President Truman and find out who is the secret invisible government which has so cleverly led our country down the road to destruction."[59] McCarthy, not to be outdone, later denounced Truman as a "son of a bitch" and blamed the firing on a White House cabal "stoned on bourbon and benedictine."[60]

Partisans like Jenner, McCarthy, and others kept up their attack long after the firing. And Truman maintained a low profile, appearing at no major public events until the opening day of baseball season at Griffith Stadium, where he was booed by the crowd. But it was remarkable how quickly MacArthur's support subsided after the initial frenzies. From the beginning many leading newspapers, including some that normally opposed Truman, had defended the right and duty of a President to punish insubordination. Among them were the *New York Times*, the *Baltimore Sun*, the *Christian Science Monitor*, and even the Republican *New York Herald-Tribune*, which commended Truman's "boldness and decision." Close observers of the crowd in New York agreed that the turnout was amazing and unprecedented but noted that many of the onlookers were more curious than anything else.[61]

By early May the emotions that had swept through the country in April were already abating. Ridgway, named to replace MacArthur, was holding the line in Korea. Congressional hearings conducted by the Senate Foreign Relations and Armed Services committees slowly and inexorably completed this process of readjustment. Led by the grave and courtly Richard Russell, a powerful Senate Democrat from Georgia, the senators elicited repeated support for the constitutional principle that had motivated Truman, as well as for his support of a limited war. The Joint Chiefs were especially effective, pointing out—contrary to MacArthur's claims—that they had never shared his views about escalation or the centrality of Asia in the grand strategy of the United States. Bradley delivered the line

58. *Newsweek*, April 30, 1951, p. 24.
59. McCullough, *Truman*, 844.
60. William Leuchtenburg, *A Troubled Feast: American Society Since 1945* (Boston, 1973), 22.
61. McCullough, *Truman*, 846–47.

that everyone remembers when he said that MacArthur's policies "would involve us in the wrong war, at the wrong place, at the wrong time, and with the wrong enemy."

ALTHOUGH TEMPERS COOLED in May and June of 1951, they revealed a stubborn fact: Americans had small patience for lingering "limited war." Most people seemed to agree with Truman that escalation of war with China and use of nuclear weapons would be costly and bloody. At the same time, however, Americans were frustrated. They had initially supported the war believing that the "credibility" of the United States and the "Free World" were at stake. But they had expected to win—America always had (they thought). As MacArthur had said, there was "no substitute for victory." As this very different and difficult conflict became a bloody stalemate after March 1951 their frustrations mounted, and "Truman's War" engendered ever-rising resentment.

Frustration of this sort suggested that democracy and prolonged military stalemate do not easily mix. Indeed, the frustration was understandable, for the fighting continued to shed a great deal of blood. Chinese and North Korean losses became staggering. Nearly 45 percent of American casualties were suffered in the last two years of fighting. This was the war of Heartbreak Ridge (September 1951), Pork Chop Hill (April 1953), bloody night patrols, ambushes, mines, flying shrapnel from artillery, sudden raids for already war-scarred real estate. Bombing and artillery denuded the landscape around the 38th parallel. More sweltering heat and rainstorms, frigid cold and howling winds, heat and rainstorms again, more cold. And no ground gained. Would the war ever end?

Beginning in July 1951 the United Nations, led by the United States, entered into peace talks with their enemies. Headlines thereafter periodically held out hope for an end to the conflict. These were cruel delusions, for the fighting lasted until July 1953. Although both sides seemed willing to accept a cease-fire that would confirm existing military realities—close to what they had been at the start of the war—they differed on the issue of repatriation of North Korean and Chinese prisoners of war. The Truman administration insisted that repatriation of such prisoners—around 110,000 in all—had to be voluntary. Those who did not want to return to North Korea or to China, estimated at more than 45,000, would not have to. China and North Korea refused to accede, arguing that many of these prisoners had been intimidated by brutal Nationalist Chinese guards who were threatening them with injury or death if they said they wanted to go home.[62] Truman probably could have given in on this point without arousing great domestic protest; most Americans did not get excited about the fate of *enemy* prisoners. But he did not, regarding the issue as one of principle. The Chinese and North Koreans, too, held firm, perhaps hoping they could get better terms once Truman was out of office after 1952. The issue was resolved only in mid-1953, when the enemy gave way. At that time 50,000 enemy prisoners, including 14,700 Chinese, refused to go home.[63]

The frustrations aroused by war and stalemate in negotiations gradually brought MacArthurite solutions back into the realm of discussion. In January 1952 Acheson contacted the British to get their approval, in case armistice talks broke down or terms of an armistice agreement were violated, for the bombing of military targets in China and for blockading of the mainland. Winston Churchill, who had become Prime Minister in 1951, demurred. He also sought assurances that the United States would not use nuclear weapons. Bradley gave him cold comfort. The United States did not plan to use atomic bombs in Korea, he said, "since up to the present time no suitable targets were presented. If the situation changed in any way, so that suitable targets were presented, a new situation would arise."[64]

Bradley's comment seemed to suggest that Truman's advisers were again willing to consider resorting to nuclear weapons. This might have been popular with the American people. In August 1950 polls had showed only 28 percent of Americans in favor of such use. By November 1951, a time of stalemate, 51 percent were willing to see the Bomb let loose on "military targets."[65] Two months later—when Acheson was exploring escalation with Churchill—Truman sat down and wrote a memo to himself spelling out a possible ultimatum to the Soviets, whom he still blamed for all that had happened in Korea. It read:

> It seems to me that the proper approach now would be an ultimatum with a ten-day expiration limit, informing Moscow that we intend to blockade the China coast from the Korean border to Indochina, and that we intend to destroy every military base in Manchuria by means now in our control—

62. Rosemary Foot, A *Substitute for Victory: The Politics of Peacemaking at the Korean Armistice Talks* (Ithaca, 1990).
63. Bernstein, "Truman Administration," 438–40.
64. Gaddis, *Strategies of Containment*, 123; Bernstein, "Truman Administration," 430.
65. Stephen Whitfield, *The Culture of the Cold War* (Baltimore, 1991), 5.

and if there is further interference we shall eliminate any ports or cities necessary to accomplish our purposes.

This means all-out war. It means that Moscow, St. Petersburg, Mukden, Vladivostok, Peking, Shanghai, Port Arthur, Darien, Odessa, Stalingrad, and every manufacturing plant in China and the Soviet Union will be eliminated.[66]

Four months later, in May, Truman returned to this idea. This time he drafted an internal memo to "the Commies": "Now do you want an end to hostilities in Korea or do you want China and Siberia destroyed? You may have one or the other; whichever you want, these lies of yours at the conference have gone far enough. You either accept our fair and just proposal or you will be completely destroyed."[67]

The President never sent such messages. As ever, he considered Europe more important to American security than Asia, and he wanted to end the fighting in Korea so that the United States could concentrate its resources in the West. His memos were ways of blowing off steam—contingent schemes that would be considered only in the event of collapse of negotiations or aggression elsewhere.[68] Still, it was clear that high American officials—Acheson, Bradley, the President—found the trials of limited war deeply frustrating as of 1952.

THE KOREAN WAR FINALLY came to a close on July 27, 1953, after the Chinese and North Koreans agreed to voluntary repatriation of prisoners of war. Why they relented after two years remains yet another debated mystery of the war. Some people point to the death of Stalin in March 1953, arguing that the new Soviet leadership pressured China to back down. Others think Eisenhower, then President, may have threatened the enemy with the use of nuclear weapons. This cannot be solidly documented. The most likely reason was that the Chinese and North Koreans were tired. Recognizing that Eisenhower and the new Republican administration were impatient and uncompromising, they decided to settle. After more than three years of fighting, an uneasy cease-fire settled on a peninsula now more implacably divided than ever. The boundaries did not differ greatly from those at the start of the fighting in 1950.[69]

66. Boyer, "'Some Sort of Peace,'" 198.
67. Ibid., 198.
68. Gaddis, *Strategies of Containment*, 123.
69. Ambrose, *Eisenhower*, 294–96, 327–30, says there was no explicit nuclear threat but that Eisenhower's reputation as a man who would go all-out to win probably influenced the enemy to deal. See also Foot, "Making Known."

How did the war affect the United States and the world?

In some ways, not very happily. Truman's failure to consult Congress set a poor precedent and helped to saddle his administration with blame when the fighting fell into apparently pointless stalemate. More important, Western "credibility" would have been stronger if the UN had stopped at the 38th parallel. The decision that provoked the stalemate—to drive further north—was indeed hard to resist: what would the American people have said if enemy troops, then on the run, had been allowed to escape and regroup behind their old borders? Still, pressing to the Yalu obviously proved costly. That decision also enabled the Chinese, who might otherwise have stayed out of the war, to establish their own "credibility" and to emerge with enhanced standing in the eyes of many "have-not" nations in the world.

The conflict in Korea also accelerated the process of globalization of the Cold War. When the fighting ended, the United States found itself ever more strongly committed to greater military support for NATO. It redoubled efforts to rebuild Japan as a bastion of capitalist anti-Communism in Asia.[70] It had to protect Rhee, a tyrant, and to station troops in South Korea for decades ahead. It also found itself more engaged in the support of Chiang Kai-shek in Taiwan and in the camp of the French in Indochina. By January 1953 America was providing 40 percent of the French effort in that little-known but highly incendiary outpost of Southeast Asia.[71]

The war had a mixed impact on the American economy. Increases in defense spending boosted the GNP, accelerating the boom psychology of the American people and promoting ever-grander expectations about personal comforts in the future. At the same time, however, defense spending sparked the flame of inflation, which incited some 600,000 steel workers to strike for better wages in April 1952. When the workers refused to settle, Truman seized the mills in hopes of forcing the strikers back to work. His dramatic move exposed the social tensions that gripped the nation during the war. It also set off a major constitutional impasse. This was resolved only in June when the Supreme Court rejected the argument that his standing as commander-in-chief justified the seizure.[72]

70. John Dower, "Occupied Japan and the Cold War in Asia," in Lacey, ed., *Truman Presidency*, 361–400.
71. McMahon, "Toward a Post-Colonial Order," 352–55.
72. Maeva Marcus, *Truman and the Steel Seizure Case: The Limits of Presidential Power* (New York, 1977); Ferrell, *Harry S. Truman*, 144–45. The case was *Youngstown Sheet and Tube Co. v. Sawyer*, 343 U.S. 579 (1952).

The increases in military spending—and the larger global commitments that America shouldered after the war—greatly changed the mix of federal expenditures in the United States. Defense spending, $13.1 billion in 1950, jumped to a wartime high of $50.4 billion in 1953 and remained between $40.2 billion and $46.6 billion for the remainder of the 1950s. By contrast, non-military spending tended at first to suffer, declining from approximately $30 billion in 1950 to a low of $23.9 billion in 1952 before rising slowly for the next several years—to $38 billion in 1958. Per capita spending for non-military ends increased hardly at all in these years. The emphasis on military expenditures boosted by the Korean War, while helpful to areas engaged in defense contracting, set public priorities that did little to promote government support for a healthy peacetime economy.

Notwithstanding these not altogether happy results of the war, it is fair to conclude that Truman, once faced with the fact of North Korean aggression, acted in the best interests of world stability. To have stood by while Kim overran the South would have been demoralizing indeed to peace-abiding nations. By intervening the United States and the UN made North Korea pay dearly for its greed. They may also have discouraged the Soviets from supporting subsequent military adventurism by client states elsewhere in the world. In these important ways Truman's decision to fight in June 1950—and his refusal thereafter to provoke a much wider war with China—not only sent strong signals against aggression but also guarded against still more dangerous escalation of the conflict.

THE WAR, FINALLY, LIFTED the Red Scare to high tide. Truman was powerless to stop the wave of anti-Communist and xenophobic feeling that washed over the country during and after the conflict. Hoover redoubled his fight against Reds in American life. Schools embarked on drills that claimed to prepare children for the horrors of atomic attack. Sales of back-yard bomb shelters seemed to increase. States and towns passed legislation that banned Communists from teaching, civil service, or office-seeking and that made the taking of the Fifth Amendment grounds for dismissal from government service.[73] Even the comic strips were affected: Buzz Sawyer went to work for the CIA; Terry now chased Communists, not pirates; Joe Palooka outwitted the Reds to rescue a scientist in Austria; Winnie Winkle was thrown into a Soviet jail; Daddy

73. Oakley, God's Country, 47, 68.

Warbucks and his friends blew up enemy planes carrying H-bombs toward America.

Fears unleashed by the Korean War helped to spread the Red Scare into
intellectual circles. This was by no means an all-encompassing development, for American intellectual life remained vibrant. Many important
books appeared during the war that were largely unaffected by concerns
about Communism: J. D. Salinger's *The Catcher in the Rye* (1951), Ralph
Ellison's *Invisible Man* (1952), Erik Erikson's *Childhood and Society*
(1950), and David Riesman and Nathan Glazer's *The Lonely Crowd*
(1950) are four of the most enduring. In the arts, architects and abstract
expressionist painters made the United States—and New York City in
particular—an international center of creative talent in the late 1940s and
early 1950s. But the threat of Communism did alarm some intellectuals.[74] Even before the war, in 1949, the British writer George Orwell
had written *1984*, a dystopian novel that was widely assumed to describe
the future under Communism. It quickly became a classic. In 1951
Hannah Arendt, a highly regarded political thinker and philosopher
known for her hostility to fascism, published *The Origins of Totalitarianism*. It tended to equate Communism and fascism by showing how
both systems relied on terror and unlimited political power.

Conservative religious leaders more eagerly joined the wartime drive
against Communism at home. During the winter of 1950–51 the evangelist Billy Graham spoke to enormous and enthusiastic crowds who heard
him warn against "over 1,100 social-sounding organizations that are communist or communist-operated in this country. They control the minds of
a great segment of our people. . . . educational [and] religious culture is
almost beyond repair." In 1952 Fulton J. Sheen, the auxiliary Catholic
bishop of New York, began attracting huge audiences to watch his new
television show, "Life Is Worth Living." His book of the same title,
published in 1953, reached number five on the best-seller list. Sheen
wore a black cassock with red piping, a scarlet cape that flowed off his
shoulders, and a large gold cross around his neck. Candles and a statue of
the Virgin Mary loomed behind him. He had piercing eyes that glowed
like coals and an apparently effortless eloquence that enabled him to speak
without notes. Sheen kept his distance from the shabbier diatribes of

74. Serge Guilbaut, *How New York Stole the Idea of Modern Art: Abstract Expression-
ism, Freedom, and the Cold War* (Chicago, 1983); and Erika Doss, *Benton,
Pollock, and the Politics of Modernism from Regionalism to Abstract Expression-
ism* (Chicago, 1991). Guilbaut, however, argues that the "freedom" celebrated by
abstract expressionists appealed to Cold Warriors in the United States.

McCarthyites. But many of his messages denounced Communism, the antithesis of Catholicism as he saw it.[75] At his peak in 1954 he reached an estimated 25 million people a week.

In the Cold War climate or the early 1950s, it was hardly surprising that television, then spreading with incredible speed to American households, should have welcomed an anti-Communist like Sheen. Indeed, the TV networks, too, felt the rising power of the Red Scare. Three days before the North Korean invasion of 1950, three former FBI agents published *Red Channels: The Report of Communist Influence in Radio and Television*. Financed by a leading supporter of the China Lobby, it included an alphabetical list of 151 people in the radio and television business along with "citations" of their involvement in various suspect organizations. These were mostly liberal associations, but *Red Channels* made them seem subversive. Sponsors grew nervous, and radio and television stations felt their pressure. Blacklisting followed. Among the so-called subversives who found it hard to get time on the air in the early 1950s were Leonard Bernstein, Lee J. Cobb, Aaron Copland, Jose Ferrer, Gypsy Rose Lee, Edward G. Robinson, and Orson Welles.[76] The left-wing singer Pete Seeger was banned from network TV until 1967. The black singer-activist Paul Robeson, an apologist for Stalinism, had his passport stripped for eight years beginning in 1950.[77] In 1954 the *New York Times* estimated that right-wing agitation had cost 1,500 radio and television people their jobs.[78]

Tensions in Hollywood also demonstrated the special anxiety that American institutions felt about Communism during the Korean War years. Thanks in part to HUAC's assaults in 1947, accommodating studios had already brought out a few films that featured explicitly anti-Communist themes. These included *The Iron Curtain* (1948) and *The Red Menace* (1949). With the coming of war, anti-Communism did a brisker business on the sets: one historian has counted around 200 such movies produced between 1948 and 1953, most of them after 1950.[79]

75. Whitfield, *Culture of the Cold War*, 80, 170–72.
76. Oakley, *God's Country*, 71; Whitfield, *Culture of the Cold War*, 166–69.
77. Martin Duberman, *Paul Robeson* (New York, 1988), 328–30, 388–89, 414–25; Whitfield, *Culture of the Cold War*, 192–201.
78. Richard Pells, *The Liberal Mind in a Conservative Age: American Intellectuals in the 1940s and 1950s* (New York, 1985), 310.
79. Peter Biskind, *Seeing Is Believing: How Hollywood Taught Us to Stop Worrying and Love the Fifties* (New York, 1983), 162. Nora Sayre, *Running Time: Films of the Cold War* (New York, 1982), 80, counts fifty such films between 1947 and 1954.

Among them were *I Married a Communist* (1950), *I Was a Communist for the FBI* (1951), *The Whip Hand* (1951), *Red Snow* (1952), and *My Son John* (1952). Many of the Communists portrayed in these films (few of which did well at the box office) were scruffy, humorless, effeminate, and sinister. They did a good deal of spying and recruiting for the party, and when necessary they murdered patriotic citizens who got in their way.[80]

Another growing genre of film, science fiction, also tried to play on anti-Communist emotions during and after the war. Following such films as *When Worlds Collide* (1951) and *War of the Worlds* (1953), "sci-fi flicks" became increasingly popular in the 1950s. Many of these movies need no deep analysis. Others, such as *Them* (1953) and *Invasion of the Body Snatchers* (1956), were truly scary, arousing fears of monsters— perhaps mutations from atomic testing. Common themes in the sci-fi movies featured "good" scientists and public officials contending with dangerous conspirators, aliens, or monsters from the "other." What one got from such themes surely varied; if we know anything from the explosion of cultural analysis in our own times it is that many individuals reach their own conclusions about what they see and read and hear. Still, some of these movies carried conservative subtexts: watch out for people who are different; things (and people) may not be what they seem; trust in authority; be careful in all that you say and do; guard against enemies and conspirators.[81]

Liberals and leftists in Hollywood had to be especially careful following the outbreak of war in Korea. Some actors, directors, and technicians lost their jobs in the war years; one estimate places the number at 350 by the mid-1950s.[82] Blacklisted screenwriters resorted to the use of pseudonyms. A particular target of Red-hunters was the British-born Charlie Chaplin, who angered conservatives both because he had been involved in a paternity suit and because he had supported a range of left-wing causes. One of his films, *Monsieur Verdoux* (1947), was withdrawn from circulation after the American Legion led protests against its pacifist message. A new film in 1952, *Limelight*, ran in only a few American cities. When Chaplin took a trip out of the United States in September 1952, the government arbitrarily rescinded his re-entry permit until he agreed to submit to rigorous examination of his political beliefs and moral behavior. Refusing

80. Les Adler, "The Politics of Culture: Hollywood and the Cold War," in Griffith and Theoharis, eds., *Specter*, 240–61; Sayre, *Running Time*, 80–99; Whitfield, *Culture of the Cold War*, 133.
81. Biskind, *Seeing Is Believing*, 102–13.
82. Pells, *Liberal Mind*, 310.

to do so, Chaplin remained in exile until returning in 1972 to receive a special Oscar. He died in Switzerland in 1977.[83]

The Justice Department's intervention in the case of Chaplin underlined the continuing importance of governmental action in feeding the flames of anti-Communism. As before, zealots on Capitol Hill proved ready with the matches. With Korea as Case Number One of Communist conspiracy, McCarthy and others spearheaded thirty-four separate probes into domestic Communist influence during the 1951–52 Congress and fifty-one in 1953–54. So politically popular was the issue of anti-Communism that 185 of the 221 GOP representatives elected in 1952 asked Republican House leaders to give them an assignment on HUAC in the new Congress.[84]

One of the most effective anti-Communists in Congress was Patrick McCarran, a conservative Democratic senator from Nevada. McCarran meant business and, as a member of the majority party before 1953, got results. In 1950 he led to passage an Internal Security Act (also called the McCarran Act) that required Communists and other "subversive" groups to register with the Attorney General. A Subversive Activities Control Board (SACB) was given broad authority to identify the groups. The law further banned individuals in such groups from holding government or defense jobs or from getting passports; it denied entrance to the United States of aliens who had ever belonged to the Communist party or other totalitarian parties or had advocated violent revolution; and it authorized detention of accused spies and saboteurs during any national emergency declared by the President. Truman vigorously opposed the legislation, calling it the "greatest danger to freedom of press, speech, and assembly since the Sedition Act of 1798."[85] Civil libertarians also denounced the act, mainly because of the sweeping authority it gave to the SACB. Still, the Democratic Congress not only approved the legislation but overrode Truman, when he vetoed it in September 1950, by votes of 286 to 48 in the House and 57 to 10 in the Senate.[86] Thanks to litigation by the

83. Richard Polenberg, *One Nation Divisible: Class, Race, and Ethnicity in the United States Since 1938* (New York, 1980), 119; Whitfield, *Culture of the Cold War*, 187–92.

84. Whitfield, *Culture of the Cold War*, 29; Richard Fried, *Nightmare in Red: The McCarthy Era in Perspective* (New York, 1990), 150–53.

85. Robert Griffith, *The Politics of Fear: Joseph R. McCarthy and the Senate* (Lexington, Ky., 1970), 118–19; John Diggins, *The Proud Decades: America in War and Peace, 1941–1960* (New York, 1988), 117.

86. Fried, *Nightmare in Red*, 116–17.

Communist party, the act was not enforced, but it placed left and liberal groups on the defensive and remained on the books for years. It exposed with special clarity the bipartisan political appeal of anti-Communism in the Korean War election year of 1950.

In 1951 McCarran struck again, this time directing his Internal Security Subcommittee to investigate "China hands" in the State Department who had "lost" China. Truman, feeling the pressure, tightened his loyalty/security procedures in April 1951 to place a greater burden of proof on government employees. John Stewart Service, an expert on China, underwent eight separate investigations before being labeled a risk and fired by Acheson in December. Another China hand, John Carter Vincent, was accused of being a loyalty risk in 1951 and quit the State Department. By 1954 most of the leading men on the China desk had been purged from government service, thereby depriving the United States government of the expertise it had previously been able to muster about the People's Republic of China.[87]

McCarran's next success, in 1952, was the McCarran-Walter Act. The law was liberal in one respect: it repealed thirty-year-old legislation that had excluded Asian immigrants from the United States, substituting instead small quotas, and it removed racial qualifications for citizenship that had also been used to discriminate against Asians. Otherwise, the McCarran-Walter Act was offensive to liberals and to Truman, who vetoed it. It maintained the existing "national origins" system of immigration by which certain groups—mainly southeastern Europeans and Jews—had been discriminated against. It also strengthened the Attorney General's authority to deport aliens who were thought to be subversive. Congress again overruled Truman's veto.[88]

The Red Scare on Capitol Hill—and elsewhere in the United States during the Korean War—exposed a final legacy of the war: it deeply damaged the Truman administration. This damage was cumulative rather than dramatic, for the Korean conflict, unlike the later quagmire that was Vietnam, was not a "living room war." People could not turn on their television sets and witness the savagery of combat. There was little in the way of organized anti-war protest: Americans either wanted to win or get out. Some 5.7 million men served in the military during the war—

---

87. Gary May, *China Scapegoat: The Diplomatic Ordeal of John Carter Vincent* (Washington, 1979); E. J. Kahn, *The China Hands: America's Foreign Service Officers and What Befell Them* (New York, 1975); Griffith, *Politics of Fear*, 133–35; Fried, *Nightmare in Red*, 145–50.
88. Polenberg, *One Nation Divisible*, 123.

about one-third the number in World War II—without much being said against the draft. But the frustrations of stalemate—and the continuing casualties—heightened the Red Scare and rendered Truman virtually powerless to control Congress or effectively to lead the country. Well before the 1952 elections it was clear that the Korean War had divided the nation and that the majority of the American people were ready for a change in leadership.

9

Ike

A poll of historians conducted by Arthur Schlesinger, Jr. in 1962 ranked President Dwight D. Eisenhower as number twenty-one among the thirty-four Presidents in American history before that time. He stood near the bottom of "average" presidents, tied with Chester Arthur and just ahead of Andrew Johnson. That changed. A poll twenty years later placed Eisenhower ninth among the ten best, between Truman and James K. Polk.[1]

Polls such as these are silly exercises that reveal more about the biases of historians (most of whom are liberals) and about the times than they do about the ability of individual Presidents. Eisenhower ranked higher in 1982 in part because some of his successors in the White House, especially Lyndon Johnson and Richard Nixon, had pursued outrageously devious and dishonest policies. By contrast Eisenhower (and Truman, too) seemed by then to be sensible and honorable. His reputation since 1982 has if anything grown more lustrous. An accomplished biographer, Stephen Ambrose, started his book in 1990 by stating that "Dwight David Eisenhower was a great and good man. . . . one of the outstanding leaders of the western world of this century."[2]

1. Vincent De Santis, "Eisenhower Revisionism," *Review of Politics*, 38 (April 1976), 196.
2. Stephen Ambrose, *Eisenhower: Soldier and President* (New York, 1990), 11. Other useful books on Ike include Herbert Parmet, *Eisenhower and the American Crusades* (New York, 1972); Robert Burk, *Dwight D. Eisenhower: Hero and Politician*

244   GRAND EXPECTATIONS

Those who have a low opinion of Ike, as he was called, tend to regard him as a career military officer with a narrow range of interests and a limited intellect. Many professors at Columbia University considered him to be wholly out of his depth in the academic world, whereupon he offered his own definition of an intellectual: "a man who takes more words than are necessary to tell more than he knows."[3] Schlesinger, a liberal Democrat, later said that Eisenhower's mind "functioned at two levels: a level of banal generality, so hortatory as to be meaningless; and a level of *ad hoc* reaction to specific events, often calm, intelligent, and decisive, but not always internally coherent. It may indeed seem in the end as it did at the time that, while Eisenhower often knew what he wanted to do at any particular moment, his larger sense of affairs was confused and con-tradictory."[4]

When Eisenhower became President in 1953, his detractors regularly made fun of his habits. Deploring his time-consuming passion for golf, they also criticized him for his taste in books—mainly western novels—and for his love of poker and bridge. This passion was indeed deep: flying back from the GOP national convention in San Francisco in 1956, Eisenhower spent eight straight hours on the plane playing bridge with his friends.[5] His critics further complained that he surrounded himself mainly with big businessmen and other rich people, sometimes at stag dinners in the White House, and cut himself off from "ordinary" folk.

Nothing gave his detractors more amusement (or concern) than Eisenhower's apparent inarticulateness. At press conferences he often seemed to stumble or to go off in all directions at once, thereby obscuring his meaning and confounding his audience. If Ike were giving the Gettys-burg Address, Dwight Macdonald once quipped, he would phrase it like this:

I haven't checked these figures, but eighty-seven years ago, I think it was, a number of individuals organized a governmental setup here in this coun-

Boston, 1986); R. Alton Lee, *Dwight D. Eisenhower: Soldier and Statesman* (Chicago, 1981); Charles Alexander, *Holding the Line: The Eisenhower Era, 1952–1961* (Bloomington, Ind., 1975); and William Pickett, *Dwight D. Eisenhower and American Power* (Wheeling, Ill., 1995).

3. Richard Hofstadter, *Anti-Intellectualism in American Life* (New York, 1963), 10.
4. Arthur Schlesinger, Jr., "The Ike Age Revisited," *Reviews in American History*, 4 (March 1983), 11. See also Richard Rovere, *The Eisenhower Years: Affairs of State* (New York, 1956), 8.
5. Ambrose, *Eisenhower*, 416.

try, I believe it covered eastern areas, with this idea that they were following up based on a sort of national independence arrangement and the program that every individual is just as good as every other individual.[6]

Critics of Eisenhower's presidency complain above all that he did not work hard at the job and that he failed to take charge. A good President, they thought, had to be strong and activist—like FDR. The liberal journalist I. F. Stone put this view well as early as January 1953 when he wrote, "Eisenhower is no fire-eater, but seems to be a rather simple man who enjoys his bridge and his golf and doesn't like to be too much bothered. He promises . . . to be a kind of president *in absentia,* a sort of political vacuum in the White House which other men will struggle among themselves to fill."[7] Objections to Eisenhower's love of golf especially punctuated his presidency. A bumper sticker read, BEN HOGAN [the best golfer of the era] FOR PRESIDENT. IF WE'RE GOING TO HAVE A GOLFER FOR PRESIDENT, LET'S HAVE A GOOD ONE. A contemporary joke had Ike asking golfers ahead of him, "Do you mind if we play through? New York has just been bombed."[8] Liberals then and later described Eisenhower as at best an appropriate President for the conservative 1950s: "the bland leading the bland." John F. Kennedy was but one of many contemporaries who, partisanship aside, believed that Eisenhower was a "non-President," with little understanding of the powers available to him.[9]

Eisenhower's defenders reply correctly that he deserves a more rounded appraisal. Ike, they emphasize, had a remarkably engaging personality and dominating presence. Although he was only five feet, ten inches tall, he carried himself with an erect military bearing and exuded physical strength and vitality. At sixty-two when he entered the White House, he was one of the oldest chief executives in American history, but even after 1955, when he had a heart attack, he remained tanned and vigorous in appearance. Although he had a hot temper, most people who came in contact with him remembered instead his bright blue, often flashing eyes and his wide, warm, and infectious smile. He radiated sincerity and openness. Field Marshal Bernard Montgomery, who clashed often with

6. James David Barber, *The Presidential Character: Predicting Perfection in the White House* (Englewood Cliffs, N.J., 1972), 161.
7. I. F. Stone, *The Haunted Fifties, 1953–1963* (Boston, 1983), 6 (Jan. 24, 1953).
8. Marquis Childs, *Eisenhower: Captive Hero* (London, 1959), 261; J. Ronald Oakley, *God's Country: America in the Fifties* (New York, 1986), 152.
9. Clark Clifford, "Serving the President: The Truman Years (2)," *New Yorker,* April 1, 1991, p. 12.

Ike during World War II, admitted, "He has this power of drawing the hearts of men towards him as a magnet attracts the bits of metal. He has merely to smile at you, and you trust him at once."[10]

Friendly writers also emphasize Eisenhower's native abilities and pre-presidential accomplishments. These had not been easy to foresee early in his life. Born in 1890 in Texas, he was raised by God-fearing pacifist parents in Abilene, Kansas, but nonetheless went to West Point, graduating in 1915. He was better known there as a football player (until sidelined by injury) than as a scholar, finishing sixty-first in a class of 164. Unlike MacArthur, he did not see action in World War I, and as a result he moved ahead only slowly in the resource-starved interwar army. In the 1920s, however, he served in Panama under General Fox Conner, a literate man who encouraged Eisenhower to read more widely in military history and the classics. Eisenhower later considered this stint as a kind of graduate education. Conner recommended him for the army's elite Command and General Staff School at Fort Leavenworth, Kansas, where Ike excelled, graduating first in a class of 275.

Eisenhower was thereafter marked as one of the ablest young officers in the army. After a tour of duty in the office of the Assistant Secretary of War, he served under MacArthur both in Washington—while Mac-Arthur was chief of staff between 1930 and 1935—and in the Philippines from 1936 to 1939. He then returned to the United States before being called after the attack on Pearl Harbor to work in Washington in the Planning Division of the War Department, where he much impressed army chief of staff Marshall. Marshall later tapped him to be supreme allied commander in Europe. Eisenhower's successful handling of D-Day, his open, democratic manner, and his ability to maintain harmony among often egotistical military and political figures made him an exceptional leader of coalition forces. He received a hero's welcome when he returned to the United States in 1945. He then became army chief of staff before leaving for Columbia in 1948 and the command of NATO forces in 1951.

By the end of World War II, when Eisenhower had become a closely watched figure, some disinterested observers appreciated his intelligence and articulateness. Steve Early, FDR's press secretary, went to one of Ike's press conferences and emerged as a strong admirer. "It was the most

10. Robert Divine, *Eisenhower and the Cold War* (New York, 1981), 6; William O'Neill, *American High: The Years of Confidence, 1945–1960* (New York, 1986), 177; Robert Donovan, *Eisenhower: The Inside Story* (New York, 1956), 3; Ambrose, *Eisenhower*, 292–93.

magnificent performance of any man at a press conference that I have ever seen," Early said. "He knows his facts, he speaks freely and frankly, and he has a sense of humor, he has poise, and he has command."[11] Another experienced journalist, Theodore White, was equally impressed while Ike was commander of NATO in 1951 and 1952: "I had made the mistake so many observers did of considering Ike a simple man, a good straightforward soldier. Yet Ike's mind was not flaccid; and gradually, reporting him as he performed, I found that his mind was tough, his manner deceptive; that the rosy public smile could give way, in private, to furious outbursts of temper; that the tangled, rambling rhetoric of his off-the-record remarks could, when he wished, be disciplined by his own pencil into clean, hard prose."[12]

Eisenhower's years in the military in fact had helped him to think and write clearly. Much of his career had involved preparing position papers and speeches, including many of MacArthur's. When speechwriters began composing for him, he proved to be a painstaking and often stern editor, seeking to rid prepared remarks of high-flown rhetoric. Although he was indeed convoluted in many presidential press conferences, he usually knew what he was doing, and he rarely said anything very damaging. Eisenhower was in fact considerably more ambitious, crafty, and egotistical than most people recognized, and he took pains to protect his image. When he decided to run for the presidency in 1952, he surrounded himself with more professional advertising and public relations experts than had any presidential candidate in American history, and by 1955 he used TV as much as he could to promote himself and his policies.[13] Samuel Lubell, a sophisticated journalist, laughed at the notion that Ike was a "five-star babe in the political woods." On the contrary, he was "as complete a political angler as ever filled the White House."[14] Ike was especially adept at a key to presidential survival: letting associates take the blame for controversial statements while appearing to remain above politics. The liberal reporter Murray Kempton later highlighted this talent in an influential article, "The Underestimation of Dwight Eisen-

11. Divine, *Eisenhower and the Cold War*, 8.
12. Fred Greenstein, "Dwight D. Eisenhower: Leadership Theorist in the White House," in Greenstein, ed., *Leadership in the Modern Presidency* (Cambridge, Mass., 1988), 77.
13. Craig Allen, *Eisenhower and the Mass Media: Peace, Prospects, and Prime-Time TV* (Chapel Hill, 1993); Robert Griffith, "Dwight D. Eisenhower and the Corporate Commonwealth," *American Historical Review*, 87 (Feb. 1982), 95.
14. Oakley, *God's Country*, 430.

hower," which is widely cited by pro-Eisenhower revisionists. Ike, Kempton concluded, was far shrewder than people realized. "He was the great tortoise upon whose back the world sat for eight years. We laughed at him; we talked wistfully about moving; and all the while we never knew the cunning beneath the shell."[15]

Revisionists such as Kempton understood that Eisenhower was more than simply cunning. Many politicians—Nixon quickly comes to mind—were as good as or better than Eisenhower at that. Ike had three other characteristics that stood him in good stead as a President and that account for the huge affection that most Americans held for him in his own time. The first was his normally prudent way of reaching decisions. When he became President, he brought with him a military way of doing things: finding loyal staff, establishing a hierarchical system of organizing it, meeting regularly with immediate subordinates, and allowing time (where possible) for contemplation before rushing into things. As critics were quick to point out, this style of decision-making tended to deprive him of the free-wheeling and sometimes innovative ideas that energized the administrations of Presidents like FDR and JFK. Over time it often insulated him from urgent public passions, such as the emerging civil rights movement. But it was orderly, and it kept him focused, his mind uncluttered, on issues he considered important. As exploited by his cautious intelligence, this administrative style encouraged prudent judgments on most (not all) matters of high public policy.

Second, Eisenhower had great self-confidence in his knowledge of foreign and defense policies. Compared to Truman, who had to learn on the job, or Kennedy, who felt he had to prove himself, Ike came to the White House with the serene self-assurance—it bordered on arrogance at times—of someone who had wide experience in these areas. He was personally acquainted with many of the world's leading statesmen and military leaders and was for the most part a wise judge of character. More important, he understood military matters and kept abreast of technological changes in weaponry. For many Americans it was comforting, amid the otherwise harrowing nuclear build-ups of the Cold War, to know that Eisenhower was in charge.

Third, Eisenhower had a sincere commitment to public service.[16] This came from a combination of forces: his upbringing in a righteous, hard-

15. Kempton, "The Underestimation of Dwight D. Eisenhower," *Esquire*, 68 (Sept. 1967), 108ff. See also Stephen Rabe, "Eisenhower Revisionism: A Decade of Scholarship," *Diplomatic History*, 17 (Winter 1993), 97–115.
16. Griffith, "Dwight D. Eisenhower," 88.

working family; his education; above all, perhaps, his career as an army officer. While he was hardly "above politics" as President, he impressed associates with his seriousness and concern for the dignity of the office. More than most world statesmen of his time, he seemed solid and sensible—at least in foreign and military affairs.

As archival materials became available to historians and political scientists in the 1970s and 1980s, these, too, seemed to confirm that Eisenhower, while ill informed on many domestic issues, was normally shrewd and prudent otherwise. They make it clear that he, not strong-willed subordinates, was the one in control. Let Schlesinger, a critic, have last words that describe Ike as at once self-promoting and politically astute. Revelations from Eisenhower's papers, Schlesinger wrote in 1983,

> unquestionably alter the old picture. We may stipulate at once that Eisenhower showed much more energy, interest, self-confidence, purpose, cunning, and command than many of us supposed in the 1950s; that he was the dominant figure in his administration whenever he wanted to be (and he wanted to be more often than it seemed at the time); and that the very genius for self-protection that led him to exploit his reputation for vagueness and muddle and to shove associates into the line of fire obscured his considerable capacity for decision and control.[17]

EISENHOWER'S GLITTERING WAR record and widespread popularity had made him attractive as a presidential possibility in 1948, but he had resisted entreaties from both parties and had stayed at Columbia. There and in Europe after 1951, however, he continued to be besieged by VIPs who wanted him to run in 1952. By the fall of 1951 non-partisan Ike clubs were springing up around the country. Truman himself in November of that year told Ike he would back him for the Democratic nomination.[18]

As a military officer Eisenhower had never registered a party affiliation or (he said) voted. (Later he said he would have voted Republican in 1932, 1936, and 1940 and Democratic—in the midst of war—in 1944.) He had held his most important assignments under Democratic Presidents and strongly supported Truman's Cold War initiatives, including the Korean War. But he was very conservative on domestic matters, believing almost passionately in the necessity of balanced federal budgets and limited gov-

17. Schlesinger, "Ike Age Revisited," 6.
18. Ambrose, *Eisenhower*, 246–48.

ernmental intervention in the social and economic life of citizens. Not for a minute did he consider running as a Democrat.

Resisting Republican entreaties proved to be more difficult. Many GOP leaders, remembering the debacle of 1948, were almost desperately eager to run the popular Ike as their candidate. It helped even more that he had taken few clear stands on domestic issues and had therefore made few enemies. Thomas Dewey, still governor of New York, began badgering him to run as early as 1949. He told Ike that only he could "save the country from going to Hades in the handbasket of paternalism-socialism-dictatorship."[19] By late 1951 Dewey, Senator Henry Cabot Lodge of Massachusetts, and other leading Republicans—most of them in the eastern, internationalist wing of the party—were developing a well-financed network of support for Eisenhower's nomination as a Republican presidential candidate in 1952.

Eisenhower, far away in Europe, managed to keep some distance from Lodge and Dewey throughout 1951 and into early 1952. He refused even to say whether he was a Republican or a Democrat. But as primary season approached he relented, agreeing to have his name placed as a Republican contender in the New Hampshire primary in March of 1952. Without leaving Europe or taking a stand on any of the issues, he won the primary by taking 46,661 votes to 35,838 for Taft, his most formidable opponent.[20]

Several considerations apparently prompted Ike to enter the political fray. One was his concern over Truman's budget message in January, which projected a sizeable deficit for the next fiscal year. Another was his distaste for Taft, who had opposed many Roosevelt-Truman era foreign policies and who had led conservative Republicans in the Senate in their support of both McCarthy and MacArthur. Taft, he told a friend, was "a very stupid man . . . he has no intellectual ability, nor any comprehension of the issues of the world."[21] Finally, Eisenhower may have been more interested in being President than he had let on. (As early as 1943 General George Patton guessed that "Ike wants to be President so badly you can taste it.")[22] Fearing the alternatives—a Democrat or Taft—

19. Robert Griffith, "Forging America's Postwar Order: Domestic Politics and Political Economy in the Age of Truman," in Michael Lacey, ed., *The Truman Presidency* (Washington, 1983), 88.
20. Dwight Eisenhower, *Mandate for Change* (New York, 1963), 54–78.
21. David Halberstam, *The Fifties* (New York, 1993), 209; James Patterson, *Mr. Republican: A Biography of Robert A. Taft* (Boston, 1972), 483–84.
22. Divine, *Eisenhower and the Cold War*, 4.

Eisenhower convinced himself in early 1952 that it was his duty to run. With characteristic self-confidence he was certain that he could do the job better than anyone else on the political horizon.[23]

His victory in New Hampshire testified strikingly to his popularity. Thereafter there was no turning back. By June he had resigned from his post at NATO and had returned to campaign against Taft and other lesser candidates, including Governor Earl Warren of California and former Minnesota Governor Harold Stassen, a near-quadrennial contender. The battle against Taft became especially intense and hard-fought, pitting the eastern Dewey-Lodge wing of the GOP against the more "isolationist" and mostly conservative middle-western wing of the party. It was a jagged rift that had long divided the party and that imperiled Eisenhower's policies throughout his subsequent political career.

Ike proved a somewhat wooden candidate at first; the hustings were new to him and made him uncomfortable. Many party regulars, moreover, deeply admired Taft, "Mr. Republican," who had led the GOP in Congress since 1939. Ike, they complained furiously, was an outsider—not even a real Republican—who had no right to enter GOP primaries, let alone to claim the nomination. Still, Eisenhower had two big assets: he seemed better informed than Taft about foreign affairs, and he was America's most popular hero. Party leaders also worried that Taft, a colorless, uncharismatic campaigner, might lose to a Democrat in November. For these reasons Eisenhower rolled up delegates in the primaries and eked out a victory over Taft at one of the most bitterly contested party conventions of modern times. It was not until September, when Eisenhower promised a conservative course, that Taft and his angry supporters agreed to support the parvenu party nominee in November.

The delegates who chose Ike sharply repudiated Truman's foreign policies, which Eisenhower had played a major role in implementing. The GOP platform denounced containment as "negative, futile, and immoral" because it "abandons countless human beings to a despotism and godless terrorism." Mindful of ethnic and anti-Communist voters, the platform went on to deplore the plight of the "captive peoples of Eastern Europe" and to call for their liberation. Republicans further pledged to "repudiate all commitments contained in secret understandings such as those of Yalta which aid Communist enslavement."[24] That the delegates

23. Ambrose, *Eisenhower*, 267.
24. Patterson, *Mr. Republican*, 509–34; Ambrose, *Eisenhower*, 270–75; Eisenhower, *Mandate for Change*, 79–110; Paul David, *Presidential Nominating Politics in 1952*, Vol. 4 (Baltimore, 1954).

could nominate Eisenhower and approve such a platform testified to the near-schizophrenic divisions within the GOP and to the unembarrassed philosophical inconsistency of American political parties.

The convention also selected Richard Nixon as Ike's running mate. Eisenhower ultimately made the choice, mainly because many of his top advisers, including Dewey, were recommending it.[25] Nixon had worked quietly but effectively to swing California delegates to Ike during the convention, thereby infuriating Warren's California loyalists. More important to ticket managers, Nixon was young (only thirty-eight in 1952), a fiercely anti-Communist partisan, and a tireless campaigner. As a Californian he brought regional balance to the ticket and was expected to help deliver the state's important electoral votes. Then and later Eisenhower remarked that Nixon seemed to have no friends; he never warmed up to him. But he seemed yoked to him for at least the duration of the campaign.[26]

Democrats and liberals professed to be shocked and appalled by the GOP ticket and platform. "T.R.B.," columnist for the *New Republic*, observed correctly that Ike actually stood to the right of Taft on some domestic issues. Eisenhower, he wrote, was "a counter-revolutionist entirely surrounded by men who know how to profit from it." The Eisenhower-Nixon team was a "Ulysses S. Grant–Dick Tracy ticket."[27] Another liberal described the convention as full of "treason screamers and poison-tongued character assassins." Most of these liberals enthusiastically backed the man whom the Democrats named to run against Ike: Governor Adlai Stevenson of Illinois. Their enthusiasm was in some ways odd, for Stevenson was hardly much of a liberal. The grandson and namesake of Grover Cleveland's Vice-President, he had grown up in Bloomington, Illinois, in a very wealthy family. His grandmother had been a founder of the Daughters of the American Revolution.[28] He had been educated at Choate School in Connecticut and at Princeton and flunked out of Harvard Law School before finishing his legal degree at

25. Sherman Adams, *First-Hand Report: The Story of the Eisenhower Administration* (New York, 1961), 34; Alexander, *Holding the Line*, 11.
26. Roger Morris, *Richard Milhous Nixon: The Rise of an American Politician* (New York, 1990), 695–736; Parmet, *Eisenhower and the American Crusades*, 102–17.
27. Alonzo Hamby, *Beyond the New Deal: Harry S. Truman and American Liberalism* (New York, 1973), 493.
28. Fred Siegel, *Troubled Journey: From Pearl Harbor to Ronald Reagan* (New York, 1984), 98. Biographies include John Bartlow Martin, *Adlai Stevenson of Illinois* (Garden City, N.Y., 1976), and *Adlai Stevenson and the World: The Life of Adlai Stevenson* (Garden City, N.Y., 1977); and Bert Cochran, *Adlai Stevenson: Patrician Among the Politicians* (New York, 1969).

Northwestern and practicing law in Chicago. But for the Democratic heritage of his family, Stevenson might well have been a Republican, like many of his wealthy friends.

In 1952 Stevenson did not differ greatly from Eisenhower. He was an ardent Cold Warrior. He opposed public housing and was ambivalent about repeal of the Taft-Hartley Act. He castigated "socialized medicine." While denouncing McCarthy, he approved of the dismissal of teachers who were Communists and supported the Truman administration's use of the Smith Act to prosecute Communist party leaders. He could sound snobbish, as when he denounced the GOP for "trying to replace the New Dealers with car dealers."[29] Like his running mate, Senator John Sparkman of Alabama, he considered civil rights to be mainly a question for the states to handle. Stevenson and Sparkman, determined not to provoke another Dixiecrat walkout, ran on a platform that was considerably more conservative concerning civil rights than the one on which Truman was elected in 1948.[30] The democratic socialist Irving Howe, casting a cool eye on liberal enthusiasm for Stevenson, later concluded that "Adlaism" was "Ikeism . . . with a touch of literacy and intelligence."[31]

For all these reasons Stevenson did not appeal much to the working-class–black–ethnic–urban coalition that Roosevelt had amassed and that Democrats needed in order to win national elections. Many party regulars found him to be aloof, for he distanced himself not only from Democratic bosses but also from the Truman administration. This behavior so irritated Truman, who had been an early backer of Stevenson, that he wrote another of his unsent letters: "I'm telling you to take your crackpots, your high socialites with their noses in the air, run your campaign and win if you can. . . . Best of luck to you from a bystander who has become disillusioned."[32]

Why then did Stevenson attract so many admirers in his time? Democrats who followed him liked his solid record as an internationalist in the 1930s and his State Department service during World War II, work that helped in the organization of the United Nations. He was a Democratic Establishmentarian with considerable experience in the realm of foreign affairs. When Illinois Democrats cast about for an honest Democrat to run as governor in 1948, they turned to Stevenson. Helped by running

29. Siegel, *Troubled Journey*, 99.
30. O'Neill, *American High*, 181–83; Hamby, *Beyond the New Deal*, 497.
31. Richard Pells, *The Liberal Mind in a Conservative Age: American Intellectuals in the 1940s and 1950s* (New York, 1985), 397.
32. Robert Ferrell, *Harry S. Truman and the Modern American Presidency* (Boston, 1983), 146.

against weak opposition, he proved to be an effective vote-getter, winning the largest plurality in Illinois history and running far ahead of Truman. He was an efficient governor who attracted able and committed people to his administration. To politicians seeking a viable presidential candidate in 1952—Truman wisely declined to run again—Stevenson, governor of an electorally significant state, was an obvious man to draft. After much hesitation he agreed to run. [33]

Liberal Democrats especially loved Stevenson—this is not too strong a verb—because he seemed to be everything that Eisenhower was not. They adored his speeches, which he spent hours practicing before delivering with a polish and vocabulary that many intellectuals considered wonderful. (Some did not: Howe noted that Stevenson was the sort of man who would call a spade an implement for the lifting of heavy objects.) There was intellectual snobbishness in this adoration; Republicans countered derisively that "eggheads" were the core of his support. But many of his speeches were indeed gusts of intelligent fresh air amid the staleness of political discourse that had often passed for campaign oratory in the 1940s and 1950s. So "eggheads" were happy to be involved. David Lilienthal pronounced that Stevenson's speeches were "simply gems of wisdom and wit and sense." The journalist Richard Rovere added, "His gifts are more imposing than those of any President or any major party aspirant for the office in this century." [34]

Stevenson ran a dignified, issues-centered campaign in which he promised to "talk sense to the American people." This, too, appealed greatly to liberal supporters and intellectuals. But as the Democratic candidate he inevitably had to contend with partisan attacks on the Truman record. Republicans hammered away at creeping corruption—the "mess in Washington," they called it—in the Truman administration after 1950. The Truman years, they cried, involved "Plunder at home, Blunder abroad." The corruption in fact was minor, mainly involving influence-peddling on a small scale, but it existed, and Truman—ever loyal to friends—had been slow to stamp down on it. Eventually Truman's appointments secretary was convicted of accepting bribes, and nine federal employees in the Reconstruction Finance Corporation and Bureau of Internal Revenue went to jail. [35]

33. O'Neill, American High, 181.
34. Hamby, Beyond the New Deal, 496–97.
35. Cabell Phillips, The Truman Presidency: The History of a Triumphant Succession (New York, 1966), 402–14; William Leuchtenburg, A Troubled Feast: American Society Since 1945 (Boston, 1973), 22; Ferrell, Harry S. Truman, 143.

More damaging to Stevenson were loud and insistent charges that the Democrats had been "soft" on Communism. The Red Scare and Korea overwhelmed other issues, including civil rights and labor controversies that had been important in 1948. As in previous years the Republican Right led this assault, often irresponsibly. McCarthy branded the Roosevelt-Truman years as "twenty years of treason." Referring to "Alger—I mean Adlai," he said that he would like to get onto Stevenson's campaign train with a baseball bat and "teach patriotism to little Ad-lie." Nixon labeled Stevenson "Adlai the Appeaser," said he had a "PhD from Dean Acheson's cowardly college of Communist Containment," and reminded voters that the country would be better off with a "khaki-clad President than one clothed in State Department pinks."[36]

Rhetoric such as this tapped into the persisting undercurrent of regional, class, and ethnic resentments that raged beneath the surface of American society in the postwar era. Like McCarthy and his allies, the *Chicago Tribune* regularly assailed liberal eastern intellectuals, on one occasion carrying the headline HARVARD TELLS INDIANA HOW TO VOTE. Its columns regularly associated virile masculinity with anti-Communism and implied that Stevenson was something less than a "real man." The *New York Daily News*, a bitterly reactionary paper, referred to Adlai as "Adelaide" and said that he "trilled" his speeches in a "fruity" voice, using "teacup words" that were reminiscent of a "genteel spinster who can never forget that she got an A in elocution at Miss Smith's finishing school."[37]

Eisenhower was uncomfortable around rabid Red-baiting such as this and avoided it himself. Most of his speeches were bland and unmemorable. When Senator Jenner, who had called Marshall "a front man for traitors," embraced him on a platform in Indianapolis, Eisenhower winced and hurried away. He told an aide that he "felt dirty from the touch of the man." But the "soft on Communism" issue dominated GOP strategy in 1952, and Eisenhower did nothing to curb the partisan zeal of other Republicans, including his running mate. Traveling into McCarthyite Wisconsin, Ike went so far as to delete from a prepared speech a paragraph that paid tribute to Marshall, who had mentored his military career. In so doing he kowtowed to McCarthy, who pumped his hand in Milwaukee when Ike gave the now sanitized speech. Reporters who had seen the original version assailed Ike for his cravenness. Eisenhower him-

36. Leuchtenburg, *Troubled Feast*, 34; Oakley, *God's Country*, 135.
37. Hofstadter, *Anti-Intellectualism*, 225, 227.

self felt ashamed. But he did not apologize for what he had done, and Red Scare rhetoric swelled throughout the campaign.[38]

Nothing fed anti-Communist feelings more than the still stalemated struggle in Korea. GOP leaders coined a symbol that stuck: K_1C_2—for "Korea, Communism, Corruption." Eisenhower himself made much of such feelings. Conceding that Stevenson could be witty, he told people that he did not find much to smile about. "Is it amusing," he asked, "that we have stumbled into a war in Korea; that we have already lost in casualties 117,000 of our Americans killed and wounded; is it amusing that the war seems to be no nearer to a real solution than ever; that we have no real plan for stopping it? Is it funny when evidence was discovered that there are Communists in government and we get the cold comfort of the reply, 'red herring'?"[39]

With polls showing Eisenhower well ahead, it seemed that nothing could derail his campaign. In mid-September, however, occurred the one great controversy of the contest: revelations in the press that Nixon had a private political "fund" donated by wealthy California supporters. This need not have been a big issue, for the fund was small (around $16,000) and legal. Most politicians, including Stevenson, had similar sources of cash. But accusatory editorials in the press unnerved Ike, who was said to have remarked at the time, "Of what use is it for us to carry on this crusade against this business of what is going on in Washington if we ourselves are not clean as a hound's tooth?"[40] Eisenhower then dawdled, refusing to offer a public defense of his running mate. Nixon grew increasingly furious as the controversy threatened to destroy his political career. Phoning Ike, he went so far as to assert, "There comes a time in matters like this when you've either got to shit or get off the the pot." Eisenhower professed to be appalled by such language and kept Nixon hanging. Nixon later complained that Eisenhower's attitude made him feel like "the little boy caught with jam on his face."[41]

With the problem dumped in his lap Nixon went on national television to defend himself. He spoke for thirty minutes, during which time he described his family's far from sizeable financial assets in detail. His wife, Pat, was deeply upset and later complained, "Why do we have to tell

38. David Oshinsky, A Conspiracy So Immense: The World of Joe McCarthy (New York, 1983), 236–38; Ambrose, Eisenhower, 282–84.
39. Childs, Eisenhower, 155.
40. Halberstam, Fifties, 237–42; Oakley, God's Country, 135–37.
41. Stephen Ambrose, Nixon: The Education of a Politician, 1913–1962 (New York, 1987), 271–300; Morris, RMN, 757–808; Ambrose, Eisenhower, 279–82.

people how little we have and how much we owe?" But Nixon marched to his own drummer. Pat, he said, did not have a mink coat (unlike the wife of a Democratic influence-peddler who was part of the "mess in Washington"), but "she does have a perfectly respectable Republican cloth coat." Nixon then told his huge audience about "the little cocker spaniel dog . . . black and white spotted" that had been sent to them in Washington "all the way from Texas" at the start of the campaign. "Our little girl—Tricia, the six-year-old—named it Checkers. And you know the kids love that dog and I just want to say this right now, that regardless of what they say about it, we're going to keep it."[42]

The Checkers speech, as it became known, was maudlin and tasteless, and many contemporaries said so. But it was also a brave performance by a determined and aggressive man who had been abandoned by many of his so-called friends. Popular reaction to his effort was overwhelmingly favorable. Many people broke into tears. Eisenhower, who had nervously watched the speech, soon recognized the favorable reaction and concluded that Nixon had saved himself. He summoned Nixon from the West to Wheeling, West Virginia, and told him, "You're my boy." This comment neatly captured the condescension that characterized Ike's feelings about his much younger running mate. Nixon was proud of what he had done, convincing himself that he was a master of television and could best anyone who tried to confront him on the screen. But he also felt bitter. He never forgot how badly he had been treated and correctly called the controversy, which could have killed his political ambitions, the "most searing personal crisis of my life."[43]

Nixon's remarkable performance had another effect: it demonstrated more than any event to that time the potential power of television in politics. This should have been apparent by then to political professionals, for TV was booming like almost nothing else in the nation. In 1951, 9 million people had watched the signing of a peace treaty with Japan. The number of households with TV sets had soared from around 172,000 in the 1948 campaign to 15.3 million in 1952. This was about one-third of American households. But the politicians had been slow to appreciate the potential for change. Stevenson was contemptuous of television, maintaining that he never watched it. He used it during the campaign, but only as a medium to display his oratorical skills. Although these were impressive, viewers did not get excited watching him reading speeches from a

42. Morris, *RMN*, 835–50; Alexander, *Holding the Line*, 18–21.
43. Ambrose, *Eisenhower*, 279–82.

studio, usually from 10:30 to 11:00 P.M. at night. Worse, Stevenson went on live, frequently speaking for more than the thirty minutes he had paid for. On several occasions, including election night, he was cut off before he finished.[44]

Eisenhower, too, started by displaying little interest in television, regarding it as a commercial medium that was for the most part beneath his dignity. He was wise enough, moreover, not to give speeches on TV: he knew that Stevenson was much the abler public speaker. But aides pressed him, especially after the Checkers speech, to let himself be televised in action. Increasingly he relented, and many of his rallies and campaign appearances were carefully scripted to convey the "I Like Ike" fervor of crowds who cheered him on. His TV advisers would cut to brief portions of his speech, but then focus again on the enthusiasm of his admirers. These were effective productions that promised to make television a force in politics for the first time.

Eisenhower also agreed to spend one evening in New York City taping "spots," as they became known. This was an amazing event. Surrounded by advertising men, Eisenhower sat in the studio and was taped giving short "answers" to questions. These "answers," mostly from phrases he had already used during the campaign, were hand-lettered on cue cards that were held before him. The lettering was big and bold so that Ike, who was near-sighted, would not have to be televised wearing glasses. After Eisenhower left the studio, having taped forty spots of twenty seconds each, the advertising experts went to Radio City Music Hall to get "everyday Americans" and bring them to the studio. There they were taped asking the questions to which Eisenhower had already given the answers. Technicians then spliced the answers to the questions.[45]

Eisenhower was at first edgy and unhappy during the process. Although he gradually warmed to the project—even writing an "answer" himself—he grumbled at one point, "To think that an old soldier should come to this." A few television executives, too, were uneasy about using them. Complicated issues, they thought, could not be boiled down to twenty seconds or a minute. And Democrats were outraged when the spots began appearing. George Ball, a young Stevenson speechwriter, accused the Republicans of selling out to the "high-powered hucksters of Madison Avenue." Stevenson added, "I think the American people will be shocked by such contempt for their intelligence; this isn't Ivory Soap versus Palm-

44. Edwin Diamond and Stephen Bates, *The Spot: The Rise of Political Advertising on Television* (Cambridge, Mass., 1992), 41–49.
45. Ibid., 50–59; Halberstam, *Fifties*, 224–36.

olive." *Reporter's* executive editor explained why his magazine was so critical of the spots: they were "selling the President like toothpaste."[46]

Criticisms like these in no way deterred those engaged in the operation. The GOP invested an estimated $1.5 million in the project, and the TV networks happily took the money. In all, twenty-eight different spots appeared on the air, many of them more than once, usually during intervals between highly popular programs. They conveyed no new information about the issues and often oversimplified them. "What about the high cost of living?" one spot asked. "My wife, Mamie," Ike answered, "worries about the same thing. I tell her it's our job to change that on November fourth." But virtually all analysts were convinced that the spots, which brought Eisenhower's luminous smile into millions of American homes, were effective. Use of spots—and of television coverage in general—henceforth became an indispensable tool in American politics.[47]

It would be too much to say, as some have, that television efforts such as Eisenhower's revolutionized campaigns and elections in the United States. Seriously contested issues, especially the Cold War, the state of the economy, and (increasingly) race relations, continued to be presented in speeches, news columns, and editorials. But the importance of TV in politics nonetheless became enormous. Other things being equal, television coverage could give a big edge to telegenic campaigners. Candidates without bundles of money—television was very expensive—operated at great disadvantage. The need for such money made politicians ever more careful about offending wealthy contributors and important special interests. Even more than in the past, money talked loudly in national politics.

The rise of television also weakened the political parties, both locally and on the national level. These were challenged by highly personal organizations of individual candidates, many of whom virtually ignored party lines and relied instead on TV to reach people directly. Voters tended increasingly to support individual candidates instead of party tickets. Others, abandoning party identifications, called themselves independents. These trends were by no means new—partisanship had been declining in the United States since the 1890s—nor were they the result of TV alone: rapidly rising educational levels, substantial geographical mobility, and advancing prosperity—which altered the ways that voters per-

46. Martin Mayer, *Madison Avenue, U.S.A.* (New York, 1958), 296–97; Halberstam, *Fifties*, 231.
47. Kathleen Hall Jamieson, *Packaging the Presidency: A History and Criticism of Presidential Campaign Advertising* (New York, 1992), 39–89.

ceived socio-economic issues—were among the many forces that underlay the growing unpredictability and independence of the electorate. Nor
was the rise of independent voting necessarily a bad thing: the highly
partisan political universe of earlier years had drawbacks of its own. Still,
the decomposition of parties—and with it the stability and reliability of
governing coalitions—became pronounced as early as the 1960s, and
television had much to do with it.[48]

By mid-October, with Nixon cleared of misconduct and the GOP TV
campaign in high gear, the election was hardly in doubt. But candidates
can never be too sure, and Eisenhower took one more arrow from his
quiver. In Detroit on October 24 he returned to the problem of Korea,
where a new Communist offensive was underway. Eisenhower proclaimed that the first task of a new administration would be "to bring the
Korean war to an early and honorable end." That task, he added, "requires a personal trip to Korea. . . . I shall make that trip. . . . I shall
go to Korea."[49] His pronouncement was a shot in the dark, for he did not
know what he would do once he got there. Would he escalate the war or
use nuclear weapons? But everyone seemed to agree that his proclamation
was a master stroke. Eisenhower, after all, was a war hero and a five-star
general. If anyone could end the awful stalemate, it was Ike.

The results of the election surprised few analysts. Stevenson took nine
southern and border states and, thanks in part to much increased turnout,
got 3.14 million more votes in a losing cause than Truman had received
while winning in 1948. But the election was a striking personal triumph
for Eisenhower, who attracted a huge following. He captured 33.9 million votes (55.4 percent of the total) to 27.3 million for Stevenson. His
total vote was nearly 12 million more than Dewey had received in 1948.
Sweeping the electoral college, 442 to 89, he even cracked the so-called
Solid South by winning Florida, Tennessee, Texas, and Virginia. Although lesser Republicans did not do so well—widespread split-ticket
voting was a sign of the decomposition of parties—they gained majorities
in both houses of Congress, 48 to 47 in the Senate and 221 to 211 in the
House. Republicans were in control of the White House and Capitol Hill
for the first time since the election of 1930. The once dominant Democratic electoral coalition, struggling to survive amid the return of good
times in the postwar era, had obviously taken a battering.

48. Walter Dean Burnham, *Critical Elections and the Mainsprings of American Poli-
    tics* (New York, 1970), 91–134; James MacGregor Burns, *The Deadlock of Democ-
    racy: Four-Party Politics in America* (Englewood Cliffs, N.J., 1963), 265–79.
49. Divine, *Eisenhower and the Cold War*, 19.

AFTER WINNING THE ELECTION Eisenhower went as promised to Korea, where he spent three days on the front. He returned convinced that the war had to be brought to a close, and he concentrated on that goal during the first six months of his administration in 1953. A combination of circumstances, including the fatigue of the enemy, led to signing of a cease-fire that took effect on July 27. This was thirty-seven anguished months after the start of the war in 1950.[50]

In securing the cease-fire Eisenhower did not win major concessions from the enemy: had the Truman administration tried to sell a similar result, it would have been attacked by the Right. Nor did the agreement end all bloodshed: border incidents killed many people over the subsequent decades that American troops remained there. But Eisenhower, a general and a Republican, escaped popular vilification. The cease-fire agreement, indeed, was probably the most important single accomplishment of his eight-year presidency, and the one he most cherished later. Giving a big boost to Eisenhower's prestige early in his presidency, it also eliminated the most acrimonious political issue of the era. A year and more of peace by late 1954 even calmed some of the passions of the Red Scare. By 1955 Americans were already trying to put the war out of mind and to concentrate on the enjoyment of the good life at home.

The ebbing of acrimony, however, took time. Especially in 1953 and 1954, Cold War passions and partisan battling continued to rend American society and culture. No President could easily have managed these controversies, and Eisenhower, who tried to avoid most of them, was no exception. A persistent political reality intensified these controversies in 1953–54: the aggressiveness and fury of the anti-Communist Right.[51] In the aftermath of the heady GOP triumph of 1952 these emotions may have been stronger—and the Left weaker—than at any time in the modern history of the United States. One scholarly overview of these years concludes: "On the coldest, darkest, and most reactionary days of the [Ronald] Reagan ascendancy [in the 1980s] there was more radical belief and activity to be seen in the United States than was present anytime in the 1950s."[52]

50. See chapter 8.
51. Ronald Caridi, *The Korean War and American Politics: The Republican Party as a Case Study* (Philadelphia, 1968), 209–45; Robert Griffith, *The Politics of Fear: Joseph R. McCarthy and the Senate* (Lexington, Ky., 1970), 186–207.
52. Maurice Isserman and Michael Kazin, "The Failure and Success of the New Radicalism," in Steve Fraser and Gary Gerstle, eds., *The Rise and Fall of the New Deal Order, 1930–1980* (Princeton, 1989), 214.

Historians, to be sure, have subsequently dug about in the byways of American culture and exposed signs of rebellion and dissatisfation with the conservative values of the early Eisenhower years. Some young people, mostly in university circles, identified with Holden Caulfield, the restless anti-hero of J. D. Salinger's novel *The Catcher in the Rye* (1951). *Mad* magazine, a zany and highly irreverent publication, began its commercially very successful career (it was number two in circulation behind *Life* by the early 1960s) in 1952.[53] I. F. Stone established his iconoclastic and liberal *Weekly* in 1953; Irving Howe started *Dissent*, an organ of left-of-center opinion, in 1954; Marlon Brando and (by 1954) James Dean stood forth as models of anti-Establishment behavior in film. But these scattered manifestations of unease and discontent did not matter much in political circles. Readers of *Dissent*, for instance, admitted that they remained a "tiny band of exiles" throughout the 1950s: circulation for the magazine approximated 4,000 at the time.[54] Writers interviewed by *Partisan Review*, another left-of-center magazine, agreed in a symposium in 1952 that they did not want to be alienated from the mainstream. Instead, they wished "very much to be a part of American life. More and more writers have ceased to think of themselves as rebels and exiles."[55] This indeed was so. The majority of left-liberal intellectuals in the early 1950s had come to perceive—some reluctantly—that American culture was dominated by a moderate-to-conservative, middle-class "consensus." Indeed, a contemporary barometer of conservatism was the founding in 1955 of the *National Review*, the anti-Communist creation of William Buckley, a young Catholic intellectual not long out of Yale. The magazine gradually became a leading organ of conservative opinion—the most successful of its kind to appear in years.

Democrats, too, seemed tame and chastened following the 1952 election. In the Congress the dominant group centered about two moderate Texans, Democratic House leader Sam Rayburn and Senator Lyndon Johnson, who became majority leader between 1955 and 1960. Among intellectual circles a few Democratic liberals, including the economist John Kenneth Galbraith, championed progressive domestic causes within the newly formed Democratic Advisory Council (DAC), a pressure group. DAC policy proposals helped to establish a liberal Democratic domestic agenda for the 1960s but did not attract great attention before then. Other

53. Todd Gitlin, *The Sixties: Years of Hope, Days of Rage* (New York, 1987), 28–29.
54. Pells, *Liberal Mind*, 384.
55. Russell Jacoby, *The Last Intellectuals: American Culture in the Age of Academe* (New York, 1987), 75–76.

liberals, including Senator Hubert Humphrey of Minnesota, covered their flanks against conservative attack by championing tough legislation to fight Communism. In 1954 Congress overwhelmingly approved Humphrey's Communist Control Act, which defined the Communist party as a "clear, present, and continuing danger to the security of the United States" and deprived the party of "all rights, privileges, and immunities attendant upon legal bodies."[56]

As Humphrey's stance made clear, anti-Communist fervor seemed politically irresistible in these years. One manifestation of this fervor was the fate of Julius Rosenberg and his wife Ethel. Accused of being part of a ring (including Fuchs) that had passed atomic secrets to the Soviets in the 1940s, they had been convicted in March 1951 of conspiracy to commit espionage, at which time Judge Irving Kaufman sentenced them to death. Their crime, Kaufman declaimed from the bench, was "worse than murder." Critics of the judgment maintained that Kaufman, the Justice Department, and the FBI had been guilty of misconduct during the trial. Other critics asserted correctly that the punishment was severe: Fuchs, a much more important figure than the Rosenbergs, was sentenced to fourteen years'imprisonment in England.[57] But campaigns on behalf of the Rosenbergs, who were Communists, failed utterly in the Red Scare of the early 1950s. As the date of their execution approached, counterdemonstrators appeared outside the White House. One held a sign, FRY 'EM well done. Another had a placard, LET'S DON'T ELECTROCUTE THEM, HANG THEM. Eisenhower refused to commute the sentence, which he thought would deter others. On June 19, 1953, a month before the cease-fire in Korea, the Rosenbergs went silently to their deaths.[58]

The Eisenhower administration moved quickly to establish its anti-Communist credentials in other ways. In April the President issued Executive Order No. 10450, which replaced the network of loyalty decrees created by Truman. The new system was broader than Truman's, including not only loyalty and security as criteria for dismissal but also "suitability," a vague and open-ended category. The order widened the power of summary dismissal, until then available only to heads of sensitive

56. Steven Whitfield, *The Culture of the Cold War* (Baltimore, 1991), 49; Griffith, *Politics of Fear*, 292–94.
57. Fuchs was paroled after nine years and lived many years thereafter as an honored figure in East Germany.
58. Morris Dickstein, *Gates of Eden: American Culture in the Sixties* (New York, 1977), 188; Richard Fried, *Nightmare in Red: The McCarthy Era in Perspective* (New York, 1990), 115–16; Whitfield, *Culture of the Cold War*, 31.

departments such as State and Defense, by giving it to chiefs of all federal departments and agencies.[59] The Eisenhower administration also continued ongoing anti-Communist efforts, including the purging of the foreign service, prosecutions of Communists under the Smith Act, deportation of Communist aliens, and exclusion of alleged subversives seeking entrance to the United States. The President supported efforts to legalize the use of wiretaps in national security cases and gave a free hand to the FBI's continuing harassment of left-wingers. In 1956 the FBI established CO-INTELPRO (counter-intelligence program), whose primary target was the Communist party.[60]

One of the most famous casualties of such governmental efforts was J. Robert Oppenheimer, "father of the atomic bomb," who was stripped of his security clearance by the Atomic Energy Commission in June 1954. He was thereupon terminated as a governmental consultant.[61] This action followed six months of investigations during which it became clear that the FBI had bugged and tapped his activities for fourteen years. Oppenheimer had many left-wing friends and relatives, including his wife, who had been a Communist; during the war he had lied to investigators about their connections in order to protect them. But this was old, known information, and many of his scientific associates, including Harvard president James Conant, were shocked at what was happening to him. They realized that Oppenheimer was being punished primarily for his opposition to development of the hydrogen bomb. And Oppenheimer's fate was sad. He lost access to scientific developments—his life and career—and was cut off from other scientists, many of whom were afraid to talk to him.[62]

The ultimate test of Eisenhower's approach to loyalty and security was of course the question of McCarthy. Once the GOP regained control of the Senate, McCarthy was in his element, for he now had a chairmanship—of the Permanent Subcommittee on Investigations—from which he launched probes that irritated the new administration. Assisting

59. Griffith, "Dwight D. Eisenhower," 113; Fried, Nightmare in Red, 133–34. Fried notes that the GOP claimed to have dismissed 2,200 federal employees by early 1954 but that these were somewhat inflated figures; many of those who left government did so for other reasons.
60. Fried, Nightmare in Red, 188–89; Jeff Broadwater, Eisenhower and the Anti-Communist Crusade (Chapel Hill, 1992).
61. His clearance was reinstated in 1963.
62. Thomas Reeves, The Life and Times of Joe McCarthy: A Biography (New York, 1982), 589–90; O'Neill, American High, 228–30; Halberstam, Fifties, 329–58.

him was a subcommittee staff headed by chief counsel Roy Cohn, a sour, troubled, and fervently anti-Communist attorney. In April 1953 Cohn and a close friend, G. David Schine, set off on a well-publicized tour of Europe in which they called for the purging of allegedly subversive literature from government libraries. The State Department panicked and issued a directive excluding books and works of art by "Communists, fellow travelers, et cetera" from United States information centers abroad. A few books were actually burned.

Eisenhower had never cared for McCarthy, and he fumed when the senator considered contesting the confirmation of Walter Bedell Smith, a close friend who had been Ike's chief of staff in the army, as Undersecretary of State early in 1953. By then Ike was becoming increasingly friendly with Taft, GOP leader in the Senate, and Taft managed to get Smith confirmed. Meanwhile, the President tried quietly to undermine McCarthy in other ways: encouraging GOP senators to oppose him; getting a reluctant Vice-President Nixon to stave off McCarthy's probes into network television; trying to prevent McCarthy from speaking at party gatherings; and suggesting (very indirectly) to publishers and other media executives that they give the rampant senator less time and space. Fred Greenstein, a political scientist, later cited these moves as evidence for what he called Eisenhower's shrewd, subtle, and effective "hidden-hand presidency."[63]

Eisenhower, however, refused to go beyond indirection or to challenge McCarthy head-on. There were several reasons for his reluctance to do battle. First, he agreed with many of McCarthy's goals. As his policies made clear, he was a staunch Cold Warrior. Second, he feared an intra-party brawl that would further imperil his shaky GOP majorities in Congress. McCarthy, after all, was a Republican, and the President was leader of the party. Third, Eisenhower recognized that a direct confrontation with McCarthy would give the rambunctious, often uncontrollable senator even more publicity—on which McCarthy thrived—than he already had. It was better, he thought, to try to ignore him and to hope that given enough rope the senator would eventually hang himself. Eisenhower, finally, was afraid that a fight with McCarthy would diminish the all-important dignity of the presidency. Why use up vital presidential resources to scrap with an alley-fighter? "I will not get into the gutter with

63. Fred Greenstein, *The Hidden-Hand Presidency: Eisenhower as Leader* (New York, 1982), 155–227; Griffith, "Dwight D. Eisenhower," 114; Ambrose, *Eisenhower*, 307–10.

that guy," he said privately. Later in the year he added—again privately—
"I just won't get into a pissing contest with that skunk."[64]

Ike's worries about the dignity of the presidential office rested on two
even deeper concerns. One was to protect his own personal popularity
with the American people. Eisenhower, while self-confident, nonetheless
craved popular approval. He generally avoided tough decisions that might
threaten it.[65] Second, Ike very much wanted to promote domestic tranquility. Throughout his presidency he feared to take actions that might
undermine what he considered to be the harmony of American society.
He also believed that his mission should be to restrain the role of government, not to force it to fulfill great goals or obligations. These
aspirations—protecting his own standing, sustaining domestic tranquility,
and curbing the activity of the State—complemented one another in his
mind and helped to explain why he often chose *not* to do potentially
controversial things: advance ambitious social programs, push for civil
rights, get involved in war in Vietnam. They also accounted for his
restrained approach to McCarthy. To wade into the arena with such a
demagogue, he thought, would endanger his popularity, incite discord,
and damage social harmony.

Whether Eisenhower should have been more bold remains one of the
most contested questions about his presidency. As it turned out, McCarthy did overreach himself and crash in mid-1954. The President thereby
stayed out of the gutter. And his personal popularity—always high—did
not suffer. On the other hand, these were in many ways dispiriting times.
Federal employees, whom Eisenhower was supposed to protect, were hurt
under his watch. If Ike had risked even a little of his immense personal
popularity and presidential prestige, he might have slowed the senator
down or hastened his demise. It could not have hurt him much to try. His
refusal to challenge McCarthy represented a major moral blot on his
presidency.

Everything finally unraveled for McCarthy in early 1954. In March
and April, Edward R. Murrow, a widely respected investigative reporter,
ran a series of programs concerning McCarthy on "See It Now," a CBS
network production. It was the first time that television—which had expanded by then to 25 million households—had exposed him in any major
way. For the most part Murrow let McCarthy's bullying words and trucu-

64. Ambrose, *Eisenhower*, 307–10; John Diggins, *The Proud Decades: America in War
    and Peace, 1941–1960* (New York, 1988), 147–50.
65. H. W. Brands, *Cold Warriors: Eisenhower's Generation and American Foreign
    Policy* (New York, 1988), 191ff; Divine, *Eisenhower and the Cold War*, 9.

lent actions speak for themselves. McCarthy finally appeared on the show in April and tore into Murrow, calling him "the leader and the cleverest of the jackal pack which is always found at the throat of anyone who dares to expose individual Communists and traitors." Scholars debate the impact of these shows, some insisting that most Americans did not watch them: "Dragnet," a popular police series, drew far more viewers at the time.[66] Others add that McCarthy was already beginning to fade before the shows started. These are valid reminders that television was hardly all-powerful. But "See It Now" did attract a great deal of attention and critical praise at the time, and it legitimated rising criticism of McCarthy from other media. If Nixon's Checkers speech indicated that TV could save a politician, "See It Now" and televised hearings that followed suggested it could also help ruin one.

What really brought McCarthy down was his ill-advised attempt at the same time to ferret out subversive activities in—of all places—the United States Army. The army responded by documenting that McCarthy and Cohn had secured special privileges for Schine, then a private in the army.[67] The Senate established a special committee headed by South Dakota Republican Karl Mundt to hear the charges and counter-charges. The "Army-McCarthy hearings," as everyone soon called them, began on April 22 and lasted thirty-six days (for a total of 188 hours) through June 17. Often sensational, they attracted more than 100 reporters and crowds exceeding 400. The hearings were televised to Americans who had little good daytime TV to divert them. Some estimates put peak audiences as high as 20 million people.[68]

Placed on the defensive, McCarthy began drinking heavily. He often slept in his office and showed up looking unkempt and unshaven. On black-and-white television he resembled a heavy from Central Casting. He spoke in a low monotone, often petulantly. Again and again he jumped to his feet to shout "Point of order," so often that the audiences ultimately broke out in laughter. He outdid himself in bullying participants and hurling abuse. Eisenhower, appalled by McCarthy's excesses, thought that McCarthy was "psychopathic" and "lawless," but again he said nothing.

What he did do, however, proved problematic indeed for McCarthy, who was demanding access to sensitive information concerning federal

66. James Baughman, *The Republic of Mass Culture: Journalism, Filmmaking, and Broadcasting in America Since 1941* (Baltimore, 1992), 49.
67. Griffith, *Politics of Fear*, 245.
68. Reeves, *Life and Times*, 595–637; O'Neill, *American High*, 199–202.

employees, in the army and elsewhere. Key Republican senators backed McCarthy's demands for such access, but Eisenhower firmly resisted them. "I will not allow people around me to be subpoenaed," he told GOP leaders on May 17, "and you might just as well know it now." Senator William Knowland of California rejoined that Congress had a right to issue such subpoenas. The President repeated, "My people are not going to be subpoenaed."[69] He then made sure that people appreciated his resolve on the issue, directing his Defense Secretary to withhold sensitive information from McCarthy and his committee. His language was sweeping in its assertion of presidential authority and dismissal of congressional rights. "It is essential to efficient and effective administration that employees of the Executive Branch be in a position to be completely candid in advising with each other on official matters." It followed that "it is not in the public interest that *any* of their conversations or communications, or *any* documents or reproductions, concerning such advice be disclosed."

This was an extraordinary claim; Arthur Schlesinger, Jr., later termed it the "most absolute assertion of presidential right to withhold information from Congress ever uttered to that day in American history."[70] Earlier Presidents had argued that discussions in Cabinet meetings were confidential, but no one had yet been so bold as to extend executive privilege to the entire executive branch.[71] Many people doubted the constitutionality of what Ike had done. It did them no good, however, for the President had spoken. Later Presidents, including Nixon during the crisis of Watergate, used Ike's precedent-setting directive to make similar assertions. More immediately, McCarthy was stymied; without access to such information he could not begin to establish the subversiveness of individuals in the army or elsewhere in the executive branch of government.

Frustrated and exhausted, McCarthy finally finished himself on the afternoon of June 9. The army's special counsel, a soft-spoken but shrewd and able attorney named Joseph Welch, was grilling Cohn on the stand.

69. Ambrose, *Eisenhower*, 365.
70. Arthur Schlesinger, Jr., *The Imperial Presidency* (Boston, 1973), 156; Carol Lynn Hunt, "Executive Privilege," *Presidential Studies Quarterly*, 16 (Spring 1986), 237–46.
71. Ambrose, *Eisenhower*, 365. He further argues, 347–52, that Ike worried above all that McCarthy might get possession of Oppenheimer's security file, then under hold pending completion of his case, and use it to tar the Eisenhower administration with keeping Oppenheimer as a consultant long after he had opposed the H-bomb.

McCarthy broke in and began accusing Welch's law firm of harboring a leftist lawyer named Fred Fisher. The charge was no surprise to Welch, for McCarthy had threatened in private to bring it up. So Welch was ready and turned to Mundt, as a personal privilege, for the chance to reply. "Until this moment, Senator," he began, "I think I never really gauged your cruelty or your recklessness." Welch then explained that he had earlier taken Fisher off the hearings because Fisher had briefly belonged to the pro-Communist National Lawyers Guild. Speaking firmly and with a tone of ineffable sadness, Welch faced McCarthy and added, "Little did I dream you could be so cruel as to do an injury to that lad," who would now "always bear a scar needlessly inflicted by you. If it were in my power to forgive you for your reckless cruelty, I would do so. I like to think that I am a gentleman, but your forgiveness will have to come from someone other than me."

McCarthy should have let the matter rest, but he bulled ahead with further attacks on Fisher. Again Welch spoke out. "Let us not assassinate this lad further, Senator. You have done enough. Have you no sense of decency, sir, at long last? Have you left no sense of decency?" When McCarthy broke in again, Welch cut him off:

> Mr. McCarthy, I will not discuss this further with you. You have sat within six feet of me, and could have asked me about Fred Fisher. You have brought it out. If there is a God in heaven, it will do neither you nor your cause any good. I will not discuss it further. I will not ask Mr. Cohn any more questions. You, Mr. Chairman, may, if you will, call the next witness.

There was a moment of silence, and the room burst into applause. Mundt called for a recess and walked out with Welch. McCarthy turned up his palms and shrugged his shoulders. "What did I do?" he asked in confusion. "What did I do?"[72]

He had destroyed himself on national television. The hearings dragged on for a few more days, but McCarthy was by then a beaten man. Senator Ralph Flanders of Vermont, a Republican, demanded that the Senate censure him then and there. Instead, the Senate moved deliberately, awaiting the findings of a special committee appointed to explore McCarthy's activities over the past few years. When it reported (after the off-year elections of 1954) it unanimously criticized him for conduct (during earlier investigations of him by the Senate) that damaged the honor of the Senate. This was the narrowest possible charge, one that ignored many

72. Reeves, *Life and Times*, 627–31; Oshinsky, *Conspiracy*, 457–71.

more reckless acts. But it virtually guaranteed a favorable response to the report. On December 2, 1954, the Senate voted to "condemn" McCarthy by a vote of 67 to 22. As had so often been the case during the Red Scare, the vote was partisan. All forty-four Democrats voting favored the resolution, as did the one Independent, Wayne Morse of Oregon. The forty-four Republicans voting divided evenly, 22 to 22.[73]

Eisenhower could relax at last. He told the Cabinet that the movement could now be called "McCarthywasism" and cut him from the list of dignitaries welcome at White House social functions. The press largely ignored him. When Nixon visited Milwaukee during the 1956 campaign, McCarthy sidled up to a seat next to him. A Nixon aide asked him to leave, and he did. A reporter found him weeping.[74]

Feeling betrayed, McCarthy also suffered from the ravages of drink. He died of a liver condition on May 2, 1957. He was only forty-eight years old and still a United States senator.

ALTHOUGH McCARTHY AND Red Scare issues threatened to obscure other political matters in the early 1950s, they were by no means the only concerns of the era. Other domestic controversies, mostly fomenting political deadlock, illuminate Eisenhower's strengths and weaknesses in these years.

Liberals who explored Eisenhower's philosophy concerning these domestic issues were sure that he was ill informed and near-reactionary. Concerning social welfare, he had snapped in 1949, "If all Americans want is security, they can go to prison."[75] He said of the Tennessee Valley Authority, "By God, I'd like to sell the whole thing, but I suppose we can't go that far."[76] Like most politicians at the time, he paid little attention to widespread rural poverty or to urban decay. Eisenhower himself conceded that he was conservative on domestic issues, admitting that Taft, who had come to favor federal aid to education and public housing, was "far more 'liberal and radical' than anything to which I could ever agree."[77]

Eisenhower seemed so vulnerable on domestic matters that liberals had a field day making fun of him. Some dubbed him "Eisen-hoover." When he tried to summarize his domestic philosophy by saying that he was

---

73. Griffith, *Politics of Fear*, 270–315; Fried, *Nightmare in Red*, 139–41. To "condemn" was less serious than to "censure," but the effect was the same.
74. Fried, *Nightmare in Red*, 135.
75. Parmet, *Eisenhower and the American Crusades*, 36.
76. Siegel, *Troubled Journey*, 103.
77. Greenstein, *Hidden-Hand*, 49; Patterson, *Mr. Republican*, 578, 590.

Churchill, Truman, and Stalin at the Potsdam Conference, Germany, July 1945. *UPI/Bettmann.*

Bess and Harry Truman, House Democratic Leader Sam Rayburn of Texas, and Margaret Truman at a party dinner, Washington, April 1947. *Acme Photo, National Archives.*

James F. Byrnes, Truman, and Commerce Secretary Henry Wallace await FDR's funeral train at Washington's Union Station, April 1945. By 1946 Byrnes, then Secretary of State, and Wallace held sharply opposing views concerning foreign policy. Abbie Rowe, *U.S. National Park Service.*

General George Marshall and Secretary of State Dean Acheson, at a meeting to celebrate progress under the Marshall Plan, April 1950. *Harris & Ewing, National Archives.*

Robert Taft (1.) and Thomas Dewey, opposing G.O.P. leaders in the 1940s, put on a cooperative pose in July 1948, during Dewey's campaign for the presidency. *Acme Photo, Library of Congress.*

Truman awards the Distinguished Service Medal to General Douglas MacArthur, Wake Island, October 1950. *Library of Congress.*

General MacArthur (r.) confers with General Matthew Ridgway (2d. from left), near the front lines in Korea, January 1951. General Courtney Whitney, a top aide to MacArthur, is at left. *U.S. Army Photo, Library of Congress.*

General Eisenhower, then NATO commander, hearing the news of MacArthur's dismissal, April 1951. *Stars and Stripes, Library of Congress.*

U.S. Marines cover a just-flushed North Korean soldier, August 1950. AP,
Library of Congress.

Jackie Robinson, who broke the color line in Major League baseball, in 1947. *Bettmann.*

Evangelist Billy Graham. *AP/Wide World Photos.*

Elvis Presley, 1957. Bob Moreland, *St. Petersburg Times, Library of Congress.*

Marilyn Monroe, posing for photographers during filming of *The Seven Year Itch* (1955). Matthew Zimmerman, *AP/Wide World Photos.*

The Consumer Culture: Automobiles. *U.S. News,* August 1952, *Library of Congress.*

The Consumer Culture: Electronics—the rise of the transistor. *Look,* July 1962, *Library of Congress.*

Senator Joe McCarthy, March 1954. UP, *Library of Congress.*

COME NOW, JOE — LETS BE REASONABLE!

DULLES

STATE DEPT.

Hunting for Reds in the 1950s. Reg Manning, March 9, 1953, *Library of Congress.*

The Warren Court at the White House, November 1953. Left to right, front row, are Justice William Douglas, Justice Stanley Reed, Warren, Ike, Justice Hugo Black, Justice Felix Frankfurter. Other justices, back row, are Robert Jackson (3d from left), Tom Clark (4th from left), Sherman Minton (5th from left), and Harold Burton (6th from left). Attorney General Herbert Brownell is far right, back row. *UP, Library of Congress.*

The Supreme Court under siege: Warren and protestors, October 1963. AP, *Library of Congress.*

Ike and Secretary of State Dulles. National Park Service, *Eisenhower Library.*

The perils of Brinkmanship. From *Herblock's Special for Today*, Simon & Schuster, 1958.

"Don't Be Afraid—I Can Always Pull You Back"

The "kitchen debate," 1959, Moscow. *AP/Wide World Photos.*

Presidential nominee Adlai Stevenson (r.) and his Democratic running mate, Estes Kefauver, in 1956. AP, *Library of Congress*.

Richard and Pat Nixon, campaigning for the presidency, August 1960.
AP, *Library of Congress.*

"conservative when it comes to money and liberal when it comes to human beings," Stevenson—ever quick with a quip—retorted, "I assume what it means is that you will strongly recommend the building of a great many schools to accommodate the needs of our children, but not provide the money."[78]

Like most wisecracks in politics, this one was a little unfair. Eisenhower was indeed uninformed on domestic matters when he became President, but he did have a fairly coherent philosophy of government. It was well described in a statement of Abraham Lincoln's that he liked to repeat: "The legitimate aim of government is to to do for a community of people, whatever they need to have done, but cannot do at all, or cannot so well do, for themselves—in their separate, and individual capacities. In all that the people can individually do as well for themselves, government ought not to interfere."[79]

What this meant in practice was what his most enthusiastic supporters called "Modern Republicanism." This was a little right of center in emphasis. Believing in limited government, Eisenhower passionately endorsed conservative fiscal policies; balancing the budget and cutting government spending—even on defense—were his highest goals.[80] Curbing spending, in turn, helped bolster his philosophical opposition to federal aid to education, a major cause of liberals in the 1950s, and to "socialized medicine." He sought to cut costly federal price supports to agriculture. He approved legislation returning "tidelands oil," which liberals argued belonged to the national government, to private interests and the states. He wanted above all to decrease the role of government because he believed that large-scale federal intervention threatened individual freedom, the ultimate good in life.

To be conservative did not mean to be reactionary. Eisenhower was emphatic about the distinction between the two. Although he sought to reduce spending, he was not a mindless slasher. The Republican Right complained that he did not cut federal expenditures enough when he took over. (A young conservative senator, Barry Goldwater of Arizona, later said that Ike ran a "Dime Store New Deal.") Like most public figures at the time, the President accepted the need for a little compensatory fiscal

---

78. William Leuchtenburg et al., *The Unfinished Century: America Since 1900* (Boston, 1973), 762.
79. Donovan, *Eisenhower*, 208.
80. Iwan Morgan, "Eisenhower and the Balanced Budget," in Shirley Anne Warshaw, ed., *Reexamining the Eisenhower Presidency* (Westport, Conn., 1993), 121–32; Pickett, *Dwight D. Eisenhower and American Power*, 145–46.

policy when the times demanded it. And keeping a close eye on spending was not an abstract end in itself but a means of warding off inflation, which seemed to him (and to many contemporary economists) to be the most worrisome problem during and immediately after the economically stimulating Korean War. This his administration did help to control, and the next few years were remarkably prosperous and stable. Even Galbraith, no friend of GOP economic policy, conceded in January 1955 that "the Administration as a whole has shown a remarkable flexibility in the speed with which it has moved away from these slogans [of balanced budgets]."[81]

The President also proved ready to accept a few moderately liberal ventures in the realm of social policy. "Should any political party attempt to abolish Social Security, unemployment insurance, and eliminate labor laws and farm programs," he warned his conservative brother Edgar, "you would not hear of that party again in our political history."[82] He thereupon signed in 1954 a broadening of Social Security. He also sought to extend the minimum wage, which covered fewer than half of the wage workers in the United States. Both programs, of course, were financed primarily by employers and workers—not by federal funds that might increase federal deficits. But Eisenhower in no way threatened the welfare state begun in the New Deal years: social welfare expenditures during his presidency rose slowly but steadily as a percentage of GNP (from 7.6 percent in 1952 to 11.5 in 1961) and (especially after 1958) as a percentage of federal spending.

Beyond these moves lay a larger vision of what the United States should be: a cooperative society in which major groups such as corporations, labor unions, and farmers would set aside their special interests to promote domestic harmony and economic stability. The State, Eisenhower believed, could serve as an arbiter of this cooperative commonwealth, acting to bring the special interests together and restraining their excess demands. As in his dealings with McCarthy, however, Ike shrank from involving the presidency in controversial questions. Better, he thought, to stand above the battle and in so doing preserve his political standing. "Partisanship," moreover, was to Ike a word every bit as dirty as "special interest." He complained that Truman had used "ward-boss, strong-arm tactics" that had not worked and that had diminished the prestige of the presidency. He added, "I am not one of the desk-pounding types that likes

81. Herbert Stein, *The Fiscal Revolution in America* (Chicago, 1969), 281–84, 462–63.
82. Griffith, "Dwight D. Eisenhower," 102.

to stick out his jaw and look like he is bossing the show. I don't think it is the function of a President of the U.S. to punish anybody for voting as he likes."[83]

In holding to this Whiggish view of the presidential role Eisenhower indeed conserved his personal prestige and popularity. If he had tried to push through major domestic legislation, he would surely have provoked determined opposition. Contemporary measures of public opinion indicated that the majority of middle-class (and politically influential) Americans in the early 1950s, especially in the aftermath of the Korean War, did not look to government for great changes. They were tired of the angry controversies of the late 1940s and early 1950s. Developing ever-larger expectations about their personal futures, they were bent on maximizing the substantial economic and educational gains that they were coming to enjoy. Pressure groups, too, resisted change that threatened their standing. It is simply ahistorical to think that Eisenhower, who was elected as a moderate, could have or should have demanded major reforms in the early 1950s.

Congress was if anything even less interested in considering major social reforms in the early 1950s.[84] This was especially so after the unexpected death of Taft from cancer in July 1953. "Mr. Republican" had hardly been a liberal, but he was not a reactionary either, and the responsibility of working for a Republican President for the first time in his congressional career had strengthened his sense of teamwork. Before his death he and Ike had become fairly good friends, even playing golf together. His passing genuinely upset the President, who held the hand of Mrs. Taft and repeated, "I don't know what I'll do without him; I don't know what I'll do without him." Thereafter the Senate Republican leader was William Knowland of California, a humorless and much more conservative figure. Eisenhower found Knowland—and the GOP right wing that henceforth dominated the party in the Senate—deaf to his "Modern Republicanism," and he gradually despaired of healing the ideological rifts in his party. On one occasion he confided to his diary about Knowland, "In his case, there seems to be no final answer to the question, 'How stupid can you get?'"[85]

83. Wilfred Binkley, *American Political Parties: Their Natural History* (New York, 1958), 354; Richard Neustadt, *Presidential Power: The Politics of Leadership* (New York, 1976), 77–79; Adams, *First-Hand Report*, 27.
84. Burns, *Deadlock of Democracy*, 192–95.
85. Ambrose, *Eisenhower*, 333–34; Patterson, *Mr. Republican*, 588–98; Adams, *First-Hand Report*, 26–28.

For all these reasons Eisenhower's first term witnessed little significant domestic legislation. Aside from the extension of Social Security, which had the support of an increasingly well organized lobby for the elderly, the only important law approved was the Interstate Highway Act of 1956. This greatly increased federal subsidies for highway-building throughout the country.[86] Many critics were appalled by this act, among them Lewis Mumford, who complained, "The most charitable thing to assume about this action is that they [the Congress] didn't have the faintest notion of what they were doing."[87] Expanded in subsequent years, the building effort had often drastic effects on air quality, energy consumption, the ecology of cities, slum clearance and housing, mass transit, and railroads.[88] But it greatly appealed to ordinary Americans, especially the expanding millions who owned cars. And it offered vast economic benefits to a wide circle of interests, including the automobile, trucking, construction, and petroleum industries, to say nothing of real estate developers, motel and restaurant chains, shopping mall entrepreneurs, engineers, and many others in virtually every congressional district. Highly popular in Congress, it promised to give something to almost everybody. It had enormous long-range importance, establishing the basis for the American transportation system for the rest of the twentieth century and beyond.[89]

Congress otherwise acted with restraint. It refused to extend coverage under the minimum wage or to curb farm subsidies that benefited large commercial operators. Surplus crops continued to pile up, and millions of small farmers and farm workers, including millions of blacks, swelled an already substantial Great Migration into overburdened cities. Congress also did little to deal with poverty, education, or mounting urban problems. It seemed especially deaf to what was soon to become the greatest domestic controversy of all: race relations.[90] Many of these issues, having been slighted, provoked ever-greater social and political divisions by the late 1950s. In the 1960s they dominated a much more activist legislative agenda.

86. Bruce Seely, *Building the American Highway System: Engineers as Policy-Makers* (Philadelphia, 1987).
87. Richard Davies, *The Age of Asphalt: The Automobile, the Freeway, and the Condition of Metropolitan America* (New York, 1975), 133.
88. Mark Reutter, "The Lost Promise of the American Railroad," *Wilson Quarterly* (Winter 1994), 10–35.
89. Griffith, "Dwight D. Eisenhower," 106.
90. For race and civil rights in the 1950s, see chapter 13.

It would be stretching things to say that Americans were especially happy with the Republican party's domestic record during Eisenhower's first term. Democrats recaptured both houses of Congress in the 1954 elections, after which interest groups continued to command center stage on Capitol Hill. But there was little doubt that Americans continued to like Ike personally; his carefully cultivated popularity persisted at extraordinarily high levels. Above all, the majority of middle-class Americans, expectant about the future, seemed more interested in private concerns than in domestic reforms after 1954. The Korean War was fading from memory; McCarthy was silenced; the economy was booming. Although Eisenhower had not done much to promote some of these developments, notably the downfall of McCarthy, he was widely credited with ending the war and with calming the roiling partisanship that had disturbed the nation in the Truman years. To millions in the mid-1950s he remained an admirable, even heroic figure.

10

World Affairs, 1953–1956

On March 1, 1954, the United States tested the world's first hydrogen bomb at Bikini Atoll in the Marshall Islands. More awesome than scientists anticipated, it proved to be 750 times more powerful than the A-bomb dropped at Hiroshima. Radioactive debris from the blast spread across 7,000 square miles of the Pacific, including inhabited islands, and enveloped a small Japanese fishing boat, *Fukuryu Maru* (the *Lucky Dragon*) that was some ninety miles east of Bikini at the time. Radioactive ash rained down on the fishermen. Some lost their appetites and grew nauseous. Their skin turned darker, and sores broke out on their fingers and necks, which had been most exposed to the radiation. When the boat got back to Japan two weeks later, twenty-three of the crew were said to be suffering from radiation sickness. Fishermen on other Japanese boats returned to port and also complained of contamination. An outcry arose, peaking six months later when Aikichi Kuboyama, a *Lucky Dragon* fisherman, died. American authorities said he had been felled by hepatitis acquired from a blood transfusion, but his organs revealed pronounced effects of radiation. Admiral Lewis Strauss, chairman of the Atomic Energy Commission, stated that the fishermen had belonged to a "Red spy outfit."[1]

1. Allan Winkler, *Life Under a Cloud: American Anxiety About the Atom* (New York, 1993), 94–96; Jeffrey Davis, "Bikini's Silver Lining," *New York Times Magazine*, May 11, 1994, pp. 43ff; David Halberstam, *The Fifties* (New York, 1993), 345; John McCormick, *Reclaiming Paradise: The Global Environmental Movement*

276

Meanwhile, the Soviet Union was conducting its own atomic experiments. On September 14, 1954, military leaders exploded a Hiroshima-sized atomic bomb in the air above 45,000 Red Army troops and thousands of civilians near the village of Totskoye. This was in the Ural Mountains, 600 miles southeast of Moscow and within 100 miles of a million people. The test, which was not disclosed until 1991, aimed at ascertaining whether troops (who were advised that the blast was an "imitation" atomic explosion) could continue to fight under such conditions. Films of the event revealed that some of the soldiers, who were less than two miles from blast center, indeed managed to struggle through maneuvers amid the smoke and dust and 115-degree heat. But many wore little or no protective clothing, and their exposure to radiation was enormous. A documentary concerning the episode later concluded that a number of soldiers and villagers fell sick, went blind, or developed cancer and other illnesses attributed to the radiation.[2]

These two events were among the most shocking of the early Cold War years. But they were hardly unique. The United States exploded at least 203 nuclear weapons in the Pacific and in Nevada between 1946 and 1961 and another ninety-six in 1962, exposing an estimated 200,000 civilian and military personnel to some degree of radiation. The Russians, the French, and the British also conducted tests. Americans near the Nevada sites were rocked and startled by the blasts and flashes of light from the explosions. Thousands of people employed in clean-up operations, as well as "downwinders" in the Pacific and in western states, claimed to suffer from the effects of radioactivity as a result of the tests.[3]

Whether scientists, politicians, and military leaders in the 1940s and early 1950s should have done more to warn the world about radiation remains a debated issue years later. Some scientists at the time were worried, not only about dangerous fallout from the tests but also about hundreds of experiments in which radiation was deliberately released into the environment and in which human beings were unknowingly dosed or injected with radioactive substances in order to learn more about bodily

---

(New York, 1989), 51–52. The United States apologized to Japan and later gave $2 million as compensation to the Japanese fishing industry.

2. New York Times, Nov. 7, 1993.

3. Winkler, Life Under a Cloud, 91–93; J. Ronald Oakley, God's Country: America in the Fifties (New York, 1986), 359; New York Times, July 24, 1993, Jan. 11, 1994. After long denying medical claims from Americans in the West who said that atomic tests had given them cancer and other ailments, the Justice Department began to relent. By 1994 it had approved medical claims from 818 (of 1,460) people.

reactions. One such scientist, in 1950, warned the AEC that such experiments had "a little of the Buchenwald touch."[4] By the mid-1950s many people were growing alarmed by what they heard and read. News items reported the presence of radioactive substances in the soil and in foods and predicted that leukemia, birth defects, possibly even horrible mutations might develop from explosions that had occurred far, far away.

Leading Eisenhower administration officials tended publicly to ignore or to dismiss such alarming reports. The evidence of danger, they said, was sketchy and debated by scientists. Many experts then believed that radioactive substances had beneficial potential: X-ray machines commonly measured foot sizes in shoe stores. Atomic testing and other kinds of experimentation, they added, were essential to national security and medical research. It is now clear, however, that these experts underestimated the dangers from the experiments. It is also clear that officials in charge of atomic testing knowingly exposed human beings to nuclear fallout. The AEC staged a well-orchestrated propaganda campaign on behalf of the peace-related blessings of atomic power, and Ike himself activated America's first commercial nuclear power plant in 1955.[5] The AEC attempted to suppress evidence of long-range fallout problems as they became more evident by the mid-1950s.[6]

Those officials who were kept apprised of what the weapons could do, however, nonetheless grew nervous. Eisenhower was one. Following a briefing in 1955 on the outcome of a hypothetical atomic war with the Russians, he estimated privately that the Soviet Union (which lagged in the nuclear arms race) would incur three times as many casualties as would the United States but that 65 percent of Americans would require medical care, most of whom would be unable to get it. He observed, "It would literally be a business of digging ourselves out of the ashes, starting again."[7]

This was indeed an unthinkable prospect—the most horrifying of the many calamities that would befall the world if the Cold War could not be contained now that the major protagonists were amassing stockpiles of

4. *New York Times*, Dec. 28, 31, 1993, Oct. 12, 22, 1994, and Aug. 20,1995. The Nazis conducted gruesome medical experiments on human beings at Buchenwald concentration camp and other places during World War II.
5. At Indian Point, New York, on the Hudson River.
6. Michael Smith, "Advertising the Atom," in Michael Lacey, ed., *Government and Environmental Politics: Essays on Historical Developments Since World War Two* (Washington, 1989), 233–62; *New York Times*, March 15, 1995.
7. John Gaddis, *Strategies of Containment: A Critical Appraisal of Postwar American National Security Policy* (New York, 1982), 174.

thermonuclear weapons. Dealing with this new world, which was considerably more frightening than the one that had confronted political leaders in the 1940s, was the most awesome task facing the Eisenhower administration. Ike's performance in the areas of foreign and defense policy could determine the fate of the earth.

EISENHOWER'S EARLY ACTIONS in these respects seemed likely to intensify the Cold War. Repeating the tough anti-Communistic messages of his election campaign, he devoted much of his inaugural address to denouncing Communism. "Freedom," he said, "is pitted against slavery; lightness against dark." In his State of the Union message he added that the United States would "never acquiesce in the enslavement of any people."[8] When Stalin died in early March, Ike made little effort to develop better diplomatic contacts with the new Soviet leadership. His neglect, which was studied, may have been unfortunate, for the new Soviet premier, Georgi Malenkov, seemed eager for contacts.[9]

Much later, when historians looked at once classified documents, it became clear that Eisenhower was wiser and subtler than his moralistic rhetoric suggested. He recognized, for instance, that world Communism was not monolithic, that the Soviet Union had severe internal problems, that Communist ideology was not the driving force behind Russian behavior, and that Soviet leaders did not intend to start a war. Conflicts between the Russians and the Chinese, he understood, were serious. Tensions with both nations must be reduced.

Eisenhower occasionally expressed these feelings to trusted aides, such as Emmet Hughes. "We are in an armaments race," he lamented in March 1953. "Where will it lead us? At worst, to atomic warfare. At best, to robbing every people and nation on earth of the fruits of their own toil." A month later he spoke out for a limit on arms and for international control of atomic energy. In December 1953 he delivered an "Atoms for Peace" speech to the UN. It called on the nuclear powers—the United States, the USSR, and the UK—to turn over some of their fissionable materials to an international agency.[10]

These efforts, however, were sporadic and not followed through.

8. Oakley, *God's Country*, 148; Marty Jezer, *The Dark Ages: Life in the United States, 1945–1960* (Boston, 1982), 60.

9. Robert Divine, *Eisenhower and the Cold War* (New York, 1981), 108.

10. Emmet Hughes, *The Ordeal of Power: A Political Memoir of the Eisenhower Years* (New York, 1963), 103–5; David Patterson, "Pacifism, Internationalism, and Arms Limitation," in Stanley Kutler, ed., *Encyclopedia of the United States in the Twentieth Century* (New York, 1995); Stephen Ambrose, *Eisenhower: Soldier and President* (New York, 1990), 323–25.

Some, such as Atoms for Peace, were at least in part propagandistic—the proposal would have weakened the Soviet Union more than the United States, which was ahead in nuclear development—and were ignored by the USSR. Eisenhower tended instead to sustain the harsh, sometimes nearly Manichean rhetoric of the campaign and of his inaugural address, especially in the first two years of his administration.

Eisenhower talked tough for many reasons. One was to reassure anti-Communist allies abroad of America's unbending resolve to stay the course. To have done otherwise, he thought, would have weakened support for NATO, which was then seeking to build up military forces and to embrace West Germany. Eisenhower also had to deal with hard-liners at home, McCarthy among them, who were stronger than ever before in Congress. Influential politicians on Capitol Hill worried not only about Soviet activity but also about maintaining the defense contracts that had become vital to the economic health of their districts during the Korean War. Many of the influential senators of the 1950s—GOP leaders William Knowland of California and Everett Dirksen of Illinois, Georgia Democrat Richard Russell (a power on the Armed Services Committee), Lyndon Johnson of Texas—ardently supported high levels of defense spending and firm foreign policy. So did important businessmen and many labor union leaders. The more than $350 billion in military spending during the Eisenhower era bolstered a host of corporations and defense workers in the country.

Above all, Eisenhower talked tough because neither he nor anyone else could be sure of Soviet or Chinese intentions. The Korean War, after all, was still killing American soldiers in July 1953. A month later the Soviets exploded their first thermonuclear device (not a bomb). In 1954 and 1955 high-level advisory committees, including the National Security Council, apprised the President of what they thought were significant increases in Soviet nuclear power. The Soviets, one such report warned in early 1955, then had the capacity to deliver a "knockout" attack on the United States.[11] Though Ike knew that America had far superior nuclear resources, he could not afford to let his guard down in such circumstances. Like all American Presidents in the Cold War era, he had to take seriously an obviously powerful adversary. In doing so he frequently felt obliged to issue dire warnings about the dangers.

The President, like most Americans after years of Cold War hostility, in

11. H. W. Brands, "The Age of Vulnerability: Eisenhower and the National Security State," *American Historical Review*, 94 (October 1989), 974. This was a report produced by a committee headed by James Killian, president of the Massachusetts Institute of Technology.

fact reflected a consensus that the Soviets were unbending and that signs of softness in dealing with them were tantamount to "appeasement." Most liberals and conservatives agreed on these apparently unchanging facts of the world order. They also believed that the United States, the world's greatest democracy, had a mission to promote democratic ideals throughout the world. For these reasons, too, Eisenhower did little, especially at first, to try to soften Cold War tensions. So while he possessed a subtler and more sophisticated knowledge of world affairs than many contemporaries, he seldom exposed his awareness to the public. He could have done more than he did during his presidency to educate the American people about the dangers of the fast-moving nuclear arms race.[12]

Eisenhower's top appointments in the fields of foreign and defense policy reflected his anti-Soviet priorities. One of these was Admiral Arthur Radford, who replaced General Bradley as Chairman of the Joint Chiefs of Staff in May 1953. Radford was a favorite of Republican conservatives, who never forgave Bradley for opposing MacArthur in 1951. A strong advocate of atomic weapons development, Radford (and other leading military advisers in the 1950s) brought a larger naval-air emphasis, especially the use of aircraft carriers, to military planning. Radford proved a ready advocate of the use of force abroad, mainly in Asia. On five occasions in the next two years (three concerning Indochina, two concerning the Nationalist-held islands of Quemoy and Matsu off the Chinese mainland) Radford pressed for American attacks, possibly including the use of nuclear weapons. Ike overruled him all five times.[13]

Charles E. Wilson, the President's choice for Defense Secretary, was another strong Cold Warrior. Wilson appealed to the President because he had been head of General Motors, the nation's largest defense contractor. Ike hoped Wilson could bring businesslike economies to the Pentagon and control the interservice rivalries that still plagued defense planning. In his confirmation hearings, however, Wilson denied that he would have a conflict of interest, even though he owned $2.5 million in GM stock and had $600,000 due him in deferred compensation. Wilson further proclaimed at the hearings that "what was good for our country was good for General Motors, and vice versa"—a comment that opponents twisted to "what was good for General Motors was good for the country."[14]

Wilson was confirmed but did not manage to control the services. (This

---

12. Gaddis, *Strategies of Containment*, 140–45.
13. Ambrose, *Eisenhower*, 379; Halberstam, *Fifties*, 396–98.
14. John Steele Gordon, "The Ordeal of Engine Charlie," *American Heritage*, Feb./ March 1995, pp. 18–22.

was never easy.) Moreover, he lost influence within the administration, in part because he could not curb his tongue. Perhaps his most memorable faux pas took place during the 1954 election campaigns, when he opposed further government aid to the unemployed by blurting, "I've always liked bird dogs better than kennel-fed dogs myself—you know, one who'll get out and hunt for food rather than sit on his fanny and yell."[15] Well before then, however, government associates found him far too blunt for his own good. At a Cabinet meeting one aide listened to Wilson, then scribbled a note to another, "From now on I'm buying nothing but Plymouths." It was said that Wilson while at GM had invented the automatic transmission so that he would always be free to drive with one foot in his mouth.[16]

Eisenhower's most important appointee, Secretary of State John Foster Dulles, seemed at first to be a sound, indeed almost inevitable choice for the position. Dulles was the grandson of John Foster, President Benjamin Harrison's Secretary of State, and the nephew of Robert Lansing, who had held the post under Woodrow Wilson. Dulles had been personally concerned with international relations for almost fifty years and had attended the Paris peace conference after World War I. He had then become an influential attorney in New York and was part of the Establishment network of well-placed lawyers and bankers who formulated postwar American foreign policy. In selecting Dulles, Eisenhower told his chief aide, Sherman Adams, "Foster has been in training for this job all his life." He reminded Emmet Hughes, "There's only one man I know who has seen *more* of the world and talked with more people and *knows* more than he does, and that's *me*."[17]

From the beginning, however, Dulles became a lightning rod for criticisms of Republican foreign policies. This was in part because he seemed extraordinarily influential. Some contemporaries, indeed, were sure that Dulles was the power behind the throne and that Ike merely acquiesced in whatever Dulles devised. This was not the case: Eisenhower made all important policy decisions himself. Indeed, the President was at times bored and irritated by Dulles, who tended to be preachy in meetings. The Secretary of State, Ike said on one occasion, had "a lawyer's mind" and tended to act like "a sort of international prosecuting attorney."[18] But

15. Ambrose, *Eisenhower*, 375. Ambrose notes that bird dogs do not hunt for food.
16. Hughes, *Ordeal*, 77; Marquis Childs, *Witness to Power* (New York), 177. The scribbler was Jerry Persons, the recipient Hughes. The quip about Wilson and automatic transmission is attributed to various journalists, usually James Reston.
17. Hughes, *Ordeal*, 251; Ambrose, *Eisenhower*, 289.
18. Richard Immerman, "Eisenhower and Dulles: Who Made the Decisions?" *Political Psychology*, 1 (1979), 21–38; Gaddis, *Strategies of Containment*, 160.

critics of Dulles were correct in recognizing that Eisenhower relied heavily on his Secretary, who was a hard worker, knowledgeable, and wholly loyal in trying to carry out the President's goals. For these reasons, and because Eisenhower did not always monitor his subordinates closely, Dulles enjoyed considerable leeway and initiative. He held office, enjoying the President's confidence, until he grew ill with cancer and had to resign in April 1959. Only then did Eisenhower step forward more boldly on his own as the spokesman for American foreign policy interests.

Critics who took aim at Dulles fired off many grievances. They emphasized first of all that he was moralistic and self-righteous. This was often true. Dulles, the son of a Presbyterian minister, was influential in national church affairs. His strong Christian faith strengthened his distaste for Communism, which he deplored as atheistic as well as unprincipled. Moreover, Dulles seemed humorless, at least on the job. Self-assured and pompous, he had a habit of looking up toward the ceiling (some critics thought toward God), hands calmly folded on his desk, while talking (critics said pontificating) at considerable length. Other critics simply described his manner as "Dull, Duller, Dulles."[19]

What most irritated liberal opponents was Dulles's apparently inflexible and ideological anti-Communism. This helped him to acquiesce in McCarthy-inspired efforts to purge the State Department of alleged subversives and appeasers. I. F. Stone, the liberal journalist, called him "McCarthy's Secretary of State."[20] While this charge was inaccurate, the critics were mostly correct in focusing on his anti-Communist zeal, for Dulles—more than most contemporary political leaders—believed that Communist ideology (rather than strategic interests) determined Soviet behavior and that the Soviet Union therefore had a grand design.[21] Perceiving issues in ideological terms, Dulles could be pickily legalistic when dealing with other political leaders. Some of these leaders were infuriated by his manner. Churchill said that Dulles was "the only case of a bull I know who carried his own china shop with him." The journalist James Reston added that Dulles "doesn't stumble into booby traps; he digs them to size, studies them carefully, and then jumps."[22]

19. Books on Dulles include Ronald Pruessen, *John Foster Dulles and the Road to Power* (New York, 1982); Frederick Marks, *Power and Peace: The Diplomacy of John Foster Dulles* (Westport, Conn., 1993); and Richard Immerman, ed., *John Foster Dulles and the Diplomacy of the Cold War* (Princeton, 1990).
20. I. F. Stone, *The Haunted Fifties, 1953–1963* (Boston, 1963), 99–101. See also 14, 15, 263 for other sour references to Dulles.
21. Gaddis, *Strategies of Containment*, 136–37.
22. Stephen Whitfield, *The Culture of the Cold War* (Baltimore, 1991), 7–9.

Analysis of Dulles's ideas and activities by historians has slightly soft-ened this acid portrait. Dulles was in fact politically shrewd. Anxious to escape the vilification from the GOP Right that had savaged Acheson, he worked hard at protecting his standing with conservatives in Congress, a very important consideration. It is also clear that Dulles was no more inflexible than Acheson—or than the Truman administration generally, which had initiated no serious negotiating with the Soviet Union (or China) in many years. Dulles's style may have seemed more rigid, but the end result was much the same: more hardening of the Cold War.[23]

These reminders are useful. Still, few contemporaries saw a flexible, subtle side to Dulles. Publicly—and in negotiations—he *was* mostly stern and unbending, with a harsh edge that not even Acheson had matched. Indeed, Dulles seemed an eager spokesman for a new administration that regularly denounced the Democrats for being "soft" on Communism. Like anti-Communist conservatives on the Hill, he seemed prepared to push the Cold War, already frigid, into a deep freeze from which it might never emerge.

The Central Intelligence Agency headed by Foster Dulles's younger brother Allen was equally anti-Communist. The agency, created in 1947, had grown slowly prior to the Korean War. But it had received authoriza-tion to conduct covert operations as early as 1948, using it to intervene at that time in Italian politics, and it grew rapidly in the early 1950s. By 1952 its budget had risen to $82 million, its personnel to 2,812 (plus an addi-tional 3,142 overseas "contract" personnel), and its number of foreign stations from seven to forty-seven. Under Eisenhower and Allen Dulles, a pipe-smoking bon vivant who was charming, popular with Congress, and well connected socially as well as politically, it grew into an important government agency.[24]

The CIA had its first significant impact early in the Eisenhower years. In the summer of 1953 it led a successful coup in Iran against Prime Minister Muhammad Mussadegh, who had earned the enmity of British leaders by nationalizing their oil interests in 1951. The coup replaced Mussadegh with the pro-Western Muhammad Reza Shah Pahlevi, who agreed to a new charter that gave British and American oil interests 40 percent each of Iranian oil revenues. The Shah received a package of

---

23. Robert Divine, "John Foster Dulles: What You See Is What You Get," *Diplo-matic History*, 15 (Spring 1991), 284–85.
24. Peter Grose, *Gentleman Spy: The Life of Allen Dulles* (Boston, 1994); Gaddis, *Strategies of Containment*, 157.

American economic aid worth $85 million.[25] In June 1954 the CIA intervened again, this time in Guatemala in an effort to help rebels overthrow Colonel Jacobo Arbenz Guzmán, the legally installed leader of the country. Arbenz Guzmán's mistake had been to promote land reform by expropriating (with compensation) significant acreage of the American-owned United Fruit Company. Unbeknownst to the American people, CIA pilots joined in bombing raids that may have helped the coup to succeed. Eisenhower, fearing the spread of Communism in Central America, was highly pleased with the result. "My God," he told his Cabinet, "just think what it would mean to us if Mexico went Communist."[26]

Because both of these coups were quickly and rather easily accomplished—and because some of the CIA's involvement remained secret—they did not attract great attention from the American press. This was unfortunate for several reasons. First, the coups exacerbated internal divisions in these countries, with disastrous long-range consequences for the people there. Second, the coups indicated the willingness of reporters at that time uncritically to accept obfuscatory CIA cover stories: it was not until the late 1950s, when a U-2 reconnaissance plane under control of the CIA was shot down over the Soviet Union, that significant numbers of reporters began to display a healthy distrust of self-serving government handouts.[27] Third, it was obvious that the coups involved well-placed economic interests. A thorough public discussion of these interests would have been useful in exposing the material forces that helped to drive America's Cold War behavior. Fourth, the coups convinced the CIA and other government officials that covert actions were easily carried out. In the next few years it conducted other such actions in Japan, Indonesia,

25. Bruce Kuniholm, "U.S. Policy in the Near East: The Triumphs and Tribulations of the Truman Age," in Michael Lacey, ed., *The Truman Presidency* (Washington, 1989), 299–338 ; Divine, *Eisenhower*, 73–78; Ambrose, *Eisenhower*, 332–33.
26. Richard Immerman, *The CIA in Guatemala: The Foreign Policy of Intervention* (Austin, 1987); Stephen Schlesinger and Stephen Kinzer, *Bitter Fruit: The Untold Story of the American Coup in Guatemala* (Garden City, N.Y., 1982); Stephen Ambrose, *Eisenhower: The President* (New York, 1984). (This is Volume 2 of Ambrose's larger biography of Eisenhower. Other references from Ambrose on Eisenhower in this chapter refer to the previously cited one-volume version [1990]).
27. Joseph Alsop, a well-connected syndicated columnist, had advance knowledge of the CIA's plan in Iran but kept quiet. *New York Times*, Jan. 23, 1994. See chapter 14 for discussion of the U-2 Affair.

and the Belgian Congo. The bravado that such efforts engendered was to prove disastrous in later years.[28]

The coups were revealing in other ways as well. Americans who read about them seemed delighted with what they were allowed to know of CIA activity. The CIA leader in Iran, Kermit Roosevelt, TR's grandson, was acclaimed as a hero.[29] Americans seemed unconcerned that the interventions violated sovereign rights. Foster Dulles was hardly challenged when he went on radio and TV following the coup in Guatemala to call it a "new and glorious chapter for all the people of the Americas."[30]

Above all, the coups indicated the power of Cold War thought and action within the Eisenhower administration. Top officials argued that Communist elements linked to Moscow were the key forces behind both Mussadegh and Arbenz Guzmán. This was not so. Though Mussadegh belatedly turned to the Iranian Communist party for help in order to bolster himself, he was fundamentally a nationalist. Arbenz Guzmán was a reformer, not a Communist. But the Dulles brothers easily convinced themselves—and many others—that Communism lay at the root of international unrest. The coups in Iran and Guatemala revealed that key figures in the Eisenhower administration, perceiving the world in black and white, had at best a dim awareness of the appeal of nationalism and anti-colonialism throughout the world. Then and later American officials would demonstrate this profound misunderstanding.

Nothing did more to sharpen the tough-minded image of the Eisenhower administration than Foster Dulles's pronouncement of a "massive retaliation" policy in January 1954. The "Free World," he said, had properly tried to contain Communism with measures such as the Marshall Plan, the Berlin Airlift, and the dispatch of troops to Korea. But these were inadequate, "emergency" reactions. Moreover, the "Free World" could not match "the mighty land power of the Communist world." Instead, it must take the initiative and rely on "massive retaliatory power." The nation should "depend primarily upon a great capacity to retaliate, instantly, by means and at places of our choosing." This would mean "more basic security at less cost." Dulles went on to say that warnings of such massive retaliation—nuclear weapons—had brought the

28. John Prados, *Presidents' Secret Wars: CIA and Pentagon Covert Operations Since World War II* (New York, 1987); *New York Times*, Oct. 9, 1994 (concerning Japan).

29. Kermit Roosevelt, *Counter-Coup: The Struggle for the Control of Iran* (New York, 1979).

30. Halberstam, *Fifties*, 387.

Chinese to heel in Korea in 1953. The Secretary seemed to be proposing that the administration brandish nuclear weapons whenever confronted by an enemy.[31]

Dulles was not simply indulging in his fondness for stern and grandiloquent phrases. On the contrary, the National Security Council, which became much more important in policy-making during the Eisenhower administration, had reconsidered defense doctrine in 1953 and had approved NSC-162/2 on October 30. This document emphasized the need for a nuclear-based strategy and for cost-cutting (mainly of ground-based forces) in defense spending. Eisenhower had read Dulles's speech in advance and had apparently penned in the key passage calling for a policy based on a "capacity to retaliate, instantly, by means and at places of our own choosing."[32] "Massive retaliation"—the "New Look," contemporaries called it—was carefully conceived administration policy.

The New Look in fact nicely complemented existing defense initiatives, which were beginning to rely heavily on the Strategic Air Command (SAC). By 1954 the SAC, still headed by the tough-talking, fiercely anti-Communist General Curtis LeMay—a kind of airborne George Patton—was replacing its propeller-driven B-36 bombers with jet-propelled B-47s. These could fly at speeds of up to 600 miles per hour and had an effective range (when refueled in the air) of nearly 6,000 miles. LeMay presided over a rapid expansion of his force between 1948 and 1955, by which time the United States had some 400 B-47s plus another 1,350 planes capable of dropping nuclear weapons on the Soviet heartland. The Soviets had perhaps one-tenth as many that could bomb the United States.[33] Given this enormous edge, it seemed only logical for the Eisenhower administration to announce a policy that rested heavily on air power and atomic weapons.

Eisenhower backed massive retaliation for two other military reasons. First, it was obvious that the Soviets possessed a very large advantage in ground forces. As Dulles pointed out, there was no way that the United States could realistically hope to catch up in that domain. Second, Eisenhower knew that missiles carrying nuclear warheads were soon to become main-line military weapons. In pursuit of such weapons he qui-

31. Samuel Wells, "The Origins of Massive Retaliation," *Political Science Quarterly*, 96 (Spring 1981), 31–52; Russell Weigley, *The American Way of War: A History of United States Military Strategy and Policy* (New York, 1973), 404.
32. Brands "Age of Vulnerability"; Divine, *Eisenhower*, 38.
33. John Diggins, *The Proud Decades: America in War and Peace, 1941–1960* (New York, 1988), 146.

etly but aggressively supported research and development of the Atlas, Polaris, and Minuteman programs, all of which were well underway by the late 1950s, and of light warheads for such missiles. America's support of bombs and warheads was intense, resulting in growth in the number of nuclear weapons available to United States forces from around 1,500 in January 1953 to 6,000 or so six years later. This was an increase of 4,500, or 750 a year, or two or more per day. The effort, which was far greater than militarily necessary, gave the United States a wide edge in missile development by the late 1950s.[34]

In supporting massive retaliation Eisenhower and Dulles adopted historically familiar American approaches to defense: faith in high technology, and aversion to large standing armies in times of peace. These were politically attractive approaches. They also had several more precise goals in mind. The policy, they believed, widened American initiative by enabling quick retaliation—nuclear if necessary—on an aggressor's own territory. The United States, for instance, could blast the Soviet Union itself instead of using troops (which were expensive to maintain and might get killed) to deter Communist trouble-making wherever it might occur— Greece and Turkey? Berlin? Korea?—throughout the world. In this sense, they thought, the new policy was both cheaper and safer than NSC-68 (1950), which had in effect called for fighting aggression wherever it occurred. Second, massive retaliation was supposed to keep an enemy guessing. Eisenhower and Dulles hoped that adversaries, like the Chinese in Korea, would think twice before defying the United States.

To Eisenhower the new doctrine promised above all to promote his vision of the good society at home. Reliance on massive retaliation would enable reductions in the size of the army, which would have been very expensive to maintain at Korean War levels, and therefore to cut costs. "More bang for the buck," contemporaries said. The President was especially anxious to balance the budget because he feared inflation, which he was sure would badly damage the economy and widen divisions in American society. These, in turn, would weaken the standing of capitalism in the global battle against Communism.

In his skepticism about the long-run capacity of the American economy

34. Ambrose, *Eisenhower*, 478; William O'Neill, *American High: The Years of Confidence, 1945–1960* (New York, 1986), 231–32, 272. The Soviets were first to launch successfully an artificial satellite, *Sputnik*, into orbit, in October 1957, but this much-noted success indicated mainly that they had the advantage in terms of thrust, a sign of their overreliance on heavy, awkward warheads. See chapter 14.

to tolerate high levels of military expenditure, Eisenhower differed sub-stantially from bullish contemporaries—and from his successors in the White House. Holding grand expectations about the potential for Ameri-can influence in the world, they felt confident that government could also promote rapid economic growth at home. They were much readier to spend generously for both defense and domestic programs. Eisenhower, too, was a Cold Warrior who wanted to lead the "Free World" against Communism. But he placed a considerably higher premium on the need for fiscal restraint, the key (he thought) to social stability. His tenacity in support of prudent financing, whether for defense or social programs, stamped a definite character on his presidency.

Eisenhower was also afraid that high levels of defense spending would give too much power to military leaders and defense contractors. The result could be a "garrison state" that distorted priorities. "Every gun that is made," he said in 1953, "every warship launched, every rocket fired signifies, in the final sense, a theft from those who hunger and are not fed, those who are cold and not clothed."[35] This did not mean that he believed in large-scale government social programs to relieve suffering; far from it, for those, too, would unbalance the budget. But he did worry that heavy spending for arms would feed what he later called the "military-industrial complex."[36]

The quest to contain costs under the New Look enjoyed modest success over the next few years. Thanks mainly to partial demobilization follow-ing the Korean War, federal spending for defense decreased from $50.4 billion in fiscal 1953 to $40.3 billion in 1956 before creeping up to $46.6 billion in 1959. After 1954 it also decreased slowly as a percentage of the federal budget and as a percentage of GNP (from 14 percent of GNP at the peak of the Korean War to around 9 percent by 1961).[37] All the armed services took cuts in personnel, especially the army, in which Eisenhower had spent most of his adult career. It lost 671,000 men and women between 1953 and 1959—a slashing that brought the number to 862,000 and that outraged many of Eisenhower's old friends and colleagues. Two of these angry generals, Matthew Ridgway and Maxwell Taylor, were army chiefs of staff in the 1950s; both wrote books in retirement that

35. Gaddis, *Strategies of Containment*, 133.
36. Sherman Adams, *First-Hand Report: The Story of the Eisenhower Administration* (New York, 1961), 154–55; Ambrose, *Eisenhower*, 311–13.
37. The percentage vis-à-vis GNP dropped a little more in the early 1960s, to around 8 percent in 1965. This reflected the considerable growth in the civilian economy in those years, not a decline in defense spending (which accelerated rapidly).

protested the reductions.[38] The cuts prompted charges that the United States would lose the flexibility to cope with local crises—"limited wars"—throughout the world. Ike, however, was determined to control costs and to curb the influence of the military-industrial complex. Sure that air and naval power offered sufficient security (especially when missiles became operational), he successfully stood his ground. Only a general with his commanding popularity and expertise could have managed this policy without severe political damage amid the Cold War fears of the 1950s.

Opponents of massive retaliation leveled other complaints at the new policy. Some insisted correctly that it amounted to a strategy of nuclear "blackmail." Ike indeed resorted to blackmail against the People's Republic of China in standoffs over the offshore islands of Quemoy and Matsu in 1955.[39] Other critics complained that massive retaliation was saber-rattling of the most dangerous sort, that it frightened allies, and that it would accelerate the arms race. More "bang for the buck" would be matched by a Soviet response of more "rubble for the ruble." Finally, they said, the policy was simply not credible. Potential aggressors, far from being deterred, would act with impunity, confident that the United States would not dare to use nuclear weapons in the vast majority of regional conflicts. The Soviets, one critic complained, would be able to "nibble the free world to death piece by piece."[40]

Eisenhower did not bend under these criticisms; he never repudiated massive retaliation or the New Look, and he avidly supported development of missiles and nuclear power. But both he and Dulles grew sensitive to the need for careful application of the policy. Eisenhower in fact came to regret the rhetoric. He told a press conference in February 1954, "I don't think that big and bombastic talk is the thing that makes other people fear." Later in 1954 he reminded Dulles that "when we talk about . . . massive retaliation, we mean retaliation against an act that means irrevocable war."[41] Dulles, writing about massive retaliation in *Foreign Affairs* in April 1954, emphasized that it was "not the kind of

38. Matthew Ridgway, *Soldier: The Memoirs of Matthew B. Ridgway* (New York, 1956); and Maxwell Taylor, *The Uncertain Trumpet* (New York, 1960).
39. Discussed later in this chapter.
40. Quote in Brands, "Age of Vulnerability," 979. These debates are also well covered in Wells, "Origins of Massive Retaliation"; Gaddis, *Strategies of Containment*, 147–51, 161, 251, 300–301; and Arthur Schesinger, Jr., "The Ike Age Revisited," *Reviews in American History*, 4 (March 1983), 1–11.
41. Gaddis, *Strategies of Containment*, 150.

power which could most usefully be evoked under all circumstances." He went on to say that he appreciated the need for other weapons.[42]

Still, the rhetoric was incendiary, and the critics were correct in lamenting it. The policy did frighten America's allies, and it did not help whatever chances may have existed following the death of Stalin for renewed dialogue with the Soviet Union, which thereupon rushed to catch up with the United States. Massive retaliation may in fact have played into the hands of Soviet hard-liners looking for reasons to accelerate their own weapons development. Finally, the rhetoric helped to further an already heated climate of domestic opinion. Massive retaliation did nothing to facilitate efforts at easing the Cold War in an age of thermonuclear capability.

It is nonetheless striking that most American political figures in the mid-1950s—Democrats as well as Republicans—adopted the same assumptions that moved Eisenhower and Dulles. Like Eisenhower, they talked as if they were certain of Soviet aggressiveness—even if privately they were not so sure about that. Though some of them, such as Adlai Stevenson, later sought to stop thermonuclear tests, most leaders tended to demand more, not less, defense spending—as well as a good deal more military flexibility. Amid the powerful anti-Communist consensus that dominated American life in the mid-1950s, voices for acting "tough with the Russians" all but silenced counsels of restraint.

NOTHING MORE CLEARLY revealed the nature of Eisenhower's conduct of world affairs than a series of crises that threatened to run out of control between 1954 and 1956. These concerned, in order, Indochina, Quemoy and Matsu, Suez, and Hungary. In several of these cases the administration seemed to toy with the idea of American military intervention. But ultimately cautious management by Eisenhower and his advisers—and good luck—enabled the United States to control its involvement. Ike's prudence under the test of these crises forms the nub of his enhanced reputation in later years.

In 1945 many native nationalists in Indochina had regarded the United States with admiration. President Roosevelt had periodically criticized colonialism and seemed prepared to put pressure on France, which had governed the area since the late nineteenth century, to surrender or soften its claims for repossession once the Japanese (who overran the region during the war) had been evicted. Although Roosevelt backed off from his

42. Wells, "Origins of Massive Retaliation," 36.

anti-colonial rhetoric in early 1945, Ho Chi Minh, the leading nationalist from Vietnam (part of Indochina), still looked to the United States for support and inspiration. When Ho's forces (working with an American intelligence unit) managed to take control of Hanoi in September 1945, he proclaimed Vietnamese independence in a message inspired by the Declaration of Independence: "We hold these truths to be self-evident. That all men are created equal." Later that day United States Army officers stood with Vietnamese patriots, listening proudly to the playing of "The Star-Spangled Banner" as American warplanes flew overhead.[43]

Ho was soon to be rudely disappointed. The French, assisted by the British, reclaimed southern Vietnam and in November 1946 shelled the northern port city of Haiphong, killing 6,000 civilians. Open warfare then broke out between the French and the Vietminh, as Ho's forces were called. Fighting was to savage Vietnam (and often neighboring Laos and Cambodia, the other parts of Indochina) for decades thereafter.

American officials during the late 1940s did not pay the warfare much attention. Some recognized that Ho was a popular nationalist leader and that the French were corrupt and often brutal. But France was needed as an ally in the developing European struggle against Communism. Moreover, although Ho was first and foremost a nationalist, he was also a Moscow-trained Communist. Then and later this basic fact was the single most important determinant of American policy toward the region. Presidents from Truman through Nixon—from the 1940s into the 1970s— insisted that Vietnam must not be allowed to fall to the Communists. As Secretary of State Marshall put the matter in February 1947, "We are not interested in seeing colonial administrations supplanted by [a] philosophy and political organization . . . controlled by the Kremlin."[44]

Communist domination of Vietnam, American officials believed, would be bad for several reasons. Militarily and economically the area was thought to provide a "natural invasion route into the rice bowl of Southeast Asia."[45] But real estate or resources were not America's major concerns. Rather, the key to American thinking was what Eisenhower, using an already worn metaphor, called in 1954 the "falling domino principle"

43. George Herring, *America's Longest War: The United States and Vietnam, 1950–1975* (Philadelphia, 1986), 3.
44. Robert McMahon, "Toward a Post-Colonial Order: Truman Administration Policies Toward South and Southeast Asia," in Lacey, ed., *Truman Presidency*, 339–65.
45. State Department team memo, 1950, cited in William Chafe, *The Unfinished Journey: America Since World War II* (New York, 1991), 257.

and what other leaders labeled "credibility." If a local communist like Ho Chi Minh could topple the domino of Vietnam, nearby dominos—Thailand, Burma, Malaysia, Indonesia, maybe even Australia, New Zealand, India, and Japan—might fall next. Such a domino effect would not only deprive the "Free World" of resources and bases; it would also demonstrate that America was a paper tiger—loud but not "credible" when a crisis arose.

For these reasons the Truman administration sided with the French, who in February 1950 established Vietnam, Laos, and Cambodia as semi-autonomous "free states" within a French Union. Bao Dai, the former emperor of Annam (a part of Vietnam within Indochina), was recognized by the United States and its Western allies as the puppet head of Vietnam. The USSR, China, and other Communist states recognized Ho. The United States stepped up military aid to the French in the area, especially after the outbreak of the Korean War, which seemed to prove the aggressive intent of Communism—Chinese as well as Vietnamese—throughout Asia.[46] The aid amounted to 40 percent of French military costs by January 1953 and totaled $2.6 billion between 1950 and 1954.[47]

The Eisenhower administration continued this policy, increasing the aid to 75 percent of the cost of the war by early 1954, and for the same basic reasons. It was anxious to get France to join the European Defense Community (EDC), a military arm of NATO, and therefore tried not to antagonize the French government. In this way, as in many others, American policy in Southeast Asia was inextricably bound to policies in Europe and to overall Cold War strategy. Far-off Vietnam, considered relatively unimportant in itself, was both a domino and a pawn on the world chessboard.[48]

The French, however, were losing badly to rebel forces led by the resourceful Vo Nguyen Giap, the Vietminh commander-in-chief. Then and later the lightly armed, lightly clad Vietminh soldiers, enjoying nationalistic support from villagers, fought bravely, resourcefully, and relentlessly—incurring huge casualties—to reclaim their country. By

46. Historic animosities divided China and neighboring Vietnam, and Ho deeply distrusted Mao. The Chinese, however, did offer Ho large stocks of weapons and sanctuary during his fight with the French.
47. Stephen Ambrose, *Rise to Globalism: American Foreign Policy Since 1938* (New York, 1985), 140–45; Herring, *America's Longest War*, 11–42.
48. David Anderson, *Trapped by Success: The Eisenhower Administration and Vietnam, 1953–1961* (New York, 1991), 154; Lloyd Gardner, *Approaching Vietnam: From World War II Through Dienbienphu, 1941–1954* (New York, 1988).

contrast, the French army was poorly led. Its commanders were contemptuous of Giap and his guerrilla forces and vastly overrated the potential of their firepower. Ike dismissed the French generals as a "poor lot." General Lawton Collins, a top American adviser, said that the United States must "put the squeeze on the French to get them off their fannies." Nothing of that sort happened, and the French, hanging on to major cities such as Hanoi and Saigon, foolishly decided in early 1954 to fight a decisive battle at Dienbienphu, a hard-to-defend redoubt deep in rebel-held territory near the border with Laos.[49]

By then various of Ike's advisers were growing anxious to engage the United States in rescue of the French. One was Vice-President Nixon, who floated the idea of sending in American ground forces. Another was chief of staff Radford, who urged massive strikes, possibly with tactical nuclear weapons, from American bombers and carriers.[50] General Nathan Twining, air force chief of staff, favored dropping "small tactical A-bombs." The result, he said, would have been to "clean those Commies out of there and the band could play the Marseillaise and the French would come marching out of Dienbienphu in fine shape."[51]

At times Eisenhower seemed tempted to involve American military forces in Vietnam. The region, he said in January 1954, was a "leaky dike." But it is "sometimes better to put a finger in than to let the whole structure wash away." On April 7, with the French in desperate straits at Dienbienphu, he gave his version of the domino theory: "You have a row of dominos set up. You knock over the first one, and what will happen to the last one is a certainty that it will go over very quickly. So you could have a beginning of a disintegration that would have the most profound consequences."[52] His remarks hinted at a decisive American commitment.

In fact, however, his domino statement was window dressing aimed mainly at reassuring domestic hard-liners and, perhaps, to make the Chinese think twice about intervening. The President never did more than toy with the idea of air strikes or use of nuclear weapons. As a general, he knew that strikes would have little military value around Dienbienphu. "I couldn't think of anything probably less effective," he later explained,

49. Herring, *America's Longest War*, 25–29.
50. Ibid., 30–32.
51. Divine, *Eisenhower*, 49; Lloyd Gardner, "America's War in Vietnam: The End of Exceptionalism," in D. Michael Shafer, ed., *The Legacy: The Vietnam War in the American Imagination* (Boston, 1990), 9–29; Ambrose, *Rise*, 143.
52. Herring, *America's Longest War*, 29–35.

". . . unless you were willing to use weapons that could have destroyed the jungles all around the area for miles and that would have probably destroyed Dienbienphu itself, and that would have been that."[53] Shown an NSC paper recommending use of atomic weapons, he exploded, "You boys must be crazy. We can't use those awful things against Asians for the second time in less than ten years. My God."[54]

Eisenhower, moreover, had already taken steps to avoid unilateral American military involvement. He and Dulles agreed that there could be no such intervention without major French concessions, including significant movement toward Vietnamese independence. Testing British reactions, he had discovered what he already suspected: Great Britain, led by Churchill, had no stomach whatever for military engagement. Neither did army chief of staff Ridgway. American intervention, Ridgway said, could entail the drafting of 500,000 to 1,000,000 additional men and the fighting of a war in a country whose people, unlike most Koreans, passionately opposed American military presence. Ridgway ridiculed the "old delusive idea . . . that we could do things the cheap and easy way." [55]

With doubts like these in mind, Eisenhower shrewdly decided to consult key congressmen, knowing that they, too, were cool to American action. On April 3, four days before his domino pronouncement, the congressional leaders told him that sentiment on Capitol Hill opposed intervention . "No more Koreas," they said, unless America's allies, notably Britain, gave firm military commitments and unless the French agreed to speed the process of Vietnamese independence. The congressional leaders, like Ike, were virtually certain that neither the British nor the French would accept such conditions. So when France two days later asked for American air strikes, Eisenhower rejected the request, pointing out that it was "politically impossible."[56]

Radford and others kept up the struggle for American military action, but the die had been cast in early April: the United States had decided *not* to intervene. Americans would *not* have to go to war. On May 7, a month after the domino statement, France's 12,000-man garrison at Dienbienphu fell in a defeat that was disastrous to French resolve and pride. France still maintained a token presence in southern Vietnam, but their

53. Divine, *Eisenhower*, 50.
54. Ambrose, *Eisenhower*, 363.
55. Anderson, *Trapped by Success*, 25–39; Herring, *America's Longest War*, 32; Halberstam, *Fifties*, 407.
56. Herring, *America's Longest War*, 33–35.

days were numbered. Ho, Giap, and the peasant-based Vietminh had won a resounding triumph against Western colonialism.[57]

The immediate aftermath of these historic events satisfied few of the contestants, who met at Geneva to work out a political settlement. Ho Chi Minh's representatives demanded a united, independent country but were pressured both by the Russians, who were trying not to drive the French into the EDC, and by the Chinese, who may have worried about American intervention, into accepting less at that time than they had fought for. Representatives of the French Union and of the Vietminh instead agreed to a temporary division of Vietnam, under separate governments, near the 17th parallel. Reunification of Vietnam, it was later specified, was to take place in July 1956 following free elections that would determine a new government. Though disappointed, Ho Chi Minh accepted the results. The North, which he was to govern, included a majority of the country's population. Southern Vietnam, by contrast, was to be governed by Bao Dai, who enjoyed French backing but virtually no popular support. It seemed certain that Ho Chi Minh, the George Washington of his country, would win elections in 1956.[58]

The United States publicly dissociated itself from these discussions and refused to be a party to the accords. Dulles visited Geneva but stayed only briefly, refusing to shake the hand of Chou En-lai, the Chinese foreign minister. A hostile biographer of Dulles said that he acted like a "puritan in a house of ill-repute."[59] Dulles instead departed for a whirl of diplomacy that led in September 1954 to creation of the South East Asian Treaty Organization, or SEATO. The signatories were the United States, Great Britain, France, Australia, New Zealand, the Philippines, Pakistan, and Thailand. All agreed to "meet common danger" in the region in accordance with each nation's "constitutional principles" and to "consult" in crisis. A separate protocol designated Laos, Cambodia, and southern Vietnam as areas which, if threatened, would "endanger" the "peace and security" of the signatories.[60] As American leaders recog-

---

57. George Herring and Richard Immerman, "Eisenhower, Dulles, and Dien-bienphu: 'The Day We Didn't Go to War' Revisited," *Journal of American History*, 71 (Sept. 1984), 343–63; Marilyn Young, *The Vietnam Wars, 1945–1990* (New York, 1991), 31–36.
58. Anderson, *Trapped by Success*, 59–67.
59. Townsend Hoopes, *The Devil and John Foster Dulles* (Boston, 1973), 222. The United States did not recognize the People's Republic, the stated reason for its refusal to participate.
60. Herring, *America's Longest War*, 45.

nized, SEATO was a weak organization. The treaty envisioned no standing armed forces like those being developed under NATO, and it required only consultation, not military action. The pact failed to get the backing of key Asian nations such as India, Burma, and Indonesia.

American officials, however, emerged hopeful from these developments. The protocol to the Geneva accords, they recognized, gave them two years in which to improve the situation. Between 1954 and 1956 the CIA, following the designs of Colonel Edward Lansdale in Saigon, harassed the North by trying to destroy their printing presses, pouring contaminants into the gas tanks of buses, and distributing leaflets predicting that the North, if it won elections in 1956, would retaliate harshly against the South.

In looking ahead to 1956 the United States relied increasingly on Ngo Dinh Diem, who assumed the premiership of the South in 1954. Diem was an ardent Vietnamese nationalist who hated the French. He was also a staunch anti-Communist and devout Catholic. A self-exile after World War II, he had settled at a Maryknoll seminary in New Jersey and developed ties with influential American Catholics such as Francis Cardinal Spellman of New York, an avid foe of Communism, and Senator John F. Kennedy of Massachusetts. These ties proved useful for cementing political support in the United States, which poured economic and military aid into the South in hopes of making Diem a viable leader.[61]

Sending large aid packages to Diem aroused sometimes splenetic protests from knowledgeable American officials, who considered Diem—accurately, as it turned out—a self-centered, stubborn, and power-hungry leader. Robert McClintock, the American chargé d'affaires in Saigon in 1954, branded Diem a "messiah without a message," whose sole policy was "to ask immediate American assistance in every form." Lawton Collins, who became American ambassador there in 1955, wanted Diem removed from office.[62] Both Foster and Allen Dulles, however, backed Diem heartily, and other American officials saw no better alternative. The aid mounted.

Until mid-1955 Diem struggled to consolidate his power in Saigon. Lacking a popular base in the countryside, he also faced sharp opposition in the cities. But he proved a tough and resourceful leader, and substantial American backing gave him a stronger hold by late 1955, when a referendum ousted Bao Dai and established Diem as President of a new republic.

61. Ibid., 57.
62. Ibid., 50.

Diem, with American approval, then made one of the most fateful decisions of the Cold War: to reject the holding of nationwide elections in
1956. The private reason for reaching this decision (in which the Chinese
and the Soviets acquiesced) was that Ho Chi Minh would easily have
triumphed. The excuses given publicly by Diem were that his government had not signed the Geneva accords and that, thanks to the authoritarian control by Ho in the North, the balloting could not have been free.
Nineteen fifty-six came and went without nationwide elections, and Vietnam remained divided, with unforeseen but ultimately terrible results for
the Vietnamese people and for American society.

How to judge the record of the Eisenhower administration concerning
events in Vietnam between 1953 and 1956? The answer is: critically. The
refusal to agree to elections in 1956, combined with rising repression by
Diem thereafter, prompted mounting nationalist rage, civil war, increasing American aid to Saigon, and—in the 1960s—full-scale American
intervention. This is not to say, as some have, that American decisions
between 1954 and 1956 (and later in the Eisenhower years) made the
American-Vietnamese War inevitable: United States leaders in the early
1960s could have dared to cut their losses. It is to say, however, that Ike's
decisions—which enjoyed bipartisan support at the time—were thereafter
perceived by American political leaders of both parties as commitments to
the protection of South Vietnam from Communism. This was a highly
dangerous legacy to leave.

In 1954–56, however, virtually no one imagined that the United States
would mire itself as deeply as it did in the 1960s. On the contrary, what
many people were pleased about in the mid-1950s was that the United
States did *not* intervene militarily in 1954. Given the pressures to do so—
from the French, from high-ranking officials like Radford, and from
others who wanted to take a stand against Communism—this was not a
wholly obvious decision to have made at the time. Other, less prudent
commanders-in-chief might have acted differently. That Eisenhower
chose to stay out did not mean that he was smarter than later Presidents
who sent in American troops: they had tougher decisions to make because
the military situation in South Vietnam grew ever more desperate over
time. Still, Eisenhower's decision not to intervene militarily testified to
his prudence. That he was able to do so with relatively few domestic
political recriminations, at a time when McCarthyism was at full tide (the
Army-McCarthy hearings did not start until April 22), revealed the respect that Washington officialdom (and the American people) had for the
general's understanding of foreign and military matters. Not becoming

directly engaged militarily, in a rigid Cold War atmosphere that tempted overreaction, was to his credit.

THE NEXT MAJOR FOREIGN POLICY controversy of the era arose from the bitterness following Chiang Kai-shek's withdrawal to Taiwan in 1949. American Asia-firsters, persistently lobbied by Chiang and his American-educated wife, still insisted that the United States had "lost" China and that the Nationalists should be helped to retake the mainland. Responding to these pressures from the Right, Eisenhower had announced that the United States would remove its Seventh Fleet from the straits between Taiwan and the mainland of China. Chiang, he implied, was now "unleashed" so that he could invade the People's Republic. This was hardly likely, given Chiang's profound military weakness, but he did manage to bomb the mainland, using American-made warplanes. In any event the symbolism of "unleashing" was of political benefit to an administration anxious to protect itself from right-wing attacks at home.

No one was more persistent on behalf of Chiang than GOP Senate leader Knowland of California. Writing in Collier's magazine in January 1954, Knowland left no doubt of his zeal. "We must be prepared," he wrote, " . . . to go it alone in China if our allies desert us. . . . We must not fool ourselves into thinking we can avoid taking up arms with the Chinese Reds. If we don't fight them in China and Formosa, we will be fighting them in San Francisco, in Seattle, in Kansas City."[63] Silly as such rhetoric sounds in retrospect—and it was indeed absurd—it came from the mouth of the Senate majority leader. If Eisenhower hoped to hold his party together in Congress, he had to play his cards carefully in dealing with Chiang Kai-shek.

So it was that a crisis of sorts arose in September 1954, when the People's Republic responded to Chiang's provocations by shelling the small and well-fortified Nationalist-held island groups of Quemoy and Matsu, two miles or so off the mainland.[64] The Nationalists shelled back.

63. William Knowland, "Be Prepared to Fight in China," Collier's, Jan. 24, 1954, p. 120; Norman Graebner, The New Isolationism: A Study in Politics and Foreign Policy Since 1950 (New York, 1956), 125. Formosa was the Portuguese name for Taiwan and was widely used in the United States at that time.
64. Matsu and Quemoy were more than 150 miles apart, off different parts of the mainland coast. References to Quemoy usually meant the main island of Quemoy, among several islands collectively called Quemoy. The Matsu group is more than 100 miles from the closest parts of Taiwan, Quemoy 150 miles from Taiwan.

Radford, again reacting sharply, advised Eisenhower to place American troops on the islands and to authorize bombing raids, using tactical nuclear weapons, on the mainland. Some of these "tactical" weapons were potentially more destructive than the bombs used against Japan at Hiroshima and Nagasaki. Other anti-Communist activists perceived the confrontation as a major test of American credibility. If Quemoy and Matsu fell, they said, China would proceed to assault Taiwan. As in the crisis over Dienbienphu, Eisenhower faced loud and partisan demands for decisive action.[65]

Eisenhower responded shrewdly. Concluding that it would be politically risky to do nothing, he reaffirmed America's commitment to protect Taiwan and the neighboring Pescadores Islands. But he was deliberately ambiguous about Quemoy and Matsu, whose strategic value and defensibility struck him and other military experts as doubtful indeed. Instead, he arranged in December a mutual defense pact with Chiang. It formalized American commitment to Taiwan in case of enemy attack but did not include one to Quemoy and Matsu. The pact also stipulated that Chiang cease unilateral raids on the mainland.

In January 1955, however, the People's Republic sent troops to one of the Tachen Islands, which the Nationalists controlled. Though these islands were 200 miles from Taiwan and of no strategic importance, their plight again aroused the Asia-first lobby. Ike decided to let the islands go, but he also determined to involve Congress (newly controlled by the Democrats after the 1954 off-year elections) in his response, and he asked lawmakers to grant him blanket authority, as commander-in-chief, to use military force in order to protect Taiwan, the Pescadores, and "closely related localities."

Congress responded quickly and enthusiastically. In so doing it ceded for practical purposes some of its constitutional authority to declare war. Few events in the history of the Cold War exposed so starkly the power of anti-Communist feelings and the way that these feelings abetted the expansion of executive power. The Formosa Resolution, as it was called, was well remembered by Lyndon Johnson, Senate majority leader in 1955, who resurrected it as a precedent in his effort nine years later to expand presidential power in dealing with Vietnam.

65. Gordon Chang, "The Absence of War in the U.S.–China Confrontation over Quemoy and Matsu in 1954–1955: Contingency, Luck, Deterrence?," *American Historical Review*, 98 (Dec. 1993), 1500–1524; Herbert Parmet, *Eisenhower and the American Crusades* (New York, 1972), 397–99; Ambrose, *Eisenhower*, 373–75.

When the People's Republic stepped up shelling of Quemoy and Matsu in March, hawks in the administration reacted still more sharply. Radford wanted to give China a "bloody nose." Dulles, in public remarks cleared by Ike, announced that the United States was prepared to use tactical nuclear weapons there. At a press conference the President then added, "In any combat where these things [tactical nuclear weapons] can be used on strictly military targets and for strictly military purposes, I see no reason why they shouldn't be used, just exactly as you would use a bullet or anything else."[66]

The threat to use nuclear weapons, even if only in a "strictly military" way, shook many people, including allies of the United States. American officials had spoken this way since the dark days of the Korean War in late 1950, at which time the mere suggestion had fanned a whirlwind of anxiety. So, too, in 1955. Who could be sure that nuclear weapons could be strictly controlled in a military situation? James Hagerty, Eisenhower's able press secretary, was so worried that he urged his boss to say nothing more if asked about the situation at a forthcoming press conference.

The President's response is central to the legend of his shrewdness as chief executive. "Don't worry, Jim," he is said to have joked to Hagerty. "If that question comes up, I'll just confuse them." The question did come up, and he managed to say nothing new.[67] Thereafter shelling from the mainland soon abated. In mid-May it stopped, and the crisis receded. Quemoy and Matsu remained in Nationalist hands.

Then and later Eisenhower's handling of the Quemoy-Matsu "crisis" struck many people as ill advised. Opponents of his management go beyond assailing him for rattling nuclear weapons. They insist that he and Dulles virtually touched off the crisis by encouraging Chiang's provocative behavior—all to placate highly partisan Asia-firsters on the domestic scene. Ike, they add, then manipulated Cold War anxieties to scare the Congress into granting him unprecedented and potentially dangerous executive power. These critics say that the United States should have worked at promoting better relations with the People's Republic, both because China was by then a well-established major power in the region and because some Chinese-American rapprochement would have widened a growing wedge between China and the USSR.[68] They question, finally, whether Ike's strong stand made much difference to the Chinese (who resumed shelling in 1958).

66. Ambrose, *Eisenhower*, 383–84; Brands, "Age of Vulnerability."
67. Ambrose, *Eisenhower*, 383–84.
68. Gaddis, *Strategies of Containment*, 194–95.

There is considerable wisdom in these criticisms. Eisenhower and Dulles did little to discourage Chiang and his American partisans, even though they recognized that the Nationalist leader was fomenting trouble. To this extent they played into Cold War passions, going so far as to use the threat of nuclear attack in defense of a few strategically insignificant islands. This was massive retaliation with a vengeance. It was mainly the restraint of the Chinese that in the end prevented more serious hostilities. [69]

Still, Eisenhower, who recognized the very limited strategic value and difficult defensibility of the islands, did manage to avoid military intervention. And to the public, who did not know about details, it seemed a well-managed business. Some later historians in fact have singled out Eisenhower's actions as a classic example of his shrewd leadership concerning foreign affairs. In particular they cite his adept handling of Congress, which gave him a useful blank check, and his capacity for ambiguity—with his comment to Hagerty as a sort of Exhibit Number One. By getting wide discretionary authority and then threatening—or so it seemed—the enemy with nuclear weapons, Eisenhower seemed to have orchestrated things in ways that he could control. In any event, the Chinese, perhaps uncertain of Ike's intent, finally stopped the shelling. [70] To Americans eager for triumphs in a dangerous Cold War, Eisenhower appeared as something of a savior, especially because he had again kept the nation *out* of war. His popularity rating soared to 68 percent when the confrontation was over. [71]

At this point the Soviet leadership under Nikita Khrushchev, by then well established at home, made a few conciliatory gestures, including the signing of a peace treaty with Austria that ended Soviet military presence there. Dulles opposed any serious negotiations with the USSR, but Eisenhower seemed eager to sit down with his adversaries. The result in July 1955 was a "summit" conference in Geneva, the first such gathering since the Potsdam conference of 1945. The "spirit of Geneva" excited hopes for a new era of "coexistence." Russian and American delegates mingled, even in the bars, and people joked about "coexistence cocktails"—"you know, vodka and Coke." The evangelist Billy Graham, a world-renowned figure, conducted a revival there and pronounced the virtues of summits. Moses, he reminded people, had had a summit parley

69. Chang, "Absence of War."
70. Divine, *Eisenhower*, 65; Ambrose, *Eisenhower*, 385.
71. Oakley, *God's Country*, 218.

and received a ten-point directive that the heads of government would do well to examine.[72]

The most dramatic development of the conference occurred when Eisenhower set forth a proposal for "open skies." Taking off his glasses, he spoke directly and from memory to Khrushchev to say that the United States was prepared to swap sensitive information concerning its armed forces with the Soviet Union. He further recommended regular and frequent aerial inspection of military installations in both countries. He concluded with a spirited appeal: "I do not know how I could convince you of our sincerity in this matter and that we mean you no harm. I only wish that God would give me some means of convincing you of our sincerity and loyalty in making this proposal."[73]

Unfortunately for Eisenhower, God did not intervene to warm up the Russians. Khrushchev was a blunt, sometimes crude diplomatist, and he made no effort to hide his contempt for Eisenhower's proposal. "In our eyes," he told the President, "this is a very transparent espionage device. . . . You could hardly expect us to take this seriously."[74] Khrushchev's reaction, while abrupt, was entirely understandable, for the open-skies idea was in part an American propaganda ploy which, having been laid on the table, Ike did little to follow up. As Eisenhower recognized, the Soviets already knew much more about American military installations—American skies were "open" to a wide range of observers— than the United States knew about Soviet sites. If Khrushchev had accepted open skies, he would have learned relatively little but would have bolstered American military intelligence. He thereby rejected the proposal, and the summit accomplished nothing of substance.[75]

As in the standoff of Quemoy-Matsu, however, Eisenhower's involvement in the Geneva conference worked well for him at home. Most Americans seemed pleased that he had made the effort to talk across the table with the Russians. They also applauded open skies, appreciating that the President was anxious to reduce the possibility of surprise attack. (Eisenhower, in fact, seems sincerely to have hoped that the Soviets

72. Richard Rovere, *The Eisenhower Years: Affairs of State* (New York, 1956), 277, 285.
73. Divine, *Eisenhower*, 120.
74. Ibid., 121.
75. David Patterson, "The Legacy of President Eisenhower's Arms Control Policies," in Gregg Walker et al., eds., *The Military-Industrial Complex: Eisenhower's Warning Three Decades Later* (New York, 1992), 217–36; Gaddis, *Strategies of Containment*, 189–96.

would consider his idea.) And although the conference accomplished nothing, it did usher in a slight thaw in the Cold War that lasted (as it turned out) for almost five years. Again, therefore, Ike's popularity soared, this time to an amazing 79 percent in Gallup poll of August 1955. The columnist James Reston commented, "The popularity of President Eisenhower has got beyond the bounds of reasonable calculation and will have to be put down as a national phenomenon, like baseball. The thing is no longer just a remarkable political fact but a kind of national love affair."[76]

A month later, while vacationing with his wife and in-laws in Denver, Eisenhower, then sixty-four, suffered a heart attack that sent him to the hospital for six weeks and forced him to recuperate carefully for a couple of months thereafter. He was fortunate, as was the United States, that no major foreign policy controversy erupted during his illness and recovery between September 1955 and February 1956. But one development did take place during this time that seemed especially promising for the future of Soviet-American relations.

That was an amazing speech that Khrushchev gave to the 1,400-odd top Soviet officials who were attending the twentieth Congress of the Communist party in Moscow in February 1956. Although the address was supposed to be secret, it soon leaked to the Russian people and to the West. Khrushchev attacked Stalin (his former boss and patron) as a paranoiac tyrant who had inflicted purges, show trials, terror, forced labor camps, and mass executions on the people of his country. Khrushchev called for the de-Stalinization of the Soviet Union and eastern Europe, maintained that capitalism and Communism were not incompatible, and seemed to welcome coexistence with the West. His speech was one of the most important documents of postwar Western history, not only because it held out the hope for a thaw in the Cold War but also because it excited hopes for reform in eastern Europe. It devastated Stalinist Communists throughout the world: the American Communist party, already weak, virtually fell apart. All these developments, of course, were welcome to Americans and to the Eisenhower administration as it readied itself for a re-election campaign later in the year.

The campaign offered few surprises. Virtually everyone was certain that Ike, still phenomenally popular, would win in a walk. But he had been ill, and that made his choice of running mate especially important. Eisenhower worried that Nixon, with whom he still had cool personal

76. Oakley, *God's Country*, 219; Ambrose, *Eisenhower*, 390–94.

relations, would harm the GOP ticket. On two occasions, the second as late as April 1956, he offered Nixon a wide choice of Cabinet posts, specifically suggesting Defense. Nixon, however, correctly perceived the offers as a way of driving him out of the vice-presidency and refused. Eisenhower, who had no politically viable alternative, acquiesced. It was to be Ike and Dick once more.[77]

The Democrats tried again with Stevenson, who ran this time with Senator Estes Kefauver of Tennessee.[78] Stevenson boldly tried to start a high-level debate about nuclear testing. But no one gave him much of a chance, and his opposition to testing commanded little backing amid the continuing anti-Communist consensus.[79] Indeed, Stevenson struck some Democratic politicians as a poor campaigner. One was young Robert F. Kennedy, who joined the Stevenson team in the fall and who professed to be appalled. Stevenson, he recalled, had "no rapport with his audience and—no feeling for them—no comprehension of what cam-paigning required—no ability to make decisions. It was a terrible shock for me." (Stevenson, equally hostile, referred to "Bobby" as the "Black Prince.")[80]

As election day approached the Democrats relied heavily on TV spots. This made sense, for by 1956 there were 35 million households with sets in the United States, as opposed to 19 million in 1952. Some of the Democratic spots introduced "negative" ads for the first time, usually aimed at raising doubts about Nixon, should some unnamed awful thing just happen to occur to Ike. One spot showed a shifty-looking, narrow-eyed, small man, over whom loomed the letters NIXON. The audio added, "Nervous about Nixon?"[81]

In the last days of the campaign two of the most frightening foreign policy controversies of the postwar era interceded to command public attention. One, especially dangerous, broke out in the Middle East, where Cold War rivalries, Arab-Israeli hostilities, and Western hunger for control of oil had long heated up an inflammable mix for all concerned,

77. Ambrose, *Eisenhower*, 400–405.
78. Kefauver bested Senator John F. Kennedy for the vice-presidential nomination in a closely contested open convention vote.
79. Robert Divine, *Blowing on the Wind: The Nuclear Test Ban Debate, 1954–1960* (New York, 1978).
80. Michael Beschloss, *The Crisis Years: Kennedy and Khrushchev, 1961–1963* (New York, 1991), 301.
81. Edwin Diamond and Stephen Bates, *The Spot: The Rise of Political Advertising on Television* (Cambridge, Mass., 1992), 82–85.

including the United States. The mix grew hotter after 1954, when Gamal Abdal Nasser, a strong Arab nationalist, secured power in Egypt. Dulles tried to steer a tight course between support of Israel and of Nasser, whom he hoped to use as a buffer against a Russian presence in the oil-rich Middle East. He therefore agreed in December 1955 to lend Nasser $56 million for construction of the Aswan High Dam on the upper Nile. The dam was a key to Nasser's dreams of breaking the poverty of his country and promoting a rise to industrialization.

In mid-1956, however, the volatile mix neared ignition. Nasser recognized the People's Republic of China and purchased arms from Czechoslovakia in the Soviet bloc. Dulles, angry at Nasser's flirtation with the Communist orbit, abruptly canceled his offer of the loan. In July Nasser then shocked the world by nationalizing the Suez Canal, which had been controlled until then by a mostly British- and French-owned canal company. Revenues from ships passing through, Nasser said, would finance the dam. Although Eisenhower and Dulles tried to work out a settlement in the months that followed, the Israelis, British, and French quietly resolved to fight. On October 29, with the election campaign in the United States entering its final stages, the Israelis attacked, smashed Nasser's ill-trained forces, and began driving toward the canal. Two days later the British and the French, in what was obviously a preconceived plan worked out with the Israelis, began bombing Egyptian military installations. They then landed paratroopers with the aim of taking the canal.[82]

News of the Israeli attack outraged Eisenhower, who stopped campaigning to take charge of the situation in Washington. Attempting at first to prevent British and French engagement, he sponsored a resolution in the UN calling on Israel to withdraw and urging other UN members to refrain from the use of force. The resolution later passed overwhelmingly, with the United States and the Soviet Union standing together against Britain and France. The British and French ignored the warning and began bombing, whereupon Nasser sank ships to block the canal.

The President was especially angry with Britain and France, for they had assured him that they would not use force. When he learned of the British bombing, he was furious with Anthony Eden, the British Prime Minister whom he had known well since World War II. "Bombs, by God," he roared. "What does Anthony think he's doing?" He telephoned

82. Charles Alexander, *Holding the Line: The Eisenhower Era, 1952–1961* (Bloomington, Ind., 1975), 172–78; Ambrose, *Eisenhower*, 421–22, 424–26, 430–33; Diane Kunz, *The Economic Diplomacy of the Suez Crisis* (Chapel Hill, 1991).

Eden and gave him a tongue-lashing that reduced the Prime Minister to tears. When Ike heard that paratroopers were about to land, he exclaimed, "I think it is the biggest error of our time, outside of losing China."[83]

The USSR then inflamed the mix by warning that it was prepared to use military power against the Israeli-British-French forces in the region. Eisenhower—it was election day—thought that the Soviets were bluffing, but he placed American military units on world-wide alert. If the Soviets intervened, he warned, the United States would send in its own troops to resist them. This was the tensest moment of the crisis and one of the most frightening of the entire Cold War. Emmet Hughes recalled that the President told him, "If those fellows start something, we may have to hit 'em—and if necessary, with *everything* in the bucket."[84]

At this flashpoint the protagonists came to their senses. The Russians did not intervene; the combatants agreed to a cease-fire and ultimately withdrew. A possible world war had been averted. But most of the major players gained little from the crisis. Nasser had become a hero to other Arab nationalists, but his army had been humiliated, and the closing of the canal for the next few months inflicted additional economic damage on his country. The Soviet Union scored a few propaganda points by posing as protector of Arab interests but did not advance its influence in the area. The Israelis had proved they were a tough fighting force but had been prevented from delivering a body blow to their enemies. The British and the French were the biggest losers by far. Having embarked on a foolish military mission, they had been isolated and forced to withdraw. They never regained their standing in the Middle East.

The United States, too, suffered a little from the crisis. Many people blamed Dulles for provoking the affair by withdrawing his offer of a loan. For this reason, and because the United States had been friendly with Israel since 1948, most Arab nations remained cool to Washington. The crisis, moreover, temporarily damaged America's relations with Britain and France. The weakening to Britain and France in the region, however, offered some benefit to the United States, which further expanded its increasingly globalistic reach by becoming the major Western power in the Middle East. Henceforth America was the most important guardian of Western oil interests there—a key to subsequent tensions in the region. In early 1957 Eisenhower stepped up military and economic aid to Middle

83. Divine, *Eisenhower*, 85; Ambrose, *Eisenhower*, 427.
84. Hughes, *Ordeal*, 222–23; Divine, *Eisenhower*, 87; Ambrose, *Eisenhower*, 431–35.

Eastern nations and made it clear that the United States would intervene if necessary to secure stability in the area.[85]

Most important in the short run, Eisenhower's conduct of the Suez Affair earned him considerable admiration both abroad and at home. He deserved it. The administration had stood strongly against the use of force in the region during the tense negotiations following nationalization of the canal: Britain, France, and Israel could have had no doubt that military action would prompt the sort of American reaction that it did. When they attacked anyway, Eisenhower moved quickly and firmly. If his opposition to British and French colonialism did not satisfy Arab nationalists, it was nonetheless a prompt and decisive response under difficult circumstances. And he had even faced down the Russians without getting in a war! Americans again had reasons to be proud of their President.

While the Suez crisis was raging, Eisenhower found himself suddenly confronted with another bloody milestone of the Cold War, this time in Hungary. The Soviet satellite nations of eastern Europe had long stirred restlessly under the Russian yoke, especially following Khrushchev's denunciation of Stalinism in February. In mid-1956 riots in Poland had forced the USSR to make some concessions, and in October discontent escalated to rebellion in Hungary. At first it seemed that Soviet diplomacy would contain the trouble, but on November 4 Khrushchev sent 200,000 troops and 4,000 tanks into Budapest and other areas to crush the opposition. This was two days before the American elections. The Soviet juggernaut did a brutally thorough job, killing some 40,000 Hungarian freedom fighters and forcing the flight of more than 150,000 refugees.[86]

The suppression of Hungary shocked the world and badly soiled the image of Communism. Was Khrushchev's rule any better than Stalin's? Still, the Eisenhower administration took some criticism for what had happened. By promoting since 1952 the goal of "liberation" of "captive peoples," it had implied that it would actively assist anti-Communist rebels. Broadcasts by the Voice of America and Radio Free Europe further encouraged foes of Soviet oppression in east Europe. The Eisenhower administration's doctrine of "liberation" appealed to dogmatic anti-Communists and to many eastern-European-Americans. But military reality in eastern Europe, which was occupied by the powerful Red Army, meant that liberation was a sham.

85. See chapter 14.
86. Ambrose, *Eisenhower*, 422–24; Alexander, *Holding the Line*, 178–81; Dwight D. Eisenhower, *Waging Peace, 1956–1963* (Garden City, N.Y, 1965), 62–69.

Eisenhower, who had spent most of his life in the army, knew this full well. Hungary, after all, was land-locked and virtually surrounded by Communist countries, including the Soviet Union, which had no toleration for rebellions next to its borders. Ike thus rejected CIA calls for the parachuting of arms and supplies to the Hungarian freedom fighters and refused to consider the dispatch of American forces. Hungary, he observed sadly, was "as inaccessible to us as Tibet." He realized—as knowledgeable observers long had known—that America's main military resource was atomic attack, or massive retaliation. This would do more to destroy Hungary than to rescue it.[87]

Eisenhower supporters nonetheless managed to derive some crumbs of satisfaction from the Hungarian revolution. The President, almost everyone acknowledged, did the sensible thing (indeed, it was the only thing) by not trying to challenge the Soviets in Budapest. Eisenhower again had used his understanding of military realities to avoid overreaction that might lead to war. Most important, perhaps, it was obvious that the major villains of the piece were not Americans but Soviets. Khrushchev's behavior had again seemed to prove the validity of two fundamental tenets of American thought about world affairs: the Soviets were tyrannical, and they must be contained.[88]

THE CRISES IN SUEZ AND HUNGARY, still unfolding on election day, probably had little effect on the voting in the United States. The results of the balloting at any rate merely confirmed what everyone already anticipated: Eisenhower won a sweeping triumph. He received 35,590,472 votes to Stevenson's 26,022,752. This was more than 57 percent of the ballots. Stevenson actually received a million fewer votes in 1956 than he had in 1952; his margin of defeat was almost 3 million greater. Eisenhower carried every state outside the South (except Missouri) and even took five states there: Virginia, Florida, Louisiana, Tennessee, and Texas. The electoral college was 457 for Ike to 73 for Stevenson.

The election was mainly a personal triumph for the President. He attracted a wide range of backers to his side, including a majority of the black vote in ten northern and twelve southern cities, a development that

87. Diggins, *Proud Decades,* 302.
88. Kenneth Kitts and Betty Glad, "Presidential Personality and Improvisational Decision-Making: Eisenhower and the 1956 Hungarian Crisis," in Shirley Anne Warshaw, ed., *Reexamining the Eisenhower Presidency* (Westport, Conn., 1993), 183–208.

threatened the viability for the future of the Democratic electoral coalition.[89] But he did not sweep his party in with him. The Democrats
retained control of Congress, gaining one seat in both the House and the
Senate. Eisenhower would have to confront a House in which Democrats
outnumbered Republicans by 233 to 200 and a Senate where they had a
majority of 49 to 47. It was the first time since 1848 that a presidential
candidate had won without carrying either house of Congress with him.

Any number of things, of course, can explain a personal victory of such
magnitude. Among them, many pundits thought, were the ineffective
candidacy of Stevenson, as well as the economy (which was flourishing
and therefore helped the incumbent). But all agreed that voters still liked
Ike. And what they especially seemed to like, aside from his attractive
personality, was his record in military and diplomatic affairs. The contrast
between the mood in 1952, when the nation had been mired in Korea and
McCarthyism, and 1956, by which time the United States had enjoyed
three years of peace, was sharp and satisfying. If Ike and Dulles had missed
chances to ease the tensions of the Cold War, if they had sometimes
pursued provocative policies, they had nonetheless avoided serious blunders. Above all, they had managed to promote prosperity and to keep the
country out of war. No wonder the voters were grateful.

89. Steven Lawson, *Black Ballots: Voting Rights in the South, 1944–1969* (New York,
1976), 256–57; Whitfield, *Culture of the Cold War*, 17.

11

The Biggest Boom Yet

Widely used words and phrases evoke the dynamism and quest for "fun" that pervaded the remarkably buoyant years of the mid-1950s, especially for the ever more numerous and steadily better-off middle classes. Hear a few: gung ho, cool jazz, hot rod, drag strip, ponytail, panty raid, sock hop, cookout, jet stream, windfall profit, discount house, split-level home, togetherness, hip, hula hoops, Formica, and (in 1959) Barbie Dolls.[1] The whole world, many Americans seemed to think by 1957, was turning itself over to please the special, God-graced generation—and its children—that had triumphed over depression and fascism, that would sooner or later vanquish Communism, and that was destined to live happily ever after (well, almost) in a fairy tale of health, wealth, and happiness.

Not everyone, of course, had these grand expectations. Poverty and discrimination still afflicted millions, especially blacks, Mexican-Americans, and Indians. Cold War concerns, including nuclear testing, remained unnerving. A recession hit the country in 1958, temporarily souring the atmosphere. By then a number of groups—blacks, some of the young, women here and there—were openly restless. Still, the mid-1950s seemed almost wonderful, especially in a material sense, to

1. See John Updike, *Newsweek*, Jan. 4, 1994; J. Ronald Oakley, *God's Country: America in the Fifties* (New York, 1986), 428. There were other, less happy new words and phrases, such as apartheid, countdown, fallout, blast-off, hard-sell, junk mail, joint, shook up, and stoned.

millions of upwardly mobile people. The Korean War was fading from memory, the Red Scare was weakening, Eisenhower stood strong in command, and the consumer culture—what a marvel it appeared to be!—seemed well on its way to softening social divisions.[2]

A few numbers tell this story, which was especially happy for the mid-decade years.[3] The GNP rose in constant 1958 dollars from $355.3 billion in 1950 to $452.5 billion in 1957, an improvement of 27.4 percent, or nearly 4 percent per year. By 1960 it had increased to $487.7 billion, or 37 percent for the 1950s as a whole.[4] By 1960 the median family income was $5,620, 30 percent higher in purchasing power than in 1950. A staggeringly high total of 61.9 percent of homes were owner-occupied in 1960, compared to 43.6 percent in 1940 and 55 percent in 1950. Thanks in part to the fiscal restraint of the Eisenhower administration, prices remained stable after the inflationary years of the Korean War (the postage stamp for ordinary letters stayed at three cents until 1958), and unemployment (save in the recession year of 1958, when it averaged 6.8 percent) was remarkably low, bottoming out at between 4.1 and 4.4 percent between 1955 and 1957.[5]

Male college and university graduates were especially blessed during

2. See chapter 3 for "booms" in the 1940s. Many writers stress the affluence of the 1950s and the special glow of the mid-1950s; book titles reveal their point of view. Among them are William O'Neill, *American High: The Years of Confidence, 1945-1960* (New York, 1986); John Diggins, *The Proud Decades: America in War and Peace, 1941-1960* (New York, 1988); and Thomas Hine, *Populuxe* (New York, 1986). David Halberstam, whose book is simply titled *The Fifties* (New York, 1993), called these years "an age of astonishing material affluence" (jacket). See also Harold Vatter, *The American Economy in the 1950s* (New York, 1963); and David Potter, *People of Plenty* (Chicago, 1954), a thoughtful evaluation of the role of affluence in United States history.

3. Save in this paragraph, most statistics in this chapter—and there are many—will be in footnotes. The source for most of them is *Statistical History of the United States, from Colonial Times to the Present* (New York, 1976). Compare chapter 3 for statistics on socio-economic developments in the 1940s.

4. The GNP grew from $227.2 billion in 1940 to $355.3 billion in 1950, an increase of 56.3 percent, but all of that occurred between 1940 and 1945; the GNP in 1950 was $355.3 billion, almost exactly what it had been in 1945. Growth in the 1960s, building heavily on technological advances in the 1940s and 1950s, turned out to be especially impressive: the GNP in 1970 was $722.5 billion, or 48.1 percent higher than it had been in 1960. Per capita GNP was $1,720 in 1940, $2,342 in 1950, $2,699 in 1960, and $3,555 in 1970. All figures here are in constant 1958 dollars.

5. The basic three-cent stamp had been in place since 1932. It rose to four cents in 1958 and to five cents between 1963 and 1968. Thereafter it jumped more quickly, up to thirty-two cents in 1995.

these years. Born in the Depression when the birth rate had fallen to a record low, these young, highly educated men were a relatively scarce and prized commodity. Corporate recruiters flocked to the campuses, sometimes making reservations a year ahead to be sure of having a place to interview. (Recruiters were not much interested in talented women, who were thought to be suited for roles as wives and mothers.) By the mid-1950s the average earnings of young men after a few years of graduation from college approached those of considerably older men.

It is doubtful that this highly favored group received a more rigorous education than earlier generations of university graduates. On the contrary, the rampant growth of schools and colleges, which accepted ever-larger percentages of high-school seniors, combined with other developments to create a long-term "dumbing down" of American secondary and higher education in many localities. That was a price of educational democratization as it accelerated in postwar America. Moreover, minority children and the poor generally received inferior schooling. Still, education flourished as one of many booming enterprises in the 1950s.[6] As never before, a college degree literally paid off.[7]

MANY FORCES POWERED this prosperity, which accelerated even more rapidly in the golden age of the 1960s: the GNP in 1958 dollars reached $658.1 billion in 1966, 35 percent higher than it had been in 1960. High among these forces were America's still substantial competitive advantage over war-damaged European and Japanese economies; the continuing availability of cheap oil, a source of energy that greatly spurred industrial and commercial growth; and ever-larger investment in research and development. R&D helped spur impressive advances in science and technology, keys to leaps in productivity and real per capita income. The 1950s witnessed especially rapid expansion of electronic and electrical firms, of tobacco, soft drink, and food-processing companies, and of the chemical, plastics, and pharmaceutical industries. IBM blossomed as a

6. Average daily attendance in schools rose from 22.3 million in 1950 to 32.3 million in 1960, and the number of teachers (and other non-supervisory staff) from 914,000 to 1.4 million (two-thirds of them women). The percentage of 17-year-olds who graduated from high school rose from 57.4 to 63.4 percent during these years (and to 75.6 percent in 1970). The total number of degree-seeking college and university students increased from 2.3 million in 1950 (14.2 percent of people aged 18 to 24) to 3.6 million (22.2 percent) in 1960 (and to 7.9 million, 32.1 percent, in 1970).
7. Richard Easterlin, *Birth and Fortune: The Impact of Numbers on Personal Welfare* (New York, 1980), 21. For criticisms of education at the time, see Diane Ravitch, *The Troubled Crusade: American Education, 1945–1980* (New York, 1983), 43–80.

leader in the computer business, soon to become a guiding star of the American economy. Transistors, having been developed after the war, emerged to commercial significance, beginning in 1953 with their use in hearing aids. The airplane and airline industries also boomed, surging past railroads in passengers carried by 1957. In 1958 Americans could fly on Boeing 707 passenger jets. Two years later Eisenhower was dazzled by the speed and comfort of Air Force One, the first presidential jet.[8]

Several other forces added to economic growth in the 1950s. One was government defense spending, which had expanded enormously during the Korean War and remained significant, Eisenhower's economizing notwithstanding, throughout the decade. In some ways, of course, spending on military goods distorted priorities, depriving civilian sectors of the economy. Still, defense contracts, which averaged around 10 percent of GNP from 1954 through 1960, stimulated many corporations and employed large numbers of workers. Thanks in part to the log-rolling of southern and western representatives and senators, defense spending especially spurred economic growth in parts of the South and West, regions that had previously lagged in the American economy.[9]

The continuing baby boom further abetted economic progress, though in some ways unevenly.[10] America's population leaped from 151.7 million in 1950 to 180.7 million in 1960. This was a growth rate of 19.1 percent, the highest of any decade (save the 1900s) in the twentieth century.[11] The population increase of 29 million was the biggest in American history before or since. Booms that had started in the late 1940s—in home-and school-building, suburban development, house-

8. Alfred Chandler, Jr., "The Competitive Performance of U.S. Industrial Enterprise Since the Second World War," *Business History Review*, 68 (Spring 1994), 1–72; John Brooks, *The Great Leap: The Past Twenty-Five Years in America* (New York, 1966); Stephen Ambrose, *Eisenhower: Soldier and President* (New York, 1990), 490.
9. Kirkpatrick Sale, *Power Shift: The Rise of the Southern Rim and Its Challenge to the Eastern Establishment* (New York, 1975), 29–35.
10. A society with large numbers of children has to direct resources at people who— not being in the labor force—are not producers. In this sense, children (like many of the elderly) can be a "burden" to an economy. Still, the baby boom children were a large market for goods. See chapter 3.
11. The decade after 1900 featured record-high immigration and a population growth from 76.1 to 92.4 million, or of 21.4 percent. Between 1940 and 1950 population had grown from 132.1 million to 151.6 million, or 14.7 percent; between 1960 and 1970 it grew from 180.6 million to 204.8 million, or 13.5 percent.

hold gadgetry, automobiles, television, children's wear and toys—expanded in the 1950s and early 1960s.

Federal agencies and private producers fostered growth by energetically encouraging people to spend their money. As in the late 1940s, the Federal Housing and Veterans administrations offered low-interest loans to facilitate home-buying and suburban expansion. Retailers and manufacturers ("Buy Now, Pay Later," GM beckoned) offered enticing installment plans. In 1950 the credit card, courtesy of the Diner's Club, made its historic arrival; these and other cards spurred huge increases in borrowing. So did advertising, which sold $5.7 billion worth of ads in 1950 and $11.9 billion in 1960. It was one of the most celebrated growth areas of the 1950s. Private indebtedness jumped from $104.8 billion to $263.3 billion during the decade.[12] Older people who had scrimped and saved, especially during the Depression, rubbed their eyes in wonder at the willingness of people to go into debt to pay for household gadgets, large new cars, swimming pools, air-conditioning, sports events, eating out, travel, and buying binges at "supermarkets," another big growth area of the age. The consumer culture surged ahead, assaulting Depression-era values of thrift and saving and enticing the upwardly mobile millions to develop ever-rising expectations about the Good Life.

Spectator sports blossomed as never before in this more affluent world. Baseball continued to be the most popular, attracting 14.3 million people to the sixteen major league parks in 1953. In 1958 the New York Giants and Brooklyn Dodgers took advantage of the growth of the West to move to San Francisco and Los Angeles, touching off subsequent westward movements of sports franchises and leading in the 1960s to dramatic expansion of the major leagues.[13] Some 2.3 million people attended pro basketball games in 1953, and 8 million watched college football. Estimated total attendance at football games at all levels in that year was 35 million. As if to anoint such growth, *Sports Illustrated*, a Luce publication, first appeared in August 1954. In sports, the magazine trumpeted, "the golden age is now."[14]

The phenomenal financial success of the Dodgers in Los Angeles depended not only on westward migration (California surpassed New York

12. These figures measure private, non-corporate debt. Private debt including corporate debt rose from $246.4 billion in 1950 to $566.1 billion in 1960. Figures here are in current dollars; the increase was a little more modest in constant dollars.
13. Expansion, plus mismanagement and the spread of television, badly hurt the minor leagues. Desegregation of baseball helped to kill the Negro leagues.
14. Oakley, *God's Country*, 250.

as the nation's most populous state in 1965) but also on the ability of people to drive to the ball park, for the growth of L.A. rested on the mega-building of multi-lane freeways. Road-building, much expanded by the Interstate Highway Act of 1956, greatly aided the petroleum, automobile, and construction industries and endowed the nation with a replicable on-the-road culture featuring motels and fast food. In August 1952 the first Holiday Inn opened, between Memphis and Nashville; by 1960 Holiday Inns had mushroomed into a hugely successful chain of franchises. In April 1955 Ray Kroc, a fifty-two-year-old businessman, built the first modern-style McDonald's—featuring the famous golden arches—in Des Plaines, Illinois. It sold hamburgers for fifteen cents (a price that did not rise until 1967, when it climbed to eighteen cents), coffee for a nickel, and milkshakes for twenty cents. A family of four could eat for $2 or less and do so in their cars if they wished. By 1960 there were 228 McDonald's franchises, with annual sales of $37 million.[15]

Automobile manufacturers profited enormously from these changes. Sales of passenger cars jumped from 6.7 million in 1950 to a record 7.9 million in 1955. In that year GM, which sold roughly half of these cars, became the first American corporation to earn more than $1 billion. GM had assets greater than those of Argentina and revenues eight times those of New York State. (Defense Secretary Wilson had had a point in saying that "what was good for our country was good for General Motors, and vice versa.")

GM and other auto manufacturers had special success in persuading Americans to turn in or dump their old models—some 4.5 million cars were scrapped annually in the 1950s—and to buy sleek, multi-colored, gas-guzzling, chrome-encrusted conveyances featuring (after 1955) sweeping and non-functional tail fins.[16] The driver behind the wheel of such garish but powerful wonders was king of the road, owner of a chunk of the American dream. By 1960 nearly 80 percent of American families had at least one car, and 15 percent had two or more. There were then 73.8 million cars registered, as opposed to 39.3 million ten years earlier.[17]

Many contemporary critics assailed the vulgarity of these automobiles.

15. Kroc bought the business in 1961 from the McDonald brothers in California. Harvey Levenstein, *Paradox of Plenty: A Social History of Eating in Modern America* (New York, 1993), 227-30. By 1992 there were 8,600 McDonald's franchises in the United States.
16. John Keats, *The Insolent Chariots* (Philadelphia, 1958).
17. Halberstam, *Fifties*, 124–27, 478–95; Stephen Whitfield, *The Culture of the Cold War* (Baltimore, 1991), 74.

The new cars, one said, looked like chorus girls coming and fighter planes going. Another likened the cars to jukeboxes on wheels. But these critics missed the point, which was that millions of Americans had fallen in love with cars, the bigger and flashier the better. Automotive design—at its most flamboyant from 1955 to the early 1960s—expressed the dynamic and materialistic mood of the era. The designs deliberately recalled the lines of jet planes and generated a streamlined, futuristic feeling—one that was emulated in many other products, from toasters to garden furniture to new kitchens that featured all manner of sleekly crafted electrical conveniences. (Many of these kitchens led into colonial-style living rooms, but Americans did not seem to mind the contrast.) The TWA airline terminal in New York, designed by Eero Saarinen, captured this buoyancy. So did Dulles Airport in Virginia, also by Saarinen, and other new buildings with soaring, butterfly roofs, bold cantilevers, and forward-leaning facades.

The "baroque bender" of contemporary design, the historian Thomas Hine explains, revealed the "outright, thoroughly vulgar joy" that many prospering Americans felt in being able to live so well. It was just this sort of pride that Vice-President Nixon expressed in 1959 when he bragged truculently about gleaming American kitchen conveniences at a trade show in Moscow—a show that the New York Times called a "lavish testimonial to abundance"—in order to remind Khrushchev (and the world) of the fantastic economic potential of the American way of life.[18]

All these developments promoted grand expectations, especially among the educated middle classes, about the potential for further scientific and technological advances. This optimistic spirit—the feeling that there were no limits to progress—defined a guiding spirit of the age and, over time, unleashed ever more powerful popular pressures for expanded rights and gratifications. Many contemporaries talked as if there were almost nothing that American ingenuity—in science, industry, whatever—could not accomplish. Engineers and scientists were perfecting weather satellites, rockets, solar batteries, and atomic submarines. Things "atomic" continued to seem promising beyond belief. Scientists at Brookhaven National Laboratory on Long Island talked of creating wonderful new hybrids of carnations in its radioactive "Gamma Garden." Researchers at Argonne Laboratory near Chicago experimented with potatoes, bread, and hot dogs

18. Hine, *Populuxe*, 3–5, 87–88, 160–68; Levenstein, *Paradox of Plenty*, 114; Elaine Tyler May, "Cold War—Warm Hearth: Politics and the Family in Postwar America," in Steve Fraser and Gary Gerstle, eds., *The Rise and Fall of the New Deal Order, 1930–1980* (Princeton, 1989), 157–58.

to show that irradiation kept foods fresh and germ-free. *National Geographic* concluded that "the atomic revolution" would "shape and change our lives in ways undreamed of today—and there can be no turning back."[19]

Biologists, medical researchers, and doctors seemed nearly omnipotent. Having developed penicillin and streptomycin in the 1940s, scientists came up with antihistamines, cortisone, and other new antibiotics in the next few years. The National Institute of Health, an insignificant government agency at its founding in 1930, received better congressional funding, expanded into an ever-larger number of disease-specific institutes, and had to be renamed (in 1948) the National Institutes of Health. In 1953 a team of researchers at the University of Cambridge, England, made a spectacular breakthrough by describing DNA (deoxyribonucleic acid), thereby stimulating unprecedented advances in genetics and molecular biology. One of the team members, James Watson, was an American chemist.[20] Physicians, who as late as the 1930s had been able to do little more than diagnose people and to console patients when they got sick, found that they now had a huge pharmacopoeia at their disposal, and they used it.[21] In 1956, 80 percent of drugs being prescribed had reached the market in the previous fifteen years. These included tranquilizers such as Milltown ("don't give a damn pills," *Time* called them), which were first introduced in the mid-1950s. Sales of tranquilizers were beginning to boom by 1960, suggesting that prosperity, for all its blessings, was associated with anxieties of its own.[22]

Medical leaders fought confidently against other scourges. Heart disease was by far the number one killer, and doctors attacked it with open-heart surgery, artificial replacement of valves, and installation of pacemakers. Two other scourges were whooping cough and diphtheria, much-feared killers of children as late as the 1930s. Vaccines greatly reduced the incidence and mortality from both by the 1950s. Researchers

19. Michael Smith, "Advertising the Atom," in Michael Lacey, ed., *Government and Environmental Politics: Essays on Historical Developments Since World War Two* (Washington, 1989), 246; and Allan Winkler, *Life Under a Cloud: American Anxiety About the Atom* (New York, 1993), 136–64. See Paul Boyer, *By the Bomb's Early Light: American Thought and Culture at the Dawn of the Atomic Age* (New York, 1985), for attitudes toward things atomic prior to 1950.
20. Francis Crick, an English physicist, was another key member of the team.
21. Lewis Thomas, *The Youngest Science: Notes of a Medicine-Watcher* (New York, 1983), 27–30; Paul Starr, *The Social Transformation of American Medicine* (New York, 1982), 338–47.
22. Oakley, *God's Country*, 313.

also developed promising leads in the effort to prevent or control mumps, measles, and rubella; their efforts began to pay off in the 1960s. Doctors were happy to take credit for these advances and for the health of the American population. People were living longer (an average of 69.7 years by 1960, as opposed to 62.9 in 1940), growing to full stature earlier (by age 20 instead of by age 25 in 1900), and becoming taller and stronger.[23]

In fact, physicians and scientists claimed too much. Better nutrition—a blessing of affluence—promoted a good deal of the improvement in life expectancy.[24] Doctors continued to be far from expert about many things. Despite innumerable claims of "breakthroughs," cancer, the number two killer, remained a mysterious, dread disease.[25] Some physicians, moreover, compromised themselves by touting cigarettes in ads, even after studies in the early 1950s had begun to demonstrate the serious health hazards of tobacco: the *Journal of the American Medical Association* still accepted cigarette advertisements at the time.[26] Medical care was so expensive that millions of Americans, lacking health insurance, continued to rely on home medications, faith-healers, or fatalistic grin-and-bear-it.

Still, growing numbers of middle-class Americans, rapidly enrolling in private medical insurance plans and enjoying easier access to care, grew enamored with the medical profession. Doctors reached the peak of their prestige and cultural status during the 1950s and 1960s, by which time they began to be celebrated in television series such as "Dr. Kildare," "Ben Casey," and "Marcus Welby, M.D." Norman Rockwell illustrations continued to lionize the friendly family doctor who came day or night, rain or shine, to heal the sick and console the dying. Men went out

23. Arlene Skolnick, *Embattled Paradise: The American Family in an Age of Uncertainty* (New York, 1991), 157. Life expectancy statistics were among the many health data that revealed the impact of poverty and racial discrimination. In 1950 life expectancy at birth for whites was 69.1, for Negroes (and "other") 60.8. In 1960 the figures were 70.6 for whites and 63.6 for Negroes. The life expectancy of women at birth in 1950 exceeded that of men by 5.5 years (71.1 years to 65.6). The gender gap in this respect slowly widened: by 1970, life expectancy of women at birth was 74.8 years, compared to 67.1 years of men—a difference of 7.7 years.

24. Thomas McKeown, *The Role of Medicine: Dream, Mirage, or Nemesis?* (Princeton, 1979).

25. James Patterson, *The Dread Disease: Cancer and Modern American Culture* (Cambridge, Mass., 1987).

26. In 1957, 52 percent of American men and 34 percent of American women over 18 smoked. It was not until the late 1970s—well after widespread scientific warnings about the dangers of tobacco—that these percentages began to fall consistently.

of their way to tip their hats (most people still wore them in the 1950s) to doctors on the street.[27]

Nothing did more to enhance the status of medical research—and to escalate already growing expectations about the capacity of science to save the world—than the fight against poliomyelitis, a deeply feared scourge of the era. Polio mainly struck children and young people, sometimes killing them, sometimes leaving them paralyzed or confined to "iron lungs" so that they could breathe.[28] Because polio was known to be contagious, especially in warm weather, many schools closed earlier in the spring or opened later in the fall. Terrified parents kept their children out of crowded places, such as movie theaters, stores, or swimming pools. Those with money rushed their children to the country. Desperate for a cure, some 100,000,000 people—nearly two-thirds of the population in the early 1950s—contributed to the March of Dimes, the major organization sponsoring research against the disease. Still, the scourge persisted. An epidemic in 1950 afflicted nearly 32,000 children; another in 1952 affected nearly 58,000 and killed 1,400.

A crash research program then paid off, especially in the lab of Dr. Jonas Salk of the University of Pittsburgh Medical School. Having developed a killed-virus vaccine against the disease, Salk (with government help) mounted a nationwide inoculation program in 1954–55. The testing operated amid relentless publicity and increasingly nervous popular anticipation. Finally, on April 12, 1955, the tenth anniversary of the death of FDR, a polio patient, Salk announced that the vaccine was effective. It was one of the most exciting days of the decade. People honked their horns, rang bells, fired off salutes, dropped work, closed schools, and thanked God for deliverance. Within a few years, by which time most American children had been inoculated, polio ceased to be a major concern. There were only 910 cases in 1962.[29]

ECONOMIC GROWTH AND AFFLUENCE, many contemporaries thought, were further eroding the class, ethnic, and religious divisions of American society. The onset of "post-industrial society," they

27. John Burnham, "American Medicine's Golden Age: What Happened to It?" *Science*, 215 (March 19, 1982), 1474–79.
28. Children with rheumatic fever or leukemia were much more likely to die than were children who got polio; polio was not usually fatal. But polio came in epidemics and was far more terrifying at the time.
29. O'Neill, *American High*, 136–39. By the 1960s live-virus vaccines favored by Dr. Albert Sabin and other researchers were preferred by the AMA.

said, was ushering in a world of relative social calm and of "consensus."[30] This notion was appealing, especially when it was used to differentiate the United States, prosperous and apparently harmonious, from the harsh and presumably conflict-ridden Soviet Union and other Communist societies. It was also a notion that was highly debatable, for affluence—great engine of change though it was—was neither all-embracing nor all-powerful.

Optimists who perceived the erosion of class distinctions pointed to undeniably significant changes in the world of work. By 1960 some of the larger corporations, such as IBM, offered their employees clean, landscaped places to work in as well as benefits such as employer-subsidized health care, paid holidays, and sick leaves. Work weeks declined a bit, to an average of around forty hours in manufacturing by 1960. The greater availability of leisure time helped to drive the boom in recreation. By the early 1960s millions of American employees could count on annual paid vacations—an unthinkable blessing for most people in the 1930s.[31]

Workers also benefited from the expansion of the Social Security program, a contributory system that paid benefits to the elderly from collection of payroll taxes on employers as well as employees. By 1951 roughly 75 percent of employed workers and their survivors had become part of the system.[32] Benefits were hardly high, averaging $42 per month for retired workers in 1950 and $70 by 1960. Retired women workers, most of whom had earned less while employed, generally received less than men, as did survivors. Still, growth in coverage and in benefits was of some help to millions of Americans. The number of families receiving Social Security checks increased from 1.2 million in 1950 to 5.7 million in 1960; in the same period the total paid in benefits rose from $960 million to $10.7 billion.

Labor unions, too, continued to secure improvements for working

30. Daniel Bell, "The End of Ideology in the West," in Bell, *The End of Ideology: On the Exhaustion of Political Ideas in the Fifties* (Glencoe, 1960), 393–407.

31. Juliet Schor, *The Overworked American: The Unexpected Decline of Leisure* (New York, 1991). Schor sees the 1950s and the 1960s as a "golden age" of relatively short work weeks in America; the "overwork" came later.

32. Those excluded included federal workers covered under civil service plans; most railroad employees, who had their own retirement plan; most household and farm workers; and many self-employed. Workers covered by Social Security (the Old Age and Survivors and Disability Insurance program, or OASDI) paid a rising percentage of their paychecks for coverage in the 1950s—from 1.5 percent in 1950 to 3 percent in 1960. Employers chipped in the same percentages. The self-employed paid higher percentages.

people. As in the past, these gains were far from universal. Some unions continued to exclude the unskilled, including large numbers of blacks or women. Labor leaders in the 1950s, moreover, largely abandoned hope of achieving governmental direction of such social policies as health insurance, concentrating instead on wringing benefits from employers. The result was that the United States continued to feature a social welfare system that was more private than those in other nations. Still, unions remained a force for many workers in the 1950s. In 1954 they represented nearly 18 million people. That was 34.7 percent of nonagricultural workers, a percentage second only to the 1945 high of 35.5 percent.[33]

Union leaders focused on bread-and-butter improvements, often succeeding in achieving higher wages, shorter hours, and better working conditions. Some secured guaranteed cost-of-living adjustments to their members; by the early 1960s more than 50 percent of major labor union contracts included them.[34] Unions also struggled to win benefits—or fringes, as they came to be called. Many managed to negotiate contracts that solidified the advantages of employee seniority and that introduced clear-cut grievance procedures, sometimes with provisions for binding arbitration. These procedures were important, for they strengthened the rule of law in the workplace, they provided much-cherished job security, and they helped management and labor avert strikes.[35] Friction hardly disappeared: a major strike in the steel industry, for instance, shook the nation in 1959. But the number of strikes (and worker-hours lost) did drop dramatically after the highs of the mid-1940s and early 1950s.[36]

As promising as these changes were, they did not fully capture the

33. Unions represented 25.4 percent of all workers in 1954, second only to the all-time high of 25.5 percent in 1953.
34. Nelson Lichtenstein, "From Corporatism to Collective Bargaining: Organized Labor and the Eclipse of Social Democracy in the Postwar Era," in Fraser and Gerstle, eds., Rise and Fall, 142.
35. David Brody, "Workplace Contractualism in Comparative Perspective," in Nelson Lichtenstein and Howell John Harris, eds., Industrial Democracy in America: The Ambiguous Promise (Cambridge, Eng., 1993), 176–205.
36. Robert Zieger, American Workers, American Unions, 1920–1985 (Baltimore, 1986), 138–57, 169; Robert Griffith, "Dwight D. Eisenhower and the Corporate Commonwealth," American Historical Review, 87 (Feb. 1982), 87–122. Work stoppages soared during the Korean War, to a record high (to that time) of 5,117 in 1952. Thereafter, however, relative peace descended: considerably fewer stoppages (fewer than 4,000 per year for most subsequent years in the 1950s) and considerably fewer workers involved in them.

higher expectations that gripped sizeable numbers of American workers at the time. These workers, to be sure, recognized that American society remained unequal, and they were far too sensible to buy into myths about progress from rags to riches. But they were delighted to have the means to buy homes, cars, and household conveniences. These gave them a larger stake in capitalist society, enhancing their dignity as individuals and their sense of themselves as citizens. Many workers also believed (at least in their more hopeful moments) that the United States promised significant opportunity and upward mobility—in short, that social class was not a hard-and-fast obstacle. Those who became parents—a commonplace experience in the baby boom era—came to expect that their children would enjoy a better world than the one they themselves had grown up in during the "bad old days," relatively speaking, of the 1930s.

Were such expectations realistic? According to census definitions of occupation, the hopeful scenario seemed to have some validity. Thanks in large part to labor-saving technology, the percentage of people engaged in some of the most difficult and ill-paid work—in mining and agriculture—continued to plummet. As of 1956, the census also revealed, there were more Americans doing white-collar work than manual labor. Millions of these upwardly mobile people swelled the migration to suburbs, thereby depopulating factory-dominated neighborhoods where working-class styles of life had prevailed. Having broken away from their old neighborhoods, many of these migrants behaved in ways that at least superficially resembled those of the middle classes. For the first time they bought new cars and major household conveniences, shopped in supermarkets instead of mom-and-pop stores, ate processed and frozen foods (which boomed in the 1950s), and dressed—at least while off the job— like many white-collar friends and neighbors.[37] Given such developments, it was hardly surprising that some contemporaries thought the United States was entering a post-industrial stage of capitalism in which class distinctions were withering away.[38]

In fact, however, nothing quite so dramatic happened either then or later. While the percentage of people defined by occupation as manual workers declined over time, the *numbers* of workers so employed continued to rise slowly but steadily (from 23.7 million in 1950 and 25.6 million in 1960 to 29.1 million in 1970). If one adds the 4.1 million farmers and

37. Levenstein, *Paradox of Plenty*, 116–18; Skolnick, *Embattled Paradise*, 55–56.
38. Daniel Bell, "Toward the Great Instauration: Religion and Culture in a Post-Industrial Age," in Bell, *The Cultural Contradictions of Capitalism* (New York, 1976), 146–71.

farm workers and the 7.6 million "service workers" (a broad category that includes janitors, maids, waitpersons, firefighters, gas station attendants, guards, and domestics) to the 25.6 million people counted as manual workers in 1960, one gets a total of 37.3 million Americans who mainly used their hands on the job. This was 10.1 million more than the number designated as white-collar at the time. "White-collar," moreover, was a misleadingly capacious category: it included 14.4 million clerical and sales workers among its total of 27.2 million in 1960.[39] Many of these people were ill paid and semi-skilled. However one looks at these numbers, two points are clear: blue-collar workers remained central to the economy of the 1950s, and classlessness—as defined by work—remained a mirage.[40]

Millions of American workers, moreover, hardly considered themselves white-collar or well-off in their jobs. Contemporary critics such as the sociologist C. Wright Mills and Paul Goodman insisted accurately that much work remained routine, boring, low-paid, and geared to the making, advertising, and selling of consumer gadgetry. People employed at such labor were often dissatisfied and angry. Absenteeism and sloppy workmanship plagued assembly lines.[41] Increasing numbers of these poorly paid employees were women, who in the 1950s were entering the labor force in record numbers.[42] Moreover, blue-collar people who moved to suburbs such as the Levittowns did not suddenly become "middle-class."[43] Rather, they tended to maintain their values and styles of life and continued to think of themselves as members of the "working classes."

39. David Halle and Frank Romo, "The Blue-Collar Working Classes," in Alan Wolfe, ed., *America at Century's End* (Berkeley, 1991), 152–78; Godfrey Hodgson, *America in Our Time* (Garden City, N.Y., 1976), 478–84; Morris Janowitz, *The Last Half-Century: Societal Change and Politics in America* (Chicago, 1978), 129–33; Richard Parker, *The Myth of the Middle Class* (New York, 1972); Daniel Bell, "Work and Its Discontents," in Bell, *End of Ideology*, 227–72.
40. Zieger, *American Workers*, 140–44, tentatively concludes that 60 percent of the American labor force was still "working-class" in 1974.
41. C. Wright Mills, *White Collar: The American Middle Classes* (New York, 1951); Paul Goodman, *Growing Up Absurd: Problems of Youth in the Organized Society* (New York, 1960).
42. See chapter 12 for discussion of women and work.
43. Herbert Gans, *The Levittowners: Ways of Life and Politics in a New Suburban Community* (New York, 1967), 417; Eli Chinoy, *Automobile Workers and the American Dream* (Boston, 1955); David Halle, *America's Working Man: Work, Home, and Politics Among Blue-Collar Property Owners* (Chicago, 1984).

The best way to describe what was happening at the time is not to trumpet the coming in the 1950s of a classless or post-industrial society. Rather, it is to observe that the United States, like other advanced industrial nations, was depending less on heavy manual labor—in factories, fields, and mines—and more on service work and office-based employment. Larger percentages of people escaped hard physical labor, earned higher real wages, and enjoyed more comfortable lives. Most, to repeat, had higher hopes than their parents had about the future. Sharp divisions in income and wealth, however, persisted. Regional differences remained pronounced; then as always the South had the highest incidence of poverty. Living standards and opportunities (especially in schools) differed greatly. At least 25 percent of Americans were "poor" in the mid-1950s.[44] From these perspectives, class distinctions, while softening when measured by census definitions of occupation, remained significant, both as facts of social life and as elements in the ways that people perceived themselves.

Unions, moreover, began to lose their potential to promote socioeconomic mobility. Falling into a slow but apparently irreversible slide after the mid-1950s, they represented by 1970 only 27.4 percent of non-agricultural workers, a percentage well below those in most other industrial countries.[45] Several forces drove these trends, among them the relatively greater growth of service work, where unions often struggled to gain a foothold. Many employers moved to the South and West, which had always been relatively hostile ground for labor organizers. Some corporations began moving their activities overseas, where cheap non-union labor offered competitive advantages.

It did not necessarily help the prospects of low-paid workers that many of the most powerful labor leaders of the early and mid-1950s became increasingly cautious and conservative. Chief among these leaders was George Meany, a strongly anti-Communist labor bureaucrat who became head of a newly merged AFL-CIO in December 1955. The merger,

44. Definitions of poverty are highly controversial. A percentage of this magnitude is most often used—by contemporaries as well as by later observers—for that era. "Poverty" by such definitions meant having insufficient income (from all sources, including government aid and charity) to live at a "decent" standard of living. That was ordinarily thought to be around $2,000 to $2,500 a year for a family of four in the mid- to late 1950s. James Patterson, *America's Struggle Against Poverty, 1900–1994* (Cambridge, Mass., 1995), 78–98. For fuller discussion of poverty in the postwar era, see chapters 17 and 18.

45. The decline of unions grew precipitous in the 1980s; by 1995 only 15 percent of non-agricultural workers in the United States belonged to unions.

indeed, symbolized the weakening state of the CIO, which became a sort of junior partner. Walter Reuther, who presided over the Industrial Union Department within the new entity, became increasingly disenchanted with Meany, who showed little interest in mobilizing the masses of unskilled and semi-skilled workers in the United States. Meany also gave up trying to advance unions in the South. Although a few unions—of retail clerks, service employees, and communications workers—managed to expand, most grew only slowly, if at all, in the 1950s and thereafter.[46]

It further hurt the cause of unions that the huge Teamsters Union under Dave Beck and Jimmy Hoffa was shown to be flagrantly corrupt and mob-dominated. Reflecting a mood of rising popular disenchantment with unions, Hollywood produced *On the Waterfront* (1954), which showed goons running a longshoremen's local. It won a number of Academy Awards. Losing patience, the AFL-CIO expelled the Teamsters in 1957. And Congress, after conducting highly publicized hearings on the internal evils of unions—evils that were said to range from Communist influence to racketeering—passed the Landrum-Griffin Act in 1959. The law opened the door to government intervention into union affairs—corporate crooks faced no comparable intrusions—and tightened restrictions against secondary boycotts and picketing.[47] Although the act probably had little effect on unionization, its passage symbolized the declining political power of the House of Labor, which never recaptured the fire that it had enjoyed in its halcyon days of the late 1930s and early 1940s. It was sad commentary on what happened thereafter that the Teamsters emerged as the largest labor organization in the country by the mid-1960s.

AMERICANS IN THE 1950S who heralded the view that ethnic consciousness was declining, like those who celebrated the advent of classlessness, could also point to developments that appeared to support their claims. By 1960 the percentage of foreign-born in the population had dropped to a new twentieth-century low of 5.4 percent (compared to 7.5 percent in 1950). The percentage of native-born people of foreign or mixed parentage also fell, from 15.6 percent in 1950 to 13.4 percent in

46. Zieger, *American Workers*, 158–62.
47. Melvyn Dubofsky, *The State and Labor in Modern America* (Chapel Hill, 1994), 217–23; Diggins, *Proud Decades*, 322.

1960. This, too, was a twentieth-century low.[48] Some trends, to be sure, suggested otherwise. Puerto Ricans, who as United States citzens were not subject to stringent immigration laws, jammed night flights from San Juan to New York City, and Mexicans began coming in larger numbers to the Southwest. Still, immigration to the United States remained insignificant in the 1950s, averaging only 250,000 a year during the decade. Ellis Island, the nation's preeminent immigration center, closed in 1955.[49]

Most "ethnic" Americans, moreover, continued to be white in skin color and of European background. In 1960 the most numerous groups of people with a foreign or mixed parentage had hardly changed since 1945. They continued to be German- and Italian-Americans (approximately 3.3 million each), followed by people with roots in Canada, Poland, Britain, Russia, and Eire. Mexican-Americans ranked next, well down the list, with 1.2 million. Asians remained an almost invisible group outside a few concentrated urban neighborhoods. The census in 1960 enumerated only 642,000 people (of a total population of 180.7 million) who had one or more parents from anywhere in all of Asia.[50]

The low level of immigration, together with the formidable patriotic fervor exerted by the Cold War, sustained an assimilationist *Zeitgeist*. The Immigration and Naturalization Service, actively promoting a more homogeneous society, abetted this spirit by aggressively deporting illegals, especially Chinese, and by pressing aliens to become American citizens. A massive effort in 1954 culminated in the induction to American citi-

48. These percentages, moreover, continued to fall in the 1960s. By 1970 the percentage of foreign-born had declined to 4.7 percent, an all-time decennial year low, and the percentage of people who were of foreign or mixed parentage had dropped to 11.7 percent. By contrast, 14.7 percent of the American population had been foreign-born in 1910, and 21.3 percent had been natives of foreign or mixed parentage in 1920. These were highs for decennial years in the twentieth century. After 1970, thanks in part to more liberal immigration laws, immigration increased—rapidly in the 1980s and 1990s. See chapter 19.
49. Bernard Weisberger, "A Nation of Immigrants," *American Heritage*, Feb./March 1994, pp. 75ff; Richard Polenberg, *One Nation Divisible: Class, Race, and Ethnicity in the United States Since 1938* (New York, 1980), 145–46. For earlier trends see Robert Divine, *American Immigration Policy, 1924–1952* (New Haven, 1957).
50. Needless to say, many people, especially from Mexico, which had a long and easily crossed border with the United States, came unnoticed by the census to the United States. Still, illegal immigration from Mexico and other countries was relatively insignificant in the 1950s; big increases happened later, mainly after 1970.

zenship of 55,000 people on Veterans Day alone, including 7,500 at the Hollywood Bowl and 8,200 at the Polo Grounds in New York City. In 1946 aliens had taken an average of twenty-three years to become citizens; by 1956 the average time had plummeted to seven.[51]

Amid long-range and apparently irreversible trends such as these it was hardly surprising that contemporaries imagined a not-too-distant, consensual time in which ethnicity would not matter much. Will Herberg, a prominent theologian, captured these expectations in 1955 in his widely hailed study of American religion and ethnicity, *Protestant Catholic Jew*. Herberg argued that religious and racial identifications remained strong but that ethnic loyalties were rapidly weakening. Like others, he was impressed by the ability of the melting pot to acculturate people into an "American Way of Life."[52]

Later developments, notably the open ethnic consciousness that arose in the 1960s and 1970s, showed that Herberg and others greatly exaggerated the heat of the melting process. It then became obvious that ethnic differences (like class distinctions) refused to boil away. Even fairly well established groups, such as Irish-Americans, often nursed old resentments and clung to neighborhood enclaves. But in the aggressively assimilationist milieu of the mid-1950s Herberg's analysis seemed persuasive. Only later did other scholars, notably Nathan Glazer and Daniel Moynihan in their perceptive book *Beyond the Melting Pot* (1963), highlight the enduring power that ethnic identifications—what one eats, who one marries, where one lives, how one votes—had in the lives of the American people.[53]

Herberg's book, among others, noted the apparently powerful role of organized religion in American society during the 1950s. Self-identification as Protestant or Catholic or Jew, he argued, had become central to the culture. Many contemporary developments bolstered this view. Even more than in the past, polls suggested, Americans openly embraced one of these faiths. The ubiquity of church-building, especially in the booming suburbs, stunned foreign visitors. Popular songs included such favorites as "I Believe," "It's No Secret What God Can Do," "The Man Upstairs," and "Vaya con Dios." The percentage of people who said they belonged to a church or synagogue increased from 49 percent in 1940 to 55 percent in 1950 to a record-high 69 percent in 1959. No other Western

51. Polenberg, *One Nation Divisible*, 145–46.
52. Herberg's subtitle was *An Essay in American Religious Sociology* (Garden City, N.Y., 1955).
53. The subtitle of Glazer and Moynihan's book was *The Negroes, Puerto Ricans, Jews, Italians, and Irish of New York City* (Cambridge, Mass., 1963).

culture was nearly so "religious" in that sense.[54] Some 66 percent of these self-identifiers proclaimed themselves Protestant, 26 percent Catholic, and 3 percent Jewish.[55]

Most Americans at the time knew little about other religions (or even their own), and they displayed small interest in ecumenicalism. Some openly disparaged other faiths and denominations. Catholic leaders clung to traditional beliefs, resisting marriage outside the Church and firmly opposing birth control.[56] Anti-Catholic feelings remained open. Former Harvard president James Conant, a prominent educational reformer in the 1950s, denounced Catholic parochial schools, even if maintained without state aid. The Catholic faith, he said, was undemocratic.[57]

Cold War concerns did much to stimulate this apparently rising religiosity, Communism, many Americans believed, was evil in part because it was Godless. Hollywood, tapping into such feelings, brought out bigselling films such as *Quo Vadis?* (1951), *The Robe* (1953), and *Ben-Hur* (1959) in which Christians starred as heroes against authoritarian and pagan Roman villains.[58] The Reverend Billy Graham explained that Communism, "a great sinister anti-Christian movement masterminded by Satan," must be battled at every turn. Eagerly promoting such views, many politicians and writers highlighted the contrast, as they saw it, between "atheistic Communism" and the spirituality of the "Free World." On Flag Day in 1954 Eisenhower signed legislation that added the phrase "one nation under God" to the Pledge of Allegiance as recited by millions of children in American schools. The new pledge, he said, would enrich a world in which there were so many people "deadened in mind and soul by a materialistic philosophy of life." The President then expressed one of his most fatuous (yet apparently popular) utterances: "Our government makes no sense unless it is founded on a deeply felt religious faith—and I don't care what it is." A year later Congress endorsed this approach by approving legislation that added the words "In God We Trust" to American currency.[59]

---

54. Leo Ribuffo, "God and Contemporary Politics," *Journal of American History*, 79 (March 1993), 1515–33.
55. Whitfield, *Culture of the Cold War*, 83–84; O' Neill, *American High*, 212–15.
56. Catholic leaders, of course, did not speak for everyone in the faith; millions practiced birth control.
57. James Hunter and John Rice, "Unlikely Alliances: The Changing Contours of American Religious Faith," in Wolfe, ed., *America at Century's End*, 310–39.
58. Nora Sayre, *Running Time: Films of the Cold War* (New York, 1982), 207–14.
59. Whitfield, *Culture of the Cold War*, 88.

In taking steps such as these, some critics said, the President and Congress were blurring the constitutionally decreed separation of church and state. But relatively few people in the 1950s seemed to notice or to care. On the contrary, these actions reflected widely held and popular feelings that fused the ideals of Christianity and "Americanism" into a firmly anti-Communist "civil religion."[60] God, many people believed, had endowed the United States with a mission to spread the sacred truths of the Declaration of Independence and the Constitution throughout the world and to destroy the diabolical dogmas of Communism.

In such a climate of opinion conservative evangelicals made impressive advances. In 1950 Billy Graham founded his Evangelistic Association. It began as a one-room office with a single secretary. By 1958 the association had a staff of 200 operating out of a four-story office building in Minneapolis. They answered 10,000 letters a week and collected and disbursed $2 million a year. Graham appeared on a weekly television show and offered a column syndicated in 125 newspapers. He told the faithful that they should believe in the literal truths of the Bible and warned that the world might soon come to an end. "Time," he said repeatedly, "is running out." Graham branded drinking, smoking, card-playing, swearing, and dancing as sins.[61]

Graham was far and away the most famous evangelist of the postwar era. A frequent visitor to the White House, he appeared to be a special friend of Eisenhower, who, though never much of a church-goer before he became President, made a show of attending thereafter. (Ike even opened Cabinet meetings with a prayer.)[62] Graham's remarkable visibility testified to the eloquence and intensity of his preaching, to the eye-catching theatrics of his massive revivals, and to the special care that he and his increasingly large staff devoted to marketing. Few postwar "personalities"—a word that reflected the media-based culture of the age—better combined old-fashioned ideas and modern packaging.[63]

Other rising evangelists, most of them deeply conservative Protestants, joined Graham in preaching against the materialism, hedonism, and

60. Robert Bellah, "Civil Religion in America," Daedalus, 96 (Winter 1967), 1–21.
61. Whitfield, Culture of the Cold War, 78–80; George Marsden, "Evangelicals and the Scientific Culture: An Overview," in Michael Lacey, ed., Religion and Twentieth-Century Intellectual Life (Washington, 1989), 23–48.
62. Graham was also close to the Nixon family. Much later (in 1993 and 1994) he conducted funeral services for Patricia and Richard Nixon.
63. William McLoughlin, Billy Graham: Revivalist in a Secular Age (New York, 1960).

secularism of modern life. They attracted millions of Americans (how many no one knows for sure), many of them relatively poor, geographically unsettled, and ill educated. These were people who felt cut off—or alienated—from the more secular world of the middle classes and who searched for consoling and unambiguous truths. Large numbers of such people followed Oral Roberts, a Pentecostal Holiness preacher and faith-healer who used television and mass-mailing techniques to build up a nationally known fundamentalist organization. By the mid-1950s Roberts had a huge ranch in Texas, a twelve-passenger plane, a television and radio show carried by 400 stations, and a university in his name. He was then drawing in more than $50 million a year in revenues.[64]

Growing numbers of Americans identified themselves as premillenarians, people who believed that an apocalypse would bring on the second coming of Christ. The Savior would subsequently cleanse the world of sin. Some of these true believers were so thoroughly alienated from contemporary American culture, including not only materialism (as they saw it) but also capitalism itself, that the word "conservative" scarcely begins to describe them. Often poor and class-conscious, they professed to be horrified by what they considered the immoral secularism of the more affluent middle classes. Their grim and pessimistic view of contemporary American culture confounded optimistic predictions about the withering away of social distinctions and belied the surface homogeneity of American society at the time.[65]

How profound were these varied manifestations of organized religious activity? The answer is unclear. On the one hand it was undeniable that many Americans—always an especially religious people—were sincere in their professions of faith. Millions joined religiously inspired voluntary organizations to engage in charitable work. Black people drew on religious faith to mount rising protests against racial injustice: no account of the emergent civil rights movement can slight the power of Christian ideals. On the other hand, church membership statistics, as always, remain at best an inexact guide to the depth of religious convictions. Herberg, among others, doubted that theological distinctions or spiritual profundity was very important to most of the people who identified with one church

64. Oakley, God's Country, 322.
65. Paul Boyer, When Time Shall Be No More: Prophecy Belief in Modern American Culture (Cambridge, Mass., 1992). Boyer estimated, 2, that these beliefs became increasingly powerful in later years. Some 40 percent of Americans, he wrote, told the Gallup poll in 1980 that the Bible was the literal word of God; another 45 percent said that the Bible was divinely inspired.

or another. People joined churches, he said, to give themselves clear *social* status in a rapidly changing, mobile culture where class and ethnic boundaries seemed a little less distinct. Church membership satisfied a need for "belonging."

The striking popularity of the Reverend Norman Vincent Peale during the 1950s further suggests the fallacy of seeing deep spiritual meaning in all aspects of religiosity at the time. Well before then Peale had been a prominent minister, at the Marble Collegiate Reformed Church on Fifth Avenue in New York City. Peale downplayed sectarian differences to reach for a mass audience. His book A *Guide for Confident Living* sold well when published in 1948. Peale's fame truly spread, however, in 1952 when he brought out another book, *The Power of Positive Thinking*. As this title suggested, Peale offered an optimistic message about the capacity of Christian teaching to evoke attractive personal qualities. Chapter titles included "I Don't Believe in Defeat," "How to Get People to Like You," and "Expect the Best and Get It."[66] Appealingly simple in its message, *The Power of Positive Thinking* essentially argued that self-confidence and faith could work wonders. The book shot quickly to the best-seller list, where it remained for more than three years. This was an amazingly durable record. And it kept on selling, to a total of 3 million copies by 1974. Behind the Bible and Charles Sheldon's novel *In His Steps* (1897) it was the most popular religious-spiritual book in the history of American publishing.

The phenomenon of Peale distressed many intellectuals in the 1950s. A few grudgingly conceded that he may have helped to moderate sectarian narrowness. A certain superficial homogenization and local good feeling may have followed. But they were otherwise appalled at what they considered his oversimplified incantation of "faith." As the conservative philosopher Russell Kirk put it, popularizers like Peale offered a "religion-in-general" that was "evacuated of content." Christianity of that sort amounted to "little more than a vague spirit of friendliness, a willingness to support churches—providing these churches demand no real sacrifices and preach no exacting doctrines."[67]

Peale's extraordinary popularity, like Eisenhower's superficial piety,

66. Donald Meyer, *The Positive Thinkers: Religion as Pop Psychology from Mary Baker Eddy to Oral Roberts* (New York, 1980), 258–95; *New York Times*, Dec. 16, 1993 (obituary); Whifield, *Culture of the Cold War*, 83–84.
67. Frederick Siegel, *Troubled Journey: From Pearl Harbor to Ronald Reagan* (New York, 1984), 113. See also William Lee Miller, "Some Negative Thinking About Norman Vincent Peale," *Reporter*, 12 (Jan. 13, 1955), 19–24.

indeed exposed the sanctimonious side of the religious "revival" in the 1950s. These phenomena did not cut deep. Still, for many—especially those in the less favored classes—fundamentalist faiths provided an anchor of consolation as well as self-identification amid the floodtides of social and cultural change. More profound than "positive thinking," they not only survived the 1950s but also spread steadily in subsequent years.

SUBSEQUENT NOSTALGIA for the mid-decade years of the "nifty fifties"—perceived in retrospect as a harmonious and mostly happy era—has conveyed the impression that social analysts had little to complain about at the time. This was not the case: much about American society in the 1950s repelled them. Intellectuals and others focused especially on the apparent flaws of suburbanization, the vulgarity of runaway materialism and consumerism, and the deterioriation of "traditional American values."

Given the suburban boom that took off after 1945, it was hardly surprising that life in the suburbs evoked considerable contemporary commentary in the 1950s. The roar of bulldozers smashing trees and overrunning farms resounded on urban fringes throughout the land. Of the 13 million homes built between 1948 and 1958, 11 million were suburban. This represented a fantastic building boom: one-fourth of *all* homes standing in 1960 had been built in the previous ten years. Some 83 percent of all population growth in the nation in the 1950s was in suburbs. The trend, already powerful, continued in the 1960s. In 1950 there had been 35 million suburbanites; by 1970 there were 72 million (in a total population of 205 million). Suburbs then housed more people than did central cities or farms, and by 1972 they offered more jobs than did the central cities.[68]

There was much about this phenomenal growth, however, for contemporary critics to bemoan. One lament was its impact on cities, which saw much of their vitality drained to the fringes. This trend was more pronounced in the United States, where the automobile revolution was by far the most advanced, than anywhere else in the world. Eleven of the

68. Landon Jones, *Great Expectations: America and the Baby Boom Generation* (New York, 1980), 38–39. Needless to say, these figures should not be taken too literally; no two definitions of a "suburb" fully agreed. The numbers used here are based on census definitions and describe people living "outside the central cities" but within "standard metropolitan statistical areas" (SMSAs). The definition of SMSAs differed from place to place and from census to census, but in general the term referred to a county or group of contiguous counties containing a city or "twin cities" with a population of 50,000 or more.

nation's twelve biggest cities (Los Angeles was the exception) lost population in the 1950s, most of it to suburban expansion. Urban playhouses, art galleries, and concert halls struggled to attract patrons. Many city newspapers folded. By contrast some suburban papers, notably *Newsday*, thrived; its circulation rose from 32,000 in 1940 to 370,000 by 1960, much of it in suburban areas on Long Island.

Hundreds of downtown movie houses either closed or let their amenities decline. Lobbies, once baroque spectacles to behold, became cluttered with popcorn, candy, and soft drink counters; uniformed ushers, who once had guided patrons down lushly padded aisles, became a vanishing breed. The decline of the movie palaces was not primarily the result of television but of the movement of millions of people to the suburbs. Drive-in theaters flourished in their place: there were 100 in 1946, more than 3,000 in 1956.[69]

The rise of suburbs badly hurt downtown hotels and retail outlets, which fought a losing battle against motels and suburban malls. The number of hotel rooms in the United States actually declined—from 1.55 million in 1948 to 1.45 million in 1964. Meanwhile, the number of motel rooms increased from 304,000 to 1 million. Malls were among the greatest success stories of the late 1950s and early 1960s. In 1955 there were already 1,000 malls in the country, whereupon a huge new surge crested. By 1956 there were 1,600, with 2,500 more in the planning phase. Urban public transit, too, shriveled in its near-hopeless struggle against automobile culture: the share of urban passenger miles carried by public buses and subways dropped from 35 percent in 1945 to 5 percent by 1965. Highways increasingly invaded central city areas themselves: by 1956 there were 376 miles of freeways within the nation's twenty-five largest cities. Cities, Lewis Mumford exclaimed, were becoming a "tangled mass of highways, interchanges, and parking lots."[70]

The decline of America's railroads in the 1950s was especially precipitous. Undercut by highways and suburbanization, many drastically curtailed schedules. Rolling stock, service, and stations deteriorated. The fate of New York City's Grand Central and Pennsylvania stations symbolized the trend. In 1961 the profit-seeking owners of Grand Central petitioned the city to lower the terminal's ceiling from fifty-eight to fifteen feet and to add three tiers of bowling alleys in the new space. Thwarted in this effort

69. James Baughman, *The Republic of Mass Culture: Journalism, Filmmaking, and Broadcasting in America Since 1941* (Baltimore, 1992), 35–36, 61–62.
70. Jon Teaford, *The Twentieth-Century American City: Problem, Promise, and Reality* (Baltimore, 1986), 98–110. Mumford cited in Russell Jacoby, *The Last Intellectuals: American Culture in the Age of Academe* (New York, 1987), 57–60.

by angry protests, the owners managed to sell the air space over the station, and the dignity of the terminal was soon dwarfed by erection of the fifty-five-story Pan American Building. Penn Station, an architectural monument, was gutted, its mighty columns felled to permit construction of offices, a sports arena, and a cramped commuter terminal.[71]

Nothing alarmed city officials more than the plight of downtown neighborhoods that decayed when upwardly mobile people left for the suburbs. In the process the cities lost still more of their tax bases. Congress had tried to deal with this problem in 1949 by approving legislation authorizing the construction, with federal subsidy, of 810,000 public housing units. The law also set aside federal money to aid local redevelopment agencies that agreed to purchase "slum" areas and to demolish old buildings. The agencies were then supposed to sell the bulldozed space to private developers, who would erect better housing for low-income people. Thus was born one of the great hopes of the 1950s: "urban renewal."

Some cities, helped by federal funds, erected splashy new areas that seemed to revitalize downtown areas: the "Golden Triangle" of Pittsburgh, the Government Center of Boston. But urban renewal had a very uneven track record insofar as housing was concerned. The process of redevelopment—identifying sites, buying land, relocating tenants, bulldozing, selling to builders, putting up new dwellings—often moved slowly, sometimes in the teeth of resistance by affected residents. In some places old buildings were razed, but new ones did not go up for years. One such area, in St. Louis, was ridiculed as "Hiroshima Flats." Detroit had its "Ragweed Acres." More serious, "urban renewal" often meant "poor removal" or "Negro removal. " Development officials tore down low-income neighborhoods only to sell the vacant land to developers who put up more expensive housing. Poor people who had been displaced while the builders tore up their homes could ill afford to move back. All too often urban renewal did two things: it shunted low-cost housing, including racial ghettos, from one part of a city to another, and it enriched builders and landlords.[72]

Meanwhile most cities, Pittsburgh and Boston among them, continued

71. Teaford, *Twentieth-Century American City*, 112; Mark Reutter, "The Last Promise of the American Railroad," *Wilson Quarterly* (Winter 1994), 10–35.
72. Kenneth Jackson, *Crabgrass Frontier: The Suburbanization of the United States* (New York, 1985), 219–30. Also Martin Anderson, *The Federal Bulldozer: A Critical Analysis of Urban Renewal, 1949–1962* (New York, 1962); Mark Gelfand, *A Nation of Cities: The Federal Government and Urban America, 1933–1968* (New York, 1975); and Thomas O'Connor, *Building a New Boston: Politics and Urban Renewal, 1950–1970* (Boston, 1993).

to lose upwardly mobile residents to suburbs. In their place flocked throngs of poor blacks (and whites). Unable to afford the suburbs (and in the case of blacks racially excluded from them), these people settled in the most broken-down areas of the urban centers. Most of these people were fleeing the farms and the small towns. The depopulation of rural America at the time, accelerated by the technological revolution that was rendering farm labor superfluous, was one of the most harrowing and large-scale demographic developments of the 1940s, 1950s, and 1960s. Because relatively few American officials attended to the problems of these people in the 1950s, the mass migrations set the stage for social and racial dynamite that exploded in the cities after 1965.[73]

If public housing had fulfilled its promise, this crisis would have been less severe. But it did not. Funding was ungenerous. Whites put up often fierce resistance to subsidized housing that accepted blacks. Builders showed little interest in such efforts, largely because public housing was considerably less profitable for them than commercial construction, much of it in the suburbs. By 1955 only 200,000 public housing units (instead of the targeted 810,000) had been constructed; by 1965 only 325,203 had. Some of these worked fairly well; it is wrong to say that public housing always failed. But many "projects" collapsed, and the subsequent reputation of public housing, like that of urban renewal, plummeted.

No public housing project received more devastating publicity than Pruitt-Igoe in St. Louis. Architects and planners hailed its design while it was being drafted in the 1950s. Its thirty-three eleven-story apartment buildings, containing a total of 2,800 units, had open galleries that tenants were expected to use as communal porches, laundries, and play areas. A "river of trees" winding between the buildings offered lots of open green space. By the mid-1960s, however, Pruitt-Igoe had deteriorated badly. The open spaces had become scrubby and covered with litter and broken glass. Muggings and rapes in corridors and elevators (when they worked) terrified residents. The architect lamented, "I never thought people were that destructive." Although renovation costing $7 million tried to pump new life into the project, it did not help. Starting in 1975 the housing authority tore down the whole thing.[74]

73. Robert Griffith, "Forging America's Postwar Order: Domestic Politics and Political Economy in the Age of Truman," in Michael Lacey, ed., *The Truman Presidency* (Washington, 1989), 57–88.
74. Teaford, *Twentieth-Century American City*, 122–26.

Critics offered widely differing opinions about what killed Pruitt-Igoe and other large urban projects. Some said that the projects were too big, too high-rise, and too sterile. Others emphasized that housing authorities needed more money to maintain the buildings and to provide proper security.[75] (All these flaws beset Pruitt-Igoe.) Others blamed racial tensions, which often flourished in those places where blacks were allowed. A common dilemma facing housing managers was whether to require upwardly mobile tenants—people who managed to exceed income limits—to move out and find housing in the private market. Telling such people to leave made sense: why subsidize people who were no longer "poor"? But their departure deprived the projects of significant numbers of stable and law-abiding families. Losing "role models," many projects also lowered "screening" regulations that had barred criminals. By the 1960s and 1970s many projects had become dumping grounds for the most troubled of the urban American poor.

Critics of suburbs, angry as they were about the deterioration of cities, did not stop there. Many deplored life in the suburbs themselves. Most suburbs, they said, were "techno-burbs" or "slurbs"—large anti-urban conglomerations that existed mainly to suit highway builders and suburban developers.[76] John Keats wrote a satire, *The Crack in the Picture Window*, which bewailed the lack of privacy there. His suburb featured a family, the Drones, whose neighbors included the Fecunds and the Amiables.[77] Another critic likened suburban life to "Disturbia," a place where the inanity of existence produced "haggard" men, "tense and anxious" women, and "the gimme kids" who, after unwrapping the last Christmas present, "look up and ask whether that is all."[78]

Easily the most common complaint about suburbs focused on "conformity." Suburbs, indeed, became symbols of what many critics imagined were the most oppressive aspects of life in the 1950s. These were not economic problems; aside from Mills and a few others, American intellectuals tended to be optimistic about the economy. Rather, critics had cultural concerns. They worried about the all-pervasive sameness, blandness, unventuresomeness, mindlessness, and threat to individualism that

75. Jane Jacobs, *The Death and Life of Great American Cities* (New York, 1961); Lewis Mumford, *The Urban Prospect* (New York, 1962); William Whyte, ed., *The Exploding Metropolis* (New York, 1958).
76. Robert Fishman, *Bourgeois Utopias: The Rise and Fall of Suburbia* (New York, 1987), 182–220.
77. (Boston, 1956).
78. Richard Gordon et al., *The Split-Level Trap* (New York, 1960), 33, 54, 142.

they thought were flowing from the onrushing materialism of middle-class life in the suburbs.

It followed, these critics thought, that people in suburbs were the ultimate "conformists." Everyone there, a hostile observer noted, "buys the right car, keeps his lawn like his neighbor's, eats crunchy breakfast cereal, and votes Republican."[79] David Riesman and Nathan Glazer, influential social thinkers, did much to encourage such criticism in a widely discussed book, *The Lonely Crowd*, that they co-authored in 1950.[80] The book argued that "American character" was changing. People were losing a more individualistic "inner-direction" that had existed in the past and were becoming more "other-directed," taking their cues from the opinions and behavior of peers. Suburbs were said to energize this development. "The suburb," Riesman wrote elsewhere, was "like a fraternity house at a small college in which like-mindedness reverberates upon itself."[81]

Other writers chimed in with similar complaints later in the 1950s. John Cheever's short stories focused on the mindless, empty lives of suburbanites. Suburbs, he wrote, "encircled the city's boundaries like an enemy and we thought of them as a loss of privacy, a cesspool of conformity."[82] Another novelist, Sloan Wilson, in a best-selling book, *The Man in the Gray Flannel Suit* (1957), skewered the soulless, consumerist lives of suburbanites and the corporate world. William H. Whyte summed up criticisms like these in a big-selling work of popular sociology, *The Organization Man*. Suburbs, he conceded, were often friendly places. Some promoted greater tolerance. But they often had a hothouse quality that highlighted "getting along" or "belonging." Whyte concluded that suburbs, together with large and bureaucratic corporations, were threatening the individualistic and entrepreneurial drives that had made America great.[83]

The conformist atmosphere of suburbs, critics added, helped to deaden political debate, thereby sustaining the middle-of-the-road, mostly con-

79. Stanley Rowland, "Suburbia Buys Religion," *Nation*, July 28, 1956, pp. 78–80.
80. With Reuel Denney, subtitled *A Study of the Changing American Character* (New Haven, 1950).
81. David Riesman, "The Suburban Sadness," in William Dobriner, ed., *The Suburban Community* (New York, 1958), 375–402.
82. O'Neill, *American High*, 23.
83. A handy summary of many of these criticisms is Richard Pells, *The Liberal Mind in a Conservative Age: American Intellectuals in the 1940s and 1950s* (New York, 1985), 232–48. Whyte's book appeared in 1956.

servative consensus that seemed to dominate, especially in the mid-1950s. No one expressed this feeling about political ideas in the 1950s more cogently than the sociologist Daniel Bell, especially in his collection of essays *The End of Ideology* (1960).[84] Bell insisted that older ideologies strong in the 1930s, notably Marxism, had lost their power to attract people. Instead, Americans focused on more private concerns and did not try to change the world. Bell did not altogether deplore this turn of events; like many contemporary thinkers, he rejoiced that the United States had avoided the bitter internal conflicts of more "ideological" societies like the Soviet Union. But he, too, was a bit wistful for a time when lively political debate had resounded in the land.

Underlying many of these criticisms of suburbia—and by extension of the "American character" in the 1950s—were deeper fears about the nation's psychological health. Buzzwords and phrases exposed these fears: "alienation," "identity crisis," "age of anxiety," "eclipse of community." The "uprooted" peopled America. "Mass society" obliterated identity and "individualism." Society was a "lonely crowd." Many of these words and phrases reflected the rising visibility of popular sociology, of psychological models, and of "experts"—whether Norman Vincent Peale on the power of positive thinking or Dr. Benjamin Spock, who soothed nervous parents with immensely reassuring advice about child-rearing. Psychiatry and psychology, like organized religion, boomed in the 1950s. The United States, it seemed, was becoming a "therapeutic culture" in which "experts" helped people to feel good.[85]

Some contemporaries thought that the rise of suburbia—and the runaway consumerism that seemed to accompany it—were undermining what they considered to be traditional American values. Mills charged forthrightly that the United States had become "a great salesroom, an enormous file, an incorporated brain, and a new universe of management and manipulation."[86] Other writers cited front-page scandals to demonstrate the apparently ubiquitous subversiveness of materialism, including the penetration by gamblers of big-time college basketball in 1951 and the

84. Daniel Bell, *End of Ideology*. Also Howard Brick, *Daniel Bell and the Decline of Intellectual Radicalism* (Madison, 1986).
85. Janowitz, *Last Half-Century*, 418–30; Skolnick, *Embattled Paradise*, 151–52; and Jonathan Imber, ed., *The Feeling Intellect: Selected Writings by Philip Rieff* (Chicago, 1990). Oscar Handlin's *The Uprooted* (New York, 1951), an account of American immigration, emphasized such psychological effects. It won a Pulitzer Prize for history.
86. Mills, *White Collar*, xv.

perversion by television of big-money quiz shows at the end of the decade. The lure of riches seemed dangerously enticing.[87]

John Kenneth Galbraith, the often iconoclastic Harvard economist, summarized and expanded on these criticisms in one of the most talked-about non-fiction books of the decade, *The Affluent Society*.[88] The title was ironic. Galbraith agreed that American society was in many ways affluent, but he emphasized that it was above all grossly materialistic. A liberal, Galbraith called for a range of public policies to improve the quality of life in America: greater spending for public education, price controls to curb profiteering, even a national sales tax to raise money for social services. Galbraith was concerned above all with the contrast, as he saw it, between private opulence and public austerity. His target, like that of many other critics in the 1950s, was as much the vulgarity of the culture as the economy.

All these denunciations of American society and culture in the 1950s revealed that contemporary critics were both lively and trenchant. Moreover, they enjoyed substantial respect at the time; writers like Whyte and Galbraith received widespread critical praise and attracted many readers.[89] Some of the criticism, however, was one-sided. Those who lambasted suburbia, for instance, tended to ignore several basic facts: the boom in building energized important sectors of the economy, providing a good deal of employment; it lessened the housing shortage that had diminished the lives of millions during the Depression and the war; and it enabled people to enjoy conveniences, such as modern bathrooms and kitchens, that they had not had before. Like the Levittowners, these people often worked hard to keep up their properties and to bring personal touches to their homes. Few suburbs, Malvina Reynolds to the contrary, were ticky-tacky and all the same. Many suburbanites also took pride in the community life that evolved around schools, churches, and other institutions: the new neighborhoods were hardly as antiseptic and isolating as Keats and others suggested. Above all, millions of suburbanites were delighted to have space—a profound human need—and to own property. A resident of Canarsie in southeast Brooklyn, then a rapidly growing (and essentially all-white) "suburban" neighborhood, remembered, "Most of us who live in Canarsie came from ghettos. But once we made it to

87. Randy Roberts and James Olson, *Winning Is the Only Thing: Sports in America Since 1945* (Baltimore, 1989), 73–92.
88. (Boston, 1958).
89. John Higham, "Changing Paradigms: The Collapse of Consensus History," *Journal of American History*, 76 (Sept. 1989), 460–66.

Canarsie, we finally had a little piece of the country." Another recalled, "It was exhilarating to own my own home. I felt like I had finally achieved something."[90]

Critics were on especially shaky ground in suggesting that suburbs— and by extension what they perceived as the materialism and "conformity" of American culture—were subverting traditional values. It was obvious, of course, that economic growth greatly boosted the consumption of goods, with much attendant waste and misdirection of resources from public needs to private display. Galbraith was on target there. Acquiring "things," however, was hardly new to American life in the 1950s—just easier because many more people had much more money. More important, people were hardly abandoning "individualism" for "conformity" or "inner-direction" for "other-direction." Traditional values—the work ethic, competing hard in order to advance in the world—seemed as vital as ever. (Immigrants at the time frequently were appalled by what they considered to be the grim and determined pace of life in the United States.) Other long-standing human aspirations, such as the quest for security and stability and the desire to live among people like themselves, also endured. People did not surrender to the tyranny of conformity. Rather, they searched understandably for whatever enabled them and their families to feel comfortable and safe.

Broad generalizations about changes in values or "national character" are often unsatisfying because they fail to capture the variety and complexity of people. Throughout American history, values of "community" (including "conformity") and "individualism" have coexisted, sometimes amid considerable tension. So, too, in the 1950s. Like Willie Loman, the "hero" of Arthur Miller's *Death of a Salesman* (1950), many Americans remained both fatuously conformist as well as driven to achieve for themselves and their families.[91] Disneyland, which opened in 1955, revealed these complexities in another way. An enormously successful business enterprise, it testified to the power of affluence and of the consumer culture. Millions of people, most (but not all) of them middle-class, traveled long distances to visit the place.[92] And Disneyland managed to have things both ways, celebrating Main Street (à la Norman Rockwell) as

90. Jonathan Rieder, *Canarsie: The Jews and Italians of Brooklyn Against Liberalism* (Cambridge, Mass., 1985), 17–18; Hine, *Populuxe*, 32–35; Gans, *The Levittowners*, xxvii, 180.
91. Skolnick, *Embattled Paradise*, 60–62, 151–52, 160–63, 174–77, 202–3.
92. George Lipsitz, "The Making of Disneyland," in William Graebner, ed., *True Stories from the American Past* (New York, 1993), 179–96.

well as Tomorrowland. Main Street evoked nostalgia for traditional small-town styles of life, while Tomorrowland appealed to the still strong yearning of Americans for the new, the dynamic, and the unknown. The positive values associated with technological progress—individualism and entrepreneurialism among them—had by no means lost their hold on the imaginations of people.

Critics of affluent excess during the boom years of the mid-1950s tended sometimes to expect human beings to deny themselves material pleasures. Yet a culture in which rising numbers of people have the luxury of fairly secure food and shelter—increasingly the case in the United States in the post–World War II era—is one in which hopes for still greater comforts will expand. The majority of Americans, their basic needs more secure, developed ever-larger expectations about life. Some, concentrating on material gain, came to crave quick personal gratification. Others, however, began to imagine a better society in which the best of American ideals could be put into practice—this, too, the nation as a whole could afford. In this way, material progress helped in time to arouse not only quests for personal gratification but also rising awareness of the needs and rights of people. It would not be long before a rights-consciousness emerged that shook the surface calm of American culture.

12

Mass Consumer Culture

To listen to commentators on American cultural life in the 1950s was often to hear a litany of complaints: the mass media were debasing public taste, sexual license was threatening traditional morality, juvenile delinquency was overrunning society, and generational change—a "youth culture"—was undermining the stability of family and community. Here are a few of the voices:

———On the mass media, especially television: "The repetitiveness, the selfsameness, and the ubiquity of modern mass culture tend to make for automatized reactions and to weaken the forces of individual resistance."

———On sexual behavior, as described by a critic of Alfred Kinsey's book *Sexual Behavior in the Human Female* (1953): The book reveals "a prevailing degeneration in American morality approximating the worst decadence of the Roman Empire. . . . the presuppositions of the *Kinsey Report* are strictly animalistic."

———On youth and juvenile delinquency: "Not even the Communist conspiracy could devise a more effective way to demoralize, disrupt, confuse, and destroy our future citizens than apathy on the part of adult Americans to the scourge known as Juvenile Delinquency."

These complainers were not all prudes or reactionaries. The critic of "modern mass culture" was Theodor Adorno, a leader of the Frankfurt School of cultural criticism, which drew on Marxian and Freudian insights to lament the commercialization of American life. The critic of Kinsey, one among many, was Henry Pitney Van Dusen, head of Union

Theological Seminary, a prestigious religious institution. Critics of mass consumer culture also included shrewd and respected intellectuals such as the sociologist Daniel Bell and the historian Daniel Boorstin.[1]

Were these Jeremiahs on target? The answer depends, of course, on the perspective of the viewer. In retrospect it is clear that they erected a number of straw men and exaggerated the curses of change. Still, they correctly placed their fingers on problems (as well as prospects) associated with one of the most profound developments of the postwar era: the dramatic expansion of a mass consumer culture. Some aspects of this culture, notably the rise of television and of "rock 'n' roll" music, struck the nation with great and sudden power. Other related developments, such as more liberalized sexuality and the advent of a "youth culture," provoked a great deal of controversy. While these changes did not stem the force of traditional values in the 1950s, they exposed undercurrents of dissatisfaction and rebellion that were to break loose more powerfully in the 1960s.

INTELLECTUALS SUCH AS Adorno, Herbert Marcuse, and other European emigrés who dominated the Frankfurt School were hardly alone in condemning what they considered to be the commodification and debasement of American cultural life. The effluvium of bad taste, they thought, not only overran the arts but also oozed throughout American culture. The United States, they said, suffered from ugly roadside "strips," rampant commercialization, mindless mass entertainment—Disneyland was often Exhibit Number One—and tasteless, fat-filled food that later made obesity a national concern.[2]

1. Adorno (1954) cited in James Gilbert, *A Cycle of Outrage: America's Reaction to the Juvenile Delinquent in the 1950s* (New York, 1986), 75; Van Dusen (1953) cited in David Halberstam, "Discovering Sex," *American Heritage*, May/June, 1993, p. 42; the alarmist about juvenile delinquency (in 1954) was Robert Hendricksen, a Republican senator from New Jersey, cited in Gilbert, *Cycle of Outrage*, 75. See also Daniel Bell, *The Cultural Contradictions of Capitalism* (New York, 1976), esp. 33–84; and Daniel Boorstin, *The Americans: The Democratic Experience* (New York, 1973), esp. 525–55.
2. For food and diets, see Harvey Levenstein, *Paradox of Plenty: A Social History of Eating in Modern America* (New York, 1993), 106–10, 119–26. General interpretations of the rise of consumerism and mass culture, especially before 1940, include David Nasaw, *Going Out: The Rise and Fall of Public Amusements* (New York, 1993); William Leach, *Land of Desire: Merchants, Power, and the Rise of a New American Culture* (New York, 1993); and Richard Fox and T. J. Jackson Lears, eds., *The Culture of Consumption: Critical Essays in American History* (New York, 1983), esp. 1–38, 101–41.

Dwight Macdonald, one of the nation's most trenchant critics, summed up many of these feelings in 1960 in a widely cited essay, "Masscult and Midcult." It took special aim at the state of the arts. "Masscult," he complained, was crassly commercial work—we know it when we see it. "Midcult, " exemplified by the Book-of-the-Month Club, Ernest Hemingway's *The Old Man and the Sea*, and middle-class magazines like the *Saturday Evening Post*, was more insidious because it "pretends to respect the standards of High Culture while in fact it waters them down and vulgarizes them."[3] "Mass society" had arrived but, alas, it was driving quality from the arts and threatening to deprive even intelligent people of their ability to discriminate between what was artistically enduring and what was merely cheap and commercial.[4]

Critics such as Macdonald identified many baleful consequences of the spread of the consumer culture in the postwar era. Techniques of sales promotion, greatly abetted by the explosion of advertising and public relations, increasingly resorted to "hype" to push a cornucopia of new products and patterns. Just as auto manufacturers and fashion designers changed their styling every year, so did other producers. Painters, sculptors, gallery owners, and curators struggled so hard to come up with what was "new" that by the 1960s they had virtually abolished the concept of avant-garde. By then pop art that replicated consumer goods, such as Andy Warhol's renditions of Campbell's Soup cans, was becoming the latest of many high-priced rages in a market-driven Art World. Warhol's intent was partly satirical. Still, he profited immensely. His visibility by the early 1960s suggested that the boundaries between High Culture and Popular Culture were becoming increasingly blurred over time.

In some ways, however, those who focused on the debasement of American aesthetic life tended to exaggerate their case. The 1950s witnessed the rise to artistic prominence of a number of essayists and novelists—J. D. Salinger, Ralph Ellison, Saul Bellow, Bernard Malamud, James Baldwin, John Updike, Philip Roth—whose work received widespread critical acclaim and remained much admired in later years.[5] It

3. Originally in *Partisan Review*, Spring 1960. Reprinted in his *Against the American Grain* (New York, 1962), quote on 37.
4. Evaluations of Macdonald and other critics include Michael Wreszin, A *Rebel in Defense of Tradition: The Life and Politics of Dwight Macdonald* (New York, 1994); Richard Pells, *The Liberal Mind in a Conservative Age: American Intellectuals in the 1940s and 1950s* (New York, 1985), 174–82, 348–49; William O'Neill, *Coming Apart: An Informal History of America in the 1960s* (Chicago, 1971), 225–26; and Gilbert, *Cycle of Outrage*, 118–20.
5. The new literary canon, however, mostly excluded women writers.

was equally unfair to dismiss the Art World as only faddish or strictly of "midcult" quality. Thanks in part to the flight of European artists and intellectuals to the United States during the 1930s and 1940s, New York City emerged after the war as a vibrant center for painters, sculptors, dramatists, actors, dancers, writers, musicians, and others in the arts. The New York School of abstract expressionist painting by Americans such as Jackson Pollock attracted international attention in the 1940s and early 1950s.

Seekers of High Culture in the United States, moreover, did not need to despair. Recordings of classical music sold well during the 1950s and later. Theater companies. art museums, and symphony orchestras, though struggling amid the decline of central cities, usually survived, even in relatively small markets. Sales of books—including classics of literature—were bolstered by the paperback revolution that exploded after 1945.[6] What these quantitative trends indicated was surely a matter of opinion; who could prove, for instance, that most of the people who went to art museums "appreciated" what they saw? Still, it was equally hard to demonstrate that taste—surely a subjective word—declined in the 1950s, especially in a country with increasing numbers of highly educated people.

The output of Hollywood in the late 1940s and 1950s exposes some of the hazards of sweeping generalizations about American culture. Movie producers, rapidly losing viewers, sought anxiously to bring people back to the theaters. Most responded, as usual, by playing cautiously to a mass audience. Westerns proliferated, many of them celebrating heroic men, submissive women, and treacherous and dirty Indians. Other movies played to Cold War obsessions by raising the specter of communism and glorifying strong military authorities.[7] Cinerama in 1952, 3-D movies in 1953, and stereo sound and CinemaScope and VistaVision in the mid-1950s resorted to technological wizardry in the hope of enticing big turnouts. *The Searchers* (1956) featured a ten-foot-tall John Wayne. Lots of films relied on lavish display, panoramic photography, vivid Technicolor (standard by the late 1950s), hackneyed boy-gets-girl scenarios, bib-

---

6. Arlene Skolnick, *Embattled Paradise: The American Family in an Age of Uncertainty* (New York, 1991), 56–57; Janice Radway, *Reading the Romance: Women, Patriarchy, and Popular Literature* (Chapel Hill, 1984), 25–30; Morris Janowitz, *The Last Half-Century: Societal Change and Politics in America* (Chicago, 1978), 351–52.

7. Nora Sayre, *Running Time: Films of the Cold War* (New York, 1982), 195–98. See also chapter 7 on the Red Scare, above.

lical extravaganzas (such as Cecil B. De Mille's *The Ten Commandments* in 1956), and predictable Hollywood endings aimed at sending people happily out of the theater.[8]

A few movies of the early postwar era, however, considered serious subjects. Among them were *Gentleman's Agreement* (1947), in which Gregory Peck coped with anti-Semitism; *Pinky* (1949), a controversial film that broke ground by showing a black woman passing for white; *The Man with the Golden Arm* (1954), on drug addiction; and *On the Waterfront* (1954). *Rebel Without a Cause* (1955) gave James Dean the chance to emote about adolescent rebellion. Hollywood even brought forth a few anti-war movies, including Stanley Kubrick's powerful *Paths of Glory* (1957), and *On the Beach* (1959), which concerned the horrors of nuclear catastrophe.[9] As these examples suggest, movies in the 1950s, while for the most part tame and unventuresome, offered a modest amount of range. Here, as elsewhere in the realm of cultural criticism, complaints about the rapid debasement of culture were excessive.

This range suggested that critics such as Macdonald tended to overlook two things. First, there was the obvious point that the mass media were necessarily commercial in nature. It stood to reason that editors and movie producers would seek to maximize profits by going after mass markets. So to a lesser extent did impresarios in the Art World. The result was predictable: much cultural production continued to be aimed at popular, not elite, taste. And it did not assail contemporary social norms, such as racial segregation. These realities existed throughout the commercialized Western world. By the late 1950s, when the people of western Europe were at last climbing out of the trough of wartime destruction, critics across the Atlantic were bewailing the "Americanization" (or "Coca-Colaization") of Europe.

Elitist critics erred also when they suggested or implied that the purveyors of mass culture commanded hegemonic power over the minds and values of people. It was true, of course, that film producers, like owners of radio and television stations, possessed vast economic resources that they might use to shape public tastes. Critics understandably worried much

8. Peter Biskind, *Seeing Is Believing: How Hollywood Taught Us to Stop Worrying and Love the Fifties* (New York, 1983), esp. 44–56, 117–20. Also see Douglas Gomery, "Who Killed Hollywood?" *Wilson Quarterly* (Summer 1991), 106–12.

9. The anti-war themes, to be sure, were cautiously packaged: *Paths of Glory*, while uncompromising, took on a relatively safe subject—the idiocy of foreign armies in World War I. Hollywood, like other creators of popular culture after 1945, mostly celebrated American involvement in World War II.

about this power. It was far less clear, however, that the producers and owners succeeded in molding popular opinions in any thoroughgoing way. For one thing, they had many "publics" to satisfy. Hedging their bets, they cooked up a smorgasbord of things for consumers to feast on. An increasing variety of films, television shows, music, drama, books, and magazines—compare a good newsstand or music store of the 1960s with one of the 1940s—arose in an effort to sate these specialized appetites.

Consumers, moreover, clung to their own tastes. As post-modern critics were later to emphasize, "texts" (whether High Culture or popular entertainment) received distinctively personal "readings" from individuals.[10] People are not easily programmed. That this is the case can be demonstrated by exploring the world of television, a dynamic force in the explosion of mass culture during the 1950s.

WAYNE COY, A MEMBER of the Federal Communications Commission, recognized in 1948 the forthcoming triumph of television. "Make no mistake about it," he said, "television is here to stay. It is a new force unloosed in the land. I believe it is an irresistible force."[11]

His prediction proved prescient, for TV grew enormously in the next few years. In 1948 it was still in its infancy; few Americans had ever seen it. People relied instead on radio, which in 1949 had 1,600 stations compared to twenty-eight for television. Then the TV boom: 172,000 American households had television in 1948, 15.3 million by 1952. In 1955 there were 32 million television sets in use, in roughly three-fourths of households. The growth continued: by 1960 some 90 percent of households, including once isolated shacks in the Deep South and elsewhere, owned at least one set; bars and restaurants catered to many other viewers. Color sets then came into vogue. By 1970, 24 million homes (38 percent of the total) had color TV.[12]

As innumerable commentators noted at the time, television stood as an

10. See George Lipsitz, *Class and Culture in Cold War America: A Rainbow at Midnight* (New York, 1981), 173–94; and John Fiske, ed., *Understanding Popular Culture* (Boston, 1989), 18–21, 159–62.
11. Joseph Goulden, *The Best Years, 1945–1950* (New York, 1976), 175.
12. Erik Barnouw, *Tube of Plenty: The Evolution of American Television* (New York, 1982); Karel Ann Marling, *As Seen on TV: The Visual Culture of Everyday Life in the 1950s* (Cambridge, Mass., 1994); Stephen Whitfield, *The Culture of the Cold War* (Baltimore, 1991), 153–54; Douglas Gomery, "As the Dial Turns," *Wilson Quarterly* (Autumn 1993), 41–46; William Leuchtenburg, *A Troubled Feast: American Society Since 1945* (Boston, 1973), 67.

icon in American homes. Millions of families dropped other activities to watch the early stars, such as the comedians Milton Berle, Arthur Godfrey, Lucille Ball, and Jackie Gleason. Water companies reported enormous increases in usage during commercial breaks. Families suspended talk during meals to watch the "tube," especially after the advent of TV Dinners in 1954. Ball, star of the immensely popular "I Love Lucy," captivated audiences in late 1952 as the date arrived for her (real-life) baby. When she featured the blessed event on January 19, 1953, some 44 million people tuned in to watch. The show received the highest TV rating (68 percent) of any in the 1950s. (A smaller number of viewers, some 29 million, watched Eisenhower's televised inauguration the next day.) By the 1960s polls reported that television was the favorite leisure activity of nearly 50 percent of the population and that TV sets were on for an average of more than three hours a day in American homes. [13]

In its early years (until 1951 or 1952) television seemed promising to people with high-brow tastes. From their perspective this was something of a Golden Age free of the crassest kinds of commercial pressures. Programs such as "Playhouse 90" and the "Kraft Television Theater" featured live drama with performers—Grace Kelly, Paul Newman, Joanne Woodward, Eva Marie Saint—starring in plays crafted by well-known writers. [14] "Meet the Press," moving over from radio, began on NBC in 1948. Edward R. Murrow's sometimes iconoclastic "See It Now" programs, such as those concerning Joe McCarthy in 1954, suggested that TV could challenge the norms of American politics.

Programming such as this, however, could dominate only so long as most viewers were relatively well-off and highly educated: in 1950 television sets still cost between $400 and $500, sums far beyond the reach of most families. As the price of sets plunged within the next few years, sponsors insisted on shows that would appeal to mass audiences. TV producers then had to be more careful about controversial material. As one advertiser put it, "A program that displeases any substantial segment of the population is a misuse of the advertising dollar." [15] CBS slowly downgraded Murrow's show to the point that it became an occasional documentary—"See It Now and Then," cynics called it by 1958. News programs, never important in scheduling decisions, scarcely mattered.

13. Janowitz, *Last Half-Century*, 337–338; Landon Jones, *Great Expectations: America and the Baby Boom Generation* (New York, 1980), 120–23.

14. James Baughman, *The Republic of Mass Culture: Journalism, Filmmaking, and Broadcasting in America Since 1941* (Baltimore, 1992), 48–54.

15. Godfrey Hodgson, *America in Our Time* (Garden City, N.Y., 1976), 148.

Until September 1963 the evening news lasted only fifteen minutes. Newscasters—such as Chet Huntley and David Brinkley, who reported the news on NBC for almost fourteen years beginning in 1956—lacked the technology of videotape in the 1950s and mostly contented themselves with showing film footage and reading a script. Marquis Childs, an advocate of serious news programming, concluded in 1956 that "the effect at meaningful and vital communication [of current events] simply has not been made."[16]

Technological developments further abetted mass production of TV offerings, thereby driving live drama from the screen. As early as 1951 Ball and her co-star–husband Desi Arnaz used edited film instead of live performance. Hollywood, seeking revenues as downtown movie theaters closed, began producing "made-for-TV" films. These could be edited and rerun innumerable times, thereby generating much higher profits for the effort involved.[17] By the mid- to late 1950s pre-staged series dominated TV prime-time, among them highly popular westerns such as "Cheyenne," "Gunsmoke," "Maverick," and "Have Gun—Will Travel," detective stories such as "Dragnet," "77 Sunset Strip," "Perry Mason," and "Hawaiian Eye," and comedies featuring performers lured from radio such as Jack Benny and George Burns and Gracie Allen.

By then critics were already assailing the banalities of the "boob tube." T. S. Eliot described TV as a "medium of entertainment which permits millions of people to listen to the same joke at the same time, and yet remain lonesome." Even Eisenhower, who became an avid watcher, said in 1953, "If a citizen has to be bored to death, it is cheaper and more comfortable to sit at home and look at television than it is to go outside and pay a dollar for a ticket."[18]

The quest for mass audiences led networks (CBS and NBC dominated at the time) to support general-interest programs. One of these was "The Ed Sullivan Show," perennially one of the most highly rated offerings of the late 1950s. Hosted on Sunday nights by the poker-faced and uncharismatic Sullivan, the show paraded the currently most popular entertainers. In 1952 NBC put on its early morning "Today Show" featuring Dave Garroway. Before then the networks had assumed that few people would tune in at an early hour of day: many channels had been blank. At first the show did not do well, but Garroway then brought on stage a chimpanzee, J. Fred Muggs. The chimp excited children, then adults,

16. Baughman, *Republic of Mass Culture*, 61.
17. J. Ronald Oakley, *God's Country: America in the Fifties* (New York, 1986), 103–10.
18. Ibid., 107; Baughman, *Republic of Mass Culture*, 74–75.

and "The Today Show" became a popular fixture. Cartoons soon dominated morning TV on weekends.[19]

The networks made special efforts to chase after the greatly expanding audiences of middle-class white suburbanites. They caught them with programs such as "Father Knows Best," "The Adventures of Ozzie and Harriet," and "Leave It to Beaver," all of which celebrated the comical but mostly triumphant experiences of middle-class, nuclear families. Most shows concerning ethnic or working-class people, such as "The Goldbergs" and "I Remember Mama," were dropped; few programs featured blacks.[20] TV producers catered carefully to sponsors (such as cigarette companies) and worked hard to reflect the norms of their viewers. They favorably portrayed businessmen and professional people such as doctors, lawyers, and scientists. Political issues were mainly off limits. So was frankness about sex: Lucy's pregnancy was mostly mentioned as an "expectancy" or some other euphemism.[21] Fathers tended to be all-knowing, mothers all-supportive (and always at home), and children (though frisky) ultimately obedient and loving. Except on the soaps, which had a grim edge not found in prime-time, nothing very bad ever happened.

By the mid-1950s the networks devoted increasing time to quiz shows. These required very little money to produce and attracted sizeable audiences who watched contestants strive to win Big Money. Although the popularity of these shows has been exaggerated, the best-known of them, "The $64,000 Question," was the nation's most widely watched program in the 1955–56 season and fourth in 1956-57. These shows came crashing to earth in October 1959, following scandalous revelations (concerning "Twenty-One") that producers had rigged the outcomes. Staging quizzes in this way assured that telegenic contestants, such as Charles Van Doren, remained on the air. Van Doren, son of a well-known Columbia University professor, stayed for an exciting run of fourteen weeks during which he was given answers in advance. He won $129,000.[22]

19. Statistics on audiences and many other matters may be found in Tim Brooks and Earle Marsh, *The Complete Directory to Prime Time Network TV Shows, 1946–Present*, 5th ed. (New York, 1992), 802–4.
20. "The Honeymooners," featuring Gleason as a bus driver, focused on working-class life and did very well. So did "The Life of Riley."
21. Oakley, *God's Country*, 104. See also James Davidson and Mark Lytle, "From Rosie to Lucy: The Mass Media and Images of Women in the 1950s," in Davidson and Lytle, *After the Fact: The Art of Historical Detection* (New York, 1992), 303–28.
22. The quiz show scandals have been descibed by many historians, including Whitfield, *Culture of the Cold War*, 172–76.

The quiz show scandals so embarrassed network executives that they resolved to promote more public service. ("Public relations" became a necessary adjunct to the consumer culture of the 1950s.) News staffs increased, current affairs documentaries made a comeback, and the networks offered to televise debates between the major presidential candidates in the 1960 campaign. In September 1963 they lengthened evening news shows to thirty minutes. Critics of TV, however, remained unimpressed. Television, they insisted, remained mindless. Programs and commercials pandered shamelessly to the money-making mania of the consumer culture. Newton Minow, chairman of the Federal Communications Commission, summed up these criticisms in 1961 by asserting, in a phrase that stuck in public consciousness, that television had become a "vast wasteland." He continued, "You will see a procession of game shows, violence, audience participation shows, formula comedies about totally unbelievable families, blood and thunder, mayhem, violence, sadism, murder, western badmen, western good men, private eyes, gangsters, more violence, and cartoons. And endlessly, commercials—many screaming, cajoling, and offending. And most of all, boredom."[23]

Minow overgeneralized, for TV was too diverse to deserve such a blanket characterization. Still, there was no doubting that television highlighted soaps, unsophisticated comedies, and violence (though not nearly so much as in later years). As a uniquely intimate medium that invaded millions of private homes, it also had to guard with special care against offending people. Far from criticizing social mores, it mainly reflected existing norms and institutions.

It remains difficult, however, to assess the impact of television on the values and beliefs of people. Debates over the cultural influence of TV, indeed, raged long after the 1950s. Those who thought this power was large argued that television strengthened violent tendencies in people, sabotaged the reading habit, stifled conversation (especially in families), and induced a general passivity of mind. *Why Johnny Can't Read*, which focused on the role of TV, became a best-seller in 1955.[24] Critics added that TV did much to debase politics, both by lowering (through "spots") the level of discussion and by increasing the costs of campaigning. It could make (Kennedy) or break (McCarthy) the careers of politicians. Critics

23. Oakley, *God's Country*, 110. David Karp added a few years later that TV was the "cheekiest, vulgarest, noisiest, most disgraceful form of entertainment since bearbaiting, dog-fighting, and the seasonal Czarist Russian pogrom" (*New York Times Magazine*, Jan. 23, 1966).
24. By Rudolph Flesch.

maintained also that television harmed radio, newspapers and magazines, and the motion picture industry, while giving enormous boosts to the advertising business and successful sponsors. When Walt Disney put on one-hour TV shows about the frontiersman Davy Crockett in 1955, he sold $300 million worth of Davy Crockett dolls, toys, T-shirts, and fake coonskin caps.[25]

Other, less hostile observers thought television did much to develop and define a more national culture. As the networks sent out nationwide messages (and commercials) to people, these analysts concluded, they helped to standardize tastes and to diminish provinciality and social division. One widely noticed version of this argument, by the Canadian critic Marshall McLuhan, went so far as to say that television was creating an interconnected "global village." McLuhan thought this globalization might be a good thing, for the "new electronic interdependence" would lead to a "single consciousness" that would link people not only throughout the country but ultimately throughout the world.[26]

Some of these generalizations about the impact of television seem irrefutable. Television indeed spurred the growth of advertising and advanced the careers of telegenic politicians such as Kennedy. Over time, mainly after 1960, it accelerated two important political trends: the rise of personalized TV campaigns and the weakening of party discipline and organization. Some magazines and newspapers also suffered a bit because of the surge of TV: *Life*, America's preeminent magazine of photojournalism, lost 21 percent of its circulation within six months in 1954; like other general-interest magazines such as *Saturday Evening Post*, *Look*, and *Collier's*, *Life* later collapsed.[27] Television advertising helped to work minor miracles for some sponsors, such as Revlon, which rode "The $64,000 Question" to striking increases in sales. There is no doubt that television reinforced the already rising consumerism that was such a prominent feature of the 1950s.

25. Marty Jezer, *The Dark Ages: Life in the United States, 1945–1960* (Boston, 1982), 132; David Farber, *The Age of Great Dreams: America in the 1960s* (New York, 1994), 49–66.
26. Among his best-known works were *The Gutenberg Galaxy: The Making of Typographic Man* (Toronto, 1962), and *Understanding Media: The Extensions of Men* (New York, 1964). A thoughtful evaluation of McLuhan's thinking is William McKibben, "Reflections: What's On?," *New Yorker*, March 9, 1992, pp. 40ff.
27. Whitfield, *Culture of the Cold War*, 153–54. See also Ben Bagdikian, *The Media Monopoly* (Boston, 1983); and William Tillinghast, "Declining Newspaper Readership: The Impact of Region and Urbanization," *Journalism Quarterly*, 58 (1981), 14–23.

Still, television was hardly all-powerful, even in the early years when it was a novelty for people. Although some magazines struggled against competition from TV, most managed all right, and a few, reaching out to specialized audiences, found growing markets. *Sports Illustrated*, starting in 1954, was but one example of a trend toward such publications, which over time heightened diversity among American magazines. Many women's magazines also flourished. Radio and films, too, found new ways to compete, often by targeting special groups: think of radio stations catering to particular tastes in music.

Television indeed lulled people into hours of sedentary viewing every day, but watchers were often far from passive; on the contrary, they frequently laughed delightedly or argued heatedly about the meaning of what they had seen. Studies of audiences suggested also that commercials did not sweep all before them. Many ads sustained sales or brand-name preferences, but it was harder to establish needs that people did not have already. Millions of people smoked and bought big cars long before the explosive rise of TV commercials. (Steadily high percentages of Americans kept on smoking even after cigarette ads were banned from radio and TV in 1971.) Viewers commonly hooted out loud at exaggerated claims for products. [28]

In the 1980s and 1990s, when "popular culture studies" became a thriving scholarly pursuit, writers were still debating the influence of television on American culture. [29] Many persisted in arguing that it was great, contributing, for instance, to long-run declines in educational test scores and to crime in the streets—developments that grew especially pronounced after 1963. Other analysts, however, doubted the strength of causal connections. Americans, they insisted, look at the "texts" of TV as they do other aspects of mass culture, in highly individualized ways. Viewers are not passive receptacles; they make choices. [30] The class, gender, religion, and ethnicity of people especially matter in affecting responses.

28. Herbert Gans, *The Urban Villagers: Group and Class in the Life of Italian-Americans* (New York, 1962), 187–96.
29. For reviews of some of this writing, see Frank McConnell, "Seeing Through the Tube," *Wilson Quarterly* (Sept. 1993), 56–65; Douglas Davis, *The Five Myths of Television Power* (New York, 1993); Fiske, *Understanding Popular Culture*, 134–36; and Gilbert, *Cycle of Outrage*, 109–26.
30. See Lawrence Levine, "The Folklore of Industrial Capitalism: Popular Culture and Its Audiences" (and rejoinders by others), *American Historical Review*, 94 (Dec. 1992), 1369–1430.

Although the jury remains out, this perspective on TV seems persuasive. Millions of American TV-watchers—in the 1950s and thereafter—remained stubbornly attached to regional, ethnic, or racial subcultures and resisted aspects of the more homogenized "outside world" that the mass media thrust at them. Nothing, it seemed, could shake the love that many Italian-Americans maintained for Frank Sinatra. Professional wrestling, a sham, nonetheless attracted large and enthusiastic audiences, especially among the working classes. The power of these personal preferences continued to divide the "global village" and to curb the capacity of TV to affect the behavior of people. [31]

MUCH OF THE HAND-WRINGING about "mass culture" in the 1950s came from the Left. From right-wing contemporaries came different laments: over the rise of sexual liberation, juvenile delinquency, and generational change.

Worries about sexual liberation were hardly new, of course, to the 1950s. Reformers and moralists had battled against prostitution and the "white slave trade" in the late nineteenth and early twentieth centuries and fretted about "flappers" and "companionate marriage" in the 1920s. Rising divorce rates had greatly alarmed Americans from 1900 onward. The social turmoil of World War II intensified such fears, as newspapers carried lurid accounts of "Victory girls," "khaki-wackies," and "good-time Charlottes" who gave themselves freely to GIs. Americans worried especially during the war about the exposure of soldiers to venereal diseases. Stories proliferated in the 1950s about "sex crimes." "Petting" seemed rapidly on the rise among unmarried people. Most of these concerns reflected long-standing assumptions about class, race, and gender, especially the double standard applied to the sexes. Lower-class white men (and blacks of both sexes), it was often said, acted like animals. What had to be nipped in the bud, traditionalists said, was greater sexuality among middle-class women, especially the young and the unmarried. [32]

It was left to Dr. Alfred Kinsey, an Indiana University entomologist, to focus these fears during the early 1950s. In 1948 he brought out his first book on American sexuality, *Sexual Behavior in the Human Male*. Pro-

---

31. Gans, *Urban Villagers*, 194–96; Radway, *Reading the Romance*, 209–22; John Fiske, *Television Culture* (London, 1987), 314–17.

32. John D'Emilio and Estelle Freedman, *Intimate Matters: A History of Sexuality in America* (New York, 1988), 260–62; George Chauncey, Jr., "The Postwar Sex Crime Panic," in William Graebner, ed., *True Stories from the American Past* (New York, 1993), 161–78.

duced with little fanfare by a medical publisher, it relied on a host of
interviews that Kinsey and his associates had been collecting for years.
The book was 804 pages long, costly ($6.50), and full of jargon, charts,
and graphs. Nonetheless, it soared quickly to the best-seller lists. His
second volume, *Sexual Behavior in the Human Female*, came out five
years later. It ultimately sold some 250,000 copies and created a minor
sensation.

It was not hard to see why, for Kinsey's books offered statistics that
staggered Americans at the time. Between 68 and 90 percent of American
males, he reported, had engaged in premarital sexual intercourse, as had
nearly 50 percent of females; 92 percent of males and 62 percent of
females had masturbated at least once; 37 percent of males and 13 percent
of females had had at least one homosexual experience; 10 percent of
males had led a more or less exclusively homosexual style of life within
the previous three years; 50 percent of men and 26 percent of women had
committed adultery before the age of 40. Kinsey added that around 8
percent of men and 4 percent of women had had some kind of sex with
animals.[33]

Kinsey's volumes encountered angry criticism from virtually all directions. A few bookstores hid the volumes. Some libraries refused to buy
them or to place them in general circulation, thereby forcing patrons to
come up to the desk and openly ask for them. Many writers disputed
Kinsey's statistics, contending that they were based on interviews with
people, including large numbers of prisoners, who spun elaborate tales
about nonexistent sexual exploits. In so doing, these reviewers said,
Kinsey painted a falsely oversexed portrait of American society—one that
thereby encouraged an "everybody's doing it" mentality. The Indiana
Roman Catholic Archdiocese announced that Kinsey's studies helped
"pave the way for people to believe in communism."[34] A minister added
that Kinsey "would lead us, like deranged Nebuchadnezzars of old, out

33. The books were published by W. B. Saunders of Philadelphia. Convenient sum-
    mations of Kinsey's data and influence can be found in Goulden, *Best Years*, 188–
    94; William O'Neill, *American High: The Years of Confidence, 1945–1960* (New
    York, 1986), 45–48; and Regina Markell Morantz, "The Scientist as Sex Cru-
    sader: Alfred C. Kinsey and American Culture," *American Quarterly*, 29 (Winter
    1979), 563–89. O'Neill called Kinsey "one of the great revolutionaries of private
    life" (48). John Burnham, *Bad Habits: Drinking, Smoking, Taking Drugs, Gam-
    bling, Sexual Misbehavior, and Swearing in American History* (New York, 1993),
    190–91, is much more censorious.
34. Whitfield, *Culture of the Cold War*, 184–85.

into the fields to mingle with the cattle and become one with the beasts of the jungle."[35]

Other reviewers, including some who distanced themselves a bit from moralists such as these, joined in the fray, which spilled over into newspapers and magazines for months after each of the volumes appeared. Reinhold Niebuhr worried that the books would encourage excessive sexual freedom. Margaret Mead predicted that Kinsey's findings might "increase the number of young men who may indulge in 'outlets' with a sense of hygienic self-righteousness."[36] Lionel Trilling, one of America's most respected literary critics, agreed that sensible talk about sex might dispel old myths and hang-ups. But he objected to the false pose of scientific objectivity that he thought Kinsey affected, and to Kinsey's simple-minded reduction of sex to physical activity, especially orgasm. Kinsey, he charged, used "facts" to celebrate an ideology of "liberation" and a "democratic pluralism of sexuality."[37] The Rockefeller Foundation, which had financed many of Kinsey's researches, bowed to rising criticism of the reports and cut off his funding in 1954.[38]

Whether Kinsey's data were accurate remains controversial years later. He was probably correct, for instance, to conclude that homosexuality was more widespread than Americans at the time wished to believe. Indeed, some gay people were beginning to organize. A group of homosexual men in Los Angeles formed the Mattachine Society in 1951 with the hope of promoting more liberal understanding and arousing opposition against the vicious harassment of "queers" by police and other authorities.[39] On the other hand, because Kinsey conducted many of his interviews with prisoners, he may have exaggerated the extent of homosexual behavior.[40]

Still, there was no doubting that Kinsey and his associates had done a great deal of research, including some 18,000 lengthy personal interviews. There were no better data at the time. Moreover, he was surely

35. O'Neill, *American High*, 47–48.
36. Goulden, *Best Years*, 194.
37. Lionel Trilling, "The Kinsey Report," in Trilling, *The Liberal Imagination: Essays on Literature and Society* (New York, 1950), 223–42.
38. Indiana University, however, continued to support Kinsey. The Institute for Sex Research there remained a center for such studies.
39. For Mattachine, see John D'Emilio, *Sexual Politics, Sexual Communities: The Making of a Homosexual Minority in the United States, 1940–1970* (Chicago, 1983), 58–86.
40. "Life Before Stonewall," *Newsweek*, July 4, 1994, pp. 78–79.

accurate in pointing out that various forms of amatory activity, especially premarital sex, had risen steadily in America during the twentieth century. Each new age cohort of young people was more sexually active— and at earlier ages—than the one that preceded it. This was as true for women as for men and for people in the middle classes as well as for those lower on the social scale. In reporting such trends Kinsey helped to demystify sex.[41]

While *Sexual Behavior in the Human Female* was still generating angry editorials, *Playboy* arrived on the newsstands, in December 1953. It was the creation of Hugh Hefner, a young college graduate who had worked for *Esquire*. Neither Hefner nor anyone else expected the issue to sell very well. But it did, in large part because its "centerfold" carried a nude photograph of Marilyn Monroe, who was already a well-known young star. The issue sold 53,000 copies, enabling Hefner (who had gone into debt to finance publication) to bring out more. Within a year *Playboy* issues were selling over 173,000 copies; by 1960 its circulation per month was over a million. By then Hefner had branched out to establish Playboy Clubs in major American cities. In 1961 his pre-tax profits from *Playboy* approached $1.8 million while his profits from the clubs were nearly $1.5 million.[42]

The rise of *Playboy* revealed the onrushing growth of America's consumer culture. Hefner deliberately and gleefully advertised a "playboy philosophy" of self-gratification. Like the expensive ads (many of them pushing cigarettes and liquor) that puffed up the size of his magazines, he equated satisfaction with hedonistic consumption. In so doing he appealed to two central fantasies of the modern consumer culture: first, that people should be free, and second, that happiness inhered in material things. The genius of Hefner, said sex researcher Paul Gebhard in 1967, was that "he has linked sex with upward mobility." Another critic added, "Real sex [for Hefner] is something that goes with the best scotch, twenty-seven-dollar sun glasses, and platinum-tipped shoelaces."[43]

41. The next major survey of American sexual behavior did not appear until 1994: Edward Laumann et al., *The Social Organization of Sexuality: Sexual Practices in the United States* (Chicago, 1994). It, too, relied mainly on interviews and provoked debate.
42. Oakley, *God's Country*, 306–7; Halberstam, "Discovering Sex."
43. Burnham, *Bad Habits*, 194–96. A provocative feminist critique emphasizing that *Playboy* especially appealed to young men seeking to be free and irresponsible is Barbara Ehrenreich, "Playboy Joins the Battle of the Sexes," in Ehrenreich, *The Hearts of Men: American Dreams and the Flight from Commitment* (New York, 1983), 42–51.

The surging circulation of *Playboy* exposed how flimsy the floodgates of traditionalism were becoming, and others soon rushed through the gap to ride on the big business of sex. One was Grace Metalious, a thirty-two-year-old New Hampshire housewife who produced *Peyton Place* in September 1956. Until then Metalious had never published a line, and her prose was overwrought. But her novel was graphic for its time about sex and made good on the claim of its jacket: to "lift the lid off a small New England town." Thanks mainly to its sexual openness (and perhaps because of a quasi-feminist subtext that appealed to female readers), *Peyton Place* sold 6 million copies by early 1958. When it surpassed 10 million it became the best-selling novel ever, overtaking Erskine Caldwell's *God's Little Acre* (1933) (which itself focused on sex).[44]

The gates kept falling. Vladimir Nabokov, a highly regarded writer, brought out *Lolita,* a book that (among many other things) highlighted the adventures of a sexy "nymphet," in 1958. It sold more than 3 million paperback copies and rose to number three on the fiction best-seller list. Heartened by the flow of profits, publishers and civil libertarians joined forces to tear down the now crumbling barriers that had protected censorship of sexually explicit materials. In 1958 they succeeded, when the unexpurgated Grove Press edition of D. H. Lawrence's *Lady Chatterley's Lover* appeared in American bookstores. By 1959 the book had risen to number five on the list of best-sellers.

To permit writing about sex was one thing; to have sex flashed on the silver screen, where it could be seen in the darkness by millions and millions of people, was another. For years states had censored movies. Since 1934 Hollywood had tried to honor a self-regulating Production Code that banned explicit handling of sensitive subjects, including homosexuality, incest, and interracial romance. Profanity was forbidden. Movies must not show the "intimate parts of the body—specifically the breasts of women." "Impure love," the Code said, "must never be presented as attractive and beautiful. . . . Passion should be so treated that these scenes do not stimulate the lower and baser element."[45]

But Hollywood, too, became more liberal in its depiction of sexuality in the 1950s. In 1953 it had explored the theme of adultery in *From Here to Eternity.* A few months after publication of *Peyton Place* it brought out *Baby Doll,* featuring the sultry presence of Carroll Baker. *Time* called the film (which was tame by later standards) "just possibly the dirtiest

44. Oakley, *God's Country,* 307–8.
45. Code as described in *New York Times,* Feb. 4, 1994. See also Gilbert, *Cycle of Outrage,* 169, 189.

American-made motion picture that has ever been legally permitted."[46] Hundreds of theater owners, worried about the reactions of traditionalists, refused to show it. Francis Cardinal Spellman of New York was so incensed that he took to the pulpit of St. Patrick's Cathedral, for the first time in seven years, in order to denounce it. Catholics who saw this "evil" and "revolting" film, the cardinal proclaimed, would do so under "pain of sin."[47]

Not even the Holy Mother Church could stem the tide, however. The Code, cut adrift under the rush of cultural change, gradually lost its force. Elizabeth Taylor, like Baker and Monroe, began to star in films that showed a good deal more female flesh than a few years earlier. The studios moved quickly to cash in on sexy books, putting out poorly made movie versions of *Peyton Place*, *Lolita*, and *Lady Chatterley's Lover* before the decade was out. They also began—very gingerly—to handle other controversial subjects: miscegenation in *Island in the Sun* (1957), homosexuality in *Compulsion* (1958), and abortion in *Blue Denim* (1959).

All these developments preceded yet another boon for sexual liberation in the United States. This was the decision of the Food and Drug Administration in May 1960 to approve the sale by prescription of Enovid, the first oral contraceptive for women. Although the decision at first attracted little attention—the *New York Times* carried the story on page 78— women reacted eagerly to arrival of "the Pill, " which promised at once to save them from unwanted pregnancies and to make them as free about sex as men had always been. The double standard, some imagined, might collapse. By the end of 1962, 2.3 million women were using the Pill. Clare Boothe Luce exulted, "Modern woman is at last free as a man is free, to dispose of her own body."[48]

The effects of the Pill, however, lay in the future. What is striking about the undoubted liberalization of sexual attitudes in the 1950s is how slowly some things seemed to change. Many institutions tried hard to hold a line: all but a few universities maintained restrictive parietal rules until the late 1960s. Their efforts suggested that parents still expected their children to conform to older norms of sexual behavior.

Other statistics suggested the enduring desire of people to find meaning in long-term, monogamous family life. Following the disruptions of wartime, divorce rates dropped sharply after 1947. Despite the passage of

46. Oakley, *God's Country*, 309.
47. Sayre, *Running Time*, 161.
48. Halberstam, "Discovering Sex," 48. See also Burnham, *Bad Habits*, 191; and O'Neill, *American High*, 48–50.

more liberal divorce laws in the 1950s, these rates remained lower throughout the 1950s (and early 1960s) than they had been since 1942. Illegitimacy rates also remained stable.[49] Marriage rates fell off a bit from the record peaks of the early postwar years but stayed very high. Despite world-record participation in higher education, Americans in the 1950s still married young (an average of around 22 for men and 20 for women) and were more likely to be married (90 percent or more at some point in their lives) than people anywhere else on the planet.[50] The baby boom did not slow until 1958, after which it declined only gradually through 1964.

Playboys (or girls) were surely not the norm. Public opinion polls revealed that young people in the 1950s unquestioningly expected to marry and raise families—what else was possible? These feelings were especially strong among women. "It is not a matter of 'want' or 'like to' or 'choice,'" one young women said about marrying. "Why talk about things that are as natural and as routine as breathing?" Another woman, asked why she hoped to marry and have children, responded, "Why do you put your pants on in the morning? Why do you walk on two feet instead of one?"[51]

Sexual liberation, driven in part by the wider liberalization of attitudes that came with the Biggest Boom Yet, and in part by the buoyant commercialization of the consumer culture, was gathering great power in the 1950s. Not until the 1960s, however, did its power really become obvious.

NO ASPECT OF LIFE in the 1950s seemed more clearly to expose the durability of traditional cultural norms than the images and status of women. Betty Friedan's *The Feminine Mystique* (1963), while overdrawn in various ways, struck a nerve by sketching the outlines of a world that assigned women to decorative and supportive roles in a rampantly materialistic consumer culture.[52]

A look at the evolution of women's fashions begins to reveal the

49. Randall Collins and Scott Cottrane, *Sociology of Marriage and the Family* (Chicago, 1991), 158.
50. Carl Degler, *At Odds: Women and the Family in America from the Revolution to the Present* (New York, 1980), 456–57.
51. Daniel Yankelovich, *The New Morality* (New York, 1974), 96–97.
52. A perceptive evaluation of Friedan's book—and other aspects of the 1950s—is Joanne Meyerowitz, "Beyond the Feminine Mystique: A Reassessment of Postwar Mass Culture, 1946–1958," *Journal of American History*, 79 (March 1993), 1455–82. An overview of women in the 1950s is Eugenia Kaledin, *Mothers and More: American Women in the 1950s* (Boston, 1984).

groundings of this world. In the war years, when unprecedented numbers of women had gone to work, it had been thought acceptable for women to wear slacks. In 1947, however, the designer Christian Dior introduced a "new look" that stressed femininity. Women's styles henceforth accentuated narrow waists to draw eyes to shapely hips, and tight tops to focus attention on the bosom. An extreme in the 1950s was the "baby doll" look that featured tightly cinched-in waists and bouffant skirts fluffed out by crinolines. Women's shoes, as one historian has said, "ushered in a bonanza for podiatrists." The toe shape got pointier and the heels so high that it seemed almost risky for women to walk. Women's fashions, largely prescribed by men who had an image of how the opposite sex should look, had hardly been so confining since the nineteenth century. [53]

The trend in fashions sent a much broader message: the place of women in society was as accessory, chiefly in the home as housewife and mother. No one made this more clear than Benjamin Spock, whose *Common Sense Book of Baby and Child Care* (1946) continued to sell extraordinarily well—almost a million copies every year of the 1950s. Spock was later chastised by conservatives for what they considered to be his "permissive" advice about child-rearing. Young people became radicals in the 1960s, they contended, because parents (mainly mothers) had failed to establish discipline in the home. That was at best an overwrought accusation. And it was ironic, for it overlooked Spock's much more traditional message: that children needed the love and care of mothers who devoted all their time to the effort, at least until children were three years of age. Fathers, Spock said, had a much smaller role in child-caring.

Women who tried to combine homemaking and career had to struggle, for "labor-saving" devices proclaimed as the deliverance of housewives stopped well short of such a rescue. The expanded availability of frozen and packaged foods, to be sure, enabled women to prepare meals more quickly. Central heating provided a good deal more comfort. Gas-powered ranges were much easier to manage than coal-burning stoves. Heavily physical demands eased for most urban and suburban housewives. But the more the conveniences, the more time it seemed to take just to keep things clean, especially for women who had (as many did) large families. Keeping the conveniences fixed and in good running order took lots of time and trips. Men, moreover, showed little inclination to take on these uninspiring chores. "Modern" housewives in the 1950s

53. Davidson and Lytle, "From Rosie to Lucy," 312.

therefore had little more time for careers, even if they had been encouraged by the culture to undertake them.

Magazines offered career women of the 1950s especially cold comfort. Women's magazines, as Friedan emphasized later, printed story after story extolling motherhood and domesticity. Titles of stories revealed the pattern: "Femininity Begins at Home," "Don't Be Afraid to Marry Young," "Cooking to Me Is Poetry." In 1954 *McCall's* left no doubt that women should be subservient: "For the sake of every member of the family, the family needs a head. This means Father, not Mother."[54] *Seventeen* advised women in 1957, "In dealing with a male, the art of saving face is essential. Traditionally he is the head of the family, the dominant partner, the man in the situation. Even on those occasions when you both know his decision is wrong, more often than not you will be wise to go along with his decision. "[55]

A major role of the wife, these stories said, consisted of helping the husband get ahead. Articles such as "The Business of Running a Home" explained that women should focus on freeing men from domestic concerns, including shopping, diapering, and other essential duties. In 1954 *McCall's* coined the neologism "togetherness," a word that nicely captured the ideal: the woman should be a helpmate so that the man could rise in the world. Mrs. Dale Carnegie, wife of a well-known expert on how to make friends and influence people, explained in 1955, "Let's face it girls. That wonderful guy in your house—and in mine—is building your house, your happiness and the opportunities that will come to your children." Split-level houses, she added, were all right, "but there is simply no room for split-level thinking—or doing—when Mr. and Mrs. set their sights on a happy home, a host of friends, and a bright future through success in HIS job."[56]

Many movies offered similar prescriptions, none more pointedly than *All About Eve* (1950), a powerful film about Eve Harrington (Anne Baxter), an aspiring young actress who feigns humility in order to ingratiate herself with Margo Channing (Bette Davis), the then first lady of the stage. By the end of the movie Eve's selfish machinations pay off; she rises to the top of her profession and receives a big award for her acting. As Margo realizes that Eve is taking her place, she becomes bitter and jealous. But then she sees the light and marries the man she loves. "A funny business, a woman's career," she muses. "You forget you'll need

54. Skolnick, *Embattled Paradise*, 71.
55. Jezer, *Dark Ages*, 247.
56. Halberstam, "Discovering Sex," 56, from *Better Homes and Gardens*.

[men] when you start being a woman again." Giving up a major role (which Eve lands), she says, "I've finally got a life to live. . . . I have things to do with my nights." The message was hard to miss: career women like Eve were evil schemers who did not understand the much more satisfying blessings of love and marriage.[57]

Within a few years career women like Eve seemed to disappear from the screen. A special example of this trend was *The Tender Trap* (1955), starring Debbie Reynolds in a characteristically feminine role. Reynolds, like Baxter, plays an aspiring actress, and she gets a much-sought-after role. Frank Sinatra, whom she really wants, congratulates her. Reynolds, however, replies, "The theater's all right, but it's only temporary." Amazed, Sinatra asks, "Are you thinking of something else? " Reynolds replies, "Marriage, I hope. A career is just fine, but it's no substitute for marriage. Don't you think a man is the most important thing in the world? A woman isn't a woman until she's been married and had children." In the end Reynolds gets her man.

It was a sign of the times that even radical intellectuals largely ignored the wider aspirations of women. One was Paul Goodman, a caustic critic of American institutions whose essays were collected in *Growing Up Absurd* in 1960. As this title indicated, Goodman was appalled by the consumerist world in which young people were trying to mature in the 1950s. But he worried only about males. "A girl," he explained in his introduction, "does not *have* to . . . 'make something' of herself. Her career does not have to be self-justifying, for she will have children, which is absolutely self-justifying, like any other natural or creative act."[58] In the essays that followed Goodman ignored the needs of women.

To amass so many traditional, anti-feminist images from the era is of course to be selective and therefore to leave the impression that American culture in the 1950s was monolithic, or even virtually misogynist, on the subject of gender relations. That would be an exaggerated conclusion. Images in television, for instance, offered some variety and provoked different reactions among viewers. "Our Miss Brooks," a popular series, featured an unmarried schoolteacher—a non-threatening stereotypical role for women. But while Brooks resorted to silly (and unsuccessful) feminine wiles to attract a male colleague, she was generally a far more

57. Written by Joseph Mankiewicz. It had an all-star cast including Gary Merrill, George Sanders, Celeste Holm, and (in the first role to gain her attention) Marilyn Monroe.
58. Paul Goodman, *Growing Up Absurd* (New York, 1960), 13; Pells, *Liberal Mind*, 214.

intelligent and worthy person than the pompous men who dominated the
school Establishment, which kept her down. Alice Kramden, the wife of
Ralph (Jackie Gleason) on "The Honeymooners," was obviously the
brains of the family. She paid little attention to Ralph's bluster or to his
crackpot ideas. Lucy, while wacky, was also shrewd: many women appeared to delight in the way that she manipulated her husband. This is
another way of repeating the obvious: people emerge with individual
understandings of what they see and hear in the media.

Movies, too, could offer a slightly more ambiguous view of the sexes
than met the eye. Some leading men—Montgomery Clift, Tony Perkins,
Dean—were shown to be soft and sensitive. Dean cried unashamedly in
East of Eden (1955). Women, meanwhile, could be decisive. Elizabeth
Taylor in *Giant* (1956) played a strong-willed eastern bride who gradually
tamed her rancher-husband (Rock Hudson). Even Debbie Reynolds, having caught Sinatra in *The Tender Trap*, let him know that she would
henceforth assert herself. "Listen to me," she told him. "From now on,
you're gonna call for me at my house, ask me where I want to spend the
evening, and you're gonna meet my folks and be polite to them and bring
me candy and flowers. . . . I've got to make a man out of you."

It is inaccurate, finally, to assume that all American women in the
1950s embraced the feminine mystique. Some were restless and unhappy.
As Friedan put it in an article in *Good Housekeeping* in 1960, "There is a
strange stirring, a dissatisfied groping, a yearning, a search that is going on
in the minds of women." What the searchers wanted, she added, was the
chance for greater self-fulfillment outside the home. "Who knows what
women can be," she asked, "when they finally are free to become themselves?" In the same month *Redbook* started a contest offering a $500 prize
for the best account of "Why Young Mothers Feel Trapped." To the
surprise of the editors, an avalanche of entries—24,000 in all—arrived at
their office. [59]

The varied messages about gender roles in the mass media and the
dissatisfaction of some mothers and housewives suggest an undercurrent
of change, especially toward the end of the decade. It is difficult, however, to know whether these manifestations of discontent amounted to
widespread feminist feeling. The people whom Friedan surveyed, for
instance, tended to be upper-middle-class white women. More highly
educated than average, they were better placed than most to act on the
deprivations that they felt. The majority of messages in the media, more-

59. Oakley, *God's Country*, 407; Jezer, *Dark Ages*, 223.

over, remained conservative. With some exceptions they did not directly challenge dominant cultural values, which assigned women to a secondary and domestic sphere.

It seems, indeed, that most American women in the 1950s did not chafe very strongly against the roles that were assigned to them. Friedan and others like her, to be sure, began to do so. But what of the millions of women who were delighted to move to the Levittowns and other suburban areas where at last they had decent housing and conveniences? Welcoming the consumer culture, they lived much more comfortably than their parents had, and they imagined that their children would do still better in life. Later, a number of developments—including still greater affluence and rising rights-consciousness, which excited expectations—helped to arouse a women's movement, especially among the young and the middle classes. As late as 1963, however, that movement remained hard to predict.[60]

It was difficult, moreover, for the majority of American women in the 1950s to expect very much in the way of advancement outside the home. Institutional barriers reflected and reinforced cultural prescriptions. Witness politics. President Eisenhower made a few highly touted appointments of women to governmental positions, notably Clare Boothe Luce as Ambassador to Italy and Oveta Culp Hobby, a Texan, to head the Department of Health, Education, and Welfare. In all he named twenty-eight women to positions requiring Senate confirmation, as opposed to twenty during the Truman years. But these appointments were token. And leading politicians of both parties had no intention of getting the Equal Rights Amendment approved. When Eisenhower was asked about ERA at a press conference in 1957, he responded, laughing, "Well, it's hard for a mere man to believe that a woman doesn't have equal rights. But, actually, this is the first time that this has come to my specific attention now, since, oh, I think a year or so. . . . I just probably haven't been active enough in doing something about it." Women activists were upset, but they should not have been surprised, to learn that the President had been paying no attention to their efforts.[61]

60. For a different argument that tends to blame Cold War culture for promoting domesticity, see Elaine Tyler May, "Cold War—Warm Hearth: Politics and the Family in Postwar America," in Steve Fraser and Gary Gerstle, eds., *The Rise and Fall of the New Deal Order, 1930–1980* Princeton, 1989), 153–81; and May, *Homeward Bound: American Families in the Cold War Era* (New York, 1988).
61. Cynthia Harrison, *On Account of Sex: The Politics of Women's Issues, 1945–1958* (Berkeley, 1988), 37, 59–61.

Witness, too, the world of education. Females, as in the past, actually stayed in high school on the average about a year longer than males—in 1950 through the tenth grade as opposed to the ninth grade for males (through the eleventh and tenth, respectively, by 1960). A slightly higher percentage of women than men graduated from high schools. But thanks in part to the blessings of the GI Bill, men were much more likely to go on to colleges and universities. In 1950 there were 721,000 women enrolled in higher education, compared to 1.56 million men. By 1960 the ratio (but not the gap) had narrowed a little: 1.3 million women attended colleges or universities compared to 2.26 million men. Only 37 percent of women graduated, compared to 55 percent of the men. Many of the women who dropped out did so, people joked, to get their M.R.S. degree and to work on their Ph.T.—"Putting Hubbie Through." Of those women who graduated, relatively small numbers went on for higher degrees, in part because graduate and professional schools had quotas limiting the percentages of women they would admit. A total of 643 women received doctorates in 1950, compared to 5,990 men. Ten years later the numbers were 1,028 for women and 8,800 for men.

Some women's colleges (which were generally headed by men) discouraged their students from taking "serious" subjects or preparing for careers. Mills College demanded a "distinctively feminine curriculum" and featured such subjects as ceramics, weaving, and flower arrangement. The president of Stephens College highlighted courses in interior decorating, cosmetics, and grooming, and added that for women "the college years must be rehearsal periods for the major performance" of marriage. Adlai Stevenson told Smith College graduates that he knew no better vocation for women than that they might assume the "humble role of housewife."[62]

Statistics concerning the world of work offered the clearest measures of the institutional barriers to gender equality. The percentage of working-age women who were in the labor force gradually increased in the 1950s, from 33.9 at the start of the decade (smaller than the 1945 high of 35.8 percent) to 37.8 in 1960. That was 23.3 million people, 9 million more than in 1940 and nearly 5 million more than in 1950.[63] This was one of the most significant social trends during the postwar era. But women

62. William Chafe, *The Paradox of Change: American Women in the 20th Century* (New York, 1991), 179–81; Leuchtenburg, *Troubled Feast*, 74; Degler, *At Odds*, 440.

63. Percentages of adult men who were in the labor force ranged in these years from 84 to 87 percent.

remained highly segregated in occupations deemed suitable for their "lesser" talents: as secretaries, waitresses, elementary school teachers, nurses, and other mostly low-paid members of the labor force. Some such women as waitresses received support from unions for better working conditions.[64] But most unions concentrated on attracting men. The median income of white female full-time workers decreased from 63 percent of the median for males in 1945 to 57 per cent in 1973.[65]

These aggregate statistics, revealing as they are, fail to catch perhaps the most significant trend in women's work in the early postwar years. This was the rapid increase in the percentage of women workers who were married. This rose from 36 percent in 1940 to 52 percent in 1950 to 60 percent in 1960—and to 63 percent in 1970. In part the increases reflected demography: higher percentages of women had husbands. But the most important cause of the trend was the desire of married women to enter the market. These were not women starting careers when they were young; they were housewives belatedly finding work, much of it low-paid, in order to help ends meet in their homes. This is what "togetherness" often meant for such families.[66]

Contemporaries and historians have disagreed over why so many of these married women entered the work force. Some have emphasized the allure of the consumer culture, which, it was thought, was creating an ever more palpable "land of desire," especially among married women in the middle classes, among whom the greatest increases in employment were taking place at the time.[67] Achieving a degree of security, these women (like many men) were thought to develop ever-greater appetites for goods. Expectations about the Good Life kept escalating. Luxuries became necessities.

64. Dorothy Cobble, *Dishing It Out: Waitresses and Their Unions in the 20th Century* (Urbana, 1991); Nancy Gabin, *Feminism in the Labor Movement: Women and the United Automobile Workers, 1935–1975* (Ithaca, 1990).
65. Degler, *At Odds*, 424–25.
66. Richard Easterlin, *Birth and Fortune: The Impact of Numbers on Personal Welfare* (New York, 1980), 66–69, 170; Collins and Cottrane, *Sociology*, 178. A total of 30 percent of women with children aged 6 to 17 worked in 1950; the percentage in 1960 was 40 percent. Percentages of women with children under 6 who worked increased in this same period from 13 percent to 20 percent. In 1960, 50 percent of women aged 45 to 54 were in the labor force, compared to 36 percent of those aged 25 to 34, 46 percent of those aged 20 to 24, and 37 percent of those aged 55 to 64.
67. Leach, *Land of Desire*; and Chafe, *Paradox*, 188, on trends among middle-class married women.

That surely happened: in this way as in many others, the rise of the consumer culture profoundly affected the behavior of people. But women were neither mindless consumers nor a homogeneous group. They divided, as always, along racial, ethnic, religious, regional, age, and class lines, and it is therefore hazardous to generalize freely, especially about complicated motivations. Women who were poor—many millions in the 1950s—could scarcely buy very deeply into the consumer culture. And many others—again the Levittowners come to mind—anxiously remembered the frightening years of the Great Depression and World War II. They wanted more consumer goods—why not?—but they also yearned for security and then more security, which they tried to advance by adding earned income to their households.[68] They cannot be accused of accumulating goods for the sake of goods.

In time many of these working women developed a heightened sense of empowerment. That was a long-range result of widespread female employment, of ascending affluence, and of the movement for civil rights, which drove ideological demands in many unanticipated ways. In the 1950s, however, most of these women were looking for jobs, not careers, and they exhibited little feminist consciousness. That arose mainly in the 1960s.

IN 1954 FREDRIC WERTHAM, a psychologist, published *The Seduction of the Innocent*, an emotional exposé of the damage that violence and brutality in comic books (which were published at a rate of more than 60 million a month by the late 1940s) were inflicting on young people. Children, he said, were learning all the wrong lessons and might become delinquents. Others would suffer from "linear dyslexia." A year later Benjamin Fine came out with *1,000,000 Delinquents*—the number of adolescents he said (accurately) would run afoul of the law in 1956. *Time* magazine, never to be outdone, then brought forth a special issue. It was entitled "Teenagers on the Rampage."[69]

Jeremiads such as these exposed a nervous underside to the surface calm of American culture in the mid-1950s, even before the oldest cohorts of baby boomers were entering their teens. The alarmists received support from a great many federal agencies such as the Children's Bureau, which—like many other concerned American institutions in the postwar era—took to heart messages from the ascendant field of psy-

68. O'Neill, *American High*, 42–44.
69. Skolnick, *Embattled Paradise*, 207; Gilbert, *Cycle of Outrage*, 5–8; Chauncey, "Postwar Sex Crime Panic."

chology and imagined that intervention by "experts" could modify personal behavior. The experts (and others) cited a host of apparently accelerating trends that purported to show an explosion of juvenile delinquency and youthful rebellion. These ranged from serious matters such as gang fights and teenaged drinking parties to more trivial matters such as the growing tendency of young "greasers" to wear cut-off T-shirts and blue jeans and to style their hair in pompadours and duck-tails. Role models such as Brando and Dean especially worried conservatives. Other alarmists blamed the apparent spate of youthful rebellion on working mothers—prisoners of the consumer culture—who left "latch-key" children behind to fend for themselves.

The United States Senate, moved to action by such concerns, responded as early as 1953 by undertaking major hearings into delinquency. These lasted off and on throughout the 1950s and attracted considerable notice after 1955 when Senator Estes Kefauver of Tennessee, a liberal Democrat, agreed to take over the investigations. (Kefauver, who had large ambitions to become President, was good at finding ways into the public eye: in 1950 he had presided over enormously popular televised hearings that sought to prove the spread of "organized crime" in America.) States and towns joined in the fight against sources of juvenile unrest. By 1955 thirteen states had passed laws regulating the publication, distribution, and sale of comic books. Leading intellectuals, including C. Wright Mills, praised Wertham's efforts.[70]

Hollywood pitched in with release in 1955 of *Rebel Without a Cause* and *The Blackboard Jungle*, both of which featured rebellious teenagers. In *Rebel* Dean defies his weak-willed, apron-clad father and domineering mother and joins a group of disaffected classmates who challenge local conventions. Brooding and surly, Dean became an idol of sorts to many adolescents. *Blackboard Jungle*, greatly aided in impact by the song "Rock Around the Clock" in its sound track, shows wild and unruly high schoolers threatening to destroy all order in the classroom. Like most apparently daring Hollywood films, both movies in fact closed with the forces of good in control. At the end of *Rebel* Dean sees the light and submits to the authority of his father, who says, "You can depend on me. Trust me." Glenn Ford, the embattled teacher in *Jungle*, manages to isolate the worst kids and regain authority. Still, these movies upset many contemporaries. Some reviewers, fearing that *Blackboard Jungle*

70. Gilbert, *Cycle of Outrage*, 64, 93–104.

would incite young people, damned it. A number of localities sought to ban it.[71]

Nothing worried traditionalists in the mid-1950s more than the impact on young people of revolutionary changes in popular music, especially rock 'n' roll. Until then "pop" music had remained fairly tame. Hits in the early 1950s had included Rosemary Clooney's "Come On-a My House," Perry Como's "Don't Let the Stars Get in Your Eyes," and Patti Page's "How Much Is That Doggie in the Window?" Even then, however, country and rhythm-and-blues tunes, some of them drawing on black musical forms, were gaining a considerable following. In 1954, *Billboard* magazine noted that rhythm and blues "is no longer identified as the music of a specific group, but one that can now enjoy a healthy following among all people, regardless of race or color."[72]

Later that year the record "Sh-Boom" came out, which in its white version by the Crew Cuts and its black version by the Chords became the fifth-best-selling song of the year. Some historians regard it as the first rock-'n'-roll hit.[73] It was quickly followed by an undoubted sensation, "Rock Around the Clock," as recorded by an all-white band, Bill Haley and the Comets. Haley's group combined country and western with rhythm and blues and featured hard-driving electric guitars and drums. "Rock Around the Clock" took off to the top of the charts and ultimately sold some 16 million recordings. Soon black rock 'n' rollers, too, like Chuck Berry ("Johnny B. Goode," 1958), Chubby Checker ("The Twist," 1960), and Fats Domino, rose to fame with a series of big-selling records.[74]

Rock 'n' roll did not supplant other forms of popular music: top hits of the late 1950s included "Tammy" by Debbie Reynolds, "Mack the Knife" by Bobby Darin, and songs by Como, Sinatra, Nat "King" Cole, Lena Horne, and other favorites. Pat Boone, a clean-cut singer, was a "pop" star. Folksinging groups such as The Weavers enjoyed a revival at the end

71. Biskind, *Seeing Is Believing*, 202–6; Sayre, *Running Time*, 110–12.
72. Oakley, *God's Country*, 272.
73. The white version was the bigger seller.
74. Carl Belz, *The Story of Rock* (New York, 1969); Ed Ward et al., *Rock of Ages: The Rolling Stone History of Rock and Roll* (New York, 1986); George Lipsitz, "'Ain't Nobody Here But Us Chickens': The Class Origins of Rock and Roll," in Lipsitz, *Class and Culture*, 195–225. Another black singing star of the era, Little Richard, reached number two on the rhythm-and-blues charts in 1955 with "Tutti Frutti." See Tony Scherman, "Little Richard's Big Noise," *American Heritage*, Feb./ March 1995, pp. 54–56.

of the decade and in the early 1960s. Jazz artists such as Ella Fitzgerald, Louis Armstrong, Duke Ellington, and many others retained loyal followings. But the rise of rock 'n' roll was nonetheless one of the most shocking cultural phenomena of the mid- and late 1950s, especially to people over the age of twenty-five. Like jazz in the 1920s, the new music seemed to separate young Americans from their elders and to usher in the beginnings of a strange and powerful "youth culture." Rock 'n' roll gave millions of young people—especially "teenagers" (a noun that came into widespread use only in 1956)—a sense of common bond: only *they* could appreciate it.[75]

No performer aroused more alarm than Elvis Presley. Elvis, twenty years old in 1955, was the son of poor Mississippi farm folk who had moved into public housing in Memphis when he was fourteen. He pomaded his hair and idolized Brando and Dean, whose *Rebel Without a Cause* he saw at least a dozen times and whose lines he could recite from memory. Presley learned to sing and play guitar while performing with local groups, often with people from his Assembly of God congregation. In 1954 he recorded "That's All Right" and a few other songs, mainly in the blues and country traditions, thereby exciting Sam Phillips, a local disk jockey, record producer, and discoverer of musical talent. Phillips loved black music and had recorded such musicians as B. B. King earlier in the 1950s. But the color line barred them from fame. "If I could find a white man with a Negro sound," Phillips is reputed to have said, "I could make a billion dollars."[76]

Presley was the man whom Phillips had been seeking. By the end of 1955 his records were hits, and his live performances, in which he affected an alienated Brando look and gyrated sexually in time with the music, were sensations. Audiences, composed mainly of young people, screeched and wailed in scenes that frightened other observers. One outraged commentator charged that Presley's performances were "strip-teases with clothes on . . . not only suggestive but downright obscene." In 1956 several of Presley's hits, such as "Hound Dog" and "Heartbreak Hotel," sold millions of records. He signed to appear in three movies. Ed Sullivan, who had prudishly announced that "Elvis the Pelvis" would never appear on his show, gave in and offered him the unheard-of sum of $50,000 to perform on three of them. (On one the cameras showed him only from the waist up.) An estimated 54 million Americans watched Elvis on one of these shows—the largest audience for anything on TV

75. Todd Gitlin, *The Sixties: Years of Hope, Days of Rage* (New York, 1987), 37–41.
76. Peter Guralnick, *Last Train to Memphis: The Rise of Elvis Presley* (Boston, 1994), 5–6, 60–65; Jezer, *Dark Ages*, 280; Halberstam, *Fifties*, 471.

until that time (and not surpassed until 67 million tuned in the Beatles, also on "Ed Sullivan," in 1964). Elvis had rocketed to fame as one of the most phenomenal stars of a decade that, thanks to the rise of the mass media, was ushering in an era of unprecedented star- and celebrity-worship. The composer Leonard Bernstein later went so far as to call Presley "the greatest cultural force in the twentieth century."[77]

The rise of the new music owed a good deal of its growth to affluence and the power of the consumer culture. By then teenagers were earning millions of dollars, often by working in fast-food places on the roadsides. Others had allowances from their parents. Many could come up with the modest sums that it took to buy record-players, as they were called at the time, and to purchase the cheap 45-rpm vinyl disks that carried the new tunes. Almost everyone could afford to feed nickels into the jukeboxes that played the music wherever teenagers assembled. Retail sales of records jumped from $182 million in 1954 to $521 million in 1960. Rock 'n' roll, like much else in the United States, quickly became com-modified—a vital part of the thriving culture of consumption.

Some adult Americans pretended not to be upset by the frenzy sur-rounding rock 'n' roll. Like many other fads, it might go away. (Elvis, they said, was not so bad—he bought houses for his parents, said his prayers, and did not smoke or drink.) But there was no doubting that the popularity of rock 'n' roll exposed the nascent rise of a sometimes restless "youth culture." And many older people openly revealed their sense of alarm. A psychiatrist, writing in the *New York Times*, proclaimed that rock 'n' roll was "a communicable disease" and "a cannibalistic and tribalistic kind of music." The racist metaphors here went unchallenged. Another critic, writing to a Senate subcommittee on delinquency, lamented that "Elvis Presley is a symbol, of course, but a dangerous one. His strip-tease antics threaten to rock-n-roll the juvenile world into open revolt against society. The gangster of tomorrow is the Elvis Presley type of today." [78]

MANY OF THESE FEARS about juvenile delinquency, rock 'n' roll, and youthful rebellion reflected contemporary confusion and anxiety amid the rapid social, demographic, and economic changes that were transforming the nation. They also addressed real phenomena, for in-creasing numbers of the young were indeed beginning to rebel against accepted ways. Some of these young people identified with Holden Caulfield, the teenage anti-hero of J. D. Salinger's novel *The Catcher in*

77. *New York Times*, June 7, 1993; Daniel Boorstin, *The Image in America: A Guide to Pseudo-Events in America* (New York, 1961), 156–61.
78. Jezer, *Dark Ages*, 279; Gilbert, *Cycle of Outrage*, 18.

the Rye (1951). Older people, Holden said, were "phonies." Others among the young—few in numbers but much noted by contemporary trackers of social trends—became "beats" who claimed to reject the materialism of the consumer culture and affected bohemian styles of life.[79] Many others identified with a peer group culture of their own—one that highlighted a new consumer-driven world of drive-in movies, fast-food hangouts, jalopies, and malls. It was small wonder that many older Americans, bewildered by the pace of social change, began to feel greatly threatened by a "youth culture."

Still, many of the "threats" to older ways of life in the 1950s were exaggerated. Statistics on juvenile delinquency (and on crime in general), while unreliable, did not show increases during the 1950s. Moreover, while many young people were restive, they saw no clear routes to collective social action. Not even rock 'n' roll, for all its liberating potential, could provide those. Instead, restless young people in the 1950s tended to rebel on a fairly small stage in which parents and neighbors remained the major impediments to gratification. Except for blacks, who grew increasingly militant in fighting against racial injustice, young people who were unhappy with the status quo did not much concern themselves with larger political or social problems. Most educators in the 1950s detected a "silent generation," both in the schools and in the burgeoning universities.

What the restless young still lacked in the 1950s was the greatly magnified sense of possibility—of open-ended entitlement—that was to give them greater energy and hope in the 1960s. Instead, they encountered still strong cultural norms that prescribed traditional roles for "growing up": "girls" were to become wives and homemakers, "boys" were to enter the armed services and then become breadwinners. Few young men, Presley included, imagined that they should avoid the draft: half of young men coming of age between 1953 and 1960 ended up in uniform, most for two years or more.

By the late 1950s millions of Americans were enjoying the bounties of affluence and the consumer culture, the likes of which they had scarcely imagined before. In the process they were developing larger expectations about life and beginning to challenge things that had seemed set in stone only a few years earlier. Older cultural norms, however, still remained strong until the 1960s, when expectations ascended to new heights and helped to facilitate social unrest on a new and different scale.

79. For discussion of the beats, see chapter 14.

13

Race

Color-consciousness has always blighted life in the United States. Light-skinned people, fearful of "pollution" from "coloreds," historically erected formidable barriers against "non-white" Americans. Between the late 1940s and 1960 these barriers opened up a bit, and a few outsiders crept through the cracks. But only for a little way. The vast majority of non-whites—including not only African-Americans but also American Indians, Asians, and many Hispanics—could not get through. [1] In matters of color and of racial consciousness the United States held firm.

THE TREATMENT OF AMERICAN INDIANS had long dem-onstrated the virtual impenetrability of such barriers. Thanks in part to military campaigns against them and mainly to the killing diseases that Europeans brought with them to the New World, the number of native

1. As in earlier chapters, I sometimes use here words—"Negro," "Indian," "Mexican"—that most Americans (whites and "non-whites" alike) used in the 1950s. (See note 12 to chapter 1.) After 1970 or so the term "Native American" became preferred by some American Indians. People conscious of having mixed backgrounds were often uncertain what to call themselves. Until the 1970s many "mixed-blood Indians" told census enumerators that they were "white." So, it is assumed, did some light-skinned people of African background, who resented being called "mulattos" or "non-white." The census did not count "Hispanics" until 1970. By the 1980s "Hispanic" came under criticism—by no means all South Americans or West Indian Americans are of Spanish background—but that debate, too, is another, later story.

people in what became the United States dropped drastically from many millions in 1600 (most estimates of these numbers now range between 4 and 7 million) to a low of around 200,000 in 1900.[2] Whites, having subdued all "Red" resistance and consigned most survivors to reservations, then further cheated Indians out of their land and tried to force them to adopt white ways. In 1924 Congress decided that native-born Indians were citizens of the United States, but neither the national government nor the states gave tangible meaning to that citizenship. Some states denied Indians the vote until the early 1950s.

Between 1934 and 1945, John Collier, Commissioner of Indian Affairs, tried to promote an "Indian New Deal" that would value native cultures, increase federal support for health and education, and provide for greater self-government on the reservations. Collier was high-handed and paternalistic, and many natives refused to cooperate with his plans. Still, his policies offered hope for a more liberal treatment of the Indians, some of whom—then and later—were encouraged in their quests for resistance and self-determination.[3]

Collier came under fire from conservative critics and resigned in 1945. His successors gradually undermined his efforts, especially after 1953, when Congress sought to end the special status that Indians had had under the law as wards of the United States. Over time, Congress then declared, Indians would be subject to the same laws, privileges, and responsibilities as other American citizens.[4] The "termination" policy, as this approach was called, aimed to cut off public aid to Indians and to get them to fend for themselves. In 1954 the Menominees, the Klamaths, and several smaller groups agreed to be "terminated" and embarked as individuals into the non-Indian world.

Termination thereafter soon lost some of its standing as official policy. A number of Indian groups joined white sympathizers to protest vigorously against the new approach, which they said was abandoning the

2. Terry Wilson, *Teaching American Indian History* (Washington, 1993), 39–42; Russell Thornton, *American Indian Holocaust and Survival: A Population History Since 1492* (Norman, 1987).
3. Robert Berkhofer, *The White Man's Indian: Images of the American Indian from Columbus to the Present* (New York, 1978), 179–88; William Hagan, *American Indians* (Chicago, 1979); Kenneth Philp, *John Collier's Crusade for Indian Reform, 1920–1954* (Tucson, 1976); and Ronald Takaki, *A Different Mirror: A History of Multicultural America* (Boston, 1993), 84–105, 228–45.
4. Francis Paul Prucha, "Indian Relations," in Jack Greene, ed., *Encyclopedia of American Political History*, Vol. 2 (New York, 1984), 609–22; Berkhofer, *White Man's Indian*, 186–90.

majority of Menominees and Klamaths to lives of neglect. Other advocates of government responsibility to native people succeeded in 1955 in
transferring Indian health facilities and programs from the Bureau of
Indian Affairs (long considered an unsympathetic and corrupt bureaucracy) to the Public Health Service, which did a better job. Congress also
provided modest financial aid to assist the growing number of individual
Indians who sought to leave their reservations and relocate. In 1958 the
government backed away from efforts to impose termination, and the
policy—in eclipse during the 1960s—was formally ended by President
Nixon in 1969. By then Indian activists were beginning to take matters
into their own hands.

Still, the termination approach signified a central fact about national
attitudes in the late 1940s and 1950s: the continuing power of white
assimilationist thinking. Indians, it was agreed, must be forced to adapt to
white ways. This thinking coexisted uneasily and in some manners incon-
sistently with white assumptions about Indian inferiority. As in the past,
whites who paid attention to Indians tended to think that native cultures
were crude and uncivilized. Most Americans, moreover, did not give
much thought to Indians, in part because the natives were mostly out of
sight and out of mind. Indeed, they remained a tiny minority: the census
enumerated 334,000 in 1940, 343,000 in 1950, and 509,000 in 1960.[5]
Relegated to reservations (or scattered in remote areas), they had little
formal education or political power. Many were poorly nourished and ill.
The vast majority lived in poverty. The miserable condition of most
American Indians in the 1950s, as throughout United States history,
testified to the continuing strength of white ethnocentrism and institutional discrimination in the country.

Asian-Americans, another group that had encountered racist treatment
in the past, fared slightly better than did American Indians in the 1940s
and 1950s, during which time new immigration laws opened a tiny bit of
space. In 1943 Congress finally repealed legislation dating to 1882 that
had barred Chinese workers from emigrating to American shores. The

---

5. The numbers in 1960 included (for the first time) people living in Alaska and
Hawaii, admitted as states in 1959. Roughly 15,000 Indians lived in these new
states, mostly in Alaska. As noted earlier, all these numbers can deceive. Many
Indians, especially prior to 1960, told enumerators that they were white. Only in
the 1960s, with the rise of Indian assertiveness and pride, did this habit begin to
change in a major way. By 1970, 793,000 people told the census that they were
Indian. This represented an increase of 450,000 over the number counted in 1950,
a jump that cannot begin to be explained by population growth.

repeal, establishing small quotas for Chinese (105 per year), reflected wartime sympathy for China, an ally against Japan. In 1946 Congress did the same for Asian Indians and Filipinos—people who in effect had been barred from emigrating to the United States since 1917. In 1952 Congress approved the McCarran-Walter Act, a major effort to recodify immigration statutes. Although it contained tough sections that widened grounds for deportation, the act loosened some restrictions. It repealed laws that had excluded Asians from settling in the United States, and it eliminated "race" as a barrier to naturalization, thereby enabling Asians to become American citizens.[6]

The practical effect of these new laws, however, was not great. Hatred and distrust of Japanese-Americans, inflamed during World War II, persisted. Moreover, the McCarran-Walter Act reaffirmed the quota system that had been enshrined in legislation of 1921 and 1924. These laws had established very low quotas for unwanted groups, especially those from southern and eastern Europe. The McCarran-Walter Act aimed to ensure that 85 percent or more of immigrants would come from northern and western Europe—areas of "Anglo-Saxon stock." It was especially hard on Asians, setting annual quotas of 185 for immigrants from Japan, 105 from China, and 100 for each of other countries within a so-called Asia-Pacific Triangle. (Family reunification provisions, however, permitted a great many more close relatives of Asian-American citizens to come in as "non-quota" immigrants, thereby boosting numbers. Some 45,000 Japanese and 32,000 Chinese thereby immigrated to the United States in the 1950s.) Although President Truman vetoed the measure, calling it restrictive and discriminatory, Congress (Democratic in both houses) resoundingly overruled him.[7] The McCarran-Walter Act formed the essence of American immigration law until 1965, when Congress approved more liberal legislation.

6. Reed Ueda, *Postwar Immigrant America: A Social History* (Boston, 1994), 42–44; Victor Greene, "Immigration Policy," in Jack Greene, ed, *Encyclopedia*, 2:579–91; Roger Daniels, *Asian America: Chinese and Japanese in the United States Since 1850* (Seattle, 1988), 195–98; Takaki, *Different Mirror*, 191–224, 246–76; and David Reimers, *Still the Golden Door: The Third World Comes to America* (New York, 1985). See also chapter 1 for further data on Asian-Americans.
7. Daniels, *Asian America*, 283–84, 305–6. Close relatives included spouses, parents, children, and siblings of United States citizens. When these newcomers gained citizenship, their close relatives, too, could come to America as non-quota immigrants. Over time, therefore, the family reunification provisions resulted in far higher immigration than legislators had anticipated. Legislation in 1948 and thereafter concerning "displaced peoples" and other war-affected refugees enabled a few more southern and eastern Europeans (and others) to come to the United

Latino-American people did not encounter nearly the same levels of exclusion and legal discrimination that marginalized Indians and Asians in the United States. Many "Hispanics," after all, were light-skinned. Puerto Ricans were American citizens who could and did enter the United States without restrictions.[8] Mexicans had been periodically lured to the Southwest, whenever American commercial farmers needed cheap labor. Primarily because of such demand for workers, Mexicans and others from Latin America had never been assigned quotas: immigration from the Western Hemisphere knew no legal limits. In 1942 Mexico and the United States approved a program whereby Mexican *braceros*— workers under government contract—might stay for specified periods of time as farm laborers in the Southwest. This arrangement was regularly extended in the 1950s. In 1959 the program peaked, admitting 450,000 workers. By 1960 *braceros* made up 26 percent of America's migrant farm labor force.[9]

But the United States was hardly a haven for Mexicans and other migrants from south of the border. On the contrary, they faced systematic discrimination, including segregation in housing and schools. The *braceros* were widely exploited by their employers. When demand for farm labor receded (as happened from time to time), American authorities rounded up Hispanic aliens (as well as some people who were American citizens) and deported them.[10] Dismal economic conditions south of the Rio Grande, however, drove ever more people across the border in the 1940s and 1950s, including "illegals" who did not want to be tied to labor contracts. Many *braceros* overstayed their presence rather than return to the even greater poverty of their native land.[11]

Many of these Mexicans, angry Anglos complained, were *mojados*, or "wetbacks who should be deported forthwith." Mexican officials, too,

States, but these numbers were not large. Americans who advocated liberalization of immigrant policies in these years remained unhappy.

8. Joseph Fitzpatrick, *Puerto Rican Americans: The Meaning of Migration to the Mainland* (Englewood Cliffs, N.J., 1971), 10–15.
9. The program was ended in 1964. See Ueda, *Postwar Immigrant America*, 32–35. Also Ernesto Galarza, *Merchants of Labor: The Mexican Bracero Story* (Charlotte, Calif., 1964); Richard Craig, *The Bracero Program: Interest Groups and Foreign Policy* (Austin, 1971); Carlos Cortes, "Mexicans," in Stephan Thernstrom, ed., *Harvard Encyclopedia of American Ethnic Groups* (Cambridge, Mass., 1980), 703; Mario Garcia, *Mexican Americans: Leadership, Ideology, and Identity, 1930–1960* (New Haven, 1989); and Takaki, *Different Mirror*, 166–90, 311–39.
10. Abraham Hoffman, *Unwanted Mexicans in the Great Depression* (Tucson, 1974).
11. Reimers, *Still the Golden Door*, 37–60.

requested a crackdown on the "illegals" in order to protect the contracts of the *bracero* program. American authorities responded to such complaints in the early 1950s with Operation Wetback, as it was called. Raiding restaurants, bars, and even private homes, they captured and deported as many as they could find. Some of those who were caught in the net suffered harsh detention and other violations of civil liberties. A number of estimates conclude that as many as 3.8 million Mexicans were rounded up and sent away between 1950 and 1955.[12]

Efforts such as Operation Wetback (which nonetheless failed to stop migrations) cannot be seen as purely "racist." Working-class Americans in the Southwest, confronting mass migrations of people who labored for next to nothing, understandably sought to stem the flow of outsiders who threatened to take away their jobs. Still, it was clear that Hispanics, whether *braceros*, "wetbacks," or others, faced substantial hostility and discrimination in the United States. Like Indians and Asians, many Anglos were saying, people from south of the border were inferior and undesirable.

IT IS DIFFICULT to generalize about the status of African-Americans in the 1950s. Much larger as a group than Indians, Asians, or Hispanics, they numbered 15.8 million people in 1950 and 19 million in 1960, or around 10.6 percent of the population. As in the 1940s, blacks continued in the 1950s (and 1960s) to flee the South in unprecedented numbers: by 1970, 47 percent of black people lived outside the South, most in the Northeast or Midwest, as compared to only 23 percent in 1940. Many of these migrants settled and resettled several times, concentrating in cities, in their restless quest for a better life.[13]

Those who left the South in the 1950s sought to escape a world of black-white relations that remained more systematically oppressive than anything experienced by other racial groups in the United States and that had changed very little over time.[14] Negative stereotyping of black people, to be sure, became a little more subtle in the 1940s and 1950s: the NAACP succeeded in driving "Amos 'n' Andy" off of TV in 1953.[15] More

12. Ueda, *Asian America*, 32–34.
13. See chapter 1.
14. Stanley Lieberson, *A Piece of the Pie: Blacks and White Immigrants Since 1880* (Berkeley, 1980), esp. 363–93.
15. "Amos 'n' Andy" remained on radio, however, until 1960. Melvin Ely, *The Adventures of Amos 'n' Andy: A Social History of an American Phenomenon* (New York, 1991).

important, whites were less likely in the postwar era to resort to violence. Reported lynchings of blacks, which had averaged twelve a year in the 1930s, fell to a total of thirteen between 1945 and 1950.[16] Thereafter the NAACP dropped federal anti-lynching bills as its first priority and focused instead on fighting other forms of racist behavior. But black people in the South during the 1950s still struggled in a Jim Crow society that segregated everything from schools and buses to bathrooms, beaches, and drinking fountains. Despite campaigns by black activists for voting rights, only a token few black people in the Deep South states were permitted to register or to vote.[17] Daily humiliations continued to remind black people of their third-class status. Whites never addressed black men as "mister" but rather as "boy," "George," or "Jack." African-American women were called "Aunt" or by their first names, never "Miss" or "Mrs." Newspapers rarely reported the names of black people but instead described them as "negro," as in "a man and a woman were killed, and two negroes." Whites did not shake hands with blacks or socialize with them on the street. When blacks encountered whites in public places, they were expected to take off their hats, but whites did not remove theirs in the same situation, or even in African-American homes.[18]

Flight to the North provided relief from some of these practices, especially in the booming manufacturing sector of the economy, where thousands of blacks found industrial work in the 1940s and early 1950s. But change came slowly.[19] As in the past, employers consistently discriminated against African-Americans in hiring, advancement, and salary. Unemployment was normally twice as high for blacks—especially males—as for whites. Poverty afflicted 50 percent or more of blacks even

16. Robert Zangrando, *The NAACP Crusade Against Lynching, 1909–1950* (Philadelphia, 1980). Lynchings of Negroes had averaged 121 a year at their peak between 1890 and 1895. A total of eleven Negroes were reported as lynched in the 1950s. There were three more reported between 1960 and 1964 and none for the remainder of the 1960s.

17. Robert Harris, Jr., *Teaching African-American History* (Washington, 1992), 51–64; Steven Lawson, *Black Ballots: Voting Rights in the South, 1944–1969* (New York, 1976), 133–39.

18. These customs had long histories, as described in Edward Ayers, *The Promise of the New South* (New York, 1993), 132. See also John Howard Griffin, *Black Like Me* (Boston, 1960), for the feelings of a white man who blackened his face and passed for black in the 1950s. The spelling by white-run newspapers of "negro," with a lower-case *n*, was another effort to demean black people.

19. Werner Sollors, "Of Mules and Mares in a Land of Difference; or, Quadrupeds All?" *American Quarterly*, 42 (June 1990), 167–90.

in the economically good years of the mid-1950s. (White rates were around 20 to 25 percent at the time.) Key unions—in construction, plumbing, sheet metal trades, and electrical work—virtually barred blacks from membership. Even the United Automobile Workers, which supported civil rights in the 1950s and 1960s, continued to be dominated by whites. It elevated no blacks to its executive board until 1962, by which time black auto workers were seething at daily humiliations on the shop floor.[20]

Talented black people who sought to break into new fields continued to face formidable obstacles. In Hollywood there seemed to be a little more room for black actors, but mainly in self-sacrificing roles. In *Edge of the City* (1957) and *The Defiant Ones* (1958) Sidney Poitier humbled himself for white friends. These roles, he said later, were "other-cheek-turners."[21] In professional sports, team owners moved slowly. The New York Yankees baseball team waited until 1955 before making Elston Howard, a gifted athlete, its first black player. The Boston Red Sox, the last major league baseball team to sign a black player, delayed taking that step until 1959, at which time only 15 percent of the 400 roster players in the major leagues were African-American. Most of them were top performers: one had to be excellent to make it.

Other sports in the 1950s remained mostly white at the highest levels. Althea Gibson broke the color bar on the tennis circuit in 1949, winning both Wimbledon and the United States championship in 1957, but few others followed her in the 1950s or early 1960s. Arthur Ashe, the first black male on the circuit, did not start top-level play until 1963. The Professional Golfers Association did not have a black player on the tour until 1961. No black golfer was invited to the Masters tournament in Georgia until 1974, when Lee Elder appeared. In part because country clubs were mostly closed to African-Americans, both tennis and golf continued to develop few black stars.[22]

Although the National Basketball Association featured a few black

20. Robert Zieger, *American Workers, American Unions, 1920–1985* (Baltimore, 1986), 174–77; Herbert Hill, "Black Workers, Organized Labor, and Title VII of the 1964 Civil Rights Act: Legislative History and Litigation Record," in Hill and James Jones, Jr., eds., *Race in America: The Struggle for Equality* (Madison, 1993), 263–341; Reynolds Farley and Walter Allen, *The Color Line and the Quality of Life in America* (New York, 1987); and William Harris, *The Harder We Run: Black Workers Since the Civil War* (New York, 1982), 123–89.
21. Nora Sayre, *Running Time: Films of the Cold War* (New York, 1982), 180–82.
22. In 1990 the Birmingham, Alabama, country club that was host for the PGA championship still excluded blacks from membership.

players when it began in 1950, notably Nat "Sweetwater" Clifton, the teams had unwritten quotas permitting only four blacks (including two starters) on their rosters throughout much of the 1950s. Talented black basketball players were thought to be appropriate for the Harlem Globetrotters, where they were expected to wear big, toothy smiles and to act like clowns. It was virtually axiomatic that blacks should not coach major teams either at the college or professional level. While Bill Russell overcame this obstacle by becoming player-coach of the Boston Celtics Basketball team in 1966, it was not until 1977 that major league baseball had a black manager (Frank Robinson) and not until 1989 that the National Football League did (Art Shell).[23]

As earlier, no aspect of racism in America cut so deeply as housing discrimination. During the 1950s, what one scholar has called a "second ghetto" arose in large northern cities such as Chicago. In the Windy City white politicians conspired with downtown businessmen and developers to prevent the central city business district from becoming "ringed" by black migrants, who were arriving from the South in record numbers. Exploiting federal funds for urban renewal, they declared downtown black neighborhoods to be "slums," tore them down, and put up commercial buildings or housing for whites in their place. African-Americans were displaced into dilapidated neighborhoods, increasingly in all-black public housing projects. Most of the 21,000 family units of public housing erected in Chicago during the 1950s were built in already black regions of the city, thereby greatly increasing the density of blacks in these areas.[24]

Black people who sought to escape the projects confronted, as in the past, unyielding and sometimes violent opposition from white property-owners elsewhere in the cities. And those who yearned for a life in the suburbs mostly dreamt in vain. When a black family sought to move into Levittown, Pennsylvania, in 1957, it was greeted with rock-throwing. Levitt dared sell a house to blacks (in New Jersey) only in 1960. More than any other thing, governmentally sanctioned housing discrimination re-

---

23. Richard Davies, *America's Obsession: Sports and Society Since 1945* (Ft. Worth, 1994), 35–61; Randy Roberts and James Olson, *Winning Is the Only Thing: Sports in America Since 1945* (Baltimore, 1989), 30–45.

24. Arnold Hirsch, *Making the Second Ghetto: Race and Housing in Chicago, 1940–1960* (New York, 1983), 242–74; Nicholas Lemann, *The Promised Land: The Great Black Migration and How It Changed America* (New York, 1991), esp. 59–107, 223–305; Thomas Sugrue, "Crabgrass-Roots Politics: Race, Rights, and the Reaction against Liberalism in the Urban North, 1940–1964," *Journal of American History*, 82 (Sept. 1995), 551-78.

vealed the power of white racist feeling against black people in the United States.

Discrimination in housing solidified an already widespread de facto segregation in northern schools. Whether this had totally bad results continued to be debated many years later, for some experts argued that all-black schools (with black faculties) offered reinforcement for African-American children that often did not exist in desegregated schools. Still, it was everywhere obvious that discriminatory housing patterns prevented black children from attending much-better-financed white schools. In short, African-American children were denied the basic right of equal opportunity. Many northern black schools, indeed, continued to be badly maintained institutions where faculty despaired of establishing elementary discipline and where little if any serious academic learning took place. High percentages of the black children in these schools came from poverty-stricken or broken homes that had no books or magazines—or even pens or pencils. These children often arrived in school without having had breakfast. The readers they studied from were usually hand-me-downs from white schools. The "Dick and Jane" stories in the books featured pink-cheeked, well-dressed children doing pleasurable things in single-family homes set in the suburbs. A host of contemporary studies suggested that the farther along black children got in these schools, the farther they fell behind white children of the same age.[25]

Underlying these and other humiliations were white attitudes that persistently downgraded African-American culture and history. While most white intellectuals had jettisoned the baggage of scientific racism, they were loath to acknowledge that blacks had developed positive traditions of their own, save perhaps in music and dance. Asserting that slavery had eradicated African-American consciousness, Nathan Glazer and Daniel Moynihan concluded as late as 1963, "The Negro is only an American, and nothing else. He has no values and culture to guard and protect."[26] It followed, other whites believed, that whites could ignore what blacks did and thought. This is what Ralph Ellison lamented in labeling his novel *Invisible Man* (1952) and what James Baldwin later meant by entitling a collection of his essays *Nobody Knows My Name* (1961). To be ignored was as bad as to be oppressed—maybe worse.

25. Diane Ravitch, *The Troubled Crusade: American Education, 1945–1980* (New York, 1983), 152.
26. *Beyond the Melting Pot: The Negroes, Puerto Ricans, Jews, Italians, and Irish of New York City* (Cambridge, Mass., 1963), 53.

OBDURATE THOUGH THESE OBSTACLES remained, they weakened a bit in the 1950s. Some of these changes, such as the disrepute of scientific racism as a result of the Holocaust, dated to the war years. The democratic ideals extolled during the war had further challenged racist practices. The rise of the Cold War obliged Truman and others to consider civil rights at home: American claims to lead the "Free World" otherwise rang hollowly. [27]

Four other forces intensified the potential for interracial progress in the 1950s: ongoing social and demographic change; rising pressure from the NAACP and other advocates of desegregation; demands from brave and determined black people at the grass-roots level; and the United States Supreme Court. These combined to ignite the modern civil rights movement, which inspired unprecedented egalitarian passion in the nation. No other movement in postwar American history did as much to arouse rights-consciousness in general—among women, the poor, and other disadvantaged groups—and to transform the society and culture of the United States.

The social and demographic changes abounded: job opportunities and military service during World War II, which had pulled millions of blacks—many of them young and impatient—out of isolated and poverty-stricken enclaves in the rural South; subsequent migrations of millions more, not only to northern industrial areas but also to growing southern cities; the ascendance in these places of better-educated young people, of a black middle class, and of resourceful leaders; greater black engagement in politics, especially in northern cities; the seductive affluence of the postwar era, which excited aspirations for a better life; and the spread of mass communications, especially television, which facilitated collective mobilization and alerted black people to the dynamic possibilities of the culture at large. African-Americans, including many who were better off in the postwar era than ever before, grew more keenly aware of their relative deprivation. Thanks to all these social and demographic changes, rights-consciousness, already rising in the 1940s, expanded for millions of American black people in the 1950s. Like whites, they were rapidly developing grander expectations. [28]

Responding to these aspirations, civil rights advocates in the NAACP, which had grown greatly during World War II, redoubled their efforts

27. See chapter 1.
28. Harvard Sitkoff, *The Struggle for Black Equality, 1954–1992* (New York, 1993), 3–18.

against racial segregation. Chief among them by the late 1940s was Thurgood Marshall, a tall, determined lawyer who led the fight against the "separate but equal" doctrine established by the Supreme Court in the 1890s. Marshall was the son of a father who was a Pullman porter and waiter at an exclusive white club in Maryland and of a mother who had graduated from Teachers College of Columbia University. After graduating from Lincoln University in Pennsylvania, an all-black college with an all-white faculty, Marshall had been denied admission, on racial grounds, to the University of Maryland.[29] He never forgot this insult. He then went to Howard University Law School, a center under the leadership of Charles Houston in the training of black civil rights lawyers.[30] By 1938, when he was only thirty, Marshall had become chief counsel for the NAACP. Marshall had a common touch and an apparently fearless determination to travel anywhere, even in dangerous areas of the South, that inspired local people to stand up against injustice.

In the 1930s and 1940s Marshall and fellow attorneys focused on ending "separate but equal" in graduate education. This was a logical strategy at the time, for most southern states could hardly pretend that they provided equality at that level, and black universities lacked the resources to fill in the gaps. No black institution then offered work leading to the Ph.D., and only two (Howard and Meharry in Nashville) provided medical education. Blacks could study dentistry, law, pharmacy, and library science in only one or two southern institutions, and they could pursue graduate work in engineering or architecture nowhere in the South.[31]

Struggling for reform in graduate education forced Marshall and his associates to litigate patiently through the various levels of the American court system. In June 1950 they had notable success with the Supreme Court; on the same day the Court rendered two important decisions. One ordered the state of Texas, which had set up a separate and inferior all-black "law school" (it had three classrooms and three faculty members), to admit a black plaintiff to its all-white school. The other decision barred

29. Mark Tushnet, *Making Civil Rights Law: Thurgood Marshall and the Supreme Court, 1936–1961* (New York, 1994); Nicholas Lemann, "The Lawyer as Hero," *New Republic*, March 13, 1993, pp. 32–37.

30. Genna Rae McNeil, "Charles Hamilton Houston: Social Engineer for Civil Rights," in John Hope Franklin and August Meier, eds., *Black Leaders of the Twentieth Century* (Urbana, 1982), 221–40; Richard Kluger, *Simple Justice: The History of "Brown v. Board of Education" and Black America's Struggle for Equality* (New York, 1976), 105–94.

31. Ravitch, *Troubled Crusade*, 121.

the state of Oklahoma from continuing to segregate facilities within its graduate school of education. Until then the school had forced the plaintiff, a sixty-eight-year-old black educator, to use separate cafeterias and library facilities and to sit alone in sections of classrooms marked RE-SERVED FOR COLOREDS.[32]

Marshall and other with the NAACP's Legal Defense and Education Fund next took on the challenge of fighting against segregated public schools, which were then attended by some 40 percent of American children in twenty-one states, ten of them outside the Confederacy.[33] States and school districts that discriminated tried to claim that the separate facilities used by blacks were equal. But their case was absurd, especially in the Deep South. South Carolina in 1945 spent three times as much per pupil on its white schools as it did on black ones and 100 times as much on transportation of white students. The value of white school property was six times that of black. Mississippi schools were even more unequal; in 1945 its white schools received four and a half times as much funding per pupil as did the black ones. Almost everywhere that segregation existed, black school years were shorter, teachers were paid less, and textbooks were dated discards from the white schools.[34]

In challenging such a vast and institutionalized core of American racism, Marshall and his allies depended greatly on grass-roots help from black people and their institutions, especially local NAACP branches. Pullman porters, who represented an elite in many black communities, often supplied local leadership. Jim Crow, ironically enough, had helped to sustain all-black institutions and communities that provided solidarity vital to protest. Many of those who assisted the NAACP were unknown outside their communities and remained largely unsung participants in the movement even in their own time. In backing Marshall and the NAACP they risked vehement white retaliation that ranged from loss of work to fear for their lives. When Levi Pearson, a black farmer in Sum-

32. These decisions, respectively, were *Sweatt v. Painter* (339 U.S. 629 [1950]) and *McLaurin v. Oklahoma State Regents. for Higher Education* (339 U.S. 637 [1950]).

33. Marshall and others, while undertaking a daunting task, thought that they would have a better chance of prevailing in public school cases, where they hoped that judges would apply Fourteenth Amendment language preventing *states* from denying people equal rights without "due process," than they would in trying to challenge *private* discrimination, as practiced by employers, restaurants, lunch counters, hotels, and other institutions. The struggle against racism required a complicated progression of legal and political battles.

34. Ravitch, *Troubled Crusade*, 121.

merton, South Carolina, dared to help the NAACP challenge school segregation, white bankers cut off his credit so that he could not buy fertilizer. White neighbors refused to lend him their harvesting machine as they had in the past, and his crops rotted in the fields. Shots were fired at his house. Pearson was luckier than some: the Reverend Joseph De-Laine, the black minister who persuaded him to bring suit, had his house burned down. DeLaine and most other blacks involved with Pearson in the case were forced out of the county.[35]

Pearson and many other black people, however, had exhausted their patience, and they stood up to be counted as plaintiffs in suits that Marshall brought against segegration in the schools. Five of these suits, including Pearson's, reached the Supreme Court by 1953, challenging school policies in Virginia, Delaware, the District of Columbia, South Carolina, and Kansas. The best-known plaintiff was the Reverend Oliver Brown, a welder in Topeka, Kansas, whose eight-year-old daughter Linda had to go to a Negro school twenty-one blocks away when there was a white school only seven blocks from her house. His suit, joined by twelve other parents, was filed in 1951 as *Brown v. the Board of Education of Topeka*.[36]

At that time the Supreme Court seemed a fragile reed for civil rights activists to hang on to. Although the Court had agreed to challenge "separate but equal" at the graduate level, it remained divided internally on many other questions. Its most prominent liberal justices, Hugo Black and William Douglas, fought openly with its most celebrated advocates of judicial restraint, Robert Jackson and Felix Frankfurter. The Chief Justice, Fred Vinson, commanded little respect from either camp. When the school cases reached the Court in 1953, Frankfurter helped to arrange a rehearing of the arguments rather than face what he thought would happen if the cases were decided then: a narrow, 5-4 decision against "separate but equal" that would invite southern resistance and destroy chances for meaningful implementation. With the new hearings set for December, Vinson died in September. "This is the first indication I have ever had," Frankfurter confided with relief to a former law clerk, "that there is a God."[37]

The new Chief Justice, Governor Earl Warren of California, emerged as proof that people matter in history. In appointing him, President

35. Joe Klein, "The Legacy of Summerton: *Brown v. Board of Education* 40 Years Later," *Newsweek*, May 16, 1994, pp. 26–30.
36. At the Supreme Court level it was 347 U.S. 483 (1954).
37. Kluger, *Simple Justice*, 656.

Eisenhower knew that the Court would soon have to decide the segregation cases—High Court decisions do not bolt from the blue—and he
should have guessed that Warren, a Republican liberal, would support
the plaintiffs. What the President did not suspect was how dramatically
Warren, a warm, gregarious, and straightforward man with a great gift for
friendship, would succeed in curbing the animosities that had polarized
the Court. Warren started on this effort immediately following his confirmation in March 1954, concentrating on securing consensus in the
school cases. Eisenhower also did not recognize how deeply Warren felt
about racial injustice. Although the new Chief Justice had helped as
California's attorney general in 1942 to intern Japanese-Americans, he
had deeply regretted this lapse in judgment, and he determined to do what
he thought was right for black children and their parents in 1954.[38] On
the school cases, as on much else during his historic fifteen-year career as
Chief Justice, Warren approached issues without worrying too much
about the niceties of legal precedent or about judicial restraint. What the
Court must do, he made clear, was to promote social justice.[39]

Warren managed to bring colleagues into line, and on May 17, 1954—
Black Monday, his critics called it—the Court electrified the nation by
unanimously overturning de jure racial segregation in the public schools.
"In the field of public education," Warren declared, "the doctrine of
'separate but equal' has no place. Separate educational facilities are inher-
ently unequal." Drawing on psychological theories raised by the plaintiffs, Warren added that segregation "generates a feeling of inferiority
[among students] as to their status in the community that may affect their
hearts and minds in a way unlikely ever to be undone."[40]

Opponents of segregation hailed the decision. The *Brown* case, said the
Chicago Defender, a leading black newspaper, was a "second emancipation proclamation . . . more important to our democracy than the
atomic bomb or the hydrogen bomb." That was an understandable and
for the most part accurate observation. The Court, which enjoyed immense standing in the eyes of the people, had spoken, and in so doing had
overturned nearly sixty years of legally sanctioned injustice. No longer, it
seemed, could segregated public schools hide behind the law. Moreover,
Americans had always placed enormous faith in the capacity of schools to

38. Dwight Eisenhower, *Mandate for Change* (Garden City, N.Y., 1963), 284–87;
    Kluger, *Simple Justice*, 657–75.
39. Anthony Lewis, *Gideon's Trumpet* (New York, 1964); James Weaver, *Warren: The
    Man, the Court, the Era* (Boston, 1967); Kluger, *Simple Justice*, 678–99.
40. Kluger, *Simple Justice*, 700–710.

promote equal opportunity and social mobility. In 1954 they imagined optimistically that racial prejudice would decline if children of different colors were brought together in the classroom. For all these reasons *Brown* conveyed profound moral legitimacy to the struggle for racial justice, not only in the schools but also in other walks of life. Activists seeking voting rights immediately redoubled their efforts, even in the Deep South. Without *Brown*, the civil rights movement would not have been quite the same.

A few southern political leaders announced that they would try to comply with the ruling. Governor "Big Jim" Folsom of Alabama, a liberal by southern standards, declared, "When the Supreme Court speaks, that's the law." The governor of Arkansas added, "Arkansas will obey the law. It always has."[41] Many border-state school districts went further, taking action to bring about substantial changes. By the end of the 1956–57 school year, 723 school districts, most of them in these areas, had desegregated their schools. In these ways *Brown* mattered: it had quick and tangible consequences for thousands of children and their families.

Whether these consequences were wholly positive was less than 100 percent clear to experts who later surveyed the impact of school desegregation over time. Virtually all agreed heartily that legally mandated segregation was wrong: equal access must be a fundamental right. They also tended to agree (though some were not so sure) that minorities who attended desegregated schools scored slightly better on standardized tests than other minority students and that they were less likely to be truants, delinquents, or dropouts. Researchers also thought (although again there were dissenters) that blacks who went to desegregated schools more frequently went on to college, succeeded there, and found work outside of all-black settings.[42]

But some observers of the decision, including the black writer Zora Neale Hurston, complained as early as 1955 that it defamed all-black schools and their teachers. Hurston, a conservative Republican in her politics, wondered why black children would want to go where they were

41. William Chafe, *The Unfinished Journey: America Since World War II* (New York, 1991), 153; Robert Norrell, *Reaping the Whirlwind: The Civil Rights Movement in Tuskegee* (New York, 1985), 72–74, 86–89; Chafe, *Civilities and Civil Rights: Greensboro, North Carolina and the Black Struggle for Freedom* (New York, 1980), 98–141; Kluger, *Simple Justice*, 724–29.
42. Julius Chambers, "*Brown v. Board of Education*," in Hill and Jones, eds., *Race in America*, 184–94; James Liebman, "Three Strategies for Implementing *Brown* Anew," in ibid., 112–66.

likely to be humiliated or threatened. Desegregation, she added, was different from integration—a (rare) situation in which people of different colors more or less willingly mix with one another. "How much satisfaction can I get," she demanded, "from a court order for somebody to associate with me who does not wish me to be near them?"[43] She called instead for strict enforcement of compulsory school laws and more funding for social workers and truant officers. Other doubters asked what would happen to all the black teachers and principals and coaches who had depended on the dual education system for employment. The answer, as it turned out, was that some did lose their jobs or had to take lesser positions doing something else.

Some critics of *Brown* also questioned a controversial premise of the decision: that black schools necessarily induced feelings of "inferiority" among African-American children. This assumption rested in large part on research done by Kenneth Clark, an eminent black psychologist. Clark had concluded, on the basis of experiments showing that black children often preferred white dolls to black dolls, that blacks had low self-regard. Desegregated schools, he thought, would counter such feelings. But this research was dubious and subject to different interpretations. Black children attending desegregated schools in the North, for instance, seemed to have lower self-esteem, as Clark defined it, than black children in segregated schools. The fact of the matter was that in 1954 there simply did not exist sufficient research that could "prove" whether any particular racial mix in schools was superior—or in what ways—to any other. The Court would have done better to avoid socio-psychological speculation, which opened it to criticism.[44]

Educational progress also involved family values and social class. These *Brown* was not asked to address, but it became increasingly clear over time that they remained central to any understanding of what schools could do. Schools in middle-class areas received considerably greater funding per pupil than did schools catering to the working classes. Moreover, parental values and the stability of neighborhoods obviously mattered a great deal: why expect schools to compensate much for disadvantages that children brought with them from their homes? Desegregating schools, it turned out, was not the deliverance that some contemporary enthusiasts, understandably swept off their feet by the moral power of *Brown*, tended to

43. Sollors, "Of Mules and Mares," 171.
44. Ravitch, *Troubled Crusade*, 124–28; Richard Polenberg, *One Nation Divisible: Class, Race, and Ethnicity in the United States Since 1938* (New York, 1980), 155.

imagine. Changing the racial character of schools could not do much to redress the larger social and economic inequality of American life.[45]

These were some of the thoughtful questions about *Brown*. From the start, however, there were openly racist responses, especially from leading southern politicians. The decision of the Court, indeed, greatly weakened racial moderates in southern politics, emboldened racists, and unleashed violent tendencies among extremists.[46] Senator James Eastland of Mississippi, a power on the Judiciary Committee (which passed on nominations of federal judges), explained that Communists were behind the ruling of the Court. "The Negroes," he said, "did not themselves instigate the agitation against segregation. They were put up to it by radical busybodies who are intent upon overthrowing American institutions."[47] Governor James Byrnes of South Carolina (once a Supreme Court justice himself, as well as Truman's Secretary of State) announced, "South Carolina will not now nor for some years to come mix white and colored children in our schools." Governor Herman Talmadge of Georgia added, "I do not believe in Negroes and whites associating with each other socially or in our school systems, and as long as I am governor, it won't happen."[48]

Leaders such as these drew heart from the tentativeness of the ruling. The Court was silent in 1954 about how and when its order should be carried out. That was because Warren and his fellow justices feared to move too far too fast. If they had said that desegregation must be carried out without delay (as they had concerning the Texas law school case in 1950), angry southern opponents might have flouted them, thereby undermining the legitimacy of the Court.

Anti-*Brown* agitators took further heart from the attitude of President Eisenhower. Like most Americans, Ike had grown up in a white world. There had been no blacks in his hometown or at West Point. He had risen in a Jim Crow army and had opposed Truman's order to desegregate the armed services in 1948. He had many wealthy southern friends who talked about the incompetence of their "darkies" and about the absolute

45. There is a vast literature on this subject. See Christopher Jencks, *Inequality: A Reassessment of the Effect of Family and Schooling in America* (New York, 1972); and Jencks and Susan Mayer, *The Social Consequences of Growing Up in a Poor Neighborhood: A Review* (Evanston, 1989), 56–65.
46. Michael Klarman, "How *Brown* Changed Race Relations: The Backlash Thesis," *Journal of American History*, 81 (June 1994), 81–118.
47. Richard Fried, *Nightmare in Red: The McCarthy Era in Perspective* (New York, 1990), 176.
48. *Newsweek*, May 24, 1954, p. 25.

need to segregate the races.[49] Conservative by temperament, he was deeply pessimistic about the possibility of significant changes in race relations and dead-set against using the federal government to force the South to mend its ways. "The improvement of race relations," he wrote in his diary in 1953, "is one of those things that will be healthy and sound only if it starts locally. I do not believe that prejudices . . . will succumb to compulsion. Consequently I believe that Federal law imposed upon our states . . . would set back the cause of race relations a long, long time."[50]

As President, Eisenhower held firm to these opinions. Where he could issue executive orders to desegregate facilities—as in federally run shipyards or veterans' hospitals—he did so. He encouraged efforts to desegregate District of Columbia schools. But he otherwise adhered to a strict constructionist view of the federal-state relationship. "Where we have to change the hearts of men," he told Booker T. Washington's daughter, "we cannot do it by cold lawmaking, but must make these changes by appealing to reason, by prayer, and by constantly working at it through our own efforts."[51] When his Attorney General, Herbert Brownell, a liberal Republican, entered an amicus curiae brief on behalf of Brown and fellow plaintiffs, the President did not stop him, but he was careful not to associate himself personally with it. While the court was considering the school cases in the spring of 1954, he invited Warren to dinner at the White House and sat him next to John W. Davis, the attorney (and Democratic presidential candidate of 1924) who was leading the legal defense team against desegregation at the time. After praising Davis as a great American, Eisenhower took Warren by the arm and privately tried to get him to understand the southern point of view. "These are not bad people," he said. "All they are concerned about is to see that their sweet little girls are not required to sit in school alongside some big overgrown Negro."[52]

When the Court issued its ruling a little later, Eisenhower was upset.

49. Stephen Ambrose, *Eisenhower: Soldier and President* (New York, 1990), 335, 406–19; Herbert Parmet, *Eisenhower and the American Crusades* (New York, 1972), 438–40; Charles Alexander, *Holding the Line: The Eisenhower Era, 1952–1961* (Bloomington, Ind., 1975), 117–18; Robert Burk, *The Eisenhower Administration and Black Civil Rights* (Knoxville, 1984).
50. Ravitch, *Troubled Crusade*, 135.
51. Robert Griffith, "Dwight D. Eisenhower and the Corporate Commonwealth," *American Historical Review*, 87 (Feb. 1982), 116.
52. Ambrose, *Eisenhower*, 367–68.

Sure that the decision would make matters worse, he became disenchanted with Warren, later grumbling privately that appointing him Chief Justice was the "biggest damn fool mistake" he ever had made. When reporters pressed him for his reaction to the Court's decision, he said that he was duty-bound to accept it. But he refused to endorse it. "I think it makes no difference whether or not I endorse it," he said. "What I say is the—the Constitution is as the Supreme Court interprets it; and I must conform to that and do my very best to see that it is carried out in this country." But "very best" did not move him to action. He told a trusted speechwriter, "I am convinced that the Supreme Court decision set back progress in the South at least fifteen years. . . . It's all very well to talk about school integration—if you remember that you may also be talking about social disintegration. Feelings are deep on this, especially where children are involved. . . . We can't demand perfection in these moral things. All we can do is keep working toward a goal and keep it high. And the fellow who tries to tell me that you can do these things by FORCE is just plain NUTS."[53]

Neither the tentativeness of the Court nor the inaction of Eisenhower accounted for the outspoken resistance of men like Eastland and Talmadge. They and others were thoroughgoing segregationists who needed little if any prodding to speak out against the Court. Eisenhower, in short, had a point in maintaining that white feelings in parts of the South ran deep on the issue—so deep that in retrospect it is hard to imagine that school desegregation could ever have been accomplished there without governmental compulsion. He also had a point in reckoning that white Americans at that time did not wish to force recalcitrant school districts to desegregate: despite widespread noncompliance with *Brown*, civil rights played only a minor role in the 1956 campaign. Still, the President's stand was both morally obtuse and encouraging to anti-*Brown* activists. Had he praised the Court for its decision and made clear his determination to enforce it with whatever it took to do so, he would at the least have driven the Talmadges and the Eastlands more to the defensive. Some of the more violent southern assaults that damaged race relations in the next few years might have been avoided.

A year after *Brown*, in May 1955, the Court turned—as it had said it would—to the question of implementation. By then, however, resistance to *Brown* in the Deep South had spread widely. Moreover, it was becom-

53. Emmet John Hughes, *The Ordeal of Power: A Political Memoir of the Eisenhower Years* (New York, 1963), 201.

ing clear that the relocation of students posed complex and time-consuming problems. For these reasons the Court again backed off from confrontation by declining to define an acceptable standard for desegregated schooling. It also refused to set a timetable for compliance. "Brown II," as the court's implementation order was called, said instead that segregated school systems must make a "prompt and reasonable start toward full compliance" and do so with "all deliberate speed."[54]

"Brown II" further emboldened southern opponents, some of whom resorted openly to violence. 1955, indeed, was an unusually violent time: eight of the eleven lynchings of blacks in the 1950s occurred in that year. Other blacks were killed for daring to assert their rights. In Belzoni, Mississippi, the Reverend George Lee was shot at point-blank range and killed for insisting that his name be kept on the voting lists. Evidence pointed to the sheriff, who was asked about the pellets found in Lee's mouth. "Maybe," he replied, "they're fillings from his teeth." No arrests were made. A few weeks later in Brookhaven, Mississippi, Lamar Smith was shot to death in broad daylight in front of the county courthouse. He, too, was a black man who had presumed to vote. As usual, no one was convicted of the crime.[55]

One of the most shocking incidents involved the killing in August of Emmett Till, a fourteen-year-old African-American boy who was visiting relatives in Tallahatchie County, Mississippi, an area that was two-thirds black and where no black person was on the rolls of registered voters or of juries. Till's "crime" was to whistle at a white woman in a grocery store. Hearing of the transgression—a tabu in much of the Deep South—the woman's husband, Roy Bryant, and his half-brother, John Milam, drove to the sharecropper shack of Moses Wright, Till's great-uncle, snatched Till, and drove off with him. Three days later Till was found dead in the Tallahatchie River. He had been shot in the head and tied to a cotton gin fan so that he would sink. His body was badly mangled. Till's mother, Mamie Bradley, had the body shipped back to Chicago, where she displayed it in an open casket for four days. Thousands of people paid their respects. National media carried the story to the country.[56]

To the surprise of many Americans who understood what Mississippi "justice" was like in such cases, Bryant and Milam were actually arrested and charged with murder. The trial, which took place before crowds of

54. 349 U.S. 294 (1955); Kluger, Simple Justice, 744–77.
55. Robert Weisbrot, Freedom Bound: A History of the Civil Rights Movement (New York, 1990), 94; Halberstam, The Fifties (New York, 1993), 430–31.
56. Weisbrot, Freedom Bound, 93.

reporters, took place in September. But it was heard before before an all-white all-male jury and was a charade and a circus. The sheriff greeted black people attending the trial with "Hello, niggers." Blacks, including reporters, were segregated in the courtroom. Wright courageously testi-fied and identified Bryant and Milam as the abductors. But the defense attorney played openly to local white prejudices, reminding the jurors in his summation, "I am sure that every last Anglo-Saxon one of you has the courage to free these men." The jury took only an hour to deliver verdicts of not guilty. "If we hadn't stopped to drink pop," a juror explained, "it wouldn't have taken that long." A grand jury, ignoring Wright's eyewit-ness account, later declined to indict Bryant and Milam for kidnapping; their bail was returned, and they went free. Wright dared not return to his shack, moved to Chicago, and never came back home.[57]

The violence and intimidation employed by southern whites, while harshest in Mississippi, broke out all over the South in 1955 and 1956. By this time the angry but scattered outbursts that had greeted *Brown* in 1954 had spread much more widely. "Massive resistance" ensued, including more violence.[58] In February 1956 Autherine Lucy, a young black woman, sought to become a student at the University of Alabama. She was almost lynched by white students, had to flee, and was formally expelled. "Bama" was not desegregated (tokenly) until 1963. In Bir-mingham a mob attacked and beat the famous black singer Nat "King" Cole when he sang at a whites-only concert in the city's auditorium. In the late summer of 1956 violence broke out in Clinton, Tennessee, where mobs of local whites, their numbers augmented to more than 2,000 by outsiders, terrorized black children seeking entry to the schools. The Highway Patrol and the National Guard used tanks and armored person-nel carriers to curb the violence and desegregated the schools. But cross-burnings, torching of Negro homes, and marches sponsored by the Ku Klux Klan disrupted the area for years thereafter. Angry whites also pre-vented blacks from enrolling in the schools of Mansfield, Texas, in 1956. The governor sent in Texas Rangers to restore order, and the local school board removed the blacks from the schools. In all these cases the federal

---

57. *Time*, Oct. 3, 1955, p. 19; I. F. Stone, *The Haunted Fifties, 1953-1963* (Boston, 1963), 107–9; Stephen Whitfield, *A Death in the Delta: The Story of Emmett Till* (Baltimore, 1988); John Dittmer, *Local People: The Struggle for Civil Rights in Mississippi* (Urbana, 1994), 54–58. Bryant and Milam later admitted their guilt, in return for cash.
58. Numan Bartley, *The Rise of Massive Resistance: Race and Politics in the South During the 1950s* (Baton Rouge, 1969).

government did nothing, maintaining that they were matters for state and local authorities to handle.

Members of the Klan, which expanded considerably in the 1950s, provoked some of this violent activity. As in the past, Klansmen incited violent intimidation and terror, including night-riding, cross-burning, and mob assaults. They claimed to be deeply religious Christians dedicated to the preservation of the Anglo-Saxon white race not only against the incursions of blacks but also of Catholics, Jews, foreigners, and all kinds of "immoral" sinners. As one Klan speaker put it in 1956, "The Ku Klux Klan is the only white Christian Protestant 100 percent American organization in America today." Another Klansman added, "We are gonna stay white, we are going to keep the nigger black, with the help of our Lord and Savior, Jesus Christ."59

Most of these overtly violent southern racists and Klansmen hailed from the lower classes of white society. Ill-educated, often almost as poor as the blacks, they whipped up a ferocious Negrophobia aimed at keeping blacks in their place. But intimidators of blacks included more than the lower classes. Thousands of more "respectable" people, including bankers, lawyers, and businessmen, openly identified with organizations such as the Citizens' Councils, which enjoyed great success at the time. The Citizens' Councils, indeed, were central in cementing a quasi-respectable facade onto massive resistance. They publicly deplored violence but condoned a great deal of it and did not act to bring perpetrators to justice. They stood foursquare for the perpetuation of Jim Crow, including racial segregation in the schools. They repeatedly denounced the Supreme Court, liberals, and northerners in general. What the North was trying to do, they complained, was to impose a "Reconstruction II" on the South. This, they charged, was more insidious than "Reconstruction I" following the Civil War.60

Southern politicians came together to present a near-united front against desegregation of the schools. In early 1956 the Georgia legislature voted to adopt as its new state flag a design that prominently featured the Confederate battle insignia. (Even in 1994, thirty-eight years later, state officials refused to take it down, protests notwithstanding, from the Georgia Dome in Atlanta when the predominantly black players of the National Football League's top two teams battled in the Dome for the Super

59. J. Ronald Oakley, *God's Country: America in the Fifties* (New York, 1956), 335.
60. Dittmer, *Local People*, 45–54; Chafe, *Civilities and Civil Rights*; Neil McMillen, *The Citizens' Council: Organized Resistance to the Second Reconstruction, 1954-1964* (Urbana, 1971), 358.

Bowl.) A much more potent sign of resistance occurred in March 1956, when nineteen of the twenty-two southern senators and eighty-two of the 106 southern representatives joined to issue the so-called Southern Manifesto. This widely noted declaration accused the Supreme Court of "clear abuse of judicial power." It promised to use "all lawful means to bring about a reversal of this decision which is contrary to the Constitution and to prevent the use of force in its implementation." The signers included every senator and representative from the states of Alabama, Arkansas, Georgia, Louisiana, Mississippi, South Carolina, and Virginia. The only southern senators who did not sign were Estes Kefauver and Albert Gore, Sr., of Tennessee and Lyndon Johnson of Texas. All three were relatively liberal and cherished hopes of running for President.[61]

The politicians opposed to desegregation did their most effective work by conjuring up a range of imaginative ruses to evade implementation of *Brown*. States cut off aid to desegregated schools, provided tuition grants to students who attended "private" all-white institutions, denied licenses to teachers who tried to work at desegregated schools, and barred members of the NAACP from public employment. "Freedom of choice" laws authorized parents to send their children to schools of their own choosing. Many opted for all-white private schools, then intimidated black parents who tried to follow suit. "Pupil placement" laws were a favorite dodge. These enabled school officials to use the results of racially biased scholastic or psychological tests as grounds for assigning students to segregated schools. In 1959 Prince Edward County, Virginia, closed all its public schools, offering children private education in their place. When blacks refused to accept what was offered them, they went without any formal schooling at all for three years while litigation ran its course. [62]

In the long run the courts, which time and again proved vital to quests for legal equality in the 1960s, stepped in to put an end to ruses such as these. But it was a long, long run that accelerated only in 1969.[63] In 1962 there were still no black children in schools with whites in the states of Mississippi, South Carolina, and Alabama. In 1964 fewer than 2 percent of blacks attended multi-racial schools in the eleven states of the Old

---

61. Ravitch, *Troubled Crusade*, 133–34.
62. David Goldfield, *Black, White, and Southern: Race Relations and Southern Culture, 1940 to the Present* (Baton Rouge, 1990), 76–87.
63. Jack Greenberg, *Crusaders in the Courts: How a Dedicated Band of Lawyers Fought for the Civil Rights Revolution* (New York, 1994); J. Harvie Wilkinson, *From "Brown" to "Bakke": The Supreme Court and School Integration, 1954–1978* (New York, 1978).

Confederacy. Many southern colleges and universities excluded blacks until the 1960s or accepted only a token few. Very few black teachers were allowed to work in white or desegregated schools. Where dual systems remained, large disparities in funding and other resources persisted.[64] Forty years after *Brown*, in 1994, Summerton, South Carolina, where Pearson had brought suit, had an all-black high school and an all-white town council. White children from the area went elsewhere to school.[65]

This was the South, which northern liberals repeatedly berated. But Americans in the North, where de facto school segregation reflected racially separate neighborhoods, could hardly claim to be color-blind. *Brown* had nothing to say about de facto school segregation, which frequently grew more pronounced as black migrations continued after 1954. So little change occurred in Topeka that the American Civil Liberties Union reopened *Brown* in 1979, asserting that thirteen of the city's schools were highly segregated by race. The suit was not settled (in favor of the ACLU) until 1993, after fourteen years of wrangling, at which point desegregation plans still awaited implementation.[66] Similar stonewalling happened elsewhere in the North: thirty-five years after *Brown* it was estimated that nearly two-thirds of minority schoolchildren in the United States attended public schools in which they exceeded 50 percent of enrollment. More than 30 percent of black children went to public schools that were at least 90 percent non-white.[67]

Developments such as these indicated that Supreme Court decisions, no matter how bold, by themselves could fail to work major changes in the behavior of people in their communities, at least in the short run. Indeed, much broader effort, including congressional action, was required to force compliance with the law. The massive resistance indicated further how deeply racial prejudice and institutionalized discrimination undercut the supposedly egalitarian ideals of the country. Conservatives, Eisenhower among them, readily accepted these dispiriting lessons; they had never had much faith in social engineering. A minority of Americans, however, persisted in demanding a racially more egalitarian world. Among them were many black people, who angrily resented southern defiance of *Brown*. The dissidents would not wait forever for courts and politicians to help them. They would take action themselves.

64. Ravitch, *Troubled Crusade*, 138, 162; John Blum, *Liberty, Justice, Order: Essays on Past Politics* (New York, 1993), 311–12.
65. Klein, "Legacy of Summerton."
66. "Segregation Persists: 40 Years After Brown," *New York Times*, May 17, 1994.
67. Chambers, *"Brown"; Washington Post*, April 11, 1992.

IN THE 1950S they acted most dramatically in Montgomery, Alabama. Briefly the capital of the Confederacy during the Civil War, Montgomery in the mid-twentieth century was a city of some 70,000 whites and 50,000 blacks. Like other southern cities, it enforced Jim Crow, segregating not only schools but virtually all public accommodations, barring most blacks from voting and limiting them mainly to menial work. Some 60 percent of employed black women were domestics, and nearly 50 percent of the employed men were domestics or laborers. The median annual income of whites in Montgomery was $1,732; of blacks it was $970. Roughly 90 percent of white homes had flush toilets, compared to 30 percent of black homes. Because few black people had cars, they had to use buses to get around.[68]

Dependence on buses galled many blacks in the city. The bus company in Montgomery hired no black bus drivers. Its white drivers enforced rules that required blacks to pay at the front of the bus, enter toward the back, and sit in the rear. Drivers often insulted and demeaned black passengers. When buses filled with whites, drivers yelled, "Niggers move back." Blacks sitting at the front of their section were expected to give up their seats and crowd to the rear.

Mrs. Rosa Parks, a forty-five-year-old Negro seamstress and downtown department store worker, regularly rode these buses. Parks was a quiet woman who wore rimless spectacles. Acquaintances knew her to be a dependable, reasonable person and a faithful church-goer. She had long chafed at Jim Crow. More than ten years earlier she had been ejected from a bus for refusing to do as she was told. A member of the NAACP, she was ready to test the bus company's policies. On December 1, 1955, she finished work and Christmas shopping and boarded a bus to go home. When white passengers filled seats in front of her, the driver yelled, "Niggers move back." Parks refused to budge. The driver hailed policemen, who booked her for violating the city's laws and told her to appear for trial four days later.[69]

E. D. Nixon, a Pullman porter who headed the local NAACP, had been waiting for an opportunity such as this and responded quickly. His

68. Adam Fairclough, *Martin Luther King, Jr.* (Athens, Ga., 1995), 23–26; William O'Neill, *American High: The Years of Confidence, 1945–1960* (New York, 1986), 257.
69. Taylor Branch, *Parting the Waters: America in the King Years, 1954–1963* (New York, 1988), 125; David Lewis, *King: A Biography* (Urbana, 1970), 46–84; Martin Luther King, Jr., *Stride Toward Freedom: The Montgomery Story* (New York, 1958).

actions, and those of the NAACP, revealed the central role played by unheralded black people in the fight for civil rights in the 1950s and thereafter. Dramatic leaders came and went, but they could do little without the sacrifices of local folk, who confronted great intimidation, including violence, on the grass-roots level. And many of these people had long been restless indeed. "The Reverend [Martin Luther King] he didn't stir us up," one young Montgomery woman told a reporter at the time. "We've been stirred up a mighty long time."[70]

Women played large roles in what followed in Montgomery and in other demonstrations to come. Jo Ann Robinson, a black English teacher, moved quickly. Hearing of Parks's arrest, she stayed up most of the night, with other members of the Women's Political Council of Montgomery, which she headed, to print protest leaflets, some 50,000 in all, to be distributed in the next few days.[71] The contribution of women like Robinson did not suggest that they were angrier than men; the mounting impatience of involved black people knew no gender boundaries. But black women were often a little less susceptible to economic pressure and to violence than were black men. Many, like Robinson, were steadfast in their goals, well disciplined, efficient, and for all these reasons vital to the cause of civil rights.

Nixon, Robinson, and other activists resolved to fight by boycotting the buses until the company agreed to their demands. These were initially very moderate: the hiring of black drivers, courtesy from drivers to black passengers, and seating on a first-come first-served basis with blacks filling up the back and whites the front. Their strategy of boycott had a considerable history: blacks had boycotted Jim Crow streetcars at the turn of the century. More recently, in 1953, a boycott in Baton Rouge had lasted a week and forced the city to let riders, regardless of race, be seated on a first-come first-served basis.[72] Starting a boycott of course demanded sacrifices: people who refused to get on the buses would have to walk or cooperate in car pools. The tactic also required widespread support; without unity among masses of black residents it would backfire. But a boycott had attractive possibilities. It would enable people to express long-pent-up feelings. It could be undertaken by blacks who acted by *not* acting and

70. Oakley, *God's Country*, 204.
71. Jo Ann Robinson, *The Montgomery Bus Boycott and the Women Who Started It* (Knoxville, 1987); Weisbrot, *Freedom Bound*, 13–15; Goldfield, *Black, White, and Southern*, 93–94.
72. Aldon Morris, *The Origins of the Civil Rights Movement: Black Communities Organizing for Change* (New York, 1984), 18–25.

who thereby risked relatively little (compared to the brazen business of trying to vote) in the way of individualized reprisals. If successful, a boycott could hit white people where it hurt, in the pocketbook. If thousands of people refused to take the buses, not only the company but also downtown merchants would suffer severe financial losses.[73]

The organizers of the boycott knew that they must rely heavily on the most important of all Jim Crow instititions: black churches. Southern blacks were perhaps the most religiously active major group in the United States. At that time the National Baptist Convention, a confederation of black churches, was the largest black organization in the United States and far and away the best supported. It had twice as many members as did the NAACP.[74] Churches in the South provided virtually the only places where large numbers of black people could meet.

Over the weekend before Parks's trial, Nixon and other organizers held long, emotional meetings in these churches. Looking for leadership, they turned to the Reverend Martin Luther King, Jr., a twenty-six-year-old pastor who had come to Montgomery in late 1954 to head the Dexter Avenue Baptist Church. Although young, King was well regarded by local people. As a relative newcomer to the city, he had not antagonized its officials. He was quickly elected president of the Montgomery Improvement Association, the organization formed to lead the boycott.

King was well educated, especially for a black man in the 1950s. Raised in Atlanta, he was the son of Reverend Martin Luther "Daddy" King, Sr., a locally famed preacher and sometimes opponent of Jim Crow. The son had graduated from Morehouse College in Atlanta, an elite Negro college, and then studied at Crozer Theological Seminary in Pennsylvania and at Boston University, whence he received his doctor of philosophy degree (after coming to Montgomery) in June 1955. Although King was hardly an intellectual, he was familiar with a number of key philosophical and theological texts, among them the teachings of non-violent protest as advocated by Henry David Thoreau and Mohandas Gandhi.

Even more important to King's thought were the writings of Reinhold Niebuhr, America's most distinguished theologian. "Niebuhr's great contribution to contemporary theology," King wrote, "is that he has refuted

73. For the boycott, see Branch, *Parting the Waters*, 137–63, 173–205; Morris, *Origins of the Civil Rights Movement*, 51–63. For studies of King, see David Garrow, *Bearing the Cross: Martin Luther King, Jr., and the Southern Christian Leadership Conference* (New York, 1986), 11–82; and Adam Fairclough, *To Redeem the Soul of America: The Southern Christian Leadership Conference and Martin Luther King, Jr.* (Athens, Ga., 1987).
74. Stephen Whitfield, *Culture of the Cold War* (Baltimore, 1991), 22.

the false optimism characteristic of a great segment of Protestant liberalism, without falling into the anti-rationalism of the continental theologian Karl Barth, or the semi-fundamentalism of other dialectical theologians." What King meant was that Niebuhr understood the profoundly sinful nature of mankind without lapsing into despair or abandoning the struggle for social change.[75] Niebuhr's Christian realism provided King with a base on which he rested growing faith in tactics of non-violence.

King's emphasis on non-violent protest, which he refined in the course of the boycott, was sincere and stubborn. Many times during his subsequent career he demanded of impatient followers that they show love, not hate, against oppressors. Even when racists bombed his home in 1956, he remained steadfast in holding to non-violent convictions. King's principled adherence to such beliefs proved inspirational, especially to the religious southern folk who most revered him. To stand up for what was right while trying to stay within the law offered his followers a moral high ground. To contest injustice while refusing to strike one's oppressors was to express the power of Christian love and forgiveness and to make one feel proud to be alive. "We got our heads up now," a black janitor in Montgomery said, "and we won't ever bow down again—no sir—except before God."[76] No approach was better suited to the mobilization of the millions of religious black people who yearned to support causes that would bring great meaning to their lives.

Non-violence, moreover, offered distinct tactical advantages in quests for civil rights. King, a thoughtful tactician as well as an inspirational moral leader, understood this. Non-violent boycotting, for instance, was reassuring, for it promised to give supporters a way of expressing themselves short of undertaking aggressive (and most likely bloody) confrontations with armed and powerful authorities. Later, anticipating violent white retaliation, King shrewdly organized protests in places (Birmingham, Selma) where volatile lawmen were in power, expecting that violence against peaceful demonstrators would promote national revulsion against white racism and elicit popular sympathy for his goals. To liberal whites, too, non-violence was reassuring, for it relieved them of the stereotype of the angry and dangerous black man. When non-violent activists were attacked—as increasingly happened—liberal whites often felt ashamed and guilty. King anticipated these profound human reactions. But he knew enough not to gloat or to reveal his more devious tactical moves. In contrast to other fiery leaders who arose to take the lead

75. Branch, *Parting the Waters*, 87.
76. John Diggins, *The Proud Decades: America in War and Peace, 1941–1960* (New York, 1988), 295–96.

in civil rights protests, he seemed moderate. Liberal whites gave money to him that they denied to those who seemed radical.

King, however, was hardly a moderate by the standards of 1955. At that time he represented a dynamic and forceful salient of what was not yet a national movement. Many white opponents called him a rabble-rouser, even a Communist. Although the demands of the Montgomery Improvement Association remained moderate, King refused to relent until the goals of the association were won. Moreover, he denounced more than Jim Crow on buses; he also challenged all aspects of racial segregation and discrimination in the United States. Desegregation, he insisted, should be accomplished non-violently, but it must be accomplished.

Above all, King was a preacher and an advocate, not a theologian or a philosopher.[77] Exposure to injustice in Montgomery and elsewhere, more than book-learning in graduate school, roused him to heights of eloquence. Like his father (and his grandfather, who had also been a minister), he rooted his thinking and his style in established, widely appreciated Negro Baptist ways.[78] In the pulpit or before a crowd he conveyed deep feeling, indeed moral passion, in a dramatic and cadenced manner that lay deep in the most powerful of African-American preaching traditions. Listeners, especially southern Christians, found him awe-inspiring as a speaker. A biographer-historian, Taylor Branch, describes this appeal: "His listeners responded to the passion beneath the ideas, to the bottomless joy and pain that turned the heat into rhythm and the rhythm into music. King was controlled. He never shouted. But he preached like someone who wanted to shout, and this gave him an electrifying hold over the congregation. Though still a boy to many of his older listeners, he had the commanding air of a burning sage."[79]

All these elements—the readiness of Nixon and the NAACP; the engagement of women such as Parks, Robinson, and others; the outstanding leadership of King; above all the willingness of rank-and-file black people to stick together—proved necessary to sustain the boycott in the difficult months to come. Parks was convicted, ordered to pay a fine of $10, refused, and was jailed. Black boycotters who could be identified were fired from their jobs. King was arrested on a trumped-up charge and sent to jail, along with 100 others, for conspiring to lead an illegal boycott. The Klan marched openly in the streets, vandalized at night, and poured

77. It was revealed long after King's death that he had plagiarized other writers in the course of preparing his doctoral thesis.
78. Keith Miller, *Voice of Deliverance: The Language of Martin Luther King, Jr., and Its Sources* (New York, 1992).
79. Branch, *Parting the Waters*, 119.

acid on vehicles used by blacks for car-pooling. The Citizens' Council distributed inflammatory handbills. Boycott opponents bombed the homes of King and other Negro leaders. President Eisenhower, still enormously popular in the country, kept his distance. "There is a state law about boycotts," he explained at a press conference, "and it is under that kind of thing that these people are being brought to trial."[80]

The solidarity of Montgomery's black citizens (a small number of whites sympathized) nonetheless prevailed. The boycott leaders brought suit, which rose slowly through the federal courts, against the policies of the bus company. Meanwhile, most blacks refused to take the buses, causing ridership to decline by an estimated 65 percent and bus company revenues to plunge. Well-organized car pools helped some of the protestors. But others walked. The most lasting anecdote of the Montgomery movement describes King stopping to ask an old woman walking on the road whether she would prefer to ride the bus. "Aren't your feet tired?" he said. "Yes," she replied, "my feets is tired but my soul is rested."

The boycott ended only after the Supreme Court ruled on November 13, 1956, almost a year after the protest had started, that the city ordinances concerning seating on the buses violated the Fourteenth Amendment. It declared that these discriminatory rules must stop as of December 20. City officials balked at first and required King to pay an $85 fine for breaking anti-boycott regulations. But they finally relented, and King and his co-leaders called off the boycott. On December 21, 381 days after the boycott had started, King sat down with a white man at the front of a bus.

One of the most sustained and coordinated black efforts in the entire history of the civil rights movement was over at last. Indeed, the boycott was a very impressive effort. It thrust King, an extraordinarily gifted leader, into a national and world spotlight. It proved that black people could come together, persevere, and suffer at great length in order to establish their dignity, and it remained an inspiring example for activists in the years to come.

Still, the Montgomery movement left some people unimpressed. Thurgood Marshall said privately at the time, "All that walking for nothing. They might as well have waited for the Court decision." King, he added, was "a boy on a man's errand." (Marshall nonetheless aided efforts by the NAACP to get King out of jail.)[81] Marshall's reaction, while ungenerous, captured an important point: it took a decision of the Su-

80. Ambrose, *Eisenhower*, 408.
81. Lemann, "Lawyer as Hero."

preme Court to force city authorities to surrender. The Court (and NAACP litigation) may have saved the boycott.

Other opponents of racial discrimination also recognized that the boycott did little to weaken the larger edifice of Jim Crow. The Montgomery Improvement Association, although very well organized and unswerving, did not change formal practices much in Montgomery. Schools, public buildings, hotels, lunch counters, theaters, and churches remained segregated. "White" and "Colored" signs confronted people entering public places. There were still no black bus drivers or black policemen. And would future boycotts be a viable strategy? It would do no good, for instance, to boycott a restaurant or park from which one was already excluded. In late 1956, when the boycott in Montgomery ended, it was far from clear what methods of protest might bring down the fortress of Jim Crow in the future.

It was also far from clear that the boycott changed white opinion outside the South. In the 1956 campaign neither Eisenhower nor Stevenson, his opponent for the presidency, paid much attention to civil rights. Both declared that they could never imagine a situation that would induce them to send in federal troops to enforce desegregation. Disgusted with the Democratic party, Harlem's Democratic congressman, Adam Clayton Powell, Jr., and many other blacks supported Ike.[82] King then established an organization, the Southern Christian Leadership Conference (SCLC), to carry on his efforts against racial discrimination. But the Conference, dominated by Negro ministers, was ill organized and aroused limited enthusiasm outside some areas of the South.

After the year-long excitement of Montgomery, militant activism for civil rights actually abated.[83] A great many black people, to be sure, had been inspired; they remained angry about discrimination and eager for change. But most white Americans had never paid much attention to the plight of minorities—whether they were Indians, Asians, Mexicans, or blacks—and for the remainder of the decade they did not much bestir themselves to improve race relations in the nation. Martin Luther King notwithstanding, they seemed more interested in enjoying the blessings of the Biggest Boom Yet. It was not until the 1960s, when a massive increase in civil rights activism arose, that they were forced to sit up and take notice.

82. Branch, *Parting the Waters*, 191–93.
83. Fairclough, *To Redeem the Soul*, 40, 53–54; Garrow, *Bearing the Cross*, 103–4, 120–21; Sitkoff, *Struggle for Black Equality*, 36.

14

A Center Holds,
More or Less, 1957–1960

Public opinion polls in the late 1950s and in 1960, one historian reminds us, reported that the American people were "relaxed, unadventurous, comfortably satisfied with their way of life and blandly optimistic about the future."[1]

A few writers, however, have perceived a more restless citizenry. The scholar Morris Dickstein, a college student at the time, recalled these years as "a fertile period, a seedbed of ideas that would burgeon and live in the more activist, less reflective climate that followed."[2] Richard Pells, a historian, adds, "Beneath the ordinary American's placid exterior, there seemed to lurk a hunger for corrosive wit, jangling sounds, disruptive behavior, defiant gestures, a revival of passion and intensity."[3]

People who perceived blandness and optimism pointed to varied phenomena: the apparently indestructible popularity of Eisenhower, the weakness of political pressure for social legislation and civil rights, the nearly total disarray of the Left. Millions of Americans still listened hap-

1. Fred Siegel, *Troubled Journey: From Pearl Harbor to Ronald Reagan* (New York, 1984), 120. Also Eric Goldman, "Good-by to the Fifties—and Good Riddance," *Harper's*, 220 (Jan. 1960), 27–29.
2. Morris Dickstein, *Gates of Eden: American Culture in the Sixties* (New York, 1977), 88.
3. Richard Pells, *The Liberal Mind in a Conservative Age: American Intellectuals in the 1940s and 1950s* (New York, 1985), 368–69.

pily to "oldies" by Pat Boone, Doris Day, and Frank Sinatra, laughed at the adventures of Ozzie and Harriet, and flocked to watch movie stars like John Wayne. As John Kenneth Galbraith complained in *The Affluent Society* (1958), Americans seemed dazzled by the glitz of the Biggest Boom Yet.

Those who challenged this serene view of American society in the late 1950s countered by citing evidence of cultural unease. "Beats" derided middle-class ways. Teenagers more than ever reveled in the rock 'n' roll of Chuck Berry and the gyrations of Elvis. The comedian Lenny Bruce, foul-mouthed and abrasive, assaulted mainstream values. Norman Mailer wrote a widely discussed essay in 1957, "The White Negro," celebrating the wonders of a loose, free, "hip" life-style. The comedian Tom Lehrer—"So long, Mom, I'm off to drop the bomb, So don't wait up for me"—attracted enthusiastic audiences of college students who roared at his brilliantly crafted songs against Cold War paranoia and nuclear overkill. Anti-war activists formed the Committee for a Sane Nuclear Policy (SANE), also in 1957.

Alienated dissenters from centrist politics could be discovered here and there. At the Highlander Folk School in the hills of Tennessee a former Socialist party organizer, Myles Horton, and Ella Baker, who was soon to be a founder of the Student Nonviolent Coordinating Committee, ran workshops for southern civil rights workers, including Rosa Parks. By 1959 Highlanders were singing a transformed gospel song as "We Shall Overcome." Hundreds of miles away in Massachusetts, Robert Welch, Jr., a retired candy manufacturer, formed the John Birch Society in 1958. The Birchers, as critics called his followers, accepted Welch's far-right view that Ike was a "dedicated, conscious agent of the Communist conspiracy." Birchers claimed to be 40,000-strong by 1963.[4]

Which observers of the late 1950s to believe? The answer depends in part on what one tries to find. Idealistic Americans, cherishing increasingly grand popular expectations about forging a new and better society, unsettled aspects of national culture and politics in these years. In the 1960s these dreamers began to shake the society.[5] As late as

4. For the Highlander School, see Todd Gitlin, *The Sixties: Years of Hope, Days of Rage* (New York, 1987), 75. For the John Birch Society see Stephen Whitfield, *The Culture of the Cold War* (Baltimore, 1991), 41–42. John Birch was an army captain and Baptist missionary killed in an encounter with the Chinese Communists soon after World War II.

5. Daniel Bell, *The Cultural Contradictions of Capitalism* (New York, 1976), esp. 33–84; and Robert Collins, "David Potter's *People of Plenty* and the Recycling of Consensus History," *Reviews in American History*, 16 (Sept. 1988), 321–35.

1960, however, they did not much divert the mainstreams of American culture and politics; between 1957 and 1960, moderates and conservatives continued to win more battles than they lost. Cold War passions remained especially intense. The moderate-to-conservative center that had dominated the United States in the early and mid-1950s, while weakening, managed to hold.

ASIDE FROM THE FOLLOWERS of rock 'n' roll, the "beats" (critics came to call them "beatniks") represented perhaps the most publicized form of dissent from mainstream culture between 1957 and 1960. By far the best-known of them were two former students at Columbia University. One, Allen Ginsberg, had graduated from the university in 1948. He was a poet, a political radical, a drug user, and a pansexual. In 1956, at the age of thirty, he rose to fame following highly public readings of "Howl," a poem he had written while under the influence of peyote, amphetamines, and Dexedrine. "Howl" foresaw a coming apocalypse: "I saw the best minds of my generation destroyed by madness / starving hysterical naked / dragging themselves through the negro streets at dawn looking for an angry fix." When police then seized *Howl and Other Poems* from a San Francisco bookstore, they ignited a sensational, widely reported trial that brought Ginsberg and beats to national attention.[6]

The other featured beat was Jack Kerouac, who was thirty-four years old in 1956. (The best-known beats were approaching middle age in the late 1950s.) In 1951 Kerouac had composed a long stream-of-consciousness-style manuscript about his restless wanderings. Rewritten many times in the next few years, it finally appeared as a book, *On the Road*, in 1957. An early, much-cited passage captured its theme: "The only people for me are the mad ones, the ones who are mad to live, mad to talk, mad to be saved, desirous of everything at the same time, the ones who never yawn or say a commonplace thing, but burn, burn, burn like fabulous yellow roman candles exploding like spiders across the stars."[7] The book sold well and brought much further notice to the beats. Then and later *On the*

6. Among the many secondary accounts of the beats are Lawrence Lipton, *The Holy Barbarians* (New York, 1958); and Bruce Cook, *The Beat Generation* (New York, 1971). See also J. Ronald Oakley, *God's Country: America in the Fifties* (New York, 1986), 397–400; Russell Jacoby, *The Last Intellectuals: American Culture in the Age of Academe* (New York, 1987), 64–67; and John Diggins, *The Proud Decades: America in War and Peace, 1941–1960* (New York, 1988), 267–69. For a sharp criticism of beat behavior see Norman Podhoretz, "The Know-Nothing Bohemians," *Partisan Review*, 25 (Spring 1958), 305–18. Ginsberg, aided by the ACLU, prevailed in subsequent legal battles and succeeded in publishing his work.

7. Kerouac, 8.

Road was a sacred text of sorts not only for the handful of self-proclaimed beats but also for many others, most of them younger than Kerouac, who responded to the message of escape from convention extolled in the book.

Some of those who were attracted to the beats became prominent a few years later. The folksinger Bob Dylan grew up in the 1950s as Robert Zimmerman of Hibbing, Minnesota. He greatly admired Ginsberg and left the University of Minnesota in 1961 to live in Greenwich Village, a center of beat and bohemian life. Tom Hayden, perhaps the best-known "New Left" leader of the early 1960s, was drawn to Kerouac and moved to San Francisco to experience the beat environment in 1960. Dr. Timothy Leary, a Harvard psychologist who claimed to be a follower of Ginsberg and other beats, began his mind-altering experiments with drugs—on Harvard students—in 1960. Ginsberg himself soon became one of Leary's acolytes.[8] All these people absorbed in one way or another a major message of the beats: Americans must reject the excesses of materialism, conformity, and the consumer culture.

The attention given the beats by the media in the late 1950s—and the fear that they aroused among conservatives—suggest that they tapped into a reservoir of discontent, especially among the young. They were a symbol of greater unrest that was to come. Still, it is a stretch to regard the beats as initiating a major cultural trend in the late 1950s, let alone to see them as comprising a "movement" that threatened a larger cultural center in the United States. Estimates of those who actually became beats range from several hundred to a thousand or more, only 150 or so of whom did any writing. They had little if anything in common with many other cultural rebels, such as fans of rock 'n' roll, or with political leftists, at the time. Leary notwithstanding, the beat phenomenon faded a bit by 1960. The media tired of it and moved on to new stories. The vast majority of Americans by then seemed either bored, disgusted, or mildly amused by beat behavior.

IN POLITICS, AS IN CULTURAL MATTERS, opponents of a conservative status quo also seemed to be gaining ground in the late 1950s, especially after a recession struck in 1958. Democrats triumphed in the 1958 elections, greatly increasing their numbers on Capitol Hill. Senator John F. Kennedy of Massachusetts cruised to victory in his quest for re-election and set sail for the presidency in 1960. Other, more liberal

8. Gitlin, *Sixties*, 51–52.

Democratic senators, such as Hubert Humphrey of Minnesota, Herbert Lehman of New York, and Paul Douglas of Illinois, took the offensive in Congress by demanding passage of federal aid to education, a public system of health insurance, and governmental assistance to "depressed areas." To combat the recession they called for expansion of federal spending on public works and for a tax cut. Their legislative initiatives set the economic agenda for Democratic presidential efforts in the 1960s.[9]

In the late 1950s, however, liberal Democrats scarcely came close to success. Congress did manage to conduct some important business, including the admission to statehood of Alaska and then Hawaii in 1959. In 1960 it approved a modest federal-state program to help the elderly with their medical costs. But Democratic moderates, led by Lyndon Johnson and Sam Rayburn, set the main agenda. This consisted primarily of acquiescing in the conservative domestic policies of Eisenhower, whose still formidable personal popularity they feared to challenge. For these and other reasons no liberal bills of significance succeeded during Ike's second term. Long-standing socio-economic problems, including poverty, continued to fester, and newer ones, such as a slowdown in manufacturing, were allowed to undermine the security of working-class people in the cities.

Nothing more clearly revealed the frustrations of liberals in the late 1950s than the state of race relations. Civil rights leaders, led by King, Roy Wilkins of the NAACP, and A. Philip Randolph, sought to stir up sentiment for racial justice by staging a "Prayer Pilgrimage" at the Lincoln Memorial on May 17, 1957, the third anniversary of *Brown*. The turnout of 25,000, however, failed to attract much attention from the national media or politicians in Washington.[10] Activists on the local level, often led by women who belonged to youth chapters of the NAACP, had slightly greater success: sit-ins in Wichita and Oklahoma City in 1958 succeeded in desegregating drugstores. These and other sit-ins between 1957 and 1960 showed that black people were ready for direct action, especially in the border states and Upper South. But the sit-ins, like the Prayer Pilgrimage, did not arouse the conscience of America's white majority to the cause of civil rights.

In the Deep South, meanwhile, racist whites continued to act with virtual impunity. Public schools remained almost wholly segregated, and

9. James Sundquist, *Politics and Policy: The Eisenhower, Kennedy, and Johnson Years* (Washington, 1968), esp. 3–250.
10. Robert Weisbrot, *Freedom Bound: A History of America's Civil Rights Movement* (New York, 1990), 77.

black colleges sympathetic to civil rights suffered cuts in state funding. A black professor at Alcorn College who tried to enter the summer session at the University of Mississippi in 1958 was arrested and placed in an insane asylum. Whites laughed, "Any nigger who tries to enter Ole Miss *must* be crazy." A black army veteran, Clyde Kennard, who sought to register at Mississippi Southern College in 1959, was harassed by police, arrested on trumped-up charges (for "reckless driving" and stealing a bag of chicken feed), and ultimately sentenced to seven years in jail. In 1959 a black Mississippian, Mack Charles Parker, was jailed on charges of raping a white woman. Two nights before his trial nine masked men snatched him from his cell at Poplarville, drove him to the Pearl River between Mississippi and Louisiana, shot him twice through the chest, and flung him into the water. His body was found nine days later. Many local people were said to know the killers, but none came forward, and no one was tried. At the impaneling of a grand jury on the case in November a circuit court judge, Sebe Dale, told the jurors that Supreme Court decisions probably caused Parker's death. The Court, said the judge, was a "board of sociology, sitting in Washington, garbed in judicial robes."[11]

Civil rights leaders hoped desperately that federal officials might help. Some of these officials tried. The FBI worked hard in a futile effort to bring Parker's murderers to justice. And Attorney General Herbert Brownell supported a civil rights bill that came to the floor of the Senate in mid-1957. Hardly a tough measure, the bill nonetheless garnered support from liberals who hoped that it could be stiffened so as to provide protection for black voters in the South. Eisenhower, however, continued to believe that progress in race relations would happen only when popular attitudes were ready for it. Clearly uninterested in the bill, he amazed reporters at a news conference in July by telling them, "I was reading that bill this morning, and there were certain phrases I did not entirely understand." Reporters, he advised, should talk to Brownell. Eisenhower also made it clear that he had no stomach for military enforcement of racial justice. "I can't imagine any set of circumstances," he said in July, "that would ever induce me to send federal troops into any area to enforce the orders of a federal court, because I believe that [the] common sense of America will never require it."[12]

Although Eisenhower's half-hearted support of the bill did it no good, the main problem facing the measure was the threat of a southern filibus-

11. Taylor Branch, *Parting the Waters: America in the King Years, 1954–1963* (New York, 1988), 257–58; Oakley, *God's Country*, 377–78.
12. Stephen Ambrose, *Eisenhower: Soldier and President* (New York, 1990), 441–44.

ter. To prevent this from happening, Senate majority leader Johnson resolved to compromise. He softened southern opponents with an amendment guaranteeing defendants charged with criminal contempt (for violating voting rights) the right of trial by jury. Many supporters of the bill hotly opposed the amendment, for they recognized that the provision for juries—of whites—would protect defendants from conviction. But Johnson, a shrewd parliamentarian, managed to persuade enough moderates, including Democrats such as Kennedy, that it was either the bill with the jury trial amendment or filibuster and no act at all. The amendment succeeded by a vote of 51 to 42. The bill then passed, 72 to 14.[13]

Some people at the time grew hopeful following this result. Whatever the faults of the bill, they said, it was the first civil rights law to get through Congress since Reconstruction. Johnson, they added, had done what he had to do and had demonstrated a statesmanship that made him a viable presidential candidate—a goal that he clearly had in mind. The New York Times called the law "incomparably the most significant domestic action of any Congress in this century."[14] All these optimists pointed to features of the act that seemed promising: creation of a Civil Rights Commission, establishment within the Justice Department of a civil rights division, and empowerment of the Attorney General's office to bring injunctions when would-be voters complained. In fact, however, the jury trial amendment, along with earlier compromises, gutted the bill of any practical impact, and the Eisenhower administration brought only a handful of suits against alleged violaters during the next three years. By 1959 the law had not added a single black voter to the rolls in the South.[15] A second voting rights law, passed in 1960, was equally ineffectual, and at the end of the Eisenhower years only 28 percent of southern blacks of voting age had the franchise. In Mississippi the percentage was 5.[16]

No racial controversy of Eisenhower's second term, however, was more discouraging to civil rights activists than the confrontation over school desegregation in Little Rock in the late summer of 1957.[17] The struggle

13. Paul Conkin, Big Daddy from the Pedernales: Lyndon Baines Johnson (Boston, 1986), 139–42.
14. Branch, Parting the Waters, 221.
15. Sundquist, Politics and Policy, 243.
16. Numan Bartley, The Rise of Massive Resistance: Race and Politics in the South During the 1950s (Baton Rouge, 1969), 7–8; Steven Lawson, Black Ballots: Voting Rights in the South, 1944–1969 (New York, 1976), 133–34.
17. Tony Freyer, The Little Rock Crisis: A Constitutional Interpretation (Westport, Conn., 1984); Harvard Sitkoff, The Struggle for Black Equality, 1954–1992 (New York, 1993), 29–33; Herbert Parmet, Eisenhower and the American Crusades

came as something of a surprise, for the city's mayor and school board had planned to comply in a token fashion with court rulings on the subject. But the governor of Arkansas, Orval Faubus, demagogically commanded 270 National Guard troops to move in around Central High School on the day before school was set to open. The troops, he said disingenuously, were needed to maintain law and order at the school. In fact, they were there to keep the nine black children assigned to Central High from entering.[18]

The crisis escalated dangerously during the next three weeks. On the first school day the black students heeded the advice of the school board and stayed home. But on the second day they were escorted to school by two white and two black ministers, only to be stopped by the Guard. They left, walking with dignity through a taunting, cursing crowd of white students and townspeople who had been stirred up by Faubus's intervention. Television cameras captured their ordeal and sent images of the event to amazed and angry viewers all over the globe.

Eisenhower now confronted a possibility that he had only two months earlier said he could not imagine: using troops to enforce desegregation orders of the federal courts. He shrank from the prospect. For the next eighteen days he tried to resolve the issue by conferring with the mayor and even with Faubus himself, who flew to the President's summer retreat in Newport, Rhode Island. Meanwhile the National Guard remained at the school, and the black children stayed home. Faubus removed the guardsmen only when a federal court ordered him to, by which time local passions had become inflamed to potentially violent proportions.

With the guardsmen gone, the nine black students came again to school on Monday, September 3—three weeks after the crisis had begun. But only 150 local policemen were on hand to protect them from a large and angry mob of whites. When the mob learned that the children had managed to get inside Central High (through a delivery entrance), leaders started shouting, "The niggers are in our school." The mob then began attacking black people on the street as well as "Yankee" reporters and photographers. The local police were clearly sympathetic to the mob; one

---

(New York, 1972), 509–12; Charles Alexander, *Holding the Line: The Eisenhower Era, 1952–1961* (Bloomington, Ind., 1975), 197–200; Dwight Eisenhower, *Waging Peace* (Garden City, N.Y., 1965), 162–76; and William Pickett, *Dwight D. Eisenhower and American Power* (Wheeling, Ill., 1995), 152–53.

18. Diane Ravitch, *The Troubled Crusade: American Education, 1945–1980* (New York, 1983), 136–38.

took off his badge and walked away. The mayor, frightened by the specter of large-scale violence, cabled the White House to ask urgently for the dispatch of federal troops. The black students were taken out of school and sent home, whereupon the mob slowly dispersed.

Eisenhower still resisted the step of sending in troops. Instead, he denounced the "disgraceful occurrences" at Little Rock and ordered people to desist and disperse. But the next day, with the black children still at home, 200 whites showed up at the school. The President then carried out what Sherman Adams, his top aide, later said was "a constitutional duty which was the most repugnant to him of all his acts in his eight years at the White House."[19] He sent 1,100 army paratroopers into Little Rock and federalized the Arkansas National Guard, thereby removing it from Faubus's command.[20] In issuing the orders Eisenhower acted not as a defender of desegregation but as commander-in-chief. Faced with Faubus's defiance and with violence, he concluded reluctantly that he had no choice. It was the first time since Reconstruction that federal troops had been dispatched to the South to protect the civil rights of blacks.

The President's actions earned him very mixed reviews. Southern politicians reviled him, with Senator Richard Russell of Georgia comparing the soldiers to "Hitler's storm troopers."[21] The lieutenant governor of Alabama, Guy Hartwick, exclaimed that "Pearl Harbor was a day of infamy. So was Eisenhower's brutal use of troops."[22] But most advocates of forceful civil rights activity were upset that Ike had been so indecisive. The President's dallying, they said, gave comfort to extremists and besmirched the image of the United States all over the world.

The soldiers arrived on Wednesday, September 25, and stayed until the end of November. The guardsmen remained for the entire academic year. Eight of the students stuck out the year, and one, Ernest Green, graduated with his senior classmates and went off to college at Michigan State. It was never easy for them, however, because a small minority of their white schoolmates regularly cursed, pushed, and spat on them. Local whites threatened to dynamite the school and to kill the school superintendent. (One such attempt was actually made but failed.) Faubus, seeking headline-grabbing fame, was enthusiastically re-elected in 1958 and for

19. Sherman Adams, *First-Hand Report: The Story of the Eisenhower Administration* (New York, 1961), 355.
20. Oakley, *God's Country*, 338–40.
21. Ibid., 341.
22. *Newsweek*, Oct. 7, 1957, p. 30.

three more terms after that. In the 1958-59 academic year he closed all the schools in Little Rock rather than see desegregation in the city. Other emboldened southern leaders followed in his footsteps and swelled massive resistance throughout much of the South in the late 1950s. By 1960 blacks were despairing about the chances for meaningful help from politicians and were resolving to take action themselves.

AMERICAN CRITICS OF Red Scare excesses derived fleeting satisfaction from a few developments of the late 1950s. But they, too, found these years frustrating.

The most significant sign of change in such matters came from the Supreme Court. By 1956–57 Warren was emerging as an advocate not only of civil rights but also of civil liberties. Others on the Court joined him to swing the tribunal toward a more liberal course. Chief among them were the two veterans appointed by FDR, Black and Douglas, as well as a newcomer named in 1956 by Eisenhower, William Brennan, Jr., of New Jersey. Before nominating him Eisenhower did not examine Brennan's views with much care—if he had, he would hardly have proposed him. Like many conservatives in the late 1950s, the President was both stunned and upset by what followed, for the Court began in 1956 to clear away some of the anti-Communist laws and regulations that had flourished during the Red Scare. It declared its civil libertarianism most clearly on June 17, 1957, which anti-Communist foes denounced as "Red Monday." Then and two weeks later a series of decisions strengthened constitutional guarantees against self-incrimination, construed the anti-Communist Smith Act of 1940 narrowly so as to guard against political trials, protected individuals from having to answer questions (from HUAC) about others, and ruled that certain defendants had rights to see reports by FBI-paid informants. As a result of these decisions, the government virtually gave up efforts to prosecute Communists under the Smith Act.[23]

Civil libertarians applauded the verdicts. I. F. Stone said that "they promise a new birth of freedom. They make the First Amendment a reality again. They reflect the steadily growing public misgiving and distaste for that weird collection of opportunists, clowns, ex-Communist crackpots, and poor sick souls who have made America look foolish and

23. The key cases were *Jencks v. U.S.*, 353 U.S. 657 (1957); *Watkins v. U.S.*, 354 U.S. 178 (1957); and *Yates v. U.S.*, 355 U.S. 66 (1957). For these cases see Whitfield, *Culture of the Cold War*, 50–51; and Richard Fried, *Nightmare in Red: The McCarthy Era in Perspective* (New York, 1990), 184–86.

even sinister during the last ten years."[24] Other Americans, however, were confused and upset by the trend. Eisenhower, asked about some of the cases at a press conference, mused, "Possibly in their latest series of decisions there are some that each of us has very great trouble understanding."[25] Justice Tom Clark, worrying about national security, complained that the Court, in affording accused people access to FBI data, provided "a Roman holiday for rummaging through confidential information as well as vital national secrets."[26]

Right-wing Americans who were already angry at the *Brown* decision were especially quick to seize on these new cases and to attack the Court. The John Birch Society, mobilizing in 1958, focused considerable resources on impeaching Warren and on curbing the authority of the Court. Prominent conservative senators led a similar, more serious assault on Capitol Hill—one that would have trimmed the Court's jurisdiction over the areas of loyalty and subversion. They employed the familiar tactic of connecting Communism and support for civil rights. The Court, said Eastland of Mississippi, was "being influenced by some secret, but very powerful Communist or pro-Communist influence."[27]

Advocates of judicial restraint offered more temperate criticisms of the Court. Among them were some of the justices themselves, notably Felix Frankfurter, who remembered how the Court's conservative activism in the 1930s had created a constitutional crisis. Judicial boldness, he thought, could foment (indeed, was fomenting) attacks on the Court in the 1950s—this time from the Right. By the late 1950s Frankfurter and John Marshall Harlan, whom Eisenhower had appointed in 1955, openly called on the Court to moderate what they considered to be its excesses of judicial activism. Edward Corwin, a Princeton professor emeritus widely regarded as one of the nation's foremost authorities on constitutional history, went so far as to write the *New York Times*, "There can be no doubt that . . . the court went on a virtual binge and thrust its nose into matters beyond its own competence, with the result that . . . it should have its aforesaid nose well tweaked. . . . The country needs a protection against the aggressive tendency of the court."[28]

24. I. F. Stone, *The Haunted Fifties, 1953–1963* (Boston, 1963), 203 (June 24, 1957).
25. Bernard Schwartz, *Super Chief: Earl Warren and His Supreme Court: Judicial Biography* (New York, 1983), 250.
26. William Leuchtenburg, *A Troubled Feast: American Society Since 1945* (Boston, 1973), 101.
27. Walter Murphy, *Congress and the Court: A Study in the Political Process* (Chicago, 1962), 89.
28. Ibid., 161.

Thanks to the efforts of Johnson and others in the Senate, Congress did not hamstring the Court. But it was a near thing; in August 1958 the conservative coalition lost a key motion to limit the Court's power by the narrow margin of 49 to 41. Perhaps aware that it risked retaliation, the Court itself seemed cautious in 1959. In the *Barenblatt* case of that year it alarmed civil libertarians by upholding, 5 to 4, a contempt citation of Barenblatt, an educator who had cited the First Amendment in refusing to cooperate with HUAC. [29] When the Court resumed its civil libertarian course in the 1960s—in retrospect, *Barenblatt* was anomalous—conservative critics erupted again. Their anger exposed a durable aspect of American thought in the postwar era: anti-Communist feelings at home remained strong indeed.

Nothing did more during these years to excite such emotions than the successful launching by the Soviet Union on October 4, 1957 of *Sputnik*, the world's first orbiting satellite. *Sputnik* was small—about 184 pounds and the size of a beach ball. But it whizzed along, going "beep beep beep" at 18,000 miles per hour and circling the globe every ninety-two minutes. A month later the Soviets launched *Sputnik II*. This weighed some 1,120 pounds and carrried scientific instruments for studying the atmosphere and outer space. It even accommodated a dog, Laika, which had medical instruments strapped to its body. [30]

Americans reacted to these dramatic accomplishments with an alarm approaching panic. The Soviets, it seemed, had far outpaced the United States in rocketry. Soon they might conquer space, perhaps to establish dangerous extraterrestrial military bases. Meanwhile, American efforts to catch up seemed ludicrous. On December 6 a nationally televised test of the Vanguard missile proved deeply embarrassing. The missile rose two feet off the ground and crashed. Press accounts spoke about "Flopnik" and "Stay-putnik." G. Mennen Williams, Democratic governor of Michigan, ridiculed the American effort:

> Oh little Sputnik
> With made-in-Moscow beep,
> You tell the world it's a Commie sky
> And Uncle Sam's asleep. [31]

29. *Barenblatt v. U.S.*, 360 U.S. 109 (1959).
30. Robert Divine, *The Sputnik Challenge* (New York, 1993); David Patterson, "The Legacy of President Eisenhower's Arms Control Policies," in Gregg Walker et al., eds., *The Military-Industrial Complex* (New York, 1992), 228–29.
31. Peter Biskind, *Seeing Is Believing: How Hollywood Taught Us to Stop Worrying and Love the Fifties* (New York, 1983), 337.

Williams at least offered a light touch. Other critics, especially Democrats, hammered at the administration for failing to keep pace with the enemy. Senator Henry Jackson of Washington spoke of "shame and danger." Senator Stuart Symington of Missouri added, "Unless our defense policies are promptly changed, the Soviets will move from superiority to supremacy. If that ever happens, our position will become impossible."[32]

A previously commissioned report by a panel of the Science Advisory Council, delivered in early November to the National Security Council, seemed to bear out these criticisms. It represented the thinking of a number of veteran (and partisan) anti-Communist Establishmentarians, including Robert Lovett, John McCloy, and Paul Nitze (who in 1950 had fashioned NSC-68 calling for major hikes in American defense spending). The Gaither Report, as the document was called, was supposedly classified, but its contents quickly leaked. It recommended enormous increases in military spending, amounting to $44 billion—to be achieved by deficit spending—during the next five years.[33] It also demanded rapid development of fallout shelters. Press accounts seized on the Gaither Report as confirmation of America's vulnerability and Eisenhower's culpability.

The Gaither Report was but one of a series of broadsides from "experts" on defense needs who were sure that the United States was falling behind the Soviets. Starting in early 1958 the Rockefeller Brothers Fund released a series of critiques along the same line.[34] Some of these came from the fertile mind of Henry Kissinger, a thirty-four-year-old student of international relations. His own book *Nuclear Weapons and Foreign Policy*, published in 1957, had already insisted that the United States needed a much more flexible and costly military posture. It became a big seller.

Documents such as the Gaither and Rockefeller reports greatly distorted the contemporary balance of terror. In fact, there was no such thing at that time—or later—as a "missile gap," a phrase that Democrats and others flung at the GOP during the 1958 and 1960 election campaigns. The *Sputnik* launches indeed demonstrated that the Soviets had an edge

32. William O'Neill, *American High: The Years of Confidence, 1945–1960* (New York, 1986), 270.
33. Named after its chair, Rowan Gaither, a lawyer and head of the Ford Foundation. For accounts of the report see John Gaddis, *Strategies of Containment: A Critical Appraisal of Postwar American National Security Policy* (New York, 1982), 184; Paul Nitze, with Ann Smith and Steven Reardon, *From Hiroshima to Glasnost: At the Center of Decision, a Memoir* (New York, 1989), 169; and Alan Wolfe, *America's Impasse: The Rise and Fall of the Politics of Growth* (New York, 1981), 120–21.
34. Subsequently collected and published in 1961 as *Prospect for America*.

in capacity for thrust—the ability to boost satellites into orbit. But in fact the Soviets lagged badly in the production of usable warheads and did not deploy an intercontinental ballistic missile (ICBM) during the Eisenhower years. In 1957 the United States had a huge advantage over the Soviet Union in the areas of military missile development and in nuclear weaponry, and it widened it during Eisenhower's second term. In the event of a Russian attack—which under the circumstances would have been suicidal for the Kremlin—the United States could devastate the military and industrial power of the Soviet Union.[35]

Eisenhower, moreover, had good reason to be confident of American nuclear superiority. Since 1956 he had benefited from extraordinary intelligence being gathered by U-2 planes, which were sleek and supersonic reconnaissance aircraft designed to fly at altitudes up to 80,000 feet (fifteen miles) and equipped with amazingly powerful cameras. Photographs could capture newspaper headlines ten miles below. Evidence from U-2 flights established without doubt in 1956–57 that the Soviets lagged far behind in development of ICBMs. The photos made it possible to lay out in sequence what the Soviets were doing and thereby gave the United States plenty of warning about Moscow's preparations, if any, for attack.

The President, self-assured about his military expertise, carefully examined the U-2 evidence, for he took great pride in his attention to American security, which he considered to be far more important than exploration of space. It was much better, he said, to have "one good Redstone nuclear-armed missile than a rocket that could hit the moon. We have no enemies on the moon." Refusing to be panicked by *Sputnik*, he insisted on adhering to the doctrine of "sufficiency": in dealing with nuclear powers the United States needed mainly to maintain sufficient military strength to survive foreign assault, with nuclear weapons in hand for a devastating counter-attack, but it need not arm ad infinitum. How many times, he asked impatiently in 1958, "could [you] kill the same man?"[36] Rejecting the fiscal extravagance of the Gaither Report, Eisenhower continued to insist on containing costs. He told his Cabinet (which needed endless reassurance) in November, "Look, I'd like to know what's on the other side of the moon, but I won't pay to find out this year."[37]

Under popular pressure Eisenhower did bend a little. In 1958 he supported establishment of the National Aeronautics and Space Agency (NASA), a civilian bureaucracy that was created to coordinate missile

35. John Blum, *Years of Discord: American Politics and Society, 1961–1974* (New York, 1991), 15.
36. Gaddis, *Strategies of Containment*, 187–88; O'Neill, *American High*, 273–75.
37. Oakley, *God's Country*, 346; Ambrose, *Eisenhower*, 453–54.

development and space exploration in the future.[38] He also recommended federal aid to promote American know how in science and foreign languages. The result, approved by Congress in September 1958, was the National Defense Education Act (NDEA). This was a historic break with twentieth-century practice, which had assigned educational spending primarily to states and localities. As its title indicated, however, the NDEA was sold as a defense measure, not as endorsement of a broader principle of federal aid to schools or universities. Individuals who received money under the act had to sign a clause affirming loyalty to the United States and to swear that they had never engaged in subversive activities.[39]

But Eisenhower otherwise hewed to his course. Indeed, he had a fundamental problem: if he spelled out in detail the nature of America's superiority in missiles and nuclear weaponry, thereby trying to relieve political pressure at home, he would have to reveal that the United States possessed super reconnaissance planes. Such a disclosure, he thought, would expose too much about American intelligence. (It was learned later that Ike need not have worried much about this; the Soviets knew of the flights in 1956 but lacked the capacity until later to shoot them down.) Full disclosure, Eisenhower further realized, would also amount to confession of high-level spying. Worse, it would publicly embarrass Khrushchev, driving him to escalate USSR defense spending. So the President did not tell all. Privately, he reassured people in the know. Publicly, he relied on trying to convince a highly nervous population (in the face of ardent partisan attacks) that he, a military expert, knew what he was doing.

As commander-in-chief Eisenhower deserves mostly good marks for his handling of concerns over *Sputnik* (and of defense policies in general during his second term). By concentrating on missile development Eisenhower presided over great gains in America's air-based and nuclear capacity relative to that of the Soviet Union. By doing so quietly, he may have allayed Soviet fears, thereby preventing the Soviets from hastening missile development of their own.[40] By refusing to panic in the face of *Sputnik* and domestic criticism he maintained a lid on defense spending.[41]

38. Ambrose, *Eisenhower*, 463.
39. Alexander, *Holding the Line*, 131–32; Oakley, *God's Country*, 346–47, 352.
40. See Michael Beschloss, *The Crisis Years: Kennedy and Khrushchev, 1960–1963* (New York, 1991), 25–26. Beschloss contrasts Eisenhower's cool command with Kennedy's boastful behavior in 1961–62—behavior that riled Khrushchev and helped foment the Cuban Missile Crisis of 1962.
41. Gaddis, *Strategies of Containment*, 164.

At the time, however, *Sputnik* probably damaged Eisenhower's politi-cal standing. Belated American missile successes—the United States finally launched its first satellite on January 31, 1958, from Cape Canaveral—did not do much to reassure American doubters. The satel-lite, carried by a Jupiter C rocket, weighed only thirty-one pounds. In October 1959 the Soviet Union landed a probe on the moon and sent back pictures of the dark side of the lunar surface. (In April 1961 it put the first man, Yuri Gagarin, into orbit.) The United States clearly lagged in this kind of effort when he left office in January 1961. It was not until February 20, 1962 that Lieutenant Colonel John Glenn became the first American to orbit the earth.

The *Sputnik* "crisis" above all revealed the enduring power of Cold War fears in the late 1950s. Although Eisenhower did his best to reassure people about American preparedness, he only half-succeeded. And this mattered politically. Americans, holding grand expectations about their "can-do" capacity, liked to think that they were the first and the best at scientific and technological innovation. If they fell behind, someone must have blundered. Reflecting these attitudes, contemporaries as varied in their politics as Chester Bowles, Dean Rusk, AFL leader George Meany, and Kissinger publicly deplored what they said was a lack of "national purpose." (What that was, they were hard-pressed to say, but obviously it was up to the President to provide it.) George Kennan com-plained in 1959 that the United States needed a greater "sense of national purpose" if it hoped to best the Soviet Union in the competition for world leadership. General Maxwell Taylor, army chief of staff in 1958–59, demanded a more flexible military posture in a book, *The Uncertain Trumpet*, that he published in 1960. Eisenhower's own Commission on National Goals warned him in 1960 that "the nation is in grave danger" and "threatened by the rulers of one third of mankind." Disgusted, the President did not release the report, *Goals for Americans*, until after the 1960 election. But claims that the nation suffered from a "missile gap" nonetheless resounded throughout the campaign.[42] When the new gen-eration took power in 1961, it zestfully accelerated the arms race.

ALTHOUGH EISENHOWER TRIED to moderate anxieties con-cerning *Sputnik* and defense spending, he otherwise pursued policies that sustained and in some ways exacerbated the Cold War during his second

42. Quotes are from Siegel, *Troubled Journey*, 116 (Kennan), and Oakley, *God's Country*, 414.

term. In his Middle Eastern policies, his use of the CIA, and above all his conduct of relations with Cuba and Vietnam, he managed (with provocative help from Khrushchev) to leave an unusually tense world situation to his successors.

His engagement with problems in the Middle East increased in early 1958 following a bloody coup that overthrew the royal family in Iraq. Nearby nations, including Jordan, Saudi Arabia, and Lebanon, grew worried that Nasser-style nationalism might engulf the region. Eisenhower attended willingly to their concerns, both because he worried about Soviet moves in the area and because he was determined to protect Western oil interests. He also welcomed the chance to take forceful action.

All these concerns led to his decision, for the only time during his eight-year presidency, to send American troops into what might become combat abroad. On July 15, 1958, marines splashed onto the beaches of Lebanon. Two days later, in what was obviously a coordinated effort, British paratroopers landed in Jordan. Fortunately for all concerned, fighting proved unnecessary, and by late October the marines were removed. The incursion accomplished nothing, for Lebanon had faced no real threat. In his eagerness to display the resolve of the United States, Eisenhower had resorted to a form of gunboat diplomacy that bestowed no honor on the nation or on his presidency.[43]

Similar criticism can be made of his continuing reliance on the CIA. During his second term he gave it still more latitude than he had earlier, when it had abetted coups in Iran and Guatemala. In 1958 the CIA attempted unsuccessfully to overthrow the government of Indonesia, and in 1959 it helped to install a pro-Western government in Laos. Eisenhower knew of and approved of such efforts. What he apparently did not know, but what his largely uncritical support of the agency encouraged, was that CIA operatives were also hatching schemes to assassinate the Congolese leader Patrice Lumumba and the new head of Cuba in 1959, Fidel Castro.[44] The CIA was becoming a rogue elephant.

In dealing with Khrushchev after 1957 Eisenhower tried sporadically to promote more amicable relations. This was hardly because he trusted the Soviets or because he thought that conciliation might lead to détente. Rather, he became ever more certain, thanks in part to the U-2 flights, that American military superiority made negotiations an increasingly safe and desirable option. In 1958 Dulles, widely considered the ultimate

43. Ambrose, *Eisenhower*, 465–70.
44. Gaddis, *Strategies of Containment*, 159; Arthur Schlesinger, Jr., "The Ike Age Revisited," *Reviews in American History*, 4 (March 1983), 1–11.

hard-liner, added his weight to such an approach by calling for reductions in defense spending. "In the field of military capacity," Dulles advised, "enough is enough."[45]

Eisenhower and Dulles pressed especially for a Soviet-American ban on nuclear tests in the atmosphere. Their support of such an effort rested in part on the advice of scientists, who by 1958 had grown more confident about their ability to distinguish a far-off nuclear test from a seismic event. It might not be necessary any more, they thought, to demand frequent on-site inspections of Russian installations—a demand that in the past had frightened the highly secretive Soviet leadership and helped to stall efforts for control of testing. Administration leaders were also growing worried about evidence linking atmospheric testing to long-range radioactive fall-out. Taking no chances, the United States ran yet another full series of tests in October 1958 and then—secure in the knowledge of American superiority—ceased atmospheric testing on October 31. A few days later the Soviets (who had conducted their own spate of tests in October) stopped, too. Although both sides kept on building bombs, chances for some kind of nuclear agreement seemed more promising than at any time in the history of the Cold War.[46]

As it happened, both sides stopped atmospheric tests for the next three years.[47] But Eisenhower failed in his quest for an agreement and presided over a Soviet-American relationship that deteriorated badly by the summer of 1960. This was mainly the doing of Khrushchev, who proved to be an erratic, confrontational, and sometimes grandstanding adversary. In November 1958 the Soviet leader sharply escalated tensions over Berlin, which remained isolated within the satellite state of Communist East Germany. Unless American troops left West Berlin by May 27, 1959, Khrushchev warned, the Soviet Union would sign a treaty with East Germany, thereby giving the East Germans a green light to deny American troops ground access to Berlin. The United States, which did not recognize East Germany, might have no recourse save to shoot its way into the city.

The early months of 1959 were difficult months for foreign policy-making in the Eisenhower administration, for Dulles was in the hospital with cancer (he died on May 24). Eisenhower, however, grabbed the reins and made it clear that the United States would stick by West Berlin. But

45. Ambrose, *Eisenhower*, 457–61.
46. Ibid., 471–79; O'Neill, *American High*, 232–39; Robert Divine, *Eisenhower and the Cold War* (New York, 1981), 127–31.
47. Or said they did; supervision and surveillance scarcely existed.

he tried not to embarrass Khrushchev by publicly calling his bluff. Instead, he resolved to keep talking about Berlin and about a test ban. When critics demanded that he increase defense spending in order to prepare for a crisis in Berlin, he grew angry and blamed selfish interests for the hulabaloo. "I'm getting awfully sick of the lobbies by the munitions," he told Republican leaders. "You begin to see this thing isn't wholly the defense of the country, but only more money for some who are already fat cats."[48]

Eisenhower's patience paid off, at least temporarily. The deadline for a treaty between East Germany and the Soviet Union passed without treaty, incident, or American concession. Khrushchev would bring up the issue again and in 1961 would erect the Berlin Wall. But for the time being he dropped his demands. Negotiations even resulted in an agreement that Khrushchev and Eisenhower would exchange visits. Khrushchev came to the United States for a whirlwind tour in September 1959. Erratic as ever, he called for Soviet-American friendship yet boasted, "We will bury you." At the end of his visit he spent three days with Eisenhower at Camp David, the presidential retreat in Maryland. There he agreed to a summit meeting at Paris in May 1960 with Eisenhower and the leaders of France and Great Britain. The "spirit of Camp David" reminded people of the "spirit of Geneva" and promoted a good deal of journalistic talk about "peaceful coexistence."

This lasted until May 1, 1960, when a U-2 reconnaissance flight by pilot Francis Gary Powers, from Pakistan to Norway, was shot down by Soviet rocket fire near Sverdlovsk, 1,300 miles inside Russian borders. The CIA, which had charge of the flights, had equipped Powers with a needle dipped in curare, a deadly poison, so that he could kill himself before being captured. But Powers bailed out and survived. His plane was found, and he was captured and interrogated. This was sixteen days before the summit conference in Paris was scheduled to open.

The U-2 Incident, as it was called, need not have torpedoed the conference. Khrushchev, though angry and embarrassed by the U-2s, could have announced right away that Powers had been captured, at which point Eisenhower could have replied that excessive Soviet secretiveness made the flights necessary. He might then have called them off, at least for a while. Khrushchev, however, resolved to show the world (perhaps mainly the Chinese, with whom Soviet relations had become dangerous) that he was tough. He therefore announced only that an American plane

48. Ambrose, *Eisenhower*, 482.

had been shot down in Soviet territory, saying nothing about a U-2 or a pilot. He hoped that the United States would spin a tissue of lies, in which case the Soviet Union could humiliate Eisenhower and claim a big propaganda victory.

The ruse worked. American officials confirmed only that a weather reconnaissance plane was missing. Khrushchev then sprung his trap on May 7, proclaiming that Powers had been captured and had confessed and that Russian officials had the plane, complete with photographic equipment that proved Powers had been spying. Powers, Khrushchev gloated, was equipped with a pistol that was noiseless. "If that gun was meant for protection against wild animals . . . then why the silencer? To blow men's brains out! The men who supplied him with the silence gun pray in church and call us godless atheists!" Powers also had with him two gold watches and seven ladies' gold rings. "What possible use could he make of all this in the upper strata of the atmosphere? Perhaps he was to have flown still higher, to Mars, and meant to seduce the Martian ladies."[49]

Eisenhower might then have kept quiet. But he was embarrassed by rumors that the mission had taken place without his authorization, and he resolved to set those rumors straight. He announced at a press conference on May 11 that he knew everything of importance that happened in his administration. The flight had been necessary, he added, because "no one wants another Pearl Harbor." To prevent such an attack the United States had to protect itself and the "Free World" by spying. Espionage activities of that sort, he concluded, were a "distasteful but vital necessity."[50]

Eisenhower then set out for Paris with the intention of going through with the summit. Khrushchev, however, showed up in no mood to settle issues such as Berlin, probably because hard-liners at home fought the idea of compromise. On the first day of the conference he arose, red-faced and angry, to demand that Eisenhower condemn U-2 flights, renounce them in the future, and punish those responsible. He also withdrew his invitation for Eisenhower to visit the Soviet Union. Eisenhower was angry but kept his temper. The flights, he said, would not be resumed. But he refused to accede to Khrushchev's other demands. When the Soviet

49. Thomas Ross and David Wise, *The U-2 Affair* (New York, 1962), 98; *Newsweek*, May 16, 1960, p. 28; Michael Beschloss, *Mayday: Eisenhower, Khrushchev, and the U-2 Affair* (New York, 1986).
50. Oakley, *God's Country*, 388–91.

leader stalked out of the room, it was clear that the summit was over before it started.

The legacy of the U-2 Affair was negative for all involved, save perhaps Khrushchev who scored the propaganda victory that he seemed to crave. American critics wondered why Eisenhower had authorized such a flight on the eve of the summit (it was supposed to have been the last until the summit was over) and were chagrined to see him caught in a lie, not only to Khrushchev but also to the American people. The debacle destroyed whatever hopes had existed (not strong) for a test ban agreement and for some understanding concerning Berlin.[51] The U-2 Incident chased away the "spirit of Camp David" and set in motion a hardening of Soviet-American relations that intensified during the next two years.

Eisenhower's foreign policies after 1956 left especially tense and unresolved situations in two other areas of the world. One crisis, in Cuba, mounted quickly after Fidel Castro staged a successful revolution against a corrupt pro-American dictatorship and triumphantly took power in January 1959. Castro at first seemed heroic to many Americans. When he came to the United States in April, he was warmly received and spent three hours talking with Vice-President Nixon. But relations soon cooled. Castro executed opponents and confiscated foreign investments, including $1 billion held by Americans. Refugees fled to the United States and told stories of Castro's atrocities. Castro signed long-term trade arrangements with the Soviet Union, denounced "Yankee imperialists," and recognized the People's Republic of China. It seemed that he was taking Cuba into the Communist bloc of nations.[52]

The Eisenhower administration responded forcefully in 1960 by cutting American economic aid and ultimately by refusing to accept Cuban sugar, a mainstay of the island's economy. An embargo was placed on American exports to Cuba. The administration encouraged the formation of a Cuban government-in-exile and authorized the CIA to promote paramilitary training of exiles in Guatemala. They hoped to stage an invasion of the island. The Soviets responded to these events by pledging to defend Cuba against attack from the United States. In Eisenhower's last days in office the United States broke off diplomatic relations with Cuba. At that time there were some 600 men being readied to attack from Guatemala.

The Eisenhower administration could hardly have preserved warm

51. Beschloss, *Crisis Years*, 149.
52. Ambrose, *Eisenhower*, 498–500, 516–17, 538–39.

relations with Castro. The Cuban leader, like many people in Latin American nations, deeply resented the economic might of the United States, whose citizens owned some 40 percent of Cuban's sugar, 90 percent of its wealth from mines, and 80 percent of its utilities.[53] The United States also controlled a bit of Cuban territory, a naval base at Guantánamo. Castro, anxious to curb American power in Cuba, indeed to promote a social revolution, could not help but provoke animosities. Still, the far-from-secret readiness of the Eisenhower administration to consider invasion—Castro was well aware of what was going on in Guatemala— added heat to the fuel of emnity. When Eisenhower left the White House, it remained only for someone to kindle the fires of war.

The legacy of Eisenhower foreign policy in Vietnam after 1956 was also grim and lasting. Having encouraged southern leader Ngo Dinh Diem to ignore the Geneva accords' call for national elections in 1956, the Eisenhower administration proceeded to step up support for his increasingly corrupt and dictatorial regime. The aid totaled some $1 billion between 1955 and 1961, making South Vietnam the fifth largest recipient in the world of American assistance during that time. By the late 1950s the United States mission in Saigon, the capital of the South, employed 1,500 people; it was the biggest American mission in the world. The economic aid helped to control inflation and to rebuild the southern economy in places like Saigon. But it did little to help villages, where more than 90 percent of the people of South Vietnam lived. Most of the aid, indeed, was military, designed to guard against invasion from Ho Chi Minh and his Communist regime in North Vietnam.[54]

Diem, having struggled earnestly against criminal elements and against various corrupt religious sects, had enjoyed some heartening successes by 1957. But he also proved to be increasingly stubborn and narrowminded.[55] Brooking no interference with his exaggerated sense of destiny, he refused to widen his circle of supporters much beyond his own extended family. With his brother Ngo Dinh Nhu as his top adviser, he tightened his autocratic rule. Local elections, a tradition in the country, were abolished. Diem filled village and provincial offices with friends, many of whom arrested local notables on trumped-up charges and forced

53. Siegel, *Troubled Journey*, 133.
54. George Herring, *America's Longest War: The United States and Vietnam, 1950–1975* (Philadelphia, 1986), 55–63.
55. Herring, "'People Quite Apart': Americans, South Vietnamese, and the War in Vietnam," *Diplomatic History*, 14 (Winter 1990), 2.

them to pay bribes in order to get released. Diem and Nhu shut down unfriendly newspapers and incarcerated many thousands of opponents. Southern opponents, feeling harassed, responded by resuming armed resistance in 1957 and by beginning a campaign of terror against Diem's supporters in the villages in 1958.

Ho Chi Minh also ruled autocratically in the North. Estimates place the number of dissidents executed there at between 3,000 and 15,000 from 1954 to 1960. Indeed, Ho was so busy centralizing his authority that he waited until January 1959 to give formal approval to Vietminh resistance in the South. Thereafter, his aid increased rapidly. The North expanded infiltration routes through Laos and Cambodia and sent increased numbers of trained agents into the South. In 1960 it was estimated that these agents, together with southern rebels, assassinated some 2,500 Diem loyalists. In December 1960 the North took a major role in the official founding of the National Liberation Front (NLF), the political arm of the struggle in the South against Diem, and revolutionary activity further intensified. Diem, christening the NLF the Vietcong (Viet-Communist), could not stem their advances in the countryside. By 1961 full-scale military operations were in progress, and the Diem regime grew steadily more precarious.[56]

It is obvious in retrospect that the Eisenhower administration erred seriously in throwing so much support to Diem. Ho, after all, was a popular liberator who had had every reason to expect that he would win national elections in 1956 and become the legitimate ruler of a united nation. By the late 1950s some administration officials recognized Ho's growing appeal and urged Diem to reform. The dictator refused, whereupon the Americans—having no viable political alternatives in the South—relented. More aid poured in, as did American military advisers, of whom there were roughly 1,000 by January 1961.[57]

While the Eisenhower administration's record in Vietnam was wrong-headed, it can be placed in context in two ways. First, Saigon was a long way from Washington. Vietnam, American aid notwithstanding, did not seem as strategically important as Berlin, Cuba, the Middle East, or even Laos (which many State Department officials in the late 1950s considered more in danger of collapse). Warnings about trouble in Vietnam were received but, for the most part, ignored amid the noise of diplomatic

56. Herring, America's Longest War, 71–79.
57. David Anderson, Trapped by Success: The Eisenhower Administration and Vietnam, 1953–1961 (New York, 1991), 199–209.

traffic. The administration simply did not pay great attention to Southeast Asia.

Second, and more important, Ho Chi Minh was a Communist. In the polarized Cold War atmosphere of the mid- and late 1950s, this alone was sufficient to make him an enemy in American eyes. South Vietnam, administration officials thought, had to have backing to protect it against the autocratic Communist rule of Ho. If Vietnam fell, moreover, it would be the domino that threatened the collapse of other non-Communist governments in Southeast Asia. Other American leaders who bothered to notice Southeast Asia agreed with this analysis. Senator John F. Kennedy, speaking passionately to the American Friends of Vietnam in 1956, delivered characteristic anti-Communist rhetoric on the subject. "Vietnam," he said, "represents the cornerstone of the Free World in Southeast Asia, the keystone of the arch, the finger in the dike." Should the "red tide of Communism" pour into South Vietnam, much of Asia would be threatened. South Vietnam, he closed, "is our offspring, we cannot abandon it, we cannot ignore its needs."[58]

AROUND 4:00 P.M. ON THE AFTERNOON of February 1, 1960, four freshmen from all-black North Carolina A&T College in Greensboro, North Carolina, entered the local Woolworth department store to stage a protest. The store was open to all—the more business the better. But its lunch counter, like those elsewhere in the South, was open to whites only. The four young men, Ezell Blair, Jr., David Richmond, Franklin McCain, and Joseph McNeil, resolved to conduct a sit-in at the counter until management agreed to desegregate it. Their bold and courageous action sparked the direct action phase of the civil rights movement in the United States.

The protestors had not entered upon their challenge carelessly. They knew a good deal about the boycott at Montgomery and had learned about tactics of resistance from a pamphlet on the subject circulated by CORE. A local white storekeeper, an NAACP member, had urged them to struggle against racial discrimination. They were understandably frightened. When they sat down, a black dishwasher behind the counter, afraid she would lose her job, loudly berated them for being "stupid, ignorant . . . rabble-rousers, troublemakers." Behind them a white policeman, confused and nervous, paced back and forth and slapped his club in his palm. No one served them, and the store closed a half-hour early at 5:00 P.M.

58. Herring, *America's Longest War*, 43.

When they went back to campus, they discovered that they were local heroes. So they remained decades later.[59]

From those modest beginnings emerged a wave of sit-ins that transformed the civil rights movement. In Greensboro the four became twenty-four by the second day. By the fourth day white women from the local University of North Carolina Women's College joined them. By then protestors, mostly black students, were starting to sit in at lunch counters elsewhere in the state. Within a week the movement had spread across the border to Hampton, Virginia, and Rock Hill, South Carolina. A week later fifty-four sit-ins were under way in fifteen cities in nine states in the South.[60] It was obvious from the way that the spark of protest jumped from place to place that black resentments, which had somehow failed to ignite other sit-ins between 1957 and 1959, had exploded.

The sit-ins of 1960 arose, as did the civil rights movement in later years, from the collective efforts of unsung local activists: they sprang from the bottom up. Many later leaders, unknown in 1960, jumped into action. One was Cleveland Sellers, then a sixteen-year-old in Denmark, South Carolina. He started leading protests there. Another was Ruby Doris Smith, who joined demonstrations in Atlanta. A "Nashville group" included such later well known activists as John Lewis, Marion Barry, ministers James Lawson and C. T. Vivian, and Diane Nash. Schooled in Gandhian principles, they believed in developing a disciplined "beloved community." An "Atlanta group" included Julian Bond. Stokely Carmichael, a student at Howard University in Washington, headed South to become involved. In New York City, Robert Moses, a twenty-six-year-old high school teacher, looked at newspaper photographs of the Greensboro youths and was inspired by their "sullen, angry, determined" look that differed from the "defensive, cringing" expression common to blacks in the South. Moses, who was to become a legendary activist, soon joined the student movement.[61]

59. William Chafe, *Civilities and Civil Rights: Greensboro, North Carolina, and the Black Struggle for Freedom* (New York, 1980), 98–141; Aldon Morris, *The Origins of the Civil Rights Movement: Black Communities Organizing for Change* (New-York, 1984), 197–215; Howard Zinn, *SNCC: The New Abolitionists*, 2d ed. (Boston, 1965); Weisbrot, *Freedom Bound*, 1–3, 19–42; Godfrey Hodgson, *America in Our Time* (Garden City, N.Y., 1976), 184.

60. Clayborne Carson, *In Struggle: SNCC and the Black Awakening of the 1960s* (Cambridge, Mass., 1981), 10–18; Branch, *Parting the Waters*, 271–73; Gitlin, *Sixties*, 81.

61. Weisbrot, *Freedom Bound*, 19.

In mid-April Ella Baker, the fifty-five-year-old executive director of the Southern Christian Leadership Conference, thought it was time to bring the student protestors together and develop strategies that would go beyond what she considered the old-fashioned approaches of Martin Luther King. The conference that she called, at Shaw College in Raleigh, attracted numbers beyond all expectations: more than 300 students. While most came from southern black colleges, those in attendance included nineteen white college students from the North. King addressed the students, but Lawson, who had engaged in the Nashville sit-ins, sparked the most enthusiasm. When the students adjourned, they had formed a new organization, the Student Nonviolent Coordinating Committee, or SNCC. Barry became its first chairman.[62]

When Baker addressed the students at the gathering, she urged them to challenge racial injustice in all walks of life, including housing, health care, voting, and employment. "The current sit-ins and other demonstrations," she said, "are concerned with something much bigger than a hamburger or even a giant-sized Coke."[63] At that time, however, SNCC was dominated by black students at southern colleges such as Morehouse and Fisk. Bent on attacking legalized racism in the South, they did not much concern themselves in 1960 with problems in the North.[64] In focusing on Jim Crow in Dixie, SNCC did not differ greatly from King and other major leaders of the movement at the time.

The protestors quickly encountered all kinds of resistance. This was expected, for the vast majority of whites in the Deep South remained implacably opposed to desegregation. On the segregated beaches of Biloxi, Mississippi, a white man shot and wounded ten black people. Even in the Upper South whites generally refused to desegregate (Greensboro did not give in until July, by which time the Woolworth store had lost an estimated $200,000 in business, or 20 percent of anticipated sales). Normally they summoned police to arrest demonstrators on charges such as

---

62. Morris, *Origins of the Civil Rights Movement*, 215–26; Branch, *Parting the Waters*, 272–93; Carson, *In Struggle*, 19–30; Manning Marable, *Race, Reform, and Rebellion: The Second Reconstruction in Black America, 1945–1980* (Jackson, 1991), 40–61. In the 1980s and 1990s Barry became a highly controversial mayor of Washington, D.C.

63. William Chafe, *Unfinished Journey: America Since World War II* (New York, 1991), 303.

64. Thomas Sugrue, "Crabgrass-Roots Politics: Race, Rights, and the Reaction against Liberalism in the Urban North, 1940–1964," *Journal of American History*, 82 (Sept. 1995), 551–78.

trespassing or disturbing the peace. Some 3,000 protestors went to jail in 1960.[65]

The activists also encountered reluctance from within black communities, which had never been monolithic. Much of this resistance reflected generational divisions, which persisted in later years. The NAACP, with some 380,000 members at that time, remained by far the largest civil rights organization in the nation, and it was dominated by older leaders who never endorsed the sit-in strategy. Local branches of the NAACP sometimes helped the protestors by providing legal services and bail money but did so on their own. Thurgood Marshall derided the Gandhian "jail-in" tactics of demonstrators as impractical and expensive and angered Nashville activists by advising them, "If someone offers to get you out, man, get out."[66] The president of Southern University in Baton Rouge, a black institution, suspended eighteen sit-in leaders from school and forced the entire student body to resign and reapply so that he could screen their applications and weed out trouble-makers. He said, "Like Lincoln, who sought to preserve the Union, my dominant concern is to save Southern University."[67]

Even so, the sit-ins spread rapidly, not only to all southern states but also to Illinois, Ohio, and Nevada. It was estimated that 70,000 demonstrators took part in them. The protestors challenged segregation at lunch counters, parks, pools, beaches, restaurants, churches, libraries, transportation facilities, museums, and art galleries. The movement for rights, once unbottled, spilled out to challenge discrimination in voting and in employment. When the campuses reopened in September, a new surge of activism indicated that pressure for change, far from abating, was unstoppable.

POLITICS AS USUAL, however, tended to persist amid this turmoil. The sit-ins still failed to arouse the conscience of the vast majority of white Americans at the time. Most whites in the North imagined that racism was a southern problem; most in the South refused to relent. Even during the fall of 1960, with sit-ins proliferating, the majority of candidates for national and state offices managed their campaigns as they always had—by saying little about race relations. This was equally true of the major presidential candidates, Vice-President Richard Nixon and Senator John F. Kennedy. Both were party regulars and centrists who

65. Chafe, *Civilities*, 136; Hodgson, *America in Our Time*, 184.
66. Weisbrot, *Freedom Bound*, 38.
67. Ibid., 31–33.

devoted relatively little attention to civil rights in 1960. The main issues at that time, as in most elections since 1945, remained the economy and the Cold War.

This focus was to be expected of Nixon, whose major crusade in life before 1960 had contested Communism. Many Democrats, indeed, had loathed him for his excesses in this effort since the 1940s. Nixon, they reiterated, had won a congressional seat in 1946 by falsely calling his opponent, the incumbent Jerry Voorhis, a Communist. Running for the Senate in 1950, he had Red-baited his liberal opponent, Representative Helen Gahagan Douglas. As Vice-President he proved to be an energetic and unfailingly partisan Cold Warrior. When he ran in 1960, he was still only forty-seven years old.[68]

Fellow Republicans admired Nixon's dogged tenacity and partisan spirit. But few of them managed to get close to one of America's most private politicians. Contemporaries remarked on Nixon's awkwardness in public and on his habit of constantly testing himself, as if in duels to the finish, against people he imagined were mortal enemies. Many of these enemies were well-connected figures: Alger Hiss, Acheson, Stevenson. Having grown up in a modest, hard-working California family, Nixon never felt secure in eastern Establishment circles, and he could scarcely conceal his resentments. Even partisan allies sometimes found him cold and excessively ambitious. Others deeply distrusted him. "Would you buy a used car from this man?" opposition posters of him proclaimed. Eisenhower noted unhappily that Nixon seemed to have no friends. Ike's secretary, Ann Whitman, agreed. "Everybody," she confided to her diary, "trusts and loves" Ike. "But the Vice President sometimes seems like a man who is acting like a nice man rather than being one."[69]

Those who opposed Nixon especially detested his style. The tasteless appeal of his Checkers speech in 1952 had appalled them. "No class," Kennedy had said. "No style, no style at all," the Washington Post's editor, Ben Bradlee (a Kennedy friend and supporter), had added.[70] Nixon's labored attempts in 1960 to project a more genial persona especially irritated detractors, who thought that he remained a phony. Stevenson sniped that Nixon had "put away his switchblade and now assumes

68. Major sources on Nixon in these years are Stephen Ambrose, *Nixon: The Education of a Politician, 1913–1962* (New York, 1987); and Roger Morris, *Richard Milhous Nixon: The Rise of an American Politician* (New York, 1990). A more favorable assessment of Nixon is Jonathan Aitken, *Nixon: A Life* (Washington, 1994).

69. Blum, *Years of Discord*, 15.

70. Morris, *RMN*, 129.

the aspect of an Eagle Scout." An editor added that the question wasn't whether there was a "new" or an "old" Nixon but "whether there is anything that might be called the 'real' Nixon, new or old."[71]

Kennedy, by contrast, seemed to promise some change. His supporters hailed him—often with hyperbole—as an accomplished writer, a hero in World War II, and a torch-bearer for a new generation.[72] Only forty-three years old in 1960, "Jack" was the youngest presidential candidate of a major party in United States history.[73] He was considered to be handsome, charming, and, with his stylish young wife, Jacqueline ("Jackie"), the epitome of cultured cosmopolitanism. A Harvard graduate, he surrounded himself with intellectuals who seemed to burst with ideas for the future, as well as with political professionals who worked effectively in the precincts. Kennedy beat Hubert Humphrey and others in a series of primaries, took the Democratic nomination on the first ballot, named Lyndon Johnson as his running mate, and promised a New Frontier for the United States.[74]

During the campaign, it became obvious that Kennedy had strong political assets, among them access to unlimited money from his wealthy family. When Humphrey lost the West Virginia primary, a key contest, he pulled out, complaining, "You can't beat a million dollars. The way Jack Kennedy and his old man threw the money around, the people of West Virginia won't need any public relief for the next fifteen years."[75] Then and later foes of Jack especially distrusted Joseph Kennedy, the cold and ruthless patriarch of the family who stopped at almost nothing where the political advancement of his sons was concerned.[76]

Many liberals, too, found it hard to summon enthusiasm for Kennedy.

71. Leuchtenburg, *Troubled Feast*, 115.
72. See Herbert Parmet, *Jack: The Struggles of John F. Kennedy* (New York, 1980), 72–78, 88–94, 100–103, 190–92, 307–10, 330–33, for solid accounts of controversial aspects of JFK's pre-presidential career, including his health and personal life.
73. Theodore Roosevelt was the youngest President, but he was older than Kennedy when he first became a presidential candidate in 1904.
74. Herbert Parmet, *JFK: The Presidency of John F. Kennedy* (New York, 1983), 3–60; David Burner, *John F. Kennedy and a New Generation* (Boston, 1988), 38–56; James Giglio, *The Presidency of John F. Kennedy* (Lawrence, 1991), 1–22; Theodore White, *The Making of the President, 1960* (New York, 1961), 150–79.
75. Thomas Reeves, *A Question of Character: A Life of John F. Kennedy* (New York, 1991), 153–67; Burner, *JFK*, 48; White, *Making . . . 1960*, 78–114.
76. See Doris Kearns Goodwin, *The Fitzgeralds and the Kennedys* (New York, 1987). Highly critical accounts include Nigel Hamilton, *J.F.K.: Reckless Youth* (New York, 1992); and Richard Whalen, *The Founding Father: The Story of Joseph P. Kennedy* (New York, 1964).

Eleanor Roosevelt and others remembered bitterly that JFK had refused to speak out against McCarthy and that he had offered only perfunctory support for progressive social causes, including civil rights. (Kennedy in fact had not been a hard-working or especially distinguished senator.)[77] The ADA had to overcome strong opposition within its ranks before agreeing to endorse him, and it never supported Johnson. Many liberals had unkind words for both nominees. Eric Sevareid of CBS complained that they were "tidy, buttoned-down men . . . completely packaged products. The Processed Politician has finally arrived." Richard Rovere wrote in *Harper's* that Kennedy and Nixon "tend more and more to borrow from one another's platforms and to assume one another's commitments."[78]

Rovere and others were correct that the candidates were packaged— each spent record sums on TV coverage and ads—and that they differed only marginally on the issues. Kennedy made much of the slow growth of the economy since the 1958 recession and of the oft-alleged lack of "national purpose" that commentators had claimed to discover in recent years.[79] He also endorsed party planks favoring federal aid to education, a higher minimum wage, and governmental medical insurance for the elderly. Support of such policies helped him attract key interest groups, notably teachers and labor unions. But Nixon, too, promised to take positive action to improve economic growth; he seemed well to the left of Eisenhower on most domestic issues. Nixon also insisted that he (unlike Ike) would be a strong and activist President. Kennedy harped on the "missile gap," even though he was frequently told by high-ranking defense officials that no such gap existed. Kennedy also seemed to support an invasion of Cuba by anti-Castro exiles. This popular line privately enraged Nixon, who dared not reveal secrets about the CIA's training of exiles in Guatemala. If anything, Nixon sounded less hawkish on the issue than did Kennedy. Overall, however, both men were solid Cold Warriors.[80]

Kennedy, of course, won the election, which attracted a record-high

77. Parmet, *Jack*, 242–87; James MacGregor Burns, *John F. Kennedy: A Political Profile* (New York, 1960), 132–52, 183–98.
78. Arthur Schlesinger, Jr., *A Thousand Days: John F. Kennedy in the White House* (New York, 1965), 65. Also Reeves, *Question of Character*, 195; Oakley, *God's Country*, 416.
79. Kathleen Hall Jamieson, *Packaging the Presidency: A History of Presidential Campaign Advertising* (New York, 1992), 122–68.
80. Beschloss, *Crisis Years*, 26–27.

turnout for a postwar presidential race (around 64 percent of the eligible electorate) and turned out to be the closest of the century. He received 49.7 percent of the vote to Nixon's 49.6 percent. His strength in populous northern states gave him a margin in the electoral college, 303 to 219.[81] But the swing of a few votes in a few big states, notably in Illinois (where fraud by Democratic retainers in the Chicago area helped him) could have turned the election the other way. The outcome was so tight that some of Nixon's advisers urged him to contest it. The Vice-President, though bitter about losing, chose not to take that divisive course (Republicans were also suspected of vote fraud) and conceded defeat. It remained for the experts to render the postmortems that would explain why he had lost.

There were plenty of diagnoses, no single one of which could explain all. Some Nixon loyalists blamed Eisenhower, who made no effort to hide his well-known ambivalence about Nixon. At a news conference in August he was asked for an example of a major idea of Nixon's that had been adopted by the administration. "If you give me a week, I might think of one," the President replied. "I don't remember."[82] The President recognized immediately how bad that sounded and apologized to his Vice-President. But damage had been done. Moreover, Eisenhower did not campaign much until the last week, at which point he spent more time praising his own accomplishments than extolling Nixon.

Other analysts tended to blame Nixon himself. Try as he might, Nixon seemed wooden in public appearances, especially in contrast to the apparently more relaxed and much more handsome Kennedy. Appearances mattered. This was especially clear in the first of a series of televised debates, a new event in American political history. Americans who listened on the radio thought that Nixon, a trained debater since high school, had "won." The majority of the millions who watched on television, however, seemed attracted to Kennedy. Nixon had hurt his knee campaigning and had a chest cold. Refusing to apply heavy makeup, he looked haggard and unshaven, and he sounded hoarse. Kennedy, by contrast, arrived tanned from California and seemed cool and controlled. While the debate may not have changed many minds—Nixon was already losing ground in the polls and may have been behind—it lifted Kennedy from his lesser-known status as challenger. It brought dismay to

81. Senator Harry Byrd of Virginia received fifteen electoral votes, including all eight from Mississippi, six from Alabama, and one from Oklahoma.
82. Ambrose, *Eisenhower*, 524; Robert Ferrell, ed., *The Eisenhower Diaries* (New York, 1981), 242.

many in the Nixon camp and great enthusiasm to Kennedy supporters, who fought thereafter with zest and high hopes.[83]

Kennedy indeed inspired fervor as the campaign neared its conclusion. At some of his rallies, groups of people—especially women—broke through barriers and ran after his car. These, *Newsweek* reported, were the "runners." Other Kennedy fans included "jumpers" who leapt up and down as his motorcade came by, "double-leapers," women who jumped together while holding hands, and "clutchers," women who crossed their arms and hugged themselves and screamed, "He looked at me! He looked at me!" Nixon usually evoked applause, but nothing like runners, jumpers, or clutchers.[84]

Many people who actually met Kennedy were almost as dazzled. They found him personally charming, magnetic, even incandescent. He was poised, handsome, youthfully energetic, and he could literally light up a room. Intellectuals and journalists responded favorably to his dry wit, irreverence, and cool, detached intelligence. Haynes Johnson, a perceptive journalist, recalled that Kennedy was "the most seductive person I've ever met. He exuded a sense of vibrant life and humor that seemed naturally to bubble up out of him."[85] Kennedy was uncomfortable indulging for long in flights of high-sounding oratory. Yet in calling for change he appealed to the idealistic hopes of people. By any standard he was the most charismatic American politician of the postwar era, especially to women and younger people.[86]

Charisma, though important, was but part of the reason for Kennedy's narrow victory. He also made tactical decisions that at least in retrospect seem politically astute, in that they strengthened the urban-ethnic-black-southern coalition of voters that Roosevelt had fashioned in the 1930s and that had held together for Truman in 1948. One such decision was his first as the nominee: naming Johnson as his running mate. This decision amazed delegates and reporters. Johnson, a presidential candidate himself, had attacked Kennedy both before and during the convention. As

83. *Newsweek*, Oct. 10, 1960, p. 23; White, *Making . . . 1960*, 279–95; Edwin Diamond and Stephen Bates, *The Spot: The Rise of Political Advertising on Television* (Cambridge, Mass., 1992), 90–95.

84. *Newsweek*, Oct. 10, 1960, p. 26; White, *Making . . . 1960*, 331.

85. Haynes Johnson, "Why Camelot Lives: JFK's Image and the Kennedys' Troubles," *Washington Post*, Aug. 18, 1991.

86. Carl Brauer, "John F. Kennedy: The Endurance of Inspirational Leadership," in Fred Greenstein, ed., *Leadership in the Modern Presidency* (Cambridge, Mass., 1988), 108–33.

Senate majority leader he had much more power than he ever would have as Vice-President. Most of Kennedy's top advisers, especially his brother Bobby, who was his campaign manager, despised Johnson, whom they considered to be a wheeler-dealer as well as a conservative. But Kennedy (and his father) decided that they wanted Johnson on the ticket in order to bolster Democratic chances in Texas, a key state, and elsewhere in the South and West. And so the offer was made.

When Johnson's friends heard of the offer, they strongly opposed the idea. But Johnson was attracted, both because he was seeking new challenges and because he was honored to be asked. Privately he accepted, and word seeped out concerning the deal. At this point liberals protested, and Bobby, thinking wrongly that Jack was reconsidering his offer, went to Johnson's suite. There he dropped hints that Johnson should pull out of the vice-presidential picture. Johnson picked up the phone to find out what Jack really wanted. Kennedy assured him that he remained the choice, and the deal later sailed through the convention. Johnson, an extraordinarily sensitive man, never forgave Bobby or forgot the insult. In the open, however, he feigned his enthusiasm for Kennedy. His presence may have protected Kennedy's margins in the South.[87]

A second key decision by Kennedy concerned his Catholicism, which led many Protestants, including Norman Vincent Peale, to question whether he ought to be President. (Martin Luther King, Sr., was another doubter.) Kennedy met the issue head-on by addressing Protestant clergymen in Houston, a center of Protestant strength. America, he said, is a nation where the "separation of church and state is absolute—where no Catholic prelate would tell the President how to act, and no Protestant minister would tell his parishioners for whom to vote." He added, "I am not the Catholic candidate for President. . . . I do not speak for my church on public matters—and the church does not speak for me."[88] The election results later suggested that voters divided sharply along religious lines and that Kennedy's Catholicism had hurt him more than it helped.[89] Still, most contemporary observers agreed that Kennedy's speech at Houston managed to soften religious rhetoric for most of the campaign. Had he evaded the issue, Protestant voters might have ensured his defeat.

87. Doris Kearns, *Lyndon Johnson and the American Dream* (New York, 1976), 290; Conkin, *Big Daddy*, 151–56.
88. Branch, *Parting the Waters*, 340; Oakley, *God's Country*, 417.
89. Richard Polenberg, *One Nation Divisible: Class, Race, and Ethnicity in the United States Since 1938* (New York, 1980), 167–68, concludes that JFK won 80

Another move that seems to have helped him politically were gestures that he and Bobby made in late October toward embattled blacks. When King was arrested in Georgia and sentenced to four months in jail on a minor traffic-related violation, Nixon tried quietly to intervene but said nothing in public. Jack, however, was persuaded by an aide to telephone King's wife, Coretta, to express his sympathy. At the same time, Bobby (unbeknownst to Jack) telegraphed the judge and requested King's release. The judge relented, and King got out of prison on bail. King then gave Jack full credit for what had happened. King Sr. came around, announcing, "I've got a suitcase full of votes, and I'm going to take them to Mr. Kennedy and dump them in his lap."[90] These efforts by Bobby and Jack may have persuaded some wayward blacks who had voted Republican in 1956 to return to the Democratic camp.[91] Whatever the cause, some 70 percent of black votes went Democratic in 1960 (as opposed to perhaps 63 percent in 1956), an increase that may have turned the tide for Kennedy in several closely contested northern states, such as New Jersey, Michigan, and Illinois, as well as in Texas (all of which Eisenhower had won in 1956).[92] Eisenhower later attributed the Republican defeat to "a couple of phone calls" by John and Robert Kennedy.[93]

So THE DEMOCRATIC COALITION survived. This center, among others, seemed to hold in 1960. Did did it make a difference?

Many Americans later imagined that it did. Kennedy, they believed, proceeded to outline a range of new frontiers and to abandon the stodgy old politics of the 1950s. It remains difficult, however, to identify very many Americans in the immediate aftermath of the election who expected that anything very dramatic was about to happen. Conservatives insisted correctly that Kennedy had no popular mandate. Eisenhower, who deeply disliked the Kennedys, also doubted that the new administration would be able to change much. In his farewell address of January 17, 1961, he predicted instead that "an immense military establishment and a

---

percent of the Catholic vote and only 38 percent of the considerably larger Protestant vote, but that Catholic votes in major northern states greatly helped him.

90. Lawson, *Black Ballots*, 256.

91. This is speculation. Blacks had voted heavily for Democratic presidential candidates since 1936 and probably returned in 1960 once Eisenhower, very popular personally, no longer headed the GOP ticket. Blacks also tended to favor liberal social policies associated with the Democratic party.

92. Lawson, *Black Ballots*, 255–58.

93. Reeves, *Question of Character*, 215.

large arms industry"—a military-industrial complex—might continue to poison the wells of international relations and to dominate domestic policy. Urgently he warned the nation to be on guard.

Liberals, of course, were pleased that Nixon had lost and that the voters had returned Democratic majorities to Capitol Hill: 65 to 35 in the Senate and 263 to 174 in the House. But they, too, had difficulty interpreting the election as a triumph for significant change. Democrats in fact lost twenty seats in the House. The fact was that Kennedy had run a pragmatic, centrist campaign in which he promised to wage the Cold War more vigorously than ever. The sit-ins notwithstanding, moral issues such as civil rights seemed no more pressing to Kennedy and his advisers at the end of 1960 than they had to the majority of white Americans throughout Eisenhower's second term. There seemed little reason to anticipate that the political center would shift very much in the days to come.

15

The Polarized Sixties:
An Overview

The year 1960, one historian writes, marked "the definitive end of the Dark Ages, and the beginning of a more hopeful and democratic period" that lasted until the early 1970s. Another historian calls the 1960s a modern Great Awakening which ignited a "Burned-Over Decade" of cultural change akin to the turbulent 1840s. William Braden, a contemporary observer, labeled the era an Age of Aquarius that heralded "a new American identity—a collective identity that will be blacker, more feminine, more oriental, more emotional, more intuitive, more exuberant— and, just possibly, better than the old one."[1]

Cultural conservatives witnessed these changes with disgust. The sociologist Daniel Bell was appalled by young people who were trying to "transfer a liberal life-style into a world of immediate gratification and exhibitionist display." The "counterculture," as it was called, "produced little culture, and it countered nothing." The columnist George Will later dismissed the decade as an age of "intellectual rubbish," "sandbox radicalism," and "almost unrelieved excess." Braden worried that Americans who were forging the "new identity" might be mistaking "vividness, inten-

1. Marty Jezer, *The Dark Ages: Life in the United States, 1945–1960* (Boston, 1982), 3; Arlene Skolnick, *Embattled Paradise: The American Family in an Age of Uncertainty* (New York, 1991), 89–99; William Braden, *Age of Aquarius: Technology and the Cultural Revolution* (Chicago, 1970), 6.

sity, and urgency for cultural sensitivity and responsible morality. They don't know what they like, but whatever they or their emotions like must be art—or must be right, and certainly righteous."[2]

Both sides of this still acrimonious debate were correct in recognizing that unusually tumultuous events shook American life in the 1960s. Cultural and social changes seemed to accelerate rapidly in the early 1960s, to reshape public policies in the mid-1960s, and to polarize the nation in the last few years of the decade. By then the thrust of activism was dramatically shifting direction: backlash mounted rapidly against the public programs and ushered in a durable age of political conservatism in America. But the tumult of the decade nonetheless had unsettled much that Americans had taken for granted before then, including vestiges of what for lack of a better word can be called "Victorian." Thereafter, people seemed much readier to challenge authority. As Morris Dickstein, a perceptive scholar put it, "The sixties are likely to remain a permanent point of reference for the way we think and behave, just as the thirties were."[3]

Dickstein's view of the sixties has much to be said for it. Signs of dramatic change were gathering force even in 1960, when the sit-ins broke out in February. SNCC came to life in April. Enovid, the birth control pill, was approved by the government in May. The Students for a Democratic Society (SDS), later to be the most prominent of many "New Left" protest groups, was born in June. In 1961 social change gathered new momentum. The civil rights movement entered a bloodier stage, with racists attacking "freedom riders" who sought to integrate interstate travel: between 1961 and 1965, twenty-six civil rights workers lost their lives in the South. More than any other development of the early 1960s, the civil rights revolution spurred the idealism, egalitarianism, and rights-consciousness that galvanized many other groups and challenged social relations in the United States.

The early 1960s witnessed publication of extraordinarily provocative and influential books that questioned conventional notions about American society and culture. In 1961 Jane Jacobs brought out *Death and Life of Great American Cities*, which skewered the grandiose pretensions of urban planners, and Joseph Heller published *Catch-22*, an unsubtle but hilarious and disturbing novel about the inanities of the military in World

2. Daniel Bell, *The Cultural Contradictions of Capitalism* (New York, 1976), 81; Wills, cited in Skolnick, *Embattled Paradise*, 78; Braden, *Age of Aquarius*, 6.
3. Morris Dickstein, *Gates of Eden: American Culture in the Sixties* (New York, 1977), 250.

War II. It sold some 10 million copies over the next thirty years, appealing especially to opponents of the Vietnam War. Two seminal books appeared in 1962. Rachel Carson's *Silent Spring* sounded an eloquent warning against pesticides and environmental pollution. Widely acclaimed, it spurred an ecological movement that gathered considerable force by the late 1960s. Michael Harrington's *The Other America* greatly dramatized the problem of poverty in the United States, adding to pressures for governmental action.[4] In 1963 James Baldwin's prophetic *The Fire Next Time* alerted Americans to the likelihood of violent racial confrontation. Betty Friedan's *The Feminine Mystique* also appeared in 1963. A huge seller, it helped to launch a renaissance of feminism.

Reflecting the anti-Establishment spirit of these books, groups of protestors began to capture public attention in the early 1960s. In Michigan Tom Hayden and other young SDS radicals crafted the Port Huron Statement in 1962, a long, sometimes contradictory, but much-cited manifesto of New Left activism.[5] In Mississippi that fall, James Meredith, an air force veteran, sought to become the first black person to attend the University of Mississippi. When segregationists retaliated with violence, President Kennedy had to send in the army. Also in 1962, César Chávez and fellow migrant workers organized the National Farm Workers Association, thereby inspiring efforts that led to highly publicized strikes and boycotts later in the decade.[6] Early in 1963 Martin Luther King staged a dramatic protest against racial discrimination in Birmingham. It provoked white violence, worldwide television coverage, and rising outrage against racism in the United States. That August King and others took part in a March on Washington that attracted some 250,000 protestors.

Other, unrelated events added to a public perception—this was important—that the times were changing with especially accelerating speed in these years. The Supreme Court shocked conservatives—and others—in 1962 by ruling that public schools in New York could not re-

4. The year 1962 also witnessed publication of *Fail-Safe* by Eugene Burdick and Harvey Wheeler. A popular novel, it centered on a nuclear disaster caused by mechanical failure. It was produced as a movie in 1964.
5. Kirkpatrick Sale, *SDS* (New York, 1973); James Miller, *"Democracy Is in the Streets"; From Port Huron to the Siege of Chicago* (New York, 1987).
6. Mark Day, *César Chávez and the Farm Workers* (New York, 1971); Juan Gonzales, *Mexican and Mexican-American Farm Workers: The California Agricultural Industry* (New York, 1985).

quire students to recite a State Board of Regents prayer in the class-room.[7] The Vatican Ecumenical Council, under the reformist leadership of Pope John XXIII, agreed to authorize use of the vernacular in parts of the Catholic mass. Traditionalists were amazed and appalled.[8] The folk-singer Bob Dylan, who had prophetically written "The Times They Are a-Changin'" in 1962, brought out "Blowin' in the Wind" in the spring of 1963. The version by Peter, Paul, and Mary, marketed in August 1963, sold 300,000 copies in two weeks and became the first protest song ever to make the Hit Parade.[9] Timothy Leary and Richard Alpert, having helped to celebrate the virtues of drugs such as LSD, were fired from their posts at Harvard University that spring but continued to beguile acolytes, espe-cially among the young.

Other blows to the familiar followed in early 1964. In January the Surgeon General of the United States issued a report by eminent scientists warning of the mortal dangers of tobacco.[10] It temporarily shook some of the millions of Americans (more than one-half of adult men, more than one-third of women) who smoked.[11] In the same month movie-goers began flocking to see Stanley Kubrick's film *Dr. Strangelove*. It featured a crazed militarist, Jack D. Ripper (Sterling Hayden), who refused to re-scind an insane attack order because he was convinced that the "Interna-tional Communist Conspiracy" was trying to "sap and impurify all of our precious bodily fluids." More effectively than any other movie of the era,

7. *Engel v. Vitale*, 370 U.S. 421 (1962). The non-denominational prayer read, "Almighty God, we acknowledge our dependence upon Thee, and we beg Thy blessing upon us, our parents, our teachers and our country." Prior to the decision, schools had had the option of using the prayer or not. Students who did not wish to recite it could leave the room. See *Newsweek*, July 9, 1962, pp. 21–22, 43–45, for vocal public reaction.

8. Jonathan Rieder, *Canarsie: The Jews and Italians of Brooklyn Against Liberalism* (Cambridge, Mass., 1985), 134–36; James David Hunter, *Culture Wars: The Struggle to Define America* (New York, 1991), 67–106.

9. Maurice Immerman and Michael Kazin, "The Failure and Success of the New Radicalism," in Steve Fraser and Gary Gerstle, eds., *The Rise and Fall of the New Deal Order, 1930–1980* (Princeton, 1989), 212–42; George Lipsitz, "Who'll Stop the Rain? Youth Culture, Rock 'n' Roll, and Social Crises," in David Farber, ed., *The Sixties: From Memory to History* (Chapel Hill, 1994), 206–34.

10. *Smoking and Health: Report of the Advisory Committee to the Surgeon General of the Public Health Service* (Washington, 1964).

11. As noted earlier, Americans cut back smoking only temporarily; by 1966 per adult consumption of cigarettes had risen to pre-report levels. See James Patterson, *The Dread Disease: Cancer and Modern American Culture* (Cambridge, Mass., 1987), 201–30.

the movie ridiculed the excesses of the Cold War. A month later the Beatles arrived in the United States from England and became an immediate sensation; a record 67 million people watched them perform on "The Ed Sullivan Show." In March Malcolm X, a charismatic black nationalist, broke with the Nation of Islam, formed the Organization of Afro-American Unity, and set about enlisting African-Americans in northern cities.

Young people seemed especially restless at the time. Gaining confidence in 1964, when a protracted "free speech" movement at the University of California, Berkeley, aroused nationwide notice, activist students— many of them veterans of civil rights protests in the South—began to demonstrate for a range of causes. [12] Some raged against poverty and racial discrimination, others (especially after escalation of American involvement in Vietnam) against American foreign policies, others against the flaws of universities themselves. By no means all campuses experienced significant unrest in the 1960s. But most of the elite colleges and universities did. These attracted many of the brightest and most privileged young people among the huge and expectant baby boom cohorts that were then swarming to the campuses. "If you are not part of the solution," the idealists believed, "you are part of the problem." Their engagement reinvigorated the political and cultural Left in the United States. [13]

These and other developments hardly added up to a coherent movement, or even a clearly visible pattern. But they came hard and fast on one another, and they received great coverage from television, which by then reached virtually all Americans, and from other sources of news that reached a more highly educated population. [14] It was in the 1960s that TV came into its own as a major force in American life, promoting a more national culture while at the same time casting its eye on profound inter-

12. W. J. Rorabaugh, *Berkeley at War: The 1960s* (New York, 1989); Seymour Martin Lipset, *Rebellion in the University* (Chicago, 1971); Kenneth Keniston, *Youth and Dissent: The Rise of a New Opposition* (New York, 1971).

13. Kenneth Cmiel, "The Politics of Civility," in Farber, ed., *Sixties*, 263–90.

14. James Baughman, *The Republic of Mass Culture: Journalism, Filmmaking, and Broadcasting in America Since 1941* (Baltimore, 1992), 91, notes that 92.6 percent of American households in 1961 had one or more TV sets and that these were on for an average of almost six hours per day in 1963. Other useful sources include Todd Gitlin, *The Whole World Is Watching: Mass Media in the Making and Unmaking of the New Left* (Berkeley, 1980), 296; Michael Schudson, "National News Culture and the Rise of the Informational Citizen," in Alan Wolfe, ed., *America at Century's End* (Berkeley, 1991), 263–82; and David Farber, *The Age of Great Dreams: America in the 1960s* (New York, 1994), 49–66.

nal divisions. Many Americans at the time indeed sensed that the times were changing, that a new if undefined *Zeitgeist*, or spirit of the times, was in the process of remaking society and culture. The restless spirit pushed with special insistence against the political center. In 1963 activists demanding racial justice forced President Kennedy to come out for civil rights legislation. Within a year and a half of Kennedy's assassination in November 1963—a shocking act that intensified the pressures for change—reformers in Congress managed to enact a spate of liberal legislation, including a "war on poverty," federal aid to education, Medicare for the aged, Medicaid for the poor, reform of immigration law, creation of the National Endowments for the Arts and the Humanities, and two historic civil rights laws that would have seemed almost unimaginable a few years earlier.

Culturally, too, the center seemed in some disarray, especially after mid-decade. Large numbers of people, most of them young, began to find common cause in seeking relief from what they considered to be the vulgarity, impersonality, and overall dullness of middle-class culture. Some of these rebels adopted New Left political opinions, but many others resisted mainstream culture, not public policies. Millions found inspiration from rock musicians, especially (it seemed) from those who were loudly and angrily anti-authoritarian. A rock concert at a farm in Bethel, New York, in 1969 attracted some 400,000 people who wallowed happily about in the rain, some in various stages of undress and drug-induced haze, for three days. Traffic jams and police barricades prevented many thousands more from attending. "Woodstock" was the culminating event of "countercultural" celebration in the 1960s.[15]

Smaller numbers of young people "dropped out" of mainstream American life to join countercultural communes. They were a tiny minority of the overall population (which rose, a little more slowly than in the 1950s, from 180.7 million to 204.9 million during the decade), but they took pride in defying conventional mores. Many openly smoked marijuana; a few experimented with harder drugs and engaged in various versions of free love. Between 1965 and 1975, when the communal movement lost momentum, some 10,000 such experiments blossomed in the country. They received lingering if sometimes snide attention from the often voyeuristic mass media.[16]

15. Named after the nearby town of Woodstock.
16. Skolnick, *Embattled Paradise*, 92–93. See Charles Reich, *The Greening of America* (New York, 1970); and Tom Wolfe, *The Electric Kool-Aid Acid Test* (New York, 1968), for contemporary accounts of countercultural activities.

Nowhere was cultural change more clear than in the realm of sexuality among young people.[17] The Pill assisted the spread of the already ascendant sexual revolution, but larger notions of personal rights and "liberation" contributed still more. So, as earlier, did agents of the consumer culture. In 1960 *Playboy* introduced its "Playboy Adviser" column, which offered explicit guidance to readers seeking new and more imaginative ways of practicing sex. (By the early 1970s the magazine was regularly reaching an estimated 20 percent of adult American men.)[18] In 1962 Helen Gurley Brown wrote *Sex and the Single Girl*, a message of female sexual liberation that she later introduced to *Cosmopolitan* magazine. In 1968 Broadway staged *Hair*, a rock musical that featured frontal nudity. Actors were paid extra for disrobing. The play became a hit in New York and in many shows on the road.

By then many university parietal rules were crumbling, often without a fight from authorities. Indeed, the sexual revolution assumed an unprecedentedly open and defiant tone, especially among women, increasing numbers of whom rebelled against the "feminine mystique" of deference and domesticity.[19] Some flaunted mini-skirts, a new style that entered the United States from France in 1965, and challenged their elders by living openly in an unmarried state with men. The mid-1960s, one survey of sexual behavior concludes, represented "perhaps the greatest transformation in sexuality [the United States] had ever witnessed."[20]

The mostly optimistic and reformist *Zeitgeist* that characterized the early 1960s weakened rapidly after mid-decade.[21] Only five days after signing of the 1965 Voting Rights Act in August, blacks began rioting in the Watts section of Los Angeles. By 1966 the interracial civil rights movement had split badly along racial lines, and advocates of "black power," among others, were renouncing non-violence. Waves of riots

17. Beth Bailey, "Sexual Revolution(s)," in Farber, ed., *Sixties*, 235–62; Edward Lauman et al., *The Social Organization of Sexuality: Sexual Practices in the United States* (Chicago, 1994).
18. John Burnham, *Bad Habits: Drinking, Smoking, Taking Drugs, Gambling, Sexual Misbehavior, and Swearing in American History* (New York, 1993).
19. Alice Echols, "Women's Liberation and Sixties Radicalism," in Farber, ed., *Sixties*, 149–74; Skolnick, *Embattled Paradise*, 85–87, 128.
20. John D'Emilio and Estelle Freedman, *Intimate Matters: A History of Sexuality in America* (New York, 1988), 302–53. Quote on 353.
21. The titles of two important histories of the 1960s stress the decentering of America after 1965. See William O'Neill, *Coming Apart: An Informal History of the 1960s* (Chicago, 1971); and Allen Matusow, *The Unraveling of America: A History of Liberalism in the 1960s* (New York, 1984). Also Hunter, *Culture Wars*.

THE POLARIZED SIXTIES 449

engulfed central cities between 1966 and 1968. American involvement in the Vietnam War, which escalated greatly between 1965 and 1968, provoked angry confrontations and demonstrations on college campuses, at draft boards, and at massive rallies in Washington and elsewhere. A number of college campuses were racked by protest and closed down at various times between 1967 and 1970. Hispanics, Native Americans, and feminists added to the air of tumult by demonstrating to promote their goals. In June 1969 homosexuals at the Stonewall Inn in Greenwich Village fought back against police harassment, igniting five days of rioting by hundreds of people and arousing greater group consciousness among the gay population.[22]

Confrontation, violence, and social disorder indeed seemed almost ubiquitous in America during the mid- and late 1960s. In 1965 protestors at Berkeley proclaimed a "filthy speech movement," a degenerate form of the free speech demonstrations a year earlier, thereby hastening a trend toward open expression of profanity in American life. Hollywood brought out *Bonnie and Clyde* in 1967 and *The Wild Bunch* in 1969, films that reveled in the choreography of killing. Rock musicians jettisoned the lyrics of Dylan and Joan Baez for "acid rock." Television shows featured more and more graphic violence. The SDS broke apart, with a few of its splinter groups practicing violent revolution. More alarming than these scattered phenomena were broader and apparently related social indicators: rates of violent crime, drug abuse, and alcohol consumption, especially among young people, rose sharply after 1963.[23] So did divorce and illegitimacy rates, which had been stable since the late 1940s.[24] Scores on Scholastic Aptitude Tests began to fall after 1964. Most shocking of all, both Martin Luther King and Robert Kennedy were assassinated in 1968. In December 1969, as if to close the decade on a specially uncivilized note, a group of Hell's Angels, acting as security at a Rolling Stones

22. O'Neill, *Coming Apart*, 269; Bruce Bawer, "Notes on Stonewall," *New Republic*, June 13, 1994, pp. 24–30; D'Emilio and Freedman, *Intimate Matters*, 318–19.
23. Stephen Ruggles, "The Transformation of American Family Structure," *American Historical Review*, 99 (Feb. 1964), 103–28; Jack Katz, "Criminal Passions and the Progressive Dilemma," in Wolfe, ed., *America at Century's End*, 390–420; James Q. Wilson, *Thinking About Crime* (New York, 1983), 23–44, 224–27, 238–40, 253–58; Charles Silberman, *Criminal Violence, Criminal Justice* (New York, 1978), 3–6, 31–33, 424–55; and Charles Easterlin, *Birth and Fortune: The Impact of Numbers on Personal Welfare* (New York, 1980), 106.
24. Diane Ravitch, *The Troubled Crusade: American Education, 1945–1980* (New York, 1983), 321–30; and Landon Jones, *Great Expectations: America and the Baby Boom Generation* (New York, 1980), 304–10.

concert in Altamont, California, beat a few concert-goers savagely with pool cues, stomped a stoned and naked young woman who tried to climb on stage, and stabbed to death a nineteen-year-old black man. The featured performers looked on uneasily but kept playing, and the cameras—making a commercial film about the Stones—kept rolling. Most rock fans in the huge audience of 500,000 seemed unaware of what had happened.[25]

DRAMATIC THOUGH THESE CHANGES WERE in the 1960s, they represented only the most widely noted aspects of an increasingly polarized era. The vast majority of Americans had little if anything to do with campus rebels, counterculturalists, or anti-war protesters. They were very much aware of the tumult—television lavished attention on it—but they went about their daily lives in familiar ways.[26] As in the 1940s and 1950s, they celebrated traditional values and institutions such as the work ethic and monogamous marriage.[27] Although ever-increasing percentages of women entered the paid work force, thus altering the dynamics of family life (and contributing to the falling off of the baby boom), most continued to do so in order to augment family resources: earning money for the home, not deep dissatisfaction with life in the two-parent nuclear family, largely explained their behavior.[28] Feminist activism, while far more visible than it had been in the 1940s and 1950s, still engaged only a minority of American women, most of them young, white, well educated, and middle-class.

As they had in the 1950s, millions of upwardly mobile Americans rejoiced especially at the ever-enlarging capacity of a thriving economy to bring material comfort to their lives. The 1960s were the longest period of

25. Michael Frisch, "Woodstock and Altamont," in William Graebner, ed., *True Stories from the American Past* (New York, 1993), 217–39.
26. A statement supported by poll data. See Daniel Yankelovich, *The New Morality* (New York, 1974), xiii.
27. Skolnick, *Embattled Paradise*, 181–91; Carl Degler, *At Odds: Women and the Family in America from the Revolution to the Present* (New York, 1980), 460–65.
28. Easterlin, *Birth and Fortune*, 60–61, 148–50. The percentage of women who had children aged 6 to 17 and who worked rose from 40 percent in 1960 to 50 percent in 1970 (and to more than 70 percent by 1990). The percentage working who had children of less than 6 years of age was 20 percent in 1960 and 30 percent in 1970—and more than 50 percent by 1995. The birth rate declined from 20 births per 1,000 population in 1960 to 18 in 1970 (and to 13 in 1990). See Randall Collins and Scott Cottrane, *Sociology of Marriage and the Family* (Chicago, 1991), 178.

uninterrupted economic growth in United States history. Per capita income (in constant 1958 dollars) rose from $2,157 in 1960 to $3,050 in 1970, an unprecedented decadal increase of 41 percent. Prices remained stable until the late 1960s. Although unemployment among 16- to 19-year-olds rose alarmingly, overall unemployment stayed low, falling to 3.5 percent in 1969.[29] Poverty as measured by the government declined rapidly, from an estimated 22 percent of the population in 1960 to 12 percent in 1969.[30]

By this time the 1950s—then the Biggest Boom Yet—seemed almost dowdy to contemporaries who remembered them. Many of the industries that had boosted that boom, such as electronics, enjoyed even more fantastic growth in the 1960s. Well-placed business and professional people came to expect as a matter of course an amazingly comfortable world that featured high-speed air travel, credit card transactions, and generous expense accounts. Architects and builders flourished, not only by catering to the explosively growing suburbs but also by designing and constructing nests of high-rise buildings in the business centers of cities. It was in the 1960s, the most glittering of times, that piles of glass and steel literally reached for the sky in urban America.

The astonishing affluence of the 1960s did much to promote the grand expectations that peaked in mid-decade. Millions of middle-class Americans—especially the youthful baby boomers—had already experienced rising levels of prosperity during the 1950s. Unaffected personally by the Depression or World War II, the boomers matured in a very different world from that of earlier, more deprived generations. Moreover, the young and the middle classes became much more numerous—and therefore more self-conscious and self-confident. The number of people aged 15 through 24 increased from 24 million in 1960 to 35.3 million in 1970, a jump of 47 percent. By then they accounted for 17.5 percent of the population, an all-time postwar high.[31] Increasingly large percentages of these young people went to colleges and universities, which also boomed as never before in the 1960s. Many came to believe that they had

29. Wilson, *Thinking About Crime*, 9–10.
30. James Patterson, *America's Struggle Against Poverty, 1900–1994* (Cambridge, Mass., 1995), 157–62; Sheldon Danziger and Daniel Weinberg, "The Historical Record: Trends in Family Income, Inequality, and Poverty," in Danziger, Gary Sandefur, and Weinberg, eds., *Confronting Poverty: Prescriptions for Change* (Cambridge, Mass., 1994), 18–50. The official poverty rate reached an all-time low of 11 percent in 1973.
31. Jones, *Great Expectations*, 80–81.

the knowledge and the resources to create a progressive, advanced society like none before in human history. Some identified themselves as participants in a "new class"—of experts in everything from engineering to social science to policy-designing. Their brimming, "can-do" certitude stimulated grand expectations about the capacity of government to solve social problems. Even more than in the 1950s, it seemed that there were no limits.

As these expectations expanded, millions of Americans began not only to anticipate ever-greater social and technological progress but also to believe that they had "rights" to all sorts of blessings, including profound psychological satisfaction. They imagined, often narcissistically, that they could achieve great personal "growth" and "self-actualization."[32] What earlier generations had considered as privileges, many in this one came to perceive as entitlements. In personal life this meant rapid gratification; in policy matters it meant deliverance from evil. Anything, it seemed, was possible in this protean time in history. People talked confidently about winning "wars" against contemporary problems, ranging from poverty to cancer to unrest in Vietnam. Some thought that they could combat not only the age-old scourges of human life—Disease and Disability—but also two others: Discontent and Dissatisfaction.[33]

These grand expectations also affected the behavior of groups. Government, many groups argued, must act to guarantee their "rights." The rights revolution that ensued engaged not only the established pressure groups—labor unions, corporations, farm organizations, blacks—but also others, including Native and Hispanic Americans ("red power" and "brown power") and feminists, who formed the National Organization for Women in 1966. Athletes, too, organized: the Major League Baseball Players Association came into being in 1966.[34] "Public interest" groups rose to demand laws to protect the environment and to improve the quality of life in myriad other ways. Elderly Americans, including militants who became known as the Grey Panthers, developed especially powerful lobbies. Even poor people got together, creating the National

---

32. Ibid., 254; Skolnick, *Embattled Paradise*, 96; Yankelovich, *New Morality*, 188, 234–38.
33. Peter Conrad and Joseph Schneider, *Deviance and Medicalization: From Badness to Sickness* (St. Louis, 1980); Renée Fox, "The Medicalization and Demedicalization of American Society," in John Knowles, ed., *Doing Better and Feeling Worse: Health in the United States* (New York, 1977), 9–22.
34. Randy Roberts and James Olson, *Winning Is the Only Thing: Sports in America Since 1945* (Baltimore, 1989), 135–39.

Welfare Rights Organization in the late 1960s and angrily denouncing Congress when it failed to meet their demands. The proliferation of these self-conscious groups, some of which (such as "seniors-only" enclaves) virtually excluded others, added to a perception by the early 1970s that the United States was becoming both a claimant society and an ever more openly balkanized culture.

The often utopian expectations stimulated in the rights revolution crashed against these and other forces by the late 1960s. Much of the rancor that thereafter roiled American life arose from the increasingly sharp disjunctures that developed between grand expectations and the more prosaic realities of American heterogeneity, notably the barriers erected by differences of class, region, gender, and race. Further rancor arose from the resentment of "ordinary" people against the special claims—many of them grandiose indeed—of the interest groups. There were limits after all. The disjunctures dominated American life for decades after the 1960s.

Still, the depth of these divisions was not altogether clear until the late 1960s, for progress before then seemed continuous and unending. Scientific and technological "breakthroughs" appeared regularly. In 1961 Haloid Xerox Corporation, started in 1959, became Xerox Corporation and transformed the ways in which institutions conducted their business. So did large, mainframe computers. Air-conditioning spread widely and promoted enormous economic growth in the South and Southwest. Television, equipped with videotape, began working wonders in its coverage of news and sports: "Wide World of Sports" appeared for the first time in 1961 and instant replay in 1963. In 1961 Dr. J. Vernon Luck, Sr., became the first surgeon successfully to reattach a severed limb—of a construction worker whose arm had been mangled in a freeway accident. Six years later a South African surgeon, Dr. Christiaan Barnard, presided over a team that managed the world's first successful human heart transplant. And the space program, set in motion by President Kennedy in 1961, captured the imagination of millions. On July 20, 1969, astronaut Neil Armstrong became the first man to set foot on the moon. Americans were thrilled to hear him proclaim, "That's one small step for a man, one giant leap for mankind." President Richard Nixon, speaking for many, boasted that the moon shot was "the greatest week in the history of the world since the Creation."[35]

35. Tom Wolfe, *The Right Stuff* (New York, 1979); Michael Smith, "Selling the Moon: The U.S. Manned Space Program and the Triumph of Commodity Scientism," in Richard Wightman Fox and T. J. Jackson Lears, eds., *The Culture of*

Many standbys of popular culture, too, offered reassuring continuities to Americans in the 1960s. Big-time sports captured ever-larger audiences, both live and on TV. Vince Lombardi, coach of the powerful Green Bay Packers football team, extolled the virtues of hard work and discipline and became something of a cult figure among Americans who proclaimed traditional values. Winning, he said, wasn't an important thing—it was the only thing. Television, too, continued to feature familiar prime-time programs along with its more violent fare. These included such hardy perennials as "The Lawrence Welk Show," "The Lucy Show," and "The Tonight Show" (which Johnny Carson took over in 1962 and stayed with for thirty years).[36] "The Adventures of Ozzie and Harriet," which had started in 1952, lasted through 1966, "Gunsmoke" from 1955 until 1975. Sitcoms such as "The Beverly Hillbillies" and "Petticoat Junction" retained solid followings for most of the decade.

The continuing popularity of other forms of popular culture also revealed the persistence of mainstream tastes. Millions of people showed little interest in rock, enjoying instead popular songs like Henry Mancini's "Moon River" (1961) and "Days of Wine and Roses" (1962).[37] In 1965 viewers flocked to see The Sound of Music, a happy, sentimental film about the singing von Trapp family. It earned more than $100 million on its first run and outdid Gone with the Wind as the all-time best-selling movie.[38] Four years later Walt Disney productions brought out The Love Bug, which became the top-grossing movie of the year, attracting far more people in that year than countercultural films such as Easy Rider and Alice's Restaurant. While attendance and sales figures do not tell the whole story, by any means, about popular tastes, they suggest an obvious continuity: millions of people still demanded non-threatening "family" entertainment. Sensational media accounts focusing on cultural "revolution" in the 1960s left a false impression of the decade: significant continuities were a feature of popular culture during the sixties.

Firmly established American attitudes toward world politics also

---

Consumption: Critical Essays in American History, 1880–1940 (New York, 1983), 175–209.

36. "I Love Lucy" became "The Lucy Show" in 1962 and ran until 1974. The Welk show ran from 1955 to 1982. For solid data on such matters, see Tim Brooks and Earle Marsh, eds., The Complete Directory to Prime Time TV Shows, 5th ed. (New York, 1992).

37. The lyrics to both were by Johnny Mercer. The songs were featured in the movies Breakfast at Tiffany's (1961) and Days of Wine and Roses (1962), respectively.

38. Baughman, Republic of Mass Culture, 139.

changed very slowly during the 1960s. While McCarthyite excesses had ebbed, a virulent anti-Communism still flourished at most levels of American politics and culture. Robin Moore's book *The Green Berets*, which celebrated the exploits of a big "Nordic type" in charge of America's Special Forces, sold 1.2 million copies within two months of its issuance in paper in late 1965. When the movie version came out in 1968, starring John Wayne, it did very well at the box office. *Patton*, which (somewhat ironically) highlighted the military exploits of "Blood and Guts" George C. Patton, was the Best Picture of the Year in 1970.[39]

Popular attitudes toward the Vietnam War especially revealed the persistent power of patriotic, anti-Communist opinion. The war sparked the most extensive protests in American history: at least 600,000 people joined "moratorium" demonstrations in Washington in late 1969. But anti-war demonstrators enraged millions of other Americans, many of them working-class people who were not necessarily pro-war but who deeply resented the fact that many of the young protestors ridiculed American institutions and avoided military service. "Here were those kids, rich kids who could go to college, didn't have to fight," a construction worker railed. "They are telling you your son died in vain. It makes you feel your whole life is shit, just nothing."[40] The anti-war protests especially angered the Cold Warriors who directed foreign policy in Washington, and until 1970 they had only limited effect on electoral politics: all three major presidential candidates on the ballot in 1968 opposed American withdrawal from the war. Significant reductions in American ground forces came only in 1969–70, by which time "cut-our-losses" realists began to coalesce effectively but very uneasily with moral opponents of the war. By then it was obvious to all but a minority of people that the United States had little chance of winning.

Political attitudes revealed other ambiguities in the 1960s. While contemporary accounts, especially in the mass media, lavished attention on the rise of the student and anti-war Left, conservative activists were also mobilizing. The Young Americans for Freedom, a right-of-center organization, was founded in 1960. It attracted as many members in the 1960s as the SDS, established in the same year. "Neo-conservative" intellectuals, regrouping to criticize the liberal programs of the early 1960s, gathered increasingly large audiences by 1970. The GOP, meanwhile, rebuilt itself after suffering serious defeats in the early 1960s; in 1966 it

39. Richard Fried, *Nightmare in Red: The McCarthy Era in Perspective* (New York, 1990), 196–97.
40. Reider, *Canarsi*, 157.

scored impressive victories, and in 1968 it recaptured the presidency. Conservatives have often controlled national politics, especially the presidency, since that time.

A final, durable continuity: America remained one of the most religious cultures in the Western World. This religiosity assumed a large variety of forms. Religious leaders and church-goers continued to contribute to the civil rights movement. Norman Vincent Peale, still preaching the message of positive thinking, prospered as a much-admired figure. So did Billy Graham, whose evangelical crusades drew millions in the United States and elsewhere in the world. Although church-going in the United States fell a bit from its peak in the 1950s, it remained high. An estimated 43 percent of Americans regularly attended services in 1968, compared to approximately 10 to 15 percent in England and France.[41]

Less noticed at the time, but obvious later, fundamentalists of varied persuasions were becoming increasingly numerous and preparing to speak out. Some were super-patriotic and politically reactionary; others were scarcely able to contain their rage at the Supreme Court and at elites—governmental, corporate, educational, scientific—that they perceived to be ruining the nation. While the fundamentalist leaders were white and upper middle-class, the followers included large numbers of poor and working-class people.[42] The appearance in 1970 of Hal Lindsey's book *The Late Great Planet Earth* suggested the depth of fundamentalist feelings in the country. This was a pre-millenarian tract that foresaw a nuclear apocalypse caused by an anti-Christ, after which Jesus Christ returned to earth and saved mankind. The book became the best-selling non-fiction book of the 1970s and sold more than 28 million copies by 1990.

What these complex trends—the changes as well as the continuities—

41. Leo Ribuffo, "God and Contemporary Politics," *Journal of American History*, 79 (March 1993), 1515–33; James Hunter and John Rice, "Unlikely Alliances: The Changing Contours of American Religious Faith," in Wolfe, ed., *America at Century's End*, 318–39; Robert Wuthnow, *The Restructuring of American Religious Society and Faith Since World War II* (Princeton, 1991). A caution concerning church attendance statistics is C. Kirk Hadaway et al., "What the Polls Don't Show: A Closer Look at U.S. Church Attendance," in *American Sociological Review* (Dec. 1993).

42. Paul Boyer, *When Time Shall Be No More: Prophecy Belief in Modern American Culture* (New York, 1992), 5; Ronald Numbers, *The Creationists: The Evolution of Scientific Creationism* (New York, 1992), 300; Stephen Bates, *Battleground: One Mother's Crusade, the Religious Right, and the Struggle for Control of Our Classrooms* (New York, 1993), 50–60.

indicate is that the 1960s were an age of increasingly open polarization and fragmentation.[43] The decade, to repeat, ushered in unprecedented affluence and escalating expectations, and it left long-range legacies, especially in the realm of race relations and in the personal behavior—much more free and anti-authoritarian—of many young people. Yet well-entrenched older values, cherished by what Richard Nixon and others called the silent majority, persisted along with these changes. The conflict between older and newer mores, contested openly on the ever-broader and more sensational stage of the mass media, sharply exposed already existing divisions in the nation, especially along lines of age, race, gender, and social class. The center that had more or less held in the late 1950s cracked in the 1960s, exposing a glaring, often unapologetic polarization that seemed astonishing to contemporaries.[44]

43. Peter Muller, *Contemporary Sub/Urban America* (Englewood Cliffs, N.J., 1981), 67–70.
44. Alan Brinkley, "The Problem of American Conservatism," *American Historical Review*, 99 (April 1994), 409–29; Leo Ribuffo, "Why Is There So Much Conservatism in the United States and Why Do So Few Historians Know Anything About It?" ibid., 438–49.

16

The New Frontier at Home

Inauguration day, January 20, 1961, was cold and bright, the sun reflecting brilliantly off new-fallen snow in Washington. The glare prevented the aging poet Robert Frost, who was invited to recite at the ceremony, from reading the poem he had composed for the occasion. He gave one instead from memory. But this was the only hitch in a memorable day. Thousands among the throng at the Capitol, and millions among those who watched the event on television, were captivated by the image of a youthful, vigorous, and eloquent Kennedy, who proclaimed his determination to advance American ideals throughout the world. Summoning the idealism and commitment of the American people, he told them: "Ask not what your country can do for you; ask what you can do for your country. . . . Ask not what America will do for you, but what together we can do for the freedom of man."[1]

1. New York Times, Jan. 22, 1961; Richard Reeves, President Kennedy: Profile of Power (New York, 1993), 35–36. Other books on the Kennedy administration include Herbert Parmet, JFK: The Presidency of John F. Kennedy (New York, 1983); David Burner, John F. Kennedy and a New Generation (Boston, 1988); James Giglio, The Presidency of John F. Kennedy (Lawrence, 1991); Jim Heath, Decade of Disillusionment: The Kennedy-Johnson Years (Bloomington, Ind., 1975); Henry Fairlie, The Kennedy Promise: The Politics of Expectation (Garden City, N.Y., 1973); and Irving Bernstein, Promises Kept: John F. Kennedy's New Frontier (New York, 1991).

Critics found Kennedy's oration to be bombastic. Yet popular reaction was generally enthusiastic, and many people never forgot his call to action. Moreover, Kennedy seemed ready to deliver on his promises. Although his selectees for top posts were hardly known as reformers—Secretary of Defense Robert McNamara, National Security Adviser McGeorge Bundy, and Treasury Secretary Douglas Dillon were Republicans—he made a show of assembling a team of highly educated and activist advisers. Many of them were academic people—"the best and the brightest"—from Harvard and other elite institutions. Dean Rusk, his Secretary of State, had been a Rhodes Scholar. In celebrating the brilliance of his team Kennedy rarely missed a chance to accentuate the difference between his presidency and that of the allegedly tired Eisenhower administration.

Kennedy's administrative style indeed differed from Eisenhower's. Where Ike had relied on a hierarchical system that he had known as an army officer, JFK sought out ideas from a corps of free-wheeling advisers. Chief among them was his brother Robert, whom he dared to name as Attorney General. McNamara, a super-efficient and dominating administrator whom JFK took from the presidency of the Ford Motor Company, was another. Serving him as advisers in the White House were Arthur Schlesinger, Jr., a Harvard history professor, and Theodore "Ted" Sorensen, an articulate young liberal.[2] Sorensen helped write many of JFK's major speeches, including the inaugural address. For political matters Kennedy relied heavily on able strategists—critics called them the Irish Mafia—such as Kenneth O'Donnell and Lawrence O'Brien. Many other Americans, most of them young and idealistic, converged on Washington to seek lesser posts in the ever-growing federal bureaucracy and to trumpet bold new ideas about town. Old hands fondly likened the atmosphere to the early days of the New Deal.

Some contemporaries, including Democrats, were appalled by what they perceived accurately as the loose administrative style of the new administration. "They've got the damnedest bunch of boy commandos running around . . . you ever saw," Adlai Stevenson told a friend.[3] And serious flaws soon revealed themselves. In April the Kennedy administration blundered impetuously into a disastrous effort to overthrow Fidel

2. Authors of early pro-Kennedy administration histories: Schlesinger, A *Thousand Days: John F. Kennedy in the White House* (Boston, 1965); and Sorensen, *Kennedy* (New York, 1965).

3. Thomas Paterson, ed., *Kennedy's Quest for Victory: American Foreign Policy, 1961–1963* (New York, 1989), 19.

Castro in Cuba. But even this debacle had no apparent effect on the young President's extraordinary popularity. Kennedy, indeed, reached out with unparallelled success to the media. He was the first President to allow his press conferences to be televised live. By May 1961 some three-fourths of the American people had seen at least one. Of these viewers, a staggering 91 percent said that they had a favorable impression of his performance, as opposed to only 4 percent who responded unfavorably.[4]

Setbacks also failed to blight the special and apparently contagious confidence that Kennedy and his advisers sustained. Many of them, like Kennedy himself, had matured during World War II, days of struggle and sacrifice that had supposedly given them the "toughness"—a favorite word of Kennedy people—to cross the new frontiers of the 1960s. Extraordinarily self-assured, they were even as young people highly conscious of their place in history. Kennedy liked to cite the words of Shakespeare in *Henry* V:

> We . . . shall be remembered;
> We few, we happy few, we band of brothers . . .
> And gentlemen in England now a-bed
> Shall think themselves accursed they were not here.

Thanks in part to this élan, Kennedy managed to bring a special aura to the American presidency. Truman and Eisenhower, to be sure, had presided over substantial growth in the size and power of the executive branch. The extraordinarily telegenic Kennedy greatly accelerated these trends by drawing popular attention to the pomp and circumstance of the office. Kennedy and his elegant wife Jackie invited a parade of famous artists, musicians, and writers to the White House. Carefully orchestrated state dinners for visiting dignitaries received wide publicity. Jackie proudly showed off the way that she redecorated the presidential home. Many reporters, themselves young and liberal, lavished attention on the high culture and taste that the Kennedys appeared to bring to government. An air of royalty was enveloping the land of the common man.

Americans began hearing more and more about the "awesome" responsibilities of the Oval Office, now regularly capitalized by credulous journalists who described the High Decision-Making taking place there and

4. Carl Brauer, "John F. Kennedy: The Endurance of Inspirational Leadership," in Fred Greenstein, ed., *Leadership in the Modern Presidency* (Cambridge, Mass., 1988), 117–18. Eisenhower's press conferences had been filmed and could be edited. Few of them had appeared on TV news, which until late 1963 lasted only fifteen minutes.

who left no doubt that the fate of the world depended on the deeds of the American President. Theodore White's popular account of the 1960 election, *The Making of the President, 1960* (1961), not only highlighted the brilliance of Kennedy and his advisers but also spoke reverently of the "hush, an entirely personal hush" that surrounded presidential activity. The hush, he added, "was deepest in the Oval Office of the West Wing of the White House, where the President, however many his advisers, must sit alone."[5]

The celebration of the American presidency, and by extension of the potential of the federal government, greatly encouraged contemporary advocates of strong White House leadership. Kennedy himself remained personally very popular throughout his presidency. Along with the booming economy, which after 1962 seemed capable of almost anything, the magnified mystique of the presidency stimulated ever-greater expectations among liberals and others who imagined that government possessed big answers to big problems. The revolution of popular expectations, a central dynamic of the 1960s, owed a good deal of its strength to the glorification of presidential activism that Kennedy successfully sought to foment.

HIGH EXPECTATIONS early gripped contemporaries who yearned for a New Frontier in the realm of domestic policies. *Newsweek* predicted following the election that Kennedy could hope for a "long and fruitful 'honeymoon' with the new Democratic 87th Congress." If Kennedy "jumps right in with a broad new legislative program," *Newsweek* added, "he will find Congress so receptive that his record might well approach Franklin D. Roosevelt's famous 'One Hundred Days.'"[6]

The magazine proceeded to list reasons why Kennedy might succeed, chief among them the support of capable Democratic leaders such as House Speaker Rayburn and Vice-President Johnson, who was to preside over a Senate that he had dominated as majority leader since 1955. Many domestic programs that adorned the Democratic domestic agenda, such as legislation to help "depressed areas," federal aid to education and to housing, and a hike in the minimum wage from $1 to $1.25 an hour, had wide support among congressional liberals. Some form of federal health insurance seemed possible.

5. (New York, 1961), 371. It sold 4 million copies. Kennedy was also the first President (so far as is known) to install hidden microphones in the Oval Office. He had this done in 1962, after which he secretly taped all sorts of meetings. William Safire, *New York Times*, Dec. 26, 1994.
6. *Newsweek*, Nov. 14, 1960, p. EE4.

Reformers had a few successes over the next three years. Working purposefully in 1961, Kennedy succeeded in enlarging the House Rules Committee, a bottleneck that had long blocked liberal efforts, and Rayburn then shepherded through a hike in the minimum wage.[7] Congress also enacted legislation providing modest public funding for manpower training and depressed areas, notably Appalachia. In 1962 it approved important (though little-noted) amendments to drug regulations; these required new drugs to be tested for efficacy as well as for safety before they could be approved for use.

Kennedy also took a few steps that later advanced the interests of women. In 1961 he named Eleanor Roosevelt to head a Presidential Commission on the Status of Women. Its report in 1963, in some ways far from feminist, advocated special training of young women to prepare them for marriage and proclaimed that motherhood was the major role of the American woman. Dominated by advocates of protective labor legislation for women, the commission also opposed the Equal Rights Amendment.[8] The President meanwhile appointed fewer women to high-level federal posts than had his predecessors: he was the only President since Hoover never to have a woman in the Cabinet.[9] Still, the commission made some difference. It called for a federal stand against sex discrimination and affirmed that women, like men, had a right to paid employment. It also stimulated formation of similar commissions on the state level. Thanks in part to the commission, Kennedy issued an executive order ending sex discrimination in the federal civil service. In 1963 he signed an Equal Pay Act that guaranteed women equal pay for equal work. Although this act excluded employees not covered by the Fair Labor Standards Act and had no provisions for enforcement, it had some effect.

7. Tom Wicker, *JFK and LBJ: The Influence of Personality upon Politics* (Baltimore, 1968), 26–148. The minimum wage increase was to be in two stages, to $1.15 in September 1961 and to $1.25 in September 1963. The minimum rate was approximately 50 percent of the average gross hourly earnings of production workers in manufacturing.

8. By then Roosevelt had died and been replaced by Esther Peterson, director of the Women's Bureau of the government. Peterson, a labor unionist and lobbyist, was a longtime Kennedy ally and was the dominant force on the commission. See Cynthia Harrison, *On Account of Sex: The Politics of Women's Issues, 1945–1968* (Berkeley, 1988), 85, 113, 139, 214–15; and Carl Degler, *At Odds: Women and the Family from the Revolution to the Present* (New York, 1980), 441.

9. Harrison, *On Account of Sex*, 75. Truman inherited Labor Secretary Frances Perkins from the FDR years and quickly replaced her with a man. He named no women to his Cabinets.

In the next ten years 171,000 employees received a total of $84 million in back pay under the act.[10] Most important, Kennedy's commission encouraged women activists on both the state and federal levels to develop networks and to talk seriously about curbing long-standing divisions within their ranks. In this way, Kennedy unintentionally aroused expectations that encouraged a much more self-conscious feminist movement after 1964.

Kennedy, who had a mentally ill sister, also moved more actively than presidential predecessors to advance the cause of mental health. In 1963 Congress passed a Mental Retardation Facilities and Community Mental Health Centers Act, which funded local mental health centers that were to provide a range of out-patient services, including marital counseling, help for delinquents, and programs for unwed mothers and alcoholics. The act sought in part to get mentally ill people out of large state hospitals, which supporters of the legislation considered to be "snake pits" of callous and inhumane treatment. Thanks to subsequent funding for this effort at deinstitutionalization, the population of mental hospitals declined from 475,000 in 1965 to 193,000 in 1975. Use of mental health services, meanwhile, exploded (six-fold between 1955 and 1980) among a populace ever more concerned about its psychological well-being.[11]

The new President took special interest in measures aimed at promoting economic growth. Some of these sought to reassure corporate leaders, most of whom had supported Republicans over the years. In 1962 Kennedy secured approval of legislation that accelerated depreciation allowances and granted businesses tax credits for investment in certain kinds of equipment.[12] The law probably enhanced corporate investment and growth. Kennedy also sought to mend political fences with business

10. Harrison, On Account of Sex, 104–5.
11. Edward Berkowitz, "Mental Retardation Policies and the Kennedy Administration," Social Science Quarterly, 61 (June 1980), 129–42; Gerald Grob, "The Severely and Chronically Mentally Ill in America: Retrospect and Prospect," Transactions of the College of Physicians of Philadelphia, 13 (1991), 337–62; Grob, The Mad Among Us: A History of America's Care of the Mentally Ill (New York, 1994); David Mechanic and David Rochefort, "A Policy of Inclusion for the Mentally Ill," Health Affairs, 2 (Spring 1992), 128–50. Other developments, notably the development of new psychotropic drugs and passage (in 1965) of Medicare and Medicaid, especially speeded up the process of deinstitutionalization.
12. Herbert Stein, The Fiscal Revolution in America (Chicago, 1969), 370; Allen Matusow, The Unraveling of America: A History of Liberalism in the 1960s (New York, 1984), 33.

that had been damaged following his heavy-handed attempt earlier in 1962 to stop leading steel companies from introducing inflationary price increases. After the companies temporarily backed down, he had been quoted as saying, "My father always told me that all businessmen were sons-of-bitches, but I never believed it till now." Angry corporate leaders had responded by wearing s.o.b. ("Sons of Business") buttons on their lapels. [13]

In 1962 Kennedy began to listen carefully to Keynesian economists, notably Walter Heller, a University of Minnesota professor whom he had named to head the Council of Economic Advisers. Heller, like many other economists in the early 1960s, was buoyantly self-confident about his discipline. "Our statistical net," he maintained, "is now spread wider and brings in its catch faster. Forecasting has the benefit of not only more refined, computer-assisted methods but of improved surveys of consumer and investment intentions." [14] Heller's enthusiasm brilliantly reflected the rapidly rising confidence that liberals, especially in the social sciences, were developing about the ability of "experts" to manage American society. This self-assurance, expanding still more in the mid-1960s, excited and energized liberal activism at the time.

Although Kennedy had to struggle to understand the theoretical arguments of Heller and other economists, he made considerable progress, and by 1962 he was ready to act on Heller's advice. For political and humanitarian reasons he was anxious to reduce unemployment and to accelerate economic growth. He also concluded that moderately higher federal deficits, which Heller foresaw, could be risked without causing serious inflation; Eisenhower, after all, had (unintentionally) run sizeable deficits during the recessionary years between 1958 and 1960. And cutting taxes always sold well with Congress and the public. So it was that he came out publicly in late 1962 for one of Heller's main goals: a cut in personal income and corporate taxes. Such reductions, it was argued, would free funds for investment and thereby promote economic expansion. [15]

13. Grant McConnell, *Steel and the Presidency, 1962* (New York, 1963); John Blum, *Years of Discord: American Politics and Society, 1961–1974* (New York, 1991), 59. Thomas Reeves, in *A Question of Character: A Life of John F. Kennedy* (New York, 1991), writes, 331, that JFK later said he was misquoted. Kennedy told a friend, "I said sons of bitches or bastards or pricks. I don't know which. But I never said anything about *all* businessmen."
14. Stein, *Fiscal Revolution*, 384.
15. James Sundquist, *Politics and Policy: The Eisenhower, Kennedy, and Johnson Years* (Washington, 1968), 34–56; Matusow, *Unraveling*, 30–59; Schlesinger,

In taking this stand Kennedy disappointed many liberals, who considered his tax cut a boon to business and upper-income interests. They called instead for tax reform, increased social spending, and investment in public works. Harvard economist John Kenneth Galbraith branded the cut as "reactionary Keynesianism" and labeled Kennedy's announcement as the "most Republican speech since McKinley."[16] Still, the President conciliated some reformers by accepting, however cautiously, the central Keynesian idea that compensatory fiscal policies, including short-run budget deficits, could stimulate economic growth. No President before him had dared publicly to assume such a position.[17] In this sense his quest for a tax cut, which ultimately passed in 1964, left an important legacy to policy-making.

Notwithstanding these varied legislative initiatives, Kennedy's record in the realm of domestic policies was hardly stellar, for three reasons. The first was his own uninspiring leadership in this area. As earlier in his career, Kennedy was a cool and unpassionate politician when he dealt with domestic issues. He identified with moderates, not with liberals, whom he disdained as "honkers." He also disdained congressional leaders, refusing to court them personally. Above all, he did not much care about domestic issues. He told Sorensen, hard at work on the inaugural address, "Let's drop the domestic stuff altogether." Sorensen did, and the address focused almost exclusively on foreign affairs, Kennedy's abiding concern. On other occasions JFK made no pretense of hiding his priorities. "Foreign affairs," he once remarked to Nixon, "is the only important issue for a President to handle, isn't it? . . . I mean, who gives a shit if the minimum wage is $1.15 or $1.25, compared to something like Cuba?"[18]

Second, Kennedy faced a line-up on Capitol Hill that he expected would defeat most major liberal initiatives. Conservative Democrats, many of them from the South, continued as since 1938 to dominate key committees and to form informal but effective coalitions with conservative Republicans. Johnson tried to overcome this coalition, but as Vice-

Thousand Days, 625–34, 644–56; Seymour Harris, *Economics of the Kennedy Years, and a Look Ahead* (New York, 1964), 66–77. By late 1963 Kennedy was also persuaded by Heller, an influential adviser, to consider programs against poverty. See chapter 18.

16. Schlesinger, *Thousand Days,* 649; Alan Wolfe, *America's Impasse: The Rise and Fall of the Politics of Growth* (New York, 1981), 68.
17. Stein, *Fiscal Revolution,* 455.
18. Michael Beschloss, *The Crisis Years: Kennedy and Khrushchev, 1960–1963* (New York, 1991), 48.

President he was far weaker on the Hill than he had been while majority leader of the Senate. Congress defeated or refused to take action on a number of Kennedy proposals, including health insurance for the aged and creation of a Department of Urban Affairs.

When liberals urged Kennedy to fight for his programs, he reminded them of his lack of mandate in 1960 and of the political realities on Capitol Hill. Citing Thomas Jefferson, he said, "Great innovations should not be forced on slender majorities." The President also hated to lose, for he recognized that presidential prestige depended in part on maintaining an aura of effectiveness. One had to conserve one's resources for major battles. "There is no sense in raising hell, and then not being successful," he said. "There is no sense in putting the office of the Presidency on the line on an issue, and then being defeated."[19]

Kennedy also failed to secure another key issue on the liberal agenda, federal aid to education. Although he supported the idea, he encountered stiff opposition from southerners such as Howard Smith of Virginia, head of the House Rules Committee. Smith proclaimed that the education bill sought to "aid the NAACP and complete the subjection of the South."[20] Religious conflicts further hurt the bill. As a Roman Catholic Kennedy was politically sensitive to charges that he secretly favored federal support of parochial schools. When he refused to endorse such aid, liberal Catholics in Congress, including John McCormack of Massachusetts (who later replaced Rayburn as House Speaker), deserted the larger cause. The bill then failed to escape the Rules Committee. The fate of the school aid bill, like that of most important measures on the Hill, reflected the continuing power of special interests in American politics, in this case the organized Protestant and Catholic churches. It also exposed persistent divisions within American society: differences along regional lines, among others, frequently cut across more obvious "liberal" versus "conservative" splits and greatly complicated policy-making.

Third, Kennedy's programs reflected the wider limitations of liberal Democratic politics. Many of his efforts manifested an interest group politics that aided influential lobbies a good deal more than the poor and the powerless. An Omnibus Housing Act passed in 1961 that offered federal support for urban renewal did more to help developers, construction unions, and Democratic activists in the cities than it did to improve housing for the poor. A Manpower Development and Training Act in

19. Schlesinger, *Thousand Days*, 709.
20. Blum, *Years of Discord*, 31–32.

1962 retained some congressional support in the next six years, but the act mainly subsidized officials and private interests who provided the training. It had at best a marginal impact on unemployment. An Area Redevelopment Act, approved in 1961, funneled federal money to Appalachia and other "depressed areas" but also had little effect, in part because it was ill funded. Worse, opponents came to perceive it as pork barrel legislation for key Democratic congressional districts. Congress refused to replenish the ARA's loan fund in 1963 and scrapped the program in 1965.[21]

Other liberal efforts were well intended but in some ways misconceived. The mental health law, for instance, assisted many not-so-sick people who were permitted to leave mental hospitals and gravitated to general hospitals, nursing homes, or community facilities. But it flung many severely and chronically ill mental patients onto the not-so-tender mercies of communities that lacked the will, the money, and the medical knowledge to care for them. In time, deinstitutionalization exacerbated social problems, including long-range drug addiction and homelessness.[22]

The tax cut especially revealed the limitations of Kennedy-style domestic policy. When finally approved in 1964 it marked a considerable change in federal tax policy. The top marginal tax rate on individuals was cut from 91 to 70 percent; the tax rate on the lowest bracket fell from 20 to 14 percent. Corporate tax rates dropped from 52 to 48 percent. The law was estimated to save taxpayers $9.1 billion dollars in 1964.[23] Heller and others were delighted, crediting the law for the extraordinary economic growth and prosperity that characterized the mid-1960s. The tax cut, they reiterated, proved their contention that social science expertise could fine-tune public policy.

In fact, however, the tax cut, as Galbraith and others had said, mainly assisted the well-off. Moreover, it probably had little to do with the great prosperity that arose by 1965. The American economy had already begun in 1962 to rebound from a largely cyclical downturn. Central to this recovery, aside from cyclical forces, were large and pre-existing developments, including the low cost of energy, such as oil; continuing technological innovation; the expansion of world trade; and heightened productivity. Certain sectors of the American economy further benefited from rapid increases in military spending during the Kennedy years.

21. Matusow, *Unraveling*, 97–107.
22. Grob, "Severely and Chronically Mentally Ill."
23. Cathie Martin, *Shifting the Burden: The Struggle over Growth and Corporate Taxation* (Chicago, 1991), 10–11.

In many areas these increases boosted expansion more than did the tax cut.[24]

It is a little unfair to single out Kennedy for the limitations of liberals' assumptions about socio-economic policy. A partisan Democratic, he largely followed the advice of others, who dominated the northern-urban wing of his party. Moreover, Kennedy naturally sought to advance popular programs, such as a tax cut, that would help him develop a larger political mandate in the 1964 election. Still, the fact remains that he devoted only sporadic attention to domestic affairs and that his administration, hamstrung by Congress, accomplished little of significance in the realm of social legislation. In this respect, as in others, his record resembled those of his predecessors, Eisenhower and Truman. There were no new frontiers there.

THE MAJOR TEST of domestic policy for Kennedy—indeed for American institutions in general at the time—was race relations. It was his fate to take office when more and more black people were losing patience with drawn-out legal strategies and were turning instead to direct action. Whites, mainly idealistic students from the North, began to join them in modest but increasing numbers. While still relying heavily on sit-ins and boycotts, the civil rights activists were developing an agenda larger than desegregating lunch counters and other facilities. Some gave thought to improving the conditions of black people in the North. In the early 1960s, however, they continued to focus mainly on ways to empower the masses of the black poor in the South.[25] Stymied by their opponents, they were very, very angry. "To be a Negro in this country and to be relatively conscious," James Baldwin asserted in 1961, "is to be in a rage all the time."[26]

Activists, many of them connected with CORE, decided to step up the battle against racism only a few months after Kennedy had taken office. The new President, they thought, would show more sympathy for their aspirations than Eisenhower had. Their strategy was to embark on "freedom rides" through the Deep South. They would board interstate buses

24. Charles Morris, A *Time of Passion: America, 1960–1980* (New York, 1984), 31–36.
25. Carl Brauer, *John F. Kennedy and the Second Reconstruction* (New York, 1977), 1–86; Giglio, *Presidency of JFK*, 159–88; Sundquist, *Politics and Policy*, 254–59; Matusow, *Unraveling*, 60–96; Burner, *JFK and a New Generation*, 118–31.
26. Cited in Morris Dickstein, *Gates of Eden: American Culture in the Sixties* (New York, 1977), 166.

and try to desegregate bus terminals wherever the buses stopped. In so doing they now had the rule of law on their side, for the Supreme Court had decided in December 1960 that segregation of interstate bus terminals was unconstitutional.[27]

The riders, including CORE leader James Farmer, fully anticipated that whites would react violently.[28] They therefore forewarned the President, Attorney General Robert Kennedy, and FBI director Hoover of their plans. Robert later said that he knew nothing beforehand about the rides, which started when seven blacks and six whites climbed on two buses in Washington and traveled through the Upper South on their way to Alabama, Mississippi, and New Orleans. In Rock Hill, South Carolina, John Lewis, a leading activist, was clubbed and knocked down when he tried to enter the white rest room. When the riders reached Anniston, Alabama, a mob slit the tires of one of the buses, smashed its windows, tossed in an incendiary device, and attacked the riders as they fled from the smoke. This was when Kennedy—and many others in the nation—awoke to the action. A dramatic new phase of the civil rights movement had begun.

The other bus rolled on to Birmingham, where a Klansman who was a paid FBI informer had alerted Hoover earlier in the week that the KKK had worked out a deal with Birmingham public safety commissioner Eugene "Bull" Connor that would allow the Klan fifteen minutes to attack the riders before Connor's police intervened. Hoover, although nominally under the control of Attorney General Kennedy, failed to inform his chief, and the riders had no federal protection when they got off the bus. There they were badly beaten by thirty-odd Klansmen wielding baseball bats, pipes, and bicycle chains. One of those attacked, a sixty-one-year-old, was left permanently brain-damaged. The battered riders then broke off their trip and were flown to safety in New Orleans.

As so often happened during the civil rights movement, the intransi-

27. *Boynton v. Virginia*, 364 U.S. 564 (1960).
28. James Farmer, *Lay Bare the Heart: An Autobiography of the Civil Rights Movement* (New York, 1986), 195–203. See also Clayborne Carson, *In Struggle: SNCC and the Black Awakening of the 1960s* (Cambridge, Mass., 1981), 34–37; Taylor Branch, *Parting the Waters: America in the King Years, 1954–1963* (New York, 1989), 419–21; Robert Weisbrot, *Freedom Bound: A History of America's Civil Rights Movement* (New York, 1990), 57–63; August Meier and Elliott Rudwick, *CORE: A Study in the Civil Rights Movement* (New York, 1973), chap. 5; Harvard Sitkoff, *The Struggle for Black Equality, 1954–1992* (New York, 1993), 88–117; Howell Raines, ed., *My Soul Is Rested: Movement Days in the Deep South Remembered* (New York, 1977), 109–30; and Anne Moody, *Coming of Age in Mississippi* (New York, 1965).

gence of whites stiffened the resolve of the activists. New riders, led by Lewis, Diane Nash, and other SNCC workers from Nashville, carried the campaign back to Alabama and Mississippi. CORE workers also returned. Some of the freedom riders, including Lewis and Kennedy aide John Siegenthaler, were savagely attacked in Montgomery.[29] Other riders, including Farmer, were arrested in Jackson, Mississippi, convicted of breaching the peace, fined $200 each, and (when they refused to pay the fines or to post bail) sent to jail for thirty-nine days before they got out on bond. Many of these activists were sent to the maximum security wing of the state penitentiary at Parchman, where guards attempted unsuccessfully to break up their unity by knocking them about with water from fire hoses, closing cell windows during the daytime to increase the already ferocious heat, and blasting them with cold air from exhaust fans at night.[30]

Segregationist southern officials such as the volatile Governor Ross Barnett of Mississippi hoped that stern measures like these would stop the freedom rides. But other activists kept coming: 328 were arrested in Jackson alone by the end of the summer. Two-thirds were college students, three-fourths were men; more than half were black.[31] The freedom rides ended only in September when the Interstate Commerce Commission (ICC), acting on an earlier request from Robert Kennedy, prohibited interstate bus and railroad companies from using segregated facilities. It had been a protracted and violent struggle.

While the freedom rides were attracting national attention, civil rights workers were busy elsewhere in the South. In the process they encountered some problems within their ranks. Intramovement tensions, especially of generation and of class, arose clearly in Mississippi, where Robert Moses and others began highly dangerous activities in August 1961. There, where racism flared as intensely as anywhere in the nation, many young blacks in their late teens and twenties had already joined the great mass migrations to the North, most frequently to the supposed promised land of Chicago.[32] Their exodus left a growing generation gap, requiring civil rights workers to recruit among older farmers, young teenagers, and their parents. These people differed from the mostly urban, middle-aged,

29. *Newsweek*, May 29, 1961, pp. 21–22.
30. John Dittmer, *Local People: The Struggle for Civil Rights in Mississippi* (Urbana, 1994), 96–97.
31. Ibid., 95.
32. Nicholas Lemann, *The Promised Land: The Great Black Migration and How It Changed America* (New York, 1991).

middle-class black people, such as preachers, porters, and educators, who had traditionally formed the backbone of the NAACP. These leaders, in turn, had risked much in their lifetimes to support the mainly legal battles against discrimination that the NAACP favored. They were often reluctant to embrace the militant tactics favored by the new and younger generation of civil rights activists.[33]

Problems in uniting local blacks became especially visible in McComb, a southwestern Mississippi town where segregationists had total control. To shake their hold SNCC leaders led by Marion Barry favored "direct action" campaigns, such as sit-ins at drugstores and other segregated facilities. Moses went along with such efforts, but he was also anxious to coalesce with local NAACP leaders whom he considered to be vital to the success of long-range change on the community level. Most of these leaders wanted to concentrate on voter registration, a traditional goal of the NAACP. Endorsing this emphasis, Moses pointed out that direct action, such as a sit-in, was often "a one-event thing, and not something the movement could sustain."[34]

By the end of 1961 Moses's worries about sit-ins proved prophetic. White authorities in McComb responded to the SNCC sit-ins by arresting and jailing demonstrators, who remained incarcerated for thirty-four days before being released on bond supplied by the Southern Christian Leadership Conference and the NAACP. Two of those released then attempted to return to their all-black school, only to be barred by their black principal. This action enraged many of their classmates, more than 100 of whom dared to march through town, carrying banners and singing "We Shall Overcome." Incredulous whites surrounded them and savagely assaulted a newly arrived SNCC worker, Bob Zellner, the only white person in the march. Police then arrested SNCC organizers and 116 students, some of whom went to jail for more than a month. Meanwhile, the principal expelled all the marchers from school, requiring them to promise not to participate in future demonstrations as the price for readmission. Most refused.

As Moses had feared, the sit-ins had the further result of dividing local black people, some of whom blamed him along with other SNCC workers for what had happened. Many African-American parents had been cool to sit-ins from the beginning. Others were appalled that SNCC "outsiders" would encourage a march by students, most of whom were under eigh-

33. Dittmer, *Local People*, 117–28, 143–58.
34. Ibid., 107–8, 118–20; Weisbrot, *Freedom Bound*, 94–95.

teen years of age, and they blamed SNCC when their children were banned from school. By the end of the year, when Moses was finally released from jail, SNCC activity in McComb virtually ground to a halt, not to revive substantially again until the summer of 1964.[35]

Violence by whites further ensured the defeat of civil rights activities in the area. In the county seat of Liberty, E. H. Hurst, a member of the state legislature, shot and killed Herbert Lee, a black farmer and father of nine who had been so bold as to drive Moses around the county to contact potential voters. An African-American eyewitness, Louis Allen, told a coroner's jury that Hurst had acted in self-defense, and Hurst was quickly cleared. Allen then told Moses that whites had coerced him into perjury—Hurst, he said, had shot Lee in cold blood—and that he would say so in public if offered protection. Moses called the Justice Department in Washington, demanding that the government provide it. The Justice Department replied that it could not possibly do so and that Hurst would be found innocent whatever it did. Allen, fearing for his life, then stuck to his original story. A year and a half later he was ambushed, shot in the face, and killed. No one was charged in the slaying.[36]

Tensions elsewhere within the movement afflicted the reputation even of King, who devoted major attention to direct action protests in Albany, Georgia, between October 1961 and August 1962. The Albany Movement, as it was called, was one of the most frustrating of all civil rights efforts of the early 1960s. Local authorities, led by police chief Laurie Pritchett, shrewdly curbed white extremists and avoided excess. King was twice jailed but each time released on bail without securing anything of significance. Arrested and jailed a third time, he came before a judge who gave him a suspended sentence. King then left town, having failed to overturn segregation in the city.

The fate of the Albany Movement brought into the open already brewing complaints by activists about King. Many younger militants, while appreciating his enormous contributions to the cause, were irritated by his style as a preacher. King, they sneered, was "de Lawd."[37] Others said he should risk jail more often, that he made key tactical errors, that SCLC was disorganized. Local black people in Albany and other places some-

35. Dittmer, *Local People*, 105–15.
36. Ibid.; Todd Gitlin, *The Sixties: Years of Hope, Days of Rage* (New York, 1987), 141.
37. David Garrow, *Bearing the Cross: Martin Luther King, Jr., and the Southern Christian Leadership Conference* (New York, 1986), 173–230; Branch, *Parting the Waters*, 550–57, 631.

times grumbled that he was above all a media star who swept into their communities, inflamed local whites, gained only token concessions (if any), and then departed, leaving them to face the angry retribution of white society.

The struggle at places like Albany accentuated other internal divisions as well. Some of these were organizational, pitting the SCLC and the NAACP against each other. King, for instance, expected the NAACP's Legal Defense and Education Fund to cover his mounting expenses. Thurgood Marshall, who headed the fund, bitterly complained, "With Martin Luther King's group, all he did was to dump all his legal work on us, including the bills."[38] Both the SCLC and the NAACP, moreover, clashed with CORE and SNCC. As in the past the NAACP mainly attracted older, middle-class blacks who believed in the efficacy of litigation. But legal action took time, and many younger civil rights workers would not wait. Committed to direct action, they pushed ahead, sometimes impetuously, without listening to their elders. Often they chose to go to jail rather than pay fines, whereupon they, too, turned to the NAACP. Roy Wilkins, head of the NAACP, complained that SNCC workers in Albany "don't take orders from anybody. They operate in a kind of vacuum: parade, protest, sit-in. . . . When the headlines are gone, the issues still have to be settled in court."[39]

WHAT WOULD KENNEDY and his brother, in the key post of Attorney General, do about the civil rights revolution in America?

Then and later they maintained that they did a good deal to help it advance along peaceful lines. In 1961 the administration created a Committee on Equal Employment Opportunities. Headed by Vice-President Johnson, it was active in probing discriminatory practices. The administration also moved to hire more blacks in the federal government: in January 1961 only ten of 950 attorneys in the Justice Department and only fifteen of 3,660 foreign service officers were black.[40] It nominated five blacks to federal judgeships. One was Marshall, who was named to the Court of Appeals. Responding to the freedom rides, it pressured the ICC to issue its ruling against segregated facilities in interstate travel. And it took special interest in voting rights efforts—these were less likely than demonstrations to provoke violence—in the South. By May 1963 the

38. *New York Times*, Jan. 31, 1993.
39. Branch, *Parting the Waters*, 557.
40. Weisbrot, *Freedom Bound*, 49.

Justice Department had become involved in voting rights struggles in 145 southern counties. This was nearly a 500 percent increase over the thirty counties so affected when the Eisenhower administration left office in 1961.[41]

For the most part, however, John and Robert Kennedy moved cautiously concerning civil rights, especially in 1961–62. Their caution rested first on political considerations. Despite the rising tide of protest, civil rights at that time still did not command great public attention or passionate popular support. JFK, a careful listener to the public pulse, could see no political gain in pressing for action, especially from a Congress that was certain to be recalcitrant. If he pushed too hard for civil rights, he risked losing southern support he hoped to get in the 1962 and 1964 elections. Kennedy worried especially about southerners in Congress, notably powers like Senator James Eastland of Mississippi, who headed the important Judiciary Committee. Accommodating Eastland, Kennedy nominated four ardently segregationist men to federal district judgeships in the Deep South. One of these men, William Harold Cox, once described black people in his courtroom as "niggers" and compared them to chimpanzees.[42]

Political concerns also led Kennedy to back off from campaign promises. Although the Democratic party platform in 1960 had indicated support for a civil rights bill, Kennedy refused to introduce one in 1961 or 1962. Upset, NAACP lobbyist Clarence Mitchell observed that the "New Frontier looks like a dude ranch with Senator Eastland as the general manager."[43] Kennedy also reneged on a campaign promise to issue an executive order banning racial discrimination in federally supported housing. Such an order, he had proclaimed, required no congressional action, just a "stroke of the pen." As 1961 and 1962 elapsed without such an order, disgruntled activists raised an "ink for Jack" campaign and sent thousands of fountain pens to the White House. Their campaign did no good. Kennedy waited until after the 1962 elections to issue the order, which he circumscribed carefully. It had little effect.[44]

Personal predilections reinforced the caution of the Kennedys. Concerning civil rights, as concerning other domestic issues, the President

---

41. William Chafe, *The Unfinished Journey: America Since World War II* (New York, 1991), 208. Also Brauer, *JFK and the Second Reconstruction*, 311–20; Parmet, *JFK*, 260–63; and Sundquist, *Politics and Policy*, 254–65.
42. Reeves, *President Kennedy*, 249.
43. Morris, *Time of Passion*, 54.
44. Branch, *Parting the Waters*, 587; Weisbrot, *Freedom Bound*, 53–54.

and the Attorney General remained cool and detached. While they believed abstractly in the goal of better civil rights, they felt no passionate attachment to the cause. The President's attitude became clear during his first week in office when black African diplomats complained that restaurants on roads to Washington were refusing to serve them. "Can't you tell them not to do it?" he asked his chief of protocol, Angier Biddle Duke. Duke tried to explain his efforts to educate the managers. But Kennedy interrupted. "That's not what I'm talking about. Can't you tell those African ambassadors not to drive on Route 40? It's a hell of a road. . . . Tell these ambassadors I wouldn't think of driving from New York to Washington. Tell them to fly!"[45]

President Kennedy worried especially that racial unrest in the United States would soil the nation's image abroad and sabotage foreign policy goals that he really cared about. While the freedom rides were starting, he was focusing intently not on race relations but on a forthcoming summit meeting with Khrushchev in Vienna. After the riders were arrested and beaten, he was worried and angry. "Tell [the riders] to call it off," he told his civil rights aide, Harris Wofford. "Stop them!" When the riders persisted, the Kennedys publicly called for a "cooling-off" period. Robert exploded to Wofford that the blacks totally failed to appreciate the need for national unity on the eve of the summit. Farmer replied that blacks "have been cooling off for 150 years. If we cool off any more, we'll be in a deep freeze."[46]

The Kennedys faced special dilemmas because of the role of FBI chief Hoover, as ever a consummate and well connected bureaucrat. Hoover had developed an overpowering hatred of Martin Luther King, whom he considered a "'tom cat' with obsessive degenerate sexual urges." He was further convinced that one of King's advisers, a New York lawyer named Stanley Levison, was a Communist. Worried about such rumors, Robert sent aides to urge King to sever his relations with Levison. (King did not.) In early 1962 Robert Kennedy authorized the FBI to tap and bug Levison's office and to tap his home telephone. In October he went further, giving Hoover the go-ahead to tap King's telephones in Atlanta and New York.[47]

Although the taps remained on King's phones for the remainder of

45. Weisbrot, *Freedom Bound*, 54.
46. Farmer, *Lay Bare the Heart*, 206.
47. David Garrow, *The FBI and Martin Luther King, Jr.* (New York, 1981); Garrow, *Bearing the Cross*, 312; Reeves, *President Kennedy*, 359–61; Gitlin, *Sixties*, 140–43.

Kennedy's presidency (and beyond), they revealed little of substance. King, they showed, liked parties and bawdy jokes and apparently engaged in a good deal of extramarital sexual activity. These discoveries, had they been publicly disclosed, would have damaged King's standing. No evidence, however, proved that Levison had associations with Communists after he had become close to King in 1956. The Kennedys nonetheless continued to sanction Hoover's obsessive and voyeuristic efforts.[48]

Why they did so remains debated. But it is obvious that they feared to challenge Hoover, who had powerful contacts on Capitol Hill and who was spreading rumors about King throughout Washington. Hoover, moreover, knew too much about President Kennedy's own reckless and irresponsible sex life. In March 1962 he apparently warned Kennedy that Judith Campbell, with whom Kennedy had been sleeping since early 1960 (and whose seventy-odd calls to the White House since January 1961 had been logged and made known to the FBI), was also the mistress of the Mafia gangster Sam Giancana. Giancana, in turn, was working with the CIA on plots to assassinate Castro. Documentation of such a network badly compromised chances of prosecuting Giancana and associated gangsters. (Robert, as Attorney General, nonetheless proceeded to do so.) President Kennedy had also exposed himself to blackmail and disgrace. Quickly he broke off relations with Campbell; so far as is known he had his last White House telephone conversation with her that afternoon. With Hoover in command of such damning information it could not have been easy for Kennedy to turn down his requests for taps on King.[49]

For all these reasons the Kennedys continued to give Hoover and the FBI wide leeway in handling racial confrontations in the South. This harmed the movement, for Hoover not only loathed King; he also feared and hated civil rights activists. Still obsessive about Communism, he was sure that Reds dominated the civil rights movement, and he amassed huge dossiers on left-wing sympathizers such as the composer Leonard Bernstein and many others.[50] The FBI hired few black agents and had none working on civil rights matters in the South. It offered movement people no protection from white violence and sometimes (as at Bir-

48. Garrow, *Bearing The Cross*, 371–82; Branch, *Parting the Waters*, 566–69, 583–86, 850–62, 903-8.
49. Beschloss, *Crisis Years*, 141–43; Reeves, *President Kennedy*, 240–41, 319–21. Campbell had visited the White House some twenty times since January 1961. Kennedy was also having sex with Marilyn Monroe, whom he saw in California two days after breaking off White House phone contact with Campbell.
50. *New York Times*, July 29, 1994.

mingham) knowingly condoned it. Nothing enraged civil rights workers more than the failure of the Kennedy administration, about which activists had once had expectations, to use federal force to shield them from attacks.

Instead, the Kennedy brothers relied on deal-making with southern politicians. Robert Kennedy spent many hours on the phone reasoning with segregationist officials such as Barnett and Eastland of Mississippi, who finally agreed that freedom riders in Jackson would be arrested peacefully. In defending this approach the administration advanced constitutional arguments, notably its exposition of "federalism." As enunciated by Assistant Attorney General Burke Marshall, Kennedy-style federalism asserted that it was the responsibility of local authorities, not the national government, to preserve order and to protect citizens against unlawful conduct. Only when local officials completely lost control should the federal government consider responding with force of its own. "We do not have a national police force," Marshall explained. "There is no substitute under the federal system for the failure of local law enforcement responsibility. There is simply a vacuum, which can be filled only rarely, with extraordinary difficulty, and in totally unsatisfactory fashion."[51]

Violent confrontation at the University of Mississippi in September 1962 revealed the dangers of this approach. This turmoil followed the efforts of James Meredith, backed by the federal courts, to enroll as the first black student at the university. Barnett, however, drew on long-discredited claims for states' rights to oppose Meredith's admission. He also whipped up a racist frenzy among students and citizens of the state. "No school will be integrated in Mississippi while I am your governor," he declared. He demanded the resignation of any state official "who is not prepared to suffer imprisonment for this righteous cause. . . . We will not drink from the cup of genocide."[52]

The Kennedys, as in the past, hoped to defuse the possibility of violence by negotiating secretly with Eastland and Barnett.[53] By the eve of Meredith's arrival they thought they had struck a deal. Barnett, they believed, would keep order on the campus. Federal presence could therefore be limited to around 500 marshals. The army would remain on call in

51. Weisbrot, *Freedom Bound*, 63; Dittmer, *Local People*, 93–94; Arthur Schlesinger, Jr., *Robert F. Kennedy and His Times* (Boston, 1978), 299–302.

52. Dittmer, *Local People*, 139–42; Weisbrot, *Freedom Bound*, 66–68; Blum, *Years of Discord*, 73–74.

53. Victor Navasky, *Kennedy Justice* (New York, 1977), 165–69, 178–81, 185–92, 231–36; Brauer, *JFK and the Second Reconstruction*, 180–204.

Memphis, sixty-five miles away. But by 7:30 on the evening of September 30, the day before Meredith was due to enroll, a hostile crowd (peaking at around 3,000 later that night) of students and outsiders gathered on the campus and began to throw bricks and Molotov cocktails at the marshals. Eight were injured, whereupon the marshals retaliated with tear gas. The Mississippi Highway Patrol, which was supposed to curb the crowd, instead withdrew; Barnett had broken his word. The crowd became a mob. Gunshots rang out in the dark, wounding marshals and bystanders. The deal between Kennedy and Barnett had ended in a riot.

By 10:00 p.m. the badly outnumbered marshals were besieged, and Robert Kennedy sent word to Memphis to bring in the first of 5,000 troops. But a series of snafus fouled up the intervention, and the men did not arrive until 2:15 A.M., nearly seven hours after the trouble had started. By that time the marshals had no more tear gas, and two bystanders had been killed and 160 wounded, twenty-eight by gunshots. The troops then restored order, and Meredith was admitted. He endured the year at the university and graduated (protected the while by federal guards) in 1963.

The rioting at "Ole Miss" did not make much difference at the time in the daily lives of the masses of black people in the United States. Meredith was an embattled, courageous token. Nor did the confrontation do much to change the strategy of the Kennedy brothers. As before, they clung to the illusion that the national administration could keep its distance. But it was increasingly clear that deal-making and "federalism" were weak reeds upon which to rely. How long could the federal government depend on others to maintain the peace?

NOT VERY LONG, for Martin Luther King determined in 1963 to force the dismantling of Jim Crow. Preparing thoroughly, he resolved to stage massive demonstrations in Birmingham. This was known as perhaps the most systematically segregated city in the South. Fifty or more racially inspired bombings of black homes and facilities had poisoned postwar race relations. Blacks held only menial jobs, even in the city's booming steel industry. Lunch counters and all public facilities were segregated. There were drinking fountains for whites only. The city even closed its parks and playgrounds rather than submit to federal orders to integrate them. It barred performances of the Metropolitan Opera because the company refused to appear before segregated audiences. Public safety commissioner Connor and his men regularly terrorized black people in the city. This was one of the main reasons why King chose it for his major effort.

Connor, he expected, would overreact and draw national attention to the civil rights movement. [54]

When King and his coalition of civil rights workers launched their boycotts, sit-ins, and demonstrations in April 1963, Connor and other officials tried at first to act with restraint. King was arrested for violating a state court order barring demonstrations and spent a week in prison, where he penned "Letter from Birmingham Jail," a widely read summation of his commitment to racial justice and non-violence. The demonstrations continued, but city authorities arrested hundreds of protestors and threatened to deplete King's available volunteers. At this point King sent out some 1,000 children from his church headquarters on a demonstration march into the downtown. Connor's forces rounded up more than 900 of them, who ended up in jail. The next day Connor ordered a new group of children not to leave the church. When some of them came out, Connor and his forces lost their heads. Firemen turned on high-pressure hoses, water from which knocked demonstrators to the pavement and cracked them against the side of buildings. Some lay bleeding and unconscious. Policemen turned on marchers and beat them with nightsticks. Other police held attack dogs on long leashes and seemed to revel in the sight of the dogs snapping at and biting the demonstrators as they fell back from the onslaught. [55]

The violence exhilarated Connor, who ultimately threw more than than 2,000 children in jail. When one of his officers held back a group of white people, he called to him, "Let those people come to the corner, sergeant. I want 'em to see the dogs work. Look at those niggers run." A few days later, with demonstrations continuing, a blast of water hit the Reverend Fred Shuttlesworth, a top King aide, slammed him into the wall of a church, and left him unconscious. When an ambulance came to take him away, Connor exulted, "I wish they'd carried him away in a hearse." [56]

Connor's activities were more than the demonstrators could tolerate. Some of them reacted by throwing stones and bottles at the police. One

---

54. *Newsweek*, Sept. 30, 1963, pp. 20–24; Weisbrot, *Freedom Bound*, 68–72; Blum, *Years of Discord*, 103-9.
55. Garrow, *Bearing the Cross*, 231–86; Raines, *My Soul Is Rested*, 139–86; Branch, *Parting the Waters*, 673–845; Aldon Morris, *The Origins of the Civil Rights Movement: Black Communities Organizing for Change* (New York, 1984), 229–74; David Lewis, *King: A Biography* (Urbana, 1970), 171–209; Adam Fairclough, *Martin Luther King, Jr.* (Athens, Ga., 1995), 71–82.
56. *New York Times*, May 8, 1963; *Newsweek*, May 10, 1963, p. 19.

waved a knife at an officer. This was the first time that a significant number of black people had broken with the non-violent mandate. On the other side the violence became much worse. Opponents of the protests bombed the Birmingham home of King's brother. Another bomb exploded in a Birmingham motel where King was thought to be staying. These bombings sent blacks to the streets in a spasm of rock-throwing. Police retaliated by beating blacks at random. As many observers recognized at the time, non-violence was losing its power as an energizing ideology. A new, more bloody phase of the civil rights movement had begun.

The Birmingham struggle was pivotal in other respects. It was the first protracted demonstration to be carried live and nationwide on television. More than any event to that time, it forced Americans to sit up and take notice. Many who saw the brutality of Connor and his forces, especially to women and children, began to speak out against racial discrimination, to write outraged letters to the editor, and to put pressure on their representatives in Congress. Larger numbers than ever before went south to participate in a wave of new demonstrations and boycotts. Birmingham did much to awaken hitherto passive people in the North.

Black people, too, were aroused by these events. Thanks to Connor's overreaction, white moderates in Birmingham recognized that they had to make some concessions. Settling, they promised to desegregate public eating facilities and to hire black salespeople. Other galling Jim Crow practices survived, however, and blacks emerged from the struggle feeling angrier than before. James Baldwin, publishing *The Fire Next Time* earlier in the year, had already concluded that desegregation would make little difference in a systematically racist society. "Do I really *want* to be integrated into a burning house?" he asked. [57] Militant civil rights workers in the South, most of them still loyal to CORE and SNCC, grew increasingly critical of King's adherence to non-violence, and they renewed protests throughout the nation. It was later estimated that more than 100,000 people took part in demonstrations over the next seven months. At least 15,000 were arrested. [58]

Kennedy, too, moved off the center. The struggle at Birmingham upset him for several reasons. He was outraged by the brutality, noting that he felt "sick" when he saw a picture of a police dog lunging at a Negro woman. He worried about the wide publicity, especially the television coverage. This had flashed about the world and damaged America's im-

57. James Baldwin, *The Fire Next Time* (New York, 1963), 127.
58. Weisbrot, *Freedom Bound*, 72; Gitlin, *Sixties*, 129; Fairclough, *King*, 71–82.

age. How could the United States claim to lead a "Free World" when it trampled on the rights of its own people? Kennedy also feared new waves of violence if he did not do something. He worried above all that he—and the government—might lose control of the dynamics of protest. Kennedy told people that he wanted to "lead," not be "swamped" by what was happening.[59]

For all these reasons Kennedy gave aides the go-ahead to prepare a civil rights bill. When Governor George Wallace of Alabama, a demagogic segregationist who had taken office earlier that year, then emulated Barnett by trying to bar two black students from the state university in June, Kennedy went on the air to announce his support of the legislation.[60] In doing so he brought an unaccustomed passion to his delivery. "The heart of the question," he said,

> is whether all Americans are to be afforded equal rights and equal opportunities. . . . If an American, because his skin is dark, cannot eat lunch in a restaurant open to the public, if he cannot send his children to the best public school available, if he cannot vote for the public officials who represent him, if, in short, he cannot enjoy the full and free life which all of us want, then who among us would be content to have the color of his skin changed and stand in his place?[61]

Kennedy's engagement marked an important turning point in the history of the civil rights movement.[62] But it was blighted that night in Mississippi. Among the many Americans who learned of his message was Medgar Evers, an activist NAACP field secretary who had devoted much of his life to civil rights activity in Mississippi. He stayed late at an NAACP meeting in Jackson, returning to his wife and three children shortly after midnight. As he climbed out of his car, a sniper shot him in the back with a bullet from a high-powered rifle. Evers staggered toward the kitchen door where his family was waiting for him and collapsed in a pool of blood. He died en route to the hospital.[63]

59. Richard Polenberg, *One Nation Divisible: Class, Race, and Ethnicity in the United States Since 1938* (New York, 1980), 188.
60. For Wallace, see Marshall Frady, *Wallace* (New York, 1970); C. Vann Woodward, "Wallace Redeemed?," *New York Review of Books*, Oct. 20, 1994, pp. 49–52.
61. *New York Times*, June 12, 1963; Brauer, "JFK," 125.
62. Brauer, *John F. Kennedy*, 204–5.
63. Dittmer, *Local People*, 165–67. Authorities arrested Byron De La Beckwith, a Citizens' Council zealot from Greenwood, on charges of murdering Evers. Beckwith avoided conviction in two trials in 1964 when white juries deadlocked. A

The martyrdom of Evers threatened to destroy Kennedy's hopes for peaceful, legislative solutions to racial conflict. In Jackson a riot was narrowly averted. Many activists, moreover, rejected Kennedy's bill as too little and too late. The measure at the time focused on curbing racial discrimination in public accommodations, a major goal of the movement. But it was crafted cautiously so as to secure the backing of congressional moderates, especially Republicans, without whose votes the bill was doomed. It authorized the Justice Department to litigate in support of non-discriminatory public accommodations only if individuals were willing to initiate the suits. Its weak voting rights section excluded elections at the state and local levels. Its sections concerning schools dealt only with de jure segregation, thus ignoring widespread de facto segregation in the North. The bill offered no answers to the problems of police brutality and of racial discrimination in employment.[64]

Moderate leaders of the civil rights movement nonetheless were encouraged by Kennedy's support—at last—of a civil rights bill. Perhaps it could grow some teeth. Led by A. Philip Randolph and Bayard Rustin, longtime activists, they resolved to stage a March on Washington for Jobs and Freedom to exert pressure on behalf of legislation and jobs for black people. As originally designed the march, which was set for August 28, would include a prolonged sit-in of thousands of demonstrators at the Capitol until Congress enacted a satisfactory law.

A demonstration such as this greatly alarmed Kennedy and his aides, who labored hard to tone down the plans. Their efforts brought results, convincing King, Roy Wilkins of the NAACP, and Urban League head Whitney Young to agree to changes. By August these advocates, supported by many white liberals, labor union officials, and church leaders, had managed to craft an agreement that would limit the march to one day. Participants would be allowed to walk from the Washington Monument to the Lincoln Memorial, where speech-making would close the event. It was further understood that there would be no sit-in on Capitol Hill and that organizers would do their best to have substantial numbers of whites at the rally. Marchers were to dress in respectable clothing. Washington area liquor stores would be closed on the day of the march, a provision that rested on the assumption that blacks would otherwise get intoxicated and rowdy. Although many of these provisions offended

---

third trial thirty years later found him guilty of murder, and Beckwith, then 73 years old, was sentenced to life in prison.

64. Sundquist, *Politics and Policy*, 263–67.

SNCC leaders, including their chairman John Lewis, they agreed to take part, in the hope that the march would give them a chance to speak their views.[65]

Administration pressure to moderate the protest continued right up to the day of the march itself on August 28.[66] When Kennedy aides and other speakers saw a draft of a fiery speech that Lewis intended to make, they pressed him to soften it. At the last minute other black leaders, feeling the pressure, got Lewis to tone it down a little. Kennedy aides stood prepared to disconnect the public address system in case things went awry. Malcolm X later observed, "There wasn't a simple logistics aspect uncontrolled," and he branded the march the "Farce on Washington."[67]

The vast majority of the hopeful and non-violent throng, however, were unaware of the angry negotiations that were taking place near the podium. It was indeed a large crowd, estimated at around 250,000 people—the biggest to that time for a political assembly in the United States. Of this number, an estimated 50,000 were white. Among the marchers were many celebrities and performers, including Joan Baez, Josh White, Odetta, Bob Dylan, and Peter, Paul, and Mary. Marian Anderson and Mahalia Jackson sang movingly during the official program at the Lincoln Memorial. But it was King who gave the most memorable speech. Finishing his prepared remarks, he seemed ready to sit down, when Mahalia Jackson called out from behind him, "Tell them about your dream, Martin! Tell them about the dream!" King obliged, setting forth his dream (which he had told on earlier occasions) in the rolling cadences that made him such a powerful orator:

> I have a dream that one day this nation will rise up and live out the true meaning of its creed: "We hold these truths to be self-evident—that all men are created equal."
>
> I have a dream that one day on the red hills of Georgia the sons of former slaves and the sons of former slaveowners will be able to sit down together at the table of brotherhood.
>
> I have a dream that one day even the state of Mississippi, a desert state sweltering with the people's injustice, sweltering with the heat of oppression, will be transformed into an oasis of freedom and justice.

65. Chafe, Unfinished Journey, 310.
66. Only a little earlier William E. B. Du Bois, long a black activist, writer, and historian, died in self-exile in Ghana. He was ninety-five. Word of his death was passed through the crowd.
67. James Forman, The Making of Black Revolutionaries: A Personal Account (New York, 1972), 377–86; Reeves, President Kennedy, 359.

> I have a dream that my four little children will one day live in a nation where they will not be judged by the color of their skin but by the content of their character. . . .

With many in the crowd in tears, King closed with a famous peroration:

> When we let freedom ring, when we let it ring from every village and every hamlet, from every state and every city, we will be able to speed up that day when all God's children, black men and white men, Jews and Gentiles, Protestants and Catholics, will be able to join hands and sing in the words of that old Negro spiritual, "Free at last! Free at last! Thank God almighty, we are free at last!"[68]

Thanks in part to King's speech, the March on Washington was celebrated by liberals at the time as a tremendous outpouring of egalitarian, interracial, and non-violent spirit. That it was. But Lewis and other activists could not forget how they had been pressured into accepting a one-day event. And black people throughout the nation, however moved they may have been by the activity, gained nothing substantial from it. As before they confronted galling daily reminders of their second-class status.

The march also failed to change opinions on Capitol Hill. Hubert Humphrey, a leading liberal, concluded ruefully that the march had not affected a single vote on the slow-moving civil rights bill. Joseph Rauh, a leading liberal lobbyist, added later, "The March was a beautiful expression of all that's best in America. But I would find it unreal to suggest that it had anything to do with passing the civil rights bill, because three months later, when Kennedy was killed, it was absolutely bogged down."[69]

Rauh was right, for the civil rights measure crept slowly through the congressional process in the next few months. Renewed violence, meanwhile, stained the South; in September a bomb blew up in a Birmingham church, killing four black little girls, and nearly sparking a riot.[70] By the end of October a weak new section was added to the bill that provided for an Equal Employment Opportunity Commission, which was to have investigatory powers. But the bill was snarled in the House and had not escaped Congressman Smith's hostile Rules Committee in late November. Although it was expected to pass the House, it was certain to encoun-

---

68. *New York Times*, Aug. 29, 1963.
69. Godfrey Hodgson, *America in Our Time* (Garden City, N.Y., 1976), 160.
70. *Newsweek*, Sept. 30, 1963, pp. 20–24. Two more blacks were killed in the next few days.

ter a filibuster in the Senate. Prospects for enactment of the bill seemed remote indeed, and the soaring rhetoric of Martin Luther King on August 28 seemed all but forgotten on the Hill.

The deadlock delaying the bill served as an apt symbol of Kennedy's larger record in the field of domestic policy between 1961 and late 1963. Indeed, his prospects in Congress (where Democrats had lost five seats in the House in 1962) seemed no better in 1963 than they had been earlier. On November 12, 1963, the *New York Times* noted, "Rarely has there been such a pervasive attitude of discouragement around Capitol Hill and such a feeling of helplessness to deal with it. This has been one of the least productive sessions of Congress within the memory of most of its members." This was a glum but accurate description of the prospects for domestic change at the time. Kennedy had aroused liberal expectations but had failed to overcome the long-entrenched power of the conservative coalition in Congress. New frontiers still stood in the distance.

17

JFK and the World

In September 1960, Kennedy gave one of his most anti-Communist campaign speeches, in Salt Lake City. It captured well the incendiary Cold War rhetoric of the era and summarized a widely held American view of the world. "The enemy," he said, "is the Communist system itself—implacable, insatiable, increasing in its drive for world domination. . . . This is not a struggle for supremacy of arms alone. It is also a struggle for supremacy between two conflicting ideologies: freedom under God versus ruthless, godless tyranny."[1]

Kennedy often spoke this way (though usually without the religious emphasis), describing a bipolar world of good versus evil. His warnings about the "missile gap" reinforced this Manichean perspective. His dramatic inaugural address, while containing conciliatory passages regarding negotiations, was best remembered for its oft-cited lines "We shall pay any price, bear any burden, meet any hardship, support any friend, oppose any foe to assure the survival and success of liberty." The oration reit-

1. Cited in Michael Beschloss, *The Crisis Years: Kennedy and Khrushchev, 1960–1963* (New York, 1991), 25. Also useful concerning JFK's foreign policies are Anna Kasten Nelson, "President Kennedy's National Security Policy: A Reconsideration," *Reviews in American History*, 19 (March 1991), 1–14; James Giglio, *The Presidency of John F. Kennedy* (Lawrence, 1991), chaps. 4, 8, 9; Stephen Ambrose, *Rise to Globalism: American Foreign Policy Since 1938* (New York, 1988), 181–200; Herbert Parmet, *JFK: The Presidency of John F. Kennedy* (New York, 1980), chaps. 6–8; and David Burner, *John F. Kennedy and a New Generation* (Boston, 1988), 72–94.

erated the grimness that he felt about the Cold War, as well as his determination to do whatever it took to stop the advance of Communism.

Some of Kennedy's alarmist statements reflected political calculations.
As a partisan presidential candidate he assailed the record of Eisenhower
even though he recognized that Ike, like most American political leaders,
was as much of a Cold Warrior as he was. In so doing Kennedy (like Ike)
missed a chance to talk some sense to the public. Indeed, JFK, who cared
above all about foreign policy, was better informed about it than he was
about many domestic affairs. He knew that the Soviet Union and the
People's Republic of China bitterly opposed each other, that neither
Communist power was ready or anxious for war, and that the restless
drives of nationalism and anti-colonialism in Asia and Africa posed perhaps greater threats to world stability than did international Communism.
To channel these drives and to tap the energy of idealistic American
volunteers he established a Peace Corps, which worked at promoting
economic development throughout the world.

His rhetoric notwithstanding, Kennedy also understood that there was
no missile gap. Secretary of Defense Robert McNamara candidly acknowledged as much to Congress early in 1961. Kennedy was realist
enough finally to understand that the United States could not and should
not try to remake the world. He hoped to pursue a slightly less ambitious
agenda: to contain Communism and to shape a balance of power more
favorable than before to the United States and its allies.[2]

To place JFK's statements in context, however, is not to argue that
foreign policies under his watch only echoed those of Eisenhower and
Truman. To the contrary, Kennedy's personal approach to foreign
affairs—combined with forces mostly beyond his control—helped in his
first two years in office to escalate tensions with the Soviet Union. These
represented the most frightening years of the Cold War.[3]

Three outside forces constrained Kennedy's freedom of action in foreign policy and further hardened the Cold War in the early 1960s. One
was the continuing power of what Eisenhower's farewell address had
labeled the military-industrial complex. Arms contractors and military
leaders, gladly reinforcing talk about a missile gap, stepped up their demands for ever-larger defense expenditures. They maintained especially
strong influence in Congress. Ike, a military officer for much of his life,

2. John Gaddis, *Strategies of Containment: A Critical Appraisal of Postwar American
   National Security Policy* (New York, 1982), 203–13.
3. Thomas Paterson, ed., *Kennedy's Quest for Victory: American Foreign Policy,
   1961–1963* (New York, 1989).

had managed to resist some of these demands. His successor had no such prestige or predilection.

The heated nature of anti-Communist public opinion in the United States was a second outside force. There was nothing new about this opinion, but the rise of Fidel Castro in Cuba and Khrushchev's truculence following the U-2 Affair had further inflamed it. So had rhetoric such as Kennedy's in Salt Lake City. Newspaper and magazine stories whipped up fears during Kennedy's first months in office. *Time* carried a major story in January arguing that the "underlying conflict between the West and communism" was erupting on three fronts, in Cuba, in Laos, and in the Congo. *Newsweek* followed with a special section on January 23. "Around the restive globe from Berlin to Laos," it began, "the Communist threat seethed, and nowhere more ominously than in Cuba." It closed by warning, "The greatest single problem that faces John Kennedy—and the key to most of his other problems—is how to meet the aggressive power of the Communist bloc."[4]

A third outside force was the provocative behavior of Khrushchev. On January 6 the Soviet leader gave a specially belligerent speech that was released two days before Kennedy's inauguration. Among other things it pledged the USSR to support "wars of national liberation." Experts on the Kremlin told Kennedy that Khrushchev had said similar things before, but the President reacted sharply, telling all his top aides to study the address with care. "Read, mark, learn, and inwardly digest it," he insisted.

On this occasion Kennedy overreacted, but he had ample reason to worry about his adversary. Khrushchev went out of his way to crow about Soviet achievements, such as the historic orbiting of the earth by Soviet cosmonaut Yuri Gagarin on April 12. He was rude to Kennedy when the two leaders met at Vienna in June, and he blustered about Soviet power during the next two years.[5] Why Khrushchev acted so provocatively is still unclear; the Soviets remained extraordinarily secretive. Perhaps he wanted to impress China with his ability to stand up to adversaries, perhaps he felt pressure from military leaders at home, perhaps he considered the youthful Kennedy to be weak. In any event Soviet behavior between 1961 and 1963 seemed unusually truculent. It induced in Kennedy and his advisers a profound unease and reinforced a toughness of their own.[6]

4. Godfrey Hodgson, *America in Our Time* (Garden City, N.Y., 1976), 8; *Newsweek*, Jan. 23, 1961.
5. Giglio, *Presidency of JFK*, 74–78.
6. Beschloss, *Crisis Years*, 60–61; Gaddis, *Strategies of Containment*, 206–8.

No President facing such outside circumstances would have found it easy to negotiate thoughtfully with the Soviet Union. Still, Kennedy brought with him assumptions and attitudes that further increased Cold War tensions. One was his belief in a defense policy of "flexible response," as it came to be called. Like many Americans, Kennedy had long deplored what he felt was the Eisenhower administration's overreliance on nuclear weapons. These, he said repeatedly, were of little use in regional conflicts. The United States must build up more conventional forces so that it could respond flexibly to circumstances. As he put it in July 1961, "We intend to have a wider choice than humiliation or all-out war."[7]

In calling for increased spending on conventional weapons Kennedy exhibited few of the fiscal concerns that had motivated Eisenhower. This was one of the major differences between the foreign policies of Democratic liberals and Republican conservatives from 1953 on. Instead, JFK listened to Establishment figures such as Paul Nitze, who chaired a pre-inaugural task force on national security. Nitze, the guiding force behind the hawkish NSC-68 in 1950, again argued that the United States could easily afford, and badly needed, increased defense spending. Economic advisers, including Walter Heller, concurred and maintained that such expenditures would stimulate the economy. This was military Keynesianism. "Any stepping up of these [defense] programs that is deemed desirable for its own sake," a pre-inaugural task force report on the economy advised, "can only help rather than hinder the health of our economy in the period immediately ahead."[8]

Kennedy's support of higher defense spending depended heavily on a man who from the start became one of his most valued advisers, Defense Secretary McNamara. A highly articulate and extraordinarily well prepared spokesman, McNamara reassured members of Congress by laying out for them a reorganization of the way the Pentagon would henceforth conduct its business. His reorganization, which featured a Planning-Programming-Budgeting System, or PPBS, promised to reduce interservice bickering, collusive bidding, and waste. As it turned out, military-industrial connections were so well entrenched that not even McNamara and his much-ballyhooed aides, employing modern methods of manage-

7. Gaddis, *Strategies of Containment*, 203.
8. Paul Samuelson, a well-known MIT economist, was chief author of this report. Nitze became Assistant Secretary of Defense for International Security Affairs in the Kennedy administration. Ibid., 203–4.

ment and computerization, could radically change things at the Pentagon. Still, in 1961 he seemed a dazzlingly competent new face.[9]

Congress, welcoming the chance to promote business growth and employment, gladly approved the new defense policies. During the next three years defense expenditures rose 13 percent, from $47.4 billion in fiscal 1961 to $53.6 billion in 1964. Despite Kennedy's criticisms of overreliance on nuclear weapons, much of this increase went for additions to the nation's already capacious nuclear arsenal, including construction of ten additional Polaris submarines (for a total of twenty-nine) and of 400 more Minuteman missiles (for a total of 800).[10] The administration, having conceded that there was no missile gap, was taking no chances.

The increases were in fact not enormous in all respects. The number of military personnel grew only gradually, from 2.5 million in 1960 to 2.7 million in 1964. In part because the GNP rose rapidly during this period, the percentage of it spent on defense actually declined a little, from 9.1 percent of GNP in fiscal 1961 to 8.5 percent in fiscal 1964. Still, the growth was considerable by contrast to the last two years of the Eisenhower administration. Moreover, Kennedy obviously assigned very high priority to defense. He gave special attention to the development of "counter-insurgency" forces, such as the Green Berets. For a while he proudly displayed a green beret on his desk.[11]

Kennedy's supporters naturally hailed these changes as improving the nation's capacity to respond in new and flexible ways. This was true to a degree. Without the resources given to the Special Forces, as they were called, the administration might have moved more carefully in places like Vietnam. At the same time, however, JFK tended on occasion to gloat about American readiness. When he let it be known in unprecedented detail in October 1961 that the United States had a huge margin of nuclear superiority over the Soviet Union, he may have caused Khrushchev deep embarrassment at home. This may have heightened Soviet fears, always profound, about the West.[12] Whether Kennedy had this impact on the Soviets cannot be proved.

Still, there was no doubting that JFK stamped a personal, activist pattern on the foreign policies of his administration. While he relied on

9. Charles Morris, A Time of Passion: America, 1960–1980 (New York, 1984), 27–29.
10. Gaddis, Strategies of Containment, 218.
11. Richard Walton, Cold War and Counterrevolution: The Foreign Policy of John F. Kennedy (New York, 1972), 176.
12. Beschloss, Crisis Years, 702.

McNamara, his brother Robert, and a few others, he made it clear that he was in command. He had little use for for the "striped-pants boys" in the State Department—he thought they dithered and shuffled papers—and he came to despair at the Buddha-like inscrutability of Dean Rusk, his loyal but bland Secretary of State. Rusk, he told Theodore White, "never gives me anything to chew on, never puts it on the line. You never know what he is thinking."[13] Kennedy preferred decisive, tough-minded advisers, and he had little time for doubters. As Chester Bowles, Kennedy's increasingly disillusioned Undersecretary of State, complained at the time, the men in Kennedy's inner circle were "full of belligerence." They were "sort of looking for a chance to prove their muscle."[14]

Kennedy's passion for decisive action in foreign policy had varied sources. One may have been his tiny margin of victory in 1960: acting boldly might rally the patriotic and expand his base of political support. In this frame of mind he announced his backing in May 1961 of the so-called Apollo program, to make the United States the first to place a man on the moon. This effort cost some $25 to $35 billion dollars before Neil Armstrong and two others reached the moon in 1969, and it produced relatively little scientific knowledge. But, as JFK anticipated, it enjoyed considerable support from proud and patriotic American people.[15] Another source of Kennedy's boldness was constitutional: as commander-in-chief Kennedy enjoyed much more discretion than he did in domestic matters, and he was temperamentally inclined to use it. A third source was his continuing fear and uncertainty about Khrushchev, whose provocative behavior aroused his competitive instincts. Finally, Kennedy, like other postwar Presidents who had been tested by World War II, heeded what he thought were the lessons of history. To show indecisiveness, as Western countries had done with Hitler in the 1930s, was to encourage aggressive behavior. Only firm and unflinching direction could preserve the all-important "credibility" of the United States, defender of the "Free World."

The President's activism in foreign affairs had a special edge of toughness about it. To critics then and later it contained a potentially dangerous

13. Ibid., 356. For Rusk, see Thomas Schoenbaum, *Waging Peace and War: Dean Rusk in the Truman, Kennedy, and Johnson Years* (New York, 1988); and Warren Cohen, *Dean Rusk* (New York, 1980).
14. Paterson, *Kennedy's Quest*, 19.
15. Walter MacDougall, . . . *The Heavens and the Earth: A Political History of the Space Age* (New York, 1985); William O'Neill, *Coming Apart: An Informal History of the 1960s* (Chicago, 1971), 59.

machismo. Some people thought this machismo arose from his upbringing in an extraordinarily competitive family. Robert displayed it, too. Others attributed it to his need to prove himself, the youngest elected President in American history, as worthy of his office. Whatever the sources, Kennedy exhibited an intense desire to demonstrate his mettle. To face crisis and to prevail was to demonstrate one's strength and to assert one's manhood. As he prepared in May for his summit meeting with Khrushchev, he said, "I'll have to show him that we can be as tough as he is. . . . I'll have to sit down and let him see who he is dealing with."[16]

NOTHING EXPOSED THESE TENDENCIES more clearly than Kennedy's attempt to overthrow Fidel Castro. The scheme that he and his zealous advisers ultimately carried out, an invasion at the Bay of Pigs on Cuba's southern coast, took place on April 17, fewer than three months after inauguration day. The invasion was one of the most disastrous military ventures in modern American history.

Kennedy had uncomplicated motives for approving the attack. Like Eisenhower, who had cut off diplomatic relations with Cuba in January, he was angered by Castro's volatility, his virulent anti-American rhetoric, and his growing rapprochement with the Soviet Union. People close to Kennedy, notably his father and Senator George Smathers of Florida, urged him to get rid of Castro before the Soviet Union set up a virtual satellite off the Florida coast. Columnists and editorial writers further raised a din for action. Moreover, Ike had authorized the training of attackers. To Kennedy, activist by temperament, it was tempting indeed to deploy them.[17]

As advisers planned an assault on Cuba, a few government officials raised doubts. Among them were liberals such as Arthur Schlesinger, Jr., and Adlai Stevenson, whom Kennedy had named as United States Ambassador to the UN. Another doubter was Bowles, who wrote Rusk, "Our national interests are poorly served by a covert operation. . . . This . . . would be an act of war." Marine Corps Commandant David Shoup warned presciently that Cuba was a large island (800 miles long) that would be very hard to conquer. Schlesinger directly conveyed his

16. Paterson, *Kennedy's Quest*, 15.
17. Thomas Paterson, "Fixation with Cuba: The Bay of Pigs, Missile Crisis, and Covert War," in Paterson, ed., *Kennedy's Quest*, 123–55. For other accounts see Beschloss, *Crisis Years*, 29–30, 100–108; Giglio, *Presidency of JFK*, 48–63; and Thomas Reeves, *A Question of Character: A Life of John F. Kennedy* (New York, 1991), 256–76.

doubts to Kennedy. So did J. William Fulbright, chairman of the Senate Foreign Relations Committee, who insisted that an American invasion would violate the charter of the Organization of American States (OAS). "To give this activity even covert support," he said, "is of a piece with the hypocrisy and cynicism for which the United States is constantly denouncing the Soviet Union."[18] Cuba, Fulbright observed, was "a thorn in the flesh, but not a dagger in the heart."[19]

Most of Kennedy's high command, however, were in no mood to listen to doubters. They were energetic, self-assured, and anxious to prove themselves. Kennedy's free-wheeling administrative style, moreover, removed institutional checks that might have curbed impetuousity. McNamara approved of a plan of attack, as did Robert Kennedy. Although the Joint Chiefs had reservations, especially about the option of landing at the Bay of Pigs, they did not oppose the venture. The CIA led the enthusiasts. CIA director Allen Dulles and his number two man, Richard Bissell, were confident that their plans to assassinate Castro could be timed and coordinated with an invasion of exiles from Central America. The United States would lend naval and air support to the effort but would do so covertly. The exiles would land, whereupon anti-Castro rebels in Cuba—presumed to be chomping at the bit—would rear up and drive the dictator from office.

What Kennedy knew about the assassination plans remains unclear. Although critics assume that Judith Campbell served as a courier who kept him informed of the plots, no documentary evidence exists to connect him with planning or knowledge of such plots prior to the invasion. On the other hand, it is doubtful that the CIA would have dared to kill a head of state on its own. It was also no secret to presidential advisers, including Dulles and Bissell, that JFK wanted Castro dead. Nothing that the President did or said deterred them from urging strong action.[20]

The operation began with an air strike on April 15 by American planes based in Nicaragua. They were painted over to look as if they were Cuban aircraft stolen by the exiles. Meanwhile United States destroyers escorted

18. John Blum, *Years of Discord: American Politics and Society, 1961–1974* (New York, 1991), 39. The OAS had been formed, under United States auspices, in 1948 and was supposed to provide for consultation among the nations of the Western Hemisphere.
19. Paterson, "Fixation," 130–34.
20. Beschloss, *Crisis Years*, 137–39; Reeves, *Question of Character*, 256–59; Ambrose, *Rise to Globalism*, 184–85.

an invasion fleet, and United States navy jets accompanied United States bombers to within five miles of the landing site. According to the original plan, planes were to launch a second air strike at the time of the invasion, thereby providing all-important cover for the amphibious operation. American frogmen, disguised as Cubans, were to be the first ashore. Kennedy and the CIA imagined that ruses such as these would hide the fact of American involvement.

Most things that could have gone wrong did. The first air strike knocked out only a few of Castro's planes and tipped him off that an attack was imminent. Indeed, well-publicized activities of Cuban exiles both in Central America and in Florida had already made it apparent that Kennedy was about to do something. News stories were full of predictions that invasion lay ahead. Pierre Salinger, Kennedy's press secretary, later observed that the invasion was the "least covert military operation in history." He added, "The only information Castro didn't have . . . was the exact time and place of the invasion." Kennedy, Salinger noted, was upset by the lack of secrecy. "I can't believe what I'm reading," he complained. "All he [Castro] has to do is read our papers. It's all laid out for him."[21]

Despite these potentially disastrous developments, the President determined to go ahead. But he altered earlier plans by refusing to authorize a second air strike to accompany the invasion. He feared that some of the planes in such a strike might be hit, thereby exposing the involvement of the United States. So it was on April 17 that the invaders, a brigade 1,400 men strong, began landing at the Bay of Pigs. Only 135 of them were soldiers by profession. A radio station on the beach, overlooked by the CIA, reported the assault. Coral reefs, also unforeseen, sank some of the landing craft. A drop of parachutists did not land near enough to cut off a main road to the beach. Castro's fighter planes, armed with rockets, tore into the ships and landing craft. Within twenty-four hours, fifty-four Soviet-made tanks were on the field of action. Castro went to the area and personally took charge of a counter-attack.

Well before then it was obvious that the venture was doomed. The Bay of Pigs, as some military advisers had recognized, proved a poor choice of site, for it was a swampy region from which soldiers, once trapped, could not melt away or find refuge. Instead, the brigade was pinned down near the beach and badly exposed to enemy fire. Desperate, they called for air support from fighter planes on the aircraft carrier *Essex*, some ten miles offshore. But Kennedy again demurred, and the invaders soon surren-

21. Reeves, *Question of Character*, 262–75.

dered. In all, 114 members of the brigade lost their lives. A total of 1,189 exiles were taken prisoner.

Recrimination arose immediately and from all sources. The covertness fooled no one: Castro, Khrushchev, and other world leaders recognized from the start that the United States had planned and supported the operation, and they assailed the President. Many critics at home blamed him especially for refusing to provide air cover. General Lyman Lemnitzer, Chairman of the Joint Chiefs of Staff, later called this decision "absolutely reprehensible, almost criminal." Eisenhower was reported privately to have described the military operations as a "Profile in Timidity and Indecision." Later he met with Kennedy and dressed him down for not using the planes.[22] Many who emphasized the potential of air support remained convinced that it would have ensured the success of the landing, whereupon the indigenous anti-Castro population would have risen to destroy his government.

Advocates of this course of action reflected a faith in the potential of air power that animated a great deal of American thought in the postwar era: again and again it was assumed—often wrongly—that air power held the key to military success. With regard to the debacle at the Bay of Pigs they were right that failure to provide air support doomed whatever chance the invaders may have had to establish themselves at the site. The critics were also correct to observe that no President would have exposed *American* men (as opposed to Cuban exiles) to devastating enemy fire without supporting them with all he had. No wonder that Kennedy and his advisers felt guilty when the attack was over.

But it is clear that the invasion of Cuba suffered from many deeper flaws of overall strategy and conception. The Bay of Pigs as a site was ill chosen. Military leaders had reservations about this choice and other matters but in their eagerness to undertake the mission mostly kept them to themselves. The CIA brought a buccaneering, can-do commitment to the project without informing others of potential problems. Its hopes for assassinating Castro were pie in the sky. Fearing leaks, the CIA also did a poor job of coordinating the invasion with agents for the anti-Castro underground in Cuba. On this as on other occasions, military leaders as well as American intelligence services poorly served the White House.

Kennedy and his advisers badly underestimated the military capacity of Castro. When the first air strike failed to knock out his planes, Castro protected them from further assaults by dispersing them. It is unlikely, therefore, that a second strike would have given the invaders full control

22. Ibid., 269, 273; Beschloss, *Crisis Years,* 144–46.

of the air. If it had, the Cuban leader had other military assets, notably a 25,000-man army and a back-up militia force of 200,000. These could easily have overwhelmed the tiny brigade of 1,400. Later writers have concluded that it would have taken at least 10,000 men and open military commitment by the United States for any invasion to have had much of a chance of triumph. And "triumph" would likely have required long-term military occupation of the island—a down-the-road probability that Kennedy and his advisers did not think through.[23]

American planners failed finally to recognize the political support that Castro enjoyed at home. Having overthrown a hated dictatorship in 1959, he remained popular among many of his countrymen. Cubans in the Bay of Pigs area, where Castro had built schools and hospitals, were especially loyal to him. Many Cubans who disliked Castro nonetheless resented the assault from the overbearing Yankees to the North and rallied to his support. The CIA, anticipating a groundswell of anti-Castro rebellion following a landing, grossly misjudged the political situation on the island. (Dulles and Bissell also guessed wrong about JFK: if push came to shove following the landings, they thought, Kennedy would commit American forces to save the venture.) It was neither the first nor the last time that United States leaders in the postwar era overestimated the potential for American military strength or underestimated the power of nationalism and patriotic fervor overseas.

The legacy of the Bay of Pigs disaster was mixed. Kennedy, having been burned, recognized that a process of unreflective "group-think" had prevailed. He gradually took steps to develop a decision-making process that included more non-military and non-CIA advisers and that required more extensive debate in advance of action. Kennedy also sought to assist social and economic reform in Latin America. The Alliance for Progress, promised during the campaign, was created to help finance such reform. It never received much money, however, and, like other Kennedy-sponsored ventures in foreign aid, such as the Agency for International Development (AID), it tended increasingly to spend for military assistance rather than for social change. In Latin America, as elsewhere in the so-called Third World, the Kennedy administration sought mainly to contain Communism, not to advance social reform.[24]

23. Paterson, "Fixation"; Trumbull Higgins, *The Perfect Failure: Kennedy, Eisenhower, and the CIA at the Bay of Pigs* (New York, 1987); John Ranelegh, *The Agency: The Rise and Decline of the CIA* (New York, 1986), 381–82.
24. Alan Wolfe, *America Impasse: The Rise and Fall of the Politics of Growth* (New York, 1981), 187–91.

The debacle in Cuba also made Kennedy think twice about American military intervention in Laos, where Communists were thought to be on the verge of taking over. Three days after the Bay of Pigs invasion he told Nixon, "I don't see how we can make any move in Laos, which is thousands of miles away, if we don't make a move in Cuba, which is only ninety miles away." In September he told Sorensen, "Thank God the Bay of Pigs happened when it did. Otherwise we'd be in Laos by now—and that would be a hundred times worse." Robert Kennedy later mused, "I think we would have sent troops into Laos—large numbers of American troops in Laos—if it hadn't been for Cuba."[25] Instead of pursuing such a course, the Kennedy administration turned to negotiation. In 1962 a fourteen-nation conference worked out a settlement for the time being.

Dealing with the Russians in the months after the invasion, Kennedy was both patient and firm. At the Vienna summit in June, Khrushchev threatened again, as he had in 1959–60, to sign a separate peace treaty with East Germany. Such a treaty would have permitted East Germany to stop a highly worrisome outflow of refugees to West Germany. It would also have encouraged the East Germans (whom the United States did not recognize) to cut off Western access to Berlin. Kennedy, like Eisenhower, refused to budge or even to negotiate. "It will be a cold winter," he told Khrushchev. He then asked Congress for another large hike in defense spending, mobilized 120,000 reservists, and called for a massive fallout shelter program.

Khrushchev responded by ordering large increases in military spending at home and in August by constructing a wall separating the two Berlins and the two Germanys. This provocative decision sparked one of the most inflammatory moments of the Cold War. Hawks in the United States urged Kennedy to challenge the Soviets by stopping the building of the wall. American and Soviet tanks and soldiers confronted each other menacingly at the borders. Kennedy, however, did not overreact. He recognized that the USSR had the right to close off its zones, and he let the construction proceed. Sending in a token force of 1,500 troops through East Germany to West Berlin, he made it clear that the United States would stand by the beleaguered city. Khrushchev then dropped his demands for a separate treaty.[26] The Kennedy administration performed more steadily and professionally in its summer war of nerves over Berlin than it had over Cuba in April.

25. Reeves, *Question of Character*, 283.
26. Blum, *Years of Discord*, 45–47; Beschloss, *Crisis Years*, 350–52; Reeves, *Question of Character*, 302–9.

In general, however, the administration did not seem to learn very much from its experiences in 1961, especially as they concerned Cuba. Kennedy continued to be enamored of the Special Forces and of secret, CIA-led efforts to undermine unfriendly governments abroad. By 1962 Kennedy had relieved both Dulles and Bissell, but he approved a CIA-sponsored counter-insurgency program in Laos. This involved the recruitment of 36,000 Meo tribesmen (later called Hmong) as well as thousands of Thai "volunteers." The CIA directed guerrilla raids against both China and North Vietnam. Air-America, a CIA-owned airline, became involved in bombing raids in Laos. This secret war in Laos continued for years until exposed in the 1970s. By then it was costing the CIA alone some $20 to $30 million a year. [27]

Kennedy above all seemed obsessed by Castro. Operation Mongoose, a highly secret, CIA-coordinated program, was developed to damage Castro's regime. "My idea," Robert Kennedy said in November 1961, "is to stir things up on the island with espionage, sabotage, general disorder." Operation Mongoose tried all these things and more. Its agents sought to contaminate Cuban sugar exports and to detonate bombs in Cuban factories, and it sponsored paramilitary raids on the island. It is estimated that Mongoose developed at least thirty-three plans to assassinate Castro between November 1961 and Kennedy's death two years later. [28]

THE BAY OF PIGS AND BERLIN CONFRONTATIONS implanted profound insecurity in both Kennedy and Khrushchev. Both men soon behaved as if their personal manhood were at stake. A *mano a mano* emotionality imparted to Soviet-American relations in 1962 a volatility that did credit to neither man as a diplomatist and that provoked the most frightening military crisis in world history. [29]

Khrushchev proceeded to act in an especially confrontational manner, resuming atmospheric atomic tests in September 1961. (Kennedy did the same seven months later.) The Soviet leader turned especially to Cuba, which he thought Kennedy was again preparing to attack. Beginning in 1961 the Soviet Union sent increasing numbers of military personnel to the island, and in the summer of 1962 it started arming Cuba with missiles. These were not defensive weapons, however, but medium-range

27. Reeves, *Question of Character*, 284. In 1963 Kennedy instructed the CIA to unseat Dr. Cheddi Jagan, the leftist head of British Guinea.
28. Beschloss, *Crisis Years*, 5–6, 375–76.
29. Barton Bernstein, "Reconsidering Khrushchev's Gambit: Defending the Soviet Union and Cuba," *Diplomatic History*, 14 (Spring 1990), 231–39.

offensive missiles designed to give the Soviets more military potential—
and diplomatic muscle—in the worldwide Cold War. They had a range of
1,100 miles, easily sufficient to hit major population centers in the
United States.[30]

By mid-October of 1962 the Soviets had almost completed their work.
As later revelations disclosed, they were then about a week away from
making operational their launching sites near San Cristóbal. Later revelations also indicated that the Soviets had in readiness nine tactical missiles
with nuclear warheads. These had a range of around thirty miles and
could have wreaked havoc on American planes or invaders. Some 42,000
Soviet military personnel—twice as many as American intelligence imagined at the time—were then on the island. Their commander, not the
Cubans (or military brass in Moscow), had authority to fire the missiles.[31]

Fortunately for the United States, the CIA grew increasingly suspicious
of Russian-Cuban activity in 1962. In September Kennedy publicly and
privately warned Khrushchev not to put Soviet missiles on Cuban soil.
Khrushchev denied that he was doing any such thing, but on October 15
photographs from U-2 reconnaissance flights showed that missile sites
were well underway in Cuba. Although the photographs failed to spot
the launchers for the tactical missiles or the nuclear warheads for the
intermediate-range missiles, they revealed twenty-four launchers for the
intermediate-range missiles. The evidence was compelling. A "missile
crisis," the most frightening confrontation of the Cold War, now faced
the world.

The deliberations of Kennedy and an "Executive Committee" of the
National Security Council that he set up to deal with the confrontation
later evoked a great deal of praise for facilitating the President's cool and
courageous "crisis management." For the next thirteen days high-ranking
officials, chief among them McNamara, Rusk, Robert Kennedy, and
National Security Adviser McGeorge Bundy, debated options far into the
nights. Hoping to avoid the group-think that pervaded the Bay of Pigs

30. Among the many accounts of the crisis that followed are Beschloss, *Crisis Years*,
esp. 525–44; Giglio, *Presidency of JFK*, 189–216; Reeves, *Question of Character*,
364–86; Arthur Schlesinger, Jr., *Robert F. Kennedy and His Times* (Boston,
1978), 499–532; Robert Kennedy (ed. by Theodore Sorensen), *Thirteen Days: A
Memoir of the Missile Crisis* (New York, 1969); Graham Allison, *Essence of
Decision: Explaining the Cuban Missile Crisis* (Boston, 1971); Dino Brugioni,
Eyeball to Eyeball: The Inside Story of the Missile Crisis (New York, 1991); and
Ambrose, *Rise to Globalism*, 192–99.
31. *Washington Post*, Jan. 14, 1992; Robert McNamara, "One Minute to Dooms-
day," *New York Times*, Oct. 14, 1992.

debacle, the "Ex-Comm" members sought advice from a range of sources, including Cold Warriors such as Dean Acheson, conciliators such as Stevenson, and military and congressional leaders. They were thoroughly scared. Further reconnaissance flights made it clear that work on the sites was progressing rapidly. Time was short. If the Ex-Comm miscalculated, they risked national safety and much more. On October 20, just before the administration decided what to do, Kennedy called his wife and children back to Washington so that they could join him in an underground shelter if necessary.[32]

By then fairly well articulated though still shifting opinions dominated the anxious discussions. Stevenson, the most conciliatory adviser, called for the demilitarization of Cuba (including the United States base at Guantánamo) and for an American promise to remove Jupiter offensive missiles that it had emplaced in Turkey, a NATO ally. These, he said, were of no use save as a first strike and were as inflammatory to the Soviets as the Cuban missiles were to the United States. Other moderates pointed out that each side already had capacity via ICBMs to inflict fearsome damage on the other. Why should America risk a nuclear war to stop the building of sites in Cuba?

Other, more militant, advisers called for much tougher American responses of one kind or another. Fulbright, a hard-liner on this occasion, recommended an invasion of Cuba. Bundy, Acheson, Vice-President Johnson, and others demanded air strikes against the missile sites. This option seemed especially popular during Ex-Comm's early deliberations. Smashing the sites might knock out the threat quickly without exposing Americans (save the pilots) to serious danger. American invasion might follow if deemed necessary.

The President and his brother refused to go along with advice such as Stevenson's, which they unkindly branded as "soft" and appeasing. Part of their reason for doing so was their profound distrust of Khrushchev. They were angry because Khrushchev's emissaries had lied to them and were continuing to lie even as Moscow directed completion of the sites in Cuba. They were also highly sensitive to charges at the time by Republicans (congressional elections were imminent) that the administration had been lax in dealing with Cuba, America's "back yard." For political reasons, they believed, the President must under no circumstances appear weak. In addition, Kennedy refused to accept the argument that the new sites added little to Soviet power. Emplacement of the missiles would give

32. Reeves, *Question of Character*, 376–80.

Khrushchev greater leverage and political prestige elsewhere in the world, as in Berlin. This would damage American credibility, then as always a key concern of American leaders in the postwar era. As the Ex-Comm continued its debates, it became clear that the President was determined not to back down from what he considered the provocative and reckless behavior of the enemy. The missiles must go.

JFK and Robert, however, also came to oppose air strikes. Some members of Ex-Comm, including Rusk and Undersecretary of State George Ball, shrank from the very idea of such attacks. "A course of action where we strike without warning," Ball said on October 18, "is like Pearl Harbor. It's the kind of conduct that one might expect from the Soviet Union. It is not conduct that one expects from the United States." Rusk added, "The burden of carrying the mark of Cain on your brow for the rest of your lives is something we all have to bear." Robert Kennedy then agreed, "we've talked for fifteen years about a first strike, saying we'd never do that. We'd never do that against a small country. I think it's a hell of a burden to carry." To launch such a strike, he added, would make his brother "the Tojo of the 1960s."[33]

The analogy to Japanese militarism infuriated hawks such as Acheson, but Robert persisted, pointing out that strikes could kill many people, including Russians and Cuban civilians, on the ground. The Soviets might well retaliate, probably in Berlin. American military leaders, moreover, weighed in with powerful practical advice. No one could be sure, they pointed out, that strikes would knock out everything that the Soviets had in Cuba. If some of the missiles survived the strikes, the Soviets might fire them off at the United States. The United States would feel obliged to retaliate, possibly with nuclear weapons of its own, and perhaps with an invasion of Cuba.

For these reasons the administration decided at almost the last moment not to begin with air strikes. Instead, JFK settled on what he considered a middle ground: the United States Navy would enforce a "quarantine" against further shipment of Soviet military equipment to Cuba. Ships that defied the quarantine would be fired on if necessary. As a contingency air strikes could also be employed.[34] On October 22, a week after U-2 evidence had reached the White House, Kennedy so informed Khrushchev, both via diplomatic communication and via a dramatic, prime-time

33. Quotes from tapes released July 27, 1994, cited in *Boston Globe*, July 28, 1994. For the reference to Tojo, see Ambrose, *Rise to Globalism*, 194.
34. The United States in fact imposed a blockade but used the word "quarantine." According to international law, a blockade was an act of war.

televised speech an hour later that let the world in on some of the behind-the-scenes activities of high American officials the previous week. He further warned against Soviet retaliation. Any missile shot at places in the Western Hemisphere, at Berlin, or anywhere else, Kennedy said, would provoke a "fully retaliatory response" upon the Soviet Union. American allies in NATO, as well as countries in the Organization of American States, had been consulted in advance and uneasily supported Kennedy's forceful but frightening stand.

Unprecedented fear and tension gripped people throughout the world in the immediate aftermath of this announcement. Billy Graham, in Argentina, preached about "The End of the World." Soviet submarines were sighted in Caribbean waters. Soviet freighters, presumed to be bringing military equipment to Cuba, were approaching the island. So were other ships carrying Soviet goods. What if they defied the quarantine and were attacked by American naval vessels? The range of potential counter-moves by the Soviets, including aggressive action in Berlin or—a worst case—the dispatch of ICBMs against the United States, was almost too horrible to think about.

Ex-Comm members, like millions of other people, waited nervously to see what would happen the next morning, when the Soviet ships would have to decide what to do. The suspense seemed nearly unbearable. Then relief. At the last moment some Soviet freighters slowly turned about. Others, carrying no munitions, agreed to be stopped and searched on the high seas. Rusk nudged Bundy, "We're eyeball to eyeball, and the other fellow just blinked."[35] It was a wonderful moment for the tired and beleaguered advisers and for peace-seeking people throughout the world.

The crisis, however, was far from over. Construction on the sites continued; very soon the missiles might be operational. U-2 photos indicated that parts of Soviet bombers were being uncrated and readied for assembly on Cuban airfields. Kennedy insisted that the missiles be removed and the sites inspected. On October 26 Khrushchev appeared to agree. In return, he said, the United States must end the quarantine and promise not to invade the island. As American officials were considering on October 27 whether to accept this settlement, a second Soviet note arrived. It added the demand that the United States remove the Jupiter missiles in Turkey, an action that presumably required NATO approval. This escalation of Soviet demands confused and antagonized American leaders. When a Soviet surface-to-air missile shot down an American U-2 over Cuba the

35. Reeves, *Question of Character*, 376–86.

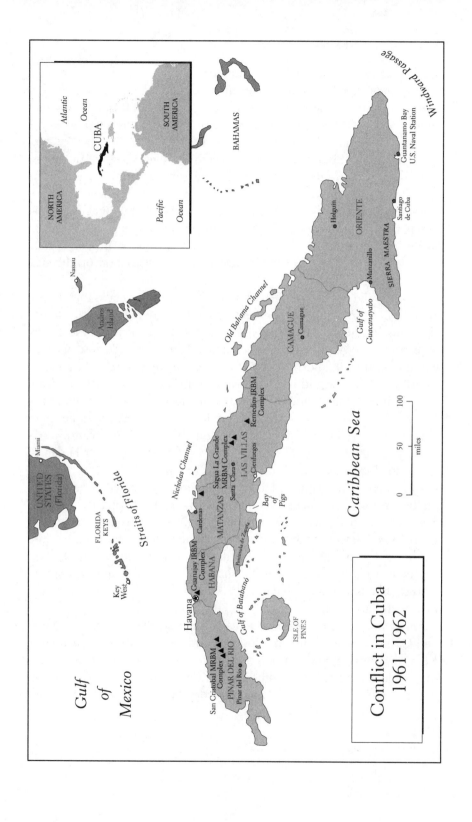

Gulf
of
Mexico

Caribbean Sea

Windward Passage

Conflict in Cuba
1961–1962

0   50   100
miles

same day, killing the pilot, the American Joint Chiefs of Staff reacted angrily and called for an immediate air strike on Cuba. A majority of Ex-Comm members supported their recommendation. In this terrifying moment it was eyeball-to-eyeball again before Kennedy himself ordered the planes to hold off for at least one more day.

At this point Robert Kennedy, drawing on suggestions from others, offered a way out of the impasse: accept the arrangement proposed in the first Soviet note and act as if the second note had never been received. President Kennedy liked the idea and told Khrushchev that he would accept the first proposal. Again employing open diplomacy, the President made his stance public. In private as well as in public he laid special emphasis on one aspect of the understanding: construction on the sites must stop immediately.[36]

Kennedy did not inform the public, however, of quiet talks that he also approved. As he was announcing his tough stance to the world, he authorized his brother to talk privately with Anatoly Dobrynin, the Soviet ambassador in Washington. Robert told Dobrynin that Moscow must commit itself by the next day to remove the missiles, in return for which the United States would later take out its missiles from Turkey and from Italy. Kennedy had planned to do so anyway, recognizing that submarine-launched Polaris missiles made the Jupiters obsolete. At the same time, JFK quietly developed a fall-back plan by which the United States, working through the UN, would support removal of its missiles from Turkey in return for Russian removal of its missiles in Cuba.

Khrushchev was so relieved to get the American note that he arose early on the morning of October 28 and personally dictated his acceptance of it. Kennedy was delighted. "I cut his balls off," he crowed privately. But Castro was outraged, both because Khrushchev had given in and because he himself had not been consulted. (He first heard the news over the radio.) Castro kicked a wall, shattered a mirror, and denounced Khrushchev as a "son of a bitch . . . a bastard . . . an asshole," and a man with "no cojones."[37]

Other interested parties were equally angry at the settlement. When anti-Castro Cuban exiles in Florida heard of it, they were irate that Kennedy would consider a promise not to invade the island. Kennedy, they said, had agreed to "another Bay of Pigs for us. . . . We are just like the Hungarians now." Later that day, when Kennedy met with

36. Beschloss, *Crisis Years*, 525–36.
37. Ibid., 542–44, 549.

American military leaders to thank them for their advice, he was staggered by their reaction. "We have been had," one said. SAC chief Curtis LeMay pounded the table. "It's the greatest defeat in our history, Mr. President. . . . *We should invade today.*" McNamara, who was present, recalled that Kennedy "was absolutely shocked. He was stuttering in reply."[38]

At that stage, however, there was little that opponents of the settlement could do: Kennedy and Khrushchev had spoken. Within the next few weeks the two leaders saw to it that most of the agreement went into effect. U-2 reconnaissance indicated that the Soviets were removing the missile launchers and missiles from Cuba. The United States stopped its quarantine and clamped down on the forays of emigrés against Cuba. By April 1963 the Jupiter missiles were taken out of Turkey and Italy and replaced by Polaris missiles in submarines.

In evaluating Kennedy's handling of the missile crisis many people have given him very high marks. Some, like McNamara, maintained at the time that the administration's prior build-up of defense forces, preparing for "flexible response," was both prescient and critical to America's ability to meet the situation. "A line of destroyers in a quarantine or a division of well-armed men on a border," he declared, "may be more useful to our real security than the multiplication of awesome weapons beyond our actual needs."[39] Other admirers of Kennedy have praised him as a cool and collected crisis-manager and lauded the much-improved decision-making process as carried out by Ex-Comm. Kennedy advisers emerged from their ordeal with great pride and self-assurance about their capacity to handle crises in the future.

Those who favorably evaluate Kennedy's actions tend to blame Khrushchev as the real villain of the piece. They surely have a point. The Soviet leader acted rashly in building up missiles in Cuba and compounded his rashness by lying about it even after it became clear that Kennedy knew he was lying. When Kennedy resisted, Khrushchev wisely backed down from a nuclear exchange. But the Soviet leader was even then reckless, having ceded to Soviet commanders in Cuba the authority to fire off missiles on their own. One commander completely surprised Moscow by doing just that, bringing down the American U-2. It was fortunate indeed that Kennedy, acting prudently, resisted the urging of his military advisers and refused to authorize a retaliatory strike. If he had, the

38. Ibid., 543–44.
39. Gaddis, *Strategies of Containment*, 216.

United States would probably have killed Russians and set in motion a series of responses that might have provoked a nuclear war.

Khrushchev's worst error was in starting something so reckless that it could not be carried out if challenged. When he turned back the ships, he accepted worldwide public embarrassment. Not only Castro but also Red China ridiculed his handling of the affair. Critics at home grew increasingly restless, finally removing him from power in October 1964 (and replacing him with leadership that was bent on still more rapid expansion of Soviet military forces). His mismanagement of the missile crisis almost certainly contributed to his ouster and helped to usher in further escalation of the arms race.

Supporters of Kennedy argue also that he acted with an admirable mix of firmness and wisdom. If he is compared to some of the hawks on Ex-Comm, that is true. To have ordered an invasion or an air strike would have been to invite nuclear disaster. If he is compared to Stevenson, praise is less clearly due, for Stevenson had a point in emphasizing that the missiles in Cuba gave the Soviets little new military potential: their ICBMs already could hit American targets. The United States, having discovered the missiles, might have quietly struck a deal. But Kennedy was correct in believing that acquiescence by the United States in Soviet missiles in Cuba—which were certain to come to public attention—would have changed international *perceptions* of Soviet strength and diplomatic boldness, thereby damaging America's credibility in the world.[40] Acquiescence might also have encouraged Khrushchev to take further liberties. Finally, it would have provoked major political recriminations in the United States. For all these reasons, Kennedy felt he had to take a firm stand. When he resisted—and forced the Soviets to back down—Americans reacted with enormous relief and fulsome praise. The journalist Richard Rovere observed in the *New Yorker* that Kennedy had achieved "perhaps the greatest personal diplomatic victory of any President in our history."[41]

Still, critics of the President's performance have many of the last words. With good cause they raise questions about actions and omissions that long pre-dated the crisis itself. If he had not followed the Bay of Pigs invasion with Operation Mongoose and other highly threatening anti-Castro activities, the Cubans might not have looked so eagerly for Soviet military help. If he had given higher priority to intelligence-gathering, he

40. Ibid., 213.
41. Beschloss, *Crisis Years*, 568.

would probably have known a little earlier of the substantial Soviet build-up on the island. If he had specifically warned the Soviets before September 1962 (perhaps following the Bay of Pigs debacle) not to bring offensive missiles to Cuba, they might not have dared to do so. Many of these American actions (and inactions) had the effect of persuading Castro and Khrushchev that another invasion—this one openly American—was soon to come. That, in turn, rattled them. The administration gave too little thought to the way in which American actions, some of them (such as Mongoose) hostile indeed, were perceived by unfriendly governments.[42]

Kennedy can also be faulted for his performance in October 1962. Angered by Khrushchev's lying, he was understandably eager to score a diplomatic victory and to humiliate the Soviet leader in the process. This he resolved to do publicly. If he had privately shown the U-2 photographs to Soviet officials and tried to negotiate—perhaps (as he ultimately did) by exchanging missiles in Cuba for missiles in Turkey—he might have managed a tense but not world-frightening diplomatic settlement. Alternatively, he might have threatened to take the strong stand that he did but have done so via more quiet diplomacy, thus enabling Khrushchev to back off in private. Instead, Kennedy felt he must face down his adversary if he hoped to avoid repeated challenges in the future, and he resorted to television. This was an unsubtle and provocative approach that required the enemy not only to give in but also to accept public humiliation. Nothing can be more risky in high-stakes diplomacy. Kennedy himself noted, "If Khrushchev wants to rub my nose in the dirt, it's all over."[43]

Kennedy partisans exaggerate considerably when they describe the decision-making process, as many did, as a cool and masterful display of the best and the brightest of American officialdom. Schlesinger later maintained that Kennedy's response revealed a "combination of toughness and restraint, of will, nerve and wisdom, so brilliantly controlled, so matchlessly calibrated, that it dazzled the world."[44] On the contrary, members of the Ex-Comm worked frantically and went without sleep. Understandably, some lost their tempers. Given the pressure of circumstances, these were predictable reactions, but they reveal that Kennedy and his advisers found themselves confronted with an unanticipated situa-

42. Paterson, ed., *Kennedy's Quest*, 140–41.
43. William Chafe, *The Unfinished Journey: America Since World War II* (New York, 1991), 204.
44. Arthur Schlesinger, Jr., *A Thousand Days: John F. Kennedy in the White House* (Boston, 1965), 841; O'Neill, *Coming Apart*, 71.

tion that demanded improvisation. What followed was not always very cool.

Nor did it reveal much mastery. No such capacity was easy to attain in complicated, fast-moving situations wherein a good deal of key information was either not known or misunderstood. Ex-Comm advisers, for instance, did not realize that the Soviets had 42,000 men on the ground in Cuba, that the Soviets might have tactical nuclear weapons ready to fire at invaders, or that decisions to shoot off missiles could be made by Soviet commanders in Cuba rather than in Moscow. Fortunately, the Ex-Comm people decided against either invasion or air strikes, which might have prompted terrific counter-force. But they did so only at the last minute and on the basis of faulty intelligence. In a word, they were lucky as well as wise.

Kennedy, meanwhile, underestimated the enemy's military capacity. Wrongly assuming that the United States had total control of the air, he kept the U-2s flying and lost one at a critical time in the negotiating process. Only the President's restraint at that point prevented major escalation. Meanwhile, Mongoose operatives were continuing to plot separately; unreachable by the CIA during the crisis, they managed to blow up a Cuban factory on November 8. What would the enemy have thought if Mongoose agents had succeeded in doing so at the height of the crisis in late October? Imponderables such as these suggest how extraordinarily difficult it is for decision-makers—whether American or Soviet—to "manage" a major crisis in the nuclear age. Officials then and later who thought they could indulged in a fair amount of self-congratulatory wishful thinking.[45]

Second thoughts about the crisis gradually sobered both Kennedy and Khrushchev, who agreed to establish a "hot-line" in 1963 in order to lessen the chances of nuclear disaster. In June 1963 the United States, the Soviet Union, and Great Britain agreed to a limited ban on nuclear testing in the atmosphere, in space, and under water. The Senate ratified it, 80 to 19, in September, and it was signed in October. The treaty did not do much to reduce tensions, for underground testing continued. Expenditures on nuclear-armed bombers, Polaris missiles, and Minutemen escalated. Other potentially important nuclear powers such as France and China refused to accede to the treaty. Still, the agreement signified a bit of a thaw in the deep freeze of Soviet-American relations. In a widely hailed speech at American University on June 11, Kennedy went so far as to suggest a reassessment of Cold War assumptions. "In the final analysis,"

45. Bernstein, "Reconsidering Krushchev's Gambit"; Paterson, "Fixation."

he said, "our most common link is that we all inhabit this small planet. We all breathe the same air. We all cherish our children's future. And we are all mortal."[46]

But the understandings that Kennedy and Khrushchev had reached over Cuba at the end of October 1962 had been rough and tentative, and serious disagreements persisted even as Kennedy talked of common humanity. American forces had had to remain on highest alert until November 20, at which point Castro had reluctantly agreed to the return of three long-range Soviet bombers to the Soviet Union. Then and thereafter, most of the 42,000 Soviet soldiers and technicians stayed in Cuba. Castro never admitted on-site inspectors, thereby promoting persistent rumors that secret installations remained. He sought to export his revolution to other parts of the Western Hemisphere, notably in Venezuela, where a Cuba-inspired plot was uncovered in November 1963. The Soviets ultimately resumed their military build-up in Cuba by installing offensive fighter bombers and starting construction of a submarine base.[47]

United States policy regarding Cuba remained equally provocative. Kennedy refused to put his no-invasion promise in writing, demanding that Cuba first agree to inspections and that it cease "aggressive acts against any of the nations of the Western Hemisphere."[48] This insistence left him—and subsequent Presidents—free to launch an invasion. As if to confirm that possibility, the Kennedy administration revived Operation Mongoose in June 1963. Only in September 1970, during the Nixon presidency, did leaders of the United States and the Soviet Union (again bypassing Castro) reach a semi-formal understanding of the settlement of 1962. At that time Nixon said the United States would not invade Cuba, and Leonid Brezhnev, the Soviet leader, agreed that Russia would stop developing offensive weapons on the island. Even this understanding, however, was a secret exchange between leaders, not a formal agreement. It was not until some years later that many officials in the American government even knew of it. Cuba remained a flashpoint of the Cold War.[49]

COMPARED TO THE DRAMATIC CONFRONTATIONS over Cuba, developments in Vietnam during much of Kennedy's tenure received little public attention. Yet they were profoundly significant. When Ken-

46. Richard Reeves, *President Kennedy: Profile of Power* (New York, 1993), 513–14.
47. Beschloss, *Crisis Years*, 564–68.
48. *New York Times*, Jan. 7, 1991.
49. Raymond Garthoff, *Détente and Confrontation: American-Soviet Relations from Nixon to Reagan* (Washington, 1985), 79–81.

nedy entered the White House, there were some 1,000 American military advisers in Vietnam. In October 1963 there were 16,732. They had been authorized to go with combat missions against the forces of the National Liberation Front and to use napalm and Agent Orange, a powerful and toxic defoliant, so as to flush out opposition. On November 1, 1963, the South Vietnamese President, Ngo Dinh Diem, and his brother Ngo Dinh Nhu, were killed in a coup. Although Kennedy had not expected them to be killed, he had in effect encouraged the effort. The assassinations left the government of South Vietnam even more disorganized and demoralized than it had been under Diem. This was the situation at the end of the Kennedy administration three weeks later.[50]

Kennedy's expansion of American commitment to Vietnam was consistent with his overall view of foreign and defense policy, a test of his inaugural promise to "pay any price" and "bear any burden" to "assure the survival . . . of liberty." Like most American political leaders at the time, he was convinced that the Soviets and the Chinese were behind the efforts of Ho Chi Minh's war of liberation. He further believed that Diem, a vehement anti-Communist, held the key to stopping enemy victory in South Vietnam. Special Forces such as the Green Berets, he thought, afforded the flexible response necessary to assist Diem against enemy guerrillas. Kennedy's chief military adviser, General Maxwell Taylor, considered Vietnam a "laboratory" for American military development. Walt Rostow, who played a major role in policy-making, was equally blunt. The United States, he said, should make aggressive use of its Special Forces. "In Knute Rockne's old phrase, we are not saving them for the junior prom."[51]

Beliefs such as these did not plunge Kennedy pell-mell into escalated involvement. On the contrary, Kennedy resisted the idea of sending American soldiers into combat. After the Bay of Pigs he became more leery of potentially costly military adventures, opting in neighboring Laos

50. For general treatments see Lawrence Bassett and Stephen Pelz, "The Failed Search for Victory: Vietnam and the Politics of War," in Paterson, ed., Kennedy's Quest, 223–52; George Herring, America's Longest War: The United States and Vietnam, 1950–1975 (Philadelphia, 1986), 73–106; John Newman, Deception, Intrigue, and the Struggle for Power (New York, 1992); Marilyn Young, The Vietnam Wars, 1945–1990 (New York, 1991), 78–83, 96–97; Guenter Lewy, America in Vietnam (New York, 1978), 18–22; Stanley Karnow, Vietnam: A History (New York, 1983), 248–53; Giglio, Presidency of JFK, 239–54; Burner, JFK and a New Generation, 95–113; and Ambrose, Rise to Globalism, 201–10.
51. Fred Siegel, Troubled Journey: From Pearl Harbor to Ronald Reagan (New York, 1984), 140; Gaddis, Strategies of Containment, 214–17, 243–45.

for negotiations and covert operations. In Vietnam he contented himself with the dispatch of 400 Special Forces.[52]

Kennedy continued to resist hawkish advice later in 1961. Taylor and Rostow led a mission to Vietnam in October and returned to call for the dispatch thereto of 8,000 American soldiers—a "logistic task force" composed of engineers, medical people, and small numbers of combat infantry to support them. Taylor explained that the men would serve as a "visible symbol of the seriousness of American intentions" and as a reserve force in case the military situation suddenly deteriorated.[53] Other advisers, however, criticized aspects of the Rostow-Taylor report. Undersecretary of State Bowles warned that the commitment of such a force would rush the United States "full-blast up a dead-end street." Kennedy was torn. On the one hand he remained very nervous about the unpredictable Khrushchev. "That son of a bitch," he said, "won't pay any attention to words. He has to see you move." As always, he worried about maintaining America's credibility in the world. On the other hand Kennedy gave only cursory attention to Southeast Asia, and he remained fearful of heavy American involvement. He noted: "The troops will march in; the bands will play; the crowds will cheer; and in four days everyone will have forgotten. Then we will be told we have to send more troops. It's like taking a drink. The effect wears off, and you have to take another."[54] Opposing the commitment of combat troops, he agreed in November 1961 to the sending of more military advisers. By the end of the year there were 3,205 in South Vietnam.[55]

In so doing, however, Kennedy was hardly giving up on Diem. In addition to increasing the number of advisers, JFK made other key decisions in November that set a course for the remainder of his administration. He rejected, first of all, the recommendations of Bowles and Averell Harriman, a special adviser, that the United States work for a cease-fire in Vietnam, to be followed in time by a negotiated agreement leading to elections that would reunite the nation. This was a course similar to the one that Harriman was then working out for Laos. Given the extent of American support for Diem since the mid-1950s, such a policy would have been difficult to sell to the American public. Moreover, Diem and Ho Chi Minh were on a collision course and would have hotly opposed it. In any event, Kennedy scarcely considered such an option. He stuck

52. Herring, *America's Longest War*, 73–78.
53. Ambrose, *Rise to Globalism*, 207–8.
54. Beschloss, *Crisis Years*, 338.
55. Herring, *America's Longest War*, 82–85.

instead with the effort to promote Diem as the non-Communist leader of a sovereign South Vietnam, even though he recognized—it was impossible not to—that Diem had become increasingly despotic and corrupt over the years. "Diem is Diem," Kennedy mused, "and the best we've got."[56]

Kennedy rejected other options suggested to him in November 1961. One, pressed by Senate majority leader Mike Mansfield of Montana, was to insist that Diem use American aid and know-how to promote social reform, especially in villages where the battle against the NLF was being lost. If Diem refused to do so, he should be cut off. Rusk, one of the strongest advocates of curbing Communism in Asia, seemed sympathetic with versions of this option from time to time. Diem, he complained, was "an oriental despot." The United States needed to take a hard line with him, else its support would be wasted.[57]

Advice such as this made sense to Kennedy—and to many others who struggled to develop a coherent policy concerning Vietnam over the years. But it was extraordinarily hard to put into practice. For one thing, Diem continued to enjoy significant political support in the United States, especially from conservative Catholics and determined Cold Warriors. Throughout his presidency Kennedy worried a good deal about the political consequences of appearing "soft" in Vietnam. Remembering that Cold Warriors (himself included) had accused Truman of "losing" China, he determined that he would not go down in history as the President who had "lost" Vietnam. For another, thanks in large part to decisions made by the Eisenhower administration, Diem *was* the government in South Vietnam. If he refused to back major reforms, what was the United States to do? The answer was that it could not do very much, except—perhaps—to look for a more pliant but equally anti-Communist successor. Indeed, JFK faced a tougher situation than the one that had confronted Eisenhower: in the late 1950s Diem had retained some plausibility as a leader, but he was losing it rapidly in the early 1960s. This created major dilemmas for Kennedy. He could and did threaten to cut Diem off, but that managed mostly to sour relations without much changing Diem's own headstrong course.

Kennedy advisers focused instead on two efforts after late 1961, "counter-insurgency" and a "strategic hamlet" program. Counter-insurgency involved using American military advisers to help Diem's armed forces engage the enemy. South Vietnamese military leaders, however, showed little interest in fighting. Many of them were instead engaged

56. Bassett and Pelz, "Failed Search"; Herring, *America's Longest War*, 85.
57. Bassett and Pelz, "Failed Search," 235.

in extortion and other forms of corruption. The United States sent in more and more military advisers: there were 9,000—as well as 300-odd military aircraft, 120 helicopters, and other heavy weapons—by December 1962. But Kennedy was busy with other concerns and did not give the effort much attention. Nor did his political advisers, who left matters mainly to General Paul Harkins, the American commander in Vietnam. Harkins tried (some 100 Americans lost their lives in the effort by December 1963) but failed to turn the tide. "Search and destroy" missions killed many villagers. Use of napalm devastated parts of the countryside. Counter-insurgency grew increasingly unpopular with South Vietnamese people in rural areas, where the NLF made rapid inroads in late 1962.[58]

The strategic hamlet program aimed to develop pockets of strength in these rural areas. American and South Vietnamese authorities were supposed to cooperate in developing civic programs, fortifications, and radio contacts in and between key villages. Moats and bamboo would protect villages from the enemy. But the program required the uprooting and moving about of people from their villages so that they would be safe in the hamlets. It also proved impossible, even when villagers were told to carry identification, to prevent subversive forces from infiltrating the hamlets. Most important, the strategic hamlet program did not promote either land reform or democratic practices. While many villagers distrusted the NLF, which was frequently brutal, they had little reason to support the corrupt and dictatorial government of Diem in Saigon. Most of them probably wanted above all to be left alone.[59]

The political situation in South Vietnam began seriously to deteriorate in 1963. Diem came to rely increasingly on Nhu, described by one leading scholar as a "frail and sinister man who tended toward paranoia and delusions of grandeur," and on Nhu's wife, "beautiful, ambitious, and acid-tongued."[60] Diem and the Nhus, who were Catholic, especially antagonized southern Buddhists, one of whom, Quang Duc, took the extraordinary step on June 11 of publicly immolating himself in protest in Saigon. Photographs of Quang Duc burning to death shocked the world, Kennedy included.[61] Other Buddhists followed suit during the summer,

58. Herring, *America's Longest War*, 85–90; Neil Sheehan, *A Bright Shining Lie: John Paul Vann and America in Vietnam* (New York, 1988).
59. Herring, *America's Longest War*, 88–90.
60. Ibid., 90.
61. June 11 in the United States was one of the most frantic days of Kennedy's presidency. It was also the day that Governor Wallace tried to stop two black

with Madame Nhu cavalierly dismissing the actions as "barbecues." She offered to supply matches and gasoline for Buddhists who wanted to burn. Diem and Nhu jailed hundreds of protestors and raided Buddhist pagodas. In August they staged a major sweep of opponents and arrested 1,400 people.

The day after these arrests Henry Cabot Lodge, Jr., arrived in Saigon as the new American ambassador to South Vietnam. Lodge was a Republican, Kennedy's opponent for the Senate in 1952 and Nixon's running mate in 1960. His appointment symbolized JFK's quest for a bipartisan political consensus in order to shore up his precarious policies in Southeast Asia. Perceiving the mood of demoralization in Saigon, Lodge had little faith in Diem's ability to rule the country. Washington, moreover, signaled to him that a coup, if likely to succeed, would be acceptable. The Kennedy administration further indicated its displeasure with Diem—and its willingness to see him removed—by cutting back aid in October. CIA officials developed close contacts with anti-Diem generals.[62]

A large and frequently bewildering increase in diplomatic cable traffic took place between Lodge's arrival in August and the coup on November 1. For the first time, Kennedy involved himself steadily in what was happening in Vietnam. Still, his top advisers, such as McNamara, Rusk, and Bundy, were not sure what to do. Harkins appeared to oppose the removal of Diem, Lodge to favor it. Offered no clear instructions, Lodge had considerable latitude and gave anti-Diem plotters—some of them in contact with the CIA—every reason to assume that the United States would not thwart a coup. When it broke out on November 1, Kennedy and other Washington officials eagerly monitored events. Neither they nor Lodge had insisted that Diem and his brother be spared, but when Kennedy heard that they had been killed, he leapt to his feet and rushed from the room in dismay. Schlesinger recalled, "I had not seen him so depressed since the Bay of Pigs. No doubt he realized that Vietnam was his great failure in foreign policy, and that he had never really given it his full attention."[63]

NOT MANY SERIOUS SCHOLARS would disagree with this assessment of "failure," which was in part one of inattention to a serious

---

students from being admitted to the University of Alabama, whereupon Kennedy went on television to announce his support of a civil rights bill. Medgar Evers was killed a few hours later. See chapter 16.

62. Herring, *America's Longest War*, 95–106; Bassett and Pelz, "Failed Search," 243–45.
63. Blum, *Years of Discord*, 128–32; Schlesinger, *RFK and His Times*, 997.

problem that demanded careful presidential engagement. From the beginning many of his military and intelligence advisers, including those in the CIA, warned repeatedly that the United States would have to make a major military commitment, at the least of 200,000 American men, if it hoped to have a good chance of victory in Vietnam. But JFK did not listen. Brushing aside such advice (the CIA, after all, had been wrong before), he persisted in hoping that the South Vietnamese, assisted by American Special Forces and "advisers," could prevail. This notion was highly questionable by 1963, but aside from dumping Diem, Kennedy never reassessed the situation, and he had no well-considered long-range plans.

Kennedy's failure stemmed also from faulty assumptions about the Cold War as it affected Southeast Asia. Like Rusk and other top advisers, Kennedy believed fervently in the domino theory. Indeed, he perceived the emerging post-colonial nations of the world as a new and vital battleground in the Cold War. If the United States acted indecisively in Southeast Asia, he thought, it would send a message of weakness to Moscow on how it would respond to insurgency elsewhere in the world. Rusk, moreover, was convinced that what was happening in Vietnam was part of a larger Communist plot. In clinging to such beliefs Kennedy did not act on what in fact he knew: the Soviets and the Chinese were bitterly opposed to each other by 1961. He also neglected to heed what some advisers (a minority, to be sure) were saying: the battle in Vietnam was a civil war, not a blueprint from the design of world Communism.

Although it is hard to find committed defenders of Kennedy's policies in Vietnam, some writers have tried to explain them in context. They reiterate that earlier decisions by Eisenhower had already committed the United States to Diem, that anti-Communist pressure at home in the United States constrained the President's freedom of action, and that Kennedy managed, despite these pressures, to forbid the direct use of American combat troops. They highlight Kennedy's increasingly well informed appraisal of the situation in the last months of his administration. "Unless a greater effort is made by the government [of South Vietnam]," Kennedy told CBS in September, "I don't think that the war can be won out there. In the final analysis, it is their war. They are the ones who will have to win it or lose it." Acting on this appraisal, it is argued, Kennedy in October ordered the return to the United States of 1,000 advisers by the end of the year. This decision, it is further claimed, was to pave the way for recall of all the advisers at some point, probably after the 1964 presidential election.[64]

64. See comments by Schlesinger, Rostow, Kennedy adviser Roger Hilsman, and others in the New York Times, Jan. 20, Feb. 15, March 23 and 29, 1992.

No one can be certain how Kennedy would have coped in 1964 or 1965 with the situation in Vietnam, which remained fluid and unpredictable. Perhaps he would have cut his losses. Kennedy did express greater doubts in private about escalating American involvement than he dared to state in public. On several occasions he reminded hawkish advisers that General MacArthur, no dove, had warned about the costs of America trying to fight a land war in Asia. It also seems possible that JFK would have reconsidered his course after the 1964 election. He told his friend Charles Bartlett in 1963, "We don't have a prayer of staying in Vietnam. We don't have a prayer of prevailing there. But I can't give up a piece of territory like that to the Communists and then get the people to reelect me."[65] It may be that if Kennedy had lived to win a major mandate in the 1964 election, he would have dared to confront Cold War passions and to extricate the United States from Vietnam.

The evidence for such a scenario, however, is sketchy. Some of those who cite Kennedy's CBS interview neglect to point out that he added, "I don't agree with those who say we should withdraw. That would be a great mistake. . . . This is a very important struggle even though it is far away. We made this effort to defend Europe. Now Europe is secure. We also have to participate—we may not like it—in the defense of Asia." Kennedy partisans also fail to note that the American advisers to be withdrawn from Vietnam in late 1963 were mostly part of a construction battalion that had finished its work. They were being brought home for Christmas and were scheduled to be replaced by others. Most of Kennedy's major advisers concerning Vietnam then and later were certain that Kennedy never intended to "withdraw" American advisers and military aid before he could be certain that the South Vietnamese could safely defend themselves. Rusk said later, "I had hundreds of talks with John F. Kennedy about Vietnam, and never once did he say anything of this sort [about withdrawal]." JFK's speech prepared for delivery in Dallas on November 22, 1963, contained a reminder about Vietnam: "We dare not weary of the task."

Kennedy's flawed policies in Vietnam do not entirely reflect his overall record in the field of foreign affairs. An assessment of this record shows that he was less of a hard-line Cold Warrior than some of his rhetoric, such as his speech in 1960 in Salt Lake City, had suggested. It further indicates that he grew a little more sophisticated and cautious as he gained experience. In the face of often considerable domestic pressures he acted

65. Paterson, *Kennedy's Quest*, 10.

prudently concerning Laos and toward the Congo and Indonesia, where his advisers also helped to broker settlements. The Peace Corps, while of limited effectiveness in coping with the enormous socio-economic problems of underdeveloped nations, earned decent marks for effort. So did the limited test ban treaty and Kennedy's growing awareness following the missile crisis of the need for negotiation with the Soviets. When he left office, the two nations had established an uneasy but promising détente. Above all, Kennedy was applauded for standing up behind Berlin and for maintaining Western defenses in Europe.

Against these accomplishments, however, may be listed troubling failures and misconceptions that plagued Kennedy and his advisers. Although he knew there was no missile gap, he eagerly increased defense spending and helped to escalate the arms race. Despite the Bay of Pigs debacle, he persisted in plans to harass and frighten Castro, thereby accentuating the provocative behavior of Khrushchev. Ignoring evidence to the contrary, he held fast to clichés—especially in public—such as the domino theory and the existence of a monolithic "international" Communism. He persisted in celebrating the capacity of the Special Forces and more broadly of military moves to solve deeper social and political problems. On many occasions, as in his policies concerning Vietnam, he proved poor at sorting out good information from bad and at developing long-range plans. Contrary to the claims of his acolytes, he did not grow very much on the job.

POLITICAL CONSIDERATIONS GAVE Kennedy little rest. Having been elected by the smallest of margins, he never stopped planning for 1964. So it was that he traveled to Texas on November 21, 1963. His wife Jackie went with him, the first time she had taken such an obviously political trip with him since the 1960 campaign. Kennedy hoped to bring peace to warring Democrats in the state and advance his political chances.

The very thought of such a trip irritated Kennedy. He was unhappy with Vice-President Johnson, who had been unable to resolve the feuds in Texas and who had pressed him to make the journey. Johnson, he knew, would get him to come to his ranch and wear "one of those big cowboy hats." Kennedy also expected opposition from right-wing activists, for Texas, like much of the South, stirred with people who despised him for his civil rights bill and for his softness, as they saw it, in foreign policy. A month earlier Stevenson had been struck and spat upon in Dallas. When JFK arrived in Dallas on November 22, an ad in the morning paper asked why he had allowed "thousands of Cubans to be jailed and sold wheat to those [the Russians] who were killing the United States in Vietnam. Why have you scrapped the Monroe Doctrine in favor of the spirit of Moscow? . . . Mr. Kennedy, WE DEMAND answers to these questions and we want them now." Kennedy observed to Jackie, "We're heading into nut country today."[66]

Hostility such as this had often caused Kennedy to reflect on how easy it would be for some crazy person to take out a rifle and kill him. But he rarely took major precautions, and he did not do so on November 22. Preparing for a ten-mile motorcade through downtown Dallas, he and Jackie climbed into an open car with Texas governor John Connally and his wife Nellie. Jackie wore white gloves and held a bouquet of red roses. As they passed the Texas School Book Depository around 12:30 P.M., gunshots rang out. One of the shots wounded Connally, and two hit Kennedy in the head and the neck. Blood spattered Jackie and the car, which sped off to a hospital. It was too late to save the President, who was clinically dead on arrival in the emergency room. The time of death was placed at around 1:00 P.M.[67]

Less than an hour and a half after the shooting, Dallas police arrested Lee Harvey Oswald, a twenty-four-year-old order-filler at the Book Depository, and charged him with the slaying of a Dallas police officer, J. D. Tippitt, who had tried to detain him in the street on suspicion of the assassination.[68] Nine hours later Oswald was charged with the murder of JFK. He vehemently denied all charges. Two days later, Jack Ruby, a strip-joint owner well known to police, approached Oswald as the alleged assassin was being transferred. Ruby pulled out a concealed pistol and killed Oswald on the spot. He explained that he acted out of grief for the President.

The killing of Oswald stimulated already rampant speculation about the assassination. Had Oswald really done it? If so, had he acted alone? Who put him up to it and why? Had Ruby been sent to silence him? This speculation grew enormously over the years and never seemed to stop: in early 1992, four books on the *New York Times* non-fiction best-seller list dealt with the events at Dallas on that unforgettable day. Much of this speculation centered on alleged conspiracies, which millions of Ameri-

66. O'Neill, *Coming Apart*, 90; Beschloss, *Crisis Years*, 670–71; Schlesinger, *RFK and His Times*, 1020–25.
67. Reeves, *President Kennedy*, 661–62; *Newsweek*, Dec. 2, 1963, pp. 20–26.
68. Known to acquaintances until then as Lee Oswald. See Max Holland, "After Thirty Years: Making Sense of the Assassination," *Reviews in American History*, 22 (June 1994), 191–209.

cans believed to have led to the killing. Many people could not believe the report that one bullet had managed to pass through both Kennedy and Connally. Others, accepting the claims of a few bystanders, were sure that there had been more than one assassin and that more than three shots had been fired.

In order to subvert these and other theories Johnson, now President, appointed a commission to offer an official government account of the assassination. It was headed by Chief Justice Warren and contained among its well-known members Richard Russell of Georgia, a power in the Senate, Congressman Gerald Ford of Michigan, a leading House Republican, and Allen Dulles, former head of the CIA. The staff of the commission concluded in September 1964 that Oswald alone had committed the crimes by firing three shots from the sixth floor of the Depository Building, whereupon he had fled into the streets, been accosted by Tippitt, and killed him, too.

The general conclusions of the Warren Report, as it was called, have satisfied most scholars. They have agreed that Oswald was a loner who had endured a deeply troubled hitch in the marines, after which he lived in the Soviet Union for thirty-two months. He was profoundly unhappy there, at one time slashing one of his wrists, and was apparently kept under surveillance by Russian officials who doubted his emotional stability. Coming back to the United States in early 1963, he remained a self-styled Marxist, identifying particularly with the virtues of Castro's regime in Cuba. At one point he tried and failed to assassinate General Edwin Walker, a right-wing leader who lived in Dallas. Oswald also moved temporarily from Dallas to New Orleans, where he opened a Fair Play for Cuba chapter. Finding no takers, he returned, embittered, to Dallas. There, the commission said, he decided to kill the President. The commission noted that Oswald's palmprint was found on the stock of the rifle, left at the Depository, that had been used to fire the fatal shot.[69]

Because Ruby killed Oswald, it is impossible to know why Oswald did what he did. But it is unsatisfactory to stop with the obvious: that he was highly unstable. Oswald was a politicized young man whose actions reflected the super-frigid context of the Cold War of the early 1960s. That was a time when governmentally sanctioned killing, whether by the CIA or the Soviet KGB, was an openly discussed and apparently viable option in the conduct of foreign policy. Castro, indeed, frequently (and accu-

69. See Gerald Posner, *Case Closed: Lee Harvey Oswald and the Assassination of JFK* (New York, 1993); Melinda Beck, "The Mind of the Assassin," *Newsweek*, Nov. 22, 1993, pp. 71–72; and "Who Shot JFK?" *Newsweek*, Sept.6, 1993, pp. 14–17.

rately) claimed that the CIA was trying to assassinate him. While in New Orleans, Oswald read in the newspapers an account of one of Castro's diatribes against the United States and apparently convinced himself that the Cuban leader wanted Kennedy dead. The assassination of JFK, in short, was not only an outrage by an unstable individual; it was also a politicized act, one of the most terrible in the history of the Cold War.[70]

The Warren Report greatly pleased President Johnson, who was espe-cially anxious to scotch rumors that the Soviets had been involved in a conspiracy to kill his predecessor. Such a finding, he knew, would gener-ate enormous and dangerous public pressures to retaliate. He was equally eager to put down other conspiracy theories. Johnson therefore hailed release of the report. So, too, did virtually all other mainstream commen-tators in the media. Worried about the fragility of American institutions in the aftermath of the shock, they wanted very much to think that only a sociopathic individual would have done such a thing.

The report, however, did not put an end to theories about conspiracy. That was in part because members of the commission had pre-judged the case. Although the commission staff interviewed some 500 people (the report was 888 pages long), its work contained flaws. Moreover, it sealed some of its evidence for seventy-five years, thereby feeding suspicions that there was something to hide. As later revelations made clear, key sources did not divulge all they knew to the commission. The CIA, for instance, hid its involvement with the Mob and with plots to kill Castro. The FBI obfuscated in order to conceal its failure to keep close tabs on Oswald, who had been known to be dangerous. Commission members them-selves, such as Russell, Ford, and Dulles, did not inform staff members, who wrote the report, about Mongoose and the broader tensions of Cuban-American relations that may have been highly pertinent in ex-plaining Oswald's behavior. Robert Kennedy, who had overseen Mon-goose since November 1961, also kept its nefarious doings secret from the commission. Such activities, these people persuaded themselves, were not relevant to the assassination. To reveal them to a commission, they thought, was to compromise the operations of the CIA.

These secrets, clearly relevant in retrospect, were scarcely suspected by conspiracy theorists in 1964. Rather, doubters at that time tended to imagine other scenarios. A few theorists, most of them on the right, believed that the commission failed to follow up leads that would have implicated Castro, the Soviet Union, or both. Others blamed the Mob,

70. Holland, "After Thirty Years."

with which both Oswald and (especially) Ruby had shadowy ties. Gangsters, they said, were furious at Kennedy for not getting rid of Castro (who had closed down their casinos in Cuba) and at Robert Kennedy for prosecuting Mafia leaders and some of their friends, such as Teamsters leader Jimmy Hoffa.

Many believers in conspiracy theories felt a deep personal loss. Idealizing Kennedy, they could not believe that one demented person could succeed in killing a President. Great events require great causes or conspiracies. Distrusting government officialdom and Establishmentarian reports, they concocted elaborate reconstructions involving Very Important People. Some pointed the finger at FBI chief Hoover, who was known to dislike Kennedy. Others said that the CIA had masterminded the killing out of fears that Kennedy would give in to the Communists and dismember the agency itself. A few people thought Johnson, Pentagon officials, or other Cold Warriors had ordered the killing of the President in order to stop the United States from pulling out of Vietnam and dismantling the military-industrial complex.[71]

So insistent were doubters of the Warren Report that a select House committee started to review it in 1976. After working for more than two years and spending some $5 million, it reported in 1979 that Oswald had shot Kennedy and that neither the Soviet Union nor Cuba nor any government agency in the United States had been involved in the killing. Most scholars have accepted this view, which echoed the major findings of the Warren Commission. But the committee otherwise challenged the commision's conclusions by asserting that there was a "high probability" that a second gunman had fired at Kennedy and missed. The committee did not identify this gunman but speculated that the Mob may have been behind this conspiracy. Its evidence for this conjecture, later challenged by other experts, was unconvincing. The committee closed some of its most sensitive records until the year 2029, thereby fueling further the rampant speculation that persisted about the case.

Although these conspiracy theories attracted a host of adherents in the late 1960s and 1970s—years of intensifying distrust of government—they were visible even in the immediate aftermath of the assassination. Prior to

71. See Giglio, *Presidency of JFK*, 277–87, for a summation of some of these theories. Also "28 Years After Kennedy's Assassination, Conspiracy Theories Refuse to Die," *New York Times*, Jan. 5, 1992; "JFK Conspiracy: Myth vs. the Facts," *Washington Post*, Feb. 28, 1992; and "The Conspiracy Theories," *Newsweek*, Nov. 22, 1993, p. 99. A film produced in 1992 by Oliver Stone, *JFK*, brought several of these theories to mass audiences.

release of the Warren Report, 52 percent of Americans told pollsters that they believed there had been some kind of conspiracy.[72] For many Americans at that time the killing was a shattering event that once and forever damaged their faith in the future. "For me," a radical student recalled, the assassination "has made all other acts irrelevant and trivial; it has displaced time with paranoia, good with evil, relative simplicity with incomprehensibility, and an ideal with dirt."[73] A perceptive historian added later that the United States "stopped being a place of infinite progress and ever-expanding promise. Instead, there were suspicions of dark and far-reaching conspiracies."[74] The belief in conspiracy reflected widespread feelings then and later, especially among people who had wanted to think that Kennedy—young, handsome, vibrant, heroic—had been the last best hope for a New Frontier. Something bright and irreplaceable had gone out of their lives, and they yearned for explanations.

The assassination, finally, made a martyr of Kennedy. While President he had enjoyed considerable political popularity. But he was never so admired as he was after his death, whereupon he frequently ranked in public opinion polls (not among historians) above Washington, Lincoln, and FDR as a "great" American President.[75] Some of this adulation stemmed from the martyrdom caused by the assassination; Lincoln, too, had become a legend after he had been shot. Some of it also depended on the response of Jackie and on the extraordinary power of television to dramatize the three days of post-assassination ceremonies that she had a major hand in designing. Millions of Americans huddled before their television sets to watch Jackie, dressed in black, walk alongside her late husband's coffin, borne by a riderless horse to Washington's St. Matthew's Cathedral for funeral services. The Kennedy children, five-year-old Caroline and three-year-old John, were at her side. Following the service John stood at attention, as he had seen the soldiers do, and saluted the coffin. The horse-drawn procession then moved in stately slowness through Washington to Arlington National Cemetery in Virginia. There Kennedy was laid to rest, his grave to be marked by an eternal flame overlooking the city.

Five days later Jackie called the journalist-historian Theodore White to the Kennedy compound in Hyannisport, where she told him a story that White placed in *Life* magazine on December 6. People should realize,

72. Holland, "After Thirty Years," 203.
73. John Diggins, *The Rise and Fall of the American Left* (New York, 1992), 191.
74. Thomas Hine, *Populuxe* (New York, 1986), 170.
75. Haynes Johnson, "Why Camelot Lives," *Washington Post*, Aug. 18, 1991.

she said, that Jack had been sickly as a boy and had spent hours reading about King Arthur and the Knights of the Round Table. In the last days of his life he had responded warmly to Lerner and Loewe's Broadway musical *Camelot*, which sentimentalized those wondrous days of chivalry and heroism. At night in his bedroom he played the recording from *Camelot* before going to sleep, and he especially loved the lines

> Don't let it be forgot,
> that once there was a spot,
> for one brief shining moment
> that was known as Camelot.

Jackie added in White's article, read by millions, that the Kennedy administration had been Camelot, "a magic moment in American history, when gallant men danced with beautiful women, when great deeds were done, when artists, writers, and poets met at the White House and the barbarians beyond the walls were held back." But "it will never be that way again. . . . There'll never be another Camelot again."[76]

If Kennedy had been alive to read this, he would probably have derided it. And rightly so, for it was myth-making of maudlin proportions. Still, it had an apparently lasting appeal to millions of people who had been shaken by the assassination and who were looking for ways to affirm the meaning of Kennedy's life. As they tried to cope with the future they were somber, to be sure. They also yearned to erect monuments to his memory.

76. *Life*, Dec. 6, 1963.

18

Lyndon Johnson and American Liberalism

Five days after Kennedy's assassination Lyndon Johnson went to Capitol Hill to address the Congress. Millions of Americans across the country watched anxiously. The new President, a tall and deliberate man, spoke slowly and clearly. "All I have," he said, "I would gladly have given not to be standing here today." He then moved to his main theme, that he would finish what Kennedy had started: "John F. Kennedy lives on. . . . No words are sad enough to express our sense of loss. No words are strong enough to express our determination to continue the forward thrust of America that he began."[1]

Kennedy, Johnson reminded his audience, had proclaimed at his inaugural in 1961, "Let us begin." Now, Johnson said, *"Let us continue."* Focusing on domestic problems (as Kennedy had not), Johnson enumerated some of the "dreams" that he said Kennedy had pursued: "education for all of our children," "jobs for all who seek them," "care for our elderly," and above all, "equal rights for all Americans whatever their race or color." Johnson stressed the issue of civil rights. "No memorial or oration or eulogy could more eloquently honor President Kennedy's memory than the earliest possible passage of the civil rights bill for which

1. *New York Times*, Nov. 28, 1963. For commentary see John Blum, *Years of Discord: American Politics and Society, 1961–1974* (New York, 1991), 135; and Doris Kearns, *Lyndon Johnson and the American Dream* (New York, 1976), 174.

he fought so long. We have talked long enough about equal rights in this country. We have talked for one hundred years or more. It is time now to write the next chapter and write it in the books of law."

It was a solemn yet uplifting oration. When Johnson finished, his audience, which yearned for leadership in the aftermath of the assassination, jumped to its feet and applauded enthusiastically. Public opinion polls suggested that Johnson had also impressed the American people. Unlike Truman, who had floundered upon taking office in 1945, Johnson, congressional aide in 1931, congressman in 1937, senator in 1949, majority leader of the Senate from 1955 through 1960, Vice-President since 1961, seemed knowledgeable and assured. The fifty-five-year-old Texan sounded like a President.

Johnson and his liberal allies, however, had to cope with a range of serious problems, the largest of which was the war in Vietnam. At home he had to preside over the resolution of major dynamics of American life that had been strengthening in recent years: the extraordinary moral power of egalitarian ideas as nourished by the civil rights movement, and rapidly rising popular expectations, many of them propelled by the promises of Kennedy. Interrelated, these dynamics peaked in the mid-1960s. They excited still grander expectations—demands, in fact, for government entitlements—that were both exhilarating and divisive. It was the peculiar fate of Johnson, a master of coalition-building on Capitol Hill, to have to deal with forces that were on their way to fragmenting the United States.[2]

JOHNSON WAS BORN and raised in the hill country of south-central Texas, the son of a rough-edged father, who struggled to make a sometimes precarious living and who served as a populistically inclined state legislator, and of a strong-willed mother, who yearned for a more genteel style of life. Strains afflicted the marriage and, biographers have surmised, left their mark on young Lyndon, their eldest. People who remembered him as a boy and a young man describe him as in many ways the son of the father: crude, boisterous, a little wild. They remember him also, however, as in awe of his mother, who sternly withheld all signs of affection when she was displeased with him. As he had yearned to win the love of his very different parents, Johnson always seemed desperately eager to make people love him. He also became adept at conciliation, a wonder-

2. Robert Divine, ed., *Exploring the Johnson Years* (Austin, 1981), 16.

ful skill for anyone who hopes to advance in politics. Throughout his life he labored long and hard to bring people together.[3]

People close to Johnson, however, also felt that his upbringing left him insecure as he battled his way up the political ladder. Unlike FDR, his role model, he lacked a patrician background. Unlike Kennedy, to whom he was often unflatteringly contrasted, he lacked inherited wealth and good looks. (Many contemporaries made fun of his ears, which were large and jutted out from his head.) Instead, Johnson had had to struggle at every turn, barely losing a senatorial primary in 1941 (his opponent probably stole it from him) and barely winning one in 1948 (he surely stole that). Following this narrow victory opponents derided him as "Landslide Lyndon." After re-election in 1954, he finally had a secure political base, which he used to run for President in 1960. But by then he had acquired a reputation as a self-aggrandizing wheeler-dealer.

Regional identification remained strong in the United States, and Johnson felt especially insecure about this aspect of his background. Among his southern and western colleagues on Capitol Hill, it was a political asset which he used to full advantage. As he grew wealthy he acquired a cattle ranch, in which he took great pride. He relished the chance to put on his boots and cowboy hat, load visitors into his Cadillac, and drive them, terrified, at ninety miles per hour about his far-flung property. But many easterners, especially well-educated people who admired the polished and stylish Kennedy, considered Johnson to be a virtual caricature of all that they associated with Texas. Many, noting his shiny, wide-lapeled suits and slicked-back hair, likened him to a riverboat gambler. These aspersions stung Johnson, a proud and vain man. Whatever he did, he came to believe, the eastern Establishment would disparage.

In the minds of Johnson and many of his admirers this Establishment had a broad reach. It consisted, they thought, of reporters and columnists for eastern media empires, such as the *Washington Post* and the *New York Times*, and of their acolytes, highly educated esthetes and snobs from

3. Major sources on Johnson's life, aside from Kearns, include Robert Dallek, *Lone Star Rising: Lyndon Johnson and His Times, 1908–1960* (New York, 1991); Paul Conkin, *Big Daddy from the Pedernales: Lyndon Baines Johnson* (Boston, 1986); Vaughn Bornet, *The Presidency of Lyndon Johnson* (Lawrence, 1983); Robert Caro, *The Years of Lyndon Johnson: The Path to Power* (New York, 1982); Caro, *The Years of Lyndon Johnson: Means of Ascent* (New York, 1989); and Alonzo Hamby, *Liberalism and Its Challengers: F.D.R. to Reagan* (New York, 1985), 231–81.

expensive eastern schools and universities. The *Times*, Johnson complained in 1967, "plays a leading part in prejudicing people against [me]. Editors won't use the word, 'President Johnson,' in anything that is good. Bigotry [against Texans] is born in some of the *New York Times* people."[4] Johnson especially identified Kennedy-lovers with the Establishment. Bewildered by the adulation that Kennedy received while he had lived, Johnson grew resentful when the "Kennedy people" did not rally unconditionally to his side after 1963. "It was the goddamnest thing," he later told a biographer. "He [Kennedy] never said a word of importance in the Senate and he never did a thing. But somehow . . . he managed to create the image of himself as a shining intellectual, a youthful leader who would change the face of the country. Now, I will admit that he had a good sense of humor and that he looked awfully good on the god-damn television screen and through it all was a pretty decent fellow, but his growing hold on the American people was a mystery to me."[5]

If Johnson had been a more reflective man, he might have understood why many Americans failed to warm up to him. For Johnson was in many ways unlikeable. Stories about his towering vanity are legion. As a senator he offered to give a calf from his ranch to parents who named their babies after him. When he became President, he ordered White House photographers to record his movements for posterity. One estimate concludes that he had 500,000 photos taken of himself. Johnson enjoyed studying these shots and regularly gave them to visitors and dignitaries. Johnson also had plastic busts made of himself, which he was known to stroke affectionately while conversing with people in the White House. Paying a visit to the Pope, he was presented with a fourteenth-century painting as a gift. In return he surprised the Pontiff by giving him a bust of himself.[6] Jokes, most of them unkind, featured Johnson likening himself to Abraham Lincoln, FDR, or—most commonly—to Jesus or God.

Johnson's vanity probably served to compensate for the insecurities that seemed central to his character. His urge to dominate may have had similar roots. Whatever the sources, his need for total loyalty among associates was legendary. Staff members came to understand that they not

4. Cited in Larry Berman, "Lyndon Baines Johnson: Paths Chosen and Opportunities Lost," in Fred Greenstein, ed., *Leadership in the Modern Presidency* (Cambridge, Mass., 1988), 144–45.

5. Robert Dallek, "My Search for Lyndon Johnson," *American Heritage*, Sept. 1991, pp. 84–88. The biographer was Kearns.

6. David Culbert, "Johnson and the Media," in Divine, ed., *Exploring the Johnson Years*, 214–48.

only had to work long hours; they also had to honor him and to surrender to his imperious will. When Johnson became President he dictated dress codes for his aides. He insisted that they be reachable at all times of the day or night. To be assured of getting in touch with Joseph Califano, one of his most trusted advisers, Johnson had a telephone installed next to the toilet in Califano's office bathroom. Worst of all for a staff member was when he was summoned to confer in Johnson's bathroom while the President was sitting on the toilet.[7]

Johnson demanded not only loyalty but also subservience from people around him. To humiliate and even to frighten others was to heighten his own sense of himself. He explained to an aide, "Just you remember this: There's only two kinds at the White House. There's elephants and there's pissants. And I'm the only elephant." A disenchanted press secretary, George Reedy, later observed that Johnson "as a human being was a miserable person—a bully, sadist, lout, and egoist. . . . His lapses from civilized conduct were deliberate and usually intended to subordinate someone else to do his will. He did disgusting things because he realized that other people had to pretend that they did not mind. It was his method of bending them to his desires."[8]

How could such a man rise so high in American politics by 1960? One reason was desire. Beginning in 1931, when he first arrived in Washington as a twenty-three-year-old congressional secretary, Johnson was obsessively ambitious. He worked extraordinarily long hours and drove himself to near exhaustion in many of his campaigns. Running for the House in a special election in 1937, he lost forty-two pounds in forty days. Seeking a Senate seat eleven years later, he lost some thirty pounds before squeaking through to victory in a primary. He was always careless of his health, bolting down food, smoking heavily, and drinking recklessly. In 1955, when he was only forty-seven, he suffered a severe heart attack and cut back on cigarettes. He remained driven and restless, unable to settle for what he had done in life.

With his eye on the future, Johnson always displayed shrewd political instincts. As a young man he sought the attention of FDR, who named him in 1935 as director of the National Youth Administration (NYA) in Texas. Johnson was then only twenty-six. Once in the House he quickly befriended Sam Rayburn, the powerful Texan who became Speaker in 1940 and held that office (except during four years of GOP control) until

7. Joseph Califano, *The Triumph and Tragedy of Lyndon Johnson: The White House Years* (New York, 1991), 26–28.
8. Berman, "LBJ: Paths Chosen," 139.

his death in 1961. Rayburn, a lifelong bachelor, treated Johnson almost as a son. When Johnson moved to the Senate in 1949, he became a protégé of Richard Russell of Georgia, leader of the southern Democratic bloc in the upper house and one of the most influential men in American government. Carefully cultivated connections such as these helped Johnson to rise with remarkable speed. When he became Democratic leader in the Senate in 1953, he was only forty-five years old, the youngest man in modern American history to hold such a position.

Johnson's ability to manipulate the political system, while unmatched, only partly explained his advancement. Equally important was his great skill as persuader and coalition-builder, especially in the Senate, where he established his national reputation in the 1950s. Johnson made it his business to know everything he could about the personal foibles and political needs of his colleagues, whom he assiduously wooed with courtesies and small favors. He knew when to flatter, when to bargain, when to threaten. He worked hard to familiarize himself with the details of legislation and counted noses carefully before going out on a limb. When he decided on a course of action, he expected to win, and he usually did.⁹

Colleagues on the Hill who resisted Johnson often received what contemporaries described in awe as The Treatment. This was Johnson the persuader at his most compelling. Senator George Smathers of Florida described The Treatment as "a great overpowering thunderstorm that consumed you as it closed in around you." It could last a few minutes or several hours. As the journalists Rowland Evans and Robert Novak described it, Johnson "moved in close, his face a scant millimeter from his target, his eyes widening and narrowing, his eyebrows rising and falling. From his pockets poured clippings, memos, statistics. Mimicry, humor, and the genius of analogy made The Treatment an almost hypnotic experience and rendered the target stunned and helpless."¹⁰

Liberals who observed Johnson in these Senate years refused to be beguiled. Most of them continued to regard him as a smooth and self-serving political operator who had engaged in vote fraud in order to enter the Senate. Some of his aides, such as right-hand man Bobby Baker, seemed even more slippery. Liberals also refused to credit him with true instincts for reform. Noting his friendship with conservative southerners like Russell—and his reluctance to criticize Eisenhower—they further

9. Conkin, *Big Daddy*, 132–36.
10. Rowland Evans and Robert Novak, *Lyndon B. Johnson: The Exercise of Power* (New York, 1966), 105; William O'Neill, *Coming Apart: An Informal History of America in the 1960s* (Chicago, 1971), 105.

denounced him for his support of Texas oil interests, his coolness to organized labor, and what they assumed was his ambivalence about civil rights.

Liberals had ample cause for questioning Johnson's claims to be one of them. Especially before 1954, he felt he had to tread carefully lest he offend politically powerful conservative business interests in Texas. Johnson, however, was more liberal than many contemporaries imagined. Obsessed though he was with advancing himself, he also hoped to advance social justice. As director of the NYA in Texas in the 1930s he had worked effectively to direct federal help to the young people who were the targets of the legislation; no other state director was so energetic or successful as Johnson. He strongly backed other New Deal social programs, especially rural electrification and public housing. When he ran for the House in 1937, he was the only contender enthusiastically to support Roosevelt's controversial plan to pack the Supreme Court. Then and later FDR remained his idol.[11]

Johnson's belief in bread-and-butter liberalism as advanced by the New Deal reflected his regional interests. He insisted that federal aid was the key to breaking down the isolation and destitution of the South, by far the nation's most poverty-stricken region. He took care while President to direct as much federal money as possible to the South and West, thereby helping to bring the Sun Belt, as it became called, into the mainstream of the American economy. But Johnson's faith in the State as benefactor of socio-economic progress transcended regional particularism. In this belief, the essence of his political philosophy, he was very much in tune with liberals of his generation.

AS JOHNSON PREPARED his domestic programs in late 1963, he knew—everyone knew—that he possessed an enormous advantage that liberal predecessors had been denied since the late 1930s: a national mood so eager for strong presidential leadership that even Congress and interest groups had to take heed. A number of developments helped to encourage such a mood. Kennedy's expansive rhetoric about a New Frontier, although ignored on the Hill in his lifetime, had kept a liberal agenda alive. Surging growth of the economy since 1962 helped enormously: the nation, it seemed, could afford expensive federal programs. And highly optimistic liberal bureaucrats, such as Walter Heller of the Council of

11. William Leuchtenburg, *In the Shadow of FDR: From Harry Truman to Ronald Reagan* (Ithaca, 1983); Hamby, *Liberalism*, 256–65.

Economic Advisers, were certain that social science expertise and computerization gave government the tools to change the world.

Forces such as these, however, paled in power compared to the impact of Kennedy's assassination. This impact, to be sure, did not move everyone in the same way. Some people thought that sinister forces had overtaken the country: reform in such a satanic society was futile. Conservatives clung to historically powerful doubts about the capacity of government to improve on the market. Special interest groups continued to resist changes that appeared to endanger them. Millions of Americans, as always, remained apathetic or apolitical. Still, Kennedy's assassination touched many people as they had not been touched before. Could the murder of so young and promising a leader be redeemed? Must his life be wasted? A palpable popular yearning for determined political leadership seemed ubiquitous.

Johnson provided it. In the months following Kennedy's death he scarcely wavered in his certainty that he, like FDR, could develop public programs to benefit society and to secure the progress of the "Free World." He was equally confident that the United States had more than enough resources to accomplish these things. Indeed, he epitomized the rise of grand expectations that captured many liberals in the 1960s. "I'm sick of all the people who talk about the things we can't do," he told an aide in 1964. "Hell, we're the richest country in the world, the most powerful. We can do it all."[12]

In pushing for such an expansive liberalism, LBJ (as many people called him) devoted unprecedented attention to Capitol Hill. "There is," he explained later, "but one way for a President to deal with the Congress, and that is continuously, incessantly, and without interruption. If it's really going to work, the relationship between the President and Congress has got to be almost incestuous. . . . . He's got to build a system that stretches from the cradle to the grave, from the moment a bill is introduced to the moment it is officially enrolled as the law of the land." He added, "A measure must be sent to the Hill at exactly the right moment. . . . Timing is essential. Momentum is *not* a mysterious mistress. It is a controllable fact of political life that depends on nothing more exotic than preparation."[13]

Johnson worked as hard at accommodating congressional sensibilities

12. Robert Collins, "Growth Liberalism in the Sixties: Great Societies at Home and Grand Designs Abroad," in David Farber, ed., *The Sixties: From Memory to History* (Chapel Hill, 1994), 19.
13. Kearns, *Lydon Johnson*, 226.

as any President in modern American history. Again and again, and at all hours, he picked up the phone to call and to flatter legislators, among them junior people who had never heard from the White House before. Leaving nothing to chance, he insisted that his political aides be on the Hill at all times. With characteristic crudeness he told them, "You got to learn to mount this Congress like you mount a woman." He told Califano never to take a congressman's vote for granted. "Don't ever *think* about those things. Know, know, *know*! You've got to *know* you've got him, and there's only one way you know." Johnson raised his right hand, made a fist, and looked at it. "And that's when you know you've got his pecker right here." LBJ then opened his desk drawer, unclasped his fist as though he were dropping something, slammed the drawer, and smiled.[14]

One of Johnson's first major efforts after Congress resumed work in early 1964 was to secure passage of the tax cut that Kennedy had supported. A version of it had previously passed the House, but a companion bill was stalled in the Senate by determined fiscal conservatives who demanded that the administration also agree to pare expenditures in the coming fiscal year. Heller and other liberals resisted this pressure and held out for a budget that would set spending at $101.5 billion. Johnson broke the stalemate by siding with the conservatives. "If you don't get this budget down to around 100 billion," he told Heller, "you won't pee one drop."[15]

Congressional conservatives, pleased by Johnson's concession, soon approved a bill. It called for cuts totaling $10 billion over the next two years, and Johnson signed it in February. As many liberals had predicted earlier, the law mainly helped upper-income groups and corporations. Loopholes persisted, unchallenged by congressional liberals or conservatives, for years thereafter. Still, Johnson's accommodating stance on taxes helped to produce a law that had seemed badly bogged down before Kennedy's death. His flexibility thereby enhanced his reputation as a legislative leader. The law, moreover, promised to benefit a good many people, pleasing members of Congress in an election year. Some economists, Heller included, hoped that the tax cut would indeed stimulate the economy and thereby demonstrate the virtues of Keynesian ideas.

Johnson's next major effort pursued a goal that he grandiloquently

14. Califano, *Triumph and Tragedy*. 142, 110.
15. Herbert Stein, *The Fiscal Revolution in America* (Chicago, 1969), 452–53; John Witte, *The Politics and Development of the Federal Income Tax* (Madison, 1985), 155–75; Allen Matusow, *The Unraveling of America: A History of Liberalism in the 1960s* (New York, 1984), 55–59.

announced in his State of the Union message in January: "This administration, today, here and now, declares unconditional war on poverty in America."[16] This idea, too, he had taken in part from Kennedy, whose economic advisers had been developing plans to battle poverty in the last weeks of his life. Heller managed to see Johnson on the evening of November 23, one day after the assassination, and to mention in a general way what Kennedy and his advisers had been thinking. "That's my kind of program," Johnson replied. "Move full speed ahead."[17]

Heller, elated, proceeded to do so. So did others in the governmental bureaucracy who had already been working on related problems. Among them were officials on the President's Committee on Juvenile Delinquency, which Kennedy had set up in May 1961. Exploring programs in various cities, the committee had become especially attracted to Mobilization for Youth, an experimental effort on the Lower East Side in New York City. Theorists for MFY, as it was called, believed that social problems such as delinquency stemmed mainly from blocked economic "opportunity" and that the remedy lay in community planning to increase such opportunities.[18] Impressed, the Kennedy administration awarded MFY a grant of $2.1 million in May 1962.[19] This idea, that poor people needed government help and guidance to promote wider opportunity, lay at the heart of subsequent liberal efforts in the 1960s to fight poverty.

Michael Harrington's *The Other America*, published in late 1962, gave added force to liberals who wanted to battle against poverty. Harrington maintained that 40 to 50 million Americans, or as much as 25 percent of the population, were deeply in need. The book attracted widespread attention, especially after it received a long and laudatory review by Dwight Macdonald in the *New Yorker* in January 1963. Kennedy apparently read the article. While Harrington had little effect on the details of subsequent policy-making—he was far to the left of administration liberals—his book did much to place the subject on the national political agenda. Poverty,

16. *New York Times*, Jan. 9, 1964.
17. Godfrey Hodgson, *America in Our Time* (Garden City, N.Y., 1976), 173; James Sundquist, *Politics and Policy: The Eisenhower, Kennedy, and Johnson Years* (Washington, 1968), 111–54; Matusow, *Unraveling*, 119–22.
18. Richard Cloward and Lloyd Ohlin, *Delinquency and Opportunity: A Theory of Delinquent Gangs* (New York, 1960).
19. Matusow, *Unraveling*, 108–12; Ira Katznelson, "Was the Great Society a Lost Opportunity?," in Steve Fraser and Gary Gerstle, eds., *The Rise and Fall of the New Deal Order, 1930–1980* (Princeton, 1989), 185–211.

having been "invisible" (in Harrington's word) for years, had now been "rediscovered."[20]

Government officials, meanwhile, were sharpening their analytical skills to produce the concept of a "poverty line." In 1964 this line was estimated to be around $3,130 for a family of four and a little more than $1,500 for a single individual.[21] By that standard, 40.3 million people were "poor," around 21 percent of the population of 192 million.[22] Certain groups, moreover, clustered below the line, including more than half of black Americans, nearly one-half of people living in female-headed families, and one-third of people over 65.

With poverty in the news, Johnson jumped at the chance to tackle it. He remembered the good that the NYA and other New Deal social programs had done in the 1930s, and he wanted to be the President who finished what FDR had started. Equally important, he shared the contemporary liberal view that the United States, a rich and resourceful country, could afford to do something. Johnson also believed unquestioningly in another liberal faith: that government had the skill to improve the lot of its citizens. What motivated Johnson to fight poverty, in short, was not the worsening of a social problem—higher percentages of Americans had been poor in the 1950s—but the belief that government could,

20. Michael Harrington, *The Other America: Poverty in the United States* (New York, 1962), 9, 171–86; Dwight Macdonald, "The Invisible Poor," *New Yorker*, Jan. 19, 1963, pp. 130ff.
21. The lines became official only in 1964, following passage of the "war" on poverty.
22. Sheldon Danziger and Daniel Weinberg, "The Historical Record: Trends in Family Income, Inequality, and Poverty," in Danziger, Gary Sandefur, and Weinberg, eds., *Confronting Poverty: Prescriptions for Change* (Cambridge, Mass., 1994), 18–50; James Patterson, *America's Struggle Against Poverty, 1900–1994* (Cambridge, Mass., 1995), 78–82. The official poverty line, which became a staple of government data thereafter, was based on estimates of food costs required for various-sized families. These costs were then multiplied by three to reach the minimum amounts that families needed to stay out of poverty. These were the lines. In 1964 it was calculated that a family of four needed $1,043 a year in order to afford a decent diet, and three times as much, or $3,130, to stay out of poverty. Liberals in later years insisted that poor people—indeed all Americans—spent considerably less than one-third of their incomes on food and that the multiplier should be higher than three. That would increase the level of the poverty line, and in so doing the number of people defined as "poor." Conservatives retorted that the lines were far too high, therefore exaggerating the numbers. The debate indicated an important point: definitions of absolute poverty are highly subjective.

and should, enter the battle. These optimistic expectations, not despair, lay at the heart of American liberalism in the sixties.

Political motivations further influenced Johnson's enthusiasm for a war against poverty. He was anxious to draw upon Kennedy's ideas and to get something passed in order to demonstrate his skills with Congress. He especially wanted to be able to point to solid accomplishments in the forthcoming presidential campaign. In February he named R. Sargent Shriver, a brother-in-law of Kennedy who already headed the Peace Corps, to develop a bill. Shriver, like Johnson, was less interested in the details of legislation than in results on the Hill, and he fashioned a hodgepodge of presumably popular prescriptions against poverty. These included programs to improve education for pre-schoolers and adults, to expand job training, and to create a domestic version of the Peace Corps that would send idealistic volunteers into poverty-stricken areas.[23]

Neither Shriver nor Johnson intended their efforts to increase governmental spending on public assistance. Both hated the very idea of long-term welfare dependency and of costly governmental outlays for public aid. "Welfare," indeed, remained a dirty word in the lexicon of liberals as well as of conservatives in the United States. Instead, Johnson hoped that a "war" on poverty would provide the "opportunity" necessary to help people help themselves. Welfare, thereby rendered unnecessary, would wither away. The goal, Shriver said repeatedly, was to offer a "hand up, not a hand out," to open "doors" to opportunity, not to establish federally financed "floors" under income. Johnson also made it clear that he would not increase taxes—after all, he had just signed a tax cut—for a large-scale program of public employment. Having promised conservatives to control spending, he intended to hold the line. From the beginning, therefore, the war on poverty had limited firepower. It was to rely mainly on educational programs and job training, mostly for young people, in order to improve their skills and their equality of opportunity. Enhancing opportunity was a profoundly American idea, rooted in a tradition dating to Thomas Jefferson and the Declaration of Independence.[24]

The poverty bill that Shriver and others fashioned in the spring of 1964

23. James Sundquist, "Origins of the War on Poverty," in Sundquist, ed., *On Fighting Poverty: Perspectives from Experience* (New York, 1969), 6–33; Sundquist, *Politics and Policy*, 111–54; Matusow, *Unraveling*, 97–107.
24. Patterson, *American Struggle*, 126–54; Mark Gelfand, "The War on Poverty," in Divine, ed., *Exploring the Johnson Years*, 126–54; Frances Fox Piven and Richard Cloward, *Regulating the Poor: The Functions of Public Welfare*, 2d ed. (New York, 1993), 248–340.

also included the idea of "community action." Those who coined the phrase did not define it with precision. Without much reflection they hoped that poverty programs would promote community development— another cherished American faith—and that local leaders would be involved at some level with formulating and carrying out the war. Only later, when the war against poverty got started, did it become clear that community action programs, or CAPs, would become the heart of the effort.[25]

A few of the draftsmen in early 1964 went further and imagined that poor people themselves should play important roles in developing community action. In the phrasing that appeared in the legislation, the poor were to have "maximum feasible participation." What the phrase meant did not become clear, and it attracted little notice during congressional debates over the bill. It was only later that a few radicals and activists tried to assign "maximum feasible participation" to themselves. In doing so they clashed sharply with local authorities, including Democratic politicians who had other ideas for the money. "Community action" and "maximum feasible participation" then became fighting words that divided the Democratic party. Such were some of the longer-range consequences of a bill that Johnson had expected to bring him praise and gratitude.[26]

These, to be sure, were unintended consequences. Neither Johnson nor Shriver foresaw them in the spring of 1964. Still, they can be faulted for their haste. Hurrying to start the war, they did not wait for studies to be done that might have told them more about the nature of poverty, which had complicated structural roots that their bill scarcely considered. The majority of poor people in the United States—as in any industrialized economy—needed more than education, job training, or the engagement of domestic peace corps workers. Millions were too old, too sick, or too disabled to benefit much from such efforts. Single mothers with young children confronted multiple problems, including the need for costly day care. Blacks, Mexican-Americans, and other minorities faced widespread discrimination in housing and jobs. Unemployment, underemployment, and low wages afflicted millions of people in the labor force. Farm and migrant workers had long been among the poorest of people; approximately 50 percent of Americans living on farms in 1960 were poor by government definition. The South, a predominantly rural region, then

25. See chapter 19 for community action after 1964.
26. Daniel Moynihan, *Maximum Feasible Misunderstanding: Community Action in the War Against Poverty* (New York, 1969).

and later suffered from the highest incidence of poor people in the nation.[27]

Radicals such as Harrington understood and highlighted the depth of these structural roots of poverty. "The entire invisible land of the other America," he wrote, "[had become] a ghetto, a modern poor farm for the rejects of society and the economy." Harrington went further in his condemnation of American society to argue that many of America's poor lived in a "separate culture, another nation, with its own way of life." He concluded, "The most important analytic point to have emerged in this description of the other America is the fact that poverty in America forms a culture, a way of life and feeling, that makes it a whole."[28]

Harrington employed the concept of a "culture of poverty" to dramatize the seriousness of the problem and to force politicians to act. Conservatives, however, made the most of the idea in subsequent debates over the nature of poverty. If poverty was rooted in the very culture of many low-income Americans, they said, then it was foolish for policy-makers to think they could do much about it.[29] Liberal efforts, it followed, at best were a waste of the taxpayers' money. At worst they were counter-productive, for they would encourage "undeserving" people—"drunks," "deadbeats," "welfare mothers"—to rely on the government. Conservatives further insisted that most "poor" people managed all right. To be "poor" in the 1960s, they said with some asperity, was to live far more comfortably than "poor" people had lived in the 1930s or at the turn of the century.[30]

Some of these conservative arguments were hard to refute. Most of the people defined by the government as poor in the 1960s had central heating, indoor plumbing, and television. Many owned cars. In much of the world they would have been considered well-off. But in comparing living standards in the 1960s to the past, conservatives failed to appreciate that escalating expectations were beginning to affect the poor as well as more

27. Danziger and Weinberg, "Historical Record"; Patterson, *America's Struggle*, 78–82, 126–54.
28. Harrington, *Other America*, 10, 15–18, 159.
29. For writing on the "culture of poverty," see ibid., 15–18, 121–26; and Oscar Lewis, "The Culture of Poverty," *Scientific American*, 215 (Oct. 1966), 19–25. Excellent evaluations of the concept are Charles Valentine, *Culture and Poverty: Critique and Counterproposals* (Chicago, 1968); and Eleanor Burke Leacock, ed., *The Culture of Poverty: A Critique* (New York, 1971).
30. A summary of such views is Charles Murray, *Losing Ground: American Social Policy, 1950–1980* (New York, 1984). See also Martin Anderson, *Welfare: The Political Economy of Welfare Reform in the United States* (Stanford, 1978).

affluent people. Television heightened these expectations by blanketing the screen with programs and commercials that advertised the affluent society. Low-income people, realizing what they lacked, developed an ever more acute sense of relative deprivation. As their expectations intensified in the next few years—in part because of hype from the "war" against poverty itself—this sense increased. It underlay much of the acrimonious social conflict that arose later in the decade.[31]

Conservative arguments about a "culture of poverty" were also flawed. To be sure, durable subcultures, most obviously among racial minorities, persisted in the United States. Notions about cultural "consensus" that had been widespread in the 1950s seemed increasingly wrong-headed in the 1960s, when blacks and other self-conscious ethnic groups reasserted their cultural roots. And some of these groups, such as blacks, Mexican-Americans, and Native Americans, had rates of poverty that were much higher than those of whites. But most low-income people, blacks included, continued also to subscribe to mainstream cultural values such as the blessings of democracy, hard work, and long-term marriage and family life. They were not isolated in a puncture-proof or inferior "culture" of their own. Where they were often most different was not in culture but in class standing. They lacked money. Lacking money, they lacked power, and they felt aggrieved. Many institutions, moreover, seemed distant or irrelevant to them. Labor unions, which had assisted upward mobility in the 1930s and 1940s, were weakening and offering relatively little help. Many low-income Americans, living in such a world, grew angry and resentful. Others remained apathetic, thereby appearing to confirm negative stereotypes about them. Feelings such as these exposed class and racial divisions that radicals like Harrington deplored.

Johnson, Shriver, and the others who developed the war on poverty, however, were not radicals. They were optimists who reflected the confidence of contemporary American liberal thought. Unlike radicals, they thought that most poor people needed only a helping hand to rise in life. Unlike conservatives, they had great faith that government could and should extend that hand. Largely unaware of rising feelings of relative deprivation, they gave little thought to the idea (which was politically unrealistic) of redistributing wealth or income. They were not much concerned about inequality. They focused instead on programs to pro-

31. Thomas Jackson, "The State, the Movement, and the Urban Poor: The War on Poverty and Political Mobilization in the 1960s," in Michael Katz, ed., *The "Underclass" Debate: Views from History* (Princeton, 1993), 403–39.

mote greater opportunity—a politically attractive goal—and pushed ahead impatiently for congressional action.

When Shriver and his advisers sent their handiwork to the Hill, Republicans and many Democratic conservatives hotly opposed it. Senate Republican leader Everett Dirksen of Illinois called the idea "the greatest boondoggle since bread and circuses in the days of the ancient Roman empire—when the Republic fell."[32] But the legislative outcome was never much in doubt, for Johnson had worked with Shriver to assure that he had more than enough loyal Democrats—and a few liberal Republicans—to pass the bill. That they did on heavily partisan but nonetheless comfortable votes (226 to 185 in the House and 61 to 35 in the Senate) in August. The final measure included most of the various ideas and programs that had surfaced earlier in the year: loans for small business and rural development, funding for work-study benefiting college students, and the domestic peace corps idea, which was entitled Volunteers in Service to America, or VISTA. The bill also authorized establishment of Job Corps centers to provide job training and of a Neighborhood Youth Corps to create low-wage jobs for young people, mainly in central city areas. Finally, it called for development of community action programs, to be worked out by local leaders in concert with Washington. The act, underlining the focus on providing opportunity, set up an Office of Economic Opportunity (OEO).[33]

To no one's surprise Shriver became head of OEO, whereupon he moved quickly to create the necessary bureaucracy and to hasten the flow of money before election time. Shriver believed sincerely in the capacity of government programs to help the needy, and he was an aggressive administrator. He proved especially adept on Capitol Hill, where he spent much of his time in the next few years trying to ensure continued support for the "war." Some of the initiatives pioneered by OEO-funded community action programs, moreover, generated fairly considerable support over time, notably Head Start and Follow Through, which aimed to improve the educational opportunity of poor children.[34] Another CAP initiative, Neighborhood Legal Services, was much more controversial

32. Richard Polenberg, *One Nation Divisible: Class, Race, and Ethnicity in the United States Since 1938* (New York, 1980), 201.
33. Patterson, *America's Struggle*, 126–41; John Schwarz, *America's Hidden Success: A Reassessment of Twenty Years of Public Policy* (New York, 1983), 20–78.
34. Debate over the long-range effectiveness of Head Start programs, however, continued decades later. See Edward Zigler and Susan Muenchow, *Head Start: The Inside Story of America's Most Successful Educational Experiment* (New York, 1992); and Diane Ravitch, *The Troubled Crusade: American Education, 1945–1980* (New York, 1983), 158–60.

but managed to offer sorely needed legal advice to welfare recipients and others. These efforts returned the problem of poverty, long neglected, to on the agenda of national policy-making.

OEO, however, was poorly funded from the first. Congress appropriated at most $800 million in new money in 1964. This was less than 1 percent of the federal budget. With at least 35 million people officially defined as poor at the time, this came to a little more than $200 per poor person per year. And very little of the money went directly to the poor. Most OEO dollars, instead, covered the salaries and expenses of administrators, professionals, and government contractors who provided services such as Head Start or job training. Some of these officials were upwardly mobile blacks who for the first time secured fairly dependable employment with the government; for them, the war on poverty was a true opportunity. But few of them had been poor. Not only radicals like Harrington but also many liberals deplored the miniscule character of the program. It was at best a skirmish, not a war. [35]

If the United States had faced a depression in 1964, larger sums for a "war" on poverty might have been acceptable on Capitol Hill. For a brief spell in the mid-1930s Congress had authorized $3 billion a year for relief and public employment, more than one-third of the federal budget. But 1964 promised great and continuing economic growth, and few people applied pressure for larger expenditures to help the poor. On the contrary, most Americans maintained historically durable and largely conservative attitudes about the subject: aside from special "categories" of the "deserving poor"—people who were disabled, widows with children, the elderly without social insurance—the needy should normally take care of themselves. In the face of such widely held faith in individualism, it was unthinkable in 1964 that Congress would do much more than it did.

Lack of money, to be sure, was a drawback if one sought truly to fight a war. But low funding was not, as some observers later claimed, solely the fault of a fiscally prudent Congress or of a subsequent rise in military spending that threatened growth in domestic programs. On the contrary, Johnson himself never sought much more than he received. This was not only because he focused on providing opportunity rather than handouts. It was also because he truly hoped to hold down spending. He came close, managing to reduce federal expenditures slightly from 1964 to 1965—from $118.5 billion to $118.2 billion—and the annual deficit from $5.9 billion to $1.4 billion. Federal government spending as a percentage of

35. Conkin, *Big Daddy*, 214; Sundquist, *Politics and Policy*, 154.

gross domestic product actually decreased a little in the early 1960s, from
18.3 percent in 1960 to 17.6 percent in 1965, before rising again to 19.9
percent in 1970.[36] Conservatives who railed at LBJ (and more generally at
liberals in the 1960s) for "throwing money at problems" somewhat distorted the realities of government fiscal policies. Indeed, increases in
federal spending for domestic purposes, while substantial by 1968, were
hardly lavish. It was not until the supposedly more conservative 1970s and
1980s that public spending for domestic programs—especially health care
and Social Security—exploded in size and created huge deficits. Some of
this increase was attributable to growth (largely unforeseen) in programs
from the Johnson years, but much also stemmed from legislation in the
early 1970s.[37]

Nor was lack of money the only drawback. As indicated, OEO-type
programs failed to attack the problems of the poor. Take job training, for
example. Some trainees managed to find work, but it was far from clear
that the training turned the tide for them. To some extent, Job Corps and
other manpower programs "creamed" upwardly mobile people who
would have found work in any event. Moreover, some of the trainees who
found employment replaced other workers, canceling out any net gains.
Otherwise, the OEO made no effort in the short run—such as providing
public jobs—to fight underemployment and unemployment. It did nothing to lessen inequality. To have placed more federal money in job
training—or in educational programs such as Head Start—would have
enhanced the life-chances of some of America's poor, but perhaps not

36. This percentage, a helpful measure of the role of federal spending in the overall
    economy, had peaked in 1945, thanks to the war, at 43.7 percent, before falling to
    16 percent in 1950, and then rising to the figure of 18.3 in 1960. After 1970 it
    stablilized at around 20 percent until 1975, when it jumped to 22 percent. It
    remained between 22 and 24.4 percent between 1975 and 1994.
37. Federal spending in current dollars rose rapidly after 1965—to $183.6 billion by
    fiscal 1969. Deficits also jumped: to $3.7 billion in 1966, $8.6 billion in 1967,
    and $25.1 billion in 1968 (after which a tax hike enacted in 1968 helped produce a
    surplus of $3.2 billion in fiscal 1969, the last federal government surplus in
    modern times). Much of this growth in spending, however, went for military
    purposes. Outlays for domestic goals also increased in these years but did not jump
    dramatically until after 1970. By fiscal 1975, federal spending had risen to $332
    billion, causing a deficit in that year of $53.2 billion. Subsequent deficits led to a
    rise in per capita federal indebtedness from $4,036 in 1980 to more than $18,000
    in mid-1995. (All numbers used here are in current dollars.) See *Statistical
    Abstract of the United States, 1994* (Washington, 1994), 297, 330–33. Also
    chapters 21, 23, and 25.

very many. In the short run, offered neither work nor welfare, they remained poor.[38]

Flaws such as these might not have mattered much if all that was at stake was a billion or so dollars a year. And indeed they did not matter in the fall of 1964, when Shriver pressed ahead excitedly with his plans. At that time, the existence of the war on poverty enhanced Johnson's reputation among liberals as a man who cared about the unfortunate and who could get Congress to do his bidding. LBJ's political timing, in which he took justifiable pride, had been good as he readied for the presidential campaign.

But the flaws did matter in the longer run—even as early as 1965. This was in part because Johnson and his aides greatly oversold their efforts, thereby raising unrealistic expectations, both among liberal observers and among the poor themselves. The struggle against poverty was hailed as the centerpiece of LBJ's liberal program. It was to be an "unconditional war." Poverty, Shriver said, could be wiped out (with sufficient funding) within ten years. Predictions such as these were an understandable and to some extent pardonable aspect of political salesmanship. But given the obduracy of destitution in all human societies, they were astonishing. When the predictions failed to materialize, mounting disillusion and cynicism undermined the liberal premise even as the expectations survived.[39]

SECURING PASSAGE OF A TAX CUT and a war on poverty helped to establish Johnson's credentials as a skilled and successful leader of Congress. But the civil rights bill that Kennedy had introduced in June 1963 represented the central test of LBJ's presidential abilities. Johnson's struggle to get it passed absorbed most of his time and effort for the first six months of 1964.

Several considerations led Johnson to throw himself into the cause. First and foremost, he believed in it. Having grown up in Texas, he had seen first-hand the viciousness of racial discrimination, and he empathized with the victims. As a congressman he had battled to ensure that federal agricultural programs treated blacks and whites equally. Milo

---

38. Gary Burtless, "Public Spending on the Poor: Historical Trends and Economic Limits," and Rebecca Blank, "The Employment Strategy: Public Policies to Increase Work and Earnings," in Danziger et al., eds., *Confronting Poverty*, 51–84, 168–204; Charles Morris, *A Time of Passion: America, 1960–1980* (New York, 1984), 106.

39. Hugh Heclo, "Poverty Policies," in Danziger et al., eds., *Confronting Poverty*, 396–437.

Perkins, a top official in the Farm Security Administration at the time, recalled that Johnson "was the first man in Congress from the South ever to go to bat for the Negro farmer." When Congress approved appropriations for public housing in 1937, Johnson persuaded officials of the United States Housing Authority to select Austin as one of the first three cities in the nation to receive funding. He then got the city to "stand up for the Negroes and the Mexicans" and to designate 100 of the 186 housing units for them.[40] Although he moved cautiously as a senator so as not to offend white supporters, he remained well to the left of most southern politicians at the time. When he singled out Kennedy's civil rights bill in his speech to Congress five days after the assassination, he made it clear that he was sincere and determined.

The new President also recognized that great effort was necessary if the bill were to become law. Kennedy's measure was certain to pass the House in some form in 1964, but to maintain credibility among liberals, many of whom deeply distrusted him, LBJ had to lead a strong bill through the gauntlet of the Senate. Johnson recalled, "If I didn't get out in front on this issue [the liberals] would get me. . . . I had to produce a civil rights bill that was even stronger than the one they'd have gotten if Kennedy had lived. Without this, I'd be dead before I could even begin."[41]

As expected the House approved a bill early in February by the comfortable margin of 290 to 130, leaving the Senate to determine the fate of the measure. When Johnson heard of the vote, he wasted no time. "All right, you fellows," he phoned aides celebrating in a House corridor. "Get over to the Senate. Get busy. We've won in the House, but there is a big job across the way."[42] Johnson and his aides knew that southern senators, led by Russell of Georgia, would try to filibuster the bill to death. Passage depended on his ability to get the Senate to vote for cloture, the only way to stop the interminable talk. Under rules at the time cloture required the votes of two-thirds of the Senate.

The key to getting two-thirds was GOP minority leader Dirksen of Illinois, whose ultimate position on the bill would guide many of the thirty-two other Republicans (one-third of the chamber) in the Senate. But Dirksen, a colorful and loquacious conservative, was on the fence. On the one hand, he seemed to favor some sort of bill. On the other, he

40. Dallek, "My Search," 88.
41. Steven Lawson, "Civil Rights," in Divine, ed., *Exploring the Johnson Years*, 99–100.
42. Robert Weisbrot, *Freedom Bound: A History of America's Civil Rights Movement* (New York, 1990), 91.

(and many other Republicans) wanted portions softened. Some sought to emasculate the measure. While Johnson was willing to consider modest changes, he knew that liberals demanded a tough bill such as the one that had passed in the House. Over the next several months he spent hours wooing Dirksen, a friend and former colleague, sometimes by inviting him to the White House, swapping stories with him, and drinking with him into the night.[43]

Johnson drove his staff members, keeping them around until late. He regularly quizzed Larry O'Brien, his congressional liaison, on exactly what various senators had said that day, and he kept long tally sheets of senators' names and columns for YES, NO, and UNDECIDED. As Califano said later, "Johnson would devour these tally sheets, thumb moving from line to line, like a baseball fanatic reviewing the box scores of his home team. It was never too late to make one more call or hold another meeting to nail down an uncertain vote."[44]

In making such an effort, Johnson had many useful allies. These included liberal labor union leaders like Walter Reuther as well as activists like Clarence Mitchell, chief lobbyist for the NAACP. Other civil rights leaders from SNCC, CORE, and SCLC pitched in. Church leaders applied pressure throughout the struggle, at one point weighing in with a prayer vigil—"coercion by men of the cloth," some observers said—at the Lincoln Memorial.[45] On the Hill, Johnson relied heavily on Senator Hubert Humphrey of Minnesota, who became floor leader for the bill. A longtime champion of civil rights, Humphrey hoped to be named LBJ's running mate in the 1964 campaign. Like Johnson, he lobbied hard with undecided senators. He joked later, "I courted Dirksen almost as persistently as I did [my wife] Muriel."[46]

The filibuster lasted three months, an all-time record, during which time opponents of the bill grew increasingly emphatic. Senator Barry Goldwater of Arizona predicted that the bill would "require the creation of a federal police force of mammoth proportions."[47] Toward the end Dirksen exacted some concessions, including one that limited sanctions against school segregation to de jure practices in the South. He then

43. Califano, *Triumph and Tragedy*, 54; Sundquist, *Politics and Policy*, 259–71, 515-18; Matusow, *Unraveling*, 92–96.
44. Califano, *Triumph and Tragedy*, 54.
45. James Findlay, *Church People in the Streets: The National Council of Churches and the Black Freedom Movement, 1950–1970* (New York, 1993).
46. Weisbrot, *Freedom Bound*, 91.
47. Sundquist, *Politics and Policy*, 270.

announced that he and most other Republicans were satisfied with the bill as it had come from the House. They voted for cloture, which was approved on June 10 by a margin of 71 to 29. Opponents were twenty-one southerners, three Democrats from outside the South, and five Republicans, including Goldwater. The bill later passed, 73 to 27. Johnson signed the 1964 Civil Rights Act on July 2.[48]

The act contained a number of strong provisions. It banned racial discrimination in privately run accommodations for the public, such as theaters, movie houses, restaurants, gas stations, and hotels, and authorized the Attorney General to eliminate de jure racial segregation in public schools, hospitals, playgrounds, libraries, museums, and other public places. The act stated that schools, as well as other federally assisted institutions, faced loss of federal funds if they continued to discriminate. It also authorized the Attorney General to bring suits on behalf of parents complaining of discrimination in the schools and declared that the government would assume their legal costs.

The law also included a section, Title VII, that forbade discrimination in employment and specified the category of sex in addition to those of race, color, religion, and national origin. The inclusion of sex as a category was originally the handiwork of House Rules Committee chairman Howard Smith of Virginia. His motive was to defeat civil rights legislation, which he vigorously opposed. If the amendment passed, he figured, liberals committed to protective legislation for women might feel obliged to oppose the entire bill, which then would fail. But Smith miscalculated, for liberals voted overwhelmingly for the final bill on the floor of the House. Title VII remained in the bill that went to the Senate and emerged unscathed in the act that Johnson approved in July. So did provision for creation of an Equal Employment Opportunity Commission (EEOC). Such was the process by which Title VII and the EEOC, later to become keys to unanticipated and unprecedented federal enforcement of gender equality, entered the law of the land.[49]

No law, of course, can work wonders overnight, and the 1964 Civil Rights Act was no exception. Voting rights remained to be protected. Many employers and unions evaded the strictures against job discrimination. De facto racial discrimination remained widespread in the North, especially in housing and schooling. Many school districts, mainly in the

48. *Newsweek*, July 13, 1964, p. 17.
49. Cynthia Harrison, *On Account of Sex: The Politics of Women's Issues, 1945–1968* (New York, 1988), 177–79; Rosalind Rosenberg, *Divided Lives: American Women in the Twentieth Century* (New York, 1992), 187–88.

Deep South, continued to employ ruses of one sort or another to avoid desegregation in public education. Until 1969, when the courts cracked down, little progress was made in this ever-sensitive area of race relations. Finally, the law did not pretend to do anything to better the mostly abysmal economic condition of black people in the United States. Like the war on poverty, it was a liberal, not a radical, measure. It aimed to promote legal, not social, equality.[50]

The civil rights act was nonetheless a significant piece of legislation, far and away the most important in the history of American race relations. Quickly upheld by the Supreme Court, it was enforced with vigor by the Johnson administration. That required a huge expansion in the reach of the State, for there were many thousands of hospitals, school districts, and colleges and universities affected by provisions of the law.[51] Although many southern leaders resisted, most aspects of enforcement proved effective in time, and the seemingly impregnable barriers of Jim Crow finally began to fall. Black people at last could begin to enjoy equal access to thousands of places that had excluded them in the past. Few laws have had such dramatic and heart-warming effects.

In retrospect, a comment by Dirksen at the time best explains why the Civil Rights Act of 1964 passed. Citing lines attributed to Victor Hugo, Dirksen mused, "No army can withstand the strength of an idea whose time has come." He added, "In the history of mankind, there is an inexorable moral force that moves us forward."[52] His point, of course, was that the drive for civil rights had acquired a momentous and a moral power by 1964 that not even filibuster rules in the Senate could withstand. The momentum, in turn, derived from the thousands of heroic efforts by civil rights activists in the preceeding years. It came from the bottom up, from the grass roots, not from the top-down strategies of VIPs in Washington.

This was true, but it was also true that Johnson had dominated the Washington stage. Some liberals still distrusted him, to be sure, as did black leaders, who suspected correctly that he, like Kennedy, was using the FBI to spy on civil rights activists. But most civil rights lobbyists on the Hill conceded his role as a star. Bayard Rustin said later that Johnson and his aides did "more . . . than any other group, any other administration. . . . I think Johnson was the best we ever had." Mitchell added

50. Hugh Davis Graham, "Race, History, and Policy: African Americans and Civil Rights Since 1964," *Journal of Policy History*, 6 (1994), 12–39.
51. Kearns, *Lyndon Johnson*, 288.
52. *New York Times*, June 20, 1964.

that LBJ "made a greater contribution to giving a dignified and hopeful status to Negroes in the United States than any other President, including Lincoln, Roosevelt, and Kennedy."[53] These were appropriate tributes to presidential leadership of an unusually high order.

WHILE JOHNSON WAS confidently commanding the ship of state in Washington, social and ideological forces elsewhere in the nation were beginning to press on the mainstream of American politics. These forces came from both the right and the left. Driving them, as so much else in the 1960s, were the imperatives of class, region, and race.

Two rising political figures especially alarmed liberals in 1964. The first to emerge as a threat to Johnson's ambitions was Governor George Wallace of Alabama. Early in his political career Wallace had directed his appeals mainly at working-class whites, paying relatively little attention to racial issues. Then and later he considered himself more of an economic populist—a spokesman for ordinary people—than a representative of the Right.[54] In 1958, however, he had lost a primary contest for the governorship to John Patterson, who Wallace felt had whipped up Negrophobia in order to beat him. Wallace was furious. "John Patterson out-nigguhed me," he was said to have cried after the primary. "And boys, I'm not going to be out-nigguhed again."[55] He wasn't. Winning in 1962, he exclaimed at his inauguration in 1963, "From this cradle of the Confederacy, this very heart of the great Anglo-Saxon Southland . . . I say, segregation now! Segregation tomorrow! Segregation forever!" As governor he continued to be liberal on economic and educational issues, but he became a hero to many southern whites when he "stood in the schoolhouse door" to block the court-ordered admission of two blacks to the University of Alabama in 1963.

Wallace was an unusually powerful public speaker. A former Golden Gloves bantamweight, he was feisty and combative. He conveyed a passion and a body language that electrified crowds who came to hear him. Charged up, his eyes glowing with intensity, he seemed scarcely able to control himself. Johnson called him a "runty little bastard and just about

53. Lawson, "Civil Rights," 94–95.
54. Marshall Frady, *Wallace* (New York, 1970); Stephen Lesher, *George Wallace: American Populist* (Boston, 1994); Kirkpatrick Sale, *Power Shift: The Rise of the Southern Rim and Its Challenge to the Eastern Establishment* (New York, 1975), 104.
55. William Leuchtenburg, *A Troubled Feast: American Society Since 1945* (Boston, 1973), 148; O'Neill, *Coming Apart*, 388.

the most dangerous person around."[56] His appeal transcended racial issues, important though those were. Again and again Wallace spoke as the champion of the common man. He attacked intellectuals, do-gooders, federal bureaucrats, radicals, Communists, atheists, liberals, civil rights workers, student protestors—soft and pampered elites all—who were threatening hard-working people. His was an appeal, reminiscent in some ways of McCarthy's, that tapped with unexpected depth into the class and regional resentments of American life.

He also had unlimited ambitions. He therefore decided to enter some Democratic presidential primaries in the spring of 1964. His purpose, he said, was to sound an alarm against the civil rights bill, which was then tying up the Senate. But he also planned to bring his larger message to a national stage. The results staggered liberals. Although Wallace had little money and no real organization, he attracted large and enthusiastic crowds, especially in white working-class areas, where his angry assaults on distant government bureaucrats aroused passionate support. In April Wallace carried 34 percent of the votes cast in the primary in Wisconsin, a normally liberal state. Later in the spring he won 30 percent of the vote in Indiana and 43 percent in Maryland. "If it hadn't been for the nigger bloc vote," he said of Maryland, "we'd have won it all."[57]

Wallace knew that he had no chance to win a presidential election, but he nonetheless unnerved political adversaries. In June he announced that he would run as an independent on a third-party ticket. By early July he had managed to get on the ballot in sixteen states. There was talk that he might receive enough electoral votes in the South to deprive Johnson of outright victory, thereby forcing the issue to be resolved by the House of Representatives.

While Wallace was rising to national prominence, conservative Republicans were developing a remarkably well organized effort on behalf of a much more ideologically pure right-wing political figure, Goldwater of Arizona. Goldwater was an affable man who had many friends on the Hill. Tall, trim and handsome, he was tolerant in his personal relationships. He belonged to the NAACP. He had been among the most consistently right-wing senators since he had come to Washington in 1953. Contemporaries labeled him a conservative, and he wrote a popular sum-

56. Califano, *Triumph and Tragedy*, 56.
57. Theodore White, *The Making of the President, 1964* (New York, 1965), 281–83; Matusow, *Unraveling*, 138–39; Jonathan Rieder, "The Rise of the Silent Majority," in Steve Fraser and Gary Gerstle, eds., *The Rise and Fall of the New Deal Order, 1930–1980* (Princeton, 1989), 243–68.

mation of his beliefs entitled *The Conscience of a Conservative* in 1960. In fact, however, Goldwater was a political reactionary who opposed virtually all efforts of the federal government to intervene in domestic social policy, including civil rights legislation. The graduated federal income tax, he believed, infringed on individual freedom, his highest value. An ardent anti-Communist, he seemed anxious to send in military force to settle overseas disputes. Goldwater's outspoken criticisms of Johnson's policies in 1964 attracted a fervent, well-financed following of conservatives and reactionaries, most of them upper-middle-class, who determined to make him the GOP presidential candidate in 1964. Diffidently, for Goldwater did not much want to become President, he agreed to run for the nomination.

It seemed in early 1964 that Goldwater would fail in this effort. Polls suggested that he had relatively little appeal, even among Republicans. Pundits observed correctly that he was too far to the right to capture moderate voters. But Wallace's successes in the primaries exposed the rage of many Americans against liberal policies and gave heart to right-wing elements in the GOP Goldwater's major contender for the nomination, moreover, had political liabilities of his own. This was Governor Nelson Rockefeller of New York, a liberal who commanded virtually no support in the conservative wing of the party. Two years earlier, when he had seemed assured of the 1964 nomination, Rockefeller had left his wife of many years and married a much younger woman. This action badly blighted his chances in 1964.

When Goldwater narrowly beat Rockefeller in a head-to-head California primary in June, it was clear that the nomination was his. A month later, in mid-July, he got the nod in a raucous GOP convention in San Francisco that exposed the bitter feelings rending the party. Goldwater delegates booed so loudly when Rockefeller arose to speak that he could not be heard. Goldwater was so angry at the rejection he received from moderates and liberals that he named Congressman William Miller of New York as his running mate. Miller was almost totally unknown but was nearly as reactionary as Goldwater himself. Goldwater closed by defending far-right organizations such as the John Birch Society in his acceptance speech. "Let me remind you," taunting his opponents, "that extremism in defense of liberty is no vice . . . and that moderation in pursuit of justice is no virtue."[58]

Goldwater's nomination, liberals exulted, was the best thing that could

58. White, *Making* . . . 1964, 230–65; Matusow, *Unraveling*, 134–38.

have happened to the Democratic party. His zealous supporters, however, were happy and hopeful. Southern conservatives figured that he might beat Johnson in the region, and they prevailed on Wallace to withdraw. Wallace did so, leaving the GOP with a good chance—for the first time since the Reconstruction era—of scoring major triumphs in Dixie. Political observers braced for a campaign centered on the politics of region and race.

ONCE THE REPUBLICANS had nominated Goldwater and Miller, Johnson and his advisers concentrated on pulling all but the Right into a big and joyous coalition that would bring him overwhelming victory. He orchestrated a Democratic convention in Atlantic City in late August that played as he directed, including the selection of Humphrey as his running mate. Before it was over, however, racial confrontations revealed fissures in American society that, while ultimately harmless to LBJ in 1964, widened later to rend his party and transform the nature of American politics.

Some of these fissures were already exposing the rise of fiercely anti-white activists associated with the Nation of Islam. The Black Muslims, as they were called, had been formed in the 1930s. Spurning Christianity as the religion of the slaveowner, they also rejected racial integration and called for black people to separate and build up their own communities. Members were to improve themselves by renouncing drink, drugs, tobacco, gambling, cursing, and sex outside of marriage. Men were to wear white shirts and suits, women long dresses, head-coverings, and no makeup. Deeply alienated from whites, the Muslims had no use for the interracial civil rights movement or for "corrupt" and "evil" white society. They were racists themselves, perceiving whites as "devils" who had been bleached in the years following creation and foreseeing a day of judgment when Allah would defeat the whites and vindicate the blacks through racial separation. Their ideas drew on versions of apocalyptic holiness religion popular among blacks as well as on historically durable traditions of black nationalism.[59]

While the Muslims enjoyed growing popularity, mainly among the most deprived people in ghetto areas of large northern cities such as Chicago, Detroit, and New York, they gained relatively few full converts.

59. Nell Painter, "Malcolm X Across the Genres," *American Historical Review*, 98 (April 1993), 432–39. Standard works include C. Eric Lincoln, *The Black Muslims in America*, rev. ed. (Boston, 1973); and C. E. Essien-Udom, *Black Nationalism: A Search for an Identity in America* (Chicago, 1962).

Estimates are that there were between 5,000 and 15,000 active Muslims in the early 1960s, 50,000 believers, and a considerably larger group of sympathizers.[60] When Malcolm X, one of the most popular Muslim leaders, dismissed Kennedy's assassination as a case of the "chickens come home to roost," he aroused bitter emotions. Elijah Muhammad, leader of the Nation of Islam, berated Malcolm for his impolitic remarks, thereby sharpening a split that had already widened between the two men. In March 1964 Malcolm X broke away from the Nation to head his own group, the Organization of Afro-American Unity.

The rise of Malcolm X (born Malcolm Little; the X was a marker for the African family name that had been lost under slavery) in the remaining eleven months of his life was impressive.[61] During this time he took two trips to Africa, exploring further the teachings of Islam. He gradually dropped the most extreme forms of anti-white racism held by the Black Muslims, gravitating instead (or so it seemed) toward a more secular, quasi-socialistic platform that foresaw some accommodation with poor and working-class whites. At the same time he remained a black national-ist, and he insisted that blacks must help themselves, using violence if provoked, if they hoped to survive the evils of white civilization. "There can be no revolution without bloodshed," he proclaimed.[62] He dismissed civil rights leaders as lackeys of the white Establishment: King was a "traitor," a "chump," and a "fool." (King retorted that Malcolm was a "hot-headed radical with a dangerous emotional appeal.") Malcolm was extraordinarily self-possessed, articulate, quick-witted, and often funny. He spoke boldly and with a controlled but obviously passionate anger. In the process he attracted growing attention and support from urban blacks in 1964.

What Malcolm X might have been able do had he lived—enemies from the Nation of Islam assassinated him in New York in February 1965—cannot be said.[63] Most politically engaged black Americans at that time remained committed to the teachings of King or to the leaders of

60. Weisbrot, *Freedom Bound*, 173.
61. Malcolm X, as told to Alex Haley, *The Autobiography of Malcolm X* (New York, 1965). In 1964 Malcolm took on the name of El-Hajj Malik El-Shabazz.
62. Hodgson, *America in Our Time*, 205.
63. Some people have argued that the FBI, which ran surveillance on the Nation of Islam and on Malcolm, knew in advance of the plot to kill Malcolm but did nothing to warn him. Others have claimed that the FBI, the police, or both were involved in some way in Malcolm's death. These remain speculations. *New York Times*, Jan. 13, 14, 1995.

other civil rights organizations. These leaders regarded Malcolm X as a wild and impractical opportunist anxious for personal glory. Would he have dared to speak so boldly if he had to live and work in the South, where activists were literally risking their lives? What, they asked, had he actually accomplished, except to stir up resentments in the cities and thereby to blight the vision of interracial progress? "What did he ever do?" Thurgood Marshall asked years later. "Name one concrete thing he ever did."[64] Other critics wondered how black people, a mostly poor and relatively powerless 11 percent of the population, could hope to advance in American life if they rejected white people and white institutions.[65]

These were good questions, and they indicate that Malcolm X was far from the inspirational figure in 1964 that he later became, in martyrdom, for large numbers of blacks in the United States. Still, Malcolm had begun to instill pride in increasing numbers of African-Americans, some of whom were angrily verging on rebellion. On July 18, 1964, a riot erupted in Harlem. Lasting a week, it featured fights between blacks and police, as well as burning and looting and attacking of whites. When it quieted down, another riot broke out upstate in Rochester. King, Randolph, Wilkins, and other black leaders called for peace, but in August riots stunned Paterson and Elizabeth, New Jersey, and Philadelphia. Compared to other urban riots in American history—and to riots later in the decade—these were relatively minor disturbances, and they stopped in September. But they exposed the coming of "the fire next time" that Baldwin had prophesied in 1963. It seemed that Johnson's civil rights legislation, primarily addressing the legal rights of blacks in the South, did little or nothing to placate the rage of blacks in the ghettos.

Racial confrontation in the South more directly threatened the President's hopes for a serene and successful campaign in 1964. Mississippi, as so often in the history of the civil rights movement, proved the major arena of conflict. There CORE, SCLC, NAACP, and SNCC had earlier formed a Council of Federated Organizations (COFO) aimed at mobilizing black people, especially to gain the right to vote. Robert Moses, still risking his life in the state, served as program director for the effort, which had intensified in 1963. Aiding him at that time was a cohort of eighty-odd white students, most of them from Yale and Stanford. They had been recruited by Allard Lowenstein, a thirty-three-year-old white activist from New York who was deeply engaged in the movement. In October 1963

64. Reported in *New York Times*, Jan. 20, 1993.
65. A critical biography is Bruce Perry, *Malcolm: The Life of a Man Who Changed Black America* (New York, 1991). It persuasively demolishes many of the "facts" in Malcolm's *Autobiography*.

they endorsed NAACP activist Aaron Henry and Tougaloo College chaplain Ed King, a white man, as nominees for governor and lieutenant governor on a Freedom party slate.[66]

Moses and others, heartened by these efforts, then began planning what became known as Freedom Summer in Mississippi, a campaign in 1964 to register blacks as voters and to establish "Freedom Schools" for black children.[67] Seeking volunteers, they finally accepted more than 900, mainly white college students who could take at least part of a summer off. Most of them underwent a week or two of training before going to Mississippi in late June. As they prepared to go they were warned by veterans of the movement that the Johnson administration opposed the venture and would offer them no federal protection. LBJ feared violence that in turn would damage Democratic party unity in the coming campaign. As the volunteers arrived in the Magnolia State, they faced enormous danger without hope of backing from the government.

What many had feared soon happened. On June 21 two white activists, Michael Schwerner and Andrew Goodman, and one black activist, James Chaney, disappeared near Philadelphia, Mississippi. When their bodies were finally found buried in an earthen dam in early August, it was revealed that Schwerner and Goodman had been shot through the head once each with .38 caliber bullets. Chaney had been shot three times.[68] Evidence implicated Neshoba County deputy sheriff Cecil Price, who had apprehended the three young men and turned them over to a Klan-dominated mob that killed them gangland style. Three years later an all-white jury convicted Price and six others, including local Klan leader Sam Bowers, of "violating the civil rights" of Schwerner, Chaney, and Goodman.[69]

66. John Dittmer, *Local People: The Struggle for Civil Rights in Mississippi* (Urbana, 1994), 200–207. Also Harvard Sitkoff, *The Struggle for Black Equality, 1954–1992* (New York, 1993), 158–67. For Lowenstein, see William Chafe, *Never Stop Running: Allard Lowenstein and the Struggle to Save American Liberalism* (New York, 1993). Also David Harris, *Dreams Die Hard: Three Men's Journey Through the Sixties* (San Francisco, 1993), 30–44.
67. Doug McAdam, *Freedom Summer* (New York, 1988); Dittmer, *Local People*, 242–71; Weisbrot, *Freedom Bound*, 95–96; Harris, *Dreams Die Hard*, 69–89.
68. Dittmer, *Local People*, 283. Chaney's body was mangled, leading to speculation that his killers had gone out of their way to disfigure him. Authorities concluded, however, that a bulldozer, used to bury the corpses, was responsible for the mangling.
69. The Schwerner family asked to have their son buried next to Chaney in Mississippi, but permission was denied; segregation in Mississippi affected cemeteries. Dittmer, *Local People*, 284, 418.

The terror in Mississippi frightened the volunteers and outraged people throughout the world. COFO leaders were angry, too—in part at themselves for lapses in their own precautionary procedures, and in much larger part at the Johnson administration for having failed to provide protection. The FBI, indeed, did not arrive on the scene until twenty hours after the three men vanished and did not take charge of the search for three days. Later in the summer, Hoover, responding to criticism, increased his forces in the state, and one of his informants made possible the discovery of the bodies and the subsequent prosecution of the killers. COFO leaders and volunteers, however, remained furious at the administration.

Violence and bloodshed further stained the efforts of that fateful summer. Between June and the end of August foes of COFO burned or bombed thirty-five homes, churches, and other buildings in Mississippi. Thirty-five volunteers were shot at (three were hit), eighty were beaten, and three more were killed. White authorities made more than 1,000 arrests of COFO workers and their allies. As Cleveland Sellers, a leader of COFO, recalled, "It was the longest nightmare I have ever had, those three months."[70]

Much of the work that summer took place in the Freedom Schools, which engaged thousands of poor, mostly rural black children and their parents in an experiment that promised much for their lives. But the most highly publicized activity was to secure equal rights for black people in state and national politics, and COFO mounted a major effort to enlarge the rolls of their Mississippi Freedom Democratic party (MFDP). Pursuing an orderly democratic process, they ultimately selected a total of thirty-four delegates and thirty-four alternates to represent the state at the Democratic National Convention in late August in Atlantic City. Four of the delegates, including Ed King, were white. Roughly three-quarters of them were small farmers.[71]

Thus was set in motion a highly contentious struggle between Johnson and liberal allies on the one hand and militant civil rights workers on the other. The Freedom party delegates did not intend that to be. Declaring their loyalty to the President and to the liberal ideals of the Democratic party, they expected to be well received at the Johnson-dominated convention. The rival white delegation chosen by party regulars, by contrast, denounced the civil rights act and explicitly opposed the party platform.

70. Ibid., 269–71; Weisbrot, *Freedom Bound*, 110–14; Gitlin, *Sixties*, 149–51.
71. Dittmer, *Local People*, 272–85.

Most of these whites were expected to support Goldwater in the fall. But Johnson, after first hoping to work out a compromise, discovered that the white Mississippians would walk out of the convention if the MFDP received any consideration whatever. Several other southern delegations threatened to join them. If that happened, Johnson faced serious electoral consequences in Dixie.

Johnson, orchestrating every detail of his nomination, had foreseen controversy and had planted FBI informants to secure intelligence about the upstarts soon to arrive at the convention.[72] He also leaned on the convention's credentials committee, which considered the rival claims, so as to ensure that it would not vote to bring the inflammatory issue before the convention floor. Still, he hoped that pressure and persuasion would help MFDP delegates see the light. Some of the nation's best-known liberals, including Reuther and Humphrey, besieged MFDP leaders with pleas and promises. Nationally known black leaders—Rustin, King, CORE leader James Farmer—also seemed ready to accept a compromise. This would have allotted the MFDP delegates two voting seats and admitted the rest with non-voting, honorary status. Johnson further indicated, although in general terms, that convention rules would be changed so that racial discrimination would not determine the selection of delegates in 1968. The credentials committee did LBJ's bidding and recommended the deal.[73]

Carrots, however, failed to move the Freedom party delegates, largely because they offered little. The two delegates to be seated, LBJ said, must include a white and a black, specifically Ed King and Aaron Henry. They were not to be seated as Mississippi delegates—to do so would be to concede the legitimacy of their claim to represent the state—but as delegates "at large." Johnson further sent word that one of the most vociferous Freedom party delegates, Fannie Lou Hamer, should not be permitted to vote at the convention.

Hamer had already attracted extraordinary attention among other delegates and reporters in Atlantic City. An eloquent spokeswoman for what it was like to be black and poor in Mississippi, Hamer was the twentieth child of impoverished, barely literate sharecroppers. In 1962 when she had dared to register to vote, she had been driven off the land where she had worked for eighteen years as a field hand. Later she had been badly beaten for encouraging other blacks to register. She emerged from these

72. Ibid., 291–93.
73. Weisbrot, *Freedom Bound*, 115–23.

ordeals as a strong and uncompromising advocate of equal rights. [74] When she appeared before the credentials committee on August 22, she described in graphic detail to spectators and to television viewers what had happened to her on that occasion:

> I was carried to the county jail. . . . And it wasn't long before three white men came to my cell where they had two Negro prisoners. The state highway patrolman ordered the first Negro to take the blackjack. . . . And I laid on my face. The first Negro began to beat, and I was beat until he was exhausted. . . . The state highway patrolman ordered the second Negro to take the blackjack. The second Negro began to beat and I began to work my feet, and the state highway patrolman ordered the first Negro who had beat to set on my feet and keep me from working my feet. I began to scream, and one white man got up and began to beat me on the head and tell me to "hush." One white man—my dress had worked up high—he walked over and pulled my dress down and he pulled my dress back, back up. All of this is on account we want to register, to become first-class citizens, and if the Freedom Democratic Party is not seated now, I question America. [75]

When Johnson saw this testimony on TV, he quickly called a presidential press conference in the hope that the networks would cover it and drive Hamer and others off the air. They did as he had anticipated, preempting among others Rita Schwerner, whose husband had been murdered in June. Johnson thought he had silenced "that illiterate woman," as he called Hamer. But the MFDP had the last word, for that evening the networks rebroadcast the hearings in prime-time, including all of Hamer's powerful testimony. Later, when asked to support Johnson's compromise, she snorted, "We didn't come all this way for no two seats." Shortly thereafter the Freedom party delegates decisively rejected the compromise.

In the short run Johnson was the winner of this struggle and the Freedom party loyalists the losers. He was nominated by acclamation, as was Humphrey, his loyal aide. He was convinced, moreover, that most white Americans—most voters—approved of his attempts to avoid identification with extremes. "Right here," he said, "is the reason I'm going to win this thing so big. You ask a voter who classifies himself as a liberal what he thinks I am and he says a liberal. You ask a voter who calls himself a conservative what he thinks I am and he says I'm a conservative. . . .

74. Kay Mills, *This Little Light of Mine: The Life of Fannie Lou Hamer* (New York, 1993).
75. Todd Gitlin, *The Sixties: Years of Hope, Days of Rage* (New York, 1987), 153.

They all think I'm on their side." And that, he said later, was "where the majority of the votes traditionally are."[76]

In the long run, however, Johnson's liberal position on civil rights could not satisfy militants at both ends of the spectrum. Given the polarization of emotions concerning race by 1964, this was no longer possible. Angry white delegates from Mississippi raged at his efforts to compromise and walked out of the convention. Some of the Alabamans also stalked out. For them and for many other southerners, Goldwater and Wallace, not Johnson, were the politicians of choice. The electoral results in November confirmed that the majority of white voters in the Deep South remained unreconstructed rebels on the subject of race.

No group was more angry than the MFPD and other civil rights militants. Long suspicious of liberals such as Kennedy and Johnson, they regarded the battle at Atlantic City as the last straw. Many never trusted white people again. Almost all refused to listen to the blandishments, as they saw them, of liberals or governmental bureaucrats. When emissaries of LBJ tried to explain to Hamer that her refusal to compromise might cost Humphrey the vice-presidency, she stared at them in disbelief. "Do you mean to tell me," she asked Humphrey, "that your position is more important to you than four hundred thousand black people's lives?" When Humphrey tried feebly to respond, she walked out in tears. When she ran into him later, she told him, "Senator Humphrey, I been praying about you . . . you're a good man, and you know what's right. The trouble is, you're afraid to do what you know is right."[77] No exchange better captured the gulf that by then was polarizing black militants and white liberals in America.

COMPARED TO THE UNREST in the cities and the turmoil at Atlantic City, the electoral campaign that followed was relatively tame. Goldwater refused to run a demagogic campaign focused on the issue of race. His position, after all, was already well suited to success in the Deep South. Instead, he tried to bring across his highly ideological opposition to Big Government as it had arisen in the United States under the New Deal, the Fair Deal, and the "Dime Store New Deal" of the Eisenhower administration. "Socialism through Welfarism," he maintained, was the greatest threat to Freedom.[78]

No presidential candidate in modern American history, however,

76. Blum, *Years of Discord*, 161.
77. Dittmer, *Local People*, 294; Weisbrot, *Freedom Bound*, 118.
78. White, *Making . . . 1964*, 375–412; Blum, *Years of Discord*, 157.

proved more impolitic. He once told the columnist Joseph Alsop, "You know, I haven't really got a first-class brain." Politically speaking, this seemed evident in the campaign, during which Goldwater went out of his way to offer voters what he called a "choice, not an echo." In the process he issued blunt statements that alienated millions of voters. He went to Appalachia to denounce the war on poverty and to the South to call for the sale to private interests of the Tennessee Valley Authority, which was highly popular in the area. He told the elderly that he wanted to scrap Social Security, and farmers that he opposed high price supports. "My aim," he insisted, "is not to pass laws but to repeal them."[79]

Some of his statements were so uncompromising that he opened himself up to ridicule. "The child has no right to an education," he proclaimed. "In most cases he will get along very well without it." American missiles were so good, he said, that "we could lob one into the men's room at the Kremlin." Angry at what he considered the arrogance of the liberal East Coast Establishment, he observed, "Sometimes I think this country would be better off if we could just saw off the Eastern Seaboard and let it float out to sea." Democrats had a field day playing with the motto of his supporters, "In Your Heart You Know He's Right." "In your guts," they quipped, "you know he's nuts."

Goldwater's bellicose statements on foreign policy left him especially vulnerable to criticism. "Our strategy," he said, "must be primarily offensive. . . . We must—ourselves—be prepared to undertake military operations against vulnerable Communist regimes." When asked earlier in the year what he would do about Vietnam, he replied that he would bomb the supply routes in the North. What would he do about trails hidden in the jungle? Goldwater answered that "defoliation of the forests by low-yield atomic weapons could well be done. When you remove the foliage, you remove the cover." Although Goldwater tried to clarify and deny these remarks—he meant tactical weapons, not atomic bombs—he was consistent on one point: let the generals have a free hand, and they would bring victory.[80]

If Americans had known what Johnson and his advisers were thinking about Vietnam at that time, Goldwater would not have seemed extreme. For the fighting in Vietnam was going so badly for Diem's forces that administration leaders were developing plans to escalate America's role. By the summer of 1964 they had decided privately that the United States

79. Frederick Siegel, *Troubled Journey: From Pearl Harbor to Ronald Reagan* ((New York, 1984), 158.
80. *New York Times*, July 18, 1964; Matusow, *Unraveling*, 148–50.

Rosa Parks being fingerprinted, February 1956, during the Montgomery bus boycott. *AP, Library of Congress*.

The desegregation of Montgomery buses: The Rev. Ralph Abernathy (front row, left), the Rev. Martin Luther King, Jr. (second row), and the Rev. Glenn Smiley, ride a bus in December 1956. The woman is unidentified. *AP, Library of Congress*.

The Greensboro sit-in, February 2, 1960. The blacks depicted are Ronald Martin, Robert Patterson, and Mark Martin. The white woman came to the counter for lunch but decided not to sit down. *UPI, Library of Congress.*

King and Hubert
Humphrey at a banquet,
May 1965. *Library of
Congress.*

Malcolm X, March 1964.
AP, Library of Congress.

"Black power" spokesman Stokely Carmichael, May, 1966. *UPI, Library of Congress.*

LBJ and JFK at the Democratic National Convention, July 1960. *UPI/ Bettmann.*

The Kennedy brothers and J. Edgar Hoover of the FBI. *John F. Kennedy Library.*

Bobby Kennedy and César Chávez. Ernest Lowe, *UPI/Bettmann.*

Secretary of Defense Robert McNamara, General Maxwell Taylor, Kennedy, and Secretary of State Dean Rusk. *AP/Wide World Photos.*

Dean Rusk and Senator J. William Fulbright, January 1962.
UPI, Library of Congress.

Left to right: National Security Adviser McGeorge Bundy, Ambassador Maxwell
Taylor, and General William Westmoreland, in Vietnam, February 1965.
UPI/ Bettmann.

Antiwar protestors, Washington, October 1967. *Library of Congress.*

Spec. 4 Ronald Abernathy of
Evanston, Ill., on patrol in
South Vietnam, March 1967.
AP, *Library of Congress.*

A wounded GI, in battle of
Hue, 1968. Donald McCullin,
Magnum Photos

LBJ, July 31, 1968. Jack Knightlinger, *LBJ Library.*

Betty Friedan, 1964.
Library of Congress.

Protest for women's rights,
1970. *UPI/Bettmann.*

Alabama Governor
George Wallace, May 1964.
Library of Congress.

Minnesota Senator
Eugene McCarthy, July 1964.
Library of Congress.

Henry Kissinger, Vice President Spiro Agnew, and President Richard Nixon, January 1970. Ollie Atkins, *National Archives.*

Kissinger, crossing his fingers and holding a whip along with a shield of peace, September 1973. Choé, *Library of Congress.*

George McGovern (r.) with campaign adviser Larry O'Brien (l.) and Senator Thomas Eagleton of Missouri, McGovern's running mate, July 1972. *Library of Congress.*

Expectations disrupted. Irene Springer, *AP/Wide World Photos.*

would sooner or later have to drop bombs on the North, and in August they manipulated naval confrontations in the Tonkin Gulf off the coast of Vietnam that gave them congressional authorization for escalation if the time should come.[81]

Having secured authority for military action in the future, Johnson posed as the peace candidate during the campaign. He denounced Republican speakers, including Goldwater, who talked of bombing the North, and said, "I want to be very cautious and careful, and use it [bombs] only as a last resort, when I start dropping bombs that are likely to involve American boys in a war in Asia with 700 million Chinese." He added,

I have not thought that we were ready for American boys to do the fighting for Asian boys. What I have been trying to do . . . was to get the boys in Vietnam to do their own fighting with our advice and with our equipment. That is the course we are following. So we are not going north and drop bombs at this stage of the game, and we are not going south and run out and leave it for the Communists to take over.[82]

From the beginning of the campaign, polls showed that Johnson had a large lead. But it was simply not in him only to win; he had to dominate. Taking no chances, he sanctioned an unprecedentedly harsh and negative series of television spots. One featured a large saw cutting through a wooden map of the United States while the narrator cited Goldwater's comment about the East Coast. Another depicted a pair of hands tearing up a Social Security card.[83] The most controversial spot characterized Goldwater as a maniac whose foreign policies would destroy the world. It showed a little girl picking petals off a daisy and counting, "one, two, . . . five—." Then the girl looked up startled and the frame froze on her eye until she dissolved into a mushroom-shaped cloud and the screen went black. While she disintegrated a man's voice, loud as if at a test site, intoned, "ten, nine . . . ." An explosion followed, where-

81. See chapter 20 for detail on Johnson and Vietnam.
82. *Public Papers of the Presidents of the United States: Lyndon B. Johnson, 1963–1964* (Washington, 1964), 1126, 1267, 1165.
83. Edwin Diamond and Stephen Bates, *The Spot: The Rise of Political Advertising on Television* (Cambridge, Mass., 1992), 122–41; Lynda Lee Kaid and Anne Johnston, "Negative Versus Positive Television Advertising in U.S. Presidential Campaigns, 1960–1988," *Journal of Communications* (Summer 1991); Kathleen Hall Jamieson, *Packaging the Presidency: A History and Criticism of Presidential Campaign Advertising* (New York, 1992), 169–220.

upon the voice of Johnson was heard. "These are the stakes—to make a world in which all of God's children can live, or go on into the dark. We must either love each other, or we must die." The spot closed with the familiar message, "Vote for President Johnson on November 3. The stakes are too high for you to stay at home." The daisy spot, as it was called, provoked a flood of protest that deluged the White House switchboard with phone calls. Embarrassed, Johnson had it pulled after one showing. But television news programs showed it repeatedly during the next few weeks. It was later estimated that 40 million Americans saw it at one time or another.[84]

Goldwater and his staff responded with some negative spots of their own. Indeed, the battle in 1964 was especially nasty. A later survey of campaigns between 1964 and 1988 concluded that a record-high 40 percent of television ads in 1964 featured negative personal jabs at the opposition.[85] But Goldwater's spots were much less negative, and they attracted less attention. Most of the GOP budget went instead to coverage of speeches and statements. These continued to feature Goldwater's uncompromising and often reactionary views, mainly on domestic issues. According to polls, they did nothing to bolster his chances in November.

SOME OF THE RESULTS of the election of 1964 signaled deep danger ahead for the Democratic party. Thanks mainly to Johnson's ardent support of the civil rights act, many southern whites showed that they wanted nothing more to do with him. Largely for this reason he lost the states of Mississippi, Alabama, South Carolina, Louisiana, and Georgia in November, as well as Goldwater's Arizona. He won narrowly in Florida. A majority of white voters in Arkansas, Tennessee, North Carolina, and Virginia rejected the President.[86]

The growth from 1964 on of the GOP in the South and the Southwest turned out to be one of the most important long-range trends in postwar American politics. Thereafter Republicans were highly competitive in presidential elections in these areas. There were many causes behind the

84. Diamond and Bates, *Spot*, 122–27.
85. Kaid and Johnston, "Negative Versus Positive."
86. Kevin Phillips, *The Emerging Republican Majority* (New Rochelle, 1969); David Reinhard, *The Republican Right Since 1945* (Lexington, Ky., 1983), 159–208; Walter Dean Burnham, *Critical Elections and the Mainsprings of American Politics* (New York, 1970), 118–20; Thomas and Mary Edsall, *Chain Reaction: The Impact of Race, Rights, and Taxes on American Politics* (New York, 1991), 32–46; and Sale, *Power Shift*, 110–15.

decline of the Democratic party in the South and Southwest, but chief among them, as Johnson himself recognized, was his open endorsement of civil rights in 1964. "I think," he told Califano when he signed the civil rights act, "we delivered the South to the Republican party for your lifetime and mine."[87]

In 1964, however, few pundits paid much attention to the travails of Democrats in the South. They focused instead on LBJ's astonishing national triumph. Voters gave Johnson 43.1 million votes to Goldwater's 27.2 million. This was 61.2 percent of the total vote, an extraordinary showing. Carrying all but six states, LBJ swept the electoral college, 486 to 52. Democratic congressional candidates coasted in on his coattails. They were slated to control the House by a margin of 295 to 140 and the Senate by 68 to 32, gains of thirty-seven and one respectively.

Notwithstanding these numbers, some observers wondered whether Johnson's mandate would last. Many who voted for him did so out of distaste for Goldwater, not out of support for the President or his programs. LBJ never aroused deep affection among voters. Still, there was no doubting that Johnson sat high in the saddle at the end of 1964. Beginning with his reassuring presence in the anxious days following the assassination, he had seemed purposeful and effective during his year in office. He had displayed great skill in dealing with Congress, getting it to approve an apparently beneficial tax cut and a "war" on poverty. He had shepherded through a historic civil rights act, presided over a year of rapid economic growth, and kept the peace. With a strongly Democratic Congress awaiting him in 1965 he was poised to lead the nation to new and unprecedented triumphs. The liberalism that he championed rode at high tide.

87. Califano, *Triumph and Tragedy*, 55.

19

A Great Society and the Rise of Rights-Consciousness

As Johnson departed from his inaugural ball in January 1965, he warned his aides, "Don't stay up late. There's work to be done. We're on our way to the Great Society."[1]

Both the phrase "Great Society" and the planning for it dated to May 1964, when Johnson addressed the graduating class of the University of Michigan. "We have the opportunity," he proclaimed, "to move not only toward the rich society and the powerful society, but upward to the Great Society." That was "where the city of man serves not only the needs of the body and the demands of commerce but the desire for beauty and the hunger for community. . . . It is a place where men are more concerned with the quality of their goals than the quantity of their goods."[2]

Some of Johnson's efforts at that time, notably the war on poverty, were already being readied to advance the Great Society. Starting in that summer, he also established the first of what ultimately became 135 "task forces" to study a wide range of social problems.[3] After the election

1. Allen Matusow, *The Unraveling of America: A History of Liberalism in the 1960s* (New York, 1964), 153.
2. *Public Papers of the Presidents of the United States: Lyndon B. Johnson, 1963–1964* (Washington, 1964), 704–6. See also James Sundquist, *Politics and Policy: The Eisenhower, Kennedy, and Johnson Years* (Washington, 1968), 361–63.
3. William Leuchtenburg, "The Genesis of the Great Society," *Reporter*, April 21, 1966, pp. 36–39; Hugh Davis Graham, "The Transformation of Federal Education Policy," in Robert Divine, ed., *Exploring the Johnson Years* (Austin, 1981), 155–

Johnson drove ahead with all of his legendary energy, for he was certain that his mandate would not last on Capitol Hill. "You've got to give it all you can, that first year," he told an aide. "Doesn't matter what kind of a majority you come in with. You've got just one year when they treat you right, and before they start worrying about themselves."[4]

In pushing for congressional action Johnson amassed very great authority in his own hands. Immediately following the election he cut back drastically on the size of the Democratic National Committee, even removing long-distance phone lines from its offices. He ordered aides and Cabinet heads to say nothing to the media about strategies—all releases would come from the White House.[5] His quest for personal control sought among other things to plug leaks and to curb the unruly federal bureaucracy. It accelerated a long-range postwar trend toward the weakening of party organizations in America and the centering of decision-making in the White House.[6] It also reflected an increasingly imperious and dissimulative manner that distressed his friends and outraged others. Journalists began writing in the spring of 1965 about a "credibility gap" emanating from the Oval Office.

In 1965, however, Johnson's imperiousness did not damage his effectiveness as a leader of domestic policy. Never had a chief executive seemed so much in control of things. Johnson sent sixty-five separate messages to Capitol Hill between January and August and never let up. Much of what he requested aimed to go beyond the bread-and-butter liberalism of the New Deal in order to create a Great Society that would be *qualitatively* better and that would guarantee "rights" and government entitlements. And Congress did his bidding, enacting the most significant domestic legislation since FDR's first term and accelerating the rights-consciousness of the people. The GOP Congressional Committee

84; Doris Kearns, *Lyndon Johnson and the American Dream* (New York, 1976), 222–23; and Paul Conkin, *Big Daddy from the Pedernales: Lyndon Baines Johnson* (Boston, 1986), 209–12.

4. Robert Divine, "The Johnson Literature," in Divine, ed., *Exploring the Johnson Years*, 11; Kearns, *Lyndon Johnson*, 216.

5. Larry Berman, "Johnson and the White House State," in Divine, ed., *Exploring the Johnson Years*, 187–213. See also Berman, "Lyndon B. Johnson: Paths Chosen and Opportunities Lost," in Fred Greenstein, ed., *Leadership in the Modern Presidency* (Cambridge, Mass., 1988), 134–63.

6. Ellis Hawley, "The New Deal and the Anti-Bureaucratic Tradition," in Robert Eden, ed., *The New Deal and Its Legacy: Critique and Reappraisal* (Westport, Conn., 1989), 77–92.

grumbled that it was the "Three-B Congress—bullied, badgered, and brainwashed." The journalist James Reston marveled that LBJ was "getting everything through the Congress but the abolition of the Republican party, and he hasn't tried that yet."[7]

COMMENTS SUCH AS THESE reflected a tendency of political observers to personalize the policy-making process. According to such analyses, legislation gets enacted—or foreign policy gets implemented—when and if a President rises above mediocrity to make his mark on the nation. As an explanation of Johnson's successful domestic leadership in 1965 this is useful up to a point. In that historic session LBJ demonstrated many of the traits that make for effective congressional action: advance preparation, thoughtful timing, amazing attention to detail, an unbending sense of purpose, and The Treatment. Behind all of these was his vision, inspiring to liberals, of a Great Society that would establish larger opportunities and entitlements for the disadvantaged.

Other, broader advantages, however, greatly facilitated his strong leadership. Chief among them, as he well recognized, was the nature of Congress in 1965. Here the disastrously impolitic campaign of Goldwater in 1964 had significant consequences. Not only was the new Congress more Democratic than at any time since 1938; it also had sixty-five freshmen, most of them young liberals who arrived in Washington ready and willing to follow their party leaders. These, majority leader Mike Mansfield of Montana in the Senate and Speaker John McCormack of Massachusetts and majority leader Carl Albert of Oklahoma in the House, were mostly loyal to LBJ.[8] The loose but normally formidable coalition of conservatives that had dominated Capitol Hill since the late 1930s was so weak in 1965 that any reasonably competent liberal President would have done well.

Pressure groups, always central to the congressional process, further aided the President. Labor unions were becoming increasingly divided on issues of race and foreign policy and were not always reliable allies of the administration, but they remained the strongest single lobby for some of LBJ's liberal measures.[9] The well-organized National Education Associa-

7. Eric Goldman, *The Tragedy of Lyndon Johnson* (New York, 1969), 334.
8. Sundquist, *Politics and Policy*, 479–84.
9. Robert Zieger, *American Workers, American Unions, 1920–1985* (Baltimore, 1986), 186; Thomas Edsall, *The New Politics of Inequality* (New York, 1984), 161–62; Ira Katznelson, "Was the Great Society a Lost Opportunity?," in Steve Fraser and Gary Gerstle, eds., *The Rise and Fall of the New Deal Order, 1930–1980* (Princeton, 1989), 185–211.

tion (NEA) worked hard for federal aid to schools. In addition, liberal "public interest groups" began to come into their own in 1965. Many of the activists who led these groups had been inspired in the Kennedy years. They were young, energetic, politically independent, suspicious of what they called "politics as usual" and of the "tired old bureaucracy" of Washington. They did not always lobby in the savviest fashion. But by 1965 they were publicizing a range of programs, including measures to clean up the environment, improve the delivery of health care, reform the schools, and crack down on what they considered the excesses of big business. They wanted to go beyond the bread-and-butter liberalism of the 1930s to improve the quality of life. Their more aggressive involvement in national affairs broadened the nature of pressure group politics in the United States.[10]

Decisions of the Supreme Court in the early 1960s gave yet another push to rights-consciousness by 1965. Starting in 1962 the Court delivered a series of landmark decisions that delighted many Americans on the left and infuriated most on the right. In that year it issued the first of a number of rulings that forced states to redraw the lines of rurally weighted voting districts—at both the state and congressional levels—so as to give urban and suburban voters a representation appropriate to their growth in numbers.[11] By 1965 states were scrambling to do so, thereby (reformers hoped) reducing the power of rural conservatives in legislatures and in Congress. In *Engel v. Vitale*, also decided in 1962, the Court held that it was unconstitutional—a violation of the separation of church and state— for public schools in New York to require children to recite a non-denominational State Board of Regents prayer. A year later it ruled against the practice of daily readings of the Bible in the public schools.[12]

These were but a few of many decisions that sought to enhance civil liberties in the early 1960s. In *Gideon v. Wainwright* (1963) and *Escobedo v. Illinois* (1964) the Court expanded the constitutional rights of alleged criminals.[13] In *Jacobellis v. Ohio* (1964) it complicated the the enforce-

10. James Wilson, "The Politics of Regulation," in Wilson, ed., *The Politics of Regulation* (New York, 1980), 357–94; David Vogel, "The Public Interest Movement and the American Reform Tradition," *Political Science Quarterly*, 95 (Winter 1980–81), 607–27; Vogel, *Fluctuating Fortunes: The Political Power of Business in America* (New York, 1989), 277–78.

11. *Baker v. Carr*, 369 U.S. 186 (1962); *Reynolds v. Sims*, 377 U.S. 533 (1964).

12. *Engel v. Vitale*, 370 U.S. 421 (1962); *Abingdon School District v. Schempp*, 374 U.S. 203 (1963).

13. *Gideon v. Wainwright*, 372 U.S. 335 (1963); *Escobedo v. Illinois*, 378 U.S. 478 (1964). The *Gideon* decision said that accused felons had a right to counsel in state

ment of laws against pornography; henceforth, the Court said, prosecutors must prove that such material was "utterly without redeeming social importance."[14] In the midst of the 1965 congressional session the Court ruled, 7 to 2, that an 1879 Connecticut statute prohibiting not only the sale but also the use (by married as well as unmarried people) of contraceptive devices violated a constitutional right of people to privacy.[15]

In *New York Times v. Sullivan* (1964) the Court unanimously overruled an Alabama court decision that had found the *Times* and four black clergymen guilty of libel because of an ad, containing errors of fact, that the clergymen had run in the paper to back a legal fight then being waged by Martin Luther King. To hold the newspaper guilty, the high court ruled, was to inhibit discussion of public issues. "Erroneous statement is inevitable in free debate," the judges said, and "must be protected if the freedoms of expression are to have the 'breathing space' that they . . . need to survive." The Court concluded by ruling that public figures and officials could recover damages for libelous statements by the news media only if they could prove that the statements were published as a result of "actual malice." Civil libertarians hailed the decision, which had great importance in later years, as a ringing defense of First Amendment freedom.[16]

In another series of cases the Court proclaimed its continuing opposition to racially discriminatory practices. In *Garner v. Louisiana* (1961) it had upheld the constitutional rights of protestors to conduct peaceful sit-ins; in *Edwards v. South Carolina* (1963) it ruled that states could not legitimately arrest civil rights demonstrators who had peacefully protested on the grounds of the statehouse; in *Shuttlesworth v. City of Birmingham* (1963) it decided that Jim Crow ordinances in the city could not be enforced; in *Heart of Atlanta Motel v. United States* (1964) it unani-

---

cases; the *Escobedo* decision said that a suspect in custody had an absolute right to the aid of an attorney during interrogation. For the *Gideon* case, see Anthony Lewis, *Gideon's Trumpet* (New York, 1964).

14. 378 U.S. 184 (1964). This and other cases managed to leave considerable confusion about the nature of obscenity and pornography.

15. *Griswold v. Connecticut*, 381 U.S. 479 (1965); *Newsweek*, June 21, 1965, p. 60. The *Griswold* decision establishing privacy as a constitutional right became the key precedent for *Roe v. Wade* (410 U.S. 113 [1973]), which legalized abortion. That decision, too, was 7 to 2. See David Garrow, *Liberty and Sexuality: The Right to Privacy and the Making of "Roe v. Wade"* (New York, 1994).

16. 376 U.S. 254 (1964). See Anthony Lewis, *Make No Law: The Sullivan Case and the First Amendment* (New York, 1991).

mously upheld the public accommodations sections of the 1964 Civil Rights Act; and in *Griffin v. County School Board of Prince Edward County* (1964) it overturned the discriminatory ruses of school segregationists in Prince Edward County, Virginia.[17] In 1964 it began a judicial assault on racist state laws by overturning a Florida statute that had forbidden cohabitation between whites and blacks in the state, and in 1965, again while Congress was in session, it overturned a Mississippi law that discriminated against blacks who wanted to vote.[18]

Many of these decisions aroused loud and abrasive controversy. Conservatives and police officials complained bitterly that the Court was coddling criminals. Southern officials resisted the decisions regarding race and civil rights. The *Jacobellis* and other rulings regarding pornography excited incredulous reactions from people across the political spectrum. The Connecticut birth control case was too much even for Justice Hugo Black, one of the most ardent defenders of civil liberties on the Court. He dissented testily, maintaining that there was no constitutional justification for the notion that people had a right to "privacy." Justices Felix Frankfurter (who left the Court in 1962) and John Marshall Harlan dissented from the majority decisions on reapportionment. Their argument, like that of many later critics of the activist Warren Court, rested on their belief in the necessity for judicial restraint. Frankfurter complained in 1962 that the Court was entering a "political thicket." It should seek "complete detachment" and abstain from "political entanglements." Harlan added that the Court "is not a panacea for every blot upon the public welfare, nor should this court, ordained as a *judicial* body, be thought of as a general haven for reform movement."[19]

No decisions unleashed more lasting controversy than those involving religion. Cardinal Richard Cushing of Boston exclaimed, "The Communists are enjoying their day." The theologian Reinhold Niebuhr observed that the *Engel* decision "practically suppresses all religion, especially in the public schools." *Engel* and other cases did more than anything else over time to arouse the religious Right from its political quietism. Other

---

17. *Garner v. Louisiana*, 368 U.S. 157 (1961); *Edwards v. South Carolina*, 372 U.S. 229 (1963); *Shuttlesworth v. City of Birmingham*, 373 U.S. 212 (1963); *Heart of Atlanta Motel v. United States*, 379 U.S. 241 (1964); *Griffin v. County School Board of Prince Edward County*, 377 U.S. 218 (1964).
18. *U.S. v. Mississippi*, 380 U.S. 128 (1965). See also John Blum, *Years of Discord: American Politics and Society, 1961–1974* (New York, 1991), 193–94, 306–41.
19. *Newsweek*, April 9, 1962, p. 30.

Americans, too, thought that the justices had lost their minds.[20]

Liberals, however, were greatly inspired by the Court. At last, they said, the judges were construing the law so as to extend the Bill of Rights and the Fourteenth Amendment to all kinds of Americans, even blacks, non-believers, Jews, and criminals. What a change in constitutional approaches to civil rights and civil liberties since the days of McCarthyism a decade earlier! The Court, moreover, did not bend under criticism; it hewed to its liberal path in the next few years. Advocates of a Great Society rejoiced that liberal programs of the 1960s, unlike those in the 1930s, were safe from judicial assault. Liberals controlled all three branches of American government.

The decisions of the Warren Court reflected and accelerated one of the major trends of the era: the rise of rights-consciousness. This, given special urgency already by the moral power of the civil rights movement, began to seem all-conquering by 1965. It was bolstered by the ever more infectious optimism of liberal social scientists who were certain that the economy, booming in the mid-1960s, could afford to sustain major policy initiatives. Liberal policy-makers, equally optimistic, were sure that the State could engineer political solutions to social and economic problems. Polls suggested that the American people had unprecedented faith in politicians and in the State.[21] The convergence of these ideas and assumptions promoted an increasingly powerful—and ultimately near-irresistible—drive for the expansion of individual rights in the United States. The drive came from the bottom up—from ordinary people demanding justice—and from the top down. A Rights Revolution was at hand.[22]

AMONG THE MANY LIBERAL ACTS of the 1965 Congress were some that in other less active sessions would have attracted considerable attention. These included a host of measures signifying the onset of a

20. Stephen Bates, *Battleground: One Mother's Crusade, the Religious Right, and the Struggle for Control of Our Classrooms* (New York, 1993), 46–52, 208–9; Richard Polenberg, *One Nation Divisible: Class, Race, and Ethnicity in the United States Since 1938* (New York, 1980), 171–72.
21. Morris Janowitz, *The Last Half-Century: Societal Change and Politics in America* (Chicago, 1978), 113.
22. Ibid., 402–3, 547; Edward Berkowitz, "Public History, Academic History, and Policy Analysis: A Case Study with Commentary," *Public Historian*, 10 (Fall 1988), 43–63.

burgeoning environmental movement: clean air legislation, establishment of parks and national wilderness areas, a law to control the spread of billboard advertising on interstate highways.[23] Congress also approved a Higher Education Act featuring guaranteed government loans for students, and it substantially expanded existing work-study programs. These widened opportunities for students, especially those from low-income families.[24] Liberals created the National Endowment for the Arts and the National Endowment for the Humanities, thereby substantially engaging the federal government in the promotion of cultural life for the first time since the New Deal. Congress considered laws to improve mine safety and consumer protection. It further increased funding for the war on poverty, although at the modest level of approximately $1.5 billion for the coming fiscal year.[25]

These and other measures, however, seemed relatively inconsequential in 1965 compared to a Big Four that passed by the end of the session: federal aid to elementary and secondary education, Medicare and Medicaid, immigration reform, and a civil rights act to guaranteee voting rights. These four laws, important by any standard of twentieth-century reform legislation, had long been on the agendas of liberal groups. They amply displayed the strengths and weaknesses of the Great Society, of Lyndon Johnson as political leader, and of modern American liberalism.

Improving elementary and secondary education ranked high among the goals of liberals in the postwar era. Some of these advocates focused on the need for higher teachers' salaries. Others, especially once the baby boomers reached school age, demanded more resources for school materials and construction. These goals had dominated debate from the 1940s, when conservatives had ignored Truman's proposals, through 1961, when conservatives and Catholics had defeated Kennedy's. Before 1965, a complicated mix of special interests—religious, racial, regional—had joined conservatives to stymie all major efforts for federal aid.

Three changes in this mix facilitated passage of aid in 1965. One involved race. Urban liberals in the past, led by Adam Clayton Powell, Jr., of Harlem, had opposed federal aid bills that would have assisted segregated schools. But the 1964 Civil Rights Act, which called for the denial of federal assistance to such schools, settled that controversy. Lib-

23. Sundquist, *Politics and Policy*, 361–81.
24. Conkin, *Big Daddy*, 228.
25. James Patterson, *America's Struggle Against Poverty*, 1900–1994 (Cambridge, Mass., 1995), 150–52.

erals, no longer worrying that aid would abet segregation, were more eager than ever to support general school assistance. The second involved the question of aid to parochial schools. Johnson, a Protestant, was not nearly so politically vulnerable among Protestants as Kennedy had been on this issue, and he resolved to satisfy both the National Education Association, representing public schools, and the National Catholic Welfare Conference, the most active lobby for parochial schools. He did so by devising a bill that would provide federally assisted educational programs directly to parochial schoolchildren rather than to parochial schools. Public schools would be expected to make these programs available to such students by way of expedients like dual enrollment, public television, and sharing of equipment.[26]

Johnson also injected the third and most important new element into the mix: a focus on poverty. Drawing on selective memories of his own life, he emphasized the benefits of education to poor people. "Compensatory education," to be extended to children from low-income families, would greatly expand his war on poverty. The bill that he endorsed trumpeted this principle while at the same time offering a formula for distributing aid that was politically attractive. It offered federal money to 90 percent of school districts in the nation and assured NEA lobbyists that local school administrators would enjoy latitude in determining how the money was spent.[27]

Thanks to these changes it was clear that aid to elementary and secondary education would pass. In the House a bill calling for $1 billion in the coming year was approved, 263 to 153, with all fifty-six freshman non-southern Democrats who voted on the bill recorded in favor. In the Senate, only eighteen voted no. Johnson signed the bill in April at the site of the one-room schoolhouse where he had learned his ABCs many years before. "As a son of a tenant farmer," he said, exaggerating the deprivation of his past, "I know that education is the only valid passport from poverty."[28]

Passage of the act dramatically increased the role of the federal government in school financing. The money, indeed, enticed members of Congress, who approved substantial sums for compensatory education in the next few years. Federal expenditures for schools rose by 1968 to ap-

26. Diane Ravitch, *The Troubled Crusade: American Education, 1945–1980* (New York, 1983), 159–61; Kearns, *Lyndon Johnson*, 227.
27. Hugh Graham, "The Transformation of Federal Education Policy," in Divine, ed., *Exploring the Johnson Years*, 155–84; Sundquist, *Politics and Policy*, 205–20.
28. Matusow, *Unraveling*, 222–23.

proximately $4.2 billion, more than ten times the amount ($375 million) spent ten years earlier. The federal share of total educational spending during the same period increased from less than 3 percent to roughly 10 percent.[29] It continued to rise in subsequent years, generating a historic change in the nature of financial support for schools in the United States.

By then, however, the principles of federal aid enshrined in Johnson's education bills came under sharp questioning. Conservatives from the start had insisted that the falling off of the baby boom in the late 1950s, combined with rapid increases in school-building, was already easing whatever "crisis" had existed in elementary schools. By the early 1960s, Goldwater pointed out, the number of students per classroom and per teacher had declined from the early 1950s. Teachers, moreover, were receiving more training. Conservatives added that while some districts needed help, there was no need for an avalanche of federal money. Most local areas had done all right in keeping pace with demographic change.[30]

Other, less partisan critics took aim at the way the act was administered in practice. The flow of federal money after 1965 did promote a little greater equality in overall per pupil spending on education in the United States. Poor states like Mississippi received a shot in the arm. But many local school administrators managed to skirt guidelines on how to spend the money, using it to cover routine administrative expenses and overhead, which did little or nothing for the poor. By the early 1970s it was an open secret that a great deal of federal educational money aimed at the poor was missing its target.[31]

Larger doubts centered about the very philosophy of compensatory education as it was established in 1965. Although Johnson's task force on education had studied conscientiously, it did not fully address the consequences of a central socio-economic trend of the postwar era: young people were staying on longer in schools. Many of these students were daughters and sons of parents with relatively little educational background. Others were blacks who had migrated from the South, where their educational facilities had been abysmal. How to instruct these masses of educationally deprived young people was a puzzle, and there was little research that would help to unravel it. Advocates of compensatory education decided in 1965 that more funding was an answer, but neither they nor teachers seemed able to use it very effectively. Money for

29. Sundquist, *Politics and Policy*, 216–17.
30. Landon Jones, *Great Expectations: America and the Baby Boom Generation* (New York, 1980), 300–302; Matusow, *Unraveling*, 106.
31. Matusow, *Unraveling*, 223–25.

compensatory education was not the same as money for good education. This, in the careful words of educational historian Diane Ravitch, would at the very least have meant using money and developing expertise to provide "intensive, individuated instruction in an encouraging, supportive environment."[32]

In 1966 there appeared a major study of American schools, *Equality of Educational Opportunity*, undertaken for the Office of Education. Named the Coleman Report after the sociologist James Coleman, its chief investigator, it questioned whether increases in per student spending for schools made much difference in the measurable educational achievement of individual students. The key to such achievement, the report concluded, appeared instead to be the background of students' families, the ethos of their neighborhoods, and the academic zeal of their classmates. For these reasons, desegregation might help, too. These conclusions surely did not mean that spending levels were irrelevant. Parents, indeed, understood that good schools required money, and they moved to districts that generously financed education. Their quest for good schools did much to accelerate suburbanization. The report did suggest, however, that dollars would go only so far and that beyond that point there was no strong correlation between per pupil spending and student achievement. Starting in 1964, average scores on Scholastic Aptitude Tests went steadily down, and per student spending went steadily up.[33]

Complicated reasons explain the much-lamented fall in scores, including the fact that higher percentages of poor and ill-prepared students took the tests, thereby driving down averages. Mindless absorption in television, others said, added to the decline in scores. Still, the faith in spending for compensatory education that had excited Johnson and others lost much of its fire in time. By the early 1970s a number of reform-minded critics had reached a different conclusion: if the government hoped to better the achievement and life-chances of "disadvantaged" children, it must pursue policies to reduce inequalities created by differences of socioeconomic class. In the context of American reform ideas, talk about class was a radical, not a liberal, notion, and it had no political prospects.[34]

The second of the Big Four reforms of 1965, Medicare and Medicaid, also emanated from a long and complicated series of postwar legislative struggles that had intensified in the late 1950s and early 1960s. From the

32. Ravitch, *Troubled Crusade*, 153.
33. Matusow, *Unraveling*, 225–26.
34. Christopher Jencks, *Inequality: A Reassessment of the Effect of Family and Schooling in America* (New York, 1972).

Roosevelt years on, the pressure of doctors in the American Medical Association, among other lobbies, had held off major changes in the American health care system. In 1965 one-half of all Americans older than 65 had no health insurance. As with the education issue, however, prospects for liberal overhaul escalated as a result of the 1964 election. Certain of chances for reform, LBJ made better medical care for the elderly his top priority and arranged to have Democratic measures introduced in January as House and Senate Bills Number One. Throughout the course of the subsequent legislative process he drove these bills ahead on mostly straight partisan votes. On one key vote in the House, fifty-eight of the sixty-five first-term Democrats voted to defeat a Republican-sponsored alternative to Medicare that might otherwise have passed.[35]

What did pass was a Medicare bill that mandated increases in Social Security taxes (paid by both employers and employees) to subsidize the costs of hospitalization for certain periods of time (in general, 100 days) for most people who were over 65 years of age. This was Plan A. The bill also offered Plan B, a voluntary program of insurance to help elderly people cover X-ray tests, up to 100 home-nurse visits, and certain doctors' and surgical fees. Plan B was to be supported by the government as well as by payments from recipients of care. The plans aided some 19 million Americans at an estimated cost in the first year (starting July 1, 1966) of $6.5 billion.[36]

Congressman Wilbur Mills of Arkansas, the influential chairman of the House Ways and Means Committee, then helped to add a Medicaid program, which few observers had anticipated and which received little sustained attention on the Hill. Like the education act, it reflected the contemporary focus on poverty. It offered federal matching grants to states that provided money to poor people already eligible for categorical welfare programs—the blind, the disabled, the needy aged not covered by Social Security, and families with dependent children (AFDC)—and to a small number of other "medically indigent" Americans who were not in these categories. Medicaid, like AFDC, was a government "entitlement": it guaranteed recipients assistance (assuming state spending) without need of annual congressional approval of appropriations. All these reforms passed with ease on partisan votes. Hailing the results, Johnson went to Independence, Missouri, to sign Medicare in the presence of Harry Truman. "No longer will older Americans be denied the healing miracle of modern

35. Sundquist, *Politics and Policy*, 317–19.
36. *Newsweek*, Aug. 2, 1965, p. 27.

medicine," he said with characteristic flourish. "No longer will illness crush and destroy the savings that they have so carefully put away over a lifetime so that they might enjoy dignity in their later years."[37]

Medicare and Medicaid considerably changed the nature of health care in the United States. Growing rapidly in the next few years, they reached one-fifth of the population by 1976.[38] Medicare helped a number of elderly people to receive health services that might otherwise have driven them into poverty. Medicaid enabled many eligible poor people to go to doctors for the first time in their lives. By 1968 it was estimated that low-income Americans consulted physicians more often than did higher-income people (5.6 visits per year as opposed to 4.9 visits).[39] Changes such as these gladdened the hearts of reformers and helped to sustain liberal support for the programs in later years.

It was not so clear, however, that there was such a thing as the "healing miracle of modern medicine." In the next few years death rates for people over 65 declined. So did rates of infant mortality. But these improvements merely sustained long-range trends, and it could not be proved that the extension of medical insurance—as opposed to other changes such as better nutrition—made much difference. "Medical miracles" were rare: the vast majority of patients with cancer, the nation's number two killer behind heart ailments, did no better after 1965. Infant mortality rates in the United States continued to be higher than those in many other nations, including some that were considerably poorer. They remained especially high among poor people and blacks, signifying the stubborn persistence of class and racial divisions in the United States.

Medicare and Medicaid, indeed, fell well short of national health insurance. They helped only the elderly and certain categories of the poor. Most Americans, including the working poor, had to contribute to employer-subsidized group insurance plans, to pay for private insurance on their own, or to do without. Those who lost their jobs often forfeited whatever coverage they may have had. And millions of people did do without. No other industrialized Western nation had higher percentages

37. Sundquist, *Politics and Policy*, 321; Matusow, *Unraveling*, 226–32.
38. Matusow, *Unraveling*, 228.
39. Theodore Marmor with James Morone, "The Health Programs of the Kennedy-Johnson Years: An Overview," in David Warner, ed., *Toward New Human Rights: The Social Policies of the Kennedy and Johnson Administrations* (New York, 1977), 173; and Karen Davis and Roger Reynolds, "The Impact of Medicare and Medicaid on Access to Medical Care," in Richard Rosett, ed., *The Role of Health Insurance in the Health Services Sector* (New York, 1976), 393.

of its people—still about 15 percent in the early 1990s—without medical insurance.

Like aid to education, Medicare and Medicaid contained widely lamented flaws. One of these involved gaps in coverage. Medicare did not pay for many things, including eye glasses, dental care, and out-of-hospital drugs, and it did not cover long-term nursing-home care or the full costs of hospitalization. Deductibles grew increasingly expensive as the costs of care rose steeply over time. Ten years after the start of the program it was estimated that beneficiaries of Medicare spent as much on the average in constant dollars as they had done in 1964.[40]

The administration of Medicaid exposed special limitations. Like many other federal-state welfare programs—AFDC was the most salient example—it depended for its support on state legislatures, some of which (mostly those in the poorer states) could not or did not appropriate much money for the federal government to match. Huge variations in benefits arose from state to state.[41] Like most services to poor people, Medicaid also tended to offer low-quality help. Many physicians shunned Medicaid patients, both because of the paperwork involved and because state governments tended increasingly to set reimbursements at a level below what doctors could receive from better-off patients. Some doctors ran so-called Medicaid mills that dispensed unnecessary services or charged for nonexistent procedures, thereby bilking public authorities and running up the costs to taxpayers. Dr. John Knowles, director of the Massachusetts General Hospital, said in 1969, "Medicaid has degenerated into merely a financing mechanism for the existing system of welfare medicine. . . . It perpetuates . . . the very costly, highly inefficient, inhuman and undignified means tests in the stale atmosphere of charity medicine carried out in many instances by marginal practitioners in marginal facilities."[42]

Critics of Medicare and Medicaid focused especially on the impact of the programs on medical costs, which rose rapidly as the programs expanded in the 1970s and 1980s. Cost increases, to be sure, had pre-dated 1965 and were rooted in the spread of private hospital insurance plans such as Blue Cross and Blue Shield after 1945. These plans had encour-

40. Marian Gornick, "Ten Years of Medicare: Impact on the Covered Population," *Social Security Bulletin*, 39 (July 1976), 19.
41. Medicaid provisions tried to minimize these state-by-state variations by stipulating that the federal government would give larger percentages of money to states with low per capita incomes, but the overall money available in each state still depended heavily on sums approved by the states.
42. Matusow, *Unraveling*, 230–32.

aged policy-owners to demand better (more expensive) care—after all, their premiums were supporting it. Doctors and hospitals were happy to offer high-tech services that other people—insurance companies—increasingly felt obliged to cover. Incentives for cost control weakened, and medical expenditures escalated. Between 1950 and 1965 hospital prices rose at the rate of 7 percent per year, compared to a rise in the general price level of 2 percent per year.[43]

Congress in 1965 might have seriously addressed these trends by imposing cost controls on doctors and hospitals providing Medicare or Medicaid services. But legislators who considered doing so faced the obdurate resistance of the AMA, which damned such controls as socialized medicine. Wilbur Mills and other congressional leaders had no stomach for a fight against such politically powerful constituents, and they approved entitlements without real controls. The law made it clear that hospitals and doctors should expect to be able to set their fees, with some exceptions. Costs then rose even more rapidly than they had in the preceding few years. During the next decade hospital prices jumped by 14 percent annually and physicians' fees by 7 percent.[44] By the 1990s the costs of health care, driven upward in part by expenditures for Medicaid and Medicare, were inciting angry debates over the role of government in American life.

As it happened, the passage of Medicare and Medicaid in 1965 represented the only major governmental changes in the American health care system during the next three decades. Americans continued to live with a medical system that led the world in its training of physicians and in technological wizardry but that was also bureaucratically complicated and far from comprehensive. This was not the intention of Johnson and fellow liberals, who did well to secure legislation that reformers had been seeking for years. In 1965 they probably accomplished all that was politically possible. The power in Congress of pressure groups—insurance companies, doctors, hospitals—overwhelmed whatever support there was at the time (and it was weak) for a governmentally operated plan of national health insurance such as those that were instituted in many other Western nations during the postwar era.

Still, the exaggerations of Johnson and other liberals at the time came to haunt them within a few years. Talk about the "healing miracle of modern medicine" and about the capacity of Medicare and Medicaid to deliver it was as utopian as talk about "wars" against poverty or the

43. Martin Feldstein, "The Welfare Loss of Excess Health Insurance," *Journal of Political Economy*, 88 (March 1973), 252.
44. Matusow, *Unraveling*, 229.

wonders of "compensatory education." Medicare and Medicaid survived to become important—and extraordinarily expensive—entitlements. But in time they raised widespread questions about the hyperbolic claims of Johnson and the wisdom of American liberalism.

Reform of the nation's immigration policies, the third of the Big Four in 1965, ranked lower among the priorities of Johnson and his advisers at the start of the legislative session. But congressional Democrats, led by Representative Emanuel Celler of New York, chose to seize upon the uniquely liberal mood to challenge existing immigration laws. These, dating from the 1920s, still employed "national origins" quotas that discriminated against southern and eastern Europeans and that sharply limited the numbers of Asians who could come to America.

By the close of the session they had succeeded, and Johnson traveled yet again, this time to the Statue of Liberty, to sign a bill that reflected the more rights-conscious spirit of the era. The legislation of 1965 did away with the old discriminatory quota system and established priorities that were expected to increase the flow of people from southern and eastern Europe—those, at least, who were not locked behind the iron curtain. It stipulated that a total of 290,000 immigrants per year (roughly the numbers then coming) could be admitted to the United States as of 1968. For the first time in United States history the law set limits on the numbers of immigrants who might be admitted from countries in the Western Hemisphere. These were to be 120,000 per year, with the rest—170,000—to come from Europe and (it was expected) in much smaller numbers from Africa and Asia. A maximum of 20,000 people might come from any single nation, except from those in the Western Hemisphere, where no such national limits were applied.[45]

At the time of its passage the new immigration law, while hailed for its repeal of the old quotas, did not seem likely to create major changes in the demography of the United States. Over time, however, it did; framers of the law failed to foresee the consequences of what they had done. This was mainly because the law also permitted the admission beyond numerical limits of close relatives of United States citizens, both native-born and naturalized. Over the next decade an average of some 100,000 were

45. Rubén Rumbaút, "Passages to America: Perspectives on the New Immigration," in Alan Wolfe, ed., *America at Century's End* (Berkeley, 1991), 212; Victor Greene, "Immigration Policy," in Jack Greene, ed., *Encyclopedia of American Political History*, Vol. 2 (New York, 1984), 579–93; Bernard Weisberger, "A Nation of Immigrants," *American Heritage*, Feb./March 1994, pp. 75ff; Polenberg, *One Nation Divisible*, 203–6; and *Newsweek*, Oct. 4, 1965, p. 35.

admitted each year in addition to the 290,000 authorized by the statute. Equally unexpected, the sources of immigration began to change dramatically after the late 1960s. Contrary to the expectations of Celler and others, the flow of immigrants from Europe declined after 1968. But the numbers from Latin America and Asia began to swell. By 1976 more than half of legal immigrants to the United States came from seven nations: Mexico, the Philippines, Korea, Cuba, Taiwan, India, and the Dominican Republic.[46]

These developments were hardly revolutionary at the time. The United States, with a population of 194 million in 1965 (and 205 million in 1970) easily absorbed the 400,000 or so legal immigrants per year who arrived in the late 1960s and early 1970s. Still, because the birth rate of other Americans had stabilized, immigrants came to compose a steadily higher percentage of the population. And the numbers of immigrants kept growing over time: by the late 1970s, more than 450,000 legal immigrants arrived each year, fewer than one-fifth of whom were Europeans. By 1980 the number of foreign-born people in the country had increased to 14 million, as opposed to 9.7 million in 1960. In the 1980s, an average of 730,000 legal immigrants came annually, of whom roughly one-tenth hailed from Europe.[47] The vast majority of the rest (as well as large numbers of illegal immigrants) came from the Caribbean, Central and South America, and Asia.

As immigration mounted in the years after 1968, scholars and politicians debated whether the results justified the reform of 1965. Decades later there was no solid consensus on the matter. Critics insisted that the flow of migrants strained schools and social services and deprived native-born Americans, including blacks, of jobs. Others, stressing that the law gave priority to immigrants with skills, retorted that the newcomers contributed to economic growth. Proponents of more liberal immigration laws further welcomed the evolution of a richer ethno-cultural mix.[48] These debates, however, focused on later, long-range consequences that had scarcely occurred to most people in 1965. What reformers saw at that

46. David Reimers, *Still the Golden Door: The Third World Comes to America* (New York, 1992), 61–91.
47. Still, the presence of foreign-born people was slight compared to what it had been early in the century. The percentage of foreign-born reached an all-time modern low of 4.7 percent in 1970, before the consequences of the 1965 reform became important. By 1980 it had risen to 6.2 percent and by 1990 to 7.6 percent. This was still well below the high of 14.7 percent in 1910.
48. Michael Fix and Jeffrey Passel, *Immigration and Immigrants: Setting the Record Straight* (Washington, 1994).

time, and were happy about, was that they had abolished discriminatory quotas and opened up the gates a little to people around the world. Like much that cleared Congress in 1965, the immigration law reflected the hopefully liberal temper of the time.

The fourth and most significant liberal accomplishment of the 1965 congressional session involved the still most divisive issue of the age: race relations. This now focused on voting rights for blacks, which the civil rights acts of 1957, 1960, and 1964 had failed to guarantee. In many Deep South counties, 90 percent or more of blacks were not registered to vote. Johnson recognized the problem and called for action in his State of the Union message in January. He also anticipated, however, another southern filibuster—one that could tie up his Great Society programs—and he did not want to jeopardize other bills in order to fight for a voting rights bill early in the session.[49]

Civil rights activists, as so often in the 1960s, set their own agendas without consulting Washington. In January, Martin Luther King amassed SCLC supporters to demonstrate against the denial of voting rights in Selma, Alabama. Activists loyal to SNCC agreed to join, in part because the cause was so compelling. Selma, a city of 29,000 people, had some 15,000 blacks of voting age, of whom only 355 were registered to vote. Its board of registrars met only two times per month and blatantly discriminated against black people daring enough to appear before it. Blacks were disqualified if they neglected to cross a *t* in registration forms or did not know the answers to obscure questions such as "What two rights does a person have after being indicted by a grand jury?"[50]

King selected Selma as his site for the same reason that he had shrewdly chosen Birmingham in 1963: he anticipated that whites would resist fiercely and violently, thereby dramatizing his cause on television and forcing the government to act. Like "Bull" Connor in Birmingham, Sheriff Jim Clark of Selma's Dallas County was expected to overreact. Clark was an unreconstructed segregationist who proudly displayed on his lapel a button, NEVER, to tell blacks that nothing would change. With the approbation of George Wallace and of white leaders in Selma, he and his

49. Steven Lawson, "Civil Rights," in Divine, ed., *Exploring the Johnson Years*, 63–90.

50. Robert Weisbrot, *Freedom Bound: A History of America's Civil Rights Movement* (New York, 1990), 127–34; David Goldfield, *Black, White, and Southern: Race Relations and Southern Culture, 1940 to the Present* (Baton Rouge, 1990), 149–73; Abigail Thernstrom, *Whose Votes Count? Affirmative Action and Minority Voting Rights* (Cambridge, Mass., 1987), 2; Sundquist, *Politics and Policy*, 271–75.

men had manhandled SNCC workers and would-be registrants in 1963 and 1964.

King judged his adversary accurately, for Clark and his deputies over-reacted in January and February 1965, arresting and jailing more than 3,000 demonstrators, including King and SNCC leader John Lewis. Deputies kicked and clubbed demonstrators and threw them into trucks that took them to jail. On one occasion Clark shoved a woman, who then knocked him down. Deputies threw her to the ground and pinned her, whereupon Clark leaned over her and smashed her with a club. Widely circulated photographs of this action appalled Americans around the country. On February 10 he arrested 165 protestors and pushed them on a three-mile forced march out of town. Electric cattle prods used by his men singed many of the demonstrators, who fell vomiting by the road-side. A few days later Clark punched the Reverend C. T. Vivian, an SCLC leader, and sent him reeling down the courthouse steps. State troopers, meanwhile, ambushed marchers in nearby Marion. When the troopers assaulted a black woman and her infirm father, her son, twenty-six-year-old Jimmy Lee Jackson, tried to intercede. He was shot in the stomach at close range by a trooper and died eight days later.[51]

King then resolved to dramatize the cause by organizing a protest march from Selma to the state capital of Montgomery, fifty-six miles to the east. There the demonstrators were to petition Governor Wallace for protection of blacks who wished to register. King hoped to start the march on Tuesday, March 9, by which time he expected that federal judge Frank Johnson would have voided an order by Wallace to ban it. Younger militants, however, determined to march on Sunday, March 7. Selma police were sure that violence would occur, but when the mayor told them that Wallace had promised there would be peace, they made no effort to stop the marchers when they moved out of their church head-quarters on the 7th. Lewis and SCLC leader Hosea Williams led the 600 black demonstrators to the Edmund Pettis Bridge at the edge of town. Many toted sleeping bags for the lengthy march to come. Across the bridge waited helmeted state troopers wearing gas masks, as well as Sheriff Clark and his men, mounted on horseback.[52]

51. Weisbrot, *Freedom Bound*, 134–38; Godfrey Hodgson, *America in Our Time* (Garden City, N.Y., 1976), 218–20.
52. David Garrow, *Protest at Selma: Martin Luther King, Jr., and the Voting Rights Act of 1965* (New Haven, 1978), 42–48; Garrow, *Bearing the Cross: Martin Luther King, Jr., and the Southern Christian Leadership Conference* (New York, 1986), 357–430; Howell Raines, ed., *My Soul Is Rested: Movement Days in the Deep South Remembered* (New York, 1977), 187–226.

There followed one of the most frightening confrontations in the history of the civil rights movement. A state police major shouted through a bullhorn, "Turn around and go back to your church." He gave the marchers two minutes to turn back. Williams asked for a "word" with the police but was told, "There is no word to be had." One minute later the troopers obeyed an order to advance. They tore forward in a flying wedge, swinging their clubs at people in the way. Lewis stood his ground, only to be cracked on the head. He suffered a fractured skull. With white onlookers cheering, the troopers rushed ahead, hitting the demonstrators and exploding canisters of tear gas. Five women were beaten so badly that they fell down near the bridge and lost consciousness. Sheriff Clark's horsemen then joined in the assault. Charging with rebel yells, they swung bullwhips and rubber tubing wrapped in barbed wire. More demonstrators fell, seventy of whom were later hospitalized. The rest were driven back to the church where they had started.[53]

The violence of Bloody Sunday, as demonstrators called it, outraged millions of Americans who saw it shown (repeatedly) on national television. Editorials in northern papers angrily denounced Wallace, Clark, and the troopers and demanded federal intervention. White supporters of civil rights, mostly from the North, began descending on Selma to help the demonstrators. From that point on it was virtually inevitable that Congress would have to take action—and soon.[54]

Bloody Sunday unleashed especial fury among militants at the scene. As Lewis left for a hospital, he cried, "I don't see how President Johnson can send troops to Vietnam . . . and can't send troops to Selma, Alabama." He added, "Next time we march, we may have to keep going when we get to Montgomery. We may have to go to Washington."[55] King, returning to Selma from Atlanta (where he had gone to preach on Sunday), started organizing a march for Tuesday, March 9. Judge Johnson, however, indicated that he wanted the march postponed pending a hearing on Wallace's request for a ban on the march. LBJ, too, applied pressure for delay. Militants, however, were eager to march on Tuesday, even if that meant violating a court order from Judge Johnson. King, caught in the middle, engaged in all-night arguments with other civil rights protestors over strategy.

Unwilling to defy a federal court, King finally accepted an arrangement devised in concert with federal mediators. He would lead a token march

53. *Newsweek*, March 22, 1965, p. 19; Weisbrot, *Freedom Bound*, 137–39.
54. Garrow, *Protest*, 72–82, 87–90.
55. *New York Times*, March 8, 1965.

across the bridge, thereby demonstrating the determination of his followers, but then would turn around and go back to his headquarters in Selma. Alabama police promised not to hurt the marchers. King told only a few trusted aides of his plan, however, and most of the 1,500 people (now including whites) who marched that day assumed they were going to challenge the authorities. When King got across the bridge, he prepared to turn about. But state troopers standing there had been informed of his plans and suddenly wheeled out of the way, leaving (or so it seemed) a clear path to Montgomery.[56]

This move of the authorities, reportedly ordered by Wallace, was obviously designed to embarrass King. Militants on the march, including James Forman and Cleveland Sellers of SNCC, had already been highly critical of King—"de Lawd." They chafed to go ahead on the now open road. But King had promised to turn around, and turn around he did, thereby aborting the march. The episode fractured already delicate coalitions within the movement. In the complicated events that followed, SCLC and SNCC barely managed to cooperate.

Militants who questioned LBJ's commitment, however, misjudged him. After Bloody Sunday the President knew he must take a stand. Overriding advisers who urged restraint, he went to Capitol Hill on Monday, March 15, to press for a strong new voting rights law. Millions of people watched on prime-time television as he spoke carefully but with great emotion. Members of Congress, outraged by the events at Selma, forty times interrupted his address with applause. Johnson closed by raising his thumbs, fists clenched, and proclaiming, "Their cause must be our cause, too. Because it is not just Negroes, but really all of us who must overcome the crippling legacy of bigotry and injustice. And, we shall . . . overcome." His speech, especially the final peroration, moved many in the movement, including King, whose eyes filled with tears.[57]

Two days later Judge Johnson sided with the demonstrators by striking down Wallace's request for a ban on a march. Johnson conceded one of Wallace's arguments—that a march on a state highway might impede traffic—but he held that the demonstrators had a right, given the "enormous" wrongs that they had suffered, to assemble and to march in a peaceable and orderly manner. The judge barred state and local officials

56. Garrow, *Protest*, 83–87; Weisbrot, *Freedom Bound*, 139–42.
57. *Public Papers of the Presidents of the United States: Lyndon B. Johnson, 1965* (Washington, 1966), 281–87; Kearns, *Lyndon Johnson*, 228–30; Conkin, *Big Daddy*, 215–17.

from interfering with the marchers. Wallace was outraged, having already called Judge Johnson a "low-down, carpetbaggin', scalawaggin', race-mixin' liar." Vindicated, King and his aides set March 21 as the starting date for the march—at last—to Montgomery.

At this point the President again intervened to assist the movement. When Wallace warned darkly that he could not guarantee the safety of the marchers, LBJ called him to Washington for three hours of The Treatment, during which he threatened to send in federal troops if necessary. The session between the President and the governor featured profane and earthy language from both parties. Wallace departed impressed by Johnson. When he was asked how LBJ compared to JFK, he replied, "Johnson's got much more on the ball." He added, "If I hadn't left when I did, he'd have had me coming out *for* civil rights."[58]

The march that began on March 21 proved to be an especially memorable event in the history of the civil rights movement. Although King did not walk the whole way, he started out at the head of a throng of thousands of people, most of them emotionally committed local blacks. Significant numbers of whites from the North joined in. Other leaders—Ralph Abernathy, John Lewis, King's executive secretary Andrew Young—circulated actively amid the marchers as they walked. Federal marshals and the Alabama guardsmen flanked and protected both sides of the highway. Helicopters circled overhead to look out for danger.[59]

After four days the marchers reached the outskirts of Montgomery, where they stopped for an evening of entertainment. The folksingers Peter, Paul, and Mary led the throng in Bob Dylan's "The Times They Are a-Changin'." The black comedian Dick Gregory delighted the crowd with jokes about Selma and the segregationist mentality. The next day King and other national leaders—Roy Wilkins of the NAACP, Whitney Young of the Urban League, A. Philip Randolph and Bayard Rustin—stood at the steps of the capitol (where the Confederate flag flew over the dome). As he had at the March on Washington nineteen months earlier, King closed an inspirational series of speeches with a powerful and uplifting oration. The crowd, now 25,000 strong, sang out the anthem of the movement, "We Shall Overcome," modifying it triumphantly, "We *have* overcome today."[60]

To an extent, they had. But that evening four members of the KKK

58. Weisbrot, *Freedom Bound*, 144.
59. Garrow, *Protest*, 114–17.
60. Weisbrot, *Freedom Bound*, 147–48.

tracked the movements of Viola Liuzzo, a white Detroit housewife who had been transporting demonstrators to and from Selma in her car. As she drove along a deserted stretch of the highway, the Klansmen drew even with her car and shot her to death. They stopped to inspect the wreckage, failing to see a young black activist who lay still in the car. Because one of the Klansmen was an FBI agent (who said he had fired in the air), the crime was solved, and convictions were later obtained. Liuzzo's killing, however, exposed the still powerful poison that contaminated race relations and left a bitter taste in the midst of satisfaction.[61]

Compared to the drama at Selma, subsequent action on Capitol Hill moved deliberately. Johnson and his aides, having derived enormous sustenance from the conflict in Alabama, applied unrelenting pressure for passage of a voting rights bill. It received strong bipartisan support save among congressmen from the South. The House approved it overwhelmingly, 333 to 85. Southerners filibustered in the Senate but lost on a vote for cloture, 70 to 30, after twenty-five days of debate.[62] The measure then passed, 77 to 19. For the signing of the bill on August 6 Johnson assembled a large audience of civil rights leaders and congressmen in the President's Room at the Capitol—the same place where Lincoln had signed the Emancipation Proclamation. "Let me say to every Negro in this country," he said. "You *must* register. You *must* vote. . . . The vote is the most powerful instrument ever devised by man for breaking down injustice and destroying the terrible walls that imprison men because they are different from other men."[63]

The 1965 Voting Rights Act greatly extended federal power in the United States. A frankly regional measure, it took aim at Deep South states by stipulating that the Justice Department could intervene to suspend discriminatory registration tests in counties where 50 percent or fewer of the county's voting-age population had been able to register. If that failed to work, the department could send in federal registrars to take over the job. The law covered state and local as well as federal elections and protected not only the right to register but also the right to vote. Two days after the bill became law, federal registrars turned up in Selma as well as in eight other counties in three southern states. Within a year the strong arm of the federal government had helped to increase the registration of eligible Negroes in the six southern states wholly covered by the law from

61. Garrow, *Protest*, 117–18.
62. Sundquist, *Politics and Policy*, 274–75.
63. *Newsweek*, Aug. 16, 1965; Steven Lawson, *Black Ballots: Voting Rights in the South, 1944–1969* (New York, 1976), 307–21.

30 to 46 percent. One of the many white office-holders defeated by the surge in black voters was Sheriff Jim Clark of Dallas County, Alabama. He was beaten in a Democratic primary in 1966.[64]

Many years later critics complained of longer-range consequences of the voting rights law of 1965. Some southern states, prevented by the law from discriminating against black voters, gerrymandered and created at-large congressional districts so as to damage the political aspirations of black candidates. In 1982 Congress amended the act so as to require that blacks and other minorities be given greater opportunity to elect one of their own to Congress and state legislatures. Reapportionment following the 1990 census gave substance to this amendment and led to the election in 1992 of sixteen new black lawmakers on Capitol Hill. These developments, the critics maintained, amounted to a special entitlement for blacks that legislators in 1965 had not intended. The result, they added, was a manipulable system of representation that catered to *groups*, or blocs, of voters, rather than a color-blind system that protected *individuals* from discrimination.[65] These developments, however, were unintended consequences of the act of 1965. They arose from a later, different politics that reflected the onward sweep and redefinition of rights-consciousness and entitlements in the United States. Long-run outcomes of the voting rights act, as of much legislation, could not be foreseen at the time.[66]

What might have been more clearly predicted were the limitations of voting rights, even in a democracy like the United States. The right to vote had a special cachet in American history dating to the eighteenth century. It was a wondrous magnet for oppressed people throughout the world. But as women had recognized after getting the suffrage in 1920, the franchise could not work wonders. Johnson exaggerated in claiming that the vote was "the most powerful instrument ever devised by man for breaking down injustice." As he intoned these words, he knew that the right to vote, however fundamental, could do only so much for black people, who faced deep socio-economic disadvantages rooted in racism

---

64. Garrow, *Protest*, 1–5, 123–32; Lawson, *Black Ballots*, 332–38.
65. Thernstrom, *Whose Votes Count?*, 1–7, 234–44; Steven Lawson, *In Pursuit of Power: Southern Blacks and Electoral Politics, 1965–1982* (New York, 1985); *New York Times*, Nov. 13, 1994.
66. These long-run consequences were not an unmixed blessing for minority candidates. Those who ran in districts created to heighten black voting power were likely to do well. Other districts, however, were more heavily white than before and were more likely to be represented by conservative whites.

and discrimination. The future proved this point. Nearly thirty years following passage of the voting rights act, the median household income of blacks in Selma was $9,615, compared to $25,580 for whites. More than half of blacks in the area lived in poverty in 1994.[67]

Daniel Moynihan, an Assistant Secretary of Labor in Johnson's administration, had already identified these economic disadvantages in a report, *The Negro Family: The Case for National Action*, that he had completed in April 1965. The Moynihan Report, as it became known, pointed to rapidly increasing rates of unemployment, family break-up, and welfare dependency among black people in the United States.[68] LBJ relied on the report as the core of a major speech on racial problems at Howard University in early June. Johnson emphasized that blacks in the United States needed not only equality of opportunity but also "equality as a fact and equality as a result." Moving beyond liberal quests for opportunity, he promised significant activity to improve the socio-economic condition of blacks, the next frontier for civil rights, later in the year.[69]

This was an extraordinary promise. By then, however, the Moynihan Report, published and widely discussed, was embroiling the administration in furious controversy. Moynihan had intended his findings to provide a "case for national action." His statistics on rising family break-up among blacks were accurate and worthy of debate. But the report linked the problems of Negro families to the heritage of slavery, thereby implying that the problems were both historical and cultural and that blacks, emasculated by slavery, could not take charge of their fate. Moynihan also used phrases such as "tangle of pathology" to describe the travail of the contemporary black family. When black militants (and white radicals) got wind of the report, they responded with outrage.[70] CORE leader James Farmer called it a "massive academic cop out for the white conscience." He added, "We are sick to death of being analyzed, mesmerized, bought, sold, and slobbered over, while the same evils that are the ingredients of our oppression go unattended."[71]

67. *New York Times*, Aug. 2, 1965.
68. For the report and subsequent angry debates about it, see Lee Rainwater and William Yancey, *The Moynihan Report and the Politics of Controversy* (Cambridge, Mass., 1967). See chapter 21 below for subsequent alarm concerning illegitimacy rates.
69. Matusow, *Unraveling*, 195–97; Patterson, *America's Struggle*, 102–5.
70. Walter Jackson, *Gunnar Myrdal and America's Conscience: Social Engineering and Racial Liberalism, 1938–1987* (Chapel Hill, 1987), 304–5.
71. Patterson, *America's Struggle*, 102.

That most white liberals in 1965 listened in embarrassed silence to the rage of activists such as Farmer showed how far the nation had moved since the late 1950s. At that earlier time few black leaders would have presumed to speak out so insultingly about white liberal allies, and few whites would have listened if they had. By mid-1965, however, black activists in the civil rights movement had acquired great moral standing among American liberals. Progressively minded whites for the most part dared not challenge them. Farmer's reaction especially indicated the intensity with which militant blacks in 1965 distrusted white liberals. The gulf between the two camps consigned the Moynihan Report to virtual oblivion and defeated whatever hopes Johnson might have had in mid-1965—or thereafter—to go beyond voting rights and seriously address the socio-economic problems of blacks in American cities. Liberalism would focus, as in the past, on expanding opportunity, not on fighting social inequality.

Notwithstanding these developments, which upset Johnson and his inner circle, there was no denying that the Voting Rights Act of 1965, like the Civil Rights Act of 1964, was a great achievement: these were the most significant of the many Great Society laws that expanded rights-consciousness in America. If most of the credit for the voting law belonged to civil rights activists, Johnson and fellow liberals deserved some praise as well. The goal of the act, after all, was to guarantee long-disfranchised black Americans the rights to register and to vote. This end the law accomplished brilliantly, thanks in large part to vigorous and unyielding federal oversight in the years ahead. By 1967 more than 50 percent of voting-age blacks had the franchise in the six most discriminatory southern states. In 1968 there were blacks in the Mississippi delegation to the Democratic National Convention. By the mid-1970s southern blacks began to win electoral office, even in Congress. The increase in black registration was so remarkable that southern white politicians, Wallace included, began by then to soften their racist rhetoric in order to capture some of the black vote. The voting rights act largely wiped out a blight on American democracy and transformed the nature of southern politics in the United States.[72]

AMERICAN LIBERALS understandably hailed the accomplishments of Johnson and the congressional session of 1965. "It is the Con-

72. Sundquist, *Politics and Policy*, 275; Conkin, *Big Daddy*, 217; Polenberg, *One Nation Divisible*, 191–92.

gress of accomplished hopes," Speaker McCormack said. "It is the Congress of realized dreams."[73] No other President cared so much as Johnson did about domestic policies or about civil rights, and none since FDR in the 1930s had come close to securing so many laws, many of them long awaited by reformers. It was a high tide of American liberalism in the postwar era.

By mid-1965, however, there were signs that the tide was about to ebb. Nothing exposed this more clearly than a riot that broke out in Watts, a predominantly black area of Los Angeles, only five days after signing of the voting rights act on August 6. Although Watts seemed to be a less squalid area than many black urban neighborhoods, it contained severe socio-economic problems: three-quarters of the adult black males living there were unemployed. The riot started following an altercation between police and a black man who resisted arrest for drunken driving.[74] Such fracases were nothing new in the history of relations between police and minorities of color (including Mexican-Americans). But urban blacks, like blacks in the South, had grown proud and angry. Charging police brutality, they rallied to the man's side. What followed was five days of rioting, sniping, looting, and burning, mostly against white-owned stores and buildings. The disturbance, ending only after 13,900 National Guardsmen came in to restore order, killed thirty-four people and injured more than 1,000, the vast majority of them blacks.[75] Damage to property was estimated at more than $35 million. Some 4,000 people were arrested. Although conservatives claimed that only a handful of "riffraff" had caused the trouble, it was obvious that the uprising commanded wide support in Watts. Some 30,000 people had participated in the rioting, and 60,000 more had stood by in support. Their hopes and expectations whetted, they had lashed out at the white world. When King walked the streets preaching non-violence, they ignored him. Clearly, the civil rights acts of 1964 and 1965 were not softening the social and economic grievances of the black masses. Maybe no liberal legislation could.[76]

73. William Leuchtenburg, A Troubled Feast: American Society Since 1945 (Boston, 1973), 142.
74. Milton Viorst, Fire in the Streets: America in the 1960s (New York, 1975), 307–42.
75. U.S. Riot Commission Report, Report of the National Advisory Commission on Civil Disorders (New York, 1968), 38.
76. See Robert Conot, Rivers of Blood, Days of Darkness (New York, 1967); Jerry Cohen and William Murphy, Burn, Baby, Burn! The Los Angeles Race Riot, August 1965 (New York, 1966); Robert Fogelson, "White on Black: A Critique of the McCone Commission Report on the Los Angeles Riots," Political Science

ALTHOUGH THE WATTS RIOT of 1965 was an extreme response, it appears in retrospect as an ominous omen of the future. One domestic crisis after another in the next few years, including even bloodier racial confrontations in the cities, shattered the optimism of social engineers and threw liberals back on the defensive. By late 1965 Johnson himself seemed close to despair. "What do they want?" he asked of his critics. "I'm giving them boom times and more good legislation than anybody else did, and what do they do—attack and sneer. Could FDR do better? Could anybody do better? What do they want?"[77] From this perspective, the first twenty months of the Johnson years stand as a shining but relatively brief era in the postwar history of American liberalism.

The rather sudden ebbing of liberal hopes caused many scholars—and contemporaries—to blame Johnson. They have a point. LBJ, unable to contain his ego, indeed wanted to outdo FDR—and every other President in history. He measured accomplishment in largely quantitative terms: the more big programs passed, the better. Some of these programs, such as OEO, were hurried into law without much research to sustain them and without much aforethought about potentially divisive political consequences. Other programs, such as aid to education, relied overoptimistically on injections of federal spending to cope with complicated social problems that, like poverty, needed more thoughtful study than they received. Getting things done quickly was not the same as getting them done well.

Many of the subsequent problems with Great Society programs stemmed from three common characteristics. One was the tendency of Johnson and his advisers to rely on highly partisan majorities. When the programs encountered difficulties down the road, Republicans and others felt free to say, "I told you so," and to denounce them at will. The civil rights acts stood as exceptions to this tendency. Thanks to the moral power of the civil rights movement—and to the overreaction of racists in the Deep South—these aroused bipartisan congressional support in the North in 1964 and 1965. Northerners of both parties, having targeted the South as the Enemy, had a stake in what they had accomplished. (After all, the acts did not much affect the North.) The goals were clear and just, the enforcement strong, the laws of lasting significance.

Quarterly, 82 (Sept. 1967), 337–67; and Jon Teaford, The Twentieth-Century American City: Problems, Promise, and Reality (Baltimore, 1986), 129.

77. Goldman, Tragedy of Lyndon Johnson, 337; Joseph Califano, The Triumph and Tragedy of Lyndon Johnson: The White House Years (New York, 1991), 305.

Second, Johnson had little stomach for taking on well-entrenched political interests. In part because he was fearful of conservatives, including corporate interests, he refused to consider creation of large-scale public employment programs, such as the WPA, that might have provided work and raised the income of the poor. Labor unions, too, feared such programs—because they threatened to endanger the jobs of the working poor. Respectful of lobbyists for the National Education Association, Johnson permitted local school administrators overwide leeway in their spending of federal money. Aware of the power of the American Medical Association, he approved health care legislation that (among other things) benefited hospitals, physicians, and insurance companies. He refused to raise taxes to pay for any of these programs.

The Great Society programs were for these reasons quintessentially liberal, not radical. Except in the area of race relations—a major exception—they made no serious effort to challenge the power of established groups, including large corporations. In no way did they seriously confront socio-economic inequality or seek to redistribute wealth. The essence of Great Society liberalism was that government had the tools and the resources to help people help themselves. It sought to advance equality of opportunity, not to establish greater equality of social condition.[78]

Oversell was a third characteristic of Johnson and the Great Society. As LBJ jetted about the country to publicize and to sign the landmark acts of his administration, he (and others) offered soaring descriptions of what he had done. The OEO could end poverty in ten years. Aid to education would provide the "only valid passport from poverty." Medicare would advance the "healing miracle of modern medicine." Voting rights were the "most powerful instrument ever devised by men for breaking down injustice." Some of these programs indeed helped people, and many others—immigration reform, governmental support of the arts and the humanities, environmental legislation—reflected noble intentions. But the Great Society did not do nearly as much to improve the economic standing of people as did the extraordinary growth of the economy. When this stopped—in the 1970s—the flaws in LBJ's programs seemed glaring. Hyperbole about the Great Society aroused unrealistic popular expectations about government that later came to haunt American liberalism.

These were indeed problems with the presidential leadership of Lyndon Johnson and more broadly with the liberal political philosophy that he

78. Katznelson, "Was the Great Society a Lost Opportunity?," 198; Conkin, *Big Daddy*, 236–42; Kearns, *Lyndon Johnson*, 218–20.

embraced. Still, it is a little unfair to harp on them. Johnson, who had an acute sense of what was possible in American politics, was correct in maintaining that he had to move quickly in 1965 if he hoped to advance the liberal agenda. After all, conservatives and interest groups had blocked it for a generation. And getting things done naturally entailed reliance on his heavily Democratic majorities. Outside of civil rights, where Republicans like Dirksen could be brought around, the GOP was neither necessary for legislative majorities in 1965 nor in much of a mood to help.

It is easy to criticize Johnson for failing to challenge interest groups or to promote redistribution of political and economic power in the country. But it is even easier to see why he did not. Interest groups had become so powerful in American politics, especially in Congress, that little significant legislation could be passed without their acquiescence. This was in part because the groups controlled major political and economic resources that could threaten a member of Congress with political defeat. It was also because other groups—the poor, minorities—remained politically very weak. Many of these people could not or did not vote, let alone find the time or money to play major roles in politics.

It was not just resources that bolstered interest groups. The groups also drew on—and deliberately aroused—substantial ideological support from politically active Americans who distrusted the State. A school aid act that contained tough federal guidelines on the spending of money would probably have incited opposition not only from teachers and school administrators—the interests, in this case—but also from thousands of parents and others who believed that education should remain a primarily local responsibility. The federal government must not "dictate" to the schools. Organized interests, for another example, led the opposition to greater State involvement in medicine, but they also had popular backing, at least among the politically influential middle classes. Many of these Americans—people who could afford doctors—believed strongly in the preservation of traditional fee-for-service medicine against the "threat" of State intervention.

The fate of the OEO revealed clearly what could happen if a government program became perceived as dangerous to well-organized interests. Although militants gained control of only a few of the community action programs, their activities raised such a storm of protest in 1965 among urban officials, most of them Democratic, that Johnson had to dispatch Vice-President Humphrey on urgent missions of mediation. Democratic powers such as Mayor Richard Daley of Chicago, however, were not to be placated unless the administration promised them control of the money. Johnson, who needed their support, quickly acquiesced, as did Democrats in Congress. Starting in 1966 Congress began amending expansive notions of "maximum feasible participation" of the poor out of existence. Direction of anti-poverty programs was returned to the alliance of urban politicians and social workers that had traditionally been in charge, but not before the war on poverty had divided the Democratic party.

Critics who lambasted Johnson for not trying hard to seek equality of social condition were also somewhat unfair. Notwithstanding rhetoric such as he employed in his speech at Howard, he did not pretend to favor the redistribution that this would have entailed. He had been elected as a liberal—as an advocate of deeply rooted American ideas about the virtues of equality of opportunity—not as a champion of major structural change in American life. Liberals, indeed, understood clearly how little political support there was in the nation for such an effort, which at the least would have called for higher taxation of the middle and upper classes. To demand equality of condition, many Americans continued to believe, was to burden the nation with taxes, regulation, and bureaucracy, to threaten prosperity, and to damage the entrepreneurial vitality and individualism that were at the heart of the American dream.

The final complaint about Johnson—that he oversold the Great Society—is both true and understandable. This is what political leaders do in order to make legislators and voters buy what they create. Johnson, a master salesman, could not restrain himself, especially when he had the now ubiquitous appliance of television to help him. Selling, often with grandiloquent flourishes, was part of the American way, not only of politics but also of the lusty commercial civilization of which it was a part.

Still, the overselling proved unfortunate for Johnson and for American liberalism. What it did was greatly strengthen powerful attitudes, notably the rise of grand expectations, that had been gathering force since the 1950s and that were starting to dominate the culture in the early 1960s. The overselling further propelled popular feelings that the United States could have it all and do it all—that there were no limits to how comfortable and powerful and healthy and happy Americans could be. This infectious optimism—about expertise, about government, about "entitlements"—stimulated a Rights Revolution that reverberated long afterwards. But the optimists were low on humility, and they underestimated the formidable divisions—of race, of class, of region, of gender—that persisted in the United States. The liberal faith of LBJ and others in the 1960s was both attractive and well-meaning, but it was destined for serious trouble ahead.

20

Escalation in Vietnam

Like many Texans of his generation, Lyndon Johnson was brought up on the story of the Alamo, where brave men had fought to the death to resist attack. As President he told people that his great-great-grandfather had died there, although there was no substance to his claim. Vietnam, he exclaimed to advisers on the National Security Council, "is just like the Alamo."[1]

This is one of many anecdotes that critics of LBJ relate about his approach to foreign policy. They portray him as ignorant about the world, imperious, devious in the extreme, and quick on the trigger. His handling of the war in Vietnam, they emphasize, demonstrated all these traits. When he sat down with Ambassador Lodge in the immediate aftermath of Kennedy's assassination, he personalized the war. "I am not going to lose Vietnam," he said. "I am not going to be the President who saw Southeast Asia go the way China went."[2] For many Americans then and later the struggle in Vietnam was simply "Johnson's War."[3]

1. William Chafe, *The Unfinished Journey: America Since World War II* (New York, 1991), 274–75.
2. Godfrey Hodgson, *America in Our Time* (Garden City, N.Y., 1976), 176.
3. Larry Berman, *Lyndon Johnson's War: The Road to Stalemate in Vietnam* (New York, 1989). Among the many other books on Johnson and the war are Neil Sheehan, *A Bright Shining Lie: John Paul Vann and America in Vietnam* (New York, 1988); Stanley Karnow, *Vietnam: A History* (New York, 1991); Frances FitzGerald, *Fire in the Lake: The Vietnamese and the Americans in Vietnam* (Boston, 1972); David Halberstam, *The Best and the Brightest* (New York, 1972);

593

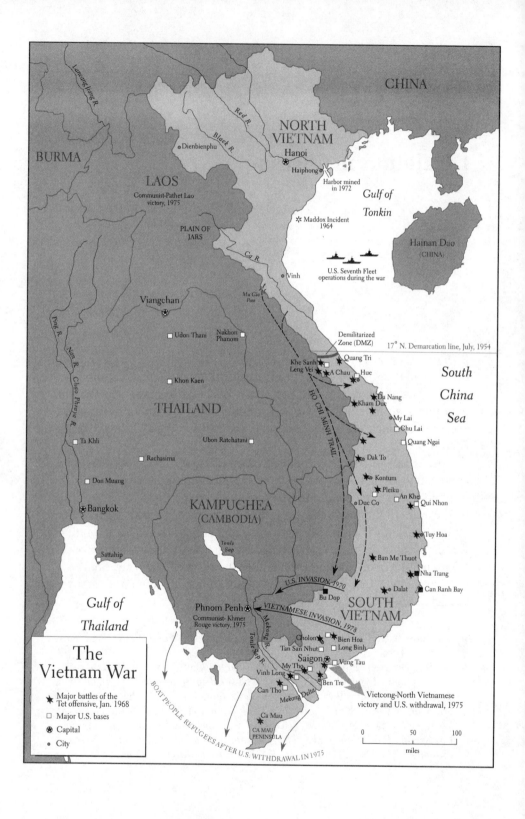

CHINA

NORTH
VIETNAM
Hanoi
Haiphong

Dienbienphu

BURMA

LAOS
Communist-Pathet Lao
victory, 1975

PLAIN OF
JARS

Viangchan

Udon Thani
Nakhon
Phanom

Khon Kaen

THAILAND

Ta Khli
Ubon Ratchatani

Rachasima

Don Muang

Bangkok

KAMPUCHEA
(CAMBODIA)

Tonle
Sap

Sattahip

Gulf of
Thailand

Phnom Penh
Communist- Khmer
Rouge victory, 1975

The
Vietnam War

★ Major battles of the
Tet offensive, Jan. 1968
☐ Major U.S. bases
✪ Capital
• City

Red R.

Black R.

Ca R.

Vinh

Mu Gia
Pass

Lancang Jiang R.

Ping R.

Nan R.

Chao Phraya R.

Mekong R.

Tonle Sap R.

Harbor mined
in 1972

✳ Maddox Incident
1964

Gulf of
Tonkin

Hainan Dao
(CHINA)

U.S. Seventh Fleet
operations during the war

Demilitarized
Zone (DMZ)

17° N. Demarcation line, July, 1954

Quang Tri

Khe Sanh
Leng Vei
A Chau
Hue

Da Nang
Kham Duc

My Lai
Chu Lai

Quang Ngai

Dak To

Kontum

Pleiku
An Khe
Duc Co
Qui Nhon

Tuy Hoa

Ban Me Thuot
Nha Trang

Dalat
Can Ranh Bay

U.S. INVASION, 1970

Bu Dop

SOUTH
VIETNAM

VIETNAMESE INVASION, 1978

Cholon
Bien Hoa
Tan San Nhut
Long Binh
Saigon
Vung Tau
My Tho
Vinh Long
Ben Tre
Can Tho
Mekong Delta

Ca Mau
CA MAU
PENINSULA

BOAT PEOPLE REFUGEES AFTER U.S. WITHDRAWAL IN 1975

South
China
Sea

HO CHI MINH TRAIL

Vietcong-North Vietnamese
victory and U.S. withdrawal, 1975

0   50   100

miles

While Johnson was more sophisticated about foreign policy than these anecdotes suggest, there was no doubt that United States involvement in Vietnam escalated enormously under his watch. At the end of 1963 roughly 17,000 American military "advisers" were stationed in Vietnam. A year later there were 23,000. Huge escalation then ensued. Following an attack on an American base in February 1965, United States planes began bombing North Vietnam. By late March American marines were regularly engaged in combat. At the end of 1965 there were 184,000 American military personnel in Vietnam; by the end of 1966, 450,000; by early 1968, more than 500,000. The number of American casualties (killed, wounded, hospitalized, and missing) increased from 2,500 in 1965 to a cumulative total of 33,000 by the end of 1966, to 80,000 by the end of 1967, and to 130,000 by the end of 1968, the peak of American involvement.[4] American planes unleashed more bombs, many of them napalm, on Vietnam between 1965 and the end of 1967 than they had in all theaters of World War II.[5] Toxic chemical defoliants such as Agent Orange, dropped to eradicate enemy cover, hit millions of acres of land in Vietnam and destroyed one-half of the timberlands in the South.[6] A motto of bombing crews read "Only You Can Prevent Forests."

Marilyn Young, *The Vietnam Wars, 1945–1990* (New York, 1991); Guenter Lewy, *America in Vietnam* (New York, 1978); Brian VanDeMark, *Into the Quagmire: Lyndon Johnson and the Escalation of the Vietnam War* (New York, 1991); Larry Cable, *Unholy Grail: The U.S. and the Wars in Vietnam, 1945–1968* (New York, 1991); Stephen Ambrose, *Rise to Globalism: American Foreign Policy Since 1938* (New York, 1985), 210–30; and especially John Gaddis, *Strategies of Containment: A Critical Appraisal of Postwar American National Security Policy* (New York, 1982); and George Herring, *America's Longest War: The United States and Vietnam, 1950–1975* (Philadelphia, 1986). Books on LBJ cited in chapters 18 and 19 are also useful, especially Doris Kearns, *Lyndon Johnson and the American Dream* (New York, 1976); and Paul Conkin, *Big Daddy from the Pedernales: Lyndon Baines Johnson* (Boston, 1986).

4. Gaddis, *Strategies of Containment*, 248; Kearns, *Lyndon Johnson*, 311. From August 1964 until the end of American involvement in January 1973, the United States suffered 47,356 battlefield deaths in the Vietnam War, compared to 33,629 in the three years of the Korean War. The number wounded was 270,000, compared to 103,284 in Korea. Other war-related American deaths in Vietnam totaled 10,795, and in Korea, 20,617. Greatly improved medical procedures held down the number of American deaths in Vietnam, which were approximately one-half, *per year*, of the number of American deaths in Korea.

5. By 1970 the total tonnage of bombs dropped by United States planes in Vietnam exceeded the tonnage dropped in all previous wars in human history.

6. Herring, *America's Longest War*, 146–54.

By the beginning of 1968 American and South Vietnamese firepower had killed, according to reasonably reliable estimates, some 220,000 enemy soldiers. This was some sixteen times the number of Americans (13,500) slain in Vietnam during the same period.[7] Civilian casualties, although smaller as a percentage of total casualties than in the Korean War (where percentages had hit record highs), were also numerous.[8] It is estimated that approximately 415,000 civilians were killed during the ten years of American involvement in the war.[9] Roughly one-third of the people of South Vietnam fled their homes as refugees at one time or another during these years.[10]

The escalation did not deter the enemy. By mid-1964 the North Vietnamese under Ho Chi Minh and his chief general, Vo Nguyen Giap, had assumed control of most military operations in the South, thereby transforming the character of the war. They relied in part on guerrilla tactics, many of them conducted by southern allies in the National Liberation Front, and increasingly as of late 1965 on conventional pitched battles in which the North Vietnamese army ordinarily had a substantial edge in manpower.[11] These engagements, as casualty figures indicated, cost them dearly: they lost approximately a million troops during the war. The United States enjoyed an enormous edge in firepower, it controlled the air, and its soldiers—contrary to later accusations of ineptitude—fought well, especially before 1969. American soldiers won the majority of such

7. Later estimates, as reported in the *New York Times*, Nov. 30, 1992, conclude that 1.5 million Vietnamese died in the war between 1964 and 1973, of whom 924,000 were North Vietnamese and their allies from the South, 415,000 were civilians, and 185,000 were South Vietnamese soldiers. Christian Appy, *Working-Class War: America's Combat Soldiers and Vietnam* (Chapel Hill, 1993), 16–17, gives somewhat higher numbers, estimating total Vietnamese deaths, 1961–75, at between 1.5 and 2 million and noting that hundreds of thousands of Cambodians and Laotians were also killed during these years, as were smaller numbers of Australians, New Zealanders, South Koreans (c. 4,000 deaths), Thais, and Filipinos (SEATO allies) who fought alongside the South Vietnamese and the Americans. An oft-cited grand total of deaths, 1961–75, is 3 million.
8. Robert Divine, "Vietnam Reconsidered," *Diplomatic History*, 12 (Winter 1988), 79–93. Civilians were 28 percent of all deaths in the Vietnam War, compared to 40 percent in World War II and 70 percent in the Korean War. There was much more bombing of large urban populations in both World War II (as of Dresden, Tokyo, or Hiroshima) and in the Korean War than in the Vietnamese conflict.
9. *New York Times*, Nov. 30, 1992.
10. Thomas Paterson, "Historical Memory and Illusive Victories: Vietnam and Central America," *Diplomatic History*, 12 (Winter 1988), 1–18.
11. Charles Morris, *A Time of Passion: America, 1960–1980* (New York, 1984), 103, 145.

battles. But Ho Chi Minh's losses were far from fatal. North Vietnam had a rapidly growing population, which in 1965 totaled around 19 million people (as opposed to 16 million in the South), and some 200,000 young men came of fighting age each year during the war. As many as were needed were thrown into the effort. The North also received substantial military aid from both the Soviet Union and China (some $2 billion between 1965 and 1968), but they relied above all on themselves and on the NLF in the South.[12] Ho Chi Minh ran a dictatorial system that he employed ruthlessly to outlast the imperialist invaders, from the world's most powerful country, and to demand nothing less than an outcome that would ultimately unite Vietnam under his control. More than any other factor, the unyielding determination of the enemy, both from the North and from Ho's supporters in the South, wore down the American commitment, which proved to be far less resolute, and foiled Johnson's efforts.

This was the longest and ultimately the most unpopular war in United States history. It had large and negative effects on American life. While it lasted the war directed vast new resources to the military-industrial complex. It further escalated the arms race and diverted the attention of Johnson and the American people from foreign problems elsewhere in the world. (The Soviet Union, building frantically, approached nuclear parity with the United States during these years.) It "Americanized" and corrupted the culture and society of South Vietnam.[13] It unnerved and at times alienated America's allies. Spending for the war created huge budgetary deficits and contributed significantly to inflation and economic instability by the early 1970s. These economic problems, worsening in the early 1970s, seriously threatened the grand expectations that had accompanied the liberal surge of the early and mid-1960s.[14]

The war helped undo Johnson's hopes for congressional expansion of Great Society programs. He understood that it would, which is one of the reasons why he put off major escalation until February 1965 and why he tried for the next few months to hide the extent of American involvement from Congress and the American people. "I knew from the start," he said

12. American soldiers commonly referred to the military side of the NLF derogatorily as the Viet Cong or Vietcong, or VC, shorthand terms for "Vietnamese Communist."
13. George Herring, "People Quite Apart: Americans, South Vietnamese, and the War in Vietnam," *Diplomatic History*, 14 (Winter 1990), 9.
14. Herring, *America's Longest War*, xi, estimated that the war cost the United States $150 billion between 1950, when Truman stepped up aid to South Vietnam, and 1975, when Congress cut back further military aid. Most of this money was spent between 1965 and 1972.

later, "that I was bound to be crucified either way I moved. If I left the woman I really loved—the Great Society—in order to get involved with that bitch of a war on the other side of the world, then I would lose everything at home. All my programs. All my hopes to feed the hungry and feed the homeless. All my dreams."[15] Still, he chose the bitch. He soon found himself dealing with military matters day and night and forsaking the domestic causes that had been closest to his heart. Care and dollars that might have gone to the Great Society were lavished on the most awesome military machine in world history.

Above all, the war polarized American society. Contrary to some accounts, substantial opposition to the war rose only slowly in the United States, and it failed before 1968 to deter Johnson and his advisers from their course. But as early as the spring of 1965 professors and students on college campuses staged "teach-ins" critical of the war. New Left organizations, small before 1965, attracted increased support and staged well-publicized anti-war demonstrations. Senator William Fulbright, chairman of the Senate Foreign Relations Committee, held televised hearings critical of the war in January 1966. To pacify critics such as these Johnson and his aides spoke of their desire for peace and promised vast American economic aid to Southeast Asia when the fighting stopped. In early 1966 he talked grandiloquently of "turning the Mekong into a Tennessee Valley."[16] But LBJ refused to agree to anything that might cause South Vietnam to fall to Communism. Rhetoric aside, Johnson departed only sporadically and slightly from this central position, which echoed those of Eisenhower and Kennedy, and he attempted no serious negotiations with the enemy—which was also inflexible—during his presidency. To offer concessions to Communists from Vietnam, he said contemptuously, was to succumb to "Nobel Prize fever."[17]

Escalation increasingly dominated his term of office. Draft calls rose rapidly after 1964, unsettling the large number of American males of the baby boom generation. A total of 11.7 million Americans served in the armed services during the nearly ten years of heavy American involvement in the war, 2.1 million of whom went to Vietnam and 1.6 million of whom saw combat.[18] This was a small minority of the 26.8 million men

15. Kearns, *Lyndon Johnson*, 251.
16. Ibid., 266–67.
17. Herring, *America's Longest War*, 46–48.
18. For data and reflections on the draft, see D. Michael Shafer, "The Vietnam Draft: Who Went, Who Didn't, and Why It Matters," in Shafer, ed., *The Legacy: The Vietnam War in the American Imagination* (Boston, 1990), 57–79; George Flynn,

who came of draft age (18 through 25) during the war; unlike the World
War II or Korean War generations, the majority of draft-age Americans
neither entered the military nor fought.[19] But millions of young men after
1964 could not be sure of their fate, especially in the Johnson years, when
calls were highest. These peaked in 1966, when 340,000 Americans were
drafted, compared to an average of only 100,000 per year between 1960
and 1964. An average of 300,000 a year were drafted in the years 1965–
1968. Four times as many other young men enlisted in order to put an
end to the mounting uncertainty posed by their draft boards or to secure
assignments that minimized their chances of combat. By 1968, one-third
of 20-year-old men in the United States were in the service.[20]

The enemy persevered, confident that the United States would lose
heart. Johnson responded by escalating American involvment. The
United States bombed and bombed and bombed, obliterating villages and
uprooting millions. American soldiers fought in jungles, forests, rice
paddies, and hundreds of villages. But it was an unusually bewildering war without a clearly demarcated battle front. Defense Secretary
McNamara and General William Westmoreland, America's new commander in Vietnam, bragged instead about enemy "body counts." Their
goal consisted of killing the Vietcong ("VC") in large numbers, thereby
depleting the enemy of fighting men and turning the tide. Many South
Vietnamese people, moreover, seemed maddeningly indifferent or,
worse, friendly with the NLF. It seemed to the bewildered and frustrated
Americans that no one could be trusted. People on both sides committed
atrocities. Casualties mounted, gradually intensifying domestic opposi-

Lewis B. Hershey, Jr.: Mr. Selective Service (Chapel Hill, 1985), 166–70; Lawrence Baskir and William Strauss, Chance and Circumstance (New York, 1978),
52–53; and Charles Moskos, "From Citizens' Army to Social Laboratory," Wilson
Quarterly (Winter 1993), 10–21.

19. Moskos, "From Citizens' Army," notes that 8 of 10 American men of draft age
entered the military during the World War II years and that roughly 50 percent of
eligible men had service during the years between 1950, the start of the Korean
War, and 1960. Roughly 40 percent served during the Vietnam years. The age
of draft registration was 18; of induction, no earlier than 18½. Seventeen-year-
olds could enlist in the marines if they had the consent of a guardian. Relatively
few 18-year-olds fought in Vietnam. A total of 250,000 women served in the
military during the Vietnam era, of whom 6,431 went to Vietnam and nine were
killed.

20. Landon Jones, Great Expectations: America and the Baby Boom Generation (New
York, 1980), 102.

tion to the war.[21] Even early in the fighting it seemed that the bloodshed would never end.

The Vietnam War ultimately enlarged widespread doubts about the capacity—indeed the honesty—of government leaders. Sapping the optimism that had energized liberals in the early 1960s, it badly damaged the Democratic party and provoked contentiousness that sharpened already significant class and racial grievances. It called into question the honor and the decency of much that Americans claimed to stand for. Nothing did more than the conflict in Vietnam to alter the course of post–World War II American society and politics or to unleash the emotions that polarized the nation after 1965.

JOHNSON DID NOT ESCALATE the war because he was temperamentally a warmonger. Unlike Kennedy, he had never demonstrated much interest in foreign policy or in the glamor of Special Forces, Green Berets, or "flexible response." Cloak-and-dagger intrigue did not appeal to him. When he heard in November 1963 about secret American efforts to sabotage Cuban installations, he ordered them stopped, even though the CIA informed him that Castro was promoting a coup in Venezuela.[22] LBJ also sought to pursue the uneasy détente that Kennedy and Khrushchev had been developing in late 1963. Throughout his administration he took great care not to make sudden moves that might provoke either the Soviet Union or China, which remained at loggerheads.

LBJ's policy of escalation in Vietnam, moreover, was neither careless nor precipitous. Contrary to some historical accounts, the United States did not get bogged down in a "quagmire" because Johnson waded into a swamp without looking where he was going. Johnson and his top advisers, to be sure, did not understand the resolve and resourcefulness of Communist revolutionaries, and they never imagined the morass that ultimately swallowed the American effort. But he was well aware of the political and military deterioriation afflicting the government of the South in 1964. By

21. As noted, American casualties during the conflict in Vietnam were relatively modest. But expectations had been high. Jones, *Great Expectations*, 102, estimates that the also 58,151 Americans who died in Vietnam (from battlefield as well as non-battlefield deaths) were offspring or parents or siblings of 275,000 people and that the 270,000 who were wounded were offspring or parents or siblings of 1.4 million people. Millions more of Americans, of course, were friends of fellow citizens who served in Vietnam.

22. Michael Beschloss, *The Crisis Years: Kennedy and Khrushchev, 1960–1963* (New York, 1991), 692–93. This order, however, did not prevent the CIA from continuing to explore ways of assassinating Castro. These explorations continued in 1965.

mid-year enemy forces controlled 40 percent of the land and 50 percent of the people in South Vietnam. The so-called Ho Chi Minh Trail, an elaborate network of roads (some of them in Cambodia and Laos) facilitating infiltration from the North, could handle trucks and other heavy equipment.[23] Johnson, deciding whether to escalate, had solid intelligence on these developments. Moreover, a few of his advisers, including Undersecretary of State George Ball and Maxwell Taylor (whom Johnson named to replace Lodge as ambassador) warned him in 1964 and early 1965 that large-scale engagement of American troops was unlikely to accomplish much. So did informal advisers such as Senate Democratic leader Mike Mansfield. CIA officials (to whom Johnson paid little attention) said much the same, as they had since 1961.[24] Unlike Kennedy, who until mid-1963 had given the situation in Vietnam little attention, Johnson understood that it demanded constant consideration and that escalation of American involvement carried with it imposing dangers. For this reason—and because he wanted nothing to disrupt his chances for election in 1964—Johnson increased American aid to South Vietnam but otherwise said little concerning the conflict during his first fourteen months in office.

Still, Johnson's personal approach to policy-making did much to promote escalation. As his comments to Lodge in November 1963 revealed, LBJ perceived Vietnam, like civil rights, as a litmus test of his ability to carry on the policies, as he saw them, of his martyred predecessor.[25] Kennedy's support of the coup against Diem, he thought, committed the United States to the preservation of successive governments in South Vietnam. From the start LBJ worked hard to retain Kennedy's foreign policy and defense team, succeeding in persuading McNamara, Rusk, and National Security Adviser McGeorge Bundy to stay on. All became top advisers on Vietnam, Rusk for the remainder of his administration. Holding firm in Vietnam would not only carry on Kennedy's policies; it would also show that Johnson could be counted on to sustain the international credibility of the United States. This, he thought, was vital. He told Lodge, "Go back and tell those generals in Saigon that Lyndon Johnson intends to stand by our word."[26]

23. Herring, *America's Longest War*, 118.
24. Gaddis, *Strategies of Containment*, 258.
25. William Leuchtenburg, *In the Shadow of FDR: From Harry Truman to Ronald Reagan* (Ithaca, 1983); Vaughn Bornet, *The Presidency of Lyndon Johnson* (Lawrence, 1983), 75.
26. Herring, *America's Longest War*, 110.

602   GRAND EXPECTATIONS

Johnson's handling of the war also reflected an inner insecurity that he felt when dealing with issues of foreign policy. Having given little attention to such matters during his political career, he relied heavily as President, especially at first, on advisers. These with few exceptions were tough-minded Kennedy men who remembered what appeasement had done in the 1930s and who demanded that the United States remain firm. Munich, symbol of appeasement, must not happen again. Some, like Rusk, believed that Ho Chi Minh was the agent of a world Communist conspiracy, in this case run by China. McNamara, who especially impressed Johnson, glowed with confidence about the technological and military capacity of the United States. Johnson shared many of these beliefs, and he easily absorbed the determination of these and other advisers in 1964. In later years he never admitted that they—and he—could have been wrong. He clung resolutely—critics said blindly—to the course that he started steering early in his presidency.[27]

Domestic political considerations figured especially heavily in Johnson's thinking about the war. Like Kennedy, he feared the backlash that might whip him if he seemed "soft" on Vietnam. Repeating "lessons" from history, he recalled:

> I knew that if we let Communist aggression succeed in taking over South Vietnam there would follow in this country an endless national debate—a mean and destructive debate—that would shatter my Presidency, kill my administration, and damage our democracy. I knew that Harry Truman and Dean Acheson had lost their effectiveness from the day that Communists took over in China. I believed that the loss of China had played a large role in the rise of Joe McCarthy. And I knew that all these problems, taken together, were chickenshit compared with what might happen if we lost Vietnam.[28]

The significance of LBJ's personal traits accounted for the growing belief, especially by anti-war activists, that Vietnam was "Johnson's War." His critics are correct in pointing to the role of these traits and in arguing that Johnson, commander-in-chief until 1969, possessed the ultimate power to stem the tide of escalation. He was the last, best, and only chance for the United States to pull itself out of the quagmire.

Critics of Johnson are also accurate in pointing to his deceitfulness about events in Vietnam. As later revelations showed, this became significant as early as his handling of the so-called Tonkin Gulf crisis of August

27. Kearns, Lyndon Johnson, 251–57; Gaddis, Strategies of Containment, 248.
28. Kearns, Lyndon Johnson, 253.

1964. Following brief fighting between the United States destroyer *Maddox* and North Vietnamese torpedo boats in the gulf on August 1, Johnson said nothing to the American people. But the engagement riled him, and he sent a second destroyer, the *C. Turner Joy*, into the gulf to help the *Maddox* resume operations. It is clear that while he was not trying to provoke another fight, he was not trying to avoid one either. When he received reports of further confrontations in the gulf on August 4, he announced that the enemy had fired on the two destroyers. In retaliation he ordered five hours of American air attacks on enemy torpedo boat bases and nearby oil storage dumps. One American airman was killed in the action.

The President also took advantage of the encounters to call on Congress to authorize him as commander-in-chief to use "all necessary measures" to "repel any armed attacks against the forces of the United States and to prevent further aggression" in the area. Congress, responding with patriotic fervor, approved the Gulf of Tonkin Resolution, as it was called, with only desultory debate. The votes were 416 to 0 in the House and 88 to 2 (Senators Ernest Gruening of Alaska and Wayne Morse of Oregon) in the Senate. The resolution, wide open in its grant of congressional power, indicated the power of the Cold War consensus in the United States. Johnson, who never asked Congress for a declaration of war, later cited it as authority for escalation far beyond anything the lawmakers could then have imagined. [29]

What really happened in the Tonkin Gulf, however, was much more mysterious than Johnson had let on. A brief naval engagement had indeed occurred in the gulf on August 1, leading North Vietnamese patrol boats to launch torpedoes at the *Maddox*. Fire from the *Maddox* and American carrier planes had badly damaged one of the boats. Events on August 4 were much less clear. As one of the destroyer commanders reported to McNamara at the time, it could not be ascertained that the enemy had fired torpedoes. Blips on radar screens—the basis for reports of attack— may have been caused by foul and freakish weather conditions. No American ships were hit, no men wounded or killed. McNamara and Johnson nonetheless chose to use the reports as pretext for a show of toughness that they had been seeking to make for some time. Johnson's goal was not to secure a resolution enabling him to wage full-scale war, either then or later. Rather, it was to put the North Vietnamese on notice that the United States was determined to fight back. It was also to show the

29. Herring, *America's Longest War*, 116–20; Ambrose, *Rise to Globalism*, 212–13.

American people that he was every bit as tough as, if not tougher than, Barry Goldwater, his opponent in the election campaign. To achieve these ends he resorted to deceit. He was to do so again and again in the ensuing fifty-three months of his administration.

In hailing the Gulf of Tonkin Resolution Johnson issued a statement that revealed much about his approach to the war in Vietnam. "Let no one doubt for a moment," he declared, "that we have the resources and we have the will to follow this course as long as it will take us." Few statements of the postwar era better expressed the grand expectations that America's liberal leaders maintained at that high tide of national optimism. Johnson, along with many of the American people, seemed to believe that the United States could build a Great Society at the same time it was fighting a war. It could afford both guns and butter. America could do it all.[30]

TO SINGLE OUT LBJ AS A VILLAIN, however, is to ignore the extraordinarily powerful cultural and political forces that long had dominated American thinking about Vietnam and foreign policy. It was not just Johnson or erstwhile Kennedy advisers such as McNamara and Rusk who demanded American firmness in Southeast Asia. As the Gulf of Tonkin Resolution indicated—it was very popular with the public— virtually all political leaders agreed with them in 1964–65. Preventing the spread of Communism, after all, remained the guiding star of American policy. Presidents Truman, Eisenhower, and Kennedy had followed it, as had their partisan opponents. All three Presidents had affirmed American support of South Vietnam and enunciated versions of the domino theory as a rationale. It was Johnson's fate to become President at a time when South Vietnam, following the assassination of Diem, was faltering badly. It was difficult for him to temporize, as his predecessors (in varying degrees) had done.[31]

Some of those who demanded resolute action thought Johnson temporized far too much. Top military figures in 1964 insisted on the need for American bombing of North Vietnam. When LBJ finally agreed to it in 1965, they chafed under his policy of calibrated incrementalism, an ap-

30. Robert Collins, "Growth Liberalism in the Sixties: Great Societies at Home and Grand Designs Aboard," in David Farber, ed., *The Sixties: From Memory to History* (Chapel Hill, 1994), 11–44.
31. John Hart Ely, *War and Responsibility: Constitutional Lessons of Vietnam and Its Aftermath* (Princeton, 1993); David Barrett, *Uncertain Warriors: Lyndon Johnson and His Vietnam Advisers* (Lawrence, 1993).

proach that involved significant escalation but that stopped short of full-scale military engagement with the enemy. General Curtis LeMay, air force chief of staff, exclaimed that the United States should bomb North Vietnam "back to the Stone Age." Admiral Ulysses S. Grant Sharp, naval commander in the Pacific, complained that America's bombing amounted to "pecking away at seemingly random targets." He added later, "We could have flattened every war-making facility in North Vietnam. But the handwringers had center stage. . . . The most powerful country in the world did not have the willpower needed to meet the situation."[32]

Other military men also blamed Johnson for timidity in his waging of the war. Westmoreland said later, "It takes the full strength of a tiger to kill a rabbit." A battalion commander added, "Remember, we're watchdogs you unchain to eat up the burglar. Don't ask us to be mayors or sociologists worrying about hearts and minds. Let us eat up the burglar our own way and then put us back on the chain."[33] Some of these critics argued that the United States should have dispatched American troops against enemy sanctuaries in Laos and Cambodia. They emphasized again and again that the United States should have done much more massive bombing of North Vietnamese military installations and infrastructure. Some urged the mining of Haiphong harbor, access point for outside aid to the North, and the bombing of Hanoi. Holding great expectations about America's capacity to control world events, they could not understand how a relatively small nation such as North Vietnam could survive against a world power like the United States.[34]

Other analysts of American military strategy grumbled that Westmoreland's emphasis, a "search and destroy" strategy that sought to chase down enemy forces in South Vietnam, entailed much wasted effort. The search and destroy approach, they noted, devastated a great many villages, making enemies of the people the United States was trying to help. Some of these critics thought it would have been better to concentrate American forces near the demilitarized zone dividing the antagonists and along

32. George Herring, "The War in Vietnam," in Robert Divine, ed., *Exploring the Johnson Years* (Austin, 1981), 27–62; Gary Hess, "The Unending Debate: Historians and the Vietnam War," *Diplomatic History*, 18 (Spring 1994), 239–64; Berman, *Lyndon Johnson's War*, 12.
33. Paterson, "Historical Memory," 5.
34. Arguments well summarized by Paterson, "Historical Memory"; Herring, "War in Vietnam"; and Divine, "Vietnam Reconsidered." See also Harry Summers, Jr., *On Strategy: A Critical Analysis of the Vietnam War* (Novato, Calif., 1982).

borders with Laos and Cambodia. The key, they insisted, was to take the war to the North, where the enemy was coming from, and pay less attention to unrest in the South. Others said it might have been better to devote greater effort to pacification, as it was called. This was a policy, intermittently employed after 1965, that depended less on search and destroy missions and more on civil-military programs to promote stability and peace in heavily populated regions of South Vietnam.

It remains extraordinarily doubtful, however, that military options such as these—many of them fine-tuned with hindsight—would have turned the tide. Hawks were correct that Johnson hamstrung the generals and the admirals: North Vietnam never had to fear the worst. But throughout the postwar era, hawkish critics had always tended to exaggerate the potential of military actions, especially bombing, to achieve political objectives. Like McNamara, they believed overoptimistically in the omnipotence of American technology. And they often presumed the superiority of white, Western ways: Asians, they thought, could not stand up for long against Western civilization. These were misplaced and ethnocentric assumptions that badly misread the situation in Vietnam. What stands out about American involvement in Southeast Asia, especially in retrospect, is the extent to which a truly enormous military commitment—both of bombs and of troops—failed to stop, let alone defeat, a much less industrialized adversary. Short of obliterating much of the North by dropping nuclear bombs on civilian centers—an option not considered at high levels—it is hard to see how greater military engagement could have achieved the goals of the United States.

Johnson's military problem, in some ways more complicated than that of his predecessors in the White House, stemmed in large part from two simple facts. First, many people in South Vietnam showed little stomach for the fight. Diem's successors, it turned out, were even less successful than he in curbing corruption and in gaining popular support. South Vietnamese leaders cooperated poorly with pacification programs, which were ill-coordinated with military strategy and often lacked even elementary security.[35] One-third of South Vietnamese combat troops deserted every year.[36] One of the most important lessons that might have been learned from the Vietnam War was that it is difficult for a nation—even a world power—to reform and protect a client state that cannot or will not

35. Herring, *America's Longest War*, 157–59.
36. Paterson, "Historical Memory."

manage itself. It may be impossible to provide protection if the state in trouble also faces widespread civil unrest and invasion, as was the case with South Vietnam.[37]

Second, the North Vietnamese were willing to fight hard and for however long it took. This was always the biggest obstacle to American chances for long-range success—however measured—in Vietnam. The longer the fighting went on—in a far-away land that seemed of little strategic value to the United States—the shorter the patience of the American people, who tolerated far fewer casualties than the enemy did.[38] One can perhaps imagine the United States massing hundreds of thousands of troops near North Vietnamese, Cambodian, and Laotian borders while trying to maintain a degree of order and safety in urban areas in parts of the South. But Ho's commanders were very good at infiltrating and at motivating their men, and the frontiers to be guarded extended more than 1,000 miles. Even by positioning 600,000 or so men in Vietnam—for who knew how many years?—the United States might not have been able to slow down infiltration from the North. Behind the border, moreover, Ho had a reserve army of 500,000 or so men. China threatened to enter the war if the United States went after Ho in the North. And who could tell about the rapidly arming Soviets, with whom LBJ hoped to maintain a modicum of détente?

Debates such as these over American military strategy animated armchair generals long after the war had ended. In 1964 and early 1965, however, Johnson firmly resisted the most hawkish advice. He feared alienating allies and provoking China and the Soviet Union. Equally important, he did care deeply about his Great Society agenda. To wage a major war in Vietnam might endanger passage of his domestic programs. Heeding the counsel of his closest advisers, he determined in late 1964 to escalate, but only after the election and in a calibrated and incremental way.

At the time this seemed like a politically sound as well as militarily promising approach to take. We return to the power of the Cold War consensus in the United States. If Johnson had done nothing, and watched South Vietnam collapse, he risked criticism not only from conservatives and hawks but also from important allies within the Democratic coalition. Labor leaders, including Walter Reuther, supported escalation

37. Sheehan, *Bright Shining Lie.*
38. Harry Summers, Jr., "The Vietnam Syndrome and the American People," *Journal of American Culture,* 17 (Spring 1994), 53–58.

in 1965-66. AFL-CIO president George Meany ardently backed it long after that. "I would rather fight the Communists in South Vietnam," he said in 1967, "than fight them down here in the Chesapeake Bay." An anti-war resolution introduced at the AFL-CIO convention in that year lost by a vote of 2,000 to 6.[39]

The majority of congressional Democrats, too, gave enthusiastic backing to Johnson's moves toward escalation. LBJ, indeed, worried more about criticism of his foreign policies from the Right on Capitol Hill than from the Left. Richard Russell of Georgia, one of the most powerful men in the Senate on defense issues, never believed that South Vietnam possessed much strategic value to American interests, and he was cool to escalation in early 1965. Once Johnson acted, however, Russell backed him to the hilt. "We are there now," he said. "If we scuttle and run, it would shake the confidence of the free world in any commitment we might make." He and his colleagues in essence abdicated congressional oversight over Vietnam policies for much of Johnson's presidency. Unprecedentedly high defense appropriation bills during these years swept through Capitol Hill with little or no debate. Even Fulbright's hearings in 1966 sparked little interest on the Hill. Until 1968 Johnson enjoyed considerable bipartisan support in Congress for his conduct of the war.[40]

Except for a few of the most anti-communist military advocates, supporters of escalation in 1964-65 were not sure—at least most of the time—that the United States and South Vietnam had to *win* the war, in the sense of destroying all hostile units on the field. Rather, they hoped to inflict so much pain on the enemy in both the North and the South that Ho would allow the war to wind down. American engagement, Johnson believed, would sooner or later force Ho Chi Minh to "sober up and unload his pistol."[41] Thanks in large part to continuing American escalation after 1965, the United States did prevent South Vietnam from falling during the LBJ years. In retrospect, however, it is clear that this policy was doomed in the long run. It reflected exaggerated expectations about the potential of American military and diplomatic muscle, as well as a misguided faith that the American people would continue to pay the price that escalation entailed. Their reluctance to do so after 1967 undercut the policy and reduced Johnson to cries of bewilderment and rage. Johnson's approach also exposed profound misunderstanding of Vietnamese cul-

39. Robert Zieger, *American Workers, American Unions, 1920–1985* (Baltimore, 1986), 171–72.
40. Herring, "War in Vietnam," 39–40; Kearns, *Lyndon Johnson*, 324–27.
41. Herring, *America's Longest War*, 142.

ture, politics, and history.[42] By 1964 the fires of revolutionary national-ism that ignited the North Vietnamese and the NLF burned far too strongly to be stopped by American arms or to be extinguished at the bargaining table.

Neither Johnson nor most of his advisers understood this in 1964 and 1965. Some, including LBJ, never did. They would have profited, as would the American people, from much wider exposure to the cultures of other nations. But the problem was only partly one of miseducation. The United States undertook escalation in 1965 because of the Cold War. In this larger calculation the nature of Vietnamese society and history was of little consequence. Johnson never even claimed that Vietnam had great strategic value. He called it a "little piss-ant country." What really mat-tered to him, as it had mattered so often to Americans in the postwar era, was the credibility of his country's commitment in its larger battle against Communism in the world. To show weakness on one part of the globe was to risk disaster on other parts. Rusk put this point clearly. "The integrity of the U.S. commitment," he said in 1965, "is the principal pillar of peace throughout the world. If that commitment becomes unreli-able, the communist world would certainly draw conclusions that would lead to our ruin and almost certainly to a catastrophic war."[43]

The focus on credibility in the fight against worldwide Communism seems both excessive and tragic in hindsight. It ignored contradictory evidence at the time, including a fact known to most of the policy-makers themselves: thanks in part to bitter Sino-Soviet rivalry, there was no such thing in the 1960s as a worldwide Communist conspiracy. Ho Chi Minh, to be sure, was a Communist, and he received aid from both China and the Soviet Union. A tyrant, he might well be expected to assault other nations in Southeast Asia once he united his own country. But Ho was above all a nationalist. And it was only speculation that he could or would overrun other nations when and if he took Saigon. The domino theory was just that—a theory. Assessing the theory in June 1964, a CIA report expressed doubts about its applicability to particular situations. If South Vietnam collapsed, it said, Laos and Cambodia might fall, too. But that was conjectural. And a "continuation of the spread of communism in the area would not be inexorable, and any spread which did take place would

42. For works dealing with cultural misunderstanding, see FitzGerald, *Fire in the Lake*; Loren Baritz, *Backfire: A History of How American Culture Led Us to Vietnam and Made Us Fight the Way We Did* (New York, 1985); and John Hellman, *American Myth and the Legacy of Vietnam* (New York, 1986).
43. Gaddis, *Strategies of Containment*, 240.

take time—time in which the total situation might change in any number of ways unfavorable to the communist cause."[44]

Thoughtful advice: CIA officials tended to be more cautious about escalation than many other American policy-makers at the time. But also unheeded advice. It competed helplessly against a Cold War consensus that continued to rule American society, politics, and culture. In deciding for escalation in early 1965 Johnson was as much a prisoner of this consensus as he was an activator of it. "LBJ," an accomplished historian of the war concludes, "appears less a fool or knave than a beleaguered executive attempting to maintain an established policy against an immediate threat in a situation where there was no attractive alternative."[45]

On February 6, 1965, enemy forces attacked an American air base at Pleiku in South Vietnam. The assault killed eight American soldiers and destroyed six helicopters and a transport plane. McNamara and Westmoreland urged LBJ to retaliate instantly. Johnson, who had been waiting for such a pretext, agreed. "We have kept our guns over the mantel and our shells in the cupboard for a long time now," he said at a two-hour meeting with top advisers. "I can't ask our American soldiers out there to continue the fight with one hand tied behind their backs." All his major advisers—Rusk, Bundy, Taylor, Chairman of the Joint Chiefs of Staff General Earle Wheeler—were consulted and agreed. Johnson thereupon ordered air attacks on targets in North Vietnam.

This response to the raid on Pleiku started the great escalation. A campaign of regular bombing attacks, Rolling Thunder, began in March and intensified over the next few months—3,600 sorties by American and South Vietnamese planes in April, 4,800 in June. Most sought to blast roads and railroads in the North so as to stem the flow of supplies to the South. LBJ, aided mainly by civilian advisers such as McNamara and Rusk, met regularly for Tuesday lunches to select bombing targets. "They can't even bomb an outhouse without my approval," he said proudly. The President also increased American troop strength. On March 8, two battalions of marines, accompanied by howitzers and tanks, came ashore to protect an American air base at Danang on the northern coast of South Vietnam. They were authorized to attack the enemy if necessary.[46]

During these early months of escalation Johnson, worried about weakening support for Great Society initiatives in Congress, tried to hide the

44. Hodgson, America's in Our Time, 237–38.
45. Herring, "War in Vietnam," 41–42.
46. Herring, America's Longest War, 126–28.

full extent of his activities in Vietnam. He did not tell the people that American forces were authorized to go on the offensive. He said he would go anywhere to talk peace. At Johns Hopkins University on April 7 he promised Southeast Asia $1 billion or more in aid—an Asian Marshall Plan—once the fighting stopped. But he also made it clear that he would never give in. "We will not be defeated," he emphasized. "We will not grow tired. We will not withdraw." The next day American bombers launched a massive raid, and LBJ authorized the dispatch of 15,000 more American troops.[47]

While Johnson was shoring up South Vietnam he found himself faced with trouble much closer to home—in the Dominican Republic. On April 24 a coup led by forces supporting Juan Bosch, a non-Communist leftist, overthrew the right-wing, pro-American government there. Bosch, however, could not establish control, and skirmishing broke out. American officials on the scene, including Ambassador Tapley Bennett, warned Johnson that Castro-inspired Communists might prevail amid the disorder. Poorly advised, Johnson was also anxious to prove that he would use American troops—in Vietnam, in the Dominican Republic, wherever seriously provoked—to beat back Communist advances. His response on April 28 was to send in 14,000 American soldiers and marines. In so doing he ignored the Organization of American States, which was supposed to meet in advance of interventions in the Western Hemisphere. The OAS, LBJ remarked, "couldn't pour piss out of a boot if the instructions were written on the heel."[48]

American intervention in the Dominican Republic proved popular in the United States. It also achieved its narrow purpose, for Bosch was prevented from ruling the country. By mid-1965 OAS troops were replacing the American forces, and in September a moderate leader, Joaquín Balaguer, took over. But LBJ had greatly overreacted. Bosch was not a Communist; indeed, he had been fairly elected President in December 1962, only to be overthrown in a coup ten months later. Nor were Communists strong amid the turmoil in April 1965. By intervening Johnson damaged whatever hopes the United States might have had of encouraging progressive forces in the Western Hemisphere: thirteen military coups took place in Latin America later in the 1960s. He demonstrated yet again

47. Ambrose, *Rise to Globalism*, 217–18.
48. Walter LaFeber, "Latin American Policy," in Divine, ed., *Exploring the Johnson Years*, 63–90; Ambrose, *Rise to Globalism*, 219–21; John Blum, *Years of Discord: American Politics and Society, 1961–1974* (New York, 1991), 222–25.

that worries about worldwide Communist conspiracies dominated American thinking about foreign policy.

If Johnson hoped that his show of force in the Caribbean would impress the North Vietnamese, he was wrong. Neither bombing nor American fighting men seemed to improve the fortunes of the South. Seeking order, the South Vietnamese military overthrew the last of many civilian governments in June. General Nguyen Van Thieu, a graduate of the United States Command and General Staff College at Fort Leavenworth, Kansas, took over as head of state. He was a self-effacing but shrewd political general with a great gift for survival in the ever-shifting intrigue of Saigon politics. Colonel Nguyen Cao Ky became premier. Ky had been trained by the French as a pilot and had later operated under CIA cover flying missions against the North Vietnamese. Gaudy and flamboyant, Ky frequently dressed in a black flying suit with purple scarf. Ivory-handled pistols jutted ostentatiously from his pockets.[49]

Top American leaders had little faith, to put it mildly, in Thieu and Ky. State Department official Chester Cooper likened Ky to a character from a comic opera. "A Hollywood central casting bureau would have grabbed him for a role as a sax player in a second-rate Manila night club." William Bundy, a State Department expert on Asia (and McGeorge Bundy's brother), observed glumly that Thieu and Ky were "absolutely the bottom of the barrel."[50] Thieu and Ky nonetheless brought some stability to South Vietnamese politics: Thieu became President in September 1967 and stayed on in that position until the North Vietnamese won the war years later. But they commanded little support among the masses of the people of South Vietnam. Following their coup, as before, the NLF made gains in the South.

With the situation continuing to deteriorate, McNamara wrote a decisive memo in late July. It laid out three options: "cut our losses and withdraw," "continue at about the present level," or "expand promptly and substantially the U.S. pressure." McNamara urged rejection of the first option as "humiliating the United States and very damaging to our future effectiveness on the world scene." The second option would "confront us later with a choice between withdrawal and an emergency expansion of forces, perhaps too late to do any good." He recommended the third. It would lead to "considerable cost in casualties and materiel" but would "offer a good chance of producing a favorable settlement in the

49. Berman, *Lyndon Johnson's War*, 17–18, 79.
50. Herring, "War in Vietnam," 3.

longer run." McNamara thought that American troop strength in Vietnam should be increased from 75,000, the number already there, to more
than 200,000.[51]

McNamara's memo, one of the most important in the history of the
war, laid out with special clarity the logic behind American policy in
Vietnam. All the main themes were there: the fear of "humiliation," the
need for American credibility on the "world scene," the faith in calibrated
escalation, the underestimation of the enemy, the grand expectation that
military pressure would ultimately force Ho Chi Minh to accept a settlement preserving the political integrity of a non-Communist South Vietnam. Like other top leaders, McNamara did not believe that escalation
would destroy all enemy forces. But he could not bear the thought of
pulling out, and he was confident that the enormously greater power of
the United States would prevail.

The five men who saw the memo were the major players: McGeorge
Bundy, Rusk, Taylor, Westmoreland, and Wheeler. Other members of
the National Security Council did not participate. Congress knew nothing
of it. All five joined McNamara in approving the memo and recommending it to Johnson. Some of them, including McNamara, wished to go
further and called on Johnson to declare a national emergency and to seek
a hike in taxes to support major increases in military spending. Military
leaders urged the President to mobilize 235,000 reservists. The administration, these advisers expostulated, was about to venture into war in a
large way and should make that clear to Congress, the American people,
and the world.

If Johnson had taken these steps, he would at least have stimulated open
debate about policies in Vietnam. But he refused to go that far. He
thought that these moves might trigger military responses from China or
the Soviet Union, that they would "touch off a right-wing stampede" at
home that would require him to escalate further, and that they would
weaken congressional support for his Great Society programs, some of
which (including the voting rights and Medicare bills) remained to be
passed. Mobilizing reservists, he feared, would provoke widespread unease among Americans who would worry about being called into combat
service. Instead, LBJ accepted the basic recommendations of the
McNamara memo without explaining their consequences to the American people. On July 28 he announced a stepping up of draft calls and the

51. Kearns, *Lyndon Johnson*, 281–83; Herring, *America's Longest War*, 138–40;
    Gaddis, *Strategies of Containment*, 242.

dispatch of an additional 50,000 men to Vietnam. Privately he committed himself to the sending of up to 75,000 more by the end of the year. That would increase American troop strength in Vietnam to around 200,000. The die had been cast.

THE WRITER DAVID HALBERSTAM, who served as a *New York Times* reporter in Vietnam in the early 1960s, later (1972) wrote one of the first large-selling books critical of the war. Its decidedly ironic title was *The Best and the Brightest*. Many writers echoed his view that liberal intellectuals and technocrats not only sparked the drive for escalation but also poured fuel on the fires after 1965. McNamara, a genius at systems analysis and rational planning, was the most ardent: critics of administration policy spoke of "McNamara's War." But the other incendiaries seemed equally bright. McGeorge Bundy had been dean of the faculty at Harvard. Rusk had been a Rhodes Scholar. So had Walt Rostow, who then had become a widely known professor of economic history at MIT before rising to the status of a top adviser. Westmoreland had been first captain of cadets at West Point and had fought in combat in both World War II and Korea. Taylor was a D-Day veteran and a liberals' general who spoke several foreign languages and wrote books.

In the next two and one-half years some of these men developed doubts about what they had done. Bundy left the administration in early 1966, to be replaced by Rostow. McNamara's doubts rose as early as December 1965, when he told Johnson that he questioned whether the American public would support the war for the long time that would be required. Johnson was intrigued enough to inquire, "Then, no matter what we do in the military field there is no sure victory?" McNamara replied, "That's right. We have been too optimistic." He then persuaded Johnson to declare a "Christmas bombing halt" in the hope that negotiations might begin. The halt lasted thirty-seven days without effect.[52]

In 1966 McNamara grew more deeply concerned. The fantastic escalation of that year was not achieving its purposes. Martin Luther King and others were denouncing him as an architect of disaster, especially for American blacks, who were dying in large numbers. By October McNamara no longer thought that further escalation of manpower or of bombing would accomplish much, and he called instead for the building of a huge infiltration barrier so as to stabilize the situation. This bizarre and expensive idea, which exposed the extent of McNamara's faith in

52. Berman, *Lyndon Johnson's War*, 12.

technological solutions, evoked little support among other advisers and was eventually dropped before construction could be completed. By 1967 McNamara was pacing about his expansive Pentagon office, staring at the large framed photograph of Defense Secretary Forrestal (who had committed suicide), and weeping. By late 1967 Johnson had given up on him. The war had savaged the self-confidence of the most certain of men.[53]

Others among the "best and the brightest," however, stayed the course until the close of Johnson's presidency in 1969. If Rostow and Rusk entertained serious doubts about the wisdom of escalation, they kept them to themselves. Both offered consistently hawkish advice. Westmoreland occasionally reported that calibrated escalation was not working, but usually to plead for additional help.[54] Most military leaders demanded more, not less, in the way of American involvement. For them, as for LBJ, the policy of escalation must not be abandoned simply because it was slow to accomplish its purposes.

Civilian advisers from time to time supported bombing halts and hoped for a negotiated settlement, but never at the cost of jeopardizing the independence of South Vietnam. Ho Chi Minh, however, insisted that any settlement must take into account the program of the National Liberation Front. This meant (at the least) NLF participation in negotiations and substantial NLF representation in a coalition government. LBJ and his advisers were certain that this would lead to a Communist-controlled regime in South Vietnam and to unification under the leadership of Ho. For this reason Johnson never encouraged serious negotiations. When a Polish diplomat seemed about to succeed in developing peace talks in December 1966, the Johnson administration spiked the idea by bombing heavily within five miles of the center of Hanoi. A similar fate met efforts for negotiations undertaken in early 1967 by British Prime Minister Harold Wilson. Johnson and Ho Chi Minh were so far apart that negotiations never had much of a chance.[55]

The failure of efforts at negotiation left the fate of Vietnam to the soldiers. In 1965 the casualties remained modest, and American morale was good in the field. Despite half-hearted fighting by many (not all) units of the South Vietnamese army, Americans usually prevailed when they could confront the enemy in head-to-head battles. But escalation on both

53. Deborah Shapley, *Promise and Power: The Life and Times of Robert McNamara* (Boston, 1993); Halberstam, *Best and Brightest*, 642–45; Robert McNamara, *In Retrospect: The Tragedy and Lessons of Vietnam* (New York, 1995).
54. Gaddis, *Strategies of Containment*, 259.
55. Herring, *America's Longest War*, 165–66, 168–69.

sides mounted in 1966, casualties increased, and fighting morale proved increasingly difficult to sustain.[56] Roughly 80 percent of American soldiers in Vietnam were from poor or working-class backgrounds. Neither in college nor in graduate school—where most students received near-automatic deferments until mid-1968—they often found themselves drafted after they got out of high school. Indeed, American combat soldiers in Vietnam were unprecedentedly young—an average of 19, as opposed to 27 in both World War II and Korea. For men so youthful, combat experience in Vietnam was especially terrifying. There was most of the time no "front" or clear-cut territorial objective. The young men—boys, mainly—felt that they were always far away "in the bush." Ordered out on patrols into dense cover, they were bait to lure the enemy into battle. Many were cut down by withering fire from the bush or were blown to bits by land mines.[57]

Morale also suffered because of the way that the military organized itself. The majority of American combat soldiers in Vietnam were draftees or "volunteers" who joined the armed forces in order to serve at a time of their own choosing. Most were required to stay there for one of their two or three years of military duty. Unlike those who had fought in Korea or in World War II, they expected to leave the battleground at a clearly specified time, whereupon someone else would take their place. Unsurprisingly, some who neared the end of their terms were not eager to take chances. Most American marines in Vietnam faced thirteen-month terms there that included up to three tours in the field of eighty days each. This was an unusually heavy and frightening exposure to combat—one that in many cases proved exhausting both physically and mentally to the men involved.[58]

For these and other reasons many American combat units found it hard, especially in the later years of the war, to develop camaraderie. In earlier wars, units had tended to stay together for the duration of the effort. Soldiers in combat grew close, sometimes dying for one another. In Vietnam, however, men often arrived in the field as individuals and were

56. John Guilmartin, Jr., "America in Vietnam: A Working-Class War?" *Reviews in American History*, 22 (June 1994), 322–27.
57. Appy, *Working-Class War*, 6–35. First-hand accounts that capture some of the terror of soldiers include Philip Caputo, *A Rumor of War* (New York, 1978); Michael Herr, *Dispatches* (New York, 1977); and Tim O'Brien, *Things They Carried: A Work of Fiction* (Boston, 1990).
58. D. Michael Shafer, "The Vietnam Combat Experience: The Human Legacy," in Shafer, ed., *Legacy*, 80–103.

placed in whatever squads needed help. Some of the men whom they fought beside had many months of service left—these might become buddies to be protected—but others were nearing the end of their term. They remained strangers, brief acquaintances who might soon go home.

Racial antagonism increasingly accentuated these problems.[59] Many American black men at first seemed ready and eager to serve in the military: through 1966 they were three times as likely as white soldiers to re-enlist when their terms were over. But they faced the same kinds of abuse and mistreatment on the front that they had faced at home. Often, it seemed, they were told to do the dirtiest jobs and pull the most dangerous duty. Between 1961, when the first American "advisers" died in Vietnam, and the end of 1966, 12.6 percent of American soldiers there were black (roughly the percentage of blacks in America's draft-age population). In 1965, 24 percent of American combat deaths were black, a high for the war.[60]

By then black leaders in the United States were becoming openly critical of American policy in the war. King criticized escalation in August 1965, and both SNCC and CORE formally opposed the war in January 1966. Blacks became less eager to serve. Many who were drafted and sent abroad to fight had imbibed racial pride from the civil rights movement, and they would not put up with second-class treatment. Made aware of these feelings, American army leaders sought to lessen racial discrimination in the field. The percentage of combat deaths that was black began to fall, to 16 percent in 1966 and 13 percent in 1968.[61] Still, sharp antagonisms persisted, reflecting not only the timeless tensions of color but also the mounting racial conflicts that were rending American society at home.

Many Americans who served in Vietnam complained bitterly—again, mainly in the later years of the war—of poor military leadership. This was hardly new in the annals of warfare, but the problem seemed especially acute in Vietnam, where "fragging" (soldiers wounding or killing their

59. Wallace Terry, *Bloods: An Oral History of the Vietnam War by Black Veterans* (New York, 1984); Peter Levy, "Blacks and the Vietnam War," in Shafer, ed., *Legacy*, 209–32.
60. Jones, *Great Expectations*, 96.
61. Blacks ultimately were 13.7 percent of all American casualties in Vietnam, a figure that was 30 percent higher than would have been the case in a color-blind situation. Shafer, "Vietnam Era Draft." Appy, *Working-Class War*, 19–21, offers slightly different figures.

own officers) became a serious matter by the 1970s. Officers who were assigned to combat normally served six-month tours. They scarcely had time to develop much rapport with men in their units. Few of those above the grade of lieutenant (the lowest officer ranks) stayed long on the front lines with the troops. They tended instead to remain in base areas, many of which were lavishly outfitted, or in the air, mainly in helicopters. Only four American generals died in combat during the years of fighting in Vietnam, three of whom crashed to their death in helicopters. (The fourth was killed by sniper fire.) Although nearly 8,000 American dead— 13.8 percent of the total—were officers, most of these were lieutenants or captains. The rest were non-commissioned officers—sergeants, corporals, and the like—and draftees and "volunteers."[62]

Men in the front lines had to worry not only about the enemy but also about the firepower from their own side. American dead and wounded from "friendly fire" were later estimated to be as high as 15 or 20 percent of all casualties. Many things accounted for these unprecedentedly high estimates: poor leadership, inadequate training, the frequency of close-in fighting in the bush, and the penchant of high-ranking officers to hurry in calls for the heaviest sort of firepower—after all, it was there—to support infantrymen in trouble.[63]

American fighting men were regularly lonely and frequently scared even when they were not in combat. They had little if any understanding of the history or the culture of Vietnam, and they did not know the language. The South Vietnamese, equally ethnocentric, were bewildered by the impatient, technologically advanced, and increasingly frustrated Americans. Personal friendships across the large gap in cultures were rare. Worse, many apparently supportive South Vietnamese seemed to have an uncanny ability to avoid the mines and booby traps that maimed United States troops. It was hardly surprising in the circumstances that many American fighting men came to believe that all Vietnamese were alike. Everyone in "Nam," it seemed, wore black pajamas and sneaked about treacherously in the night. "They say, 'GI number one' when we're in the village," an American soldier complained, "but at night the dirty rats are VC." Another wrote, "During the day they'll smile and take your money. At night they'll creep in and slit your throat."[64] More and more, American combat veterans thought that no South Vietnamese civilian deserved

62. Ibid.; "Vietnam: Who Served and Who Did Not?," *Wilson Quarterly* (Summer 1993), 127–29; James Fallows, "What Did You Do in the Class War, Daddy?" *Washington Monthly*, Oct. 1975.

63. *Newsweek*, April 25, 1994.

64. Herring, "War in Vietnam," 90.

a second thought before being shot in the midst of a fire fight in a village. "The rule in the field," one explained, "if it's dead, it's VC."[65]

All these problems and fears helped to account for the savagery of the fighting in Vietnam. To some this seemed even more gruesome than in other wars. NLF and North Vietnamese combatants, battling against alien invaders, fought tenaciously, resorting to sabotage and stab-in-the-back assaults on American soldiers in the night. Americans, with awesome firepower at their disposal, dropped ever more napalm and Agent Orange and poured bombs on villages and enemy installations. Commanders in the field often called in the planes, helicopters, and heavy artillery wherever they thought the enemy might be. As an American major explained in 1968 following the obliteration by American firepower of Ben Tre, a village in the Mekong Delta, "We had to destroy the village in order to save it."

The savagery of combat intensified in 1966–67, by which time it was becoming increasingly difficult for many of the soldiers to understand why the United States was fighting. McNamara, Westmoreland, and their own commanders, after all, were measuring progress only in body counts. When soldiers returned from patrol, they were asked one thing: how many did you kill? One soldier exclaimed, "What am I doing here? We don't take any land. We don't give it back. We just mutilate bodies. What the fuck are we doing here?"[66]

Increasing numbers of America's combat soldiers were revulsed by the slaughter, and many thousands suffered from serious personality disorders thereafter. Others, however, grew callous and cruel. That happens in wars. William Broyles, a marine lieutenant, later described the experiences of his unit:

> For years we disposed of the enemy dead like so much garbage. We stuck cigarettes in the mouths of corpses, put Playboy magazines in their hands, cut off their ears to wear around our necks. We incinerated them with napalm, atomized them with B-52 strikes, shoved them out the doors of helicopters above the South China Sea. In the process did we take down their dog-tag numbers and catalog them? Do an accounting? Forget it.
>   All we did was count. Count bodies. Count dead human beings. . . . That was our fundamental military strategy. Body count. And the count kept going up.[67]

65. Shafer, "Vietnam Combat Experience."
66. Ibid.
67. *Newsweek*, Feb. 14, 1994, p. 31. Appy, *Working-Class War*, 320, says some 500,000 American veterans of Vietnam suffered from post-traumatic stress disorder, which often lasted for decades.

A COMMON ARGUMENT of supporters of American escalation is that the war was lost not in the bush but on the home front in the United States. They emphasize the damaging role of two institutions: the media, especially television, for presenting an overly negative picture of American goals and accomplishments; and the universities, for succoring draft-dodging student activists against the war. The anti-war stance of the media and of the students, they say, encouraged a broader-based disenchantment that ultimately reached Capitol Hill.

In the long run the war indeed increased the skepticism—indeed the suspicion and testiness—of the media, which understandably grew disenchanted by glowing government handouts and public relations efforts. Their rising suspiciousness, which grew confrontational by the early 1970s, was a very important legacy of the war—one that reflected and intensified broader popular dissatisfactions with government. Prior to 1968, however, most newspapers and magazines supported Johnson's policies. Reporters necessarily relied heavily on handouts from American military and political leaders, and newspapers printed thousands of stories carrying greatly inflated statistics about enemy body counts and other supposed accomplishments of the military effort. As in the era of Joe McCarthy, journalists normally felt obliged to accept what office-holders and top military leaders—those who commanded relevant information—had to say.

Television news programs, too, tended either to soft-pedal the horrors of the war (footage of such horrors was hardly pleasurable to watch at dinnertime) or to go along with the administration until 1968. Walter Cronkite, the avuncular, much-admired anchorman for CBS news, came close to applauding Johnson's actions at the time of the Tonkin Gulf "crisis." Henceforth, he told viewers, the United States was committed to "stop Communist aggression wherever it raises its head."[68] Then and for the next three years television newscasts not only had little criticism to offer of American troop build-ups but also presented a great many unflat-

68. Michael Delli Carpini, "Vietnam and the Press," in Shafer, ed., Legacy, 125–56; Chester Pach, "And That's the Way It Was: The Vietnam War on the Network Nightly News," in Farber, ed., Sixties, 90–118; James Baughman, The Republic of Mass Culture: Journalism, Filmmaking, and Broadcasting in America Since 1941 (Baltimore, 1992), 111–14; Kathleen Turner, Lyndon Johnson's Dual War: Vietnam and the Press (Chicago, 1985); and William Hammond, "The Press in Vietnam as Agent of Defeat: A Critical Examination," Reviews in American History, 17 (June 1989), 312–23.

tering portrayals of anti-war activists, "hippies," and other critics of American institutions.

Although the conflict in Vietnam was America's first "living room war," newscasts devoted little of their evening broadcasts (only twenty-two minutes long excepting commercials) to actual fighting. With some notable exceptions there was not much footage of heavy combat that showed dead or wounded American soldiers; one later estimate found a total of seventy-six such cases in 2,300 newscasts surveyed between 1965 and 1970. In part because many battles were fought at night and in locations far from television cameras, TV coverage of combat tended to show soldiers leaping out of helicopters and dashing off into the bush. Sounds of artillery boomed in the background. By the late 1960s the presentation of such scenes probably accentuated an already widespread popular conviction that the war had no clear goals and that it was going badly. But it also showed the courage of American soldiers, not the most bloody aspects of the fighting. To a degree, such coverage may have helped to sanitize a sanguinary struggle. [69]

The influence of college and university students on American attitudes toward the war is a somewhat more complicated story. Clearly, the potential for student influence was enormous, for it was in the mid-1960s that the coming of age of baby boomers began to swell the already rising numbers of American young people on the campuses. The number of Americans aged 18 to 24 rose from 16.5 million in 1960 to 24.7 million in 1970, a jump of almost 50 percent. By then approximately one-third of this age group (or 7.9 million) was studying at least part-time at institutions of higher education. The explosive leap in the numbers of college-educated young people, one of the most salient demographic trends of the decade, promoted increasing amounts of talk by the mid-1960s about "youth culture," "youth rebellion," and "generation gap."

These demographic developments, moreover, interacted dynamically with changes in attitudes, notably the rise of liberal optimism, escalating expectations, and rights-consciousness. Millions of young people, especially those with the resources to pursue higher education, were swept up in the hopes for change that civil rights activists and others had done

---

69. Lawrence Lichty, "Comments on the Influence of Television on Public Opinion," in Peter Braestrup, ed., *Vietnam as History: Ten Years After the Paris Peace Accords* (Washington, 1984); Michael Arlen, *Living-Room War* (New York, 1982); and Hodgson, *America in Our Time*, 150–51. See also Ralph Levering, "Public Opinion, Foreign Policy, and American Politics Since the 1960s," *Diplomatic History*, 13 (Summer 1989), 383–93.

much to unleash and that Kennedy and Johnson had appealed to in order to promote liberal programs. As Bob Dylan had prophesied in his song "The Times They Are a-Changin'," many young people thought they had the potential to transform American life. Rarely before had two such interrelated and powerful trends—demographic and ideological— occurred at the same time. Reinforcing each other, they underlay much of the turmoil that distinguished the 1960s from the 1950s.[70]

A minority of these young people—again, mostly those with higher education—turned well to the left in the early 1960s. Some grew rebellious upon arriving on campuses that had become enormous and impersonal in recent years. Anger at bureaucracy and at the seeming indifference of authority figures fueled a good deal of the discontent of young Americans in the 1960s. Other students railed at what they perceived as complacency among older, affluent people who seemed to ignore social problems. Accumulating material goods, they said, might have been all right, indeed necessary, for their elders. But it was unsatisfying, even cold-hearted, to some in the younger generation. Many of the young, indeed, felt guilty that they were well-off while others were not. They believed above all that they must act—"put our bodies on the line"— against injustice. They would work to smash racism and poverty and transform the consciousness of the nation.[71]

Few of these young people were in open rebellion against their elders. On the contrary, they tended to be the children of relatively affluent and indulgent parents and to have gone to schools and colleges where they had been encouraged to think for themselves. (Nixon later grumbled that they were a "Spock-marked generation.") In general, they were more likely than others in their age group to have grown up in politically liberal families. Some were sons and daughters of radicals—or "red-diaper babies," as critics named them. Many displayed a special self-consciousness and self-confidence. Having grown up in a world that differed significantly from that of their parents, they imagined that their elders just did not understand. As Dylan put it:

70. Morris Dickstein, *The Gates of Eden: American Culture in the 1960s* (New York, 1977), 188; Maurice Isserman and Michael Kazin, "The Failure and Success of the New Radicalism," in Steve Fraser and Gary Gerstle, eds., *The Rise and Fall of the New Deal Order, 1930–1980* (Princeton, 1989), 212–42.

71. For the rise of the Left see John Diggins, *The Rise and Fall of the American Left* (New York, 1992), 173–90; Todd Gitlin, *The Sixties: Years of Hope, Days of Rage* (New York, 1987), 83–85; and Winnie Breines, "Whose New Left?," *Journal of American History*, 75 (Sept. 1988), 528–45.

Come mothers and fathers
Throughout the land
And don't criticize
What you can't understand
Your sons and your daughters
Are beyond your command
There's a battle
Outside and it's ragin'
It'll soon shake your windows
And rattle your walls,
For the times they are a-changin'.

Among these young activists in the early 1960s were some who derived their ideas from radical thinkers of the 1950s, such as the sociologist C. Wright Mills, the free-thinking Paul Goodman, and Michael Harrington.[72] Most of the philosophically aware young activists of the early 1960s liked especially to distinguish themselves from representatives of the "old" or Marxist Left, which they thought was doctrinaire. Labor unions, too, seemed to them stodgy and conservative, and liberals could not always be trusted. Most of the young activists considered themselves members of a new generation and of a New Left, a term that came into general usage in 1963. They would rid themselves of dogmatism and employ democratic procedures to generate sweeping social change.[73]

From the beginning, members of the New Left differed a good deal among themselves, and their major causes changed considerably over time. Some who joined the best-known protest group of the time, Students for a Democratic Society, focused in the early 1960s on developing community action against poverty. When the war on poverty began in 1964, they redoubled their efforts to revitalize the ghettos and other poor pockets of American society. Most SDS workers—and other young activists—were especially inspired by the civil rights movement. Student

72. Important books for these people included Goodman's *Growing Up Absurd: Problems of Youth in the Organized Society* (1960); and Mills, *The Power Elite* (1956). Mills wrote an essay, "Letter to the New Left," in the October 1960 issue of *Studies on the Left* that called on the young to become the vanguard of change. See Priscilla Long, ed., *The New Left: A Collection of Essays* (Boston, 1969), 14–25.

73. Maurice Isserman, *If I Had a Hammer: The Death of the Old Left and the Birth of the New Left* (New York, 1987); Gitlin, *Sixties*, 4–6. The Students for a Democratic Society, for instance, broke away from the Student League for Industrial Democracy in 1960 in large part because the students considered SLID to be caught up in tired sectarian battles against Communism.

protestors who demonstrated in the highly publicized movement for free speech at the University of California at Berkeley in 1964 included key leaders who had been radicalized by the racism that they had witnessed as civil rights volunteers in the South.[74]

Perhaps the most prominent young radical in the early 1960s was Tom Hayden, a University of Michigan student who had worked with SNCC in 1961. Raised a Catholic, Hayden was a serious thinker with a commitment to elevating the spirit and improving human relationships in the United States. In 1962 he emerged as chief author of a major position paper of the SDS, the Port Huron Statement.[75] This manifesto, widely cited (if not widely read) by fellow activists in the early 1960s, lamented the alienating affluence of American civilization and the consequent "estrangement" of modern man. Although it did not renounce capitalism, it echoed Mills (whom Hayden much admired) by asserting that "in work and in leisure the individual is regulated as part of the system, a consuming unit, bombarded by hard-sell, soft-sell, lies, and semi-true appeals to his basest drives." Hayden denounced what he deemed to be the intellectual aridity of American universities and called on college students to "consciously build a base for their assault upon the loci of power." The statement was vague about specific goals, welcoming, for instance, both decentralization of government and a larger welfare state. But it clearly aimed at harnessing the moral idealism of young people to humanize capitalism. This was to be done through the process of "participatory democracy," a rallying cry for many who joined the New Left in the 1960s.

Although the SDS and other radical groups attracted rising attention in the media during the early 1960s, they gained very few members at the time. In October 1963 SDS had six chapters (plus thirteen more on paper) and a total of 610 paid-up members. In January 1965, before LBJ escalated the war on a large scale, SDS had a few dozen chapters but still only 1,500 paid-up members.[76] Most of them were still concentrating on the

74. W. J. Rorabaugh, *Berkeley at War: The 1960s* (Berkeley, 1989).

75. So called because it was drafted at an SDS gathering at Port Huron, Michigan. The best source on Hayden and the early SDS is James Miller, *"Democracy Is in the Streets": From Port Huron to the Siege of Chicago* (New York, 1987). The Port Huron Statement is reprinted there, 329–74. See also Tom Hayden, *Reunion: A Memoir* (New York, 1988); Gitlin, *Sixties*, 112–26; Diggins, *Rise and Fall*, 198.

76. Todd Gitlin, *The Whole World Is Watching: Mass Media in the Making and Unmaking of the New Left* (Berkeley, 1980), 22–31. Gitlin, an activist for peace in the early 1960s, became president of SDS from mid-1963 to mid-1964. See also Kirkpatrick Sale, *SDS* (New York, 1973).

issue of poverty. Although some of its members had been active in demonstrations against nuclear testing, they had paid little notice before 1965 to foreign policies or to events in far-off Asia.

Johnson's escalation soon transformed the SDS and other New Left organizations. Some of the activists, to be sure, continued to concentrate on problems of the ghetto—Hayden was busy mobilizing the poor in Newark—and others focused on reforming the universities. But many new young people flocked to SDS and other New Left groups out of fear and anger at LBJ's foreign policies. As early as April SDS dramatized the anti-war cause by sponsoring a March on Washington. By 1966 SDS membership was three times what it had been a year earlier. Equally important, SDS and other student-dominated organizations were developing considerably more anti-war sympathy among otherwise apolitical university students, many of whom were frightened by the threat of the draft. No other group of Americans was nearly so vocal against the war— or so widely assailed by advocates of escalation—as the student New Left and its much larger band of student sympathizers.[77]

Some of the students were attracted to speakers at university teach-ins who emphasized the strategic fallacies of escalation. Hans Morgenthau, a prominent political scientist, repeatedly emphasized that Vietnam had little geopolitical importance and that the United States, like France in the 1940s and 1950s, was sacrificing resources as well as international prestige by tying itself down in the war. Other well-known American students of foreign policy, including George Kennan and the columnist Walter Lippmann, made similar points. They were realists who deplored the distortions that Johnson's military escalation was creating in foreign and defense policies.[78]

Most youthful opponents of the war, however, approached the issue from a moral point of view. Here they differed sharply from many of their elders, such as Johnson, who had lived through the rise of fascism and the struggles of World War II. The young, with a much more present-minded view of things, did not worry nearly so much about Communism or the Cold War. Where LBJ and his advisers applied the "lessons" of Munich and of appeasement, the young activists were sure that the conflict in

77. Kenneth Heineman, *Campus Wars: The Peace Movement at American State Universities in the Vietnam Era* (New York, 1993); Nancy Zaroulis and Gerald Sullivan, *Who Spoke Up? American Protest Against the War in Vietnam, 1963–1975* (Garden City, N. Y., 1984).

78. William O'Neill, *Coming Apart: An Informal History of America in the 1960s* (Chicago, 1971), 142.

Vietnam was a civil war, not a Communist conspiracy. They were appalled by the immorality, as they saw it, of American actions, which they equated with racism in the American South. Placards proclaimed their position: STOP THE KILLING; UNCONDITIONAL NEGOTIATIONS YES—KILLING VIETNAMESE CHILDREN NO; WAR ON POVERTY, NOT ON PEOPLE; ONE MAN ONE VOTE—SELMA OR SAIGON; ESCALATE FREEDOM IN MISSISSIPPI.[79]

"Eve of Destruction," a popular song by P. F. Sloan, captured the anxiety that many of these young people felt in 1965. Sung in a brooding, surly way by Barry McGuire, it was accompanied by a pounding drumbeat and occasional whines of the harmonica. The lyrics were apocalyptic. "When the button is pushed there's no runnin' away / There'll be no one to save with the world in a grave." Racial injustice was rampant:

> Handful of Senators don't pass legislation
> And marches alone can't bring integration
> What human respect is disintegratin'
> This whole crazy world is just too frustratin' . . .
> Look at all the hate there is in Red China
> Then take a look around at Selma, Alabama.

And Vietnam:

> The Eastern world, it is explodin'
> Violence flarin', bullets loadin'
> You're old enough to kill but not for votin'
> You don't believe in war but what's that gun you're totin'
> And even the Jordan River has bodies floatin'.

By contrast to many of the hit songs of 1964 (the Beatles' "A Hard Day's Night," the Supremes' "Baby Love," and the Beach Boys' "Little Deuce Coupe"), "Eve of Destruction" was bitter and discordant. It symbolized a move toward a much angrier, louder—and sometimes deliberately unintelligible—form of rock music that rose in popularity during the late 1960s. Some people thought "Eve of Destruction" subversive. Many stations refused to play it. It zoomed to the top of the charts five weeks after its release in July, the fastest-rising song in the history of rock.[80]

After Johnson committed himself in mid-1965 to much-increased escalation, a minority of American activists virtually disowned the govern-

79. Kearns, *Lyndon Johnson*, 328–32; Gitlin, *Whole World*, 54.
80. George Lipsitz, "Youth Culture, Rock 'n' Roll, and Social Crises," in Farber, ed., *Sixties*, 206–34; Gitlin, *Sixties*, 195–96.

ment. Hayden, the pacifist Staughton Lynd, and the Communist Herbert Aptheker journeyed to Hanoi at the end of the year and returned to celebrate the "rice-roots democracy" of the North Vietnamese state. Hanoi, they said, was not abetting the fighting in the South. The travelers ignored well-documented evidence, available since the late 1950s, that North Vietnam had killed thousands of uncooperative peasants and had imprisoned thousands of others in forced-labor camps. Outraged by the war, the radicals were credulous visitors.[81]

Other militant foes of the war shared the rage that animated Hayden and his cohorts. Demanding immediate American withdrawal, they attended rallies in 1966 waving NLF flags and wearing buttons, VICTORY TO THE NATIONAL LIBERATION FRONT. They chanted, "Ho, Ho, Ho Chi Minh / The NLF is gonna win." They adopted as heroes not only Ho Chi Minh but also Mao Tse-tung and Ché Guevara, the Marxist guerrilla strategist who had helped Castro stage his revolution in Cuba. Some were so psychologically alienated as to feel virtually homeless in the United States. They called themselves the "NLF behind LBJ's lines."[82]

Colorful and quotable, the student radicals received considerable coverage in the media. Pro-war Americans were quick to blame them for impeding the war effort. This was inaccurate, however, in important respects. Nothing that the students said or did at that time weakened Johnson's resolve to escalate the war. Anti-war activists enjoyed strength, moreover, on only a few campuses in the United States. The more elite the college or the university, the more likely it was to harbor visible opponents of the war in Vietnam. The state universities of Michigan, Wisconsin, and California-Berkeley were three such places. Prestigious private institutions such as Harvard and Columbia also proved relatively receptive to radical dissent. Most campuses, however, remained quiet until the late 1960s: one careful estimate has concluded that only 2 to 3 percent of college students called themselves activists between 1965 and 1968, and that only 20 percent ever participated in an anti-war demonstration.[83] It was highly inaccurate to lump together the widely different institutions of higher education in the United States and accuse them, especially in 1965–66, of succoring anti-war activity.

Class differences in opinions about the war further divided young people (and others) in America. This was a complicated matter, for "attitudes" toward the war were neither unidimensional nor consistent over

81. Gitlin, *Sixties*, 261–62; Diggins, *Rise and Fall*, 208.
82. Gitlin, *Sixties*, 262–69.
83. Terry Anderson, *The Movement and the Sixties* (New York, 1995), preface (n.p.).

time. Many people who told pollsters that they were opposed to the war meant that they disagreed with the way that the administration was handling it; some wanted more escalation, not less. Surveys that tried to get at class attitudes, however, did not support the contention that the poor and the working classes were the strongest advocates of war. Instead, they indicated that backing for the fighting was in fact highest among the young (people aged 20 to 29) and the well educated. Over time it became especially weak among the elderly, blacks, women, and the poorly educated. It is therefore inaccurate to argue that poor and working-class people—for the most part, non-students—were the most avid advocates of escalation or that middle- to upper-middle-class Americans, including university students, were the greatest advocates of restraint. To the dismay of anti-war professors, high levels of education did not translate readily into opposition to the fighting in Vietnam.[84]

Most university students who expressed sympathy with anti-war activity in the mid-1960s, moreover, were hardly radical in their general views. Only a small percentage of them joined SDS or other such groups. The much larger group of anti-war sympathizers on the margin criticized the immoral course of the government, sometimes with inflammatory rhetoric, but hoped that teach-ins and non-violent demonstrations would help Washington see the light. Although they feared and avoided the draft, they were slow to resist it actively—that came later, mainly after 1966. While they pressed the United States to stop the bombing, they were not so likely to demand total American withdrawal, and they hoped for some form of power-sharing arrangement between South Vietnam and the NLF. Given the bloody history of Vietnam, this was a forlorn hope. Many anti-war sympathizers nonetheless clung to it and continued long after 1965 to believe in the possibility that reasonable men could find a reasonable solution.[85]

Notwithstanding all these qualifications, there was no doubt that radical anti-war activity in the United States was nowhere more visible in the mid-1960s than on some of the most elite campuses. These tended to enroll the most privileged, academically oriented, and highly motivated young people in American life. Why this vanguard of students came to

84. Hodgson, *America in Our Time*, 385–92; Charles De Benedetti, and Charles Chatfield, *An American Ordeal: The Antiwar Movement of the Vietnam Era* (Syracuse, 1990).

85. Diane Ravitch, *The Troubled Crusade: American Education, 1945–1980* (New York, 1983), 223; Gitlin, *Sixties*, 296. Baskir and Strauss, *Chance and Circumstance*, 4–6, estimate that there were 570,000 "apparent draft offenders" during the ten war years—of 26.8 million who came of draft age at the time.

oppose the war remains debated. Feelings of guilt, strongest among many of the most privileged, had something to do with it. So did particular campus traditions, such as the power of pre-existing peace movements and other liberal ideas. Jewish students, many of whom came from liberal or radical families, were relatively well represented on these campuses and in the leadership of some of the campus protests. Finally, many of the most privileged students had the greatest of expectations, both about their own futures—which were threatened by the draft—and about the future of the nation, which seemed destined to disgrace itself if fighting continued.[86]

Pro-war Americans in 1965 and 1966 were therefore correct to single out university students, whether radical leaders who belonged to SDS, followers who turned out for demonstrations, or young men who feared the draft, as the most determined leaders of protest against the conflict in Vietnam. And because enrollments at places like Berkeley had become so huge—it had 27,000 students in the mid-1960s—opponents of war did not have to muster a very high percentage of students on such campuses in order to bring out a good-sized crowd. Indeed, during the early years of the war university students (and organizations such as SNCC) seemed to offer the *only* substantial opposition to the fighting, for the conflict then remained more popular than people later cared to remember. Many important institutions—labor unions, corporations, Congress, the media, many (not all) of the churches—either supported escalation or went along with it through 1965 and 1966. Their attitudes reflected the continuing power of anti-Communist opinion in the United States, as well as the willingness of most people to believe what their leaders were telling them. Americans had not become very cynical about their public officials—yet. The "credibility gap" was widening in 1965 and early 1966, but it had yet to become a chasm.

IN 1967 THE GAP WAS BECOMING profound. By then Johnson had been escalating the conflict for two years. Military spending was exploding in size—from $49.6 billion in 1965 to $80.5 billion for the fiscal year of 1968—and creating ever-larger deficits.[87] Draft calls had

86. Arlene Skolnick, *Embattled Paradise: The American Family in an Age of Uncertainty* (New York, 1991), 82–84.
87. Deficits in the Johnson years rose from $1.41 billion in fiscal 1965 to $3.70 billion in 1966 and to $8.64 billion in 1967. In 1968 they ballooned to $25.1 billion. A tax increase in 1968 helped to bring in much greater revenue, and the federal government ran a small surplus (of $3.2 billion) in 1969. (See chapter 23 for later figures.)

peaked. Some 450,000 American soldiers were in Vietnam. Most important, casualties had risen alarmingly since 1965. Nothing did more than the casualty figures, which LBJ could not conceal, to advance the tide of popular concern about the war in Vietnam. This did not yet engulf Congress or other American institutions. But it was spreading beyond the student Left. Casualty reports—and the sense of futility surrounding the fighting—had done the same during the war in Korea. By the end of 1967 the tide seemed threatening indeed to the captains of state in Washington.

Some of the anti-war protestors then began to turn from a strategy of protest to one of more active resistance. Most of this remained non-violent, but some of it was confrontational. Resisters began engaging in street theater and guerrilla actions. They taunted policemen ("pigs"), soldiers, and other uniformed symbols of authority, tossed bottles, and sat in at draft headquarters. Other resisters turned in or burned draft cards, thereby breaking the law, and were prosecuted. In October 1967 the Reverend Philip Berrigan, a Catholic priest, doused draft records in Baltimore with blood from ducks.[88] Some young people eligible for the draft went underground, and others fled to Canada and elsewhere. A highly publicized siege of the Pentagon in October 1967—some 20,000 demonstrators took part—provoked brutal suppression by authorities as McNamara watched nervously from the roof.[89] Police attacked draft resisters on "Bloody Tuesday" in Oakland in the same month. Prominent opponents of the war, among them Dr. Spock and the Reverend William Sloane Coffin, openly counseled young men to resist the draft and were indicted. Some of the militants were devising strategies of protest that would literally "stop America in its tracks."[90]

This was the radical fringe of draft resistance in the United States. More common, however, were draft avoiders. Ever-larger numbers of young people—most of them college students—sought to avoid the draft by getting married and fathering children, staying as long as possible in college and graduate school, joining the military reserves or the National

88. Flynn, *Lewis B. Hershey*, 176.
89. Vividly captured by Norman Mailer, *Armies of the Night* (New York, 1968).
90. Gitlin, *Sixties*, 250–56, 291–93; Flynn, *Lewis B. Hershey*, 172–87. It is estimated that 10,000 people ultimately went underground and that 60,000 to 100,000 fled the country. Approximately 9,000 Americans were convicted of draft violations, most from 1967 on; 3,250 of them went to prison. These were not huge numbers, of the 26.8 million who were eligible, but many were defiant, and their activities were widely publicized. Jones, *Great Expectations*, 94; Shafer, "Vietnam Era Draft."

Guard, or finding friendly family doctors (including psychiatrists) who would say that they were too sick to be inducted. A few claimed to be homosexual, grounds for avoiding military service. The effort to escape induction, which meant eluding the call of local draft boards, could be all-consuming, for men were eligible until the age of 26. "My whole life style, my whole mentality was cramped and distorted, twisted by fear of the United States government," one high school graduate recalled. "The fear of constantly having to evade and dodge, to defend myself against people who wanted to kill me, and wanted me to kill."[91]

Many of those who sought to escape the draft managed to do so in the late 1960s. Some lived in areas where enlistment rates were high, thereby obviating the need for their draft boards to dig deep into their pool of eligibles. Most of the others who escaped the draft used every imaginable ploy to achieve their ends. Those with good connections—doctors, friends on the draft boards—managed much better than those who did not. Those who had deferments because they attended college or graduate school did best of all, especially if they remained on campus beyond the age of 25. For these reasons the Vietnam-era army (unlike the armies that had fought in World War II or Korea) consisted disproportionately of the poor, minority groups, and the working classes. They were getting drafted and killed while others—many of them university students who were loudest against the war—stayed safely at home.[92]

No one had intended selective service to work in this way. Supporters of the system, which had been created in the 1940s, argued that local draft quotas were based on reliable counts of eligible manpower and that the local boards—more than 4,000 in all—would do a better and fairer job than a far-off bureaucracy of "channeling" young men in their communities. Some young men would be drafted; others with special qualifications that made them more useful in civilian life would be spared. But the baby boom changed the situation by creating a huge available pool in the mid-1960s. Local boards, with bigger pools to choose from, called far smaller percentages of young men than had been the case during World War II and Korea. So the students and the privileged, manipulating the system, generally escaped service. By 1967 the unfairness of the process was obvious to all. It was a scandalous state of affairs that increasingly enraged working-class young men, their parents, their relatives, and their friends.

91. Jones, *Great Expectations*, 94–95.
92. Shafer, "Vietnam Era Draft."

It also upset Congress, which authorized changes in July 1967, and Johnson, who began to implement them in early 1968. He did not attempt to stop the deferment of college students. But he said that starting in the spring of 1968, graduate students (excepting those studying divinity, dentistry, or medicine) who had not completed two years of study would be eligible for the draft. So would college seniors graduating in 1968. Occupational deferments were also tightened. Johnson made it clear that he expected draft boards to focus on people who had graduated from college. "Start at age twenty-three," he said. "If not enough, go to twenty-two, then twenty-one, then twenty, and lastly nineteen." As it happened, these moves did not change things dramatically, for many draft boards by mid-1968 had filled their quotas, and a lottery system was introduced in late 1969. But his directions greatly unsettled students and their parents. For the first time, it seemed, the college-educated sons of the middle classes might have to face the terrors of the bush.[93]

LBJ's ideas about the draft reflected visceral feelings that he harbored about anti-war activists. A conventional patriot, he was infuriated by draft card burnings and other studied insults to the American way. Militants who attacked him personally outraged him. On one occasion a demonstrator confronted him with a sign, LBJ, PULL OUT LIKE YOUR FATHER SHOULD HAVE DONE. He complained bitterly to Joseph Califano, a trusted aide, that the "thickness of daddy's wallet" offered protection to the privileged and to the hypocritical young men who were pontificating in order to save their skins.[94]

Johnson's rising rage against anti-war activists convinced him by 1967 that they included Communists acting on directions from abroad. Determined to stop the conspiracy, he told the CIA to spy on them. This was a violation, as he well knew, of the CIA's charter, which was supposed to prevent the agency from conducting surveillance at home. This program, which ultimately compiled information on more than 7,000 Americans, was later known as CHAOS. Johnson also encouraged the FBI to infiltrate

93. Joseph Califano, *The Triumph and Tragedy of Lyndon Johnson* (New York, 1991), 196–203. One who was affected was J. Danforth Quayle, later to be Vice-President from 1989 to 1993. He joined the National Guard and avoided Vietnam. Another was Bill Clinton, later President of the United States. Clinton graduated from college in the spring of 1968, and he worried deeply about the draft. After he was called for a draft physical (receiving a classification of 1-A) in February 1969, he manipulated the system so as to remain in a university and to avoid being called. He then entered the first lottery in December 1969. He received a very low number and never was called.

94. Califano, *Triumph and Tragedy*, 200–203.

and disrupt the peace movement. The Counter-intelligence Program (COINTELPRO) that had been begun in the 1950s to fight domestic Communism now directed "black bag" jobs of surreptitious entry into private residences of anti-war activists. When the CIA failed to come up with evidence connecting anti-war people with Communism, LBJ leaked information to right-wing congressmen that such connections existed. The congressmen then charged, as he had assumed they would, that "peaceniks" were being "cranked up by Hanoi."[95]

Reactions such as these exposed the emotional stress that afflicted Johnson because of the Vietnam War by 1967. He was obsessed by the carnage, often unable by then to think of anything else. Earlier he had ordered the installation in the Oval Office of three large-screen television sets. He did the same in a den next to the Oval Office, in his bedroom at the White House, and in his bedroom at the LBJ Ranch, his home in Texas. He also placed Associated Press and United Press International news tickers in the Oval Office. When he sensed that important news was coming off the tickers, he left his desk, ripped off a sheet, and read through the reports. When he stayed at his ranch—a source of solace in these trying times—he placed a transistor receiver in his ear while he walked about his property.[96]

The rise of opposition to his policies intensified this obsessiveness. While polls in mid-1967 showed that a majority of the people still supported the war, they also revealed widening public dissatisfaction with Johnson's performance as President. Approval of his handling of the war dropped to a low of 28 percent in October.[97] Despairing, Johnson agonized over news of American casualties. He wept when he signed letters of condolence. He found it hard to sleep, and he paced the corridors of the White House at night. Frequently he invaded the operations room to check casualty figures at 4:00 and 5:00 A.M. Sometimes he slipped off in the night, accompanied only by agents of the Secret Service, to pray and read scriptures with monks at a small Catholic church in southwest Washington.[98]

95. Herring, America's Longest War, 182; Gitlin, Sixties, 264.
96. Larry Berman, "Lyndon B. Johnson: Paths Chosen and Opportunities Lost," in Fred Greenstein, ed., Leadership in the Modern Presidency (Cambridge, Mass., 1988), 134–63; David Culbert, "Johnson and the Media," in Divine, ed., Exploring the Johnson Years, 214–48; Califano, Triumph and Tragedy, 167–68.
97. Herring, America's Longest War, 174; Berman, Lyndon Johnson's War, 60, 86.
98. Berman, "LBJ: Paths Chosen," 144; Califano, Triumph and Tragedy, 250. By then Johnson had two sons-in-law in the military, one of whom, Marine Captain Charles Robb, was slated to go to Vietnam in 1968.

At the same time, Johnson lost tolerance for disagreement or even debate. Reporters and columnists became enemies. "I feel like a hound bitch in heat in the country," he said in a characteristic sexual metaphor, "If you run, they [the press] chew your tail off. If you stand still, they slip it to you."99 "Newspapermen," he added, "are the only group in the country who operate without license. Reporters can show complete irresponsibility and lie and mis-state facts and have no one to be answerable to."100 Johnson came to loathe public figures—Robert Kennedy, by then a senator from New York, Martin Luther King, Fulbright (or "Halfbright," as Johnson called him)—who dared to question his policies. Those who raised doubts at White House meetings, such as McNamara, got nowhere. Johnson distrusted even advisers on the National Security Council. Some of these people, he complained later, were "like sieves. I couldn't control them. You knew after the National Security Council met that each of those guys would run home to tell his wife and neighbors what they said to the President."101 By late 1967 Johnson felt fully comfortable only with a handful of aides, such as Rusk and Rostow, who joined him for the Tuesday noon luncheons where they picked bombing targets. He dominated them, virtually demanding sycophancy. The deft and solicitous legislative conductor who had orchestrated the Great Society had become a frightened and high-strung maestro who shuddered at the sound of a dissonant note.

What especially frustrated the President was the trap that the military situation had snared him in. Military leaders such as Westmoreland assured him that the United States could prevail, but Westmoreland also kept asking for authorization to carry the ground war into Laos, Cambodia, and North Vietnam. He also requested more in the way of American manpower and materiel. In April 1967 he called for 95,000 more soldiers, which would bring the total to 565,000. He added that another 100,000 would be needed in fiscal 1969. He told Johnson that "last month we reached the crossover point. In areas excluding the two northern provinces, attrition [of the enemy] will be greater than additions to the force." Johnson, however, opposed extending the ground war, and he did not dare announce expansion of troop numbers on such a scale. "Where does it all end?" he replied. "When we add divisions, can't the enemy add divisions? If so, where does it all end?" Westmoreland had no good answer to this question.102

99. Berman, *Lyndon Johnson's War*, 183; Kearns, *Lyndon Johnson*, 313–16.
100. Berman, "LBJ: Paths Chosen," 144–45.
101. Kearns, *Lyndon Johnson*, 319–20; Conkin, *Big Daddy*, 185–87.
102. Berman, "LBJ: Paths Chosen," 155; Herring, *America's Longest War*, 176–79.

Westmoreland's reports tended to be overoptimistic. In June he estimated enemy capability at 298,000 men, compared to CIA numbers that
ranged between 460,000 and 570,000. "Westy's" estimates excluded
"self-defense" forces of 200,000 or more that the NLF could deploy if
necessary.[103] These numbers were too much for McNamara, who had
become increasingly skeptical of upbeat reports from the field. He then
authorized a wide-ranging review of the war that later became known as
the Pentagon Papers. Even Westmoreland admitted that defining the size
of the opposition depended on how one counted the foe. Guessing the
size of enemy manpower, he conceded, was like "trying to estimate
roaches in your kitchen."[104]

Most advisers, moreover, knew that Johnson detested pessimistic reports, and they shielded him from bad news. The discouraging CIA
estimates stopped at the desk of Wheeler, Chairman of the Joint Chiefs of
Staff. Wheeler was especially frightened that requests such as Westmoreland's might somehow leak to the media. He told Westmoreland, "If
these figures should reach the public domain, they would, literally, blow
the lid off of Washington. Please do whatever is necessary to insure that
these figures are not repeat not related to News Media or otherwise exposed to public knowledge."[105]

For political reasons Johnson dared not adopt the more expansive ideas
of military advisers in mid-1967. In August he limited troop increases to
45,000. These would bring the total to around 515,000. But the President
was hardly deceived by being shielded from bad news, for by then he had
studied hundreds and hundreds of documents and reports. He had "big
ears" for intelligence. Optimistic reports from the field notwithstanding,
he knew full well that the war was going badly for the United States and
South Vietnam. The bombing was not deterring the enemy; missions to
search and destroy, and pacification, were not working; Communists
controlled vast areas of the South Vietnamese countryside.

The President nonetheless stayed his course. Having approved escalation for nearly three years, he refused to believe that the South could not
be saved. Summoning Westmoreland to Washington in November 1967,
he set in motion a blitz of happy-sounding publicity. Coached by LBJ,
Westmoreland gave the press what was by then a tired cliché: there was
"light at the end of the tunnel." He added, "I am absolutely certain that
whereas in 1965 the enemy was winning, today he is certainly losing.

103. Berman, Lyndon Johnson's War, 38–39.
104. Berman, "LBJ: Paths Chosen," 155–60.
105. Berman, Lyndon Johnson's War, 27–30.

There are indications that the Vietcong and even Hanoi know this. . . .
It is significant that the enemy has not won a battle in more than a
year."[106]

The public relations blitz of November 1967 was characteristic Johnson: bold, carefully orchestrated, optimistic, grand in its expectations.
Although he agonized in private, he would not give in. Like the Texans
who had fought at the Alamo, he would stand tall in defense of his beliefs.

106. Ibid., 116.

21

Rights, Polarization, and Backlash, 1966–1967

Although the war in Vietnam shook the optimism of liberals who had imagined that a New Frontier and a Great Society would transform the United States, it was but one of many sources of backlash, as it came to be called, that gathered strength after 1965 and polarized the nation. The roots of the backlash, while varied, sprang mainly from divisions of race and class. They were so deep that they would almost surely have fragmented the nation even in the absence of military escalation.

The extraordinary affluence of postwar American society, a blessing to millions of people, did much to exacerbate these divisions. By 1967-68, when the economy peaked, it had worked wonders in living standards. Even more than earlier, people took for granted conveniences that would have seemed luxurious to earlier generations. Millions of Americans in the ever-larger middle classes now assumed that their children could go to a college or a university and fare better than they had. They also imagined that modern science and technology would radically reduce suffering and endow their lives with comfort. It was only a matter of time, they were certain, before the United States, the world's greatest nation, would place a man on the moon.

Rising living standards, however, also expanded expectations, particularly among the huge and self-assertive cohorts of young people—the baby boom generation—who had grown up in the abundance of the postwar world. By the late 1960s these expectations also affected Americans of less favored socio-economic backgrounds. Blacks, having achieved legal protection under the civil rights legislation of 1964 and 1965, were quick to demand social and economic equality. Poor people, including welfare recipients, insisted on their "right" to better benefits. Ethnic groups, notably Mexican- and Native Americans, grew increasingly self-conscious and, like blacks, turned to direct action to achieve their goals. Women, too, raised the banner of equal rights. In 1966 expectant leaders formed the National Organization for Women.

Most of these groups turned eagerly to government, especially Washington, for redress of their grievances. Like liberals, whom many protestors came to disdain, they hoped that they could influence the system and expand the Great Society. Their expectations, however, had grown grand indeed, and they were impatient. Increasingly, they sought not only benefits but also guarantees and entitlements. The rise of rights-consciousness, having flourished in the early and mid-1960s, became central to the culture by 1970.

The Rights Revolution received special inspiration from the civil rights movement, which in turn had drawn on one of the most enduring elements of the American creed: belief in the equal opportunity of individuals.[1] As passage of the civil rights laws had suggested, large majorities of people supported this ideal, and they persisted in doing so amid the backlash that ensued: the laws continued to matter. As events after 1965 showed, however, Americans were much less sympathetic when people demanded the "right" to *social* equality or special entitlements for *groups*. That was taking the Rights Revolution too far.

Those who resisted the expectant demands of rights-conscious groups in the mid- and later 1960s controlled many of the levers of power, including Congress, in the United States. Some championed "rights" of their own, including the right to run their lives (or their businesses) without government regulation. As LBJ and liberals lost influence after 1965, these conservatives won most of the battles. In so doing they frustrated the new claimants, driving them to new levels of fury and sometimes to more confrontational forms of protest. These protests, in turn, fanned a backlash that whipped out at a wide variety of targets, including

1. William Galston, "Practical Philosophy and the Bill of Rights: Perspectives on Some Contemporary Issues," in Michael Lacey and Kurt Haakonssen, eds., *A Culture of Rights: The Bill of Rights in Philosophy, Politics, and Law* (New York, 1991). An important philosophical defense of rights, drawing on the ferment of the 1960s, is John Rawls, *A Theory of Justice* (Princeton, 1971).

anti-war demonstrators, the New Left, "welfare bums," and "uppity" blacks. By 1968 life in the United States had grown far more contentious than it had been in the hopeful days of the Great Society. Fragmentation and polarization, much of it displayed on TV, seemed to endanger the give-and-take of political compromise and to threaten the social stability of the nation.

AMID THIS ONGOING POLARIZATION liberals (and others) took heart from a few developments in 1966 and 1967. One, of course, was the continuing advance of prosperity. To be sure, there were problems to worry about, including widespread poverty among minority groups, agricultural workers, and female-headed families. Moreover, early seeds of inflation, fueled by the boosts in military spending, began to worry economists.[2] Still, the economy seemed extraordinarily healthy in these years. High levels of productivity continued, bolstering uneven but real economic growth. Unemployment overall fell from 4.5 percent in 1965 to 3.5 percent in 1969, the lowest it had been since the Korean War.[3] Income taxes, having been reduced by the tax cut of 1964, remained relatively moderate, considerably lower than they were in other industrialized nations.[4] Inequality of personal assets actually declined slightly.[5] Notwithstanding the concerns among economists—concerns that turned out to be justified in the 1970s—most Americans remained confident during the Johnson years about the future performance of the economy. This confidence did much to inflate already substantial expectations.[6]

Advocates of liberal programs in these years observed with satisfaction that the federal government managed to sustain spending on human resources even amid the escalation in Vietnam. Indeed, such expendi-

2. Paul Kennedy, *The Rise and Fall of the Great Powers: Economic Change and Military Conflict from 1500 to 2000* (New York, 1987), 384; Alan Wolfe, *America's Impasse: the Rise and Fall of the Politics of Growth* (New York, 1981), 161; David Calleo, *The Imperious Economy* (Cambridge, Mass., 1982), 201–16; Godfrey Hodgson, *America in Our Time* (Garden City, N.Y., 1976), 245–54. See chapters 23 and 25 for discussion of these economic problems.
3. Calleo, *Imperious Economy*, 201–10.
4. John Witte, *The Politics and Development of the Federal Income Tax* (Madison, 1985).
5. Carol Shammas, "A New Look at Long-Term Trends in Wealth Inequality in the United States," *American Historical Review*, 98 (April 1993), 412–31.
6. Paul Conkin, *Big Daddy from the Pedernales: Lyndon Baines Johnson* (Boston, 1986), 204-6; Doris Kearns, *Lyndon Johnson and the American Dream* (New York, 1976), 297.

tures, while small compared to defense spending, rose substantially between 1965 and 1970, and much more rapidly than in the New Frontier–Great Society years between 1961 and 1965.[7] As a percentage of gross domestic product they increased from 5.2 percent in 1960 to 5.4 percent in 1965 to 7.7 percent in 1970. The United States, it seemed, had the wealth to spend for butter (though not so much as reformers would have liked) while also handing out very large sums for guns.

Liberals rejoiced especially at the performance of the Supreme Court, which continued as a bulwark of support for the Rights Revolution until Chief Justice Warren's retirement in 1969. In 1966 it endorsed efforts to bring federal might into play in prosecuting the people who had murdered Chaney, Schwerner, and Goodman in 1964; in 1967 it unanimously struck down a Virginia law that had banned interracial marriage.[8] Civil libertarians especially hailed the majority opinion of *Miranda v. Arizona* (1966), in which the Court extended, 5 to 4, earlier decisions that had broadened the rights of criminal suspects.[9] In a voice charged with emotion Warren declared from the bench that an alleged criminal "must be warned that he has a right to remain silent, that any statement he does make may be used in evidence against him, and that he has a right to the presence of an attorney, either retained or appointed." Police might interrogate an accused who waived these rights, but even then the suspect might stop the questioning by demanding to see a lawyer or invoking his or her constitutional rights.

As it turned out *Miranda* did not much change the practices of law enforcement: police and prosecutors managed to figure out ways of maneuvering around the decision. At the time, however, the decision aroused heated controversy, for "crime in the streets" was becoming an

7. "Human resources," as defined by the government, include such matters as Social Security, income security (such as AFDC and food stamps), Medicare and Medicaid, public funding for education and training, and benefits for veterans. The sums expended for these purposes increased in current dollars from $26.2 billion in 1960 to $36.6 billion in 1965 to $75.3 billion in 1970 (and to $173.2 billion in 1975). (Most of these increases were for Social Security and Medicare, not for means-tested programs for the poor.) Defense spending rose from $48.1 billion in 1960 to $50.6 billion in 1965, jumped to $81.7 billion in 1970 (and, after dropping a little in the early 1970s, increased slightly to $86.5 billion in 1975). *Statistical Abstract of the United States, 1994* (Washington, 1994), 330–32. See also chapters 18, 23, and 25.
8. *Loving v. Virginia*, 388 U.S. 1 (1967).
9. 384 U.S. 436; John Blum, *Years of Discord: American Politics and Society, 1961–1974* (New York, 1991), 212–13.

inflammatory national issue. Enraged conservatives began to mobilize into increasingly self-conscious lobbies. Justice John Marshall Harlan pounded the table in registering dissent. "This doctrine has . . . has no sanction, no sanction. . . . It's obviously going to mean a gradual disappearance of confessions as a legitimate tool of law enforcement." Justice Byron White, a Kennedy appointee, added, "In some unknown number of cases the Court's rule will return a killer, a rapist, or other criminal to the streets . . . to repeat his crime whenever it pleases him." The executive director of the International Association of Chiefs of Police grumbled, "I guess now we'll have to supply all squad cars with attorneys."[10]

These important decisions pointed to a key trend that was to accelerate in coming years: the tendency of people to join groups in defense of their rights and seek legal redress in order to advance them. More and more, Americans turned to litigation. The legal profession, both guiding and reflecting this trend, grew rapidly in size and attracted many of the nation's most ambitious and rights-conscious college graduates. Its most sucessful practitioners joined a "new class" of highly trained, well-paid "experts"—technocrats, consultants, medical specialists, scientists—that flourished in these dynamic years. Many of the lawyers, stimulated by the example of the highest court in the land, trumpeted the seductive language of rights to advance the claims of their clients.

Some of these lawyers undertook liberal causes, volunteering as counsel for aggrieved and oppressed people, often under the auspices of the Legal Services programs of the OEO. In the process they widened the rights of welfare recipients, assisted minorities seeking fair housing, and abetted the efforts of the poor. Much of the support at the top for the Rights Revolution after the mid-1960s came not from legislation—Congress balked at further moves—but from the courts and from government bureaucracies composed of liberal activists, many of them in agencies expanded by such Great Society landmarks as the OEO and the aid to education and civil rights acts. Then and later, opponents of these activists complained bitterly that the courts and federal agencies were overriding a more conservative popular will and giving unreasonably favorable political treatment to adamant groups, especially blacks.[11]

The drive for rights proved particularly compelling in Washington,

10. *Newsweek*, June 27, 1966, pp. 21–22.
11. Paul Gewirtz, "Discrimination Endgame," *New Republic*, Aug. 12, 1991, pp. 18–23; Nathan Glazer, *Affirmative Discrimination: Ethnic Inequality and Public Policy* (New York, 1975).

where newly created bureaucracies such as the Office of Federal Contract Compliance and the Equal Employment Opportunity Commission became increasingly responsive to protest groups. New notions about "affirmative action" began slowly to secure a hold in the mid-1960s. In September 1965 LBJ issued Executive Order No. 11246, which (as amended to include sex in 1967) stated that contractors should "take affirmative action to ensure that applicants are employed, and that employees are treated during employment, without regard to their race, color, religion, sex, or national origin." In May 1968 Johnson's Department of Labor elaborated on this order, requiring contractors to prepare ethnic censuses of potential workers and to develop "specific goals and timetables for the prompt achievement of full and equal employment opportunity." The department designated groups to be counted as "Negroes," "Orientals," "American Indians," and "Spanish Americans." These were guidelines that still focused on enhancing employment opportunity for disadvantaged individuals. They did not establish quotas for groups, and they were yet to be implemented. Still, a trend toward governmental enforcement of equal rights seemed clear. Many liberals were encouraged by it; others, foreseeing "reverse discrimination," grew alarmed.[12]

THE RISE OF ORGANIZED MOVEMENTS among previously marginalized groups was indeed contagious in these years. One was composed of farm workers, especially in California. In 1965 César Chávez, a thirty-eight-year old Mexican-American who had picked crops most of his life, took the lead in merging the predominantly Mexican and Mexican-American membership of the National Farm Workers Union with Filipino workers to create the United Farm Workers Organizing Committee. Chávez, a frail, shy, and almost monastic man, possessed a charisma that gave him legendary status as a leader. Uniting workers as never before, the UFW joined a strike of grape-pickers in Delano, California, in 1965, thereby setting off a much wider action of workers, La Huelga ("the strike"), throughout the San Joaquin Valley. La Huelga, dramatized na-

12. Diane Ravitch, *The Troubled Crusade: American Education, 1945–1980* (New York, 1983), 282–84; Robert Weisbrot, *Freedom Bound: A History of America's Civil Rights Movement* (New York, 1990), 292–95; Hugh Davis Graham, *The Civil Rights Era: Origins and Development of National Policy, 1960–1972* (New York, 1990), 469–70, 550–65; Graham, "Race, History, and Policy: African Americans and Civil Rights Since 1964," *Journal of Policy History*, 6 (1994), 12–39; and Paul Burstein, *Discrimination, Jobs, and Politics: The Struggle for Equal Employment in the United States Since the New Deal* (Chicago, 1985).

tionally by a boycott in 1968 against producers of table grapes in the valley, lasted for five years. At the peak of the boycott some 17 million Americans stopped buying grapes, thereby costing the growers millions of dollars. In 1970 the growers agreed at last to sign union contracts.[13]

The contagion of rights-consciousness especially attracted women, who grew more politically engaged than they had been since the achievement of women's suffrage in 1920. Their restlessness had begun to flourish openly in 1963, thanks in part to the publication in that year of Betty Friedan's highly popular *The Feminine Mystique*. Friedan had graduated Phi Beta Kappa from Smith College in 1942 and had been bent upon going to graduate school, but instead became a wife and mother of three in suburban New York. Increasingly unhappy, she discovered at her Smith reunion in 1957 that many of her classmates shared her discontent. The home, she wrote later, had become a "comfortable concentration camp." Friedan did not use the word "feminist" in the book, writing instead about the "problem that has no name." This was the sense of frustration and lack of fulfillment that some well-educated, middle-class women such as herself had experienced as round-the-clock housewives.[14]

Friedan said little that was new, but she wrote effectively out of personal experience, and hundreds of thousands of women identified with her. Her book came at an especially opportune time. Kennedy's Commission on the Status of Women was already drawing attention to some of the problems that Friedan complained about. After concluding its deliberations in October 1963, the commission issued some 83,000 copies of its final report in the next year.[15] More important, Friedan's book appealed because it reflected two of the most powerful social trends of the era. One

13. Mark Day, *César Chávez and the Farm Workers* (New York, 1971); Juan Gonzales, *Mexican and Mexican-American Farm Workers: The California Agricultural Industry* (New York, 1985). The victory in 1970, however, was a high point for the union and for Chávez, whom some union leaders considered authoritarian and inefficient. Many other developments, including grower opposition, mechanization, and rising immigration, further damaged union efforts. By the late 1980s fewer than 10 percent of the grapes in the Delano area were harvested by members of the UFW. See *New York Times*, April 24, 1993.

14. Carl Degler, *At Odds: Women and the Family in America from the Revolution to the Present* (New York, 1980), 443; William Chafe, *The Paradox of Change: American Women in the 20th Century* (New York, 1991), 195–97; Arlene Skolnick, *Embattled Paradise: The American Family in an Age of Uncertainty* (New York, 1991), 114–15; David Halberstam, "Discovering Sex," *American Heritage*, May/June 1993, pp. 39–58; and Cynthia Harrison, *On Account of Sex: The Politics of Women's Issues, 1945–1968* (Berkeley, 1988), 172.

15. Harrison, *On Account of Sex*, 172.

was the ever-increasing number of women—many of them housewives—who were working outside the home. In 1963, 36.1 percent of American women over the age of 14, or 24.7 million people, were in the labor force. This was an increase of 6.3 million over 1950.[16] More than three-fifths of them, or 15.4 million, were married. The other current was the civil rights movement. It sensitized millions of Americans—men as well as women—to inequality in the United States.[17]

The optimistic, prosperous spirit of the time further advanced Friedan's message. Relatively few of the women who responded to *The Feminine Mystique*, surveys suggested, were from minority groups or the blue-collar classes. Many of these people, after all, had always worked outside the home, ordinarily in low-paid and sex-segregated jobs, and they found little that was liberating in her talk about careers. Those among them who were mothers of small children knew they could never afford day care that would enable them to seek employment with a future. For these and other reasons the women's movement after 1963 was slow to attract wives and mothers in the working classes. But many middle-class, well-educated women responded eagerly to Friedan's book. Although they had not yet developed a collective, strongly feminist view of their world, they sensed that there was more to life than cleaning house and bringing up children.

Moreover, women in general were developing increasingly high expectations, which they were passing on to their children. Public opinion polls revealed how rapidly these expectations expanded during the 1960s. A Gallup survey in 1962 indicated that only about one-third of American women considered themselves victims of discrimination. Eight years later the proportion had risen to a half, and by 1974 to two-thirds. By any standard these were striking measures of social and cultural change.[18]

The sense of being discriminated against grew especially acute among young women who worked in SDS or in civil rights groups. Men in these groups may have been a little more sensitive to the rights of women than were men in American society at large. But they, too, could be domineering, patronizing, and chauvinistic. When Stokely Carmichael, a leader of

16. In 1950, 31.4 percent of American women over 14 had been in the labor force. These trends accelerated in the mid- and late 1960s: by 1970 there were 31.2 million American women in the work force, or 42.6 percent of women over 16.

17. Skolnick, *Embattled Paradise*, 72–73, 106–10; Jo Freeman, *The Politics of Women's Liberation: A Case Study of an Emergent Social Movement and Its Relation to the Policy Process* (New York, 1975); Landon Jones, *Great Expectations: America and the Baby Boom Generation* (New York, 1980), 168–72.

18. Chafe, *Paradox*, 211.

SNCC, was asked what should be the role of women in the civil rights movement, he replied (in jest), "The only position for women in SNCC is prone." Anti-war activists opposed the draft with the slogan "Girls say yes to guys who say no." Women were one-third of SDS membership but had only 6 percent of the seats on the SDS executive council in 1964. At the SDS convention in 1965 a male delegate assessed the situation. Women, he said, "made peanut butter, waited on tables, cleaned up, got laid. That was their role." When a woman arose at the gathering to discuss the "women's issue," she was greeted with catcalls. One delegate shouted, "She just needs a good screw."[19]

By then some of the female activists began to react in outrage. As one woman later complained of the men who called themselves radicals, "They had all this empathy for the Vietnamese, and for black Americans, but they didn't have much empathy for the women in their lives; not the women they slept with, not the women they shared office space with, not the women they fought at demonstrations with. So our first anger and fury was directed against the men of the left."[20] Two of these angry women, Casey Hayden (Tom Hayden's wife) and Mary King of SNCC, reacted by writing and circulating widely a memo on "Sex and Caste" in late 1965. It drew a parallel between the subordination of blacks and of women. Both groups, they wrote, "seem to be caught in a common-law caste system that operates, sometimes subtly, forcing them to work around or outside hierarchical structures of power that may exclude them. . . . It is a caste system which, at its worst, uses and exploits women."[21]

The memorandum of King and Hayden did not do much to change males in the anti-war and civil rights movements. Attitudes about gender roles die hard. But growing numbers of many young women on the left came to share their anger. By 1966 and 1967 some of them were calling not only for equal rights but also for "women's liberation." Joining "consciousness-raising" groups, they shared with other women their understanding of sexual exploitation and injustice. They read and circulated memoranda such as "The Myth of the Vaginal Orgasm" and "The Per-

19. William Chafe, *The Unfinished Journey: America Since World War II* (New York, 1991), 334; John D'Emilio and Estelle Freedman, *Intimate Matters: A History of Sexuality in America* (New York, 1988), 311.
20. Maurice Isserman, "The Not-So-Dark and Bloody Ground: New Works on the 1960s," *American Historical Review*, 94 (Oct. 1989), 1000–1001.
21. Sara Evans, *Personal Politics: The Roots of Women's Liberation in the Civil Rights Movement and the New Left* (New York, 1979), 97–101; Chafe, *Paradox*, 198; Todd Gitlin, *The Sixties: Years of Hope, Days of Rage* (New York, 1987), 365–70.

sonal Is Political." They were a vanguard in support of a "sexual politics" that was to demand reproductive freedom, the legalization of abortion, changes in family dynamics, and—in time—lesbian rights. The rise of women's liberation exposed as clearly as any development of the mid-1960s the tendency of protestors to assert the rights of groups and to employ the language of rights and entitlements.[22]

Most of the women who led the drive for women's rights in the mid-1960s, however, were not primarily advocates of sexual liberation. Rather, they were middle-aged political leaders like Congresswoman Martha Griffiths of Michigan and professional women like Friedan who demanded legal equality for the sexes. Their primary goal was to get the EEOC to enforce Title VII of the 1964 Civil Rights Act outlawing sexual discrimination in employment. Friedan, called to Washington to assist the cause in June 1966, grew frustrated by the timidity of the commission—some of its members thought Title VII was a joke—and by the lethargy, as she saw it, of other women lobbyists for the cause. At a luncheon she sat down and scribbled on a napkin the words calling for a National Organization for Women.[23]

Four months later, NOW was formed, with Friedan as its first president. It had only a small membership, most of whom were white, middle-class women who advocated legal equality. Then and later the organization had difficulty attracting blue-collar or minority women to its ranks, and it was cool, especially at first, to demands for sexual liberation. Still, it was an assertive organization, calling forthrightly for gender equality. Its statement of purpose emphasized that "a true partnership between the sexes demands a different concept of marriage, an equitable sharing of the responsibilities of home and children and of the economic burdens of their support."

NOW also managed to be heard. It was in part its lobbying that induced Johnson to add "sex" in October 1967 to the phrase "race, creed, color, or national origin" that he had used in 1965 in his Executive Order 11246 against discrimination in employment. NOW also endorsed the Equal

22. Alice Echols, *Daring to Be Bad: Radical Feminism in America, 1967–1975* (Minneapolis, 1989); Echols, "Nothing Distant About It: Women's Liberation and Sixties Radicalism," in David Farber, ed., *The Sixties: From Memory to History* (Chapel Hill, 1994), 149–74. Contemporary feminist manifestos include Kate Millett, *Sexual Politics* (New York, 1969); and Germaine Greer, *The Female Eunuch* (New York, 1971).
23. Rosalind Rosenberg, *Divided Lives: American Women in the Twentieth Century* (New York, 1992), 188–90; Harrison, *On Account of Sex*, 189–95, 207–8.

Rights Amendment to the Constitution, thereby jettisoning the long-held focus of female lobbyists on protective legislation. In the new climate of equal rights, the courts gradually struck these protective laws down in coming years. In 1968, after further lobbying, NOW and other advocates even succeeded in reaching the EEOC. Before then the commission had insisted that complaints about discrimination first be heard by the courts. Henceforth, it now decided, it would agree to hear them itself. The EEOC also opposed airline regulations that had authorized the firing of stewardesses who married or reached the age of 32, as well as want ads that appeared in sex-segregated newspaper columns.[24]

Thanks to the activity of NOW and other new women's organizations after 1966, the media gave great attention to the rise of "feminism" in the United States. From its modest beginnings in the mid-1960s the women's movement, though hampered as before by racial and class divisions within its own ranks, grew to establish one of the most durable legacies of the decade. The rise of the women's groups, moreover, exposed with special clarity the proliferation of well-organized lobbies in American politics, as well as the focus of these lobbies on legal action and administrative procedures. Much of the pressure for equal rights in America after the mid-1960s emanated from the courtrooms and the offices of the government bureaucracy.

That this was so, however, also revealed the political limitations of organizations such as NOW. Like many other pressure groups for expanded rights in the mid- and late 1960s, NOW found Congress cool to its efforts: only in late 1970 did ERA escape the House Judiciary Committee. Congressional resistance, in turn, reflected wider popular doubts, among women as well as among men, about many feminist causes. Polls indicated that while women were growing increasingly sensitive about gender discrimination, only a small minority of liked to be called "feminists." The majority of housewives, indeed, told pollsters that they were largely content with their lives. Many resented being told by "elitists" that raising families was boring. Other women questioned the emphasis that NOW placed on the satisfactions of working outside the home. Most American women in the 1960s, as earlier, had been socialized in traditional ways. They anticipated marrying, having children, and, if necessary to support their families, working. For this reason—not to compete with men— millions did take jobs outside the home. But many of them had derived little satisfaction from the jobs that were open to them, which notwith-

24. Harrison, *On Account of Sex*, 201–8.

standing the efforts of feminists and others continued to be low-paid and sex-segregated, and they often returned to the home if family budgets so permitted.[25]

Attitudes such as these indicated that powerful cultural continuities persisted in the United States amid the turmoil of change in the 1960s. Activists for sexual liberation encountered especially fierce resistance from traditionalists. But even those women, like the leaders of NOW, who concentrated on legal rights, faced formidable opposition. Frustrated, some of them grew confrontational and assailed the "sexism"— soon to be a rallying cry—of American men, indeed of American society. The "sexists," bewildered and hurt, shot right back. The war of words expanded, firing up the thunderous language of rights and sparking in time an explosion of litigation. By the late 1960s a "battle of the sexes" had become one of many polarizing struggles that were rending American society.

WHATEVER SATISFACTIONS that rights-conscious activists derived from the Warren Court, the grape-pickers' strike, and the women's movement were shaken by the blows that battered Johnson's Great Society in the last three years of his presidency.[26] LBJ tried to move ahead on the domestic front, especially in 1966, when he called on Congress to approve a sizeable agenda. It included truth-in-packaging and truth-in-lending legislation to benefit consumers, an increase in the minimum wage, and laws to improve safety in mines and on the highways. Other goals were equally important: improving child nutrition, cleaning up rivers, reforming the bail system, providing rent supplements for the poor, increasing support for urban mass transit, and expanding the availability of low-cost housing. Johnson sought especially to stop the racially discriminatory practices of developers and landlords. The list of his objectives highlighted a special concern of liberals in mid-decade: to improve the quality of life in the cities.[27]

Congress approved a few of these measures in 1966, finishing its work in October with passage of what LBJ and other liberals hailed as the Model Cities Act. In the next two years the act offered more than $900

25. Chafe, *Paradox*, 156, 196; Richard Easterlin, *Birth and Fortune: The Impact of Numbers on Personal Welfare* (New York, 1980), 60–61, 148–50.
26. Sidney Milkis, *The President and the Parties: The Transformation of the American Party System Since the New Deal* (New York, 1993), 195–218.
27. Joseph Califano, *The Triumph and Tragedy of Lyndon Johnson: The White House Years* (New York, 1991), 118–36, 257–59; Conkin, *Big Daddy*, 231–34; Wolfe, *America's Impasse*, 102–4.

million in federal matching grants (80 percent federal, 20 percent local) to cities for a range of programs aimed at improving housing, education, health care, crime prevention, and recreation. Congress, however, refused to approve an "open housing" measure that Johnson had urged to combat discrimination in the real estate market. And the effect of laws that did pass was modest. The Model Cities Act, the most highly touted measure of the session, aided some people, including upwardly mobile blacks who found jobs in the expanding bureaucracy necessary to operate the programs. [28] But, like urban renewal legislation in the past, it did little for the poor. Some of the projects amounted to little more than pork barrel efforts that shifted federal money to Democratic urban machines. Others were badly managed. [29]

So it was that the 1966 session turned out to be at most a feeble last hurrah for liberalism on Capitol Hill. Throughout the session conservatives harped on the flaws of Great Society laws passed earlier, notably those that were part of the war on poverty. [30] They especially assailed the Warren Court, which they blamed for a rise of crime and racial unrest in the cities. Gerald Ford of Michigan, the House Republican leader, asked, "How long are we going to abdicate law and order—the backbone of any civilization—in favor of a soft social theory that the man who heaves a brick through your window or tosses a fire bomb into your car is simply the misunderstood and underprivileged product of a broken home?" [31]

Advocates of rights for the disadvantaged scored a few victories in the 1966 elections, including the triumph of Walter Mondale, a young liberal who replaced Humphrey in the Senate. Massachusetts voters sent to the Senate Edward Brooke, a moderately liberal Republican who became the first black person in American history to be popularly elected to the

28. The growth of government in the 1960s was an important source of employment in general. Paid civilian employment of the federal government increased from 2.4 million in 1960 to 2.53 million in 1965 to 2.98 million in 1970 (remaining near that level in the 1970s). These numbers had been 3.81 million in 1945, the last year of World War II; had dropped to 1.91 million by 1950; and then had risen to 2.4 million by 1955. State and local government employment increased more steadily and more rapidly, from 3.2 million in 1945, to 4.3 million in 1950, to 5.1 million in 1955, to 6.4 million in 1960, to 8 million in 1965, and to 10.1 million in 1970.

29. Jon Teaford, *The Twentieth-Century American City: Problem, Promise, and Reality* (Baltimore, 1986), 136–38; Conkin, *Big Daddy*, 231–32.

30. See Daniel Moynihan, *Maximum Feasible Misunderstanding: Community Action in the War Against Poverty* (New York, 1969), 129–38.

31. James Sundquist, *Politics and Policy: The Eisenhower, Kennedy, and Johnson Years* (Washington, 1968), 285.

upper chamber.[32] But conservatives did better. Californians chose Ronald Reagan governor over two-term incumbent Edmund "Pat" Brown, conqueror of Nixon in 1962. Alabamans, prevented by law from re-electing Wallace, installed his wife Lurleen instead: Wallace remained governor in fact. Lester Maddox, a vehement racist who had gained notoriety for banning blacks from his restaurant, succeeded to the governorship in Georgia. Republicans, most of them conservatives, replaced forty-seven Democratic incumbents in the House and three in the Senate.

A chastened Johnson recognized the conservative handwriting on the wall. His State of the Union message in 1967 was so modest that the columnist James Reston ridiculed it as a "guns and margarine" address. LBJ continued, however, to press for urban reforms, and in 1968 Congress authorized $5.3 billion in federal money over the next three years to subsidize low-cost private housing. It was hoped that the law would lead to the construction of 1.7 million homes or apartments within the next ten years. The housing law did result in an upsurge of building. But many developers, as in the past, managed to evade quality controls and to build shoddy structures. Scandals erupted, draining congressional support. The program was largely suspended by 1970.[33]

LBJ also kept up pressure for his open housing bill. In April 1968 Congress turned around and approved it. The measure was ambitious, banning discriminatory practices affecting 80 percent of housing in the nation. Congress acted, however, only because it was frightened by rioting following the assassination at the time of Martin Luther King. The law generally required complainants to bear the burden of proof and provided for only weak enforcement. Given the widespread refusal of whites to live near blacks, the open housing act, a landmark piece of civil rights legislation on paper, did virtually nothing in practice to promote residential desegregation in the cities.[34]

The 1968 Congress also revealed its continuing fear and anger about

32. *Newsweek*, Nov. 21, 1966, p. 37.
33. Conkin, *Big Daddy*, 233–34.
34. Weisbrot, *Freedom Bound*, 271–72; Graham, "Race, History, and Policy"; Allen Matusow, *The Unraveling of America: A History of Liberalism in the 1960s* (New York, 1984), 206–8; Steven Lawson, "Civil Rights," in Robert Divine, ed., *Exploring the Johnson Years* (Austin, 1981), 93–125. Other sections sought to help American Indians, by prohibiting tribal governments from making or enforcing laws that violated specified constitutional rights and by prohibiting states from assuming civil or criminal jurisdiction over Indian areas without the consent of the tribes affected. See chapter 22 for discussion of King's assassination and social turmoil in 1968.

urban unrest, which had mushroomed since 1965. The open housing law contained tough provisions against people who crossed state lines with intent to incite riots: penalties of up to five years in jail and fines of $10,000. Two months later Congress approved an Omnibus Crime Control and Safe Streets Act that called for the spending of slightly over $100 million to upgrade law enforcement. Lawmakers also included in the bill provisions that authorized local and state as well as federal law enforcement officials to engage in wiretapping and bugging in a number of situations. For a while the President considered vetoing the measure, which took aim at the *Miranda* and other Supreme Court decisions. But the bill was highly popular not only on the Hill but also, it seemed, among constituents. Government, people were saying, must get "tough" on crime. In the end Johnson unhappily signed the bill, the last and in many ways the most conservative piece of domestic legislation of his presidency.[35]

Johnson's difficulties with Capitol Hill, striking though they were in contrast to his successes in 1964 and 1965, were nothing new in modern United States political history. All his liberal predecessors, including FDR, had lost battles to pressure groups and to the conservative coalition in Congress. Newly troubling to reformers, however, was the rise of well-articulated doubts about the capacity of government to remedy social problems. These doubts seemed especially strong among a number of once liberal intellectuals and policy-makers who subjected Great Society programs—aid to education, Medicare and Medicaid, above all the war on poverty—to close examination after 1965. Some of these people wrote for *Commentary*, a magazine edited by Norman Podhoretz, who turned from left-of-center in the early 1960s to conservative after 1968. Others published in a new journal, *The Public Interest*, that first appeared in 1965. It attracted some of the nation's best-known intellectuals in the social sciences, including Daniel Moynihan, the sociologists Nathan Glazer, James Wilson, and Daniel Bell, and the political philosopher Irving Kristol.[36]

Some of these writers refused to be typed as "conservative." Bell described himself as a socialist in economics, a liberal in politics, and a conservative only in culture. He thought of himself as a "skeptical Whig," someone who believed in progress but doubted that government could do much to enhance it. But many of the others, notably Kristol and Edward

35. Califano, *Triumph and Tragedy*, 305.
36. Peter Steinfels, *The Neoconservatives: The Men Who Are Changing America's Politics* (New York, 1979).

Banfield, a student of urban problems, emerged as leaders of a "neo-conservative" intellectual surge in the United States. Along with libertarians such as Milton Friedman, an especially influential economist, they pointed with alarm at the dramatic increase in the number of government programs, bureaucracies, and employees that the Kennedy and Johnson administrations had created. Government, they said, was swelling to elephantine proportions. In so doing it was greatly stimulating popular expectations. Yet the gulf between expectation and accomplishment had widened, thereby inciting dangerously high levels of frustration and polarization. Meanwhile, the number of federal regulations had multiplied seemingly beyond belief, threatening to immobilize American institutions in a tangle of red tape.[37]

Intellectuals, of course, have ordinarily had little impact on the course of public policy in the United States. That was true in the mid- and late 1960s as well. Few Americans—or members of Congress—read *The Public Interest*. Still, the rapid rise of the "neo-cons" to intellectual respectability was revealing. And their complaints, especially about the "dead hand of bureaucracy," epitomized a new mood of doubt. After all, the government had obviously oversold its expertise. The war on poverty was at best a skirmish. Worse, the "best and brightest" of liberals had blundered badly in Vietnam. For these reasons conservatives successfully took the offensive in public debate. Liberals, so confident and optimistic just a few years earlier, seemed tired and unsure of themselves. This sudden and largely unexpected development was one of the most lasting legacies of the mid-1960s.[38]

AS THE BATTLES over open housing legislation in Congress revealed, racial conflict continued to be the most divisive issue in American politics and society between 1966 and 1968. During these years "black power" replaced interracialism as a guiding principle in the civil rights

37. Books that display doubts about the virtues of governmental expansion and regulation in the 1960s include Moynihan, *Maximum Feasible*, on the war on poverty; James Wilson, ed., *The Politics of Regulation* (New York, 1980); and Ravitch, *Troubled Crusade*, on education. Ravitch noted, 312–20, that there were ninety-two federal government regulations affecting education in 1965 and almost a thousand in 1977. A Bible of sorts for conservatives is Charles Murray, *Losing Ground: American Social Policy, 1950–1980* (New York, 1984).
38. The trend to the right, of course, greatly assisted the Republican party, which grew much stronger in the United States after 1966. See Jerome Himmelstein, *To the Right: The Transformation of American Conservatism* (Berkeley, 1990), 63–94, for comments on the 1960s.

movement, agitation for racial justice and "rights" mounted in the North, and race riots shattered the cities. Even more than usual, these were turbulent years in postwar race relations.

There was a good deal of irony surrounding this turbulence. Black people in the United States had never gained so much as they had between 1963 and 1966. The civil rights acts had succeeded, at last, in providing for legal equality. Litigation, plus the threat of losing federal aid to education, began to force school districts (save in the Deep South) to admit blacks to white schools: in 1964 only 2 percent of southern blacks attended biracial schools, a percentage that rose to 32 by 1968.[39] Robert Weaver, the first black person to hold a Cabinet position—in newly created HUD (Housing and Urban Development, 1965)—led administration attempts to improve housing opportunities for blacks and the poor. Blacks also made unprecedented use of their political potential. In 1967 voters in Gary elected Richard Hatcher, a black man, as their mayor. Voters in Cleveland chose Carl Stokes, another black. These were the first African-Americans to become big-city mayors in American history. In June of the same year Thurgood Marshall became the first Negro to be confirmed as a justice of the Supreme Court. Blacks even fared better economically, thanks mainly to the overall increase in prosperity (not to the Great Society). Commentators observed happily that a significant black middle class was coming into being. In fact, the United States was on its way to becoming perhaps the least racist white-majority society in the world.[40]

Polls indicated, moreover, that most blacks were optimistic about their personal futures, as well as positive about the virtues of desegregation. Only a minority seemed attracted to radical causes. Millions, indeed, remained deeply attached to their churches, whose memberships far outnumbered those of civil rights organizations. Many of these churches were led by conservatives. Only a very small number of African-Americans, as before, joined or openly identified with the Nation of Islam or

39. Ravitch, *Troubled Crusade*, 167.
40. Bart Landry, *The New Black Middle Class* (Berkeley, 1987), 67–70 estimates that the percentage of blacks in "middle-class" occupations doubled between 1960 and 1970, growing from approximately 13 percent to 25 percent. The percentage of whites in such occupations increased also, but at a slower pace, from 44 percent in 1960 to 50 percent in 1970. His definition of "middle-class" occupations included sales and clerical workers. See also Teaford, *Twentieth-Century American City*, 148–50; James Smith and Finis Welch, *Closing the Gap: Forty Years of Economic Progress for Blacks* (Santa Monica, 1986); Orlando Patterson, "Race, Gender, and Liberal Fantasies," *New York Times*, Oct. 20, 1991.

other militantly black organizations. Martin Luther King, who struggled non-violently for open housing laws in Chicago and other American cities during these years, remained by far the most admired black leader in the country.

The irony, of course, was that for militant blacks, especially the young, these improvements in race relations fell far short of what they were coming to expect. Their expectations whetted by the acquisition of political and legal rights, they demanded social and economic rights: open housing, better schools, decent jobs, everything that middle-class whites were enjoying. Social and technological change further drove these expectations. Thanks in part to the massive south-to-north migrations of blacks in the postwar era, blacks were far less isolated than they had been in the past. Having escaped the rigidities of Jim Crow in the South, they felt a liberating sense of possibility in the North. More than their elders, they could see what they were missing in life. Television especially sharpened the sting of relative deprivation. Available to virtually all people by the mid-1960s, it flashed to viewers the ever more fantastic wonders of the affluent society. The relative deprivation of blacks, moreover, was in some ways growing. While the median family income of black people was increasing more rapidly than that of whites in these years (rising slightly from 57 percent of white income in the late 1940s to 61 percent in 1970), the gap in their *absolute* earnings was widening.

Many black people had other grievances in these years. One was the war in Vietnam, which grew steadily more unpopular with African-Americans as casualties mounted in 1966. Some of the black veterans who finished their tours of duty in "Nam" returned angry at the discrimination that they had experienced in the service and determined to fight at home for justice. "I ain't coming back playing, 'Oh, Say Can You See,'" one such veteran exclaimed. "I'm whistlin' 'Sweet Georgia Brown,' and I got the band."[41]

Black intellectuals, too, stressed that African-Americans must take pride in themselves, their race, and their history. James Baldwin wrote bitterly of the psychological damage that afflicted blacks who internalized the inferiority, thereby hating themselves, that whites ascribed to them. The psychologist Kenneth Clark, whose research had influenced *Brown v. Board of Education*, wrote an introduction in 1966 to a widely read collection of essays, *The Negro American*, in which he warned that whites

41. Richard Polenberg, *One Nation Divisible: Class, Race, and Ethnicity in the United States Since 1938* (New York, 1980), 235–37.

could not be relied on to do much more for the cause of racial justice. "The new American Dilemma," he wrote, "is power." Blacks must take it and act for themselves. "Ideals alone . . . do not bring justice," he said. "Ideals, combined with necessity, may."[42] Some radical black intellectuals began to deny that blacks and whites could ever understand each other. The playwright and essayist LeRoi Jones wrote in 1965 that whites could not appreciate black jazz music. The difference between the white listener and black listener, he said, is "the difference between a man watching someone have an orgasm and someone having an orgasm."[43] In 1968 Jones changed his name to Imamu Amiri Baraka. Like others of his persuasion, he looked to Africa as a dominant and positive source of black cultural forms in the United States. An impassioned chauvinist, he also celebrated the superior capacities of African-American people.

With racial polarization developing, it required only an incident to undermine interracialism in the civil rights movement. That happened when James Meredith, the loner who had tried to integrate Ole Miss in 1962, determined in June 1966 to make a 220-mile pilgrimage from Memphis, Tennessee, to Jackson, Mississippi. Black people, Meredith hoped, would be inspired and emboldened to register to vote.[44] Two days into his trek, however, a white man in Hernando, Mississippi, pumped three rounds of birdshot from a 16-gauge automatic shotgun at Meredith and a handful of fellow marchers on the highway. Meredith fell bleeding to the ground and was rushed to a hospital in Memphis.

When Meredith started his march, few people had paid much attention. No civil rights organizations were involved. The shooting changed all that. Top civil rights leaders—King, Floyd McKissick, who had taken over in January from James Farmer as head of CORE, and Stokely Carmichael, who had replaced John Lewis in May as the leader of SNCC— quickly determined to resume the march. Meredith, whose injuries proved to be superficial, was a little bewildered and worried about what these leaders were proposing to do. But he gave his assent to carry on the

42. For Baldwin, see Morris Dickstein, *Gates of Eden: American Culture in the Sixties* (New York, 1977), 169. For Clark, see Talcott Parsons and Kenneth Clark, eds., *The Negro American* (Boston, 1965), xviii. See especially Walter Jackson, *Gunnar Myrdal and America's Conscience: Social Engineering and Racial Liberalism, 1938–1987* (Chapel Hill, 1987), 305–7.
43. Werner Sollors, "Of Mules and Mares in a Land of Difference; or, Quadrupeds All?" *American Quarterly*, 42 (June 1990), 183–84.
44. *Newsweek*, June 20, 1966, pp. 27–31; John Dittmer, *Local People: The Struggle for Civil Rights in Mississippi* (Urbana, 1994), 389–407.

march. Without the benefit of planning aforethought, a dramatic new protest—the first on such a scale since Selma fifteen months before—was thrown into high gear.

For the next ten days the marchers, numbering between 30 and 250, walked toward Jackson without major incident. But in Greenwood, police arrested Carmichael and two other SNCC workers on charges of violating a local ordinance against putting up tents on the grounds of a local black school. Carmichael, a West Indian who had gone to Howard University, was a proud and fiery young man. While he had deferred to the non-violent philosophy of King, he did not believe in it, and he had already profoundly antagonized Roy Wilkins of the NAACP and Whitney Young of the Urban League, who had come to Memphis to join in strategy sessions. Both leaders angrily returned to New York. Carmichael had also demanded that whites be excluded from the march, backing off only under pressure from King. When Carmichael got out of jail on bail, he seized the occasion by firing up an agitated crowd of 600 at the scene.

His speech became a milestone on the road away from interracial cooperation in the civil rights movement. "This is the twenty-seventh time I have been arrested," he announced. "I ain't going to jail no more." Carmichael then shouted five times, "We want black power!" Each time the crowd cheered more enthusiastically, and Carmichael warmed to the response. "Every courthouse in Mississippi," he shouted, "ought to be burned tomorrow to get rid of the dirt." Repeatedly he asked his listeners, "What do you want?" The crowd called back, louder each time, "Black power! Black power! BLACK POWER!"[45]

Carmichael's call was neither spontaneous nor new in the long history of black protest in the United States. The trend toward black direction of civil rights activities, even if it excluded whites, had been relentless since the travail of the Mississippi Freedom Democratic party at the Democratic National Convention in 1964. SNCC, in choosing Carmichael over Lewis a month before, had signified its readiness for it. Willie Ricks, a SNCC advance man on the march, had tried out the phrase at rallies and urged Carmichael to use it. Even so, the response of blacks in Greenwood, many of them local people, to the slogan of "black power" was striking. Militant blacks thereafter talked about "black power" all the

45. Dittmer, *Local People* 395–96; Weisbrot, *Freedom Bound*, 199–200, estimated the size of the crowd at 3,000 and attributed slightly different language to Carmichael.

time, sometimes to the accompaniment of virulently anti-white rhetoric.[46]

What the phrase actually meant became a matter of sometimes agitated debate in the next few weeks and months.[47] Confusion over the matter was understandable, for "black power" was more a cry of rage than a systematic doctrine. Some who used it were activists who recognized the superior resources of the white-dominated political system. They considered black power a temporary strategy of solidarity similar to that which other once powerless ethnic groups had employed to gain a foothold in pluralistic democratic politics. "Before a group can enter the open society," one manifesto said, "it must first close ranks."[48] Other advocates of black power, however, were radical and/or black nationalists who believed in excluding whites from black institutions and who rejected not only the premises of integration but also white society in general. Carmichael, one of these, also advanced Marxist social ideas from time to time. Still other militants, echoing LeRoi Jones, saw black power as a way to resurrect black pride and highlight African-American culture, which they regarded as more free and less "up-tight" than that of whites.[49]

In essence, however, the quest for black power was a more or less inevitable result of the dynamics of civil rights protest in the 1960s. At its core for most advocates was a not terribly complicated idea. Carmichael explained it in his speech at Greenwood when he said, "We have to do what every group in this country did—we gotta take over the community where we outnumber people so we can have decent jobs."[50] Proponents of black power insisted simply that whites could not be trusted to help them. It followed that black people had to control their own political and economic institutions. If whites felt abandoned, that was too bad. If they grew violent, blacks must be ready to defend themselves.

For these reasons black power posed a challenge to King and other advocates of interracialism and non-violence. They remained committed,

46. Manning Marable, *Race, Reform, and Rebellion: The Second Reconstruction in Black America, 1945–1990* (Jackson, 1991), 86–113; David Colburn and George Pozzetta, "Race, Ethnicity, and the Evolution of Political Legitimacy," in Farber, ed., *Sixties*, 119–48.
47. Stokely Carmichael and Charles Hamilton, *Black Power: The Politics of Liberation in America* (New York, 1967), tried to explain.
48. Ibid., 44.
49. Fred Siegel, *Troubled Journey: From Pearl Harbor to Ronald Reagan* (New York, 1984), 166–68.
50. Dittmer, *Local People*, 397.

as Carmichael was not, to cooperative strategies with white liberals and to a goal of racial integration. A. Philip Randolph condemned black power as a "menace to peace and prosperity." He added, "no Negro who is fighting for civil rights can support black power, which is opposed to civil rights and integration."[51] For a while King considered pulling SCLC, his organization, out of the march. Rejecting such a move, he nonetheless issued a statement saying, "the term 'black power' is unfortunate because it tends to give the impression of black nationalism. . . . black supremacy would be as evil as white supremacy." Wilkins went further, writing in *Life* magazine that black power was "the reverse of Mississippi, a reverse Hitler, and a reverse Ku Klux Klan." Although Wilkins later retracted the statement, it was obvious that Carmichael's approach frightened him. White liberals, too, were upset. Vice-President Humphrey probably spoke for many when he declared, "Racism is racism—and there is no room in America for racism of any color."[52]

The divisions separating Carmichael from King attracted great coverage in the media as the marchers, now more numerous, resumed their trek toward Jackson. Most blacks on the scene were attracted to the slogan of "black power," which in a loose and general way they had been seeking for years. They continued to cheer Carmichael and other militants at rallies. Most also revered King, however. Throngs of people flocked to the roadsides just to have the chance to see him and crowded to his platforms to be able to say that they had touched him. As the leaders struggled for control of the march, many black followers sought to avoid a clear choice between them.

It took ten more days after Greenwood for the marchers to reach Jackson. En route King and twenty others split off to hold in Philadelphia (Mississippi) a second anniversary commemoration of the killings of Chaney, Goodman, and Schwerner. There they were attacked by a group of twenty-five whites with clubs while police and FBI agents looked on. King was aghast, calling Philadelphia "a terrible town, the worst I've seen." Meanwhile, in Canton, state police sought to bar the main body of marchers from pitching their tents on a black schoolground. When the marchers refused to disperse, sixty-odd troopers in full battle gear fired tear gas into the campers, some 2,500 strong. The police then charged in, stomping and gun-butting the men, women, and children who were weeping and blinded by the gas. When the police riot was over, the field

51. William O'Neill, *Coming Apart: An Informal History of America in the 1960s* (Chicago, 1971), 174.
52. Dittmer, *Local People*, 397; Weisbrot, *Freedom Bound*, 200.

resembled a war zone. Dr. Alvin Poussaint, on the scene as a representative of the Medical Committee for Human Rights, set up a makeshift clinic. "We were up that whole night treating the victims," he recalled.[53]

By contrast to these violent events the rest of the march was anticlimactic. Meredith rejoined the marchers, who numbered around 2,000, on the final leg of the trek that ended at the state capitol in Jackson. This was Sunday, June 26, three weeks after Meredith had begun his journey in Tennessee. That afternoon King and others addressed a jubilant throng of 15,000 and proclaimed the march "the greatest demonstration for freedom ever held in the state of Mississippi." Supporters cried, "Freedom now." Carmichael's supporters chanted, "Black power," in competition. Carmichael received big applause when he told the people that blacks "must build a power base in this country so strong that we will bring [whites] to their knees every time they mess with us."[54]

For blacks in Mississippi the march had indeed been a modest success. Some 4,000 blacks had registered to vote, and 10,000 had marched at least part of the way. The violence of whites had further focused liberal attention on racism in the South and revealed that the civil rights acts of 1964 and 1965, while historic, fell far short of protecting black people in Mississippi. Reporters covering the action, however, continued to be virtually obsessed with the meaning of "black power" and with the conflicts between King, Carmichael, and other civil rights leaders. It was clear to them and to many American people that King, while remaining an extraordinarily gifted and much beloved leader, would have to share the stage, as he had at Jackson, with rivals for leadership. The civil rights movement was fragmenting.

The rise of black power seemed contagious in the next few months. In Oakland, California, two young militants, Huey Newton and Bobby Seale, adopted the symbol of a black panther that had been used in 1965 by SNCC workers in Lowndes Country, Alabama. Newton and Seale called their organization, founded following the killing by San Francisco police of an unarmed sixteen-year old black youth, the Black Panther Party for Self-Defense. It set up free health clinics, ran educational programs, and offered free breakfasts for schoolchildren. The emphasis, however, was military. Panthers liked to dress in uniforms of black trousers, light blue shirts, black leather jackets, black berets, and dark sunglasses. They armed themselves with (then legal) unconcealed weapons, formed

53. Dittmer, *Local People*, 399–400.
54. Ibid., 402.

disciplined patrols aimed at "policing the police," and preached a blend of black nationalism and socialism. They were fond of citing Ché Guevara, Ho Chi Minh, and Chairman Mao, who had said, "Power grows out of the barrel of a gun."[55]

Reflecting their Marxist leanings, Panther leaders professed their willingness to work with radical whites. Still, it was mainly a black nationalist organization, one of the most confrontational that whites had yet come across. In May 1967 Newton, Seale, and thirty-odd followers armed themselves with shotguns and M-16 rifles and marched into the California state legislature in Sacramento in order to protest against a bill that would have made it illegal for people to carry unconcealed weapons. As they had anticipated, this dramatic move gave them great play in the media. The Panthers also proved ready to engage in gunplay, which during the next few years killed at least nineteen Panthers in the course of conflicts with police, the FBI, each other, and various black revolutionaries.[56] Newton shot and killed an Oakland police officer in October. Arrested on charges of murder, he was eventually convicted of voluntary manslaughter and spent three years in jail before his conviction was reversed on a judge's error.

The Panthers were not so much a political group as they were angry young men attracted to Third World revolutionary ideology, paramilitary activities, and in some cases to violence. Newton grew fond of ever more high-sounding military titles, including Minister of Defense, Supreme Commander, and Supreme Servant. He and some of his followers later ran an extortion racket. Another top Panther, Minister of Education Eldridge Cleaver, was a convicted rapist who had bragged about his conquests of white women. In early 1968 he and other top Panther leaders were arrested following a shootout with Oakland police.

By 1969 the Panthers had virtually broken up, save in Oakland where they remained a presence. Seale and Newton were in jail, and Cleaver had fled the country to avoid imprisonment. Still, for a while after 1966 they were front-page news. The FBI devoted great resources to campaigns against them of infiltration, harassment, and destabilization. So did police departments: Chicago police in 1969 shot and killed a well-known Panther, Fred Hampton, while Hampton was asleep at his apartment.

55. Blum, *Years of Discord*, 265; Gitlin, *Sixties*, 348–50; Hugh Pearson, *The Shadow of the Panther: Huey Newton and the Price of Black Power in America* (New York, 1994); and William Van Deberg, *New Day in Babylon: The Black Power Movement and American Culture, 1965–1975* (Chicago, 1992).

56. *New York Times*, Jan. 3, 1991.

Perceived as heroic by many young blacks in the ghettos, as well as by the most extreme of white radicals, the Panthers achieved some success in spreading their mission in 1966 and 1967. By 1968 they may have had 5,000 members in a dozen large cities. Their mystique went farther. Cleaver's autobiographical *Soul on Ice*, which appeared in March 1968, received praise from many liberal whites, even though it contained a good deal of hate. A famous poster of Newton, sitting with a spear in one hand and a rifle in another, adorned the rooms of blacks in the ghettos as well as of young radicals in the universities. [57]

While the Panthers attracted an especially large amount of publicity, they were far from the only blacks to react angrily against what they considered the injustices of white America. Many attacked the war. King, an opponent since 1965, escalated his rhetoric against it in early 1967, calling the United States "the greatest purveyor of violence in the world today." Muhammad Ali, the heavyweight boxing champion of the world, announced in 1966, "I ain't got nothing against them Vietcong." Faced with the draft in 1967, he claimed exemption as a Muslim minister and conscientious objector, whereupon he was summarily stripped of his boxing title and prevented from fighting important bouts. He was also convicted of draft evasion and spent the next three years struggling in the courts (ultimately successfully) to reverse a sentence of a $10,000 fine and five years in prison. [58] Carmichael went further, all the way to Hanoi in the summer of 1967. There he announced, "We are not reformers. . . . We want to stop cold the greatest destroyers, . . . American leadership." [59]

Other blacks, including many who were neither radicals nor Muslims, began questioning the assumed virtues of school desegregation. For integrationists who considered *Brown v. Board* a virtually sacred text, this was heresy. But the doubters gathered strength in the climate of rising black pride. Citing the performance of African-America students in some all-black schools, they quarreled with the claims of white liberal "experts" who were saying that black students would founder unless they attended schools with whites. Why, they asked, was a predominantly white school better than a properly supported and predominantly black one? [60] They also began to demand local—-meaning black—control of their schools, a

57. John Diggins, *The Rise and Fall of the American Left* (New York, 1992), 226.
58. Randy Roberts and James Olson, *Winning Is the Only Thing: Sports in America Since 1945* (Baltimore, 1989), 167–79; Weisbrot, *Freedom Bound*, 227.
59. Weisbrot, *Freedom Bound*, 251–52.
60. Ravitch, *Troubled Crusade*, 164–74.

course of action that often pitted them against white teachers and tax-payers. A furious fight by blacks for control of schools in the Ocean Hill section of Brooklyn in 1967–68 badly polarized racial attitudes in New York.[61]

The most militant civil rights leaders, meanwhile, rejected white co-leaders. In December 1966 the executive committee of SNCC voted formally to expel the five whites who still remained on it. In July 1967 CORE, an interracial organization since its founding in 1942, eliminated the term "multi-racial" from the membership clauses of its constitution. Moves such as these killed off whatever chance SNCC and CORE still retained of reaching out to moderate blacks and white liberals. Their funding dried up, and their white support declined drastically. But their leaders, issuing ever more strident statements, did not seem to care. By early 1968 SNCC was dying so fast that it was ridiculed as the "Non-Student Violent Non-Coordinating Committee." Looking for help, it sought an alliance with the Black Panthers. Carmichael became "Prime Minister of the Panthers."

A few other black leaders made Carmichael look tame. One was Hubert "Rap" Brown, who replaced him as head of SNCC in May 1967. Like many other young militants by that time, Brown grew an African-style ("Afro") hair-do, and he went around in a blue denim jacket and dark sunglasses. He made a name for himself by inciting resistance to whites, whom he called "honkies," and to police, whom he branded as "pigs." "Violence," he said, was "as American as apple pie." In Cambridge, Maryland, long the site of racial tensions, Brown climbed atop a parked car in August 1967 and declared, "Black folks built America, and if America don't come around, we're going to burn America down." A few hours later a fire erupted in a ramshackle school for black children. When white firemen were slow to put out the blaze—they feared snipers—the fire spread and burned out the heart of the black district of the city. Brown was later arrested on charges of riot and arson, got free on bail, and shrugged off the charges. The real blame for the unrest of blacks, he said, lay with Lyndon Johnson, "a wild, mad dog, an outlaw from Texas."[62]

ALL THESE CONFRONTATIONS PALED before the outbreak of urban riots in 1966 and 1967. There were thirty-eight riots in 1966, the most serious in Chicago, Cleveland, and San Francisco. They killed

61. Weisbrot, Freedom Bound, 238–42.
62. Newsweek, Aug. 7, 1967, p. 28.

seven people, injured 400, and led to 3,000 arrests. Burning and looting caused an estimated $5 million in property damage. The turmoil reached all-time highs in 1967. During the first nine months of the year there were 164 insurrections; thirty-three of these were frightening enough to require the intervention of the state police, and eight brought in the National Guard. The two biggest riots, in Newark and Detroit in July, lasted for almost a week in each instance and left twenty-three and forty-three people dead. Hundreds were injured. Thousands of buildings were burned or looted, and thousands of people were left homeless. When the riot finally died down in Detroit, Mayor Jerome Cavanaugh remarked, "It looks like Berlin in 1945."[63]

Many Americans, including prominent conservatives like Governor Reagan of California, reacted angrily to the violence by blaming radical agitators and "riffraff" who engaged in looting just to rip off stores.[64] Others laid the blame at the feet of unsettled new migrants, many of them from the South. Television coverage came in for a good deal of condemnation: black people, critics thought, observed on TV that police and bystanders stood around and did nothing while looters and rioters went freely about their business. Emboldened, the TV-watchers then went out and emulated the rampages of others. There was some validity in these critiques. An almost carnival-like atmosphere pervaded some of the riots. Scenes on television did little if anything to discourage people from joining in. Looters were central to the disturbances. Unlike the "community" riots that had erupted following World War I, which had involved a good deal of fighting between blacks and whites, the disturbances of the 1960s were "commodity" riots featuring stealing and burning. This change resulted in part from the great expansion over time in the size of virtually all-black ghettos: whites in the 1960s were much less likely than they had been in 1919 to live or work among blacks, and they feared to walk through their neighborhoods. Interracial fighting was therefore rare—save between blacks and police—in 1967.

Other observers emphasized poverty and class grievances as a source of the outbreaks. This was true, but mainly in a relative sense. Save for the most down-and-out in the ghettos, urban black people lived more com-

63. *Report of the National Advisory Commission on Civil Disorders* (New York, 1968), esp. 1–34; Robert Conot, *Rivers of Blood, Years of Darkness* (New York, 1968); *Newsweek*, July 24, 1967, pp. 21–23 (on Newark), and Aug. 7, 1967, pp. 18–26 (on Detroit); Matusow, *Unraveling*, 362–65.
64. Charles Morris, *A Time of Passion: America, 1960–1980* (New York, 1984), 117–28.

fortably on the average than they had in the 1950s or early 1960s, when the central cities had been relatively quiet. Moreover, there was no clear correlation between the incidence or depth of poverty and disorder: the black ghetto in Detroit (as in Watts) was generally better-off than those in other cities. Other desperately poor ethnic groups, such as Puerto Ricans in New York, did not riot. Still, there was no doubting the anger fueled by the sense of relative deprivation. Most of the rioters were poor or working-class; non-rioting blacks were likely to be better-off. In Detroit, angry class feelings seem to have driven some poor whites as well as blacks. Of the 8,000 who were arrested in the Motor City, 700 were white.[65]

Still another liberal view held that rage against racial discrimination inspired the riots. This argument found its most enduring statement in the report of the National Advisory Commission on Civil Disorders named by President Johnson. Known as the Kerner Report after its chairman, Governor Otto Kerner of Illinois, it called in March 1968 for massive expansion of government programs to soften the bitter heritage of "white racism" in employment, housing, welfare, education, and every other walk of life. Research by the commission and by others in the aftermath of the riots confirmed the key role of racial feelings, noting that thousands of black people, not just the poor or a handful of agitators, took part. Most were young. They were likely to be high school dropouts but nonetheless better educated than the average for blacks in the central cities, and to be long-term residents—not "riffraff" or recent migrants. High percentages—40 percent in some places—were unemployed and angry about job discrimination. Looters tended to converge purposefully on white-owned stores, leaving black-owned establishments unscathed. Most of the major disturbances occurred following incidents between white police and local black residents. These incidents were mostly minor—the sorts of confrontations that happened every day. But that was the point. News of the incidents, distorted by rumors or spread by outraged local residents, tore like wildfire through a population harboring explosive resentments against police and escalated into riots when the police handled situations badly.

The racial tension that blew up into riots had deep urban roots, especially in residential discrimination. Confrontations between blacks seeking housing and whites, mostly working-class, who were determined to keep them out, had long afflicted race relations and urban politics in many northern cities and had intensified in the wake of black migrations

65. Sidney Fine, *Violence in the Model City: The Cavanaugh Administration, Race Relations, and the Detroit Race Riot of 1967* (Ann Arbor, 1989).

in the 1940s and 1950s. Many of them had even then erupted in violence, which nervous white politicians and business leaders had tried to keep out of the news. These antagonisms reached a breaking point by the mid-1960s, infuriating many blacks and frightening local whites.[66] Although such conflicts most commonly pitted blacks against whites, they were also beginning to involve Hispanic-Americans as well. In San Antonio the Mexican-American Youth Organization and in Los Angeles the Brown Berets, a Chicano group, were mobilizing vigorously against residential discrimination—and by extension school segregation.[67]

By the time the biggest waves of riots swelled forth in 1966–67, Americans generally were growing edgy about ever-higher levels of unrest, disorder, and violence in the culture. The national homicide rate doubled between 1963 and 1970—from 4.6 per 100,000 people to 9.2 per 100,000.[68] The carnage in Vietnam, never far from people's minds, provided a gory backdrop for violent behavior. *Bonnie and Clyde*, a film celebrating violence, appeared in August 1967 and appeared to reflect a contemporary mood. Violence was rising especially rapidly at the time in black central city areas, in part because of demographic changes: the coming of age of the baby boom generation and the peaking of south-to-north migrations. For these reasons the ghettos by the mid-1960s were home for an unprecedentedly huge population of young men, increasing percentages of whom came from broken homes without strong discipline. Young men, moreover, are always the group most susceptible to crime and violence.[69] The decline in the number of manufacturing jobs in many of these areas further exacerbated racial tensions. The special demographics and economics of the mid-1960s, and the incapacity of hard-pressed cities to cope with them, were generating what Americans were later to call a black "underclass" and creating extraordinarily unsettling conditions.[70]

66. Arnold Hirsch, *Making the Second Ghetto: Race and Housing in Chicago, 1940–1960* (New York, 1983), 256–58; Thomas Sugrue, "Crabgrassroots Politics: Race Relations and the Fragmentation of the New Deal Coalition in the Urban North, 1940–1960," paper delivered at Brown University, April 1994.
67. Teaford, *Twentieth-Century American City*, 132–33.
68. *New York Times*, Oct. 23, 1994. By 1980 it had risen further, to 10.1 killings per 100,000.
69. James Wilson, *Thinking About Crime* (New York, 1983), 98–124; Charles Silberman, *Criminal Violence, Criminal Justice* (New York, 1978), 118–19, 163.
70. William Julius Wilson, *The Truly Disadvantaged: The Inner City, the Underclass, and Public Policy* (Chicago, 1987); Christopher Jencks and Paul Peterson, eds., *The Urban Underclass* (Washington, 1991); Jencks, *Rethinking Social Policy: Race, Poverty, and the Underclass* (Cambridge, Mass., 1992), esp. 143–203.

Many black and poor people, moreover, had become politicized by the civil rights movement and had begun to develop higher expectations from life. Some joined a newly formed National Welfare Rights Organization (NWRO). Following in the path blazed by civil rights workers, NWRO leaders engaged in sit-ins and other forms of non-violent direct action. Later, NWRO activists took over welfare offices. Other local residents organized to protest against urban redevelopment that threatened to disrupt their neighborhoods.

Most of the blacks who took part in the riots of 1966 and 1967 apparently did not expect much in the way of tangible results. Fired up by conflicts with the police, they started disturbances that exploded suddenly, raged out of control, and then stopped before participants could develop much of a program. Rather, the rioters rampaged so as to express themselves against long-simmering injustices and to be noticed, at last. Although they damaged their own already blighted neighborhoods, they often felt that the rampage had been worth it. As one rioter had said in Watts, "We won, because we made them pay attention to us."[71] Congressman Adam Clayton Powell, Jr., of Harlem added in 1967, "Violence is a cleansing force. It frees the native from his inferiority complex."[72]

Some Americans reacted to the riots by calling on LBJ and Congress to do something to improve life in the ghettos. Atlanta Mayor Ivan Allen, Jr., insisted, "This is a great national problem. Congress can't wait for local governments to try to cope with problems that can be met only on a national level." The New York Times editorialized, "There are at least two conclusions to be drawn. . . . One is that if Detroit is an example of America's best efforts to solve the racial and other problems confronting the cities, the best is not nearly good enough. The other is that even if progress is achieved on a broad front, the United States must be prepared to contend with serious turbulence in its cities for a long time to come."[73]

The New York Times, like the Kerner Commission, was correct that the ills lay deep. Liberals, however, did not have a very clear idea of what to do to cure them. Great Society programs, after all, had failed to ensure peace. And hostilities across the color line seemed implacable. What then to do? Should programs attempt to pump money into the ghettos, perhaps by aiming it at black entrepreneurs, perhaps by offering tax breaks or other incentives to businesses that would agree to locate there? Should resources be set aside to increase the recruitment of blacks into police forces, so as to

71. Hodgson, America in Our Time, 180–81.
72. Newsweek, Aug. 7, 1967, p. 31.
73. Ibid.

temper the incendiary feelings that ignited confrontations? Or should government and private initiative focus on ways of helping blacks to get *out* of the ghettos? That would involve truly tough actions to eradicate housing and job discrimination in white neighborhoods. It might also require the spending of greatly increased sums on job-training and education programs. These, in turn, would surely take a long time to make much difference. Neither in 1967 nor later was there a consensus among liberals in the United States on what to do. Instead, there was widespread bewilderment and panic.

Conservatives, by contrast, scored political points by denouncing agitators. Reagan noted accurately that "the first victims . . . are the good, responsible members of the Negro community." He added, however, that rioters were "the lawbreakers and the mad dogs against the people." Eisenhower hinted at the existence of conspiracy: "A lot of people think there is definitely a national planning system because they [the riots] seem to follow such a definite pattern."[74] Polls indicated that many Americans agreed with him. One showed that 45 percent of whites blamed outside agitators (including Communists); another poll indicated that only one-sixth of whites acknowledged the existence of police brutality.[75] Sustained by popular responses such as these, conservatives in Congress mobilized to attack an administration effort then pending to exterminate rats in the ghettos. One denounced the measure as a "civil rats bill." Another suggested that the President "buy a lot of cats and turn them loose." Others wondered if Johnson would next seek legislation to do away with snakes, squirrels, bugs, and blackbirds.[76]

It was not only white conservatives, however, who were appalled at what had happened. Wilkins and Young, called to the White House for advice, were shocked and baffled. Bayard Rustin proclaimed, "The rioting must be stopped. Whatever force is necessary should be used. . . . If the rioting continues, an atmosphere will be created in which the established civil-rights leadership will be robbed of standing. . . . The movement could be destroyed and the leadership passed over to the hands of the destructive elements of the ghettos."[77]

President Johnson, absorbed by the war in Vietnam, had mixed reactions. Unlike others who blamed a few zealots for the troubles, he understood the pain that discrimination had wrought in black people, and he

74. Ibid.
75. Kearns, *Lyndon Johnson*, 307.
76. Califano, *Triumph and Tragedy*, 211–12.
77. *Newsweek*, Aug. 7, 1967, p. 31.

persevered with his rat control program (which ultimately passed) and his open housing bill.[78] Still, he was angered by the lawlessness and hurt by the barrage of criticism that people were flinging at him. Stung by assertions that the government had not done enough to alleviate racial problems in the cities, he ignored the Kerner Commission's report when it appeared in 1968.

THE ESCALATING DEMANDS for rights after 1965, and especially the riots, did more than bewilder people. They also aroused significant backlash, the most vivid of the many reactions that arose amid the polarization of the era. It long outlasted the 1960s.[79]

Some of those who lashed back carried with them the fervor of religious commitment. These included a few blatant racists who belonged to groups such as the KKK. Much more numerous, however, were rising numbers of white people who followed evangelists like Billy Graham; disciples of fundamentalist Protestants such as Oral Roberts, who had established his own university and medical school in Oklahoma; and adherents of student organizations such as the Campus Crusade for Christ. Devout and politicized Catholics also denounced the outrageous excesses, as they saw them, of blacks and other minority groups.

Many of these angry whites could hardly be called "conservative" in a traditional sense. They included millions of struggling, often class-conscious people who raged with almost equal fervor at what they perceived as the special privileges of corporate elites, Establishment priests and ministers, wealthy medical practitioners, liberal school boards, permissive bureaucrats and judges, and "experts" in general. They displayed a mounting unease with much that was "modern," including the teaching of Darwinian theories in the schools, and with much that "know-it-all" social engineers told them to believe. They were disturbed by feminists, sexual liberation, radicals, and anti-war demonstrators, and they were outraged by the "idolatrous" and "criminal-coddling" Warren Court. Many perceived a conspiracy that was masterminded by an eastern Establishment. Threatened by the insouciance of the younger generation, they especially resented the contempt that they received from more secular Americans. By 1966 they were beginning to take part in politics as never before, especially in the South and the Southwest, where population growth was explosive. Reagan was but the best-known of the anti-liberal

78. Kearns, *Lyndon Johnson*, 305–7.
79. James Button, *Black Violence: Political Impact of the 1960s Riots* (Princeton, 1978).

office-holders who actively politicized their concerns and benefited from what became identified in the 1970s as a new and powerful Religious Right.[80]

Many other Americans in the mid- and late 1960s, whether religious or not, were repulsed by the attitudes and behavior of the counterculture, as it was called. This, a kin of the student Left, had attracted some earlier attention, especially after 1965. But it reached a new efflorescence in January 1967, when some 20,000 people, most of them young, gathered at Golden Gate Park in San Francisco to celebrate their "hippie" style of life. Timothy Leary was on hand to hail the wonders of LSD, a synthetic hallucinogen, and other mind-enhancing drugs. He urged the young to "turn on, tune in, drop out." Allen Ginsberg chanted Hindu phrases. Jerry Rubin, a radical who had headed a militant Vietnam Day Committee in Berkeley, appealed for bail money. People in the audience passed each other flowers, LSD tablets, and sandwiches, provided free for the occasion by the "acid chemist," Augustus Owsley Stanley III. Reporters and television cameras conveyed the colorful action to the world.[81]

The Human-Be-In, as organizers called the occasion, was by far the best publicized but not the last of such celebrations. The nearby Haight-Ashbury neighborhood of San Francisco continued to thrive as one of a number of settings for congregations of "flower people" in the late 1960s. Many of the hippies burned incense, did drugs, created psychedelic art, and walked about garbed in outlandish clothing. Hippie men let their hair grow long; some of the women dispensed with bras. Like millions of other young people, they listened to acid rock, as it was called. (The phrases "acid test" and "freak out" dated from the mid-1960s.) Groups like the Doors, the Rolling Stones, and the Grateful Dead developed great cult followings. Even the Beatles recorded their music after 1965 while under the influence of drugs, especially marijuana and LSD. Other counter-culturalists moved to communes, where they ate organic or macrobiotic food and (the media liked to emphasize) practiced various free versions of sex.

Young people who participated actively in such communal experiences

80. Kirkpatrick Sale, *Power Shift: The Rise of the Southern Rim and Its Challenge to the Eastern Establishment* (New York, 1975), 90–103; Stephen Bates, *Battle-ground: One Mother's Crusade, the Religious Right, and the Struggle for Control of Our Classrooms* (New York, 1993), 212–14; and Paul Boyer, "A Brief History of the End of Time," *New Republic*, May 17, 1993, pp. 30–33.
81. Gitlin, *Sixties*, 210; O'Neill, *Coming Apart*, 242; Diggins, *Rise and Fall*, 210; Matusow, *Unraveling*, 275–307.

were never more than a small minority of their age group at the time. Like the beats, whom they resembled in some ways, they tended to be apolitical. They sought freedom from authority, escape from conventional middle-class conventions, and satisfaction from levels of personal intimacy that they despaired of finding in mainstream society. Most did not attain these blessed states, and the counterculture faded after 1970. Their unconventional ways, moreover, made them easy targets for ridicule by conservative politicians. Governor Reagan delivered perhaps the most famous one-liner. A hippie, he said, is someone who "dresses like Tarzan, has hair like Jane, and smells like Cheetah."

Still, the hippies and "dropouts" attracted considerable media attention while they held the stage, especially between 1967 and 1970.[82] And the rise of a self-conscious acid-rock culture beguiled larger numbers of people, most of them young. The apparent rejection by many young people of conventional middle-class styles of life, the assault to the senses of acid rock, and the open use of drugs offered a noisy and discordant counterpoint to a mainstream culture that, thanks in part to divisions widened by the war, seemed increasingly cacophonous by 1967.

"Long-haired hippies," while irritating to many Americans, aroused less backlash than did young radicals who resisted the draft and otherwise impugned the symbols of American patriotism. Many Americans who railed at these "spoiled rich kids" had tired of the war. But they were often deeply patriotic and class-conscious. One expostulated, "The college types, the professors, they go to Washington and tell the government what to do. Do this, they say; do that. But their sons, they don't end up in the swamps over there, in Vietnam. No sir. . . . I think we ought to win that war or pull out. . . . I hate those peace demonstrators. . . . The sooner we get out of there the better." His wife added, "I'm against this war, too—the way a mother is, whose sons are in the army, who has lost a son fighting in it. The world hears those demonstrators making noise. The world doesn't hear me."[83]

Those who joined the backlash were equally appalled by the personal behavior of people, especially blacks, whom they blamed for what they considered to be the rampant moral degradation of the era. It was then, following in the wake of the Moynihan Report, that the media devoted increasing attention to statistics about "family break-up" in the United States. These statistics were indeed troubling for those Americans—the

82. Skolnick, *Embattled Paradise*, 92–93.
83. Polenberg, *One Nation Divisible*, 228.

vast majority—who imagined that the two-parent nuclear family anchored national stability. Contemporary reports indicated that illegitimacy rates began to rise rapidly after 1963. The rates among blacks were around 23 percent in 1963, compared to around 2 percent among whites, and jumped to 36 percent in 1970, compared to 3 percent among whites.[84] Divorce rates also increased (after declines from 1946 through 1958), from 9.2 per 1,000 married couples in 1960 to 11.2 in 1968.[85] The trends seemed relentless, threatening the stability of neighborhoods, disrupting the schools, and setting loose a tidal wave of laments about social "crises" that crested in later years.[86]

The sources of these trends, which (along with rapidly rising rates of violent crime) developed suddenly and at much the same time, were complex. One, however, was the coming of age of the baby boomers, many of whom had developed more permissive feelings about sex, illegitimacy, and divorce: the young did much to accelerate the ongoing sexual revolution, which burst into the open at the time.[87] Another source was the migration of millions of young people from rural or small-town settings, where neighborhood sanctions had dominated, to the relative freedom and anonymity of cities. A source of rising divorce rates was the ever-higher percentage of women who worked: more women now had the resources, however meager, to break away from unhappy marriages and strike out on their own. Underlying all these changes was the peaking at the same time of the *Zeitgeist* of freedom, expectations, and rights. These, expanded by the unparalleled affluence of the 1960s, did much to challenge traditional mores.

The Graduate, an Oscar-winning movie that appeared in late 1967,

84. Rates measure the percentage of all live births delivered by single women. The rates continued to rise, to 48 percent for blacks and 11 percent for whites in 1980 and to 67 percent for blacks and 22 percent for whites in 1993. This was an overall percentage of approximately 30 percent. The increasing rates did not reveal a rise in fertility; birth rates generally (save among unmarried women under 19) went down as the baby boom declined after 1957. Rather, they indicated that higher percentages of women having babies were unmarried. See James Patterson, *America's Struggle Against Poverty, 1900–1994* (Cambridge, Mass., 1995), 240; Murray, *Losing Ground*, 125–28.

85. Randall Collins and Scott Cottrane, *Sociology of Marriage and the Family* (Chicago, 1991), 157. The divorce rate shot up more after 1968, to 20 per 1,000 married females in 1974.

86. Concerning problems in schools, see Charles Silberman, *Crisis in the Classroom: The Remaking of American Education* (New York, 1990).

87. Beth Bailey, "Sexual Revolution(s)," in Farber, ed., *Sixties*, 235–62.

dramatized these changes. It featured a young man (Dustin Hoffman) who was in no way a hippie, a user of drugs, or a political radical. But he seemed unconnected to traditional values. Alienated from many things, he felt no kinship with fraternity men at his university or with materialistic adults of the older generation. The sound track, featuring the song "Sounds of Silence" by Paul Simon and Art Garfunkel, emphasized the cleavages (silences) that separated people from one another. Like James Dean in *Rebel Without a Cause* twelve years earlier, the character played by Hoffman seemed to epitomize those young Americans who felt cut off from conventional American civilization.

Family disorganization excited special confusion and alarm among middle-aged and older people. Some felt the whole world was spinning away under their feet, flinging their cherished values into a black hole. They spoke with fear and disgust about the lack of "family" among the younger generation, especially among blacks. "You see, we were poor," an Italian-American parent explained to an interviewer. "But we helped one another. But the colored don't have that family life like we do. I don't know what it is with them. I can't understand these people." A Jewish parent added, "In East New York the Jewish values were passed on from generation to generation. We never dreamed of doing other than those values, and we were aghast when people didn't do as they should. We still talk about those things. A girl having an illegitimate child, or a girl getting a divorce, it was unheard of, you just didn't have it. You knew all on the block, the street was like a small town; you knew who was doing this and who was doing that."[88]

Many of those who railed against family break-up assailed the rise of welfare, especially the Aid to Families of Dependent Children (AFDC) program, which mainly aided low-income divorced, separated, or single women and their children. The rolls of AFDC, like divorce and illegitimacy, rose rapidly in the 1960s, from 3.1 million recipients in 1960 to 4.4 million in 1965 to 6.1 million in 1968. Costs of the program, which was supported by both the federal and state governments, increased during the same eight years from $3.8 billion to $9.8 billion. Like the rise in divorce and illegitimacy, the upward course of welfare rolls seemed both rapid and escalatory: by 1970 the number of recipients shot up to 9.7 million, and the cost to $14.5 billion.[89]

The sources of this surge in welfare dependency, while related in part

88. Jonathan Rieder, *Canarsie: The Jews and Italians of Brooklyn Against Liberalism* (Cambridge, Mass., 1985), 63–66, 139.
89. Frank Levy, *Dollars and Dreams: The Changing American Income Distribution* (New York, 1987), 173.

to the rise in family break-up, were rooted especially in the rights-consciousness of the era. The incidence of poverty, which declined throughout the 1960s, was not the primary cause. Rather, the rolls grew because much higher percentages of eligible people demanded to be helped. They expanded also because activists working in community action programs informed potential recipients of their rights and offered them legal aid; this was one area where the war on poverty had significant (though unintended) effects. Some recipients, swept up in the quest for rights, not only insisted on help-from AFDC but also joined the NWRO and demanded a more generous and less intrusive administration of benefits. They proclaimed especially that they and their children had a "right" to a decent life. Their behavior indicated that the stigma of being on welfare, which had been powerful throughout American history—even in the Depression—had lost some of its force. So had the tendency of poor people to defer to people in authority. These were among the most profound and lasting developments of the 1960s.[90]

The costs of supporting this surge of recipients were never very high even in the late 1960s: federal outlays for AFDC were generally about one-ninth of federal expenditures for social welfare and about 2 to 3 percent of all federal spending. Nor was life on welfare a very enticing prospect for poor people: average aid per family on AFDC rose from $108 per month in 1960 to $168 in 1968, sums that left recipients well below the official poverty lines set by the government. These lines, moreover, were not high. Although they rose to reflect increases in living expenses (the line for a family of four reached $3,350 in 1967), they remained a good deal lower than activists wanted. Still, spending for welfare, supported by taxpayers, was mounting. So by 1970 were costs for a food stamp program, which had been an insignificant federal effort as recently as 1966. And black people, while never a majority of welfare recipients, were disproportionately on the rolls because they had especially high rates of poverty and of family break-up. Backlash against AFDC, much of it racial in nature, grew fierce by the late 1960s.[91]

90. Patterson, *America's Struggle*, 157–70; Robert Plotnick and Felicity Skidmore, *Progress Against Poverty: A Review of the 1964–1974 Decade* (New York, 1975), 82; Kirstin Grønbjerg et al., *Poverty and Social Change* (Chicago, 1978), 72–88; Edward Berkowitz, *America's Welfare State: From Roosevelt to Reagan* (Baltimore, 1991), 91–152.

91. Kathryn Hyer, "The Measurement and Meaning of Poverty," *Social Problems*, 22 (June 1975), 652–62; Mollie Orshansky, "How Poverty Is Measured," *Monthly Labor Review*, 92 (Feb. 1969), 37–41; Richard Cloward and Richard Elman, "Poverty, Injustice, and the Welfare State," *Nation*, 202 (March 7, 1966), 264–66.

Opponents of welfare, moreover, focused on far more than costs. In-deed, they expressed visceral feelings on the subject, complaining indig-nantly about "leeches," "cheaters," and "welfare bums." Cherishing the work ethic, they grew irate when they thought about blacks and other "loafers" on the dole. One city worker later exploded, "These welfare people get as much as I do and I work my ass off and come home dead tired. They get up late and they can shack up all day long and watch the tube. . . . I go shopping with my wife and I see them with their forty dollars of food stamps in the supermarket, living and eating better than me. . . . Let them tighten their belts like we do."[92]

Emotions such as these gripped millions of Americans, most of whom were willing to support public assistance only for "deserving" people such as widows, orphans, the elderly poor, and the disabled. By the mid-1960s, however, relatively few AFDC mothers were widows; most were younger women whose marriages had collapsed or who had had children out of wedlock. They did not seem "deserving" at all. Working-class people, many of them poor, seemed especially upset by welfare. Many, to be sure, had themselves gone on relief during the 1930s or when tragedy, such as the loss of a major breadwinner, had hit their families. But they had tried to manage stoically, enduring the intrusions of social workers and the restrictions on possessions—no telephones, no linoleum—that the strin-gently run system had required. As soon as they had been able to find work, they did so; the young had often left school in their early teens to support their families. "Welfare helped us, and it was right and just that it did," one New Yorker remembered. But "then we could shift for our-selves."[93]

What especially bothered many of those who joined the backlash against welfare was that their own children seemed to be tempted by the same "degraded" values that were luring the "leeches" on the dole. Many Americans in the younger generation, they believed, were demanding instant gratification. Worse, they thought, these young people were insub-ordinate, lacking either appreciation for or understanding of the sacrifices of their parents. A Jewish businessman remembered, "My old man gave me a handtruck when I was nine, down in the garment district. He said to me, 'Here, go to work!'" Not so with young people in the 1960s, he complained; "nobody wants to work or wait for anything."[94]

Fear of violent crime greatly intensified these feelings of backlash. Polls increasingly showed that Americans considered "crime in the streets" to

92. Rieder, *Canarsie*, 102.
93. Ibid., 104.
94. Ibid.

be the nation's number one problem. Family break-up, illegitimacy, and crime, they were certain, went hand-in-hand.[95] Blacks, the vast majority of whom were law-abiding, were among the Americans who worried about these trends. They were affected more than any other group, for most violent crime in the cities was black-on-black. A resident of a crime-ravaged low-income apartment complex in Washington articulated such feelings: "I would like to say I'm black and I'm proud. But I can't say that so easily because I'm not proud of what black people are doing to each other in this building." She added, "When we first moved in, I would go down the hall and wash off things that had been written on the walls. Now I'm afraid to go out into the halls."[96]

White working-class people, however, seemed loudest and angriest about crime. Many, of course, lived near the most run-down sections of cities. Like the black woman in Washington, they feared for their safety. They normally blamed young black men—whose rates of arrest for violent crimes and drug-dealing were far higher than those of whites—so much so, indeed, that the rates could not be entirely explained away as reflections of racial prejudice on the part of police: a homicide, after all, was a homicide. Some of these whites labored to control their feelings. A Jewish woman explained, "I guess I don't really hate the blacks. I hate that they make me look over my shoulder." Other whites, however, were more open. "You can't walk . . . anywhere," a resident of Brownsville in Brooklyn exploded. "It's because these people don't know how to live. They steal, they got no values. They say it's history, but that's bullshit. It's not history, it's the way they live. They live like animals."[97]

Whites like this angrily rejected the argument that black people deserved special consideration because of their long history of being oppressed. Many did not consider themselves to be prejudiced. They insisted that they supported the right of all people to equal opportunity, still a most hallowed American political ideal. But they hotly resented being dismissed as "racists" by privileged integrationists—"limousine liberals"—who lived in lily-white suburbs. And they drew a firm line against special treatment to protect or advance minority groups as groups.[98]

95. Wilson, *Thinking About Crime*, 64–65.
96. Silberman, *Criminal Violence*, 161.
97. Rieder, *Canarsie*, 177, 26.
98. Colburn and Pozzetta, "Race, Ethnicity." Two books describing such feelings in Boston, a hotbed of conflict over such issues in the 1970s, are J. Anthony Lukas, *Common Ground: A Turbulent Decade in the Lives of Three American Families* (New York, 1986); and Ronald Formisano, *Boston Against Busing: Race, Class, and Ethnicity in the 1960s and 1970s* (Chapel Hill, 1991).

Liberal-dominated agencies such as the EEOC, they complained, were moving toward "reverse discrimination." One white man asked, "Who will pay the Jews for two thousand years of slavery? Who will compensate the Italians for all the ditches they dug?" Another exclaimed, "What happened four hundred years ago, all those whites who whipped them and beat them, are we responsible for it? I don't even have anything to do with slavery. What's past is past."[99]

THESE MANIFESTATIONS OF BACKLASH—against family break-up, illegitimacy, welfare, crime, riots, black activists, anti-war demonstrators, long-haired hippies, government programs that favored minorities, elitists, liberals generally—exposed a major development of the mid-1960s: rapidly rising polarization along class, generational, and racial lines. The backlash represented considerably more than white racism, which polls suggested was less intense than in the past. It also affirmed the behavior and the moral standards of traditional ways. It exposed a fragmentation of society and culture that seemed if anything to grow in the next thirty years.

The rising numbers of people who became part of the backlash did not much perceive themselves as part of an organized movement. Particularly at first, they tended to express local grievances arising from tensions in their neighborhoods. But they also worried about larger forces that threatened them. Increasingly, they used the word "squeeze" to capture their plight. From the bottom they felt squeezed by blacks and other minorities who were demanding special rights and privileges. From the top they felt pressed by the more affluent and powerful, including their superiors at work. Public employees, indeed, went out on strikes in record numbers in 1967. Other workers felt "blue-collar blues": there were more work stoppages in 1968 than in any year since 1953.[100] These feelings of squeeze provoked an often bitter rage that rested in part on unabating class and ethnic identifications.[101]

Backlash also threatened the Democratic party. This had been apparent as early as the 1964 and 1966 elections, and it grew more ominous as the presidential election of 1968 approached. Many Americans blamed John-

99. Rieder, *Canarsie*, 111; Jonathan Rieder, "The Rise of the Silent Majority," in Steve Fraser and Gary Gerstle, eds., *The Rise and Fall of the New Deal Order, 1930–1980* (Princeton, 1989), 254–55.

100. Robert Zieger, *American Workers, American Unions, 1920–1985* (Baltimore, 1986), 169–70.

101. Hodgson, *America in Our Time*, 483–86; Rieder, *Canarsie*, 98.

son and the Democratic party not only for mismanaging the Vietnam War but also for creating the social turmoil that disturbed the nation after 1965. They especially resented liberals—permissive, patronizing, hypocritical, and sanctimonious do-gooders who reproved them for their resistance to the claims of minorities and assorted trouble-makers. (A conservative, it was said, was a liberal who had been mugged; a liberal was a conservative who hadn't been mugged—yet.) In an increasingly fragmented and polarized society these angry people were a political force to be reckoned with.

22

The Most Turbulent Year: 1968

January 30, 1968, was the first day of Tet, a festive holiday in Vietnam that marked the beginning of the lunar year. Americans in Vietnam hoped for some respite from the fighting. But at 2:45 that morning a team of NLF sappers blasted a hole in the wall that surrounded the American embassy in Saigon. Racing into the compound, they tried but failed to smash through the heavy door at the entrance to the embassy. They then took cover behind large concrete flowerpots and assailed the building with rockets. Military police fired back at them in a fight that lasted until 9:15 in the morning. All nineteen of the enemy were either killed or badly wounded. Five Americans and a South Vietnamese civilian employee lost their lives. A reporter described the scene as "a butcher shop in Eden."[1]

The attack on the embassy formed part of a much broader military plan, elements of which had already been launched outside of Saigon, that came to be known as the Tet offensive. Starting in late 1967 Hanoi had intensified pressure on towns and bases in the central highlands of South Vietnam and along the demilitarized zone, and especially on the marine garrison at Khe Sanh near the border with Laos. Top units of the

1. George Herring, *America's Longest War: The United States and Vietnam, 1950–1975* (Philadelphia, 1986), 186. Also William Chafe, *The Unfinished Journey: America Since World War II* (New York, 1991), 345; *Newsweek*, Feb. 12, 1968, pp. 27–29.

NLF began at the same time to infiltrate major cities. American and South Vietnamese troops fought back and inflicted heavy casualties on the enemy forces. General Westmoreland devoted special effort to protecting Khe Sanh, a beleaguered outpost which he feared might otherwise become a second Dienbienphu. Hanoi launched these assaults in part to make the United States and South Vietnam reduce their forces in Saigon and other major cities, thereby exposing themselves to the attacks at the opening of Tet.

Within a few hours of the battle at the embassy, the enemy forces assailed a large number of targets in South Vietnam, including five major cities, sixty-four district capitals, thirty-six provincial capitals, and fifty hamlets. President Thieu declared martial law, thus conceding that there were then no secure areas in the South. Johnson tried to minimize the dangers, likening the situation to the previous year's riot in Detroit—"a few bandits can do that in any city."[2] But the United States and South Vietnam had to stage a major counter-offensive to overcome the enemy. Battles raged for the next three weeks, killing perhaps 12,500 civilians and creating a million refugees. Some of the fighting was bloody indeed. It took twenty-five days of heavy artillery and air bombardment to reconquer the old Vietnamese capital of Hue, which was reduced to a "shattered, stinking hulk, its streets choked with rubble and rotting bodies." Fighting there killed some 5,000 enemy troops, untold numbers of civilians, 150 American marines and 350 South Vietnamese soldiers. When Americans re-entered the city, they found 2,800 bodies buried in mass graves. They were people slaughtered by the enemy as suspected collaborators with the South. The victors retaliated by assassinating suspected Communists.[3]

When the battles subsided, Westmoreland declared that the United States and the South Vietnamese had inflicted devastating losses on the attackers. "The enemy exposed himself by virtue of his strategy and he suffered heavy casualties," he said. This was in fact the case. The Tet offensive failed in the end to take cities away from South Vietnamese control or to unleash a general revolt (as Hanoi may have hoped to do) against the government in Saigon. Thanks to heavy bombing, the United States also managed to repel the attacks on Khe Sanh, killing many thousands of enemy soldiers in the process. Rough estimates of overall casualties during the three weeks following Tet conclude that North Viet-

2. Larry Berman, *Lyndon Johnson's War: The Road to Stalemate in Vietnam* (New York, 1989), 147; *Newsweek*, Feb. 12, 1968.
3. Todd Gitlin, *The Sixties: Years of Hope, Days of Rage* (New York, 1967), 298; Herring, *America's Longest War*, 187; Chafe, *Unfinished Journey*, 346.

namese and NLF battle deaths reached 40,000, compared to 2,300 South Vietnamese and 1,100 Americans. It took Ho Chi Minh and General Giap more than two years to compensate for the fearful losses that they had sustained in Tet and its aftermath.[4]

Rarely, however, has "victory" been so costly. The initial thrusts of the Tet offensive, especially the breaching of the embassy wall, convinced already skeptical Americans that Johnson, Westmoreland, and other administration officials had been lying all along. Critics of Johnson were all the angrier about being deceived because of the publicity blitz that LBJ and Westmoreland had conducted in late 1967. After Tet it was as clear as could be that there was no "light at the end of the tunnel," as Westmoreland had maintained at that time.[5] Indeed, the shock from Tet greatly intensified an adversarial relationship that had been developing between the media and the State since the mid-1960s. For many in the media the credibility gap, a chasm after Tet, was never to be bridged thereafter.[6]

The reaction of CBS anchorman Walter Cronkite, the most widely admired television newsman in the nation, served for many at the time. Until then Cronkite, like other newscasters, had tried to maintain an "objective" stance. That had normally required him to report what Johnson, Westmoreland, and other administration officials released, without overt editorial comment. When Cronkite heard of the Tet offensive, however, he was furious, the more so because he sensed that television reports on the war had misled the American people. "What the hell is going on?" he is supposed to have snapped. "I thought we were winning the war!" Cronkite journeyed to Vietnam to see the situation for himself. When he returned, he reported on February 27, "It seems more certain than ever that the bloody experience of Vietnam is to end in a stalemate."

4. George Herring, "The War in Vietnam," in Robert Divine, ed., *Exploring the Johnson Years* (Austin, 1981), 50.
5. Peter Braestrup, *The Big Story: How the American Press and Television Reported and Interpreted the Crisis of Tet in Vietnam and Washington* (New York, 1978); John Mueller, *War, Presidents, and Public Opinion* (New York, 1973).
6. Michael Delli Carpini, "Vietnam and the Press," in D. Michael Shafer, ed., *The Legacy: The Vietnam War in the American Imagination* (Boston, 1990), 125–56; David Culbert, "Johnson and the Media," in Divine, ed., *Exploring the Johnson Years*, 214–48; William Hammond, "The Press in Vietnam as Agent of Defeat: A Critical Examination," *Reviews in American History*, 17 (June 1989), 312–23; Kathleen Turner, *Lyndon Johnson's Dual War: Vietnam and the Press* (Chicago, 1985); and Lawrence Lichty, "Comments on the Influence of Television on Public Opinion," in Peter Braestrup, ed., *Vietnam as History: Ten Years After the Paris Peace Accords* (Washington, 1984), 158.

Cronkite was but one of many in the media—and elsewhere—who were disbelieving. The columnist Art Buchwald, a humorist, went further. Westmoreland's claim of American victory, he wrote, was like Custer saying at Little Big Horn, "We have the Sioux on the run. . . . Of course we still have some cleaning up to do, but the Redskins are hurting badly and it will only be a matter of time before they give in."[7]

The disillusion of newsmen like Cronkite later led pro-war commentators to blame the press for misinterpreting what had happened after Tet and for failing to make it clear that American and South Vietnamese forces had prevailed in the field. They are right that America's successful military retaliation seemed to get lost amid the domestic recriminations that followed Tet. They are also correct to observe that some reporting during this anxious period was both disturbing and shocking. None was more so than the vivid coverage of an execution by the chief of the South Vietnamese police of an enemy officer on a Saigon street. An AP photographer and two television camera crews captured the execution, a slightly sanitized version of which was thereupon shown on two television networks in the United States. Anti-war Americans pointed to the gory killing as proof of their contention that the war in Vietnam was immoral.[8]

Critics were wrong, however, to maintain that the media were thereafter heavily negative about the war. The credibility gap notwithstanding, many reports on the conflict in Vietnam continued to convey a sense of progress in the fighting. Pro-war observers were also in error to argue that the media following Tet greatly altered public opinion in the United States about the war. What the reports probably did do at that time was to reflect and intensify popular doubts that had been rising, primarily because of discouraging casualty figures in 1966 and 1967, for some time.[9] These doubts centered not so much on the wisdom of fighting the war, which a small majority still seemed to believe in, as on LBJ's handling of it. Public approval of Johnson's conduct of the conflict, already low at 40 percent following his public relations blitz in November, fell to 26 percent in the immediate aftermath of Tet.[10]

7. Herring, *America's Longest War*, 188–89.
8. Culbert, "Johnson and the Media," 234. A little later, in March, American soldiers massacred more than 100 civilians at the village of My Lai. This, the only well-documented case of murders on such a scale by Americans, remained unreported until 1969.
9. See especially Hammond, "Press in Vietnam;" Mueller, *War, Presidents, and Public Opinion*; and Chafe, *Unfinished Journey*, 358.
10. Herring, *America's Longest War*, 200–202.

What Tet did more generally was to deepen the mood of gloom that was already growing profound in the United States. It was the first of a near-numbing series of blows in 1968 that bashed what hopes remained for healing the fragmentation and polarization that had been widening since 1965. After 1968, in many ways the most turbulent year in the postwar history of the United States, there was no turning back to the higher hopes that liberals had had in 1964 and early 1965.

THE RESPONSE TO THE TET OFFENSIVE by Wheeler, Westmoreland, and other military leaders in Vietnam was to ask for 206,000 additional American troops, half of them to be sent by the end of the year, in addition to the 525,000 or so who were already there. Such an effort was thought to require mobilization of the reserves. The details of the request, of course, were secret, but rumors of military calls for further escalation seeped out. Johnson, although shaken by the reaction to Tet, was not moved by popular disgruntlement to back off from American commitments. Nor, however, did he warm to the idea of further increases in American troop levels, which would have been perilous politically. He therefore turned the request over to advisers.

At this point, however, he encountered serious doubts from within his administration, and especially from Clark Clifford, who replaced McNamara as Defense Secretary on March 1. Truman's former aide, since 1949 a Washington attorney and unofficial adviser to Democratic Presidents, had opposed escalation of the war in 1965 but had then—like virtually all American experts on foreign and military policies—supported Johnson's course as the best way to preserve a non-Communist South Vietnam. When Clifford received the request for more troops, he ordered a review of the war. He asked Pentagon officials, "Does anyone see any diminution in the will of the enemy after four years of our having been there, after enormous casualties and after massive destruction from our bombing?" No one perceived any lessening of the enemy's will to fight. Clifford, moreover, was persuaded by leading Establishment figures, among them Dean Acheson and Averell Harriman, that Westmoreland's request would severely threaten America's financial standing in the world. Leading business figures at the time questioned the capacity of the nation, which was then facing a gold drain, to escalate further. Many of these doubters were coming to believe that the war was damaging America's ability to meet its strategic commitments in Europe, where national security interests were paramount. For all these reasons Clifford counseled against significant escalation.

Instead, he recommended to Johnson on March 4 that the United States send over a token force of 22,000 and that LBJ call up an undetermined number of reserves. Clifford further urged that Thieu and Ky be pushed into assuming greater responsibility for the war. This was a recommendation for what later became known as Vietnamization: the South Vietnamese would carry more of the burden, the United States less. Johnson, reassured by the ferocious American counter-attack at that time, was inclined to accept these cautious recommendations but delayed taking action for the time being.[11]

By then the volatile state of popular opinion at home may have begun to affect the President, who as always avidly followed the polls. Especially worrisome to him was the outcome of the first presidential primary, in New Hampshire on March 12. Although Johnson's name was not on the ballot, party regulars had launched a write-in campaign for him. When the votes were counted, however, he had won only 49 percent of the Democratic turnout. Senator Eugene McCarthy of Minnesota, a strong opponent of the war who had challenged him for the Democratic presidential nomination in January, not only received 42 percent, a stunning figure against an incumbent President, but also won more delegates than Johnson did to the Democratic National Convention that summer. A majority of McCarthy's voters, polls later discovered, were hawks who blamed Johnson for not winning the war. At the time, however, the vote was interpreted as a sign of left-of-center anti-war sentiment. It was surely anti-Johnson. When New York senator Robert Kennedy, whom Johnson loathed, announced his own candidacy on March 16, pressure intensified on LBJ to make some conciliatory moves. This pressure, combined with the gold crisis and the improving military situation, seems to have induced Johnson on March 22 formally to accept Clifford's relatively moderate recommendations.

Clifford, meanwhile, kept seeking more data. On March 26 and 27 he convened many foreign policy experts—dubbed "the wise men" by contemporaries—to help him with further recommendations. The wise men included Acheson, until then a renowned hawk on the war, Maxwell Taylor, another hawk, George Ball, McGeorge Bundy, Matthew Ridgway, Henry Cabot Lodge, and others. Most of them had been called in to advise the administration in November, at which time they had supported the President. This time they looked more carefully at the new data. Some, like Taylor, emerged from this process to urge backing for West-

11. Berman, *Lyndon Johnson's War*, 179; Herring, *America's Longest War*, 194–95.

moreland. Most of the wise men, however, were deeply upset by what they found. They concluded that the enemy could match whatever force the United States threw into the arena of battle. Bundy observed that Vietnam was a "bottomless pit." Clifford commented, "There are grave doubts that we have made the type of progress we had hoped to have made by this time. As we build up our forces, they build up theirs. . . . We seem to have a sinkhole. . . . I see more and more fighting with more and more casualties on the US side and no end in sight to the action."[12]

Johnson bridled when Clifford and others relayed the pessimism of the wise men to him. "These establishment bastards have bailed out," he is supposed to have said.[13] Confronted with such opinion, however, he made some conciliatory moves that he had been considering earlier. Within the next few days he accepted the advice of Rusk, who had recommended that the United States call a partial bombing halt in North Vietnam. Johnson also signified his readiness to engage in peace talks with the North Vietnamese and chose Harriman as America's representative in the event that the North Vietnamese agreed to talk.

On March 31 Johnson appeared on prime-time television to announce these decisions. They were important in that they marked—after three years—a tacit, though not binding, admission of the failure of continued escalation. But they did not represent much change in the policies he had been pursuing since late 1967. Rather, they were tactical moves aimed primarily at calming down domestic dissent. When Hanoi surprised him by responding favorably to the idea of peace talks, Harriman was sent to Paris, which was chosen as the site for the effort, in May. But an impasse quickly developed. Hanoi demanded that the United States cease all bombing. Johnson, fearing that this would jeopardize American troops, insisted that Hanoi agree to cut back its military activity in the South. Bombing continued, and talks went nowhere. Peace in Vietnam seemed farther away than ever.[14]

United States military effort actually intensified during this time. Indeed, Johnson cut back the bombing in the North only because he had become convinced that it was not doing much good and because bad weather was in any case expected to hamper missions over the northern part of North Vietnam in the near future. Meanwhile, the United States stepped up bombing of enemy resources in the South. In March and April the United States conducted the largest search and destroy missions in the history of the war. It then launched an Accelerated Pacification Program

12. Berman, *Lyndon Johnson's War*, 180.
13. Herring, *America's Longest War*, 206.
14. Ibid., 204–8.

to secure as much of the countryside as possible in the event of serious negotiations. Finally, it greatly increased military aid to Saigon, pushing up the force level of the Army of South Vietnam from 685,000 to 850,000.

Johnson did add a big surprise in his televised speech on March 31. Waiting until the end of his address, he paused, then added. "There is division in the American home now . . . and holding the trust that is mine, as President of all the people, I cannot disregard the peril to the . . . prospect for peace. . . . I do not believe that I should devote an hour a day of my time to any personal partisan course. . . . Accordingly, I shall not seek, and I will not accept, the nomination of my party for President."

Why Johnson took this step, which astonished many who heard him, was not altogether clear. But it was not because he feared to lose his fight for renomination. Although McCarthy and Kennedy were challenging him boldly, they stood little chance of securing the party's endorsement against an incumbent President. Rather, Johnson seems to have decided not to run again because he was tired, both physically and emotionally, and because he knew he had lost his political capacity to get things done. By stepping aside he hoped to bring a little more harmony to a populace that was already badly fragmented by disputes over the war, race relations, and the many other contentious issues that had inspired backlash in the previous two years.

ONLY FOUR DAYS LATER, on April 4, a high-powered bullet from a sniper's rifle badly damaged whatever hopes Johnson and others still retained for a softening of racial polarization in the United States. It shattered the jaw of Martin Luther King as he stood on a motel balcony in Memphis, Tennessee, where he had been supporting striking black sanitation workers seeking recognition of a union. King's presence in Memphis was characteristic of the still non-violent efforts he had been making since 1965 to promote economic justice for the masses of the black poor in the cities. The bullet, which killed him, did much to destroy chances for non-violent leadership on behalf of social justice for blacks.[15]

News of King's murder frightened Congress into passing Johnson's open housing bill within the week.[16] But nothing could arrest the rage that overwhelmed many black Americans. On the night of the killing

15. Robert Weisbrot, *Freedom Bound: A History of America's Civil Rights Movement* (New York, 1990), 266–70.
16. See chapter 21.

rioting erupted in Washington, where looters and arsonists destroyed white-owned stores (and black tenements above them) in black sections of the city. Nine people were killed in rampaging that followed. Riots damaged more than 130 other cities, causing property damage estimated at more than $100 million. Police arrested 20,000 people. A total of forty-six people, all but five of them blacks, died in the wave of violence.[17]

Violence and extremism, indeed, seemed ubiquitous in April and May of 1968. Overseas it afflicted Paris, where left-wing students took over the Latin Quarter, collaborated with factory workers, and ultimately brought down the government of Charles de Gaulle. In Czechoslovakia rebels of the "Prague Spring" revolted against Communist rule, only to be overwhelmed in August by Soviet tanks. Riots engulfed the Free University in West Berlin. Bloody confrontations between students, workers, and authorities convulsed Tokyo, Bologna, Milan, and Mexico City, site of the Olympic games that fall. The widespread outbursts, most of them student-inspired, and the violent repression that they often evoked from police, unnerved political leaders throughout the industrialized world.[18]

In the United States militant students also threatened the status quo, mainly on a few of the most prestigious university campuses. What happened at Columbia University in late April set the stage. Mark Rudd, a passionate admirer of Ché Guevara who had recently returned from a trip to Cuba, led defiant student protestors against a variety of alleged misdeeds by a rather hapless university administration. These sins included the support of classified war research and indifference to the needs of nearby black residents of Harlem. Dispensing with civility, Rudd wrote an open letter to university president Grayson Kirk in which he quoted LeRoi Jones, "Up against the wall, motherfucker, this is a stick-up." In the prolonged confrontations that followed, approximately 1,000 students (of 17,000 in all at the university)—some of them spearheaded by Rudd and the SDS, others by militant blacks—managed to seize five university buildings and to rifle files in the president's office. They ran up red flags atop two of the buildings and festooned the walls of offices with portraits of Marx, Malcolm X, and Ché Guevara. After six days of occupation, police were called in at 2:30 in the morning. Their response exposed the backlash that energized working-class Americans who raged at the protests of the privileged. Swinging clubs, the police tore into the students. More

17. Weisbrot, *Freedom Bound*, 270.
18. John Diggins, *The Rise and Fall of the American Left* (New York, 1992), 221.

than 100, plus some police, were injured. A total of 692 people were arrested. The university virtually shut down, to the dismay of thousands of non-demonstrating students and faculty.[19]

The conflict at Columbia was the most violent and uncivil battle to that time; it received enormously wide publicity; and it encouraged a host of struggles at other campuses, most of them between 1968 and 1970. It has been estimated that there were 150 violent demonstrations (and many more that were non-violent) on American campuses, including many of the most prestigious ones, in the 1968-69 academic year alone. A few of the protests rivaled Columbia's, notably a demonstration at Cornell in 1969, where black students brandishing guns and sporting bandoliers of ammunition forced concessions from the university administration. Other student protestors jostled faculty and staff, vandalized libraries, disrupted classes, and—in a second episode at Columbia in the spring of 1968—burned years of research notes by a faculty member.[20]

Like the demonstrators at Cornell, students "won" some of these battles. President Kirk, for instance, resigned, and most of the student protestors there were not disciplined for the damage or disruption they had caused. Universities introduced changes giving student groups a somewhat larger role in decision-making on campus. Curricula were broadened, usually to the accompaniment of greater choice of courses and fewer requirements. Black Studies programs began to proliferate. Most important, the unrest of 1968 greatly heightened the rights-consciousness of students. From then on, many university administrators and faculty moved with caution lest they provoke campus uprisings.

Whether the curricular changes were "reforms," of course, sparked loud and lasting debate. Many parents and professors lamented the decline of "general education." A number of new courses introduced to pacify protestors lacked academic rigor. Other Americans, including a few professors who were themselves members of the Left, were appalled at what they considered the arrogance and play-acting of the demonstrators, who seemed to be emulating the theatrics of on-the-street agitators and Third World revolutionaries. The historian Eugene Genovese, a leading Marxist scholar, branded the students as "pseudo-revolutionary middle-

19. Terry Anderson, *The Movement and the Sixties: Protest in America from Greensboro to Wounded Knee* (New York, 1995), 193–98; William Leuchtenburg, *A Troubled Feast: American Society Since 1945* (Boston, 1973), 176; Gitlin, *Sixties*, 306–9; Diggins, *Rise and Fall*, 219.
20. Leuchtenburg, *Troubled Feast*; Diane Ravitch, *The Troubled Crusade: American Education, 1945–1980* (New York, 1983), 200–205.

class totalitarians."[21] William O'Neill, another historian, observed wryly that many universities prior to the rise of student unrest had at least required hard work and discipline—training for life in the real world. In some of the post-protest universities, he lamented, "The Protestant ethic gave way to the pleasure principle in college but not in life."[22] Reactions such as these reflected a widespread sense among Americans that the students were spoiled brats.[23]

Black activism off the campuses aroused equally contentious emotions. Following King's assassination, the Reverend Ralph Abernathy, King's most trusted aide, tried to pick up the fallen banner by pressing ahead with a plan that King had endorsed before his death—a Poor People's March on Washington. Abernathy hoped to stimulate national action against poverty among blacks. The result of the march, however, proved embarrassing to Abernathy and co-leaders. Attempting to dramatize the plight of the poor, organizers built a shantytown, Resurrection City, on the mall in Washington. But construction was hurried and shoddy, leaving early arrivals in mid-May without adequate power, water, or sanitation facilities. Heavy rains created seas of mud. The number who braved conditions to live there never exceeded 2,500 and usually totaled around 500. Picketers of government buildings aroused little attention. Activists representing Mexican-Americans and Indians—the march was to be a multi-ethnic effort—clashed with Abernathy and other black organizers, whom they accused of trying to dominate the proceedings. Some of the marchers smashed windows and tossed one another into fountains.

The debacle of the Poor People's March ended only in late June when police dispersed the last few residents of Resurrection City. By then virtually everyone involved was glad that the struggle had ended. The failure was partly one of disorganization. But it was mainly a reflection of the time. Many whites had responded enthusiastically in 1963 to the March on Washington, which had dramatized the goals of the civil rights bill then under consideration. By 1968, however, the black agenda focused much more directly on poverty and racial discrimination in the North. Whites were much less supportive of demands such as these, especially amid the backlash following rioting in the cities. Reflecting such feelings, Congress did nothing.[24]

21. Ravitch, *Troubled Crusade*, 227.
22. William O'Neill, *Coming Apart: An Informal History of America in the 1960s* (Chicago, 1971), 302.
23. Ravitch, *Troubled Crusade*, 223–24.
24. Weisbrot, *Freedom Bound*, 272–75.

Black militants were so divided and demoralized for the remainder of 1968 that they commanded little attention, especially by contrast to the previous few years, in the media. Eldridge Cleaver, having published *Soul on Ice* early in March, remained sporadically in the news as a presidential candidate for the California-based Peace and Freedom party, but once he fled into exile, surfacing for a time in Cuba, he attracted support from only a few on the fringes. With SNCC, CORE, and the Black Panthers in virtually total disarray, no black organization—not even the still active NAACP—came close to filling the void left by the assassination of King.

By far the most widely noticed black protest in these months broke out at the Olympic games in October. Two of America's many top athletes at Mexico City were Tommie Smith, gold medal winner in the 200 meters, and John Carlos, who finished third in the race. Both, like many on the national track team, were African-Americans. Before they ascended the stand to receive their medals, they rolled up their sweat pants to reveal black socks, and they displayed protest buttons on their chests. On the stand they bowed their heads and raised black-gloved fists in a black-power salute. Their gesture of defiance, televised throughout the world, brought international attention to the cause of racial justice. For many athletes the protest became a defining moment; they could never again ignore the political and racial dimensions of sports.[25]

Smith and Carlos, however, lost in the short run. Officials of the United States Olympic Committee suspended them from the team and banned them from the Olympic village. White politicians denounced them for their lack of patriotism. Moreover, some black athletes feared to stand with them. This was in part because they had a lot to lose if they defied white America—look what had happened to Muhammad Ali! The boxer George Foreman, a black man, walked around the ring waving a small American flag after knocking out a Russian challenger to win the gold medal in the heavyweight division at Mexico City. In the United States, O. J. Simpson, winner of the Heisman Trophy as college football's best player, had refused to join the Black Student Union at the University of Southern California, the predominantly white school where he had played. Asked for his reaction to the defiance of Smith and Carlos, he commented, "I respect Tommie Smith, but I don't admire him."[26]

25. *Newsweek*, Oct. 28, 1968, p. 74.
26. *Newsweek*, Aug. 29, 1994, p. 44.

ALL THESE EVENTS OF 1968—anguished responses to Tet, the assassination of King, riots in the cities, confrontations on campus, the further spread of black-power ideology and of ethnic consciousness— heightened the fragmentation and polarization that had come to light in the previous two years. They also inflamed a presidential campaign that became in many ways the most heated of the twentieth century.[27]

The most fiery struggles badly burned the Democratic party. McCarthy, whom Allard Lowenstein and other liberal activists had persuaded to challenge Johnson in January, took an early lead among anti-war Americans, especially students.[28] In New Hampshire and subsequent Democratic primaries thousands of young people, "Clean for Gene," canvassed energetically to help him. They respected his intelligence, his wit, his care in developing positions on the issues, his commitment to opening up party processes to new groups of people, and his refusal to pander to audiences. Above all, they admired his courage, rare among established politicians, in challenging an apparently invulnerable President of his own party.

McCarthy was indeed an unusual politician. As a young man he had spent a nine-month novitiate at a monastery before abandoning thoughts of becoming a monk. He had then taught at Catholic colleges, where he also wrote poetry. Among his friends was Robert Lowell, perhaps America's most distinguished poet. An ardent supporter of Adlai Stevenson in 1960, McCarthy had backed LBJ, not JFK, for the presidential nomination after Stevenson gave up. He then became known as a fairly liberal senator and as a reliable supporter of LBJ (who dangled the vice-presidential nomination before him in 1964) before revolting against the President's policies in Vietnam. His strong showing in the New Hampshire primary inspired his followers to hope that he might win the nomination.[29]

From the beginning, however, McCarthy left many people cold. He was often arrogant with supporters, including his own staff, and contemptuous about the glad-handing rituals of democratic political campaigning.

27. Theodore White, *The Making of the President, 1968* (New York, 1970); Lewis Chester et al., *An American Melodrama: The Presidential Campaign of 1968* (New York, 1969).

28. William Chafe, *Never Stop Running: Allard Lowenstein and the Struggle to Save American Liberalism* (New York, 1993), 262–314.

29. For McCarthy, see John Blum, *Years of Discord: American Politics and Society, 1961–1974* (New York, 1991), 291; and Allen Matusow, *The Unraveling of America: A History of Liberalism in the 1960s* (New York, 1984), 407–9; Leuchtenburg, *Troubled Feast*, 203; O'Neill, *Coming Apart*, 376–77; and Gitlin, *Sixties*, 297.

He made few efforts to cultivate the press. When energized, he could be an inspiring public speaker, but more often than not he made no apparent effort to reach out to listeners. Some observers wondered if he really wanted to win. McCarthy seemed especially uncomfortable trying to cope with the passionate emotions aroused by race. He avoided speaking in ghettos and in other places where blacks were numerous. When King was murdered, he said nothing. Although liberal opponents of Johnson respected McCarthy, many yearned for someone who could excite the masses of blacks and working-class Democrats.[30]

That someone, of course, was Robert Kennedy. For a while after his brother's assassination in 1963, "Bobby" had seemed traumatized. He nursed deep and abiding resentments against Johnson, with whom he had clashed bitterly during the 1960 convention, and whose presence in the White House—in place of his brother—was a rankling reminder of what might have been. These bitter feelings never abated; if anything they intensified over time. But some of the old ruthlessness, for which opponents had both feared and hated him in the early 1960s, seemed to soften. Supporters said he had grown. Even enemies sensed that he had mellowed.[31]

Liberal political organizers, led by Lowenstein, had worked hard in late 1967 to get Kennedy to challenge Johnson. They knew that he would be a charismatic campaigner and that he possessed a unique asset: the mystique and the magic of the Kennedy name. Kennedy was sorely tempted, both because he loathed Johnson and because he had grown increasingly critical of the war. But he had refrained from breaking openly with LBJ. Moreover, many of the political professionals to whom he turned for advice counseled him against running. They pointed out what seemed to be the obvious: Johnson, as President, could not be denied the Democratic nomination. Better to wait until 1972.

When Kennedy reluctantly agreed with this analysis, many people were both upset and angry. Lowenstein responded, "The people who think that the future and the honor of this country are at stake because of Vietnam don't give a shit what Mayor Daley [of Chicago] and Governor Y and Chairman Z think. We're going to do it, and we're going to win, and it's a shame you're not with us because you could have been President."[32] Lowenstein then turned to McCarthy, who boldly took the plunge that

30. Jeremy Larner, *Nobody Knows: Reflections on the McCarthy Campaign of 1968* (New York, 1969); Matusow, *Unraveling*, 407–11.
31. Arthur Schlesinger, Jr., *Robert F. Kennedy and His Times* (New York, 1978); O'Neill, *Coming Apart*, 364, 373–74; Chafe, *Unfinished Journey*, 352–53.
32. Chafe, *Never*, 271; O'Neill, *Coming Apart*, 361.

Kennedy had shied away from. When Johnson's popularity dwindled in early 1968, especially after Tet, many liberals openly expressed their contempt of Kennedy. They carried placards, BOBBY KENNEDY: HAWK, DOVE, OR CHICKEN?[33]

When Kennedy finally jumped into the contest—after the New Hampshire primary exposed Johnson's vulnerability—he infuriated many liberals who had committed themselves to McCarthy earlier in the year. They complained, often bitterly, not only that Kennedy was "chicken" but also that his candidacy would split the liberal and anti-war camps that opposed Johnson's policies. The result, they predicted after LBJ's withdrawal on March 31, would be that Vice-President Humphrey, Johnson's surrogate, would win the Democratic presidential nomination. In 1960 or even in 1964 many liberals would have welcomed that outcome, for Humphrey had been a committed supporter of civil rights and other social programs. But as Vice-President he had swallowed doubts about the war and had supported Johnson's policies. He was anathema to many Democratic liberals in 1968.

Kennedy, despite these handicaps, gradually cut into the base of McCarthy's anti-war and liberal support. This was not because he was more fervently anti-war than McCarthy. On the contrary, while both candidates called for an end to American bombing and for allowing the National Liberation Front a role at the peace table and in the subsequent political life of South Vietnam, McCarthy was willing to endorse in advance a coalition government including the NLF, and Kennedy was not. Kennedy indicated that he would maintain America's commitment to South Vietnam and would support "retaliatory action" against the North if necessary. Nor was it because Kennedy had necessarily better answers for inner-city problems. He favored expansion of public and private spending to build up black areas in the cities. (He was himself contributing a great deal of his own money to such efforts in the Bedford-Stuyvesant area of New York.) This approach commanded only lukewarm support from many people concerned with urban racial problems. A program of "gilding the ghettos," they said, ran counter to the goal of most people who lived there—to escape. If enriching the ghettos worked, which critics doubted, it would reinforce racial separation. McCarthy, emphasizing the goal of integration, denounced Kennedy's stance and called instead for the construction of "new towns" on the edge of cities, so blacks could move out and live where the jobs were.[34]

33. Leuchtenburg, *Troubled Feast*, 204.
34. Matusow, *Unraveling*, 408.

Kennedy's campaign caught fire, instead, because he seemed much more engaged and eloquent, especially on the subject of race relations, than McCarthy. When Kennedy heard that King had been killed, he ignored advisers who warned him to stay out of the exploding inner cities. Instead, he braved the black center of Indianapolis—he was then running in the Indiana primary—where he climbed on top of a car to speak movingly of his support for racial equality. He was so intense, so obviously shattered by the killing, that the once seething crowd grew attentive and respectful. Later he spoke in the impoverished black sections as well as the white working-class wards of Gary. He delivered the same candid and unpatronizing message wherever he went: assailing racial prejudice, denouncing riots, deploring the rise of welfare, celebrating the virtues of hard work. In particular he appealed to the idealism and consciences of people in the middle classes. He thereby forged coalitions of supporters that cut across race and class lines and that brought him victory in the primary. Thousands of liberals, recognizing McCarthy's weaknesses, swarmed to the Kennedy cause.

In the remaining few weeks of the primary season Kennedy solidified his appeal as a champion of poor and working-class Americans. In Oklahoma he deplored the poverty of Indians on the reservations; in California he befriended Chávez; in New York he identified with the plight of Puerto Ricans. Although he lost a primary to McCarthy in Oregon—the only time any Kennedy had failed to win an election—he attracted huge and sometimes frighteningly responsive crowds almost everywhere else he went. Throngs of people heaved against him and his worried bodyguards; women lunged to touch his hair. More than once he emerged from crowds with torn clothes and with hands bleeding from the hundreds of squeezes and slaps that besieged him. Veteran political observers were astonished and shaken by the powerful emotions that Kennedy aroused.

Kennedy capped his exciting run with a close but decisive victory over McCarthy in the key California primary in early June. In his moment of triumph, however, he was fatally shot by Sirhan Sirhan, a deranged Arab nationalist, in a hallway of a Los Angeles hotel. The killing and its aftermath unleashed vivid memories of JFK's murder more than four years earlier. When Bobby's body was carried on a train from New York to Washington, where he was to be buried near his brother, throngs of weeping and waving Americans stood alongside the tracks. In Baltimore thousands sang "The Battle Hymn of the Republic" even before the train appeared. The death of Robert Kennedy further smashed the already beleaguered forces of American liberalism and devastated people who had looked to him as the only remaining hope to heal a fragmented nation.

Could Kennedy have won the nomination if he had lived? That became one of the most frequently asked questions in the history of modern American politics. When he was killed, he needed 800-odd additional delegates to win the nomination. Some of them might have come from McCarthy—if McCarthy, an unpredictable man, proved willing to let them go. Others might have abandoned Humphrey, whose chances then seemed hopeless for November.[35] Still, the Johnson-Humphrey forces maintained a firm hold on the party machinery, which they manipulated without qualm at the convention. Johnson hated Kennedy as much as Kennedy hated him. All these political realities would have worked strongly against Kennedy's chances for the nomination.

The Democratic convention that took place in Chicago in late August turned out to be such a wild and bloody affair that the first-ballot nomination of Humphrey, by then foreordained, was scarcely noticed.[36] Chicago Mayor Richard Daley had long anticipated some sort of confrontation. The mayor, indeed, reflected the complex feelings of many who joined the backlash of the late 1960s. By then he had lost enthusiasm for the war effort, mainly because he had concluded that it could not succeed. But Daley, like many of the working-class people who were the source of his power, was revulsed by anti-war demonstrators, whom he regarded as elitist, pampered, sanctimonious, and unpatriotic. He was equally hostile to unruly blacks: in April, amid rioting in Chicago following the assassination of King he had ordered his police to "shoot to kill" arsonists and "shoot to maim or cripple" looters. By the time the convention opened Daley had barricaded the site and had amassed a formidable force of 12,000 police (plus 5,000 National Guardsmen and 6,000 federal troops in readiness nearby) to quell the slightest disturbance. When demonstrators arrived, he denied them permits to sleep in public parks, to march, and otherwise to engage in meaningful protest. He was eager for an excuse to put them down.

Many anti-war activists, forewarned that there would be serious trouble at Chicago, stayed home. The numbers who came from out of town were therefore relatively small; estimates placed them at around 5,000. Another 5,000 or so from the Chicago area joined them on occasion during the five days of protests that followed, but most demonstrations, scattered as they were over seven miles of Chicago shorefront, were

35. Joseph Califano, *The Triumph and Tragedy of Lyndon Johnson: The White House Years* (New York, 1991), 323; Gitlin, *Sixties*, 182.
36. David Farber, *Chicago '68* (Chicago, 1988); Farber, *The Age of Great Dreams: America in the 1960s* (New York, 1994), 221–24.

small—police ordinarily outnumbered protestors by three or four to one. Many of those who journeyed to Chicago were pacifists and advocates of non-violence who belonged to the National Mobilization Committee to End the War in Vietnam, or Mobe, as it was called. Mobe, however, was a sprawling coalition of groups, some of which seemed ready and willing to provoke violence in order to advance their cause. Tom Hayden, a key leader, was one of these. By the summer of 1968, following the assassinations and rioting that had raised the level of turbulence in the country, it was clear that many of the activists who arrived in Chicago were anticipating a fight.[37]

A smaller though much more colorful group of demonstrators called themselves Yippies, or members of the Youth International Party. The Yippie phenomenon—one could hardly call it a movement—was largely the creation of two incredible characters, Abbie Hoffman and Jerry Rubin. Both were veterans of hippie and anti-war activities, including the march on the Pentagon in 1967. They had a wonderful affinity for the absurd, a gift for theatrics, and a keen awareness of the way in which the zaniest antics attract attention from the media. They expected and welcomed violent retaliation from Daley's police, and they very much wanted to be noticed.[38] They announced that the Yippies would dress up as bellboys and seduce the wives of delegates, and give out free rice on the streets. They proposed to nominate a pig, Pigasus, for the presidency. The Yippie slogan declared, "They [the Democrats] nominate a president and he eats the people. We nominate a president and the people eat him."[39]

Some of the violence that disrupted Chicago erupted as early as Sunday August 25, the eve of the convention, when Yippies seeking to camp in Lincoln Park three miles north of the convention site taunted police. "Pig, pig, fascist pig," they chanted. "Pigs eat shit!" When the Yippies disobeyed an order to leave the park area at 10:30 P.M., police chased after them through the streets of the city, clubbing them as they fled. Those who refused to leave, perhaps 1,000 in all, suffered the same fate. The police also assaulted reporters and photographers from *Newsweek*, *Life*, and the Associated Press. The battle at Lincoln Park continued sporadi-

37. Matusow, *Unraveling*, 411–22; Blum, *Years of Discord*, 306–10; O'Neill, *Coming Apart*, 382–85; Gitlin, *Sixties*, 320–26.
38. Marty Jezer, *Abbie Hoffman: American Rebel* (New Brunswick, 1992); Abbie Hoffman, *Revolution for the Hell of It* (New York, 1970); Jerry Rubin, *Do It: Scenarios of the Revolution* (New York, 1970).
39. Matusow, *Unraveling*, 412–13. Rubin was also Cleaver's running mate on the Peace and Freedom ticket in 1968.

cally and violently for the next two nights. Confrontations also arose outside the convention hotel, the Hilton, where protestors chanted, "Fuck you, LBJ," "Dump the Hump," "Sieg Heil," and "Disarm the Pigs."

On Wednesday, August 28, the angriest struggles ravaged the city. That was the day on which allies of Johnson secured a strongly pro-war plank (by a vote of 1,567 to 1,041) and on which Humphrey, having accepted the plank, was later nominated on the first ballot. Although much of the oratory in the hall was boring, tempers heated up by the evening, especially after televised accounts of violence outside reached the delegates. At one point Connecticut senator Abraham Ribicoff stood on the podium to nominate Senator George McGovern of South Dakota, a candidate who represented many former supporters of Kennedy. Ribicoff stared down at Daley, twenty feet away in the audience, and exclaimed, "With George McGovern we wouldn't have Gestapo tactics on the streets of Chicago." Infuriated delegates from Illinois jumped up shouting and waving their fists. Daley was purple with rage and shouted back with words that, while drowned out in the bedlam, were lip-read by many in the national television audience: "Fuck you, you Jew son of a bitch you lousy motherfucker go home."[40]

The violence against demonstrators outside the Hilton and near the hall at the time was indeed shocking. When protestors tried to march to the hall, thousands of police, acting on orders from Daley, determined to stop them. Taking off their badges, they charged, tossed tear gas, cracked people with clubs, and shouted, "Kill, kill, kill!" All who strayed into their path—demonstrators, bystanders, medics, reporters and photographers—became targets. Hundreds were bloodied, though no one—miraculously enough—was killed. It was dark, but television lights illuminated some of the scenes, and a national audience, listening to protestors shouting, "The whole world is watching," looked at outbursts of graphic violence. Later, police staged a pre-dawn raid on McCarthy's fifteenth-floor headquarters at the Hilton, clubbing young volunteers whom they accused of lobbing urine-filled beer cans at police lines below.[41]

During these stunning confrontations the Johnson-Humphrey forces remained unrepentant. Humphrey, winning at last the presidential nomination that he desired so much, the next day chose Edmund Muskie, a senator from Maine, as his running mate and defended the actions of Daley and his police force. The mayor, he said, had done nothing

40. Gitlin, Sixties, 334. Matusow, Unraveling, 421, has a slightly different reading.
41. Newsweek, Sept. 9, 1968, pp. 24, 41.

wrong: "The obscenity [of the demonstrators], the profanity, the filth that was uttered night after night in front of the hotels was an insult to every woman, every daughter, indeed every human being. . . . You'd put anybody in jail for that kind of talk. . . . Is it any wonder that the police had to take action?"[42]

Many Americans did wonder. They argued that Daley need not have worried much about the demonstrators, whose numbers were modest. He could have let out-of-towners sleep in the park and have been more generous in setting guidelines for marches and demonstrations. He could surely have restrained his police. Instead, he encouraged them to run amok. In so doing he played into the hands of Hoffman, Rubin, and other demonstrators—seven of whom ("the Chicago Seven") federal authorities then proceeded to prosecute for conspiring to riot. Rubin later commented, "We wanted exactly what happened. . . . We wanted to create a situation in which the . . . Daley administration and the federal government . . . would self-destruct."[43]

It was no wonder, however, that Humphrey reacted as he did. An earnest, well-meaning man, he was appalled by the often juvenile behavior of some of the demonstrators. Millions of Americans agreed with him: polls suggested that a majority of the American people defended the riotous behavior of the Chicago police under the circumstances. Still, the disorder at Chicago hurt Humphrey and the Democratic party, which limped out of Chicago more badly wounded than ever. McCarthy refused to appear with Humphrey or to endorse him. The nominee, reconsidering his defense of Daley and the police, soon admitted that disaster had occurred. "Chicago," he conceded two days later, "was a catastrophe. My wife and I went home heartbroken, battered, and beaten."[44] He recognized, as did the vast majority of political pundits when the convention finally ended, that it would take some sort of miracle to resuscitate the Humphrey-Muskie ticket and the Democratic party. How far the mighty—the liberal Democrats who had swept to victory in 1964— had fallen!

As HUMPHREY RALLIED to begin his campaign he understood that he had to deal with two formidable foes, George Wallace of Alabama, who had announced in February as presidential candidate under the

42. Gitlin, Sixties, 338.
43. Blum, Years of Discord, 309. Bobby Seale of the Black Panthers was tried separately. Following wild courtroom scenes, five of the seven were found guilty of incitement, but their convictions were subsequently overturned.
44. Matusow, Unraveling, 422.

banner of the American Independent party, and Richard Nixon, whom the Republicans had nominated for President three weeks before the Democratic debacle at Chicago.

Wallace was indeed a frightening force in 1968. Although he knew he could not win the election, he hoped to capture enough southern and border states so as to throw a close race into the House of Representatives. Surprising many political observers, he managed to get on the ballot in all fifty states, and his popularity climbed steadily, to 21 percent in the immediate aftermath of the Democratic convention.[45] As in the past, Wallace commanded enthusiastic support among southern segregationists. Most far-right organizations, including the KKK, Citizens' Councils, and the John Birch Society, openly aided his operations.[46] Much of the power driving his campaign came from the South, exposing more sharply than ever the regional rifts that had widened during the Goldwater-Johnson contest in 1964.

The appeal of Wallace in 1968, however, transcended regional lines, important though those were. It rested also on his evocation of backlash in many working-class areas of the North. Wallace was an energetic, aggressive, caustic, sneering, often snarling campaigner. Eschewing openly racist oratory, he called for "law and order" in the streets and denounced welfare mothers who he said were "breeding children as a cash crop." He gleefully assailed hippies, leftists, and radical feminists, some of whom picketed the Miss America pageant in Atlantic City just after the Democratic convention, dumped what they called objects of female "enslavement"—girdles, bras, high-heeled shoes, false lashes, and hair curlers—into a "freedom trash can," and earned the label forever after of "bra-burners."[47] Wallace took special pleasure in attacking anti-war demonstrators, often with thinly veiled references to violent retribution that excited many of his followers. "If any demonstrator ever lays down in front of my car," he proclaimed, "it'll be the *last* car he'll ever lay down in front of." He also set forth an economic program designed to appeal to blue-collar working folk. It included support for a federal job-training program, safeguards for collective bargaining, a higher minimum wage, and better

45. Blum, *Years of Discord*, 310.
46. Kirkpatrick Sale, *Power Shift: The Rise of the Southern Rim and Its Challenge to the Eastern Establishment* (New York, 1975), 103.
47. John D'Emilio and Estelle Freedman, *Intimate Matters: A History of Sexuality in America* (New York, 1988), 301–2. The protestors draped a sheep in yellow and blue ribbons and crowned it queen. Parading it along the boardwalk, they sang, "There she is, Miss America."

protection for people who lost their jobs or could not afford adequate medical care.[48]

Wallace seemed most passionate in attacking know-it-all federal bureaucrats and self-styled experts who tried to tell honest working-class folk what to do. "Liberals, intellectuals, and long-hairs," he cried, "have run the country for too long." His audiences cheered when he denounced "over-educated, ivory-tower folks with pointed heads looking down their noses at us." These were "intellectual morons" who "don't know how to park a bicycle straight." He added, "When I get to Washington I'll throw all these phonies and their briefcases into the Potomac."[49]

In driving for the presidency in 1968 Wallace made some mistakes, among them his selection in early October of General Curtis LeMay as his running mate. LeMay, who had directed fire-bomb raids on Japan in World War II, remained a fierce, plain-spoken advocate of air power, including nuclear weapons. He was widely believed to have been the model for the mad general in Stanley Kubrick's *Dr. Strangelove* (1964), a devastating satire of Cold War military zeal. In a disastrous early press conference following his selection as running mate LeMay told reporters, "I don't believe the world would end if we exploded a nuclear weapon." Despite all the tests in the Pacific, "the fish are back in the lagoons, the coconut trees are growing coconuts, the guava bushes have fruit on them, the birds are back."[50] Humphrey soon started calling Wallace and LeMay the "bombsy twins."

When Wallace heard comments like this, he was dismayed. Like LeMay, he had supported the war, but by 1968 he knew that it was a divisive issue, and he had no clear ideas about resolving the stalemate. Like many Americans in 1968, he was not so much pro-war as anti-anti-war. This remained a key theme of his campaign: stimulating popular anger at privileged and "unpatriotic" young folk who went about ridiculing the military while dodging the draft.

For all his appeal, however, Wallace remained a third-party candidate.

48. Walter Dean Burnham, *Critical Elections and the Mainsprings of American Politics* (New York, 1970), 143–58; Richard Polenberg, *One Nation Divisible: Class, Race, and Ethnicity in the United States Since 1938* (New York, 1980), 221.
49. Leuchtenburg, *Troubled Feast*, 211. For Wallace, see also Jonathan Rieder, "The Rise of the Silent Majority," in Steve Fraser and Gary Gerstle, eds., *The Rise and Fall of the New Deal Order, 1930–1980* (Princeton, 1989), 243–68; Mickey Kaus, *The End of Equality* (New York, 1992), 37–38; and Thomas Edsall, "Race," *Atlantic Monthly*, May 1991, pp. 53–86.
50. Herbert Parmet, *Richard Nixon and His America* (Boston, 1990), 526.

Political pundits did not expect him to win any states outside of the Deep South. The biggest worry of Humphrey and his advisers was the candidacy of Nixon, who held what appeared to be an insurmountable edge over the Democrats following the Chicago convention. The former Vice-President, fifty-five-years old in 1968, had seemed politically doomed following his defeat in a race for governor of California at the hands of Pat Brown in 1962. A very poor loser, he had lashed out at the press at the time, "You won't have Nixon to kick around any more." In 1968, however, he managed to win the GOP nomination on the first ballot, in part because he was a centrist in the party and in part because he had doggedly supported Republican candidates throughout the 1960s while lining up backing for himself. He then surrounded himself with a coterie of advertising, public relations, and television experts and ran a carefully scripted campaign that stressed his experience, especially in the field of foreign policy. Some contemporary observers, hoping for the best, speculated that there was once again a "new Nixon."[51]

Nixon still nursed bitterly the many slights and injustices that he imagined had been his lot. Life, he thought with enduring self-pity, consisted of a series of "risks" and "crises."[52] He retained the same passion to succeed in politics that had driven him in the past to excesses, sometimes vicious, of partisanship and personal invective. Uncomfortable before crowds, he remained an uninspiring campaigner. His speeches, as in the past, bordered at times on the mawkish. His movements, notably a hands-over-the head gesture of triumph, seemed studied and phony. John Lindsay, a liberal Republican from New York City, observed that Nixon looked like a "walking box of circuits."[53]

Nixon's running mate, Governor Spiro Agnew of Maryland, was of doubtful help to him. Agnew had won his governorship in 1966, one of a number of Republicans who swept into office in the reaction against Johnson and the Democratic party that year. He had then seemed to be a moderate, backing Governor Nelson Rockefeller of New York, a Republican liberal, for the presidency in 1968. But Agnew, a Greek-American

51. Stephen Ambrose, *Nixon: The Triumph of a Politician, 1962–1972* (New York, 1989), 133–222; Joe McGinniss, *The Selling of the President, 1968* (New York, 1969); John Judis, *Great Illusions: Critics and Champions of the American Century* (New York, 1992), 182–85; Parmet, *Richard Nixon*, 20–21.

52. Joan Hoff-Wilson, "Richard M. Nixon: The Corporate Presidency," in Fred Greenstein, ed., *Leadership in the Modern Presidency* (Cambridge, Mass., 1988), 164–98.

53. Ibid., 180; O'Neill, *Coming Apart*, 380.

whose father had been a poor immigrant, was yet another political leader stung into backlash, especially by riots in Baltimore following the murder of King. At that time Agnew gained notice by damning the "circuit-riding, Hanoi-visiting, caterwauling, riot-inciting, burn-America-down type of [black] leader."[54] Nixon, seeking a running mate who would be strong for "law and order," chose Agnew without looking too carefully into his background. He had reason to doubt his choice during the campaign when Agnew dropped ethnic slurs, speaking of "Polacks" and calling a reporter a "fat Jap. " Resurrecting the tactics of Joe McCarthy, Agnew called Humphrey "squishy soft on communism." He observed, "If you've seen one slum, you've seen them all." The *Washington Post* concluded that Agnew was "perhaps the most eccentric political appointment since the Roman Emperor Caligula named his horse a consul."[55]

These problems aside, Nixon ran a well-calculated, very well financed campaign. The GOP spent enormous sums on radio and television—estimated at $12.6 million as opposed to $6.1 million for the Democrats—and pioneered the practice of what became known as "demographic marketing." This involved hiring "media specialists," who assumed unprecedented influence in the campaign. They undertook market research to determine the concerns of special groups and then directed particular ads, most of them "spots," at blocs of voters. They also did much with new TV technology—videotape, zoom shots, split screens—to make their spots lively. It was the first truly high-tech television campaign in American history.[56]

On the issues Nixon tried not to rock the boat; after all, he began the fall campaign with a huge lead over Humphrey. Concerning Vietnam, the driving political issue of the era, he said only that he had a "secret plan" to end the war. On domestic matters Nixon echoed Wallace, but in a more genteel fashion, by catering to the contemporary backlash. (Humphrey mocked Nixon as a "perfumed, deodorized" version of Wallace.) This meant celebrating "law and order," denouncing Great Society programs, rapping the liberal decisions of the Supreme Court, and deriding hippies and protestors. He lambasted the "busing" of children, then being applied

54. Fred Siegel, *Troubled Journey: From Pearl Harbor to Ronald Reagan* (New York, 1984), 228; Parmet, *Richard Nixon*, 510.
55. Parmet, *Richard Nixon*, 524–25.
56. Edwin Diamond and Stephen Bates, *The Spot: The Rise of Political Advertising on Television* (Cambridge, Mass., 1992), 142–46, 175; Bernard Yamron, "From Whistle-Stops to Polispots: Political Advertising and Presidential Politics in the United States, 1948–1980," Ph.D. Thesis, Brown University, 1995, 144–210.

in places as a means of desegregating the schools. Much of his campaign, like his choice of Agnew, reflected what pundits later called a Southern Strategy, which aimed to corral the backlash white vote in the South (and elsewhere). "Working Americans," he declared, "have become the forgotten Americans. In a time when the national rostrums and forums are given over to shouters and protestors and demonstrators, they have become the silent Americans. Yet they have a legitimate grievance that should be rectified and a just cause that should prevail."[57]

Only one issue seemed likely to derail Nixon as the campaign progressed: Vietnam. If Humphrey could win over McCarthy, McGovern, and onetime Kennedy supporters, he might patch up the badly tattered Democratic party. This, of course, was an extraordinarily difficult task, because he did not want to alienate the President, who adamantly refused to stop the bombing. But Humphrey knew he had to try. On September 30 he bought television time for a well-advertised speech in Salt Lake City, which he gave at a lectern without the vice-presidential seal. In the speech Humphrey proclaimed his willingness, under certain conditions, to "stop the bombing of North Vietnam as an acceptable risk for peace because I believe it could lead to the success in the negotiations [in Paris] and thereby shorten the war." His statement was hedged and pleased neither Johnson, who fumed that Humphrey had gone too far, nor McCarthy, who waited another month before endorsing the party ticket—and then only weakly. At first the speech made no difference in polls of voters. But by mid-October political observers sensed a change in the mood of people, especially liberals and opponents of Johnson's conduct of the war. They had never had any use for Nixon, let alone Wallace, and they yearned for a reason to return to the Democratic fold.[58]

Humphrey, too, seemed rejuvenated by the shift in popular mood. Until then a somewhat demoralized campaigner, he became ever more passionate in his backing of social programs, including civil rights, that liberal Democrats had traditionally supported. Labor union leaders, who had always admired Humphrey, redoubled their efforts to draw workers out of the clutches of Wallace and back into the party of FDR and Truman. The Democratic coalition that had been a key element in American politics since the 1930s seemed to be coming together again. Wallace, meanwhile, started to sink in the polls, partly because of LeMay, partly because voters recognized in the end that they would be wasting their votes on him. The polls showed Humphrey's resurgence. In

57. Rieder, "Rise of the Silent Majority," 260–61; Matusow, *Unraveling*, 432.
58. Matusow, *Unraveling*, 431–32.

late September, Gallup had given Nixon 43 percent, Humphrey 28 percent, and Wallace 21 percent. By October 24, two weeks before the election, Humphrey still lagged, but the numbers—44 percent for Nixon, 36 percent for Humphrey, 15 percent for Wallace—showed movement. What had seemed in early September to be an open-and-shut election was growing exciting.[59]

At this point rumors flew about that the Johnson administration was reaching an agreement that might cut the carnage in Vietnam: a total bombing halt by the United States and reciprocal though unspecified military restraint by North Vietnam. Ho Chi Minh, it seemed, was willing to negotiate directly with South Vietnam, which he had always refused to recognize as a legitimate state. In return the United States would no longer balk at involvement in the talks of the National Liberation Front. On October 31 LBJ announced that the United States would stop the bombing. This softening of American policy infuriated Nixon, who accused the administration of playing politics with the war.

South Vietnamese president Nguyen Van Thieu, however, then spiked hopes for negotiations by announcing that he would not take part in peace talks in Paris if the NLF was involved. Why he did so caused great political controversy in the United States. Johnson, high-handedly ordering the FBI to wiretap the South Vietnamese embassy, blamed Anna Chan Chennault, the Chinese-born widow of General Claire Chennault of Flying Tigers fame. She was then vice-chairman of the GOP National Finance Committee. The tapes revealed that Chennault, who had some access to Nixon, called the embassy on November 2 and urged that Thieu hold his ground. Nixon, she added, would offer South Vietnam a better deal. Johnson angrily phoned Nixon, who denied any involvement in Chennault's machinations.

Nixon was indeed deeply interested in the negotiations that Harriman and others were frantically conducting in Paris. Henry Kissinger, a Harvard professor of government who was supposedly helping the administration in Paris, was in fact double-dealing. Eager to secure a high-ranking job in Washington, Kissinger sought to curry favor with Nixon (who he thought would win), by secretly relaying classified information about diplomatic developments to the GOP.[60] Nixon was thus well briefed on all that was happening and in a good position to take actions that would delay progress.

There was no solid evidence, however, that Nixon instigated or knew of

59. Ibid., 433–34.
60. Parmet, *Richard Nixon*, 519; Chafe, *Unfinished Journey*, 391.

Chennault's actions. Most likely, Thieu, who was a shrewd politician, needed no prodding from Chennault or others to anticipate a better relationship with a Republican administration in Washington. He also faced strong pressure from political allies in South Vietnam who understandably feared any lessening of American resolve. He therefore refused to relent under American pressure. Hanoi, he said, must formally agree to de-escalate the war and must negotiate directly with South Vietnam. He did not back down from this position until two weeks after the election, following still stronger pressure from the United States. The South Vietnamese did not join the negotiations in Paris until mid-January.[61]

Humphrey, too, was unhappy with Johnson, in part because LBJ had done little to support the campaign, in part because the President did not keep him well informed about the negotiations. By election day the two men were barely on speaking terms. Still, the President's decision to halt the bombing greatly encouraged advocates of de-escalation and further boosted Humphrey's now fast-moving effort for the presidency. Campaigning enthusiastically, he surged forward. Final polls indicated that the election was a toss-up, too close to call.

NIXON WON, BUT ONLY BARELY. He received 31,785,480 votes to 31,275,166 for Humphrey and won the electoral college by a margin of 301 to 191. Wallace captured 9,906,473 votes, taking Alabama, Arkansas, Georgia, Louisiana, and Mississippi and scoring 46 in the electoral college.[62] Nixon received only 43.4 percent of the vote, Humphrey 42.7 percent, Wallace 13.5 percent. If Wallace and Humphrey between them had managed to secure thirty-two more electoral votes, they would have denied Nixon an electoral college majority, throwing the election into the heavily Democratic House of Representatives.

Democrats could take some solace from the fact that Humphrey had managed to make things close. Indeed, some of the groups that made up the Democratic coalition held firm. This was especially true of black voters, an estimated 97 percent of whom backed Humphrey. The coalition further proved its endurance in races for the House of Representatives, where the party maintained a margin of 245 to 187. In the Bedford-Stuyvesant district of New York City they elected Shirley Chisholm, the first black woman ever to win a House seat, over James Farmer. Although they lost seven seats in the Senate, they still could look forward to a

61. Herring, *America's Longest War*, 217–18; Matusow, *Unraveling*, 435–36.
62. One elector from North Carolina also voted for Wallace. The Peace and Freedom ticket of Cleaver and Rubin won 36,563 votes.

margin of 57 to 43 in 1969. Among the Democratic senators re-elected were Ribicoff, McGovern, and others who would uphold the liberal cause.

Those who had feared Wallace also derived some satisfaction from the election returns. Outside the Deep South he did much less well than it had seemed he would earlier in the year. Strom Thurmond, running on the States' Rights ticket in 1948, had won thirty-nine electoral votes, almost as many as Wallace did twenty years later. Although Wallace made inroads among working-class people, especially among Catholics and Italian-, Irish-, and Slavic-Americans, he failed to break the grip that Democrats had on the majority of blue-collar Americans. It was estimated that Wallace received around 9 percent of the votes of white manual workers in the North. Most blue-collar Americans, like most blacks, seemed to remain reliable members of the Democratic coalition.[63]

It was obvious, however, that the election of 1968 marked a huge turnabout from 1964. In that year 43.1 million voters had opted for LBJ, almost 12 million more than Humphrey received (in a turnout that was 3 million higher) four years later. Even Kennedy, winning slightly fewer than 50 percent of the ballots in 1960, had attracted 34.2 million voters, 3 million more than Humphrey did. Between them Nixon and Wallace won 57 percent of the vote.[64] No amount of wishful thinking could obscure the fact that the 1968 election boded ill for the future of the Democratic party.[65]

This did not mean that voters were elated to see Nixon in the White House. He had run a banal, lackluster campaign and had watched an enormous lead dwindle to almost nothing. He would be a minority President with Democrats in control of both houses of Congress. Having offered little in the way of positive programs, he had no mandate for much of anything, except perhaps to dismantle things that Johnson had erected. As Samuel Lubell, a student of elections, put it, Nixon was "little more

63. *Newsweek*, Nov. 11, 1968, p. 36; Rieder, "Rise of the Silent Majority," 251, 261; Matusow, *Unraveling*, 438.

64. Pundits were divided on whether Wallace hurt or helped Nixon. On the one hand, he appealed to millions of voters who might otherwise have voted for the GOP. On the other hand, it was widely assumed that many of those who voted for Wallace were at least nominally Democratic. See Matusow, *Unraveling*, 438, and Parmet, *Richard Nixon*, 526, for differing intepretations. However one interpreted these matters, the key point remained: Humphrey did not come close to a majority.

65. As emphasized at the time by Kevin Phillips, *The Emerging Republican Majority* (New Rochelle, 1969).

than a convenient collection basket, the only one available into which [voters] were depositing their numerous discontents with the Johnson administration."[66]

Still, reflections on the 1968 campaign and election showed how deep the trouble had become for the Democrats. The course of events in that unusually turbulent year first of all exposed a further decomposition of the political parties. This was a major legacy of the McCarthy movement, which indicated that a largely unheralded candidate—especially if he had a powerful issue—could come out of nowhere and destabilize a major party organization. The political coming of age of television, well advanced by 1968, further abetted political figures who ran more as individuals than as party regulars. Although it was not fully appreciated at the time, the election of 1968 also presaged the sharp decline of a system of nominating presidential candidates that had featured the role of party bosses and state conventions. Henceforth presidential nominations—and campaigns—depended much more on the ability of individuals to tap into grass-roots feelings, to exploit primaries, which proliferated after 1968, and to sound impressive on the sound bites of television. These changes affected Republicans, but they especially changed things for the Democrats, who were more divided and unruly. After 1968 the Democratic party became less and less a purposeful political organization when it came to presidential politics and more and more a loose coalition of freewheeling individuals.[67]

The campaign and election of 1968 also revealed the persistent power of regional differences in the United States. This was especially obvious, of course, in the Deep South, where racial hostilities drove political life. Wallace, like Strom Thurmond in 1948 and Barry Goldwater in 1964, exposed the weakness of Democratic presidential candidates in the area. As Johnson himself had predicted following passage of the 1964 civil rights bill, the identification of northern Democrats with civil rights destroyed the hold that the party had once had in the region. Humphrey, indeed, fared poorly not only in the Deep South but also in the Upper South and the border states, where Wallace and Nixon divided up most of the votes. Nixon carried Kentucky, Tennessee, Virginia, Missouri, Okla-

66. Siegel, Troubled Journey, 202–3.
67. Arthur Schlesinger, Jr., "The Short Happy Life of Political Parties," in Schlesinger, The Cycles of American History (Boston, 1986), 256–76; David Broder, The Party's Over: The Failure of American Parties (New York, 1972); Burnham, Critical Elections, 135–74.

homa, and (thanks in part to Thurmond, who had switched to the GOP) South Carolina.

Nixon also did extraordinarily well in states west of the Mississippi. His strength there did not attract much notice at the time, but it accelerated a trend that, like the rise of the GOP in the South and border states, became central to American presidential politics for the next several decades. The only states that Humphrey won west of the Mississippi were Minnesota, Texas, Washington, and Hawaii. All but Texas were historically liberal states. The other western states, however, including the key state of California, almost always went Republican after 1968. From then on GOP presidential candidates could run confidently in most states outside of the Middle West and Northeast. This was a comfortable feeling.

The sources of Republicanism in the Plains and West were not so obvious as those that affected politics in the South. They also varied, for there were significant differences between the politics of a state such as Arizona and those of North Dakota or California. Local issues often figured in major ways. In general, however, the rise of the GOP in the West reflected the resentments that Goldwater in 1964 and Wallace and Nixon in 1968 tried to exploit: distrust of far-off government bureaucrats, especially liberals, who tried to tell people what to do. Some of those with the most vocal resentments in the West represented powerful special interests—oil and mining companies, large-scale farmers, ranchers, real estate developers—who raged at federal efforts in support of the environment, Indians, or exploited farm workers. "Rights," they said, meant freedom from government interference. Others were fundamentalists— they were especially numerous in the South and the West—who perceived liberals, such as the majority on the Warren Court, to be heretical. But the rise of Republicanism in the region also affected millions of other people who were neither conservative on economic issues nor fundamentalist. Save in Hawaii, where Asian-Americans were an important voice, in enclaves such as Watts, and places in the Southwest where people of Hispanic background were developing a political voice, the mass of voters west of the Mississippi were non-Hispanic whites. Most of them lived in the country, in small towns, in suburbs, or in cities of modest size. They were increasingly repelled by the ethnic, big-city eastern world that seemed to them to consist of ghettos, riots, crime, welfare, and family break-up. Their feelings of backlash, exacerbated by the polarizing events of the mid-1960s, seemed if anything to grow over time.

Backlash was indeed the dominant force in the exciting campaign and election of 1968. Much of this continued to reflect racial antagonisms, the most powerful determinant of electoral behavior in the 1960s.[68] Humphrey received less than 35 percent of the white vote—a remarkable statistic—in 1968. But backlash transcended racial divisions, fundamental though those were to an understanding of American society in the late 1960s. It stemmed also from rising dissatisfaction with Johnson's policies in Vietnam and from doubts—well exploited by Wallace and Nixon—about the liberal social policies that the Great Society bureaucrats had oversold after 1964. The backlash represented a powerful reaction against liberalism, a major casualty of the 1960s.[69]

The reaction, as it turned out, was lasting. It affected not only Humphrey in 1968 but also liberals generally thereafter. Polls from the mid-1960s on showed that smaller and smaller percentages of the American people had faith in their elected officials or in the ability of government to do things right.[70] Turnouts, while subject to varying interpretations, suggested the same. The turnout of eligible voters in presidential elections, having reached a postwar high in the closely contested election of 1960, thereafter dropped consistently. It fell to 60.6 percent in 1968, as compared to 64 percent in 1960 and 61.7 percent in 1964.[71] Those who studied these trends, moreover, concluded that the falling off in voting—and more generally in political involvement—was disproportionally serious among poor, working-class, and lower-middle-class people, most of whom had been Democrats. Never very influential in politics, they seemed to feel ever more alienated in the 1960s and thereafter. Their disaffection with politicians contributed further to the decomposition of the parties and to the ability of special interests to play major roles in the governance of the country.[72]

68. Edsall, "Race."
69. Thomas Edsall, with Mary Edsall, *Chain Reaction: The Impact of Race, Rights, and Taxes on American Politics* (New York, 1992); E. J. Dionne, Jr., *Why Americans Hate Politics* (New York, 1991).
70. Morris Janowitz, *The Last Half-Century: Societal Change and Politics in America* (Chicago, 1978), 113. Gallup periodically asked people, "If you had a son, would you like him to go into politics as a life's work?" The percentage that said yes rose from 20 percent in 1950 to 27 percent in 1955 to 36 percent in 1965, and then plummeted to 23 percent by 1973. See also Daniel Yankelovich, *The New Morality* (New York, 1974), 95.
71. And to 55.1 percent in 1972 and 52.6 percent in 1980. See chapters 23 and 25.
72. Thomas Edsall, "The Changing Shape of Power: A Realignment in Public Policy," in Fraser and Gerstle, eds., *Rise and Fall*, 269–93; Chafe, *Unfinished Journey*, 458–59.

The turbulent year of 1968 was therefore a pivotal year in the postwar history of the United States. The social and cultural antagonisms that rent the country, having sharpened appreciably since 1965, widened so significantly following the shattering experiences of Tet and the assassinations of King and Kennedy that they could not thereafter be resolved. The social and political history of the United States in the next few years witnessed mainly an extension—sometimes an acceleration—of the conflicts that reached their peak in 1968. There seemed to be no turning back.

23

Rancor and Richard Nixon

Eight months after Richard Nixon assumed office a gigantic "happening" transformed the 600-acre farm of Max Yasgur in Bethel, New York. Promoters hailed the event as "An Aquarian Exposition" and named it the Woodstock Music and Art Fair, Woodstock being the name of a nearby town. It was conservatively estimated that 400,000 people, most of them in their teens and twenties, flocked to "Woodstock" for the three-day event on August 15–17. Mammoth traffic jams prevented hundreds of thousands more from taking part in the revelry.

It rained intermittently while the young people were there, but no one semed to mind. Those who were close enough heard some of the most popular rock musicians of the era, including Janis Joplin, Jimi Hendrix, and Jefferson Airplane. Others settled as best they could in the mud and reveled in their freedom. Some took off their clothes and wandered about in the nude. A few had sex in the open. The vast majority smoked marijuana or used other drugs. While hundreds suffered from bad trips caused by low-grade LSD, most of those who attended remembered a wonderfully free and mellow occasion. Rollo May, a psychoanalyist, described Woodstock as a "symptomatic event of our time that showed the tremendous hunger, need, and yearning for community on the part of youth."[1]

Woodstock was indeed a mellow event, a high point of countercultural

---

1. "The Message of History's Biggest Happening," *Time*, Aug. 28, 1969, pp. 32–33; Michael Frisch, "Woodstock and Altamont," in William Graebner, ed., *True Stories from the American Past* (New York, 1993), 217–39.

expression in the United States.[2] It was one of a number in 1969, a year of many ironies. In Washington, Richard Nixon, exemplar of traditional values, struggled to preserve the older ways. Elsewhere, demonstrations, protests, and rebellions against these ways seemed to flourish. *Easy Rider*, a low-budget movie starring Peter Fonda, Dennis Hopper, and Jack Nicholson, celebrated the joys of getting high on drugs and motorcycling on the open road. It drew large audiences and made substantial profits. Rock musicians like Joplin, Jim Morrison, and the Rolling Stones stimulated wild enthusiasm among many of the young.

Significant cultural changes that had already advanced apace in the postwar era were indeed accelerating at that time, branching out to affect American society in general. In 1969, 74 percent of women believed that premarital sex was wrong; four years later, only 53 percent did.[3] In 1965, 26 percent of Americans had opposed abortion, even when the pregnancy represented a serious risk to the health of the woman; the years later, fewer than 8 percent felt that way. The more open depiction of sex, especially in magazines and in film, startled contemporaries. By 1973 Americans could watch Bernardo Bertolucci's *Last Tango in Paris*, in which (among other things) Marlon Brando grabbed a stranger, kissed her, tore off her underpants, and had sex with her standing up.

Changing attitudes toward sex may even have begun to promote slightly greater tolerance toward homosexuals. In June 1969 homosexuals at the Stonewall Inn in Greenwich Village, tired of harassment by police, fought back and set off five days of confrontations. Their activism did much to arouse group consciousness among the gay population. In 1973 the National Organization for Women, which had previously disdained lesbians (Friedan referred to them as the "lavender menace") endorsed gay rights.[4] In 1974 the American Psychiatric Association removed homosexuality from its list of psychological disorders.[5]

2. Terry Anderson, *The Movement and the Sixties: Protest in America from Greensboro to Wounded Knee* (New York, 1995), 241–92. For accounts of the counterculture, see also David Farber, *The Age of Great Dreams: America in the 1960s* (New York, 1994), 167–89; Allen Matusow, *The Unraveling of America: A History of Liberalism in the 1960s* (New York, 1984), 275–307; and Theodore Roszak, *The Making of a Counterculture* (Garden City, N.Y., 1969).
3. Rosalind Rosenberg, *Divided Lives: American Women in the Twentieth Century* (New York, 1992), 201.
4. William Chafe, *The Paradox of Change: American Women in the 20th Century* (New York, 1991), 210–12.
5. John D'Emilio and Estelle Freedman, *Intimate Matters: A History of Sexuality in America* (New York, 1989), 324–25; Martin Duberman, *Stonewall* (New York, 1993).

The hastening of the sexual revolution was perhaps the most important cultural change to emerge from the 1960s. Yet it was but one manifestation of the grand expectations and rights-consciousness that more than ever were defining contemporary understandings of the Good Life.[6] Increasing numbers of Americans—especially those who were young and relatively well-off—sensed by the Nixon years that they had an unusually wide set of choices in life. They expected to enjoy greater freedom than their parents had. They need not defer gratification or sacrifice themselves to a job—or even to a spouse. Many yearned for "self-fulfillment" and "growth." This meant the chance to pursue all sorts of leisure activities, to be creative, to seek adventure, to embrace the "joy of living." Charles Reich's jejune *The Greening of America*, which treated these quests for transcendence as "rights," was serialized in the *New Yorker* and became a best-seller in 1970.[7]

Searchers for a "meaningful" and "satisfying" life—these were key adjectives at the time—often demanded the "right" to good health and empowerment in achieving it. Many people, among them feminists, struck out against orthodox practitioners and the American Medical Association, which they said was monopolistic, racist, and sexist.[8] Other Americans insisted that the government do more to fight disease, forcing through in the process funding for a "war on cancer" in 1970. Still others assailed the tobacco industry, succeeding by 1970 in requiring cigarette companies to place on packages the label "Warning: The Surgeon General has determined that cigarette smoking is dangerous to your health." Smoking ads were banned from television and radio in January 1971.[9]

It was yet another golden age for advocates of diet reform, regarded by many as a key to better health. Counterculturalists, among others, spread the gospel of organic gardening and hailed the virtues of oats, dates, sunflower seeds, prunes, and raisins. It was especially important, many

6. Dennis Wrong, "How Critical Is Our Condition?," *Dissent* (Fall 1981), 414–24.
7. Daniel Yankelovich, *The New Morality* (New York, 1974), 4–5, 38–39; Charles Reich, *The Greening of America: How the Youth Revolution Is Trying to Make America Liveable* (New York, 1970).
8. Jonathan Imber, "Doctor No Longer Knows Best: Changing American Attitudes Toward Medicine and Health," in Alan Wolfe, ed., *America at Century's End* (Berkeley, 1991), 298–317.
9. James Patterson, *The Dread Disease: Cancer and Modern American Culture* (Cambridge, Mass., 1987), 201–30. These were token victories: opponents of smoking wanted more explicit warnings (later achieved); the ban on TV and radio ads saved the companies a good deal of money that they had previously spent for such purposes. Per capita cigarette smoking increased until the late 1970s.

advocates thought, to avoid the adulterated fare produced by industrial civilization: better to eat "natural" food as peasant people did in Third World cultures. Thanks in part to the efforts of Ralph Nader, a free-wheeling and indomitable critic of corporate misbehavior in American life, broader "food scares"—about the dangers from additives, meat-packing, and processing—also arose in the late 1960s and early 1970s. Robert Finch, Nixon's Secretary of Health, Education, and Welfare, issued an order in 1969 banning the sale of cyclamate, an artificial sweet-ener thought to be associated with cancer. DDT, a pesticide, was banned in the same year. The American Heart Association stepped up its warn-ings about excessive intake of cholesterol. By the early 1970s it was esti-mated that some 50 percent of Americans regularly took vitamins or other food supplements.[10]

Good health required "fitness." Dr. Kenneth Cooper's book *Aerobics* (1968) sold 3 million copies over the next four years. Health clubs prolif-erated, and the sale of sports equipment, including a variety of specialized sneakers, boomed.[11] Fitness, moreover, came to require spiritual and psychological growth as well as physical well-being. Promoters of the "Human Potential Movement," therapists from the Esalen Institute and other meccas, and gurus of indescribable exoticism popped up like genies to lighten the burdens of America's stress-ridden civilization. Experts touted confrontational "encounter," gestalt therapy, bioenergetics, "sen-sitivity training," meditation, massage, breathing, drugs, and even easy recreational sex. Any or all would bring out the inherent spirituality of the self, enlarge human potential, and light up the dawn of the New Age.[12]

Many who embarked on the road to "self-actualization" in these turbu-lent times carried heavy anti-Establishment baggage with them. Indeed, assaults against authority, which had already engaged activists of all sorts in the 1960s, gained ever-greater ground. Critics exclaimed that large and bureaucratic institutions—schools, universities, hospitals, governments, corporations—threatened the natural goodness of human beings. Cru-saders for "free schools" denounced the educational Establishment and spearheaded an "open education movement" that would be informal,

10. Harvey Levenstein, *Paradox of Plenty: A Social History of Eating in Modern America* (New York, 1993), 160–77, 204–6.
11. Randy Roberts and James Olson, *Winning Is the Only Thing: Sports in America Since 1945* (Baltimore, 1989), 222–24; Landon Jones, *Great Expectations: America and the Baby Boom Generation* (New York, 1980), 257–59.
12. Todd Gitlin, *The Sixties: Years of Hope, Days of Rage* (New York, 1987), 425; Aristides, "Incidental Meditations," *American Scholar* (Spring 1976), 173–79.

based on "discovery," and above all "child-centered."[13] Other challenges to authority seemed ubiquitous. Soldiers confronted officers, reporters balked at editors, patients challenged doctors, artists denounced curators, graduate students protested against requirements and formed unions. High school students resisted teachers and defiantly wore forbidden buttons. Native Americans and Mexican-Americans raised banners of "red power" and "brown power." A few priests and nuns even broke with the church to marry, sometimes each other. Rights-conscious litigants sustained further expansion of the legal profession. No institutions could shut off the rhetoric of rights that rang throughout the land.

Feminists emerged as among the most vociferous of these advocates of change. Between 1969 and 1973, peak years of feminist activism, they formed hundreds of organizations including caucuses within professional groups, campus collectives, centers for women's studies, and political action committees. The Women's Equity Action League, developed in 1968, led efforts against discriminatory practices in industry, education, and other institutions. By 1974 there were scores of new feminist periodicals, most of them created in 1970 and 1971. Ms., begun in January 1972 as a one-time publication, sold 250,000 copies within eight days.

Some of the most visible feminists were radicals who professed to despise men: A bumper sticker read DON'T COOK DINNER TONIGHT, STARVE A RAT TODAY.[14] Much more numerous, however, were those women, including millions of housewives who did not call themselves feminists, who demanded an end to unequal treatment. They were better organized and more powerful than ever before in their quest for rights. By early 1972 they affected Congress, which approved the Equal Rights Amendment by overwhelming majorities. As of mid-1973, twenty-eight states had ratified it.[15]

The Supreme Court, as in the 1960s, extended an especially strong hand to the Rights Revolution. In 1969 it struck down state laws requiring

13. Diane Ravitch, The Troubled Crusade: American Education, 1945–1980 (New York, 1980), 235–37, 250–56; Gitlin, Sixties, 429–30.
14. Fred Siegel, Troubled Journey: From Pearl Harbor to Ronald Reagan (New York, 1984), 209.
15. Richard Polenberg, One Nation Divisible: Class, Race, and Ethnicity in the United States Since 1938 (New York, 1980), 270; Arlene Skolnick, Embattled Paradise: The American Family in an Age of Uncertainty (New York, 1991), 103–5, 117–18; Carl Degler, At Odds: Women and the Family in America from the Revolution to the Present (New York, 1980), 446–47; Ravitch, Troubled Crusade, 293. For varied reasons, the surge later ebbed, and ERA finally failed of ratification, lacking three states, in 1982.

a year's residence before people could receive welfare benefits.[16] Earl Warren retired later in the year, to be replaced by Warren Burger, a moderate Minnesotan named by Nixon. In many ways, however, the departure of Warren seemed to make little difference. In 1970 the Burger court ruled that statutory entitlements of welfare and disability benefits could not be taken away from people without due process.[17] In 1971 it unanimously supported tough new federal guidelines concerning job discrimination, and a year later it determined in *Furman v. Georgia* that capital punishment was "cruel and unusual punishment" and therefore in violation of the Eighth Amendment.[18] This decision, decided by a vote of 5 to 4, contained loopholes that permitted states to sanction capital punishment if it was not arbitrary or capricious; twenty-nine states quickly did so. Moreover, the Burger Court proved tough concerning crime, reaching a number of judgments that weakened the *Miranda* rules. Still, through early 1973 it showed few signs that it would impose a new conservatism on the land.

In 1973 the Burger Court announced, in *Roe v. Wade*, that women had a constitutional right to abortion.[19] Using as a precedent the "right to privacy" that the court had enunciated eight years earlier in striking down Connecticut's law against contraception, the decision surprised a great many Americans.[20] Abortion, after all, had long been driven underground in the United States. For the next two decades, *Roe v. Wade* aroused extraordinarily powerful emotions, which further polarized a nation that was already fragmented along the lines of race, gender, and social class.

Some of those who demanded more from life during the Nixon years were easily defrauded. "Clients" who looked to gurus for therapeutic

16. *Shapiro v. Thompson*, 394 U.S. 618 (1969), by a margin of 6 to 3. Warren was a dissenter.
17. Mary Ann Glendon, "Rights in Twentieth-Century Constitutions: The Case of Welfare Reform," in Hugh Davis Graham, ed., *Civil Rights in the United States* (University Park, Pa., 1994), 140–50. The case was *Goldberg v. Kelly*, 397 U.S. 618 (1969).
18. *Griggs v. Duke Power Co.*, 401 U.S. 424 (1971)(to be discussed later in this chapter); and *Furman v. Georgia*, 408 U.S. 238 (1972).
19. 410 U.S. 113 (1973), by a vote of 7 to 2. See *Newsweek*, Feb. 5, 1973, pp. 27–28; David Garrow, *Liberty and Sexuality: The Right to Privacy and the Making of "Roe v. Wade"* (New York, 1994); Norma McCorvey, with Andy Meisler, *I Am Roe* (New York, 1994).
20. The Connecticut case was *Griswold v. Connecticut*, 381 U.S. 479 (1965). See chapter 19.

deliverance often received "touchy-feely" sensations and little more. Other Americans seeking a New Age were simply hedonistic, caught up in the excesses of a consumer culture that seemed to know no limits. Worse, some of those attracted to the wilder excesses of drugs and acid-rock culture came to ruin. The stomping of concert-goers by Hell's Angels at the Altamont, California, rock festival in December 1969 was one of the most alarming signs that the spirit of Woodstock could go very badly amiss.[21] In 1970 both Joplin and Hendrix died of drug overdoses. Jim Morrison also died, probably of excessive consumption of alcohol. All were twenty-seven years of age.

New Left organizations also fell victim to excess in 1969–70. SDS, having grown rapidly in the mid-1960s, splintered badly. This process, rooted in an ever-growing rage and self-indulgence that drove true believers to extremes, became most obvious at the national convention of SDS in June 1969, where the organization broke apart along racial and ideological lines.[22] Some of the splinters—the self-named Motherfuckers and the Crazies, for example—were largely nihilistic. Another dissident faction became known as Weatherman, after a Dylan song, "You Don't Need a Weatherman to Know Which Way the Wind Blows." The Weathermen stomped out of the convention chanting, "Ho, Ho, Ho Chi Minh," and supported violence in support of foreign and domestic liberation movements. "Our whole life," Weatherman proclaimed, "is a defiance of Amerika. It's moving in the streets, digging sounds, smoking dope . . . fighting pigs."[23] In October the group staged "Four Days of Rage" of "trashing" in Chicago that stopped only when police arrested 290 of the 300 people involved. A total of seventy-five police were hurt in the melees.[24]

Although the vast majority of radicals remained non-violent, a few zealots in the Weatherman and other sects resorted at that time to violence on a scale rarely experienced in American history. Between September 1969 and May 1970, there were at least 250 bombings linked to white-dominated radical groups in the United States. This was an average

21. See chapter 15.
22. Theodore Draper, *The Rediscovery of Black Nationalism* (New York, 1970), 108.
23. Maurice Isserman and Michael Kazin, "The Failure and Success of the New Radicalism," in Steve Fraser and Gary Gerstle, eds., *The Rise and Fall of the New Deal Order, 1930–1980* (Princeton, 1989), 212–42.
24. William O'Neill, *Coming Apart: An Informal History of America in the 1960s* (Chicago, 1971), 295; John Diggins, *The Rise and Fall of the American Left* (New York, 1992), 231; Gitlin, *Sixties*, 399–401.

of almost one per day. (The government placed the number at six times as high.) Favorite targets were ROTC buildings, draft boards, induction centers, and other federal offices. In February 1970 bombs exploded at the New York headquarters of Socony Mobil, IBM, and General Telephone and Electronics. The spate of bombings slowed down only in March, when three members of Weatherman accidently killed themselves while preparing explosives in a Greenwich Village townhouse. Ironically, they were the only people killed in the bombings to that time.[25]

The collapse of SDS did not signify the end of anti-war activity in the United States. Far from it—foes of war in Vietnam, reaching beyond students, gathered ever-greater strength in 1969 and 1970. Nor did the death of SDS mean that radicals left no legacy.[26] Many on the left moved on in later years to work for social change as advocates of the poor and as community organizers. Still, the fall of SDS distressed many radicals both then and in later years, for it had been the largest and most visible New Left organization in postwar United States history. Moreover, many of the blows that sank it—and other groups of radicals at the time—were self-inflicted. The SDS and other New Left groups, mostly composed of students, never established strong institutional bases off-campus. Disdaining unions, the students aroused increasingly vehement backlash from labor leaders and from the working classes. They failed for the most part to reach the poor, especially in black ghettos. The radicals, individualistic and sectarian, also resisted organizational discipline. Many became petulant, intolerant, millenarian, and extreme.

The splintering of the New Left exposed broader social and economic forces that affected American society in the Nixon years. Most of the youthful protestors and demonstrators of the early and mid-1960s had grown up in a prosperous era that had encouraged dreams of social transformation. In particular, the cause of civil rights had inspired them. Losing many of their battles, however, they lashed out ever more fiercely against Authority. The internecine brawls that tore them apart in 1969 and 1970 were microcosms of larger frustrations that were accumulating as crusaders for greater rights encountered increasingly unyielding opposition. The struggles between these antagonists symbolized and reflected the very contentious home front that Richard Nixon had been elected to preside over.

25. Gitlin, *Sixties*, 399–401; Diggins, *Rise and Fall*, 231.
26. Winifred Breines, "Whose New Left?," *Journal of American History*, 75 (Sept. 1988), 528–45; Isserman and Kazin, "Failure and Success."

NIXON WAS A PROFESSIONAL POLITICIAN with no interest in challenging authority. It was impossible to imagine him going to Woodstock or signing up for therapy at the Esalen Institute. The President was an avid fan of professional football, which had enjoyed a boom during the 1960s: the first Super Bowl took place in 1967. He watched *Patton* (1970), which celebrated military discipline, many times. Otherwise he never seemed to relax: even on vacation he appeared always to be garbed in suit, white shirt, and tie. Nixon not only disliked hippies and anti-war activists; he also seized countless opportunities while President to arouse backlash against them. Like LBJ, he set the FBI to work to destroy the Black Panthers and others whom he presumed to be revolutionaries.

Nixon would have sneered at the notion that it was his job to promote personal liberation or transcendence. He remained a wary, humorless, tightly controlled man who felt the need constantly to stiffen himself for crises and to protect himself against enemies, especially intellectuals and journalists. A hard worker, he put in twelve- to sixteen-hour days, in part because he was unable to delegate authority, even on trivial matters. Much of the time he isolated himself. Weeks went by without press conferences. It proved extraordinarily difficult for people, even representatives and senators, to get beyond his top aides, John Ehrlichman and H. R. Haldeman. Neither of these men had had political experience outside of campaigns. Wholly loyal to their boss, they were curt, humorless, and ruthless in their quest for efficiency. Critics soon referred to the Ehrlichman-Haldeman team as the Berlin Wall.[27]

When Nixon took office in 1969, it seemed highly unlikely that much in the way of domestic legislation would emerge. Having denounced the Great Society during the campaign, he had no desire to expand social programs. Like Kennedy, he had little interest in domestic policies. He

27. Among the many sources on Nixon in these years are Garry Wills, *Nixon Agonistes: The Crisis of the Self-Made Man* (Boston, 1970); Stanley Kutler, *The Wars of Watergate: The Last Crisis of Richard Nixon* (New York, 1990); H. R. Haldeman, *The Haldeman Diaries: Inside the Nixon White House* (New York, 1994); J. Anthony Lukas, *Nightmare: The Underside of the Nixon Years* (New York, 1976); Herbert Parmet, *Richard Nixon and His America* (Boston, 1990), esp. 535, 565, 572, 610; Joan Hoff-Wilson, "Richard Nixon: The Corporate Presidency," in Fred Greenstein, ed., *Leadership in the Modern Presidency* (Cambridge, Mass., 1988), 164–98; Joan Hoff, *Nixon Reconsidered* (New York, 1994); Alonzo Hamby, *Liberalism and Its Challengers: From FDR to Bush* (New York, 1992), 282–338; and especially Stephen Ambrose, *Nixon: The Triumph of a Politician, 1962–1972* (New York, 1989). Ambrose's book is one of three volumes that he wrote on Nixon.

said later, "I've always thought this country could run itself, without a president. All you need is a competent Cabinet to run the country at home."[28] It therefore came as something of a surprise to contemporary observers that Nixon signed a fair amount of significant legislation between 1969 and 1972. This was mainly the doing of Democrats in Congress, who pursued a more liberal course. Nixon often went along with them, in part because he did not care a great deal about domestic matters, in part because he recognized the political gains to be harvested by supporting generous social legislation, and in part because he was, in fact, a moderate himself. Throughout his career he had been a Republican centrist, closer in fact to liberals like Rockefeller than to conservatives like Taft or Goldwater. Nixon was easily the most liberal Republican American President, excepting Theodore Roosevelt, in the twentieth century. In 1971 he even called for passage of a comprehensive national health insurance plan—one that would have combined private with expanded public initiatives to provide coverage for all. No President since Truman had gone so far; none until President William Clinton in 1993 would try again.[29]

The list of domestic legislation signed during Nixon's first term was indeed fairly impressive, some of it bolstering the quest for rights that lay at the heart of the culture. It included extension for five more years of the Voting Rights Act of 1965, funding for the "war on cancer," and greater federal spending for medical training and the National Endowments for the Arts and Humanities.[30] Nixon also signed what later became an important measure (so-called Title IX) in 1972 banning sex discrimination in higher education. Title IX became vital in subsequent efforts by women to counter patterns of gender bias in colleges and universities.[31]

Although Nixon ultimately reorganized the bureaucratic apparatus of

28. Alan Wolfe, *America's Impasse: The Rise and Fall of the Politics of Growth* (New York, 1981), 73.
29. Hoff, *Nixon Reconsidered*, 137–38. The proposal, of course, did not pass, and Nixon knew that it would not. Still, he seems to have been sincere in support of it.
30. Congress also sent to the states the Twenty-sixth Amendment to the Constitution, ratified in 1971, which gave 18-year-olds the right to vote.
31. Jeremy Rabkin, "Office for Civil Rights," in James Wilson, ed., *The Politics of Regulation* (New York, 1980), 304, 314–16, 436–37; Ravitch, *Troubled Crusade*, 292. Nixon was cool, however, to the larger agenda of women's rights: in 1972 he vetoed a bill that promised to set up a national system of day-care centers, arguing that the bill would establish "communal approaches to child-rearing, over against the family centered approach." See John Blum, *Years of Discord: American Politics and Society, 1961–1974* (New York, 1991), 411.

the war on poverty, he did not dismantle the programs themselves. Head Start, publicly financed job training, and other OEO efforts continued to be funded at modest levels. Thanks to the urging of Daniel Moynihan, a Democrat who became chief adviser concerning social policies, Nixon also introduced a Family Assistance Plan (FAP) that would largely have replaced AFDC with guaranteed annual incomes for poor families— whether working or not. Minimum incomes for families of four would be $1,600 a year, plus $800 or more worth of food stamps. The plan, which Nixon hoped would reduce the role of welfare bureaucrats, got caught in a political crossfire: the National Welfare Rights Organization protested that the payments were far too low, liberals objected to requirements that some recipients work, conservatives denounced the idea as a giveaway, Democrats in general distrusted the President, and Nixon himself—never deeply engaged in the issue—lost interest. In the end, FAP did not pass. If it had, it would not have addressed the roots of poverty; there was still no consensus on how to do that. Still, FAP was a bold new approach. It would have made life a little more comfortable for many poor families, especially in the South where welfare benefits were low, and it might have simplified the bureaucratic apparatus of welfare. Nixon's association with the plan, while fleeting, suggested that he was open to ideas for streamlining the system.[32]

Nixon also indicated his willingness to support unprecedentedly high levels of spending that his rights-conscious Democratic Congresses approved for domestic purposes. In this respect he and Congress exercised much less fiscal restraint than federal officials—including the supposedly extravagant liberal Democrats in the Johnson years—had shown in the past. During Nixon's first term, means-tested federal spending per person in poverty rose by approximately 50 percent. Much of this increase represented congressionally approved hikes in entitlements such as food stamps and Medicaid; much of it also went to provide welfare to women, now exercising their rights to aid, who (with their children) swelled the rolls of AFDC.[33] In 1972 Congress also enacted a Supplemental Security In-

32. Daniel Moynihan, *The Politics of a Guaranteed Income: The Nixon Administration and the Family Assistance Plan* (New York, 1973); Vincent Burke and Vee Burke, *Nixon's Good Deed: Welfare Reform* (New York, 1974); Hoff, *Nixon Reconsidered*, 123–37; and Charles Morris, *A Time of Passion: America, 1960–1980* (New York, 1984), 140–43.

33. The number of recipients of AFDC rose from 7.4 million in 1970 to 11.1 million in 1975, stabilizing thereafter until the recession of 1991–92. Federal benefits for AFDC increased in current dollars from a total of $2.2 billion in fiscal 1970 to

come (SSI) program, which replaced existing federal-state assistance to the indigent aged, blind, and disabled with national (and therefore uniform) payments. Benefits under SSI, which began in 1974, were considerably higher than they had been and were indexed to keep pace with inflation.[34] At the same time Congress approved substantial raises in Social Security benefits and indexed them, too. Total government outlays for social insurance, which as always were much higher than spending targeted at the poor, jumped from $27.3 billion in 1969 to $64.7 billion in 1975. In time, as more and more Americans reached retirement age, these increases proved to be extraordinarily important.[35]

These and other changes—and especially continuing economic growth—helped in the short run to cut poverty from 12.8 percent of the population in 1968 to 11.1 percent in 1973, a low in modern United

---

$4.6 billion in fiscal 1975. States added $1.4 billion in 1970 and $3.8 billion in 1975. See Committee on Ways and Means, U.S. House of Representatives, *Overview of Entitlement Programs: 1992 Green Book* (Washington, 1992), 654, 660.

34. AFDC, the largest of the "categorical assistance" programs established in 1935, remained a federal-state program, and therefore one that varied considerably from state to state. AFDC payments were not indexed, and the real value of benefits fell considerably after 1973. The different approach of Congress to the indigent aged and the disabled on the one hand and "welfare mothers" on the other suggested the continuing sense among Americans that some poor people (such as the aged and disabled) were more "deserving" of aid than others.

35. Gary Burtless, "Public Spending on the Poor: Historical Trends and Economic Limits," in Sheldon Danziger, Gary Sandefur, and Daniel Weinberg, eds., *Confronting Poverty: Prescriptions for Change* (Cambridge, Mass., 1994), 51–84. Overall federal spending between 1969 and 1975, years in which defense expenditures increased only slightly (from $82.5 billion to $86.5 billion), jumped from $195.6 billion to $332.5 billion. Deficits accumulated between the fiscal years 1970 and 1975 totaled $123.4 billion, compared to total deficits of $50.3 billion between 1945 and 1950 (almost all of it in the war year of 1945), $9.1 billion between 1950 and 1955, $11 billion between 1955 and 1960, $22.9 billion between 1960 and 1965, and $44.6 billion between 1965 and 1970. Per capita federal debt rose between 1970 and 1975 from $1,814 to $2,475. (It had been $1,849 in 1945, $1,688 in 1950, $1,572 in 1960, and $1,613 in 1965.) After 1975, federal spending, deficits, and the debt rose to levels that would have been unimaginable prior to 1975. By 1990 federal spending was $1,252 trillion, the deficit $220 billion, and per capita federal debt $13,100. Expenditures for interest on the debt by then came to 21.1 percent of federal outlays, a percentage roughly twice as high as those between 1950 and 1975. All these figures are in current dollars. See *Statistical Abstract of the United States, 1994* (Washington, 1994), 330–33.

States history, and in the long run to guarantee more generous entitlements to the indigent aged and disabled.[36] Nixon, anxious (like Congress)
to win the support of the elderly—by then a well-organized, highly rights-
conscious lobby—was happy to sign these measures into law in an election year.[37]

The President hoped that these increases in federal expenditures for
social welfare could bolster what he called a New Federalism.[38] His
"revenue-sharing" plan, approved in 1972, earmarked block grants of
federal money—a proposed $16 billion between 1973 and 1975—to states
and localities, which were given greater freedom to spend it as they
wished. Cutting back on the federal bureaucracy, indeed, greatly appealed to Nixon, who distrusted Washington officialdom.[39] Revenue-
sharing appealed to many governors. Thanks in part to Nixon's political
problems during his second term, however, it did not greatly change the
balance of power between Washington and the states.

Some Native American groups, too, gained modestly during Nixon's
terms of office. By then they had definitely caught the fever of rights-
consciousness. In 1969 a group captured national attention by seizing the
island of Alcatraz, which they said was Indian land, and by proclaiming
their intention to turn it into an Indian cultural center. Four years later
activists in the American Indian Movement (AIM) forcibly occupied
Wounded Knee, South Dakota, site of a massacre of Sioux in 1890,
forcing a seventy-one-day stand-off with United States marshals before the
government agreed to reconsider the treaty rights of the Oglala Sioux.
Other Indians overran and wrecked the Bureau of Indian Affairs in Washington. Congress, with Nixon's support, responded to moderate demands.

36. The poverty rate had been 21.9 percent in 1961 and 17.3 percent in 1965. Thanks
to problems with the economy after 1973 (see chapter 25), to a rise in the number
of poor female-headed families, and to the declining real value in the 1970s and
1980s of AFDC benefits, poverty thereafter increased, ranging between 11.8 and
15.3 percent of a larger population between 1978 and 1992. In 1992, when the
rate was 14.5 percent, there were 36.9 million Americans defined by the government poverty line as poor, compared to 33.6 million in 1965, 26.3 million in
1969, and 23.2 million in 1973. See Sheldon Danziger and Daniel Weinberg,
"The Historical Record: Trends in Family Income, Inequality, and Poverty," in
Danziger et al., eds., *Confronting Poverty*, 18–50.
37. James Patterson, *America's Struggle Against Poverty, 1900–1994* (Cambridge,
Mass., 1995), 197–98.
38. Hoff, *Nixon Reconsiderd*, 65–73.
39. Otis Graham, *Toward a Planned Society: From Roosevelt to Nixon* (New York,
1976), 188–263; Hoff-Wilson, 175–78.

In 1970 it agreed to return the sacred Blue Lake and surrounding lands in New Mexico to Taos Pueblo. In 1971 it approved the Alaska Native Claims Settlement Act, thereby resolving long-standing controversies to the satisfaction of most Alaskan Indians. In 1973 it formally reversed the "termination" policy of the 1950s by re-establishing the Menominees as a federally recognized tribe and providing for the return of their common assets to tribal control. An Indian Education Act, passed in 1972, authorized new and federally supported programs for Indian children. These and other efforts—mainly to resolve land claims—barely made a dent in the poverty and isolation afflicting most Native Americans, especially those on reservations. Still, they represented signs that officials were beginning to recognize the history of exploitation by whites.[40]

One of the most surprising acts of Nixon in the field of domestic policy concerned racial discrimination in employment. With his approval, Secretary of Labor George Shultz established in October 1969 the so-called Philadelphia Plan. This required construction unions in Philadelphia employed on government contracts to set up "goals and timetables" for the hiring of black apprentices. In 1970 this mechanism was incorporated in government regulations governing all federal hiring and contracting—thereby involving corporations that employed more than one-third of the national labor force. In so doing the Nixon administration transformed the meaning of "affirmative action." When Congress approved Title VII of the 1964 Civil Rights Act to ban job discrimination, it had affirmed a meritocratic and color-blind principle: hiring was to be done without regard to race, religion, sex, or national origin. Although executive orders issued in the Johnson years had called on employers to pursue affirmative action in order to counter discrimination against *individuals*, the orders had not demanded "goals and timetables" or "set-asides" that would protect *groups*. After 1970, however, many American institutions—corporations, unions, universities, others—were required to set aside what in effect were quotas, a process that engaged the federal government as never before in a wide variety of personnel decisions taken in the private sector. This dramatic and rapid transformation of congressional intent took place as a result of executive decisions—especially Nixon's—and court interpre-

40. William Hagan, *American Indians* (Chicago, 1979); Francis Paul Prucha, "Indian Relations," in Jack Greene, ed., *Encyclopedia of American Political History*, Vol. 2 (New York, 1984), 609–22; Peter Carroll, *It Seemed Like Nothing Happened: The Tragedy and Promise of America in the 1970s* (New York, 1982), 104–6; Hoff, *Nixon Reconsidered*, 27–42.

tations. Affirmative action of this sort never had the support of demo-cratically elected representatives.[41]

The Supreme Court, moreover, indicated that it would support such approaches to curb discrimination. The most important case, *Griggs v. Duke Power Co.* (1971), involved intelligence tests (and other expedients) administered by employers to determine the eligibility of workers for certain positions. Could such tests be used if the results differentiated between blacks and whites in such a way as to bar substantial numbers of blacks from better jobs? The Burger Court ruled unanimously that em-ployers would henceforth have to demonstrate that such differentiating tests were essential, irreplaceable, and directly related to the jobs in ques-tion. If not, the tests were discriminatory. Even neutral tests, the Court ruled, "cannot be maintained if they operate to 'freeze' the status quo of prior discriminatory practices." After *Griggs* and other decisions, em-ployers could protect themselves against charges of discrimination only by managing to show a statistical parity between the racial composition of their work forces and that of local populations.[42]

In supporting the Philadelphia Plan, Nixon seems to have acted in part to get even with unions, most of which had opposed him in 1968, in part to promote "black capitalism" that might attract African Americans to the GOP, and in part on the assumption that a key to progress in race relations rested in employment. This was a sound insight, for blacks had been shut out of many areas of the industrial revolution in the United States. If they ever hoped to make use of their legal rights (as established in the 1964 and 1965 civil rights acts), they had to have equal social and economic oppor-tunity as well. Moreover, the new regulations in time made a difference, especially in government employment, where increasing numbers of blacks found places. For these reasons, the regulations aroused lasting controversy. Many employers and white workers denounced them as reverse or affirmative discrimination. Even the NAACP, perceiving the Philadelphia Plan as a political ploy aimed at breaking up alliances be-

41. Hoff, *Nixon Reconsidered*, 90–92; Seymour Lipset, "Affirmative Action and the American Creed," *Wilson Quarterly*, 16 (Winter 1992), 52–62; Ravitch, *Troubled Crusade*, 282–84; Thomas Edsall, "Race," *Atlantic Monthly*, May 1991, pp. 53–86; and Hugh Davis Graham, *The Civil Rights Era: Origins and Development of National Policy, 1960–1972* (New York, 1990), esp. 302-8.
42. Paul Gewirtz, "Discrimination Endgame," *New Republic*, Aug. 12, 1991, pp. 18–23. The Supreme Court much later backtracked on issues such as those involved in the *Griggs* case. Especially notable in this regard was the case of *Wards Cove Packing Co. v. Atonio* in 1989.

tween blacks and trade unions, opposed it. It was ironic that such far-reaching definitions of affirmative action took root in a Republican administration.[43]

IT WAS ALMOST AS IRONIC that a movement to protect and sustain the environment enjoyed special legislative successes during the Nixon years. This movement had been building steadily for some time, especially since the publication in 1962 of Rachel Carson's eloquent *Silent Spring*. Carson, an experienced marine biologist, trained her sights against agribusiness interests that polluted the environment with toxic pesticides such as DDT. Dangerous chemicals, she wrote, built up in human fat, invaded water and mothers' milk, and became "elixirs of death." The springs would soon be silent. More broadly, Carson presented an ecological perspective, positing the interconnectedness of human beings and all of nature. People, she warned, must refrain from activities that upset the delicate integrity of ecological systems.

The many people who joined the legions of environmental activists in the next few years included some, such as Nader, who attacked corporate devils—in his case General Motors and the gas-guzzling automobiles that wasted resources and polluted the air. But they embraced a vision that went well beyond conserving natural resources and assailing corporate capitalism. It was a vision that combined a range of causes: enhancing opportunities for outdoor recreation, preserving the natural beauty of things, protecting the public health, battling pollution, stopping atomic tests, safeguarding wildlife and endangered species, regulating commercial development, curbing dam-builders, stemming population growth, and counteracting the drive for production and consumption that environmentalists blamed for blighting the planet. Some environmentalists were hostile to industrial growth itself. By the late 1960s the many Americans who embraced environmental visions were engaging in a broad-based popular movement for the first time in United States history. The number of people who belonged to the twelve top environmental groups jumped from 124,000 in 1960 to 819,000 in 1969 to 1,127,000 in 1972. Polls indicated that millions more supported the goals of these organizations.[44]

43. Hugh Davis Graham, "Race, History, and Policy: African Americans and Civil Rights Since 1964," *Journal of Policy History*, 6 (1994), 12–39.
44. See Michael Lacey, ed., *Government and Environmental Politics: Essays on Historical Developments Since World War Two* (Washington, 1989), especially the following essays: Thomas Dunlap, "The Federal Government, Wildlife, and

The sources of the movement were varied, but they rested in part—like so much agitation for change at the time—on the unprecedented affluence of the postwar era. Prosperity greatly increased the number of people with the resources, education, and leisure to be concerned with such issues. Most environmentally active leaders were well-educated, middle-class people who were optimistic about the capacity of modern science and of governmental regulation to bring about change. Many, indeed, were driven by the same moral passions, high expectations, and rights-consciousness that inspired the civil rights and feminist movements. They dreamed of a world, apparently affordable and within reach at last, in which the quality of life would be enhanced for all.

Environmentalists had enjoyed modest successes during the New Frontier–Great Society years: a Clean Air Act in 1963, a Wilderness Act in 1964, a Clean Water Act in 1965, and an Endangered Species Act in 1966. In 1967 movement leaders coalesced to form the Environmental Defense Fund, a key lobby thereafter. In 1968 Congress approved a Wild and Scenic Rivers Act and a National Trails Act. Senator Edmund Muskie of Maine, Humphrey's running mate in 1968, was especially persistent in pushing for greater spending to aid municipal sewage treatment operations and in enlarging the role of the federal government in the control of water pollution.[45]

By the time Nixon reached office the environmental cause had grown stronger than ever, thanks in part to media attention given to Malthusian prophets of doom. Paul Ehrlich, a professor of biology at Stanford, published *The Population Bomb* (1968), which foresaw the starvation of hundreds of millions of people throughout the world during the 1970s and 1980s if population growth were not controlled. The book sold some 3 million copies in paperback. Barry Commoner, another biology pro-

Endangered Species," 209–32; Samuel Hays, "Three Decades of Environmental Policies: The Historiographical Context," 19–79; Robert Mitchell, "From Conservation to Environmental Movement: The Development of the Modern Environmental Lobbies," 81–113; and Joseph Sax, "Parks, Wilderness, and Recreation," 115–40. See also Hays, *Beauty, Health, and Permanence: Environmental Politics in the United States, 1955–1985* (New York, 1987); and Thomas McCormick, *Reclaiming Paradise: The Global Environmental Movement* (Bloomington, Ind., 1989), ix, 11, 54–56. Estimates of numbers in environmental groups is from Mitchell, "From Conservation," 96.

45. David Vogel, *Fluctuating Fortunes: The Political Power of Business in America* (New York, 1989), 93–112; Robert Gottlieb, *Forcing the Spring: The Transformation of the American Environmental Movement* (Washington, 1993), 124–61.

fessor, reaped fame at the same time by issuing apocalyptic jeremiads about the coming of nuclear disaster. In 1970 he was hailed on a *Time* magazine cover as the "Paul Revere of ecology." A year later he published *The Closing Circle*, an impassioned book that warned of the dangers of environmental pollution. In 1972 the Club of Rome, a loose association of scientists, technocrats, and politicians, produced *The Limits to Growth*. Employing computers to test economic models, the authors concluded that the world would self-destruct by the end of the century unless planners figured out ways to limit population and industrial growth and to expand supplies of food and energy. *The Limits to Growth* sold 4 million copies in thirty languages by the late 1970s.[46]

Nixon had little interest in environmental problems—indeed, he was bored by the issue—but he was savvy enough not to swim against the tide of reform, especially after a huge oil spill in Santa Barbara in early 1969 aroused national alarm. The result in the next few years was that he accepted a spate of bills, many of them passed by large bipartisan majorities. The most important of these laws, signed in January 1970, was the National Environmental Policy Act, which set up the Environmental Protection Agency (EPA). The EPA was empowered to require the filing of acceptable environmental impact statements before federal projects could be approved and more generally to enforce a range of guidelines. In February *Time* gushed that the environment "may well be the gut issue that can unify a polarized nation," and on April 22 throngs of people, including 10 million schoolchildren, gathered in communities to celebrate the nation's first Earth Day. Some 10,000 people flocked to the Washington Monument for twelve hours of revelry marking the occasion.[47]

Congress also passed other environmental legislation, including a measure that led to creation of the Occupational Safety and Health Administration (OSHA) in 1970, a Clean Air Act in 1970, a Federal Water Pollution Control Act in 1972, and an Endangered Species Act in 1973. (Nixon, fearing the costs of regulation, vetoed the water act but was overridden.) The air and water acts were tough against polluters, providing for specific goals and timetables and permitting little discretion in administration. The EPA, entrusted with the task of enforcement, was not to consider the costs of clean-up. The Endangered Species Act broke sharply from past practice, mandating protection of everything (except

46. McCormick, *Reclaiming Paradise*, 70–82.
47. Ibid., 67.

pest insects) above the microscopic level. By 1992 there were 727 species listed as endangered, half of them plants.[48]

The quest for environmental reform, as for affirmative action and other causes at the time, relied not only on federal bureaucracies but also on the courts. In time, environmental groups often managed to avoid the tedious and expensive business of litigation at the state level, bringing instead a single action in a federal court. Environmental groups benefited from two other key developments. First, judges tended to grant them "standing" in the courts, thereby permitting them to litigate even when they could not show themselves to be directly injured. Second, the Internal Revenue Service cleared the way for many environmental groups to receive non-profit tax status. Prospects for lobbying and litigation—and for the practices of attorneys engaged in environmental law—brightened considerably.[49]

The surge of interest in environmental causes managed even to curb the onrush of irrigation projects and dam-building that the Bureau of Reclamation and the Army Corps of Engineers had enthusiastically undertaken, especially since the 1920s. LBJ had been a strong advocate of such projects, which were popular among commercial farmers and which brought huge amounts of water to California and other areas of the West. As early as the 1950s, however, opponents of such projects had begun to mobilize more effectively, citing the disastrous consequences—to flora and fauna, to water tables, to places of beauty, to native cultures, in short to the ecology of the Mountain West—of the enormous diversions of water involved. By the late 1960s they had developed significant political power. Thanks to the protests of environmentalists and others, Congress refused after 1972 to approve large new irrigation projects in the West. This was a remarkable turnabout from previous policy.[50]

After 1972 the surge of environmentalism receded a bit. Critics of doomsayers like Ehrlich and Commoner counter-attacked, citing errors in their assumptions and predictions. One hostile review of *Limits to Growth* was entitled "The Computer That Printed Out W*O*L*F*."[51]

48. Alfred Marcus, "Environmental Protection Agency," in James Wilson, ed., *The Politics of Regulation* (New York, 1980), 267–303; *New York Times*, May 26, 1972; Dunlap, "Federal Government."
49. Mitchell, "From Conservation," 100–103.
50. Donald Worster, *Rivers of Empire: Water, Aridity, and the Growth of the American West* (New York, 1985), 259–84, 322–26. Congress did reauthorize funding for existing projects.
51. McCormick, *Reclaiming Paradise*, 79–81.

Other critics characterized environmentalists as privileged elitists whose self-interested opposition to growth and development would harm the working classes. "Some people," the black leader Vernon Jordan observed later, "have been too cavalier in proposing policies to preserve the physical environment for themselves while other poor people pay the costs." A popular labor union bumper sticker read, IF YOU'RE HUNGRY AND OUT OF WORK, EAT AN ENVIRONMENTALIST.[52] Charges such as these manifested—yet again—the class and regional divisions that fragmented the nation.

Efforts to curb pollution also ran into serious obstacles, especially from business and corporate leaders. Indeed, the rise in regulatory activity and in paperwork incited greatly growing hostility to Big Government in the 1970s.[53] More than 2,000 firms contested EPA standards within the next few years. The EPA, moreover, had been given an enormous regulatory burden: to oversee some 200,000 potential polluters. Economists and scientists working for the agency considered some of the goals and timetables to be unrealistic and resented being pressured into rapid action. Regulation moved slowly, fought at every step in the courts. By the late 1970s crusaders for pollution control were on the defensive, and Congress approved amendments that postponed deadlines for air and water standards.[54]

For all these reasons advocates of environmentalism, like many other crusaders for social change, did not achieve their high expectations. But they had hardly failed. On the contrary, the environmental movement, rooted as it was in the fertile soil of postwar affluence and concern for the quality of life, not only survived the counter-attacks of the mid-1970s and 1980s but also enjoyed considerable success in some ways—notably in improving the quality of air and water in the United States. Although embattled, it stood out as a legacy of the reform spirit of the 1960s.[55]

52. Siegel, *Troubled Journey*, 215.
53. Jonathan Rauch, "What Nixon Wrought," *New Republic*, May 16, 1994, pp. 28–31.
54. Marcus, "EPA," 285–97; Hays, "Three Decades," 44–47; Malcolm Baldwin, "The Federal Government's Role in the Making of Private Land," in Lacey, ed., *Government and Environmental Politics*, 183–207; and James Wilson, "The Politics of Regulation," in Wilson, ed., *Politics of Regulation*, 387–89. Economic stagnation after 1973, to be discussed in chapter 25, gave special impetus to corporate arguments against environmental regulations.
55. Gregg Easterbrook, *A Moment on the Earth: The Coming Age of Environmental Optimism* (New York, 1995); "The Good Earth Looks Better," *New York Times*, April 21, 1995.

WHILE LIBERALS COULD DERIVE some satisfaction from the rise of feminism, affirmative action, and environmentalism between 1969 and 1972, they remained as hostile as ever to Nixon. Indeed, political polarization not only persisted but also sharpened under the watch of the new administration. Some of this was predictable, given the implacable social and political divisions that had arisen in the 1960s. Some of it, however, stemmed from activities of the Nixon administration, which proved highly partisan in many ways. Nowhere was this more clear than in the area of race relations.

On racial issues Nixon and his gruff and conservative Attorney General, John Mitchell, were moved mainly by political considerations. Despite rhetoric to the contrary, they were less interested in moderating interracial fears than they were in protecting themselves against the appeal of George Wallace, who was expected to run again in 1972. This meant placating conservative white voters in the South and border states and bringing disaffected Democrats—those who had shown enthusiasm for Wallace in 1968—into the GOP fold.

Nixon and Mitchell sought especially to win over southerners who were bucking desegregation of the schools, an issue that rose again to the center of public debate after 1968. In that year the Supreme Court had indicated, in *Green v. County School Board of New Kent County, Va.*, that it had finally lost patience with southern resistance. Striking down so-called freedom-of-choice plans, which perpetuated segregation, it placed the burden of proof on schools to come up with workable plans for change. "A dual system," it said, "is intolerable."[56] Mitchell, however, sought to abet southern resisters by deferring guidelines, created by the Johnson administration, that would have terminated federal funding to segregated schools. Mitchell also opposed extending the Voting Rights Act of 1965, due otherwise to expire in 1970, on the implausible grounds that it was no longer needed. In August 1969, with schools about to open, HEW Secretary Finch sided with Mississippi segregationists who sought to postpone court-ordered desegregation. These and other actions highlighted the politically motivated Southern Strategy of the new administration and infuriated proponents of desegregation, including civil rights lawyers in the Department of Justice.

Nixon's efforts to put off desegregation of the schools ran up against determined and mainly successful opponents. The NAACP's Legal Defense and Education Fund brought suit that halted federal aid to segre-

56. 391 U.S. 430 (1968); Blum, *Years of Discord*, 315–18.

gated schools. Meanwhile, the Supreme Court rejected further obstruction of desegregation. In October 1969 it ruled unanimously in *Alexander
v. Holmes Board of Education,* "The obligation of every school district is
to terminate dual school systems at once and to operate now and hereafter
only unitary schools."[57] The enunciation of "at once" and "now" brought
a few teeth, at last, to the doctrine of "all deliberate speed" that "Brown II"
had set forth fourteen years earlier.[58] In the 1968–69 school year 32
percent of black schoolchildren in the South had attended schools with
whites. By 1970–71 the percentage jumped to 77 per cent, and by 1974–
75 it was 86 percent. Nationally, the change was less significant. School
districts continued to figure out ways to maintain defacto segregation. So
did state universities, especially in the South. Still, the changes in Nixon's
first term seemed promising: between 1968 and 1972 the percentage of
students attending schools that were 90 to 100 percent minority in enrollment decreased from 64.3 percent to 38.7 percent.[59]

Having been rebuffed by the judges, Nixon resolved to counter-attack
by naming a southerner to the Supreme Court when a vacancy arose in
late 1969. His nominee, Clement Haynsworth, was a South Carolina
federal judge. Liberals, however, fought back by airing reports that Haynsworth had a record of hostility to unions and civil rights. They also cited
conflicts of interest in some of his rulings. Nixon fought tenaciously but to
no avail. When the nomination reached the floor of the Senate, seventeen Republicans joined the majority of Democrats to defeat the nomination, 55 to 45. Mitchell snapped, "If we'd put up one of the twelve
Apostles it would have been the same."[60]

The President then tried again, this time in January 1970, by nominating to the High Court G. Harrold Carswell, a former Georgia state legislator who had become a federal circuit judge in Florida. Carswell, however,
suffered from greater liabilities than Haynsworth. As a legislator in 1948

57. 396 U.S. 19 (1969).
58. Nathaniel Jones, "Civil Rights After *Brown*: The Stormy Road We Trod," in
    Herbert Hill and James Jones, Jr., eds., *Race in America: The Struggle for
    Equality* (Madison, 1993), 97–111.
59. Gary Orfield, "School Desegregation After Two Generations: Race, Schools and
    Opportunity in Urban Society," in Hill and Jones, eds., *Race in America,* 234–
    62. Thereafter, progress toward school desegregation stalled, in large part because
    of "white flight" from cities and because of court decisions. In 1988, 32.1 percent
    of students attended schools that were 90 to 100 percent minority in enrollment.
    See discussion below of *Milliken v. Bradley.* For state universities, see *New York
    Times,* May 18, 1995.
60. Ambrose, *Nixon: Triumph,* 314–16; Weisbrot, *Freedom Bound,* 282–83.

he had said, "Segregation of the races is proper and the only practical and correct way of life. . . . I have always so believed and I shall always so act." Some of his judicial decisions had reaffirmed these beliefs. Nixon's advisers warned him that Carswell did not have a chance. The President nonetheless persisted until early April, when the Senate defeated him again, this time rejecting Carswell, 51 to 45. Pursuing the Southern Strategy to the end, Nixon called a press conference and stoutly defended his nominees. "When you strip away all the hypocrisy," he said, "the real reason for their rejection was their legal philosophy . . . and also the accident of their birth, the fact that they were born in the South."[61]

Having made his case, Nixon then nominated Harry Blackmun, a moderate from Minnesota, to the opening. The Court, however, persisted in pursuing a liberal course on matters of race. In March 1971 it rendered the *Griggs* decision that toughened affirmative action guidelines. A month later it decided, again unanimously, in favor of county-wide, court-ordered busing of students in and around Charlotte, North Carolina, as a means of achieving desegregation in the schools.[62] The decision affected 107 schools and many thousands of students, of whom 29 percent were black. Many liberals were delighted, hoping that busing would compensate for racially separate housing patterns. Busing indeed helped Charlotte to maintain one of the most desegregated school districts in the nation.[63]

Elsewhere, however, court-mandated busing became one of the most controversial issues of the 1970s, provoking passionately contested reactions, especially in the North. Many of those who protested busing had moved to all-white or mostly white neighborhoods in part to ensure that their children would not have to go to class with lower-class blacks. (Some wanted to avoid blacks of any class.) Cherishing the creed of "neighborhood schools," they were outraged that judges and government bureaucrats—some of them people without children in the public schools—were telling them what to do. Indeed, the majority of Americans rejected court-ordered busing, damning it as a desperate and divisive approach to complicated problems. A Gallup poll in October 1971 revealed that whites opposed busing by a ratio of 3 to 1. Even black people disapproved, by a margin of 47 percent to 45 percent. The issue of "forced" busing,

61. Ambrose, *Nixon: Triumph*, 330–31, 337–38.
62. *Swann v. Charlotte–Mecklenburg County Board of Education*, 402 U.S. 1 (1971).
63. *New York Times*, May 18, 1994.

already volatile before 1971, enormously abetted public backlash thereafter, fomenting violence in Boston and other cities.[64]

Nixon was philosophically opposed to court-ordered busing. Moreover, he was quick to recognize the political advantage of catering to popular resistance. For these reasons he clamped down hard when officials in HEW and the Department of Justice tried to hasten desegregation. He wrote Ehrlichman, "I want you personally to jump" on those departments "and tell them to *Knock off this Crap.* I hold them . . . accountable to keep their left wingers in step with my express policy—Do what the law requires and not *one bit more.*" Publicly, he declared his determination to "hold busing to the minimum required by law." Congress, he said, should "expressly prohibit the expenditure" for school desegregation of "any . . . funds for busing."[65] In March 1972 he called for a moratorium on all new busing orders by federal courts until legal issues, then under appeal, could be resolved. Vigorous oppostion to court-ordered busing became a main theme of his campaign for re-election in 1972.[66]

Nixon's stand helped to tie up efforts for court-ordered busing between 1971 and 1974. Meanwhile, the retirements in late 1971 of Supreme Court justices John Marshall Harlan and Hugo Black enabled the President to add two more judges of his choice to the Court. His nominees were Assistant Attorney General William Rehnquist, a Goldwater Republican, and Lewis Powell, a Virginia attorney who had been president of the American Bar Association. Both were confirmed, thereby pushing the ideological bent of the Court to the right.

In 1973 and 1974, with popular feeling high against busing, the new Court rendered two key decisions that gladdened conservatives and depressed liberals. In the first in 1973, *San Antonio Independent School District v. Rodriguez*, the Court affirmed by a vote of 5 to 4 the widespread American practice of local financing of schools—a practice that

---

64. J. Anthony Lukas, *Common Ground: A Turbulent Decade in the Lives of Three American Families* (New York, 1986); Ronald Formisano, *Boston Against Busing: Race, Class, and Ethnicity in the 1960s and 1970s* (Chapel Hill, 1991); and Harvey Kantor and Barbara Brenzel, "Urban Education and the 'Truly Disadvantaged': The Historical Roots of the Contemporary Crisis, 1945–1990," in Michael Katz, ed., *The "Underclass" Debate: Views from History* (Princeton, 1993), 366–402.

65. Ambrose, *Nixon: Triumph*, 460–61.

66. Jones, "Civil Rights," 101.

resulted in large disparities in per student spending. The "right" to an education, the judges said in rejecting Mexican-American complaints, was not guaranteed by the Constitution.[67] The second, *Milliken v. Bradley,* was announced less than a month before Nixon left office in August 1974. It involved schools in Detroit and its suburbs and was also decided by a vote of 5 to 4. In this case, as in the *Rodriguez* decision, all four of Nixon's appointees were with the majority.[68] The judges overruled a lower court ruling in 1971 that had ordered the merging of school districts so as to promote metropolitan desegregation of predominantly black Detroit and fifty-three suburban districts, most of them white-dominated, outside the city. The suburban districts, Burger reasoned, had not willfully segregated or violated the Constitition. District lines, there-fore, could be sustained, separating Detroit from its environs.

The *Milliken* decision was pivotal in the postwar history of race rela-tions, for it badly hurt whatever hopes reformers still maintained of over-turning de facto segregation of the schools and of slowing a dynamic that was accelerating in many American urban areas: "white flight" of familes to suburbs.[69] Flight in turn eroded urban tax bases, further damaging schools and other services in the cities. A "white noose" was tightening around places like Detroit. Justice Thurgood Marshall, appalled by the Court's decision, declared, "Unless our children begin to learn together, there is little hope that our people will ever learn to live together. . . . In the short run it may seem to be the easier course to allow our great metropolitan areas to be divided up into two cities—one white, the other black—but it is a course, I predict, our people will ultimately reject. I dissent."[70]

Marshall was prophetic about the further racial polarization of urban areas in the future. Many black ghettos grew even more desperate, vir-

67. 411 U.S. 1 (1973). The wealthiest school districts in the San Antonio case spent $594 per pupil, the poorest $356.
68. 418 U.S. 717 (1974). His appointees were Burger, Blackmun, Powell, and Rehn-quist. A second *Milliken* decision in 1977 (433 U.S. 267) established various compensatory mechanisms for schoolchildren in Detroit, as ordered to be devel-oped by the *Milliken* decision in 1974.
69. For developments in urban life in these years, see Peter Muller, *Contemporary Sub/Urban America* (Englewood Cliffs, N.J., 1981), 179–81; and Jon Teaford, *The Twentieth-Century American City: Problem, Promise, and Reality* (Bal-timore, 1986), 136–43.
70. Richard Kluger, *Simple Justice: The History of "Brown v. Board of Education" and Black America's Struggle for Equality* (New York, 1976), 773.

tually isolating an "underclass" that lived there. By 1974, however, the judges were hardly alone in abandoning ghettos to their fate. The *Milliken* decision reflected the backlash that had grown since the mid-1960s and that Nixon, Mitchell, and others in his administration had done much to stimulate. For these reasons the cause of racial desegregation continued to stall in the early 1970s. Black leaders who seemed threatening, such as the Black Panthers, were silenced, sometimes violently. Other black leaders remained divided and demoralized. The civil rights revolution, which had inspired grand expectations in the 1960s, reeled on the defensive in the 1970s and thereafter.

RACIAL POLARIZATION WAS but one cultural division that widened in the Nixon years. Ethnic and class conflicts also seemed to grow sharper, defying prophecies about the heat of the melting pot and resulting (many thought) in greater residential segregation of the social classes in the 1970s.[71] Here, too, deliberate actions by Nixon exacerbated tensions. Far from trying to muffle the resentments of many in the working and lower-middle classes, Nixon fanned their anxieties in the hope of drawing them, along with southern whites, to the Republican party. Hard-working and patriotic people, he said in 1969, were the "great silent majority" of Americans.[72]

No one was happier to pursue this strategy than Spiro Agnew, who emerged in late 1969 as one of the most visible Vice-Presidents in modern American history. Four days after a massive anti-war protest in October, he starting firing away at a wide range of enemies: "A spirit of national masochism prevails, encouraged by an effete corps of impudent snobs who characterize themselves as intellectuals." People promoting peace demonstrations were "ideological eunuchs." The press, he added two weeks later, were "a tiny and closed fraternity of privileged men" who engaged in "instant analysis and querulous criticism." Agnew, relying on speechwriters, relished high-sounding alliteration. He denounced opponents as "nattering nabobs of negativism" and as "hopeless hypo-

71. Mickey Kaus, *The End of Equality* (New York, 1992), 53–54. This is a disputed point, in part because of the ever-complicated problems involved in defining "class." Contemporary observers, however, generally agreed that class and ethnic consciousness seemed sharp in the early 1970s. See Michael Novak, *The Rise of the Unmeltable Ethnics: Politics and Culture in the Seventies* (New York, 1972).
72. Ambrose, *Nixon: Triumph*, 310. The phrase "great silent majority," enunciated by Nixon in November 1969, initially sought to rally support for his policies in Vietnam but from the start had a broader applicability.

chondriacs of history." He lamented that "a paralyzing permissive philosophy pervades every policy they [anti-war demonstrators] espouse."[73]

Although Agnew's primary targets were opponents of the war, he scattered shot more widely about, at one point taking aim at *Easy Rider* and Jefferson Airplane. Like Wallace in the 1968 campaign, he delighted in attacking self-styled experts, whom he blamed for encouraging permissiveness in American life. His targets included a host of people: opponents of school prayer, advocates of busing, hippies, counterculturalists, radical feminists, pushy blacks, spoiled university students, and intellectuals.

Outraged liberals and journalists countered Agnew's barrage with attacks of their own. But Agnew, egged on by Nixon, did not run for cover. Indeed, Nixon was developing tough tactics of his own. He was determined to destroy the most critical of his enemies, groups such as Weatherman and the Black Panthers. In mid-1970 he approved a plan devised at his instigation by Tom Huston, a young aide. The so-called Huston Plan would have increased funding for the CIA and the FBI and authorized these and other agencies to engage in a range of illegal activities, including covert opening of mail and much greater bugging and wiretapping. Only the opposition of FBI chief Hoover, who feared that the plan would damage the reputation of his agency (and who anticipated competition from Huston), prevented Nixon from going ahead with the plan. It was ironic indeed that Hoover, who had employed high-handed investigatory tactics for years, should have been the one to face down the President. Perhaps only Hoover, still a brilliantly self-protective bureaucrat, could have managed to do so.[74]

Nixon, although thwarted in this effort, persisted in his overall political strategy: to portray his enemies in as violent and unpatriotic a guise as possible. In so doing he raised the level of public acrimony to new highs in 1970. Worried about losing strength in Congress in the off-year elections, Nixon pumped money into close races. In October he took to the road himself to defend "law and order" and to assail his enemies as purveyors of "violence, lawlessness, and permissiveness." He went out of his way to anger demonstrators in the expectation that they would resort to extremes of vulgarity and violence. When they did—in San Jose, protestors stoned his armored car—he seized the chance to

73. Jonathan Rieder, "The Rise of the Silent Majority," in Fraser and Gerstle, eds., *Rise and Fall*, 243–68; Carroll, *It Seemed*, 6–7; Parmet, *Richard Nixon*, 575, 584.
74. Ambrose, *Nixon: Triumph*, 367–69.

denounce what he called "the viciousness of the lawless elements in our society."[75]

The partisan efforts of Nixon and Agnew in 1970 were among the most aggressive and divisive in the history of postwar political campaigning. *Time*'s Washington bureau chief, Hugh Sidey, observed that "Nixon's campaign was an appeal to narrowness and selfishness and an insult to the American intelligence. He diminished the presidency."[76] The effort, however, did not work. Democratic candidates for the House received 4.1 million votes more than their Republican challengers, a margin that was 3.4 million higher than in 1968. They widened their majority in the lower chamber by nine seats. Republicans increased their numbers by two seats in the Senate but remained a minority. Democrats also gained eleven governorships (but lost in New York and California, where Rockefeller and Reagan were re-elected). The results of the election did not bode well for the GOP in 1972.[77]

There were a number of explanations for these results, including popular revulsion at GOP tactics as well as the normal mid-term reaction against incumbents. The uncertain state of the economy, however, may also have played a role. Although overall economic growth seemed healthy, signs of economic instability that had become apparent in 1968 grew more worrisome in the next two years. The unemployment rate rose between 1968 and 1970 from 3.6 to 4.9 percent—a jump of more than 33 percent. The consumer price index increased by roughly 11 percent in the same period. Analysts of the economy coined a new and memorable term for what seemed to be happening: "stagflation." Larry O'Brien, John F. Kennedy's former campaign manager who headed the Democratic National Committee, popularized another new term, "Nixonomics." "All the things that should go up," O'Brien maintained, "—the stock market, corporate profits, real spending income, productivity—go down, and all the things that should go down—unemployment, prices, interest rates— go up."[78]

Like most partisan explanations of economic change, O'Brien's was simplistic. The causes of instability were considerably more complex and structural. Much of the inflation stemmed from the build-up of federal spending under the Johnson administration, a good deal of which went to support the war after 1965. The huge federal deficit of 1968 ($25.1 bil-

75. Ibid., 390–97.
76. *Time*, Nov. 16, 1970, p. 16.
77. Ambrose, *Nixon: Triumph*, 396.
78. Carroll, *It Seemed*, 128.

lion) exceeded the total of all deficits between 1963 and 1967 and combined with high levels of consumer spending to heat up the economy. Some of the unemployment stemmed from weaknesses in manufacturing and chemical companies, which proved less competitive than in the past against technologically superior overseas rivals, especially from the revived and booming economies of Germany and Japan.[79] By 1971 the United States had an unfavorable balance of international trade for the first time since 1893. A leveling off in defense spending further threatened jobs and added to popular anxieties about the future.

Less apparent at the time, but in many ways more problematic, were deep-seated structural developments in the work force. By the late 1960s millions of baby boomers were already crowding the job market. Ever-higher percentages of women were also looking for employment outside the home. A rise in immigrant workers, made possible after 1968 by the immigration law of 1965, did not affect most labor markets but further intensified popular unease. These developments combined to hike the numbers seeking work by 10.1 million between 1964 and 1970, or 1.6 million per year. Many of these people landed in the service sector of the economy—as employees in fast-food chains, discount retail outlets, hospitals, and nursing homes—or as clerical or maintenance workers. Most of these jobs tended to be part-time, offering low pay and benefits.[80]

Americans who found more promising employment also seemed edgy and uncertain. Millions of workers, eager to take advantage of the consumer culture, complained of long hours that created high levels of stress and left little time for leisure.[81] Young people confessed to feeling "crowded." Products of rapidly growing schools and colleges, they were keenly aware of the baby boom bulge that was sharpening competition for work and careers. They were troubled above all by what they sensed was the decline in their prospects compared to those of people who were just above them—those who had entered the work force in the 1950s and Golden 1960s. Having grown up in an age of enormous expectations,

79. Ruth Milkman, "Labor and Management in Uncertain Times," in Wolfe, ed., *America at Century's End*, 131–51.

80. David Calleo, *The Imperious Economy* (Cambridge, Mass., 1982), 185; Thomas Edsall, "The Changing Shape of Power: A Realignment in Public Policy," in Fraser and Gerstle, eds., *Rise and Fall*, 269–93; Bennett Harrison and Barry Bluestone, *The Great U-Turn: Corporate Restructuring and the Polarizing of America* (New York, 1990), vii–xxviii, 3–20.

81. Juliet Schor, *The Overworked American: The Unexpected Decline of Leisure* (New York, 1992), 80–82.

they found themselves in a world where the future seemed less auspicious than it had been. They put off marriage, child-bearing, and home-buying. Some of the extraordinary optimism that had accompanied the unprecedented economic boom of the mid-1960s—and that had given such a special spirit to that dynamic era—was abating.[82]

Labor-management relations, too, seemed to deteriorate by the early 1970s. Increasing numbers of younger blue-collar workers, expecting high levels of personal satisfaction, chafed under the routine of assembly line labor.[83] Others dared to disrupt public services: a strike of 180,000 postal workers in the summer of 1970 caused the calling out of the National Guard, which struggled for a while to deliver the mail.[84] Many unionized workers, moreover, felt increasingly insecure. For some time their unions had had trouble mobilizing new members, especially in the service industries and in the South and West, the fastest-growing areas of employment. Unions also failed to proselytize aggressively among women, high percentages of whom worked only part-time.

For these reasons the union movement continued to falter. While the total number of unionized workers had risen a little in the 1960s, the percentage of employees who belonged to unions maintained its long-range downhill slide. George Meany, still heading the AFL-CIO, had grown increasingly conservative and non-confrontational over the years, as had many leaders of constituent unions: they had become haves rather than have-nots. As the power of unions fell off, employers became bolder in their demands. Rank-and-file workers, many of them agitated already about court-ordered busing, affirmative action, and other divisive social issues, grew visibly more restless.[85]

82. Richard Easterlin, *Birth and Fortune: The Influence of Numbers on Personal Welfare* (New York, 1980), 23–28. Jones, *Great Expectations*, 152–57, notes that high school graduates began abandoning the notion, so central to attitudes of teenagers in the 1960s, that college would help them find a good life: the percentage of high school graduates who went on to college decreased between 1973 and 1976 from 62 percent to 54 percent. (The stagnant economy after 1973 furthered this trend.)

83. Carroll, *It Seemed*, 66–67. The most dramatic manifestation of this rebellion broke out in early 1972, when General Motors assembly line workers in Lordstown, Ohio, struck to gain better control of working conditions. The strike cost GM $150 million, but the corporation ultimately triumphed.

84. Robert Zieger, *American Workers, American Unions, 1920–1985* (Baltimore, 1986), 166.

85. Thomas Edsall, *The Politics of Inequality* (New York, 1984), 155. Union membership rose from 18.1 million in 1960 to 20.7 million in 1970, but fell as a

An enormously popular television sitcom, "All in the Family," high-lighted the feelings of many of these working-class Americans when it first invaded the nation's living rooms in January 1971. The rise of the show, indeed, exemplified an important trend of the era: the omnipresence of the mass media, especially TV, and the blurring of the line between "news," current events, and popular entertainment. Archie Bunker, pro-tagonist of the series, was a middle-aged, blue-collar father. He was blunt, bigoted, and xenophobic—much more openly so than any television character to that time. Most liberals who watched the series thought that it satirized the world-view of bigots such as Archie. That it did, but gently. Many working-class people told interviewers that they identified with him. As one worker told *Life*, "You think it, but ole Archie he says it, by damn."[86]

By this time the worrisome economic trends were jolting Nixon into new and unorthodox approaches. In January 1971 he startled the news-man Howard K. Smith by telling him, "I am now a Keynesian in eco-nomics," and in August he jolted the nation by announcing a New Economic Policy. This entailed fighting inflation by imposing a ninety-day freeze on wages and prices. Nixon also sought to lower the cost of American exports by ending the convertibility of dollars into gold, thereby allowing the dollar to float in world markets. This action transformed with dramatic suddenness an international monetary system of fixed exchange rates that had been established, with the dollar as the reserve currency, in 1946. Seeking further to aid American producers, Nixon placed a tempo-rary 10 percent surcharge on imports. Four months later, the dollar having fallen, he accepted a 13.5 percent devaluation of the dollar against the West German mark and a 16.9 percent devaluation against the Japa-nese yen. The surcharge on imports was then discontinued.[87] That Nixon, who had fiercely opposed controls throughout his political life,

---

percentage of non-agricultural employment from 31.4 percent to 27.4. In 1945, a postwar high, 35.5 percent of non-agricultural workers had belonged to unions; in 1954, a high thereafter, 34.7 percent had. The decline of unions continued after 1970, especially in the 1980s. By 1995 around 15 percent of non-agricultural workers in the United States belonged to labor unions.

86. Polenberg, *One Nation Divisible,* 225; Carroll, *It Seemed,* 62–66; James Baugh-man, *The Republic of Mass Culture: Journalism, Filmmaking, and Broadcasting in America Since 1941* (Baltimore, 1992), 145–47.

87. William Leuchtenburg, *A Troubled Feast: American Society Since 1945* (Boston, 1973), 225; Calleo, *Imperious Economy,* 62–65, 94–96, 105–7; Morris, *Time of Passion,* 158, 176; Carroll, *It Seemed,* 128; Siegel, *Troubled Journey,* 236.

would resort to such moves indicated his flexibility (foes said his inconsistency) as well as his alarm at what was happening to the economy. Smith quipped that Nixon's conversion to Keynesianism was "a little like a Christian crusader saying, 'All things considered, I think Mohammed was right.'"[88]

In the long run the New Economic Policy did not help much.[89] In 1972, however, it papered over a few cracks in the economy. Ending fixed convertibility gave a short-run boost to United States companies selling abroad. America's balance of payments deficit dropped by the end of 1972. The wage and price controls put a lid of sorts for the time being on inflation at home. By election time, as Nixon had hoped, the economic picture was a little rosier than it had seemed in 1971, and in January 1973 he temporarily relaxed the controls.

THREE CHARACTERISTICS DOMINATED President Nixon's handling of domestic matters. The first was flexibility. A Republican centrist for all of his political life, he had none of the ideological fervor of a reactionary like Goldwater or of a principled conservative such as Taft or Eisenhower. Although highly partisan when it came to his own political survival, he did not waste energy trying to stop every liberal idea that came from the Democratic majorities in the Congress. Some of their proposals, such as changes in policy affecting Native Americans, struck him as worthwhile; in any event they did not cost much. Others, such as environmental reforms, enjoyed popularity among the people: it was not worth it, Nixon thought, to oppose them. Still other policies, such as increasing spending for social insurance, had the support of influential lobbies, such as the elderly. Having no strong feelings about issues such as these, Nixon went along with many of them.

The second characteristic was Nixon's especially keen instinct for political survival. In this sense he was one of the greatest politicians of the postwar era. Those who counted him out, as many had following his 1960 and 1962 campaigns, were rudely surprised. In his handling of domestic policies as President he managed most of the time to do what it took—no matter how inconsistent he had to be—to assure himself of re-election in 1972. His support of the elderly, his Southern Strategy, and his New

88. Ambrose, *Nixon: Triumph*, 404; Carroll, *It Seemed*, 128. See also Ellis Hawley, "Challenges to the Mixed Economy: The State and Private Enterprise," in Robert Bremner, Gary Reichard, and Richard Hopkins, eds., *American Choices: Social Dilemmas and Public Policy Since 1960* (Columbus, Ohio, 1986), esp. 168–72.
89. See chapter 25.

Economic Policy were all in differing ways political at the core. A healthy concern for political consequences, of course, is necessary among those who hope to hold high office. Still, Nixon calculated with special persistence, regularly exhibiting a readiness—as in his appeals to class and racial feelings—to do whatever it might take, regardless of its cost in national divisiveness, to advance his interests. Political maneuvering was the great game of Richard Nixon's life. He played it grimly and with pride in his expertise at it. He had no other hobbies.

The third characteristic exposed a special abrasiveness, ruthlessness, and win-at-all-costs mentality. Nixon had always been shy and somewhat awkward, especially for a politician, and he remained as President a lonely, unhappy, often beleaguered man. He was so uncomfortable giving orders to people, even his top aides, that he communicated increasingly via memoranda. Many of these were in the form of marginalia on daily News Summaries, as they were called, that conservative aides such as Patrick Buchanan prepared for him. Some of the Summaries were forty or fifty pages long, but Nixon, isolated in the Oval Office or in private quarters, read them carefully. His comments on them revealed revengeful, aggressive, and violent feelings about people who seemed threatening. "Get someone to hit him," he would write of an opponent. "Fire him," "cut him," "freeze him," "dump him," "fight him," "don't back off."

Outbursts such as these, penned day after day, may have been therapeutic in some ways, but they did not help him relax: unlike all other postwar Presidents, Nixon never seemed to enjoy the job. His jottings exposed the emotions of a self-pitying, humorless, confrontational, and deeply suspicious public official. Nixon had a siege mentality that was contagious, encouraging aides to develop a "Freeze List" and an "Opponents List" of people who were never to be invited to the White House and then a long "Enemies List" of reporters, politicians, and entertainers. From the start of his administration he thought that only he could protect the "silent majority" of patriotic and hard-working Americans from the conspiratorial clutches of liberals and leftists who wielded unwarranted power in the press and the universities. His excess in trying to expand his power ultimately brought him down.[90]

90. Ambrose, Nixon: Triumph, 408–12.

24

Nixon, Vietnam, and the World, 1969–1974

Nixon, a student of international relations, was confident that he could allay world tensions. Although he had been one of the nation's most partisan Cold Warriors, he had gradually softened his rhetoric in the 1960s. He came to the presidency with the hope of bringing about better relations, later billed as détente, with the Soviet Union, and of opening dialogue with the People's Republic of China. [1]

During his years in office Nixon made a little progress toward these goals, enhancing greatly his image in time for the presidential election campaign in 1972. Triumphantly returned to power, he managed to reach a cease-fire in Vietnam in January 1973. In the process, however, he pursued policies—notably concerning Vietnam—that prolonged and sharpened rancor in the United States. Many of his efforts in foreign affairs, like those on the home front, were designed to win personal political objectives, not to break decisively with policies of the past. When he left office in August 1974, the Cold War—a constant of the years since 1945—remained about as frigid as ever.

1. Lloyd Gardner, *The Great Nixon Turnaround* (New York, 1973); Robert Litwak, *Détente and the Nixon Doctrine: American Foreign Policy and the Pursuit of Stability, 1969–1974* (New York, 1974); Joan Hoff, *Nixon Reconsidered* (New York, 1994), 147–207.

THE PRESIDENT'S ALTER EGO in foreign policy-making was Henry Kissinger, his National Security Adviser. Kissinger, a Jew, had been forced to flee his native Germany in the late 1930s. After serving in the American army during World War II he became a brilliant student and then professor of government at Harvard University. By the 1960s he was chafing for public office. He was a gregarious, arrogant, and extraordinarily egotistical self-promoter who carefully cultivated good relations with journalists and who was ready to work for almost anyone who would give him access to power. Having ingratiated himself with Nixon—he had great talent at obsequiousness—he got his chance as security adviser in 1969.[2]

Kissinger embraced a "realistic" view of international relations. Rejecting what he considered excessively moralistic approaches to policy, he admired statesmen who sought instead to broker a stable and orderly balance of power in the world. An early book, A World Restored (1957), lauded the efforts of Metternich, Castlereagh, and other conservative proponents of Realpolitik who designed the post-Napoleonic agreements at the Congress of Vienna in 1815.[3] The wise and realistic architect of foreign policy, Kissinger believed, must not try to change the internal systems of other nations; he must not be sentimental; he must accept limits and work within them. Kissinger hoped to promote a manageable relationship among the United States, the USSR, and the People's Republic of China, as well as a balance of power in the non-Communist world among the United States, western Europe, and Japan. With the big powers in place, the rest of the world could be stabilized.

Nixon, having moderated his moralistic anti-Communism by 1969, had come to share this approach. In July 1969 he enunciated what became known as the Nixon Doctrine, the essence of which was that the United States must first consider its own strategic interests, which in turn would shape its commitments—rather than the other way around. Other nations must normally expect to assume primary responsibility for their own defenses. Although the Nixon Doctrine did not change much in practice, it signaled that the new administration would not try to save the world. What mattered was carefully defined strategic interests, not moralistic attachments. Nixon, like Kissinger, liked to think of himself as

2. Walter Isaacson, Kissinger: A Biography (New York, 1992); Robert Schulzinger, Henry Kissinger: Doctor of Diplomacy (New York, 1989); Seymour Hersh, The Price of Power: Kissinger in the Nixon White House (New York, 1983).
3. John Judis, Grand Illusions: Critics and Champions of the American Century (New York, 1992), 189–91.

tough and analytical. To be sentimental in dealing with other nations, he thought, was foolish.[4]

Nixon also shared Kissinger's passion for secrecy and intrigue. Kissinger, as suspicious a man as Nixon, so feared leaks that he authorized unconstitutional wiretaps on members of his own staff. Both men were contemptuous of government bureaucrats, even in the National Security Council itself. They had scant respect for Congress, which they thought played to the voters when it concerned itself with world affairs. Nixon had special scorn for so-called experts in the State Department: they were the very sort of eastern Establishment people who had sneered at him all his life. For these reasons Kissinger and Nixon deliberately bypassed Secretary of State William Rogers, a friend of the President's who had little knowledge of foreign problems.[5] To manage these evasions Kissinger and Nixon set up a maze of secret "back channels" connecting them to loyalists in various offices and embassies around the world. Through these channels they could conduct elaborate negotiations and hide them from the State Department bureaucracy. These channels remained in place after Kissinger replaced Rogers as Secretary of State in 1973.[6]

Needless to say, this was a cynical and high-handed way of managing foreign relations. In evading official channels—and largely ignoring Congress—Nixon and Kissinger narrowed their scope of advice, and they further entrenched already centralized procedures in the making of policy. An imperial presidency, ascendant under Kennedy and Johnson, had arrived. On many occasions Nixon's reach for personal control sabotaged ongoing negotiations undertaken by Rogers and others at State.[7] Moreover, Kissinger and Nixon deeply distrusted each other. Kissinger was sometimes contemptuous (behind Nixon's back) of the President. He called Nixon "our drunken friend," a "basket case," or "meatball mind." Kissinger was also given to fits of temper. After one of these tantrums Nixon confided that he might have to fire Kissinger unless he got psychological help. Nixon apparently added later, "There are times when Henry

4. John Lewis Gaddis, *Strategies of Containment: A Critical Appraisal of Postwar American National Security Policy* (New York, 1982), 298.
5. Henry Kissinger, *White House Years* (Boston, 1979), 26.
6. Joan Hoff-Wilson, "Richard M. Nixon: The Corporate Presidency," in Fred Greenstein, ed., *Leadership in the Modern Presidency* (Cambridge, Mass., 1988), 164–98; Judis, *Grand Illusions*, 180–83; Harold Hongju Koh, "Reflections on Kissinger," *Constitution* (Winter 1993), 40–41.
7. For a critical view, see Stephen Ambrose, "Between Two Poles: The Last Two Decades of the Cold War," *Diplomatic History*, 11 (Fall 1987), 371–79.

has to be kicked in the nuts. Because sometimes Henry starts to think he's president. But at other times you have to pat Henry and treat him like a child."[8]

This volatile personal chemistry nonetheless survived a range of acid tests and produced what seemed to be tangible results, especially with the Soviet Union. In September 1970 Nixon and Leonid Brezhnev, the Soviet leader, reached an understanding concerning Cuban issues that had festered since the missile crisis of 1962. The Soviets agreed to stop building a submarine base in Cuba and to refrain from arming Castro with offensive missiles; the Americans promised in return that they would not invade. Characteristically Nixonian, the agreement was reached in secret; even after it was made virtually no one in government even knew of it. It therefore had no legal standing. Still, it indicated the search by both men for common ground on an inflammatory issue. In September 1971 the two leaders also accepted a four-power agreement that lessened tensions over Berlin, another of the world's flashpoints. Although these moves toward détente did not stop the Cold War, they moderated hostility to some degree.[9]

By 1972 the administration's *Realpolitik* seemed to be working wonders. In February, Nixon, his way paved by secret journeys that Kissinger took in 1971 to Peking, made a lavishly televised week-long visit to the People's Republic of China, thereby dramatizing his commitment to better relations with one of America's most determined foes. That Nixon, a life-long Cold Warrior who had assailed Truman for "losing China," could and did make such a journey staggered and excited contemporaries. The rapprochement promised to soften hostilities between the two countries and to enable the United States to play off China against the Soviet Union, whose relations with Mao Tse-tung remained unfriendly. The warm reception that Nixon received in Peking also suggested that the Chinese might turn a blind eye or two if the United States resorted to escalation in North Vietnam.

No act of Nixon's presidency was more carefully staged. None better

8. *Newsweek*, Sept. 7, 1992.
9. Raymond Garthoff, *Détente and Confrontation: American-Soviet Relations from Nixon to Reagan* (Washington, 1985), 38–54, 76–106. See also Paul Kennedy, *The Rise and Fall of the Great Powers: Economic Change and Military Conflict from 1500 to 2000* (New York, 1987), 406–10. In 1970, Nixon also placed in force a treaty on the non-proliferation of nuclear weapons that the United States, the USSR, Great Britain, and China had signed in 1968. By 1995, when the treaty was extended in perpetuity, 178 nations promised to adhere to it.

demonstrated the flexibility that made him such a formidable politician.[10] It is doubtful, moreover, that any Democratic leader could have taken such a trip without suffering severe political recriminations, for Cold War feelings remained intense. The Vietnam War still raged. The President, moreover, made it clear that the United States would reduce its military presence in Taiwan; later that year Taiwan was voted out of the UN. Nixon dared to take these steps because he knew that his reputation as a Cold Warrior would insulate him against assaults from the Right. He told Mao, "Those on the right can do what those on the left only talk about." Mao nodded cheerfully, "I like rightists."[11]

Three months later, in May, Nixon made another well-publicized journey, this time to Moscow for a summit meeting with Brezhnev. There the two leaders put the finishing touches on earlier negotiations that had led to a Strategic Arms Limitation Treaty (SALT I). The treaty placed upper limits on the future buld-up of ICBMs for five years. The two leaders also signed a treaty restricting the deployment by both sides of anti-ballistic missile (ABM) systems. Both agreements received senatorial approval later in 1972. Experts who followed the complicated, highly technical negotiations concluded that the agreements did not amount to much. SALT I did not stop the building of ICBMs already underway or prevent the situating on missiles of MIRVs (multiple independently targeted re-entry vehicles). The ABM treaty left plenty of room for the development of sophisticated defensive systems. Soviet as well as American arms build-ups continued at a rapid pace, especially of MIRVs and long-range bombers. Still, the agreements demonstrated Nixon's willingness to talk with an old enemy. They had large symbolic and political value for the White House.[12]

Not all of Nixon's foreign policies during this time evoked praise. By focusing so intently on great power relationships the President proved himself as blind as his predecessors in the White House to much of the rest of the world. Even rising powers such as Japan felt slighted. The

10. Gaddis, *Strategies of Containment*, 296–98; John Blum, *Years of Discord: American Politics and Society, 1961–1974* (New York, 1991), 380, 395; Garthoff, *Détente and Confrontation*, 199–247; Hoff, *Nixon Reconsidered*, 187–91, 201–3.
11. Gaddis, *Strategies of Containment*, 284–85.
12. Stephen Ambrose, *Nixon: The Triumph of a Politician, 1962–1972* (New York, 1989), 546–48, 614–16. Also Gaddis, *Strategies of Containment*, 320–26; Garthoff, *Détente and Confrontation*, 127–36, 184–98, 289–318; Peter Carroll, *It Seemed Like Nothing Happened: The Tragedy and Promise of America in the 1970s* (New York, 1982), 77–79.

concentration on what the Soviets and Chinese were doing caused special neglect of regional conflicts. This was obvious in South Asia, where Nixon and Kissinger were overeager in 1971 to court Pakistan as a conduit for their secret approaches to China. For this and other reasons (they thought the Soviets were masterminding Indian opposition to Pakistan) they sided with Pakistan's brutal suppression of Bengalis, who sought to secede. It was later estimated that Pakistan killed as many as a million people. Nixon's highly secretive policy, rooted in notions about great-power balance, ignored the expertise of State Department specialists in the region. It caused lasting ill feeling with the Bengalis and with India.[13]

The administration also revealed itself to be considerably more ideological than its professions of *Realpolitik* suggested. In Chile, Nixon and Kissinger encouraged covert American action to keep a Marxist, Salvador Allende, from gaining power. When Allende nonetheless won a democratic election in the fall of 1970, they continued to authorize the CIA to destabilize his regime, which was overthrown in 1973. Allende was assassinated in the uprising. Although there was no direct evidence linking the United States with the coup, Nixon and Kissinger rejoiced over it. American actions in Chile—as in Vietnam, Angola, Iran, and other places where Communism seemed to threaten—remained as uncompromising and ideological in the Nixon years as they had been since 1945.[14]

Some critics at the time grumbled that Nixon's foreign policies primarily reflected calculations of domestic political gain. Was it accidental, they asked, that the trips to Peking and Moscow took place in an election year? There was a great deal of truth to this complaint, for Nixon and Kissinger carefully timed their moves. Moreover, Nixon and Kissinger did not change the overall direction of American foreign relations. Even the opening to China was mostly symbolic; diplomatic recognition did not occur until 1978. Détente, though a worthwhile goal, did not transform Soviet-American relations, which grew especially rigid in Nixon's second term. Nixon and Kissinger, however, acted and spoke as if they were breaking dramatically and successfully with the past. "This was the week that changed the world," Nixon proclaimed in a toast at Peking. There, as in Moscow and other places, he and his aides took special pains to accommodate television, which captured his every move. When it came to foreign policy-making, the President was a master of political timing and of the art of public relations.[15]

13. Gaddis, *Strategies of Containment*, 262–88.
14. Ibid., 329–38.
15. Ambrose, "Between Two Poles"; Garthoff, *Détente and Confrontation*, 8–9, 29–33.

That mastery mattered much in 1972, the election year. Those who complained about what he had done in South Asia or Chile, who denounced him for his secretive and duplicitous ways of operating, and who exposed his exaggerated claims could scarcely be heard over the applause that followed his trips in early 1972 to Peking and Moscow. At last, it seemed, the United States had a man of experience and vision in the Oval Office. Nothing did more to advance Nixon's political prospects at that time than the reputation that he had managed to cultivate as an advocate of détente.

THE MOST IMPORTANT TEST of Nixon's foreign policies, however, involved Vietnam. In dealing with this divisive conflict, the President and Kissinger juggled two not always compatible goals: de-escalation of American troop commitments and escalation of military support for the South Vietnamese. His efforts prolonged the war and failed to save South Vietnam. They also provoked greatly increasing domestic opposition— which he and his aides tried to stifle in whatever ways they could. Indeed, Nixon's sensitivity to domestic dissent about the war—a sensitivity that bordered on paranoia—led him to whip up popular backlash against "unpatriotic" advocates of American withdrawal. It poisoned his administration and led to many of the excesses that ultimately destroyed his presidency. Like many of his policies, however, his course of action regarding Vietnam was conducted with clever political timing, so that the end of war seemed imminent in the last few weeks of the 1972 election. No policy of his presidency better exhibited Nixon's political skills, at least in the short run.[16]

When Nixon took office, Americans—and others—waited for him to unveil the "secret plan" that he had said would end the war. In fact, however, he had none, save the hope that efforts toward détente might encourage the Russians to put pressure on Hanoi. He also thought that he could scare the enemy—as Eisenhower was thought to have done to the North Koreans in 1953—into believing that they risked unimaginable American retaliation if they did not agree to settle. Apparently he confided his faith in this approach to Haldeman in early 1969 by touting it as his "madman theory" of ending the conflict: "I want the North Vietnamese to believe I've reached the point where I might do *anything* to stop the war. We'll just slip the word to them that, 'for God's sakes, you know Nixon is obsessed about communists. We can't restrain him when

16. Hoff, *Nixon Reconsidered*, 208–42; Stanley Karnow, *Vietnam: A History* (New York, 1983), 567–612; Schulzinger, *Henry Kissinger*, 29–51.

he's angry—and he has his hand on the nuclear button'—and Ho Chi Minh himself will be in Paris in two days begging for peace."[17]

Whether Nixon ever said this—he denied doing so—is unclear. But he did hope to frighten the enemy into a settlement within a year of taking office. The trouble with this approach was that it misread the "lessons" of history, which rarely repeats itself. The North Vietnamese, unlike the North Koreans in 1953, remained determined to prevail at whatever cost. As in the Johnson years, they refused to consider any agreement that permitted the United States to stay in Vietnam or that allowed Thieu, leader in the South, to take part in a coalition government in South Vietnam. Although Nixon stepped up military pressure by bombing heavily and greatly expanding South Vietnamese forces, the enemy did not bend. Nor did détente with the Soviets assist the American cause: Moscow continued to send military aid to Hanoi. Nixon's threats made no difference to the fundamental reality of the Vietnam War: the North and the NLF would fight to the end to win, and the South would not.[18]

Moreover, Nixon, like Johnson, personalized the issue. "I will not," he said in late 1969, "be the first President of the United States to lose a war." To back down, Nixon thought, would be to invite political assaults from the Right, to diminish the prestige of the presidency, and to tarnish the all-important "credibility" of the United States. Again and again he asserted that he would win "peace with honor." For these reasons, Nixon refused until 1972 to consider any settlement that would have permitted the North Vietnamese to keep troops in South Vietnam or that would have given any diplomatic standing to the NLF. Because Hanoi insisted on such terms, peace talks, which Kissinger and others conducted in Paris from 1969 on, went nowhere.[19]

Instead, Nixon persisted in the attempt to blast the enemy into talking, authorizing far greater bombing than Johnson had. The bombing increased already severe ecological damage to the countryside and uprooted masses of civilians from their homes. Nixon and Kissinger widened the war geographically as well by attacking neutral Cambodia, where North

17. William Chafe, *The Unfinished Journey: America Since World War II* (New York, 1991), 38; Gaddis, *Strategies of Containment*, 300. Nixon denied ever having outlined such a scenario; see Hoff-Wilson, "RMN," 187–89. Ho Chi Minh died in September 1969, but new leaders in Hanoi carried on his policies.
18. Michael Lee Lanning and Dan Cragg, *Inside the VC and the NVA: The Real Story of North Vietnam's Armed Forces* (New York, 1992).
19. George Herring, *America's Longest War: The United States and Vietnam, 1950–1975* (Philadelphia, 1986), 224–25, 256; Neil Sheehan, "The Graves of Indochina," *New York Times*, April 28, 1994.

Vietnamese troops maintained sanctuaries. This further escalation of the war, for which he never sought congressional support, began in March 1969 with a highly secret campaign of bombing raids. When the *New York Times* printed a story about it nine days after the start of the bombing, Nixon and Kissinger enlisted the aid of the FBI to wiretap staff members of the National Security Council—their own advisers—in hopes of uncovering the leak. Meanwhile the bombing continued; in the next four years B-52s dropped more than a million tons of explosives on Cambodia. When a pro-American government led by Lon Nol staged a successful coup in Cambodia in March 1970, Nixon tried to bolster Nol's regime by authorizing a joint South Vietnamese–American invasion that he said was aimed at the enemy's sanctuaries. These interventions, which proved to be of limited military value, unleashed great anti-war protest in the United States and badly destabilized Cambodia, which later fell victim to fratricidal civil war.[20]

While escalating the war in these ways, Nixon also started the process of cutting back on aspects of the strictly American contribution to it. This was part of a process that sought to divide the anti-war movement, about which he and Kissinger were almost obsessive. In May 1969 he announced his support of a plan to change the selective service system so as to move from an oldest-first to a youngest-first order of call. This meant that local boards would pick 19 year-olds first and that older males (save those who left or graduated from college) would no longer be threatened. A lottery would determine which young men would be chosen. In September Defense Secretary Melvin Laird said that the call for October would be spread over three months and that there would be no draft calls for November and December. The total drafted in these months, some 30,000, was one-tenth the number called per month at the peak of escalation under the Johnson administration. In November Congress approved the lottery system, and on December 1 the first drawing took place. The lottery did not greatly democratize the process of raising manpower, for student deferments remained (until 1971), and physical exemptions continued to be relatively easy to get. But it did seem a little fairer. The falling off of calls was especially reassuring; no one with a number higher than 195 (out of 365) was ever called.[21]

20. William Shawcross, *Sideshow: Kissinger, Nixon, and the Destruction of Cambodia* (New York, 1979); Marilyn Young, *The Vietnam Wars, 1945–1990* (New York, 1991), 245–49.
21. Ambrose, *Nixon: Triumph*, 264–66; Christian Appy, *Working-Class War: American Combat Soldiers and Vietnam* (Chapel Hill, 1993), 29. The draft itself was ended in 1973, after American soldiers had been called home from Vietnam.

Nixon was able to lower draft calls because he was pursuing a policy of what by late 1969 became known as Vietnamization.[22] This was more or less the same approach that Johnson had begun to employ after Tet. It involved pouring money and arms into the military of the South Vietnamese, increasing the size of their army (from 850,000 to a million), and getting it to bear a greater brunt of the fighting. As early as June 1969 Nixon announced the withdrawal of 25,000 American combat troops from Vietnam. The policy of Vietnamization upset Thieu and South Vietnamese military leaders, who sensed that the United States was pulling the rug from under them. As it turned out, they were correct, but Nixon denied it at the time. The President insisted instead that the United States remained committed to the anti-Communist cause.

Vietnamization unintentionally helped to damage morale among Americans in combat. In May 1969 American troops struggled for nine days, absorbing great losses, to take an enemy position. This became known as Hamburger Hill. Ordered to try again, they almost mutinied, then finally succeeded. Having achieved their goal, they were told that the hill had no military value, and they withdrew. The battle of Hamburger Hill aptly summarized the bloody, apparently pointless nature of the war on the ground.[23]

As the near mutiny revealed, American troops were growing weary of this sort of effort. Before 1969 they had fought with great bravery and discipline. But when it became clear that Nixon intended to cut back on American troop strength, many wondered why they should pay the price. Enlisted men increasingly refused to carry out orders. "Fragging" of officers became serious: more than 1,000 incidents were reported betwen 1969 and 1972. Racial conflicts tore units apart. Desertions increased, to an average of 7 per 100 soldiers. More than twice that average were reported as AWOL. By 1971 it was estimated that 40,000 of the 250,000 American men then in Vietnam were heroin addicts.[24] A vocal minority of American troops came home enraged and ready to protest against the war. In April of that year some 1,000 veterans camped out on the mall in Washington. Calling out the names of their dead buddies, they flung their medals on the Capitol steps.[25]

---

22. Guenter Lewy, *America in Vietnam* (New York, 1978), 166–89; Hoff, *Nixon Reconsidered*, 163–66.
23. Carroll, *It Seemed*, 4; "Hamburger Hill" was hardly a new phrase; the Korean War, too, featured one.
24. Ambrose, *Nixon: Triumph*, 418.
25. Todd Gitlin, *The Sixties: Years of Hope, Days of Rage* (New York, 1990), 417–19; and Thomas Paterson, "Historical Memory and Illusive Victories: Vietnam and

Vietnamization, which moved slowly at first, did little to dampen anti-war dissent. Draft resisters continued their efforts in 1969 and 1970, staging vehement protests against companies such as Dow Chemical and General Electric. Draft card burnings and turn-ins increased. Much more significant, however, was the swell of peaceful actions between mid-1969 and early 1971. Anti-war activists organized massive demonstrations, one of which, Mobilization Day in Washington and other cities on November 15, 1969, attracted crowds estimated at between 600,000 and 750,000 people. Demonstrations such as these, by far the largest in the history of the war, indicated that the collapse of organizations such as SDS was relatively insignificant. On the contrary, the anti-war movement by 1969 was moving well beyond the campuses and into the neighborhoods of America. It embraced an uneasily diverse coalition of people: draft resisters, students, anti-war veterans, blacks, working-class people, parents, the elderly, women for peace, and many others. Among people who opposed continuation of the fighting were mounting numbers of the "silent majority" who did not think, as many students did, that the war was "immoral." But they had come to believe that the war could never be won and that it must be ended before it tore up the United States.[26]

Nixon professed to be unmoved by anti-war activity. He made a show of telling people that he watched a Washington Redskins football game during one of the big demonstrations in Washington. "We've got those liberal bastards on the run now," he told his staff. "We've got them on the run and we're going to keep them on the run."[27] Ever more fearful of anti-war protestors, he actively fomented backlash against demonstrators. Moreover, millions of Americans—they may in fact have been a silent majority—still hoped that the United States could end the war in Vietnam without losing. Many of these people (as well as some who opposed the war) were offended by the antics of the radical few, like Jerry Rubin and Abbie Hoffman, who continued to seek—and to get—widespread media coverage. On Mobilization Day, Rubin and Hoffman marched to the Justice Department, raised an NLF flag, built barriers, and set fires, thereby distracting attention from a much larger peaceful march to the Washington Monument. Divisions within the anti-war movement,

---

Central America," *Diplomatic History*, 12 (Winter 1988), 10. See also Ronald Spector, *After Tet: The Bloodiest Year in Vietnam* (New York, 1993).
26. Herbert Parmet, *Richard Nixon and His America* (Boston, 1990), 570–76; Gitlin, *Sixties*, 379; Blum, *Years of Discord*, 356–59.
27. Carroll, *It Seemed*, 6.

which were sharp along lines of age and class, created problems for the cause.[28]

Still, the rise of anti-war activity could not be ignored, never more so than following announcement of the American invasion of Cambodia on April 30, 1970. Before going ahead with the assault Nixon steeled himself by watching *Patton* at his retreat at Camp David. As rumors circulated of American involvement he went on a national TV hook-up to defend his actions. In this widely noted speech, he explained that the North Vietnamese sanctuaries had to be wiped out and that Americans would leave Cambodia when that limited objective was accomplished. But Nixon was otherwise belligerent, defiantly asserting his toughness. The speech, indeed, outlined as clearly as any President ever had the rationale for the involvement of the United States in Cold War ventures such as Vietnam. This was the credibility of American commitments. "If, when the chips are down," Nixon explained, "the world's most powerful nation, the United States of America, acts like a pitiful, helpless giant, the forces of totalitarianism and anarchy will threaten free nations and institutions around the world." He added, "I would rather be a one-term president and do what I believe was right than to be a two-term president at the cost of seeing America become a second-rate power."[29]

The invasion of Cambodia set off a wave of protests, especially on the campuses. The next night students at Kent State University in Ohio flung bottles at police cars and smashed store windows, and the night after that persons unknown fire-bombed the university's ROTC building. The governor of the state, James Rhodes, sent in National Guardsmen to keep order. On May 4, however, some 500 protestors gathered, some of whom threw rocks at the guardsmen, who retaliated with tear gas. Although the closest demonstrators were sixty feet away, some of the guardsmen then opened fire, killing four students and wounding nine. None of the four who were killed was a radical; two of them were women walking to class. A presidential Commission on Campus Unrest later assailed the "indiscriminate firing" as "unnecessary, unwarranted and inexcusable."[30]

28. Herring, *America's Longest War*, 173; Gitlin, *Sixties*, 394–96; Kenneth Heineman, *Campus Wars: The Peace Movement at American State Universities in the Vietnam Era* (New York, 1993).
29. Blum, *Years of Discord*, 367; Herring, *America's Longest War*, 233–35; William Leuchtenburg, *A Troubled Feast: American Society Since 1945* (Boston, 1973), 244.
30. Leuchtenburg, *Troubled Feast*, 244; Joseph Kelner, "Kent State at 25," *New York Times*, May 4, 1995.

News of the killings at Kent State inflamed already super-heated college campuses. News a few days later that Mississippi policemen had killed two and wounded eleven black college students at Jackson State College ignited further protest. Campuses that until then had experienced little antiwar activity were convulsed by violence; three students were stabbed at the University of New Mexico. Student strikes that May affected some 350 campuses; demonstrations engaged some two million students, or 25 percent of university students in America—by far a record high. Thirty ROTC buildings were burned or bombed. The National Guard had to be called out in sixteen states and on twenty-one campuses. More than seventy-five colleges and universities had to close down for the remainder of the academic year.[31]

As in the 1960s, however, reactions on university campuses offered but a partial glimpse into the kaleidoscope of American public opinion about that most divisive and long-lasting of wars. A *Newsweek* poll taken a few days after the killings at Kent State discovered that 58 percent of respondents blamed the students, and only 11 percent the National Guard. Some 50 percent approved of the invasion of Cambodia, compared to 39 percent who opposed it. Amid the volatility following announcement of the invasion and the tragedy at Kent State it was difficult to know what to conclude from survey data such as these. Still, they suggested that something like a silent majority, while tired of the war, was sticking to some of its guns.

Events in New York City indicated that a few of these people were willing to take violent action. When New York City mayor John Lindsay, a liberal, decided to set aside a day in memory of the victims at Kent State, hundreds of people, many of them students, gathered peacefully for the occasion in the financial district. Just before the noon lunch hour, some 200 construction workers suddenly descended on the commemorators. As police stood by, the workers swung their hard hats as bludgeons at the students, beating as many heads as they could. The workers then marched on City Hall, bringing a mob of sympathetic bystanders with them. There a uniformed postal worker raised the American flag, which had been dropped to half-staff in commemoration of Kent State. A Lindsay aide reacted by relowering the flag. This set the crowd aflame. The mob swarmed past police who again did little to intervene, across the tops of parked cars, and up the steps of City Hall. Chanting "All the way with the

31. Terry Anderson, *The Movement and the Sixties: Protest in America from Greensboro to Wounded Knee* (New York, 1995), preface (n.p.), 350–52; Gitlin, *Sixties*, 409.

USA," they returned the flag again to full staff. The workers then pushed into nearby Pace College and knocked more heads before melting away—it was 1:00 P.M.—to return to work. A total of seventy people were bloodied in the assault. Only six workers were arrested. Six days later Peter Brennan, leader of the local construction workers' union, traveled to the White House to present Nixon with an honorary hard hat. Nixon accepted it as a "symbol, along with our great flag, for freedom and patriotism to our beloved country."[32]

Nixon, buoyed by support such as this, held to his course. Indeed, he further broadened the war by sending the South Vietnamese army into Laos in February 1971, an effort that turned into military disaster. But large-scale dissent also persisted, both on the campuses and elsewhere. In April 1971 a radical group of militants, calling themselves Mayday, vowed to "shut government down" in Washington. They performed "lie-ins" on bridges and major avenues in the District. Mobs roamed the streets and broke windows. Police retaliated violently, provoking one of the worst riots in Washington history. Some 12,000 were arrested.[33]

Other war-related controversies further rent the nation that spring. Lieutenant William Calley, charged with murdering many of the civilians at My Lai in 1968, was found guilty in late March by a military court and sentenced to life in prison at hard labor. His conviction enraged many pro-war advocates, who were sure that he had been made a scapegoat, and further inflamed domestic debate. Nixon reacted by releasing Calley from the stockade and confining him to his quarters at Fort Benning, Georgia, while the conviction was reviewed. The trial and its aftermath, which lightened Calley's punishment, aroused more bitter feelings and served as a nasty reminder of the viciousness of the conflict.[34]

Equally divisive was the start of publication in mid-June by the *New York Times* of the so-called Pentagon Papers. This was a 7,000-page collection of documents, originally commissioned by McNamara in 1967, concerning America's conduct of the war to that time. Many of the documents had been classified top secret. Inasmuch as the papers focused on the war under Kennedy and Johnson, Nixon might well have paid no heed. But Kissinger worried about the safety of his secret back channels and argued that publication of the papers was a violation of national

32. *Newsweek*, May 18, 1970, p. 50; Carroll, *It Seemed*, 57.
33. Herring, *America's Longest War*, 240–42.
34. Lewy, *America in Vietnam*, 356–59; Ambrose, *Nixon: Triumph*, 428; Thomas Boettcher, *Vietnam: The Valor and the Sorrow* (Boston, 1985), 390–93. Calley was paroled in November 1974.

security. More important, he was outraged—Haldeman said he put on "one of his most passionate tirades"—because the man who leaked the documents to the *Times* was Daniel Ellsberg, whom Kissinger had placed on the National Security Council staff as a consultant. Nixon agreed with Kissinger and sought a court injunction to stop further publication. Two weeks later the Supreme Court stymied Nixon's effort, ruling by a vote of 6 to 3 that publication of the documents did not violate national security. Justice Black stated that the effort to stop publication constituted "prior restraint" and was a "flagrant, indefensible" violation of the First Amendment. The judgment was a milestone in judicial interpretation of the rights of the press.[35]

The decision infuriated Nixon and Kissinger, who determined to get even with Ellsberg and plug the leaks. They turned first to Hoover and the FBI for help, but Hoover dragged his feet—in part, apparently, because he was a friend of Ellsberg's father-in-law. They then resolved to set up their own operation, the White House Special Investigations Unit. Nixon told Ehrlichman, "If we can't get anyone in this damn government to do something about [leaks], then, by God, we'll do it ourselves. I want you to set up a little group right here in the White House. Have them get off their tails and find out what's going on and figure out how to stop it." Ehrlichman organized a staff in the Executive Office Building, with a sign on the door reading PLUMBERS. They were ordered to do what it took to plug leaks.

Among the people who eagerly helped the plumbers were Nixon aide Charles Colson, a zealous and unprincipled loyalist, G. Gordon Liddy, a former FBI agent with an irrepressible attraction for derring-do, and E. Howard Hunt, a former CIA agent. Within a month of the establishment of the plumbers' office Ehrlichman authorized them to break into the office of Ellsberg's psychiatrist, Dr. Lewis Fielding, in Los Angeles, in order to dig up information. Whether Nixon knew of the covert action remains unclear, but he had pressed Ehrlichman to get the plumbers moving. On Labor Day weekend Hunt, Liddy, and three Cuban exiles recruited by Hunt staged the break-in, only to find nothing of interest. The administration's obsessiveness about leaks, having led to criminal activity, was to have catastrophic consequences later on.[36]

35. *New York Times v. the United States, United States v. the Washington Post,*
    403 U.S. 713 (1971). The dissenters were Harlan and Nixon's two appointees to
    that time, Burger and Blackmun. See Blum, *Years of Discord,* 388; Ambrose,
    Nixon: Triumph, 446–47.
36. Ambrose, *Nixon: Triumph,* 447–49, 465–66.

Meanwhile, Nixon steadily pursued Vietnamization, forcing the South Vietnamese army to do more and more of the fighting. By March 1972 there remained only 95,000 American troops in the country, compared to more than 500,000 when he had taken office in 1969. At this point the North Vietnamese staged another major military attack, the so-called Easter offensive, and Nixon retaliated with massive bombing of the North as well as of its strongholds in the South. "The bastards have never been bombed like they're going to be bombed this time," he said. Nixon also imposed a naval blockade on North Vietnam and mined the North Vietnamese harbor of Haiphong—steps that Johnson, fearful of direct Chinese intervention, had never dared to take.

If Nixon had responded this fiercely in 1969 or 1970, he might well have unleashed massive domestic protests. In mid-1972, however, these did not occur—in part because by then the South Vietnamese were suffering most of the casualties. Moreover, neither the Russians nor the Chinese sent in combat troops. Both nations by then seemed more interested in improving relations with the United States than in backing North Vietnam forever and anon. Brezhnev, indeed, welcomed Nixon to Moscow for the summit in May 1972 even though bombs had damaged four Soviet ships. The ferocity of the American military response, killing an estimated 100,000 North Vietnamese troops, managed also to stem the enemy assault. South Vietnam held on to its major cities, and Thieu remained in power.[37] The "peace with honor" that Nixon had promised still seemed far away in the summer of 1972, but he nevertheless could claim that America was staying the course.

BY THIS TIME NIXON was focusing ever greater attention on the business of getting himself and Agnew, once again his running mate, re-elected. This turned out to be a relatively easy but also divisive and ultimately damaging effort.

His greatest asset was the weakness of his opposition. One of his fears had been that Senator Edward "Ted" Kennedy of Massachusetts would oppose him. But Kennedy then self-destructed on July 19, 1969 (only hours before America landed its men on the moon), when a car he was driving plunged off a bridge at Chappaquiddick Island in Massachusetts. Kennedy escaped from the car, swam to shore, and went to bed at his hotel. He reported the accident only the next morning, at which point it was discovered that a passenger in the car, twenty-eight-year-old Mary Jo Kopechne, had drowned in the accident. Kennedy's irresponsible behav-

37. Herring, *America's Longest War*, 246–48; Chafe, *Unfinished Journey*, 399; Young, *Vietnam Wars*, 254–80.

ior did not hurt him with the star-struck voters of Massachusetts, who repeatedly re-elected him to the Senate. But it badly damaged his aspirations to be President. Luck had smiled on Nixon.

Another tragedy, this one on May 15, 1972, further helped the President's chances for re-election. On that day a deranged young man by the name of Arthur Bremer shot and severely wounded George Wallace, who had been running strongly in Democratic presidential primaries, winning in Florida, Tennessee, and North Carolina, and finishing second in northern states such as Wisconsin, Indiana, and Pennsylvania. Although it was clear that Wallace would not win the Democratic nomination, it seemed likely that he would ultimately bolt the Democrats and run as an independent. If so, he might drain millions of votes away from Nixon. The shooting changed all that. Wallace won primaries in Maryland (where he had been shot) and Michigan on the day after Bremer wounded him. But the bullet had penetrated his spinal column and paralyzed him from the waist down. In chronic pain and in a wheelchair, Wallace was forced to withdraw.

Nixon then had the pleasure of watching the Democrats tear themselves apart. Edmund Muskie, an early leader, had already dropped out, the victim in part of his own ineptitude as a campaigner, in part of "dirty tricks" (later exposed during the Watergate controversy) that Nixon had ordered to undermine his candidacy. The battle for the Democratic nomination then featured Hubert Humphrey, still eager to be President, and George McGovern of South Dakota. Entering the decisive California primary in June, McGovern had what seemed to be a safe margin. But Humphrey campaigned vigorously, focusing popular attention on his opponent's hostility to defense spending—in a state where thousands of jobs depended on it. Humphrey also heaped scorn on a "demogrant" proposal that McGovern had proposed. A rehashed and ill-considered version of Nixon's Family Assistance Plan, it would have given $1,000 in tax money to millions of Americans. Humphrey also managed to depict McGovern as a radical, associating him with the "Three A's"—of Acid, Abortion, and Amnesty (for Vietnam draft evaders). McGovern favored the third of these, but not the first, and he had taken no clear position on the second. Still, Humphrey's charges seemed to hit home. McGovern staggered through to a narrow victory in California, but he was a wounded candidate. Nixon relished the thought of repeating Humphrey's attacks against McGovern in the campaign ahead.[38]

38. Fred Siegel, *Troubled Journey: From Pearl Harbor to Ronald Reagan* (New York, 1984), 247; Ambrose, *Nixon: Triumph*, 554.

In mid-July McGovern won the Democratic nomination but paid another high price in the process. New party rules that a Commission on Party Structure (which he had headed) had developed in the years following the contentious party convention in 1968 set aside greatly increased percentages of delegate seats for women, blacks, and young people in 1972. The new rules, indeed, accelerated the trend toward grass-roots politics pioneered by McCarthy in 1968 and revolutionized the nature of electoral procedures in the party.[39] At the convention, 38 percent of the delegates were women (compared to 13 percent in 1968), 15 percent were black (compared to 5 percent), and 23 percent were younger than 30 (compared to 2.6 percent). Many of these people were enthusiastic liberals who had had little experience in national politics; a few pressed for radical causes. Urban bosses, labor leaders, and representatives of white ethnic groups—keys to the Democratic electoral coalition—were outraged. The crowning insult to these party faithful—and symbol of how far to the left the party had moved in four years—came when the delegates voted to exclude Mayor Richard Daley and his followers. They seated instead a delegation led by the Reverend Jesse Jackson, a young black preacher who had been an ally of Martin Luther King. Only one of the Jackson slate of fifty-nine was an Italian-American; only three were Polish-American. Frank Mankiewicz, a McGovern spokesman, conceded wryly, "I think we may have lost Illinois tonight."[40]

The disarray of the convention seemed only to grow as the spectacle careened to a close. McGovern had trouble finding a vice-presidental nominee, finally settling on Senator Thomas Eagleton of Missouri, a relative unknown. But the delegates then proceeded to advance thirty-nine additional candidates for the number two slot, including Mao Tse-tung, Archie Bunker, and Martha Mitchell, the outspoken wife of Nixon's campaign manager. By the time McGovern gave his acceptance speech it was 2:30 in the morning. Americans who watched the convention were shocked by the disorder that had overtaken the party.[41]

Nixon then got still one more unexpected bit of luck. Ten days after the end of the convention Eagleton admitted that he had earlier in his life

39. Byron Shafer, *Quiet Revolution: The Struggle for the Democratic Party and the Shaping of Post-reform Politics* (New York, 1983).
40. Carroll, *It Seemed*, 82; Siegel, *Troubled Journey*, 248; Thomas Edsall, *The Politics of Inequality* (New York, 1984), 158; Edsall, "Race," *Atlantic Monthly*, May 1991, pp. 53–86.
41. Edwin Diamond and Stephen Bates, *The Spot: The Rise of Political Advertising on Television* (Cambridge, Mass.), 187–88.

undergone electroshock therapy for depression. McGovern at first stood "1,000 per cent" behind his running mate. But controversy over Eagleton's mental health intensified, and McGovern backed down, thereby displaying inconsistency and indecisiveness. McGovern found a replacement, Sargent Shriver, only after embarrassing rebuffs from Muskie, Kennedy, and Humphrey. Nixon, who was scornful of Shriver, was delighted at the vulnerability of the ticket that now confronted him. Discussing how to belittle Shriver in the campaign, he told Haldeman, "Destroy him . . . kill him." The McGovern-Shriver ticket, he confided, was "a double-edged hoax."[42]

McGovern was by all accounts a decent man. As a bomber pilot in World War II he had won the Distinguished Flying Cross. He had then earned a doctoral degree in American history, taught at Dakota Wesleyan, and served as both congressman and senator from South Dakota. He was one of the first senators to call for pulling out of the Vietnam War, supporting an amendment (with Senator Mark Hatfield of Oregon) in 1970 that would have required withdrawal of all American forces there by mid-1971. Getting out of Vietnam was his chief issue in 1972. He also tried to blame Nixon for a break-in on June 17 at Democratic National Committee headquarters in the Watergate building of Washington. Nixon, he said, ran the "most corrupt Administration in history." But McGovern was an uninspiring speaker and a poor organizer. Well-meaning and liberal, he seemed to learn about the issues only as he went along. "Every time he opened his mouth," one early supporter said at the end of the campaign, "it came out irresponsible. Starting with the Eagleton affair, I just felt that this was a man who was not sure. So I voted for Nixon with no enthusiasm."[43]

Comments such as this indicated that popular perceptions of McGovern—as an earnest but bumbling and left-of-center liberal—badly hurt him. The AFL-CIO executive council, with but three dissenters, refused to endorse him for the presidency. So did leaders of individual unions, especially in the building trades.[44] Many working-class Democrats, traditionally the heart of the party, were also disaffected. Some

---

42. Ambrose, *Nixon: Triumph*, 582–83.
43. Leuchtenburg, *Troubled Feast*, 256; Blum, *Years of Discord*, 419; *Newsweek*, Nov. 13, 1972, p. 31.
44. Herbert Hill, "Black Workers, Organized Labor, and Title VII of the 1964 Civil Rights Act: Legislative History and Litigation Record," in Hill and and James Jones, Jr., eds., *Race in America: The Struggle for Equality* (Madison, 1993), 263–341.

opposed McGovern because they thought his strong opposition to the war both demeaning and unpatriotic. Others regarded him as a leftist who spoke for middle-class liberals and intellectuals, not blue-collar people. McGovern, indeed, in many ways represented a political culmination of protest movements and causes of the 1960s; he was the most left-of-center presidential candidate of any major political party in United States history. Millions of onetime Democrats either voted for Nixon in November or refused to vote at all.[45]

Nixon was disdainful of McGovern, whom he considered preachy. He was anxious to attack him. But the opposition was so badly divided that he scarcely had to fight. Moreover, he enjoyed substantial support from the press. Outside of the *Washington Post*, the media largely ignored McGovern's charges about the break-in at Watergate during the campaign. Nixon, playing the role of high-minded statesman, managed to keep all but friendly reporters at some distance and isolated those few who had the temerity to be critical: although journalists had grown more skeptical as a result of the credibility gap over Vietnam, they were not yet as adversarial as they were to become in the next few years. And publishers overwhelmingly backed the incumbent. Of the 1,054 dailies surveyed by *Editor and Publisher*, 753, or 71.4 percent, endorsed him; only fifty-six backed McGovern.[46]

The expectation of victory, however, did not stop Nixon from doing all he could to pile up support. As in the 1970 election, he posed as the defender of "law and order" and assailed the opposition as "soft" on crime and big on "forced" busing. He took special pains to fine-tune the economy so that it would peak in November. Early in the year he approved a $5.5 billion space-shuttle project, not because he thought it had great scientific promise—experts told him it did not—but because he understood the political gains to be made from such a venture.[47] During the campaign he substantially stepped up federal spending. As Melvin Laird, his Defense Secretary, recalled, "Every effort was made to create an economic boom for the 1972 election. The Defense Department, for example, bought a two-year supply of toilet paper. We ordered enough trucks . . . for the next several years."[48] Congress, too, propelled

45. Jonathan Rieder, *Canarsie: The Jews and Italians of Brooklyn Against Liberalism* (Cambridge, Mass., 1985), 247–52; Edsall, *Politics of Inequality*, 145–78.
46. James Baughman, *The Republic of Mass Culture: Journalism, Filmmaking, and Broadcasting in America Since 1941* (Baltimore, 1992), 177.
47. Ambrose, *Nixon: Triumph*, 498.
48. Blum, *Years of Discord*, 408.

election-year spending by its approval of sharp hikes in Social Security benefits: some $8 billion in extra checks went out in October. Veterans' benefits also increased, as did federal grants to state and local governments under Nixon's revenue-sharing plan.

Leaving nothing to chance, Nixon took pains to gain the support of potent interest groups, especially in the corporate sector. In 1971 he agreed to increase federal supports for milk prices, in return for which the dairymen, a strong lobby, contributed $2 million to his campaign. Early in 1972 he killed an anti-trust suit against ITT, which donated $400,000 in support of the GOP convention. The GOP campaign in 1972 was virtually a textbook demonstration of the huge role that big money, especially corporate money, had come to play in American presidential elections. The list of firms that made large contributions (some of them illegal) to Nixon's re-election effort read like a Who's Who of regulated industries, including airlines, bankers, and truckers.[49]

Nixon and Kissinger labored especially hard in the summer and fall of 1972 to manage the situation in Vietnam so that it would appeal to American voters. With the elections near at hand in the summer of 1972, they redoubled their efforts at the peace table in Paris, conceding for the first time a key demand of the North Vietnamese: that they be allowed to keep troops in the South after a cease-fire. Nixon also backed away from America's commitment to Thieu, agreeing to establishment of a tripartite electoral commission composed of Saigon, the NLF, and neutralists. It would have the task of arranging a settlement after the cease-fire went into effect. Led by chief delegate Le Duc Tho, the North Vietnamese also compromised, agreeing that Thieu might stay on in control of the South until later settlements could be reached. By mid-October an agreement had been hammered out along these lines. United States troops would leave South Vietnam within sixty days of a cease-fire, and North Vietnam would return American POWs. The tripartite commission would step in to arrange a political settlement, administer elections, and take responsibility for implementing the results.[50]

Kissinger, who was almost desperately eager to settle the war by election time (and take credit for the result), thought he had made a breakthrough. On October 31 he stated publicly, "Peace is at hand." But he had failed to consult Thieu, who refused to accept an agreement that would permit

49. James Wilson, "The Politics of Regulation," in Wilson, ed., The Politics of Regulation (New York, 1980), 388; Ambrose, Nixon: Triumph, 434–35, 502–5; Blum, Years of Discord, 414–18.
50. Herring, America's Longest War, 250–53.

North Vietnamese troops to remain in the South or that would acknowledge NLF sovereignty. Such concessions, Thieu feared, spelled doom for him and his government. Nixon, moreover, tended temporarily to side with Thieu. Confident of winning the election, peace or no peace, he held out hope of better terms—something like the "peace with honor" that he had promised the people. For these reasons the election passed without a settlement. Still, the administration's well-publicized efforts helped to display him as a man seeking peace and to undercut the anti-war opposition. As his trips to China and the USSR had indicated, the President was a master at making symbolic moves in foreign policy that advanced his political fortunes at home.

No one was surprised when Nixon won resoundingly in November. He received 47.1 million votes, 60.7 of the total cast, to McGovern's 29.1 million, and carried every state except Massachusetts and the District of Columbia. His total vote was 15.3 million higher than it had been in 1968 (and 5.4 million higher than the combined vote for him and Wallace in 1968). McGovern got 2 million fewer votes than Humphrey had received four years before. This was one of the most one-sided presidential elections in modern American history—as overwhelming for Nixon as the election of 1964 had been for Johnson.

What the election results said about American politics was not quite so clear. Although Democrats lost twelve seats in the House, they maintained control of the chamber. Thanks in part to the Voting Rights Act of 1965, Barbara Jordan, elected in Texas, and Andrew Young, a winner in Georgia, became the first black southerners to go to the Hill since the era of Reconstruction.[51] Democrats gained one seat in the Senate. It was obvious that two long-range trends of postwar American politics, the rise of split-ticket voting and the decline of party organizations, were accelerating. The election, moreover, was more a rejection of McGovern than it was a sign of voter affection for Nixon. Voter participation, which had been dropping in the 1960s, fell further in 1972—to the lowest levels since 1948. Polls suggested the power of yet another troubled legacy from the 1960s: rising distrust of national politicians and doubts about the capacity of government to do the right thing most of the time.

Still, there was no doubting the magnitude of Nixon's victory or the

---

51. Other newcomers, later to become well-known Washington figures, who were first elected to Congress in 1972 included Pat Shroeder, a liberal Democrat chosen to the House from Colorado; Jesse Helms, an ultra-conservative elected to the Senate from North Carolina; Sam Nunn, a Democratic senator from Georgia; and Joseph Biden, a Democratic senator from Delaware. Those re-elected included Senators Edward Brooke of Massachusetts, Walter Mondale of Minnesota,

remarkably rapid fall of the Democratic coalition, at least in presidential elections. This decline had first become sharp in 1966, when sizeable numbers of white working-class voters either had refused to vote or had deserted to the GOP. In 1968 Nixon and Wallace had attracted millions of such voters, and in 1972, with Wallace sidelined, Nixon won them for himself. The backlash that had whipped through American life since the mid-1960s reverberated more strongly than ever in 1972.

SAFELY RE-ELECTED, NIXON FELT FREE to unleash the awesome destructiveness of American air power on the North Vietnamese. He told Admiral Thomas Moorer, Chairman of the Joint Chiefs of Staff, "I don't want any more of this crap about the fact that we couldn't hit this target or that one. This is your chance to use military power to win the war, and if you don't, I'll consider you responsible." The twelve-day "Christmas bombing" that followed was indeed intensive, blasting the city of Hanoi and arousing storms of protest around the world. The *New York Times* called it "Stone Age barbarism." The 36,000 tons of explosives dropped during that time were more than had been used between 1969 and 1971. They killed some 1,600 civilians. The North, by then equipped with surface-to-air missiles, knocked down fifteen B-52s and eleven other American planes, causing the death or capture of ninety-three American airmen.[52]

On December 26, the eighth day of the bombing, the North Vietnamese (who had run out of missiles) indicated that they were ready to return to the peace table when the bombing stopped. Nixon called it off on December 30, and the two sides soon resumed talks in Paris. On January 14, Kissinger and Tho reached an agreement that was essentially the same as the one they had worked out in October.[53] This time Nixon imposed it on Thieu, sweetening the taste a little by promising unilaterally (and without congressional input) that he would continue to provide him with military support and would "respond with full force" if the North Vietnamese violated the agreement. The cease-fire started at

---

Strom Thurmond of South Carolina, and James Eastland of Mississippi. Margaret Chase Smith, a Republican senator from Maine, lost after twenty-four years of service. *Newsweek*, Nov. 13, 1972, p. 36.

52. Herring, *America's Longest War*, 253–55; Parmet, *Richard Nixon*, 625; Stephen Ambrose, *Nixon: Ruin and Recovery, 1973–1990* (New York, 1991), 38–58.

53. For their efforts Kissinger and Le Duc Tho won the Nobel Peace Prize, an award that amazed many contemporaries.

midnight following January 27, five days after Lyndon Johnson suffered a fatal heart attack at his ranch.

In announcing the cease-fire Nixon told the world five times that it represented the "peace with honor" that he long had promised. But Americans were skeptical. A Gallup poll showed that two-thirds of the people did not believe that Nixon was telling the whole truth. Times Square, which had been mobbed on V-J Day in 1945, was deserted. "There's nothing to celebrate," an American Legion commander told *Newsweek*, "and nobody to celebrate with."[54]

Americans were right to be skeptical, for it was clear that the agreement was far from a "peace with honor." The United States had conceded the most important demand of the North—that its troops be permitted to stay in the South—and had gained nothing more than it could have had in October. The bombing had accomplished nothing. Suppose, critics asked, that Nixon had been willing to grant this demand in 1969? If he had, they said, the North might have dropped its insistence (as it did in 1972) that Thieu be removed immediately from power. A cease-fire might have been reached at that time. Instead, critics emphasized, there had been four more years of carnage. Between January 1969, when Nixon took office, and January 1973, when the cease-fire went into effect, the United States lost 20,553 servicemen—or more than a third of the 58,000 who died during the war. Official estimates conclude that 107,504 South Vietnamese troops were killed during these four years, along with half a million troops of the North Vietnamese and NLF.[55]

The cease-fire hardly stopped the fighting. As Nixon and Kissinger knew (but did not explain to the American people), the political future of South Vietnam would have to be settled by force. The United States continued to pour military aid into South Vietnam. It dropped 250,000 tons of bombs on Cambodia in the next seven months, more than the tonnage used against Japan in all of World War II. Congress, however, stiffened, cutting off appropriations for such bombing as of August 15, 1973. In November it overrode a presidential veto to pass a War Powers Act. This required American Presidents to inform Congress within forty-eight hours of deployment of United States forces abroad and to bring the troops home within sixty days unless Congress explicitly endorsed what the President had done.[56]

54. *Newsweek*, Feb. 5, 1973, p. 16.
55. Herring, *America's Longest War*, 256; Hoff-Wilson, "RMN," 189.
56. Herring, *America's Longest War*, 257–68; Carroll, *It Seemed*, 94. Neither Nixon nor subsequent Presidents acknowledged the constitutionality of the law, which had little effect in the future.

By then the Nixon administration was fighting to defend itself against Watergate-related charges of illegal activities. It was losing political muscle that it needed to direct its foreign policies. But Congress would have refused to help Thieu in any event, and it cut back aid. It nearly eliminated it after Nixon left office in August 1974. Lon Nol fell from power in Cambodia in April 1975, replaced by a brutal Khmer Rouge regime led by Pol Pot. The Khmer Rouge killed an estimated 2 million people during the next three years, at which point the North Vietnamese went to war and chased them into hiding. Thieu, overwhelmed by a North Vietnamese military offensive, was forced to resign on April 21, 1975. As his loyalists scrambled desperately to climb aboard United States helicopters, Hanoi ran up its flag in Saigon on May 1 and renamed the capital Ho Chi Minh City. South Vietnam was a state no more.

NIXON AND KISSINGER WERE NOT, of course, responsible for America's original involvement in Vietnam. Nor were they really to blame for the fall of Thieu. That was mainly the result of North Vietnam's unwavering determination to take over the country, of South Vietnam's inability to resist, and of America's fatigue with the fighting. From the beginning American leaders had underestimated the will to fight of the North, overestimated the staying power of the South, and misjudged the endurance of the American people. There are events in the world that not even the greatest military powers can control.

Some people have argued, moreover, that Nixon did a decent job of managing the war, especially given the terrible circumstances that he confronted on taking office. His escalation in Cambodia, they point out, followed earlier incursions by the enemy—the North Vietnamese were the first to destabilize that unfortunate country—and seemed to make sense from a strictly military point of view. It was hardly fair, they add, to blame the United States for all the blood that later stained the Cambodian landscape. Some people also think that Nixon's overtures to Mao and Brezhnev gave him greater freedom than Johnson had had to unleash bombing attacks, thereby enabling him to pound a little sense into the North Vietnamese in Paris. In any event, the President's defenders insist, the bombings were not reckless in the sense of risking wider world war, for Nixon had first assured himself that Peking and Moscow would tolerate them. Indeed they did, even the Christmas bombing of 1972. Nixon's rapprochement with the Chinese and the Russians may also have increased feelings of isolation in Hanoi by 1972, thereby helping a little (although not nearly so much as American concessions) to induce the North to agree to the temporary continuation in power of Thieu.[57]

Nixon and his defenders emphasize above all that it would have been very difficult before 1972 to secure a good settlement, for two reasons. First, the North Vietnamese were stubborn and wily negotiators. More than once during the fruitless years of talks they seemed forthcoming, but mainly to court world opinion, whereupon they dug in their heels. Second, Nixon and Kissinger emphasized that they dared not compromise much before 1972. If they had reached an agreement that seemed to pave the way for North Vietnamese victory, they would thereby have abandoned all that the United States had been fighting for, including its allies in Saigon. This would damage America's "credibility" in the world and encourage powers like the Soviet Union and China to underwrite "proxy" wars in the future. Such an agreement, Nixon realized, would also have exposed him to criticism from the millions of Americans who still hoped in 1969 (and later) for a peace with honor. Instead, he pursued Vietnamization, a policy that reduced American involvement, strengthened South Vietnam's military capability, and ultimately (once concessions were made) got the United States out of the war. Carefully managed Vietnamization, Nixon maintained, produced a settlement that was acceptable politically in the United States, thereby enabling the nation to moderate potentially disastrous domestic recriminations.

Still, Nixon's policies in Vietnam look as politically motivated in hindsight as they did to critics at the time. Total Vietnamization, given the well-demonstrated corruption and political instability that rent South Vietnam, was almost certainly doomed to failure. Without huge and apparently endless infusions of American support, neither Thieu nor anyone else in Saigon could withstand the relentless drive of the enemy. Hoping for the best, Nixon persisted anyway, using Vietnamization as a fig leaf to reduce American casualties and as a means of enabling Thieu to hold out for a while. That would create an interval—politically crucial to Nixon—between the withdrawal of United States forces and South Vietnamese defeat. Much of the bloodshed that occurred on his watch might have been averted if he had tried harder to compromise. When he finally did, in the election year of 1972, he secured an agreement that was politically more palatable than it would likely have been in 1969, when support for the war had been stronger. But it was no better than what

57. Gaddis, *Strategies of Containment*, 299; Garthoff, *Détente and Confrontation*, 259; Charles Morris, *A Time of Passion: America, 1960–1980* (New York, 1984), 146–48.

might have been achieved at that time, and it was accomplished only after four more years of slaughter. It was not a peace with honor.

The Vietnam War taught Americans a few lessons, chief among them the dangers of large-scale military intervention in strategically marginal areas of the world. In future years American politicians and military leaders were more likely to set limits before sliding into quagmires like the one that had swallowed so much humanity in Southeast Asia. "No more Vietnams," they warned. Given the grand expectations that Americans had had until then about their ability to shape the world, this was a shift of historic importance—one that demarcates the postwar era. As Maxwell Taylor explained later in the 1970s, "We [the United States] certainly had a feeling after World War II that we could go almost any place and do almost anything. Well, we did many things at enormous cost, but henceforth we're going to have trouble feeding and keeping happy our population just as every other nation is. This is not a time for our government to get out on limbs which are not essential."[58]

This useful lesson, however, was learned only after extraordinary costs that Nixon's policies helped to escalate. After 1969 the war further savaged and badly destabilized Vietnam and Cambodia. As before, it diverted the attention of American foreign policy-makers from serious problems elsewhere, especially in Latin America, Africa, and the Middle East. Continuing fixation on Vietnam also left the United States relatively weaker vis-à-vis the Soviets, whose arsenal of missiles and delivery capacity reached parity with that of the United States by the 1970s.[59] At home the war provoked serious economic difficulties, especially inflation, by 1973. It accelerated the rise of an imperial presidency and contributed powerfully—thanks to Nixon's quest for control—to the constitutional crisis of Watergate.

More generally the war undercut the standing of political elites. Nothing did more than Vietnam to subvert the grand expectations that many Americans had developed by 1965 about the capacity of government to deal with public problems. Popular doubt and cynicism about "the system" and the Washington Establishment lingered long after the men came home.

The war above all left an abiding sourness in the United States. Vet-

58. Lloyd Gardner, "America's War in Vietnam: The End of Exceptionalism," in D. Michael Shafer, ed., The Legacy: The Vietnam War in the American Imagination (Boston, 1990), 28. See also Morris Dickstein, Gates of Eden: American Culture in the Sixties (New York, 1977), 271.

59. Kennedy, Rise and Fall, 406–8.

erans of the war tended to feel this with special intensity. Unlike the servicemen who had returned to parades and celebrations in 1945, those who came back after 1968 encountered an increasingly weary and cantankerous nation. Dumped into civilian life after surviving the terrors of the bush, they experienced staggering problems, including unemployment, guilt, depression, rage, and a sense of rejection. Hundreds of thousands suffered from "post-traumatic neurosis," flashbacks, and nightmares. Suicide rates among the veterans were much higher than in the population at large.[60]

Sourness in America extended well beyond the reception of veterans, neglectful though that was. Wider recriminations that had risen during the war persisted for years thereafter. Many people, including political leaders like Ronald Reagan, never stopped insisting that the war need not have been lost. They regarded anti-war activists and draft avoiders with a fury and contempt that did not abate with time.[61] Other people, including many who had once supported the war, raged at the military and political leaders who had dragged the country into the conflict and at Nixon for his Machiavellian maneuvers. A vocal few still insisted many years later that North Vietnam—contrary to what Nixon maintained in March 1973—had not turned over all American prisoners-of-war or soldiers said to be missing in action. America's longest war inflicted wounds that time was very slow to heal.

60. Larry Berman, *Lyndon Johnson's War: The Road to Stalemate in Vietnam* (New York, 1989), 5; Landon Jones, *Great Expectations: America and the Baby Boom Generation* (New York, 1980), 102.
61. Angry debates over the alleged draft-dodging of GOP vice-presidential nominee J. Danforth Quayle of Indiana in 1988, and of Democratic presidential candidate William Clinton of Arkansas in 1992, roiled the election campaigns in both years.

25

End of an Era?
Expectations amid Watergate
and Recession

Even as the United States was pulling the last of its soldiers out of combat in Vietnam, the scandal known as Watergate was beginning to wreck Nixon's second administration. From then until he resigned under fire in August 1974 the President found himself driven ever more into a corner. As he ran for safety he shed advisers, including Haldeman and Ehrlichman, who were involved in his cover-up of the break-in at the Watergate complex in Washington. But one bizarre development after another, including the revelation in July 1973 that the President secretly taped conversations in the Oval Office, gave him no rest. By itself the break-in was petty, but Nixon's attempts to cover it up were crude, cynical, and illegal acts to obstruct justice. The prolonged but often exciting events of the Watergate scandal ultimately created a constitutional crisis and further polarized the nation.[1]

1. Key sources for what follows are Stephen Ambrose, *Nixon: Ruin and Recovery, 1973–1990* (New York, 1991); Ambrose, *Nixon: The Triumph of a Politician, 1962–1972* (New York, 1989), 420–22, 501–5, 543–44, 558–63, and passim; Stanley Kutler, *The Wars of Watergate: The Last Crisis of Richard Nixon* (New York, 1990); John Blum, *Years of Discord: Politics and Society, 1961–1974* (New York, 1991), 421–75; and James Neuchterlein, "Watergate: Toward a Revisionist View," *Commentary*, Aug. 1979, pp. 38–45.

In its broadest sense, the scandal of Watergate arose from the tumultuous and destabilizing trends of the 1960s, especially the war in Vietnam
and the deviousness and power-grabbing associated with the rise of an
imperial presidency.[2] It became the constitutional impasse that it did in
large part because of Nixon's special passion for vengeance and control. In
its narrower sense, however, the scandal began with Gordon Liddy, the
irrepressible "plumber" who had broken into the office of Ellsberg's psychiatrist. In February 1972 Liddy was working as an espionage expert with
Nixon's Committee to Re-elect the President (CREEP), headed by
Nixon's friend and former Attorney General, John Mitchell. Liddy recommended to Mitchell that CREEP tap the phones of Democratic National Committee chairman Lawrence O'Brien at the Watergate complex. Mitchell and his top aide, Jeb Stuart Magruder, approved the idea,
and undercover agents of CREEP broke in to tap O'Brien's phone on May
27. Something went wrong with the tap, however, whereupon an undercover team of three Miami-based Cuban exiles, Frank Sturgis, and James
McCord, chief of security for CREEP, returned to fix it on June 17. A
watchman caught them breaking into the office and called police, who
arrested them.

It remains disputed why CREEP resorted to such illegal activity. The
most widely accepted theory, however, holds that CREEP wanted to learn
all it could about Democratic electoral strategy.[3] More specifically, it may
have hoped to discover what O'Brien, a former lobbyist for the reclusive
tycoon Howard Hughes, might know about possibly embarrassing connections among Hughes, Nixon's brother Donald, and various criminal
figures. Some people have also speculated that Nixon sought to find out if
O'Brien had evidence linking him to assassination plots against Fidel
Castro.[4]

Whether Nixon knew in advance of the plans to break into the Democrats' national headquarters is another unresolved question. He always

2. See Arthur Schlesinger, Jr., *The Imperial Presidency* (Boston, 1973), for historical
background.
3. See Joan Hoff, *Nixon Reconsidered* (New York, 1994), 309–12, for discussion of a
different theory for what happened at the Watergate complex—that the break-ins
sought to uncover evidence about a call-girl ring that might embarrass the Democrats.
4. Michael Beschloss, *The Crisis Years: Kennedy and Khrushchev, 1960–1963* (New
York, 1991), 135–37. CREEP may also have hoped to find out what O'Brien knew
about Nixon's earlier decision to drop the government's anti-trust suit against
ITT—a decision that O'Brien made much of in the campaign. Nixon had accepted
large and secret campaign contributions from Hughes.

denied that he did, and his press secretary, Ron Ziegler, dismissed reports of presidential involvement by branding the affair as a "third-rate burglary attempt." Nixon may have been telling the truth in denying foreknowledge; thanks to years of effort by his lawyers, who brought suits to prevent release of relevant documents and tapes, it has been impossible to know. But Nixon's super-loyal aides could have had no doubt in 1972 about his partisan zeal to embarrass "enemies," especially in an election year. Like their boss, they were contemptuous of democratic procedures and of the niceties of constitutional protections. After all, the President had already authorized wiretapping of his own advisers, tried (via the Huston Plan) to involve the FBI and CIA in illegal surveillance activities, encouraged creation of an "Enemies List," and ordered establishment of the "plumbers" who burglarized the office of Ellsberg's psychiatrist. Mitchell, Magruder, and others, moreover, were well aware of Nixon's special dislike of O'Brien. In the siege mentality that Nixon incited among his aides, something like Watergate was probably an excess waiting to happen.

Unfortunately for Nixon, address books taken from two of the burglars contained the name of E. Howard Hunt, a onetime "plumber" then working for CREEP.[5] Both he and Liddy had been in the Watergate building the night of June 17; both were later arrested as co-conspirators behind the break-in and tapping. At this point Nixon might have fired any and all aides who were involved, thereby clearing some of the air. But it was the middle of an election campaign, and he decided on a cover-up. "Play it tough," he ordered Haldeman. "That's the way they play it, and that's the way we are going to play it."[6] Within a few days of the break-in he arranged to provide hush money for the accused. Sums for this purpose ultimately approached $500,000.[7] On June 23 he ordered Haldeman to have the CIA stop an FBI investigation of the affair. This probe had been started by L. Patrick Gray, acting head of the FBI, who had been named to this post after Hoover had died on May 2. Nixon's order was an illegal use of the CIA and a deliberate obstruction of justice.[8]

5. Blum, *Years of Discord*, 421–27. Hunt had recruited the three Cuban exiles for the job.
6. *Time*, Aug. 19, 1974, p. 27.
7. "Nixon's Endgame," *Newsweek*, Aug. 8, 1994, pp. 50–51.
8. The reason given to stop the FBI was that its investigation would compromise CIA operations necessary to national security. This was not the case, but both the CIA and Gray, a compliant Nixon loyalist eager for confirmation, went along with it. Gray later cooperated with Nixon's plans to the extent of destroying relevant documents. He was not confirmed as head of the FBI.

Nixon's decision to cover up was his fatal error. One revelation after another frustrated his efforts, many of them illegal, to sit on the lid. But it is easy in retrospect to see why he tried. Gathering with Haldeman, Ehrlichman, and others immediately after the break-in, he learned (if he did not know already) of CREEP's involvement. To confess to such activity risked not only admission of administration culpability concerning Watergate but also revelations of other clandestine efforts, such as the earlier break-in—conducted by Hunt and Liddy—at the office of Ellsberg's psychiatrist. Nixon, moreover, prided himself on being able to fight his way out of a hole. He considered his life to be a series of crises, all of them plotted by implacable conspirators. Watergate was such a crisis, and he would overcome it, too.

Developments as early as the autumn of 1972 suggested that the cover-up might not succeed. In late September a grand jury indicted the burglars, as well as Hunt and Liddy, who were scheduled to be tried in the court of federal district judge John Sirica. Washington Post reporters Carl Bernstein and Robert Woodward followed the case closely, printing stories describing the "slush fund" that CREEP was amassing. They also established links between CREEP and presidential aides such as Haldeman and White House Special Counsel Colson, who had originally hired Liddy and Hunt as plumbers. In compiling their stories, which later were turned into a best-selling book on the scandal, Bernstein and Woodward relied on a hidden source who became known as Deep Throat.[9] The identity of this person is yet another mystery surrounding the affair. Best guesses are that the reporters relied on tips from a highly placed and disaffected FBI agent. Angry at being prevented from investigating the break-in, the agent is thought to have leaked information to the press.[10]

During the election campaign, however, Nixon was fortunate. To the dismay of McGovern, who had been assailing the break-in, the trials of the conspirators were postponed until after the election, and Sirica ordered people involved to say nothing about the case. Most reporters, moreover, paid little attention to the burglary: compared to the possibility of peace in Vietnam it seemed an insignificant story. The role of the press in the unfolding of the scandal, both then and later, was less important

9. Carl Bernstein and Robert Woodward, All the President's Men (New York, 1974); Kutler, Wars of Watergate, 458–59.
10. "Two Decades After a Political Burglary, the Questions Still Linger," New York Times, June 15, 1992.

than journalists were to claim. Probing by judges and politicians mattered a good deal more.[11]

This digging approached pay dirt in early 1973. In January the five burglars, plus Hunt and Liddy, were found guilty. Judge Sirica announced that there was more to the break-in than met the eye and threatened tough sentences. McCord, fearful of taking the rap for everyone else, came forth in March to implicate higher-ups in the administration. He talked also to members of a select Senate investigating committee, headed by folksy Sam Ervin of North Carolina, that had started to look into the affair. McCord's revelations started a scramble for safety by others involved in the scandal, including John Dean, Nixon's counsel, and Magruder. Both talked to a grand jury that had been convened on the matter. By late April Nixon himself was feeling the pressure, and he forced Dean to quit. Richard Kleindienst, who had replaced Mitchell as Attorney General, also resigned. Nixon even forced out Haldeman and Ehrlichman. The Berlin Wall around the White House had fallen.[12]

Ervin's committee opened televised hearings on the matter in May, enabling the public to watch McCord accuse Dean and Mitchell of foreknowledge of the break-in, and Mitchell and Magruder of authorizing it. Testimony before the Ervin Committee that summer, especially by Dean, broke further news of the plumbers, of the Huston plan to abuse the powers of the FBI and the CIA, of presidential wiretapping, and of Nixon's authorization of hush money in order to seal the cover-up. Americans were stunned to learn from Alexander Butterfield, a former aide to Haldeman, that the President secretly taped conversations in the Oval Office. By the autumn of 1973 all concerned—Sirica, the Ervin Committee, and Archibald Cox, a Harvard Law School professor whom Nixon had been forced by pressure to name as an independent special prosecutor—were battling the President and his lawyers for release of the tapes.[13]

Nixon fought gamely on, in October ordering his new Attorney Gen-

11. James Baughman, *The Republic of Mass Culture: Journalism, Filmmaking, and Broadcasting in America Since 1941* (Baltimore, 1992), 177–78; Kutler, *Wars of Watergate*, viii, 615. In the long run, official lies about Watergate, like lies about the war in Vietnam, helped to make the media more suspicious and confrontational about public leaders. But that is mainly a later story. See chapter 20 for the media and Vietnam.
12. Ambrose, *Nixon: Ruin*, 81–136; Kutler, *Wars of Watergate*, 290–320.
13. Ambrose, *Nixon: Ruin*, 179–228.

eral, Elliot Richardson, to fire Cox. Richardson, however, resigned rather than carry out the order. William Ruckelshaus, next in command at Justice, also resigned. Finally Nixon prevailed on Solicitor General (by then Acting Attorney General) Robert Bork to do the firing. Critics of these actions, which took place on October 20, called them the Saturday Night Massacre. Nixon then agreed to name another prosecutor, Leon Jaworski of Houston, and surrendered some of the tapes to Sirica. But it was obvious that the President was not ready to give in. As before he cited executive privilege as grounds for not releasing all the tapes. One of the tapes he did turn over to Sirica, moreover, contained an eighteen-and-one-half-minute gap—erased by accident, Nixon's secretary volunteered—of a key conversation between the President and Haldeman on June 20, three days after the break-in. The erasure aroused a further storm of suspicion.[14]

Other walls also crumbled in late 1973. Vice-President Agnew was shown to have accepted kickbacks from contractors while governor of Maryland and even while Vice-President. On October 10 he was forced to resign after pleading no contest to a charge of tax evasion, whereupon Nixon named House GOP leader Gerald Ford of Michigan as Agnew's replacement.[15] Investigations into Nixon's financial affairs at the time were even more damaging to the administration. They revealed that his lawyers had backdated his signature on a deed of gift of his papers to the National Archives, so as to qualify for an income tax deduction in 1969. Examination of his tax returns, which had also been prepared by lawyers, indicated that he had failed to declare taxable improvements made by the government to his substantial personal properties at Key Biscayne, Florida, and San Clemente, California. Nixon responded with a memorable statement, "I have never profited . . . from public service. . . . I have never obstructed justice. . . . I am not a crook." He promised to pay his back taxes. It was obvious, however, that he had amassed great wealth in the course of his life and that he had fudged his tax returns.[16]

Jaworski, meanwhile, proved to be as tenacious a prosecutor as Cox, and a grand jury to which he had presented evidence responded on

14. Ambrose, *Nixon: Ruin*, 229–62; Kutler, *Wars of Watergate*, 383–414.
15. Some observers have wondered if the House of Representatives would have proceeded with impeachment proceedings in 1974 if Agnew had remained as Vice-President. Many representatives would have preferred Nixon, with all his faults, to Agnew.
16. Blum, *Years of Discord*, 451–65; Ambrose, *Nixon: Ruin*, 263–88.

March 1, 1974, by indicting seven members of the CREEP and White House staffs, including Haldeman, Ehrlichman, and Mitchell, on charges of obstruction of justice and impeding the investigation of Watergate. Nixon was named as an unindicted co-conspirator. The grand jury instructed Sirica to turn over tapes in his possession to the House Judiciary Committee, which by then was considering whether to impeach the President. Both Sirica and the House Committee subpoenaed the White House for release of the many tapes still in Nixon's possession.[17]

Nixon continued to resist. Instead of surrendering the tapes themselves, he released on April 30 some 1,300 pages of edited transcripts from the tapes. These, he explained, had "expletives deleted." Defending his action on television, the President said that the transcripts "included all the relevant portions of the subpoenaed conversations . . . the rough as well as the smooth. . . . The President has nothing to hide." In fact, the transcripts had been sanitized. Even so, the transcripts were damaging to the President, for they showed that he had discussed with Dean possible payments to Hunt and that he had ordered aides to perform "dirty tricks" on political opponents. The phrase "expletive deleted," occurring with great frequency on the transcripts, further undermined the President's moral standing.

Sirica, who had listened to the tapes involved, knew that the transcripts had been sanitized. One of the tapes, for instance, had recorded the President as telling aides to say nothing to the grand jury in 1973. "I don't give a shit what happens," he had said. "I want you all to stonewall it, let them plead the Fifth Amendment, cover-up or anything else, if it'll save it—save the plan." The transcripts that he released revealed none of this, expletives included.[18] Sirica, Jaworski, and the House Committee therefore insisted again that Nixon turn over the tapes themselves. Nixon still refused to do so, citing executive privilege, and took the case to the Supreme Court. It promised to hear the matter in July.

On July 24 the Court decided unanimously that executive privilege did not apply in the Watergate case, a criminal matter, and ordered Nixon to surrender all the tapes to Sirica.[19] After hesitating for a few hours, Nixon agreed to do so. It was too late, however, for the House Judiciary Commit-

17. This and the following paragraphs are drawn from Ambrose, *Nixon: Ruin*, 289–445, and Kutler, *Wars of Watergate*, 443–550.
18. Blum, *Years of Discord*, 464.
19. *U.S. v. Nixon*, 418 U.S. 683 (1974). The decision, however, did for the first time give claims for executive privilege a constitutional standing, thereby enabling later Presidents (notably Reagan in the Iran-*contra* affair) to hide behind it.

tee was already concluding nationally televised discussions on articles of impeachment. Between July 27 and July 30 it voted to impeach the President for obstruction of justice concerning the Watergate investigation, for violation of constitutional rights concerning illegal wiretaps and misuse of the FBI, CIA, and IRS, and for violating the Constitution by resisting the committee's subpoenas. The votes, 27 to 11 and 28 to 10, revealed that seven or eight Republicans joined all twenty Democrats on the committee in the decisions to impeach.

While the committee was voting these articles of impeachment, the President's lawyers were listening to the tapes. What they heard staggered them, especially the tape of Nixon's order on June 23, 1972—six days after the break-in—in which he ordered the CIA to stop the FBI from investigating. This was the "smoking gun" that for many people cinched the case against the President. Republican leaders, sensing disaster, began calling upon Nixon to resign from office. These included Kissinger, General Alexander Haig, who had replaced Haldeman as Nixon's top adviser, Senator Barry Goldwater, and Republican national chairman George Bush. When the tapes were released to the public on August 5, confirming many of the worst suspicions about the President's behavior, pressure on Nixon to resign became overwhelming.

It was a mark of Nixon's tenacity—indeed his desperate passion for control—that he still tried to stay in office. But it was clear that he could count on no more than fifteen members of the Senate, which would sit as a court to try him once the House as a whole impeached him. Rather than prolong the inevitable, he gave an unapologetic speech on the evening of August 8 and told the American people that he would resign. Gerald Ford was sworn in as President the next morning.

COMPARED TO THE TEN-YEAR TRAVAIL caused by war in Vietnam, the Watergate break-in was trivial. Moreover, some of the illegal things that the Nixon administration did, such as wiretapping, had been tried by other Presidents, notably Kennedy. Nixon's profanity, of which much was made at the time, was common among politicians, as many who professed to be appalled well knew.[20] If he had quietly destroyed the tapes, it is conceivable that he could have hung on in the White House, for few politicians, even Democrats, were eager to wade into the un-

20. Neuchterlein, "Watergate." It is hard to imagine any administration (or any institution) benefiting from public access to tapes of frank and uncensored private conversations. That is why people have private conversations.

charted waters of impeachment unless they had solid evidence. Why Nixon preserved the tapes, thereby inciting a feeding frenzy of suspicion that led to a constitutional crisis, is yet another of the mysteries surrounding the whole business.[21]

Commentators who seek to avoid the doom-and-gloom postmortems that followed the scandal like to add that the controversy had some beneficial consequences. Among them of course was the forced departure from government of Nixon and many of his aides, who had endangered constitutional liberties and consistently lied to the public: fourteen top officials, including two from the Cabinet, were fined or went to jail.[22] Celebrating this outcome, some people said that the affair revealed the stability and strength of America's political institutions, especially the press, the judiciary, and Congress. The resignation of Nixon, *Time* exulted, was an "extraordinary triumph of the American system."[23] A Brooklyn man added, "It's great. I can tell my senator to go to hell, the rights are mine. I'm proud of a country that can throw out a president. . . . Democracy was strengthened by Watergate. It proved the Constitution works. The political system passed the test."[24]

It is hard, however, to derive great satisfaction from the resolution of the scandal. While it is true that the "system"—Congress, the courts, the press—helped to topple Nixon, it is also true that the President did much to hurt himself. If he had not insisted on preserving the tapes, he might well have survived. It required a good deal of luck for the "system" to bring him down.

In struggling to save his presidency, Nixon repeatedly claimed that opponents who harped on Watergate were undermining his domestic and foreign policies. There was a little bit of truth in this, for the declining political standing of his administration in 1973 and 1974 emboldened his foes. Defense Secretary James Schlesinger, who was hostile to Kissinger, openly undercut administration efforts to negotiate new SALT agree-

21. Theories about this range widely. Some argue that he did not dare, others that he thought some of the tapes might exonerate him, others that he was either stupid or stubborn.
22. These included Mitchell, Kleindienst, Colson, Dean, Haldeman, Ehrlichman, Magruder, Hunt, and Liddy. All but Kleindienst, who pleaded guilty to a misdemeanor for refusing to testify fully and accurately before a Senate committee investigating the ITT case, served time. Kleindienst received a $100 fine and a suspended one-month jail sentence.
23. *Time*, Aug. 19, 1974, p. 9.
24. Jonathan Rieder, *Canarsie: The Jews and Italians of Brooklyn Against Liberalism* (Cambridge, Mass., 1985), 160.

ments in the winter of 1973-74. Opponents of détente in the House of Representatives, allying with friends of Israel, coalesced in December 1973 to pass the so-called Jackson-Vanik amendment, which sought to deny most-favored-nation status to the Soviets unless they agreed to major concessions concerning Jewish emigration.[25] The possibility that this amendment might be approved in the Senate irritated the Soviets and soured relations with the United States. When Nixon journeyed to Moscow again in June 1974, he managed to accomplish nothing of significance.[26]

But it is difficult to argue that Watergate much changed policy possibilities in Washington. Nixon started his second term with no new programs in mind, save to cut back sharply on the number of federal bureaucrats (many of whom did all they could to oppose the President after 1972) and defiantly to impound congressional appropriations, especially for control of water pollution.[27] These efforts greatly heightened partisan controversy with the Democratic Congress from the very beginning of his term and virtually ensured that little legislation of consequence would be approved. In 1974 he called for expansion of a guaranteed student loan program to aid college and university students, and he reiterated earlier requests for expanded health insurance and welfare reform. But, as in the past, he did little to follow up on his declarations of intent, in part because he knew that Democrats in Congress would by then block almost anything that he sought.[28]

Détente, moreover, had always been oversold by Nixon. The Soviets had supported it only when it suited their interests, reserving the right (as did the United States) to go their own way. In October 1973 Moscow did not inform the United States that Syria and Egypt were about to wage war on Israel. In February 1974 it expelled the novelist Aleksandr Solzhenitsyn, leading the physicist Andrei Sakharov to go on a hunger strike on the second day of the summit in June. In claiming that controversy over Watergate endangered American foreign policy, the President and his defenders greatly exaggerated. Nixon himself later admitted that "the military establishments of both countries" stopped progress toward arms

25. Congress later approved this amendment, in December 1974. It did nothing for the Jews and harmed what remained of détente.
26. Kutler, *Wars of Watergate*, 603–7.
27. Hoff, *Nixon Reconsidered*, 25–27; Kutler, *Wars of Watergate*, 133–37; Louis Fisher, *Presidential Spending Power* (Princeton, 1975), 175–201.
28. Ambrose, *Nixon: Ruin*, 297–99, 596–97.

control. "These problems," he added, "would have existed regardless of Watergate."[29]

It is tempting to think that the Watergate scandal trimmed the imperial presidency down to more manageable size and that it revitalized moral considerations in the conduct of government business. Whenever later presidents appeared to exceed their authority or to resort to acts of doubtful constitutionality—as Ronald Reagan did in the Iran-*contra* affair in 1987—the misdeeds of Nixon were resurrected. Iran-*contra* became "Irangate."[30] Still, the extraordinary popularity of Reagan to that time— as of other presidents since 1974 who have asserted large prerogatives in foreign policy matters—suggests that Americans have continued to admire bold executive leadership. While the backlash against Nixon's highhandedness (and Johnson's) has moderated the imperious temptations of his successors, it has not altered the constitutional balance of American government, which until the end of the Cold War remained heavily tilted toward 1600 Pennsylvania Avenue on military and foreign policy concerns.

Politically, the squalid business of Watergate had significant partisan results, at least in the short run. Democrats scored major triumphs in the 1974 elections and sent Jimmy Carter to the White House in 1976. Conservatives in the GOP, leaping into the vacuum left by Nixon and his centrist allies, gradually established control of the party and blocked whatever hopes may have remained for serious consideration of such unresolved issues as welfare reform and health insurance. In 1980 they elected Reagan to the presidency. But policy-making from 1974 through 1976 did not differ much from what it would likely have been with Nixon in command. President Ford, after pardoning his predecessor, avoided new frontiers. Instead, he struggled much of the time to deal with already existing problems: tensions over court-mandated busing and affirmative action, highly emotional debates over abortion, and above all the stagnation of the economy.

The central issue raised by Watergate, finally, was not resolved. This was how to make American government, especially the President, more accountable to the people. A raft of legislation in 1973–74, including the

29. Raymond Garthoff, *Détente and Confrontation: American-Soviet Relations from Nixon to Reagan* (Washington, 1985), 409–37; John Gaddis, *Strategies of Containment: A Critical Appraisal of Postwar American National Security Policy* (New York, 1982), 310–15; Kutler, *Wars of Watergate*, 607.
30. Kutler, *Wars of Watergate*, 710–11.

War Powers Act of 1973, a law to regulate campaign financing and spending (1974), a Freedom of Information Act (1974), and a Congressional Budget and Impoundment Act (1974), tried to promote such accountability, but these laws for the most part failed to accomplish what they set out to do, largely because Presidents and other politicians figured out ways of evading them. As acts by subsequent Presidents made clear, White House high-handedness could and did happen again in the future.[31]

Well before these later abuses occurred, many Americans had feared as much. Watergate, they believed, proved—yet again—the deviousness and arrogance of government officials who claimed to serve the public interest. First, Lyndon Johnson and exaggerated claims about a Great Society. Then lies upon lies about Vietnam. Now, Watergate and many more lies. A teacher expressed the feeling of many Americans:

> After Watergate, it's crazy to have trust in politicians. I'm totally cynical, skeptical. Whether it's a question of power or influence, it's who you know at all levels. Nixon said he was the sovereign! Can you believe that? I was indignant. Someone should have told him that this is a democracy, not a monarchy.[32]

CRITICAL FEELINGS SUCH AS THESE remained powerful in the United States after 1974. Together with abiding popular resentments about other domestic issues—busing, affirmative action, abortion, crime, welfare dependency—they sharpened social divisions and stymied liberal reformers. Conservatives maintained the initiative in Washington for much of the next two decades. While grand expectations about "rights" at home, as well as grand designs for America's role in the world, did not disappear after 1974—these were lasting legacies of the postwar era— many people seemed anxious and contentious. This was not because they were worse off absolutely—most people managed as well or a little better economically, especially in the mid- and late 1980s—but because their very high expectations became frustrated. The United States, so powerful for much of the postwar period, seemed adrift, unable to reconcile the races (or the classes or the sexes) at home or to perform as effectively on the world stage. A woman exclaimed, "Sometimes you get the feeling nothing has gone right since John Kennedy died. We've had the Vietnam War, all the rioting. . . . Before then you were used to America

31. Hoff, *Nixon Reconsidered*, 329–38; Kutler, *Wars of Watergate*, 574–603.
32. Rieder, *Canarsie*, 250.

winning everything, but now you sometimes think our day may be over."[33]

Most people, however, did not go through the day thinking about Big Political Issues. They tended, rather, to concentrate on concerns closer to home: their families, their neighborhoods, their work, their economic well-being, their futures. These private spheres seemed very different to people after 1973–74. Poll data suggested that more and more Americans after that time were losing faith in the capacity of the nation to move ahead in the future.[34] The very high hopes of the previous decades—a key to the drive, the optimism, the idealism, and the rights-consciousness of the era—were becoming harder to achieve.

Nothing did more to generate these anxieties than the downturn of the economy in 1973–74. Signs of trouble, of course, had appeared earlier, leading Nixon to impose controls in 1971. But controls, while of some use in 1972, were a band-aid that could not stop the bleeding. All the structural problems that economists had warned about coalesced after 1973–74 to jolt American life. These included sagging productivity, declining competitiveness in world markets, accelerating inflation, rising unemployment, especially among minorities and the millions of baby boomers now seeking work, and a slowing down in the creation of good-paying, career-enhancing jobs outside of the increasingly dominant service sector.[35]

Nixon's New Economic Policy, resumed after the 1972 election, failed to curb inflation, mainly because it did not attack the underlying problem—too much spending, both public and private—and prices mounted in 1973. Nixon reimposed controls in June and relaxed them in August, but the cost of living kept moving up. The NEP, moreover, did nothing to deal with the deep-seated problems of American manufacturing. The automobile industry, already sluggish, was especially hard-hit, mainly by competition from abroad. Trying to repair the damage, Nixon devaluated the dollar again in early 1973. It did not help: American car sales dropped by 11 million in 1973, and unemployment—most worrisome in manufacturing—rose in 1974 to 7.2 percent, the highest since

33. Barry Bluestone and Bennett Harrison, *The Deindustrialization of America* (New York, 1982); Landon Jones, *Great Expectations: America and the Baby Boom Generation* (New York, 1980), 255.

34. Daniel Yankelovich, *New Rules: Searching for Self-Fulfillment in a World Turned Upside Down* (New York, 1981), 181–84.

35. Jones, *Great Expectations*, 158–60. Cutbacks in defense spending in 1973–74 also hurt some regions.

1960. The AFL-CIO complained (with some exaggeration) that the United States had become "a nation of hamburger stands, a country stripped of industrial capacity and meaningful work . . . a service-economy . . . a nation of citizens busily buying and selling cheese-burgers and root-beer floats."[36]

On top of these problems came yet another blow—one of the most traumatic to befall the United States in the postwar era. This was the "energy crisis." Some observers had earlier warned that the United States was vulnerable if oil prices increased, but neither Nixon nor Congress heeded them. The energy crisis then shook Americans following the "Yom Kippur War" between Israel and Arab enemies in October 1973. The war exposed the limitations of détente, for the United States and the Soviet Union, rallying to their allies, seemed ready to face off mili-tarily in the Middle East. Kissinger, acting while Nixon was asleep, im-petuously called a day-long, worldwide high alert of American forces, including the Strategic Air Command. While it lasted, the alert was profoundly unnerving. Elizabeth Drew of New Yorker called it "Strange-love Day."[37]

A more enduring consequence of the war was the impact that it had on Arab leaders. Upset by Nixon's devaluations of the dollar, the currency normally used to pay for oil, they had already raised their charges, helping to cause worldwide inflation. Arab leaders further resented the United States for its longtime support of Israel. Retaliating, they imposed an embargo on shipments of oil to the United States. Two months later, in December, the Organization of Petroleum Exporting Countries (OPEC) raised its charges to $11.65 per barrel, a price that was 387 percent higher than before the Yom Kippur War.[38]

The effects of the embargo, which lasted until March 18, 1974, and of higher prices of oil thereafter, seriously upset hosts of people in the United States. Cheap oil had been a key to American prosperity and economic growth in the postwar era, enormously benefiting major industries such as automobiles and utility companies, accelerating large-scale social trans-formations such as the spread of suburbanization, and stimulating the consumerism that lay at the heart of postwar American culture.[39] With

36. Peter Carroll, It Seemed Like Nothing Happened: The Tragedy and Promise of America in the 1970s (New York, 1982), 129–30.
37. Garthoff, Détente and Confrontation, 360–85, 404–7. Kissinger called the alert in the early morning hours of October 25, five days after the turmoil engendered by Nixon's "Saturday Night Massacre" of October 20. Efforts by the UN and others brought about a cease-fire and averted a confrontation of world powers.
38. Blum, Years of Discord, 457.

6 percent of the planet's population in 1973, the United States consumed a third of the oil produced in the world.

The postwar availability of cheap oil from abroad, especially the Middle East, had also led the United States to rely on overseas sources. In 1960, 19 percent of oil consumed in America had come from abroad; by 1972 the figure had risen to 30 percent. Then, and suddenly, a more complicated world had arrived. As demand in the United States outstripped supplies, domestic oil prices (and profits for American producers) increased substantially. Still, shortages persisted. Desperate car owners waited hours for gasoline—lines in New Jersey were as long as four miles—sometimes fighting among themselves or attacking gas station attendants. By the time the embargo ended, prices for heating oil and gasoline had risen in places by as much as 33 percent.

In some ways the effects of the embargo were exaggerated; oil and gasoline costs in the United States still remained a good deal lower than they were in most industrialized nations. But they were nonetheless traumatic for many Americans, because the embargo reinforced the sense of national vulnerability that had already arisen amid the frustrations of the Vietnam War. First, the United States was humbled by North Vietnam, the land that Johnson had called a "piss-ant country." Now it was shaken by the actions of Arabs, some of whom (as the Saudis) were supposed to be allies. The wonderful American Century that Henry Luce had foreseen in 1941 seemed to be collapsing early. As one scholar put it, the embargo represented a "watershed, sharply dividing the second half of the twentieth century into two elongated quarter-centuries—the twenty-seven-year period extending from the end of World War Two to 1973 (the postwar quarter-century) and the other twenty-seven-year period extending from 1973 to the end of the century."[40]

The impact of the embargo and of other structural flaws in the economy touched off unsettled times that lasted until 1983.[41] In 1974 alone retail prices increased by 11 percent and wholesale prices by 18 percent. Unemployment continued to rise, reaching a postwar high of 8.5 percent by 1975. Real gross national product fell more than 2 percent in 1974 and nearly 3 percent in 1975. By the end of the decade the United

39. David Calleo, *The Imperious Economy* (New York, 1983), 112–13.
40. Yankelovich, *New Rules*, 164–65. Carroll, *It Seemed*, 117–18, argues that the embargo caused "the most revolutionary shift of world power in the twentieth century." For a sweeping study of oil in world politics, see Daniel Yergin, *The Prize: The Epic Quest for Oil, Money, and Power* (New York, 1991).
41. Frank Levy, *Dollars and Dreams: The Changing American Income Distribution* (New York, 1987), 62–65.

States was confronting a "stagflation" that featured double-digit inflation as well as double-digit unemployment.

The ill health of the American economy resisted whatever cures policy-makers tried to prescribe, thereby further stimulating dissatisfaction with government and with "experts" in general. Nixon conceded as much, telling Americans that "we are heading toward the most acute shortage of energy since World War II." Desperate for quick fixes, Nixon, Congress, governors, and state legislators called for thermostats to be lowered, cut back air travel, dropped speed limits, and accelerated the licensing of nuclear power plants. Funding increased to promote exploration of energy sources at home. The private sector did its part: factories reduced hours to conserve fuel; colleges canceled mid-winter sessions; commuters formed car pools or tried mass transit. Nothing, however, seemed to make much difference. And the profits of oil companies continued to skyrocket, hastening the already considerable shift of national wealth and power to the Southwest, but otherwise intensifying class and regional antagonisms. THINGS WILL GET WORSE, a newspaper headline warned, BEFORE THEY GET WORSE.[42]

IT IS CORRECTLY OBSERVED that Americans have a special tendency to dissect their culture, as if uncertain that their great experiment in democracy can hold together. Many of the would-be Tocquevilles who searched for the essence of the United States in the mid-1970s—and later—were almost as pessimistic as the headline-writer above. Americans, they said, had become discontented, fractious, alienated, and divided into ever more self-conscious groups that identified themselves narrowly by region, gender, age, religion, ethnicity, and race. Pundits thought that people were abandoning social concerns and becoming less willing to defer gratification. Tom Wolfe said in 1976, the bicentennial of the United States, that the seventies were the "Me Decade" in America. Christopher Lasch wrote a little later that the United States had become a "culture of narcissism."[43]

42. Carroll, *It Seemed*, 118. For energy policy after World War II, see Richard Vietor, *Energy Policy in America Since 1945: A Study of Business-Government Relations* (New York, 1984).
43. Arlene Skolnick, *Embattled Paradise: The American Family in an Age of Uncertainty* (New York, 1991), 134-37. Lasch popularized this phrase, introduced by him in the *New York Review of Books* in 1976, in a much-discussed book, *The Culture Of Narcissism: American Life in an Age of Diminishing Expectations* (New York, 1979).

Jeremiads such as these left the false impression that American society and culture were suddenly falling apart. On the contrary, many features of American life in the post–World War II years persisted after 1974. As before, the United States remained one of the most stable societies in the world. Most Americans still held strongly to long-established values, including commitment to the Constitution, respect for the law, belief in the necessity of equal opportunity, and confidence in the utility of hard work. No Western culture was more religious. Not even the travail of Vietnam dimmed the certainty of Americans that Communism must be contained and the Cold War carried on, in Asia as well as elsewhere. These were among the many attitudes and values that flourished from 1945 to 1974 and that remained alive and well thereafter.

It was also wrong to assume, as some pessimistic liberals did, that Americans were ready to jettison the progressive social policies that had gained ground since 1945, especially in the 1960s. Racial tensions notwithstanding, the majority of people remained committed after 1974 to the civil rights statutes that had dramatically transformed the legal status of minorities since 1945. These were the most significant legislative accomplishments of the era. Americans also continued to support higher levels of domestic spending for health, education, and Social Security—in real dollars per capita and as a percentage of GNP—than they had in the 1940s and the 1950s. The rise of big government and of an expanded welfare state were major legacies of the period: social insurance payments for the elderly and disabled mushroomed after 1975.[44] A third liberal legacy of the postwar era, advances in civil liberties, also survived the 1970s. Americans who had fought McCarthyism were highly pleased by what the Warren Court and political activists had done to breathe life into Bill of Rights freedoms of press, speech, and religion.

Some manifestations of postwar rights-consciousness, to be sure, had taken hold less firmly between 1945 and 1974, and they encountered damaging backlash thereafter. Advocates of women's rights, for instance, had done much before 1974 to change the nature of thinking about gender relations: henceforth, discussions of "sexism" remained at the

---

44. See footnote 35, chapter 23. America's social welfare state, however, continued to be less comprehensive and less generous than that of most other industrialized nations; means-tested, non-indexed programs such as AFDC fared relatively worse than they had before 1974; and social insurance continued to be financed by regressive payroll taxes. These reached 15.3 percent for Social Security and Medicare by the mid-1990s.

center of public debates.[45] But feminism lost some of its drive thereafter: ERA, which had seemed certain of ratification in 1973, failed in the next decade.[46] And many minority groups, while enjoying larger legal rights than earlier, continued to face widespread discrimination, especially in jobs, schooling, and housing. They also suffered disproportionally from poverty, as well as from levels of violent crime and drug usage that escalated frighteningly in the 1970s and 1980s.

Nonetheless, many goals of postwar American liberalism, notably the dismantling of Jim Crow and the rise of federal standards in social policy, especially for the disabled and the elderly, were far closer to realization in 1974 than they had been in 1945, and they endured thereafter in a political culture that, while more conservative than it had been in the mid-1960s, continued to support social programs. These standards were considerably more generous than people in the 1940s and 1950s might have imagined possible. Backlash did not kill everything.

Another goal of many Americans in the postwar era, greater personal choice, also advanced with special speed between 1945 and 1974, especially amid the fantastic affluence of the 1960s. Age-old stigmas seemed to collapse. By the early 1970s people had much more freedom to dress, to wear their hair, and to socialize as they pleased. They traveled about much more widely and were exposed to a considerably broader world. They had far greater choice of film, television, and music, not to mention the means to exercise it. The variety of paperbacks and of magazines that was available after 1974 would have been unimaginable in 1945. These were not trivial matters. Thanks in part to the Pill—and, more important, to the rise of permissive attitudes in the culture at large—women had much greater sexual freedom. After 1973 they had a constitutional right to an abortion. Whether all these developments were "good" or "bad" obviously depended on the perspective of the viewer.[47] Still, liberating forces in personal life had been rapid and dramatic, more so than in most eras of comparable length. These, too, survived very well after 1974.

Doomsayers after 1974, preoccupied with stagflation, sometimes

45. David Farber, *The Age of Great Dreams: America in the 1960s* (New York, 1994), 264–65.
46. Mary Frances Berry, *Why the ERA Failed: Politics, Women's Rights, and the Amending Process* (New York, 1986).
47. By the 1980s it became especially clear that the sexual revolution, among other things, had helped to promote extraordinarily high rates of illegitimacy (which in turn greatly increased the numbers of children in poverty). Sexual "liberation" also helped to sustain the rise of AIDS after 1980—but that is another story.

tended to overlook the continuing strength of these important postwar developments and to focus instead on contemporary problems. Like many other Americans, they also grew corrosively critical of "experts"—whether in government, medicine, law, or business—whose "answers" to the problems had been oversold. The media, much more suspicious than earlier, exacerbated these popular concerns. Polls in the unsettled mid-1970s, however, suggested that Americans were about as content with their daily lives in the present, including the quality of their work, as they had been in the "nifty fifties" or in the booming sixties. They continued to derive special satisfaction from their roles as spouses and parents. As in most eras of human history, these things changed slowly if at all.

Still, the sputtering performance of the economy understandably alarmed people. As stagflation persisted, Americans expressed more and more open doubts about their ability to get ahead and about the chances that their children would do as well as or better than they had. Faith in upward social mobility—so central to the American dream—seemed to weaken. Some of the special vibrancy and energy of postwar American culture—qualities that had often astonished and delighted newcomers to the United States—seemed in decline.[48]

These economic difficulties were more damaging to the fuller realization of the still grand expectations of Americans than the Vietnam war or Watergate, important though those had been.[49] For economic progress had been a tonic for millions of people between 1945 and the early 1970s. Gains in science and technology had helped to promote impressive advances in productivity, as well as a phenomenal spread of home-ownership, great access to higher education, and widespread enjoyment of consumer goods. These were goals that had only been glimmering in 1945. For all its vulgarity, the consumer culture that exploded in the postwar era greatly improved comfort for the majority of Americans. In so doing it partially obscured persistent inequalities of race, class, region, and gender.[50]

48. Nicholas Lemann, "How the Seventies Changed America," *American Heritage*, July/August 1991, pp. 39–49; Yankelovich, *New Rules*, 24–25. In fact, a degree of upward mobility remained, but perceptions—conditioned by high expectations in earlier years—were often otherwise.

49. As noted, Vietnam had much to do with the rise of these economic difficulties—and with the suspicion and divisiveness of the culture. Economic problems were deeply interrelated with others.

50. Economic inequality, as measured by shares of national income possessed by various slices of the income pyramid, declined slightly between 1945 and 1974 and appeared to rise considerably thereafter, especially in the late 1970s and

The vibrancy of the economy in the postwar era had indeed stimulated unprecedented and, by the late 1960s, near-fantastic expectations about the Good Life. These expectations, in turn, had combined with the civil rights revolution—a moral cause of transcendent power—to accelerate a rights-consciousness that had always been inherent in American democratic culture. When economic growth declined, especially after 1974, it did not kill these expectations or destroy the quest for rights: individuals and groups, having been empowered in a rights-conscious culture, continued to demand a wide range of freedoms, entitlements, and gratifications. But the sluggishness of the economy widened the gulf between grand expectations and the real limits of progress, undercutting the all-important sense that the country had the means to do almost anything, and exacerbating the contentiousness that had been rending American society since the late 1960s. This was the final irony of the exciting and extraordinarily expectant thirty years following World War II.

1980s. By 1990 the wealthiest 20 percent of families in the United States had 44.3 percent of aggregate income, and the poorest 20 percent had 4.6 percent— compared to percentages in 1950 of 42.7 and 4.5, in 1960 of 41.3 and 4.8, in 1970 of 40.9 and 5.4, and in 1980 of 41.6 and 5.1. Sheldon Danziger, "The Historical Record: Trends in Family Income, Inequality, and Poverty," in Danziger, Gary Sandefur, and Daniel Weinberg, eds., *Confronting Poverty: Prescriptions for Change* (Cambridge, Mass., 1994), 18–50.

Bibliographical Essay

As footnotes in the text suggest, the literature concerning postwar United States history is vast. This brief bibliographical essay mentions only those books that proved most useful to me. It begins by identifying general interpretations of the era as well as sources concerned with various themes and topics: race relations, religion, the economy, and so on. The bibliography then follows the chronological organization of the chapters, referring to books (for articles, see footnotes) that deal with particular time periods and controversies, beginning with the Truman era and concluding with sources on the early 1970s.

General interpretations: Among the best books that seek to make sense of this era are William Chafe, *The Unfinished Journey: America Since World War II* (New York, 1991), an especially well written and well argued survey; John Blum, *Years of Discord: American Politics and Society, 1961–1974* (New York, 1991); John Diggins, *The Proud Decades: America in War and Peace, 1941–1960* (New York, 1988); Steve Fraser and Gary Gerstle, eds., *The Rise and Fall of the New Deal Order, 1930–1980* (Princeton, 1989), a collection of articles focusing on labor and politics; Godfrey Hodgson, *America in Our Time* (Garden City, N.Y., 1976); William Leuchtenburg, *In the Shadow of F.D.R.: From Harry Truman to Ronald Reagan* (Ithaca, 1983); Leuchtenburg, *A Troubled Feast: American Society since 1945* (Boston, 1973); Alonzo Hamby, *Liberalism and Its Challengers: From F.D.R. to Bush* (New York, 1992), a book of informed essays on major political figures; Marty Jezer, *The Dark Ages: Life in the United States, 1945–1960* (Boston, 1982), a critical account; William O'Neill, *American High: The Years of Confidence, 1945–1960* (New York, 1986), which presents a very different view from Jezer's; James Sundquist, *Politics and Policy: The Eisenhower, Kennedy, and Johnson Years* (Washington, 1968), a still useful analysis of public programs; Morris Janowitz, *The Last Half-Century: Societal Change and Politics in America* (Chicago, 1978), which is informative on social trends; Alan Wolfe, *America's Impasse: The Rise and Fall of the Politics of Growth* (New York, 1981); Wolfe, ed., *America at Century's End* (Berkeley, 1991), an especially strong collection of topical essays on trends since World War II; and Frederick Siegel, *Troubled Journey: From Pearl Harbor to Ronald Reagan* (New York, 1984). Important books offering interpretive surveys of postwar foreign policies are John Gaddis, *Strategies of Containment: A*

Critical Appraisal of Postwar American National Security Policy (New York, 1982); Stephen Ambrose, *Rise to Globalism: American Foreign Policy Since 1938* (4th rev. ed., New York, 1988); and Daniel Yergin, *The Prize: The Epic Quest for Oil, Money, and Power* (New York, 1991).

Thematic Books: Among these sources are Diane Ravitch, *The Troubled Crusade: American Education, 1945–1980* (New York, 1983), a balanced treatment of a controversial subject; Charles Silberman, *Crisis in the Classroom: The Remaking of American Education* (New York, 1970); James Baughman, *The Republic of Mass Culture: Journalism, Filmmaking, and Broadcasting in America Since 1941* (Baltimore, 1992), a brief, careful account; Walter Dean Burnham, *Critical Elections and the Mainsprings of American Politics* (New York, 1970); Kathleen Hall Jamieson, *Packaging the Presidency: A History and Criticism of Presidential Campaign Advertising* (2d ed., New York, 1992); John Diggins, *The Rise and Fall of the American Left* (New York, 1992), which interprets twentieth-century trends; Thomas Edsall, with Mary Edsall, *Chain Reaction: The Impact of Race, Rights, and Taxes on American Politics* (New York, 1992), which focuses on the years since 1960; Fred Greenstein, ed., *Leadership in the Modern Presidency* (Cambridge, Mass., 1988), a collection of scholarly essays on presidents; Harvey Levenstein, *Paradox of Plenty: A Social History of Eating in Modern America* (New York, 1993); Charles Silberman, *Criminal Violence, Criminal Justice* (New York, 1978); and James Wilson, *Thinking About Crime* (rev. ed., New York, 1983).

Some topics have stimulated especially substantial coverage by historians and others. For up-to-date surveys of aspects of **race relations**, see Robert Weisbrot, *Freedom Bound: A History of America's Civil Rights Movement* (New York, 1990); and Harvard Sitkoff, *The Struggle for Black Equality, 1954–1992* (New York, 1993). Other broadly framed books concerning race include Steven Lawson, *Black Ballots: Voting Rights in the South, 1944–1969* (New York, 1976); Lawson, *In Pursuit of Power: Southern Blacks and Electoral Politics, 1965–1982* (New York, 1985); August Meier and Elliot Rudwick, *CORE: A Study in the Civil Rights Movement, 1942–1968* (New York, 1973); Aldon Morris, *The Origins of the Civil Rights Movement: Black Communities Organizing for Change* (New York, 1984); Reynolds Farley and Walter Allen, *The Color Line and the Quality of Life in America* (New York, 1987), a valuable survey of socio-economic data; Bart Landry, *The New Black Middle Class* (Berkeley, 1987); and Herbert Hill and James Jones, Jr., eds., *Race in America: The Struggle for Equality* (Madison, 1993), a collection of scholarly articles focusing on labor, education, and the law. Also David Goldfield, *Black, White, and Southern: Race Relations and Southern Culture, 1940 to the Present* (Baton Rouge, 1990); Manning Marable, *Race, Reform, and Rebellion: The Second Reconstruction in Black America, 1945–1990* (Jackson, 1991); Walter Jackson, *Gunnar Myrdal and America's Conscience: Social Engineering and Racial Liberalism, 1938–1987* (Chapel Hill, 1990); and Nicholas Lemann, *The Promised Land: The Great Black Migration and How It Changed America* (New York, 1991), a well-written book on a key development of the postwar era. Valuable sources on race relations also include the Kerner Commission, *Report of the National Advisory Commission on Civil Disorders* (Washington, 1968); and Kenneth Clark and Talcott Parsons, eds., *The Negro American* (Boston, 1966), a collection of thoughtful articles. Other, more focused books concerning race relations and civil rights are noted below in the chronologically organized part of this bibliography.

Ethnic relations: A very intelligent starting point is Richard Polenberg, *One Nation Divisible: Class, Race and Ethnicity in the United States Since 1938* (New York, 1980). Also well done are Jonathan Rieder, *Canarsie: The Jews and Italians of Brooklyn Against Liberalism* (Cambridge, Mass., 1985), which focuses on backlash after 1970; Reed Ueda, *Postwar Immigrant America: A Social History* (Boston, 1994), a brief interpretive survey; Ronald Takaki, *Strangers from a Different Shore: A History of Asian Americans* (Boston, 1989); Roger Daniels, *Asian America: Chinese and Japanese in the United States Since 1850* (Seattle, 1988); Mario Garcia, *Mexican Americans: Leadership, Ideology, and Identity, 1930–1960* (New Haven, 1989); David Reimers, *Still the Golden Door: The Third World Comes to America* (New York, 1992); Terry Wilson, *Teaching American Indian History* (Washington, 1993), a useful historiographical guide; and Nathan Glazer and Daniel Moynihan, *Beyond the Melting Pot: The Negroes, Puerto Ricans, Jews, Italians, and Irish of New York City* (Cambridge, Mass., 1963).

Religion: A fine collection of essays is Michael Lacey, ed., *Religion and Twentieth-Century Intellectual Life* (Washington, 1989). It complements a solid survey, Robert Wuthnow, *The Restructuring of American Religion* (Princeton, 1988); and James Davison Hunter, *Culture Wars: The Struggle to Define America* (New York, 1991), which stresses the growing importance, especially since 1960, of divisive religious/ moral world views in the postwar era. For more specialized accounts, see Paul Boyer, *When Time Shall Be No More: Prophecy Belief in Modern American Culture* (New York, 1992); and William McLoughlin, *Billy Graham: Revivalist in a Secular Age* (New York, 1960). A key primary source is Will Herberg, *Protestant, Catholic, Jew: An Essay in American Religious Sociology* (Garden City, N.Y., 1955).

Urban/suburban history: Among the most helpful books in this area are Kenneth Jackson, *Crabgrass Frontier: The Suburbanization of the United States* (New York, 1985), a broad historical survey; Arnold Hirsch, *Making the Second Ghetto: Race and Housing in Chicago, 1940–1960* (New York, 1983), an excellent monograph; Mark Gelfand, *A Nation of Cities: The Federal Government and Urban America, 1933–1965* (New York, 1975); and two brief but interpretive surveys, Peter Muller, *Contemporary Sub/Urban America* (Englewood Cliffs, N.J., 1981); and Jon Teaford, *The Twentieth-Century American City: Problem, Promise, and Reality* (Baltimore, 1986). An influential critique of urban planning is Jane Jacobs, *The Death and Life of Great American Cities* (New York, 1961). Two brilliant books by Herbert Gans are *The Levittowners: Ways of Life and Politics in a New Suburban Community* (New York, 1967); and *The Urban Villagers: Group and Class in the Life of Italian-Americans* (New York, 1962).

Labor: In addition to coverage in books mentioned above, see Robert Zieger, *American Workers, American Unions, 1920–1985* (Baltimore, 1986), a fine brief survey. Other relevant sources are Melvyn Dubofsky, *The State and Labor in Modern America* (Chapel Hill, 1994); Christopher Tomlins, *The State and the Unions: Labor Relations, Law, and the Organized Labor Movement in America, 1880–1960* (New York, 1985); Gary Gerstle, *Working-Class Americanism: The Politics of Labor in an Industrial City, 1914–1960* (New York, 1989), on Woonsocket, R.I.; David Halle, *America's Working Man: Work, Home, and Politics Among Blue-Collar Property-Owners* (Chicago, 1984); William Harris, *The Harder We Run: Black Workers Since the Civil War* (New York, 1982); and Juliet Schor, *The Overworked American: The Unexpected Decline of Leisure* (New York, 1991). See also sources below for women's history.

Economic and business trends: A good starting point is Frank Levy, *Dollars and Dreams: The Changing American Income Distribution* (New York, 1987), a clear and well-written account of historical trends. Wolfe, *America's Impasse*, noted above, is valuable for its focus on the role of economic growth in American culture and politics. Three books emphasizing economic decline are Bennett Harrison and Barry Bluestone, *The Great U-Turn: Corporate Restructuring and the Polarizing of America* (New York, 1990), which is very critical of corporate decisions; David Calleo, *The Imperious Economy* (New York, 1983), which pays considerable attention to international events; and Paul Kennedy, *The Rise and Fall of the Great Powers: Economic Change and Military Conflict from 1500 to 2000* (New York, 1987), a sweeping interpretation. More specialized sources are Bruce Schulman, *From Cotton Belt to Sunbelt: Federal Policy, Economic Development, and the Transformation of the South, 1938–1980* (New York, 1991); David Vogel, *Fluctuating Fortunes: The Political Power of Business in America* (New York, 1989); Robert Collins, *The Business Response to Keynes, 1929–1964* (New York, 1981); Herbert Stein, *The Fiscal Revolution in America* (Chicago, 1969); Cathie Martin, *Shifting the Burden: The Struggle over Growth and Corporate Taxation* (Chicago, 1991); John Witte, *The Politics and Development of the Federal Income Tax* (Madison, 1985); and James Wilson, ed., *The Politics of Regulation* (New York, 1980). John Kenneth Galbraith, *The Affluent Society* (Boston, 1958), is an important contemporary source.

Poverty and welfare: These subjects have evoked a great deal of writing. Helpful places to begin are Jeffrey Williamson and Peter Lindert, *American Inequality: A Macroeconomic History* (New York, 1980); Sheldon Danziger, Gary Sandefur, and Daniel Weinberg, eds., *Confronting Poverty: Prescriptions for Change* (Cambridge, Mass., 1994), an authoritative collection of essays by social scientists; Edward Berkowitz, *America's Welfare State from Roosevelt to Reagan* (Baltimore, 1991); and John Schwarz, *America's Hidden Success: A Reassessment of Twenty Years of Public Policy* (New York, 1983). See also James Patterson, *America's Struggle Against Poverty, 1900–1994* (Cambridge, Mass., 1995); and Margaret Weir, Ann Orloff, and Theda Skocpol, eds., *The Politics of Social Policy in the United States* (Princeton, 1988). A conservative interpretation of welfare programs is Charles Murray, *Losing Ground: American Social Policy, 1950–1980* (New York, 1984). A view from the left is Frances Fox Piven and Richard Cloward, *Regulating the Poor: The Functions of Public Welfare* (updated ed., New York, 1993). Sources—among many—that deal with inner-city poverty, mainly of African-Americans, include William Julius Wilson, *The Truly Disadvantaged: The Inner City, the Underclass, and Public Policy* (Chicago, 1987); Michael Katz, ed., *The "Underclass" Debate: Views from History* (Princeton, 1993); and Christopher Jencks, *Rethinking Social Policy: Race, Poverty, and the Underclass* (Cambridge, Mass., 1992). Other useful books include Carol Stack, *All Our Kin: Strategies for Survival in a Black Community* (New York, 1974); Daniel Moynihan, *Maximum Feasible Misunderstanding: Community Action in the War on Poverty* (New York, 1969); and James Sundquist, ed., *On Fighting Poverty* (Boston, 1969), a collection of essays by scholars. Michael Harrington, *The Other America: Poverty in the United States* (New York, 1962), did more than any other book to place the problem of poverty on the national agenda in the 1960s.

Legal developments: The role of the courts is large in this history. A handy reference to Supreme Court cases is Congressional Quarterly, *Guide to the U.S. Supreme Court*

(2d ed., Washington, 1990). A solid survey is Paul Murphy, *The Constitution in Crisis Times, 1918–1969* (New York, 1972). Blum, and Hill and Jones, cited above, offer knowledgeable commentary on key legal and constitutional developments. Other important sources include Anthony Lewis, *Gideon's Trumpet* (New York, 1964), an incisive treatment of *Gideon v. Wainwright* (1963); Richard Kluger, *Simple Justice: The History of "Brown v. Board of Education" and Black America's Struggle for Equality* (New York, 1976); J. Harvie Wilkinson, *From "Brown" to "Bakke": The Supreme Court and School Integration, 1954–1978* (New York, 1979); David Garrow, *Liberty and Sexuality: The Right to Privacy and the Making of "Roe v. Wade"* (New York, 1994); and Stanley Kutler, *The American Inquisition: Justice and Injustice in the Cold War* (New York, 1982).

Women: Here, too, the literature is vast, reflecting the rise of feminism since the 1960s. Betty Friedan, *The Feminine Mystique* (New York, 1963), did much to accelerate the women's movement. The best histories include William Chafe, *The Paradox of Change: American Women in the 20th Century* (New York, 1991); and Carl Degler, *At Odds: Women and the Family in America from the Revolution to the Present* (New York, 1980). See also Susan Hartmann, *The Home Front and Beyond: American Women in the 1940s* (Boston, 1982); and Rosalind Rosenberg, *Divided Lives: American Women in the Twentieth Century* (New York, 1992). More focused accounts are Jo Freeman, *The Politics of Women's Liberation: A Case Study of an Emerging Social Movement and Its Relation to the Policy Process* (New York, 1975); Sara Evans, *Personal Politics: The Roots of Women's Liberation in the Civil Rights Movement and the New Left* (New York, 1979); Wini Breines, *Young, White, and Miserable: Growing Up Female in the Fifties* (Boston, 1992); Alice Echols, *Daring to Be Bad: Radical Feminism in America, 1967–1975* (Minneapolis, 1989); Cynthia Harrison, *On Account of Sex: The Politics of Women's Issues, 1945–1968* (Berkeley, 1988); and Janice Radway, *Reading the Romance: Women, Patriarchy and Popular Literature* (Chapel Hill, 1984). Barbara Ehrenreich, *The Hearts of Men: American Dreams and the Flight from Commitment* (New York, 1983), is a spirited collection of feminist essays focusing on the attitudes and behavior of men.

Family/demography/social trends: I rely heavily throughout the book on the following: Arlene Skolnick, *Embattled Paradise: The American Family in an Age of Uncertainty* (New York, 1991); Landon Jones, *Great Expectations: America and the Baby Boom Generation* (New York, 1980); Richard Easterlin, *Birth and Fortune: The Impact of Numbers on Personal Welfare* (New York, 1980); and Randall Collins and Scott Cottrane, *Sociology of Marriage and the Family* (3d ed., Chicago, 1991). See also Elaine Tyler May, *Homeward Bound: American Families in the Cold War Era* (New York, 1988). Books on sexuality include John D'Emilio and Estelle Freedman, *Intimate Matters: A History of Sexuality in America* (New York, 1988); D'Emilio, *Sexual Politics, Sexual Communities: The Making of a Homosexual Minority in the United States, 1940–1970* (Chicago, 1983); and Edward Lauman et al., *The Social Organization of Sexuality* (Chicago, 1994), an attempt to expand on the insights of Alfred Kinsey in the 1940s and 1950s.

Cultural trends: Two widely read collections of scholarly essays are Richard Fox and T. J. Jackson Lears, eds., *The Culture of Consumption: Critical Essays in American History, 1880–1980* (New York, 1983); and Lary May, ed., *Recasting America: Culture and Politics in the Age of Cold War* (Chicago, 1989). The work of Daniel Bell, a sociologist, has had considerable influence; see especially *The Cultural Contradictions of Capitalism* (New York, 1976); and *The End of Ideology: On the Exhaustion of Political Ideas in the 1950s* (Glencoe, 1960). Other relevant books include Russell Jacoby, *The Last Intellectuals: American Culture in the Age of Academe* (New York, 1987); Richard Pells, *The Liberal Mind in a Conservative Age: American Intellectuals in the 1940s and 1950s* (New York, 1985); Stephen Whitfield, *The Culture of the Cold War* (Baltimore, 1991), an especially perceptive interpretation; and George Lipsitz, *Class and Culture in Cold War America: A Rainbow at Midnight* (New York, 1981). Important contemporary intepretations of social and cultural trends include David Riesman et al., *The Lonely Crowd: A Study of the Changing American Character* (New Haven, 1950); David Potter, *People of Plenty: Economic Abundance and the American Character* (Chicago, 1954); William Whyte, *The Organization Man* (New York, 1956); C. Wright Mills, *The Power Elite* (New York, 1956); Paul Goodman, *Growing Up Absurd: Problems of Youth in the Organized Society* (New York, 1960); and Christopher Lasch, *The Culture of Narcissism: American Life in an Age of Diminishing Expectations* (New York, 1979). Among many special studies are Bruce Cook, *The Beat Generation* (New York, 1971); James Gilbert, *A Cycle of Outrage: America's Reaction to the Juvenile Delinquent in the 1950s* (New York, 1986), which is more wide-ranging than the title suggests; Peter Guralnick, *Last Train to Memphis: The Rise of Elvis Presley* (Boston, 1994); and Ed Ward et al., *Rock of Ages: The Rolling Stone History of Rock and Roll* (New York, 1988). Thomas Hine, *Populuxe* (New York, 1986), is a beautifully illustrated, very readable treatment of trends in art, design, and other aspects of popular culture, mainly in the 1950s.

For studies of the **media** and **film,** a good place to start is Baughman, *The Republic of Mass Culture,* noted above. Sources concerning movies include Nora Sayre, *Running Time: Films of the Cold War* (New York, 1982); and Peter Biskind, *Seeing Is Believing: How Hollywood Taught Us to Stop Worrying and Love the Fifties* (New York, 1983). Three thoughtful interpretations of the media, especially television, are Todd Gitlin, *The Whole World Is Watching: Mass Media in the Making and Unmaking of the New Left* (Berkeley, 1980); Edwin Diamond and Stephen Bates, *The Spot: The Rise of Political Advertising on Television* (Cambridge, Mass., 1993); and Karal Ann Marling, *As Seen on TV: The Visual Culture of Everyday Life in the 1950s* (Cambridge, Mass., 1994). See also Robert Sklar, *Prime-Time America: Life On and Behind the Television Screen* (New York, 1980); Erik Barnouw, *Tube of Plenty: The Evolution of Modern Television* (New York, 1975); John Fiske, ed., *Understanding Popular Culture* (Boston, 1989); and Fiske, *Television Culture* (London, 1987). A much-discussed contemporary work is Marshall McLuhan, *Understanding Media: The Extensions of Man* (New York, 1964).

Sports and popular culture are the subject of Richard Davies, *America's Obsession: Sports and Society Since 1945* (Ft. Worth, 1994); and Randy Roberts and James Olson, *Winning Is the Only Thing: Sports in America since 1945* (Baltimore, 1989).

Although **environmental history** is a relatively young field, several good books inform my treatment of the subject. See Michael Lacey, ed., *Government and Environmental Politics: Essays on Historical Developments Since World War Two* (Washington, 1989); Samuel Hays, *Beauty, Health, and Permanence: Environmental Politics in the United States, 1955–1985* (New York, 1987); William McKibben, *The End of Nature* (New York, 1989); and John McCormick, *Reclaiming Paradise: The Global*

Environmental Movement (Bloomington, Ind., 1989). Donald Worster, *Rivers of Empire: Water, Aridity, and the Growth of the American West* (New York, 1985), is an environmentalist call to arms. Rachel Carson, *Silent Spring* (New York, 1962) is in many ways the bible for American environmentalists.

The books mentioned below indicate the sources that helped me the most in my chronologically organized approach to the events of the postwar era.

The late 1940s and Harry Truman: A sprightly, thoughtful interpretation of the postwar mood is Joseph Goulden, *The Best Years, 1945–1950* (New York, 1976). Alan Brinkley, *The End of Reform: New Deal Liberalism in Recession and War* (New York, 1995), sets the stage for liberal ideas in the postwar era. Michael Lacey, ed., *The Truman Presidency* (Washington, 1989), includes first-rate essays on aspects of the Truman years. Alonzo Hamby, *Beyond the New Deal: Harry S. Truman and American Liberalism* (New York, 1973), is an outstanding treatment of the subject. Robert Donovan's two volumes, *Conflict and Crisis: The Presidency of Harry S. Truman, 1945–1948* (New York, 1977), and *The Tumultuous Years: The Presidency of Harry S. Truman, 1949–1953* (New York, 1982), are solid accounts. David McCullough, *Truman* (New York, 1992), is a well-written biography. Robert Ferrell, *Harry S. Truman and the Modern American Presidency* (Boston, 1983), also readable, is much briefer. Intelligent special studies include Steven Gillon, *Politics and Vision: The ADA and American Liberalism, 1947–1985* (New York, 1987); Barton Bernstein, ed., *Politics and Policies of the Truman Administration* (Chicago, 1970); Maeva Marcus, *Truman and the Steel Seizure Case: The Limits of Presidential Power* (New York, 1977); Allen Matusow, *Farm Policies and Politics in the Truman Administration* (Cambridge, Mass., 1967); William Berman, *The Politics of Civil Rights in the Truman Administration* (Columbus, 1970); Donald McCoy and Richard Ruetten, *Quest and Resistance: Minority Rights and the Truman Administration* (Lawrence, 1973); and Ingrid Winther Scobie, *Center Stage: Helen Gahagan Douglas* (New York, 1992). See also James Patterson, *Mr. Republican: A Biography of Robert A. Taft* (Boston, 1972). For cultural histories of the atomic age, see Paul Boyer, *By the Bomb's Early Light: American Thought and Culture at the Dawn of the Atomic Age* (New York, 1985), which carries the story to 1950; and Allan Winkler, *Life Under a Cloud: American Anxiety About the Atom* (New York, 1993), a briefer interpretation that moves beyond the Truman era.

Foreign affairs in the early Cold War years: The origins of the Cold War have stimulated a great deal of historiographical controversy. I have tried to fashion an interpretation drawn from varied sources, notably John Gaddis, *Strategies of Containment*, noted above; Gaddis, *The United States and the Origins of the Cold War, 1941–1947* (New York, 1972); and Melvyn Leffler, *A Preponderance of Power: National Security, the Truman Administration, and the Cold War* (Stanford, 1992). Also helpful are Walter Isaacson and Evan Thomas, *The Wise Men: Six Friends and the World They Made: Acheson, Bohlen, Harriman, Kennan, Lovett, McCloy* (New York, 1986); Thomas Paterson, ed., *Cold War Critics: Alternatives to American Foreign Policy in the Truman Years* (Chicago, 1971); Paterson, *On Every Front: The Making of the Cold War* (New York, 1979), a revisionist interpretation; Robert Pollard, *Economic Security and the Origins of the Cold War, 1945–1950* (New York, 1985); Michael Hogan, *The Marshall Plan: America, Britain, and the Reconstruction of Western Europe, 1947–*

1952 (New York, 1987); and Daniel Yergin, *Shattered Peace: The Origins of the Cold War and the National Security State* (Boston, 1977). Other relevant sources include Wilson Miscamble, *George Kennan and the Making of American Foreign Policy, 1947–1950* (Princeton, 1992); Gaddis Smith, *Dean Acheson* (New York, 1970); Lloyd Gardner, *Architects of Illusion: Men and Ideas in American Foreign Policy, 1941–1949* (Chicago, 1970), which presents critical essays on individual policy-makers; and Dean Acheson's detailed memoir, *Present at the Creation: My Years in the State Department* (New York, 1969).

The Korean War: Among the many books on the war are Dorothy Foot, *The Wrong War: American Policy and the Dimensions of the Korean Conflict, 1950–1953* (Ithaca, 1985); Burton Kaufman, *The Korean War: Challenges in Crisis, Credibility, and Command* (Philadelphia, 1986); William Stueck, *The Road to Confrontation: American Policy Toward China and Korea, 1947–1950* (Chapel Hill, 1981); and Callum MacDonald, *Korea: The War Before Vietnam* (New York, 1986). Military events receive good coverage in Robert Leckie, *Conflict: The History of the Korean War, 1950–1953* (New York, 1962); and David Rees, *The Limited War* (Baltimore, 1964). Bruce Cumings, *The Origins of the Korean War* (2 vols., Princeton, 1981, 1990), is highly critical of American policies. For General Douglas MacArthur, see D. Clayton James, *The Years of MacArthur, 1945–1954* (Boston, 1985); and Michael Schaller, *MacArthur: Far Eastern General* (New York, 1989).

McCarthyism and the Red Scare: Especially sound interpretations of these events are Stephen Whitfield, *Culture of the Cold War*, cited above; and Richard Fried, *Nightmare in Red: The McCarthy Era in Perspective* (New York, 1990). Robert Griffith, *The Politics of Fear: Joseph McCarthy and the Senate* (Lexington, Ky., 1970), concentrates persuasively on political manifestations. Griffith and Athan Theoharis, eds., *The Specter: Original Essays on the Cold War and the Origins of McCarthyism* (New York, 1974), contains well-researched articles on aspects of the Red Scare. David Caute, *The Great Fear: The Anti-Communist Purge under Truman and Eisenhower* (New York, 1978), offers an incredulous survey. Two biographies are David Oshinsky, *A Conspiracy So Immense: The World of Joe McCarthy* (New York, 1983); and Thomas Reeves, *The Life and Times of Joe McCarthy: A Biography* (New York, 1982). See also Richard Rovere, *Senator Joe McCarthy* (New York, 1959), for a sharply critical study. More specialized books include Ellen Schrecker, *No Ivory Tower: McCarthyism and the Universities* (New York, 1986); Allen Weinstein, *Perjury: The Hiss-Chambers Case* (New York, 1978); and Richard Freeland, *The Truman Doctrine and the Origins of McCarthyism* (New York, 1971), which criticizes the Truman administration. Daniel Bell, ed., *The Radical Right* (New York, 1964), includes essays rooting the Red Scare in social tensions.

The 1950s and Eisenhower politics: I have relied extensively on J. Ronald Oakley, *God's Country: America in the Fifties* (New York, 1986), a well-written survey; and on Stephen Ambrose, *Eisenhower: Soldier and President* (New York, 1990). Other useful treatments of the 1950s include Hine, *Populuxe*, cited above; Charles Alexander, *Holding the Line: The Eisenhower Era, 1952–1961* (Bloomington, Ind., 1975); and David Halberstam, *The Fifties* (New York, 1993). Among the many books focusing on Ike are Craig Allen, *Eisenhower and the Mass Media: Peace, Prosperity, and Prime-Time TV* (Chapel Hill, 1993); Jeff Broadwater, *Eisenhower and the Anti-Communist Crusade* (Chapel Hill, 1992); Robert Burk, *Dwight D. Eisenhower: Hero and Politician*

(Boston, 1986), a brief biography; William Pickett, *Dwight David Eisenhower and American Power* (Wheeling, Ill., 1995), also brief; Herbert Parmet, *Eisenhower and the American Crusades* (New York, 1972), a longer study; Fred Greenstein, *The Hidden-Hand Presidency: Eisenhower as Leader* (New York, 1982), a pro-Ike revisionist account; Shirley Anne Warshaw, ed., *Reexamining the Eisenhower Presidency* (Westport, Conn., 1993); and Emmet John Hughes, *The Ordeal of Power: A Political Memoir of the Eisenhower Years* (New York, 1963), an engaging account by one of Ike's speechwriters. See also Stephen Ambrose, *Nixon: The Education of a Politician, 1913–1962* (New York, 1987); and I. F. Stone, *The Haunted Fifties, 1953–1963* (Boston, 1963), which collects many of Stone's acerbic essays about events of the era.

Foreign affairs in the Eisenhower era: Robert Divine, *Eisenhower and the Cold War* (New York, 1981), introduces the subject in stimulating essays. Ambrose, *Eisenhower*, offers a favorable interpretation. Other relevant studies include Divine, *The Sputnik Challenge: Eisenhower's Response to the Soviet Satellite* (New York, 1993); David Anderson, *Trapped by Success: The Eisenhower Administration and Vietnam, 1953–1961* (New York, 1991); Lloyd Gardner, *Approaching Vietnam: From World War II Through Dienbienphu, 1941–1954* (New York, 1988); Peter Grose, *Gentleman Spy: The Life of Allen Dulles* (Boston, 1994); and David Wise and Thomas Ross, *The U-2 Affair* (New York, 1962).

Race relations in the 1950s: In addition to Kluger, *Simple Justice*, and surveys of race relations noted above, I recommend beginning with Taylor Branch, *Parting the Waters: America in the King Years, 1954–1963* (New York, 1988), a sweeping, well-written history. Excellent treatments of Martin Luther King, Jr., include David Garrow, *Bearing the Cross: Martin Luther King, Jr., and the Southern Christian Leadership Conference* (New York, 1986); Adam Fairclough, *To Redeem the Soul of America: The Southern Christian Leadership Conference and Martin Luther King, Jr.* (Athens, Ga., 1987); and David Lewis, *King: A Biography* (Urbana, 1970). Specialized studies include Jo Robinson, *The Montgomery Bus Boycott and the Women Who Started It* (Knoxville, 1987); and Tony Freyer, *The Little Rock Crisis: A Constitutional Interpretation* (Westport, Conn., 1984). For reactions of southern whites the best books are Numan Bartley, *The Rise of Massive Resistance: Race and Politics in the South During the 1950s* (Baton Rouge, 1969); and Neil McMillen, *The Citizens' Council: Organized Resistance to the Second Reconstruction, 1954–1964* (Urbana, 1964). C. Vann Woodward, *The Strange Career of Jim Crow* (rev. ed., New York, 1974), provides essential historical context.

General works on the 1960s: Three critical books are Allen Matusow, *The Unraveling of America: A History of Liberalism in the 1960s* (New York, 1984), which emphasizes failings of liberal domestic policies; Charles Morris, *A Time of Passion: America, 1960–1980* (New York, 1984), which contains many acute observations; and William O'Neill, *Coming Apart: An Informal History of America in the 1960s* (Chicago, 1971), an often sardonic account. Todd Gitlin, *The Sixties: Years of Hope, Days of Rage* (New York, 1987); and Terry Anderson, *The Movement and the Sixties: Protest in America from Greensboro to Wounded Knee* (New York, 1995), devote considerable attention to protest activities. Books focusing on the Left include James Miller, *"Democracy Is in the Streets": From Port Huron to the Siege of Chicago* (New York, 1987); Kirkpatrick Sale, *SDS* (New York, 1973); Maurice Isserman, *If I Had a Hammer: The Death of the Old Left and the Birth of the New Left* (New York, 1987); W. J. Rorabaugh, *Berkeley at*

War: The 1960s (Berkeley, 1989); and Michael Wreszin, *A Rebel in Defense of Tradition: The Life and Politics of Dwight Macdonald* (New York, 1994). Cultural events in the 1960s receive lively and intelligent coverage in Morris Dickstein, *Gates of Eden: American Culture in the Sixties* (New York, 1977). See also William Braden, *The Age of Aquarius: Technology and the Cultural Revolution* (Chicago, 1970); David Farber, ed., *The Sixties: From History to Memory* (Chapel Hill, 1994), an up-to-date collection of scholarly essays; and Jim Heath, *Decade of Disillusionment: The Kennedy-Johnson Years* (Bloomington, Ind., 1975).

John F. Kennedy: Among the many studies of Kennedy and his presidency, four fair-minded accounts merit special mention. They are Herbert Parmet, *JFK: The Presidency of John F. Kennedy* (New York, 1983); Richard Reeves, *President Kennedy: Profile of Power* (New York, 1993); David Burner, *John F. Kennedy and a New Generation* (Boston, 1988); and James Giglio, *The Presidency of John F. Kennedy* (Lawrence, 1991), a very judicious volume. A highly critical study is Thomas Reeves, *A Question of Character: A Life of John F. Kennedy* (New York, 1991). Arthur Schlesinger, Jr., *1000 Days: John F. Kennedy in the White House* (Boston, 1965); and Theodore Sorensen, *Kennedy* (New York, 1965) are lengthy, pro-Kennedy accounts by advisers. Other relevant books include Schlesinger, *Robert Kennedy and His Times* (Boston, 1978); Victor Navasky, *Kennedy Justice* (New York, 1971), which looks critically at the Kennedy administration's handling of issues related to the Attorney General's office; and Gerald Posner, *Case Closed: Lee Harvey Oswald and the Assassination of JFK* (New York, 1993), a thoroughly researched treatment of an unendingly controversial matter. Theodore White, *The Making of the President, 1960* (New York, 1961) is the first of his four books on presidential elections (through 1972). All, focusing on personalities and political strategies, are fast-paced. Key sources on JFK's foreign policies include Michael Beschloss, *The Crisis Years: Kennedy and Khrushchev, 1960–1963* (New York, 1991); and Thomas Paterson, ed., *Kennedy's Quest for Victory: American Foreign Policy, 1961–1963* (New York, 1989), a critical study. For the space program see Walter McDougall, . . . *The Heavens and the Earth: A Political History of the Space Age* (New York, 1985); and Tom Wolfe, *The Right Stuff* (New York, 1979), a highly ironic look by a clever writer.

Civil rights: This, too, is a crowded field of scholarship and popular writing. In addition to books on race relations noted above, see Carl Brauer, *John F. Kennedy and the Second Reconstruction* (New York, 1977); Clayborne Carson, *In Struggle: SNCC and the Black Awakening of the 1960s* (Cambridge, Mass., 1981); Hugh Davis Graham, *The Civil Rights Era: Origins and Development of National Policy, 1960–1972* (New York, 1990), a careful, well-researched effort; William Chafe, *Civilities and Civil Rights: Greensboro, North Carolina, and the Black Struggle for Freedom* (New York, 1980), a thoughtful book that ranges beyond its title; and John Dittmer, *Local People: The Struggle for Civil Rights in Mississippi* (Urbana, 1994), an outstanding history. Anne Moody, *Coming of Age in Mississippi* (New York, 1968), is a moving memoir by a civil rights activist, while Howell Raines, ed., *My Soul Is Rested: Movement Days in the Deep South Remembered* (New York, 1977), contains well-chosen reminiscences. More specialized books are David Garrow, *The FBI and Martin Luther King, Jr.: From "Solo" to Memphis* (New York, 1981); Garrow, *Protest at Selma: Martin Luther King, Jr. and the Voting Rights Act of 1965* (New Haven, 1978); Stokely Carmichael and Charles Hamilton, *Black Power: The Politics of Liberation in America*

(New York, 1967); Malcolm X (with Alex Haley), *The Autobiography of Malcolm X* (New York, 1965); E. U. Essien-Udom, *Black Nationalism: A Search for an Identity in America* (Chicago, 1962); Lee Rainwater and William Yancey, eds., *The Moynihan Report and the Politics of Controversy* (Cambridge, Mass., 1967), which reprints the highly debated report and much illuminating commentary about it; and Robert Conot, *Rivers of Blood, Years of Darkness: The Unforgettable Classic Account of the Watts Riot* (New York, 1968).

Lyndon Johnson: Books that focus on LBJ and his domestic policies include Paul Conkin, *Big Daddy from the Pedernales: Lyndon Baines Johnson* (Boston, 1986), a balanced study; Bruce Schulman, *Lyndon Johnson and American Liberalism: A Brief Biography with Documents* (New York, 1995); Doris Kearns, *Lyndon Johnson and the American Dream* (New York, 1976), which explores personal sources of LBJ's policies; Joseph Califano, *The Triumph and Tragedy of Lyndon Johnson: The White House Years* (New York, 1991), a sympathetic account by a former adviser; and Robert Divine, ed., *Exploring the Johnson Years* (Austin, 1981), which brings together well-researched essays. Robert Caro's two volumes of biography, *The Years of Lyndon Johnson: The Path to Power* (New York, 1982), and *The Years of Lyndon Johnson: Means of Ascent* (New York, 1989), offer a very hostile treatment of LBJ from his birth through 1948. Robert Dallek, *Lone Star Rising: Lyndon Johnson and His Times, 1908–1960* (New York, 1991), carries Johnson's story to 1960.

Vietnam: George Herring, *America's Longest War: The United States and Vietnam, 1950–1975* (2d ed., Philadelphia, 1986) is a remarkably disciplined general history. Larry Berman, *Lyndon Johnson's War: The Road to Stalemate in Vietnam* (New York, 1989), focuses critically on Johnson's policies. Neil Sheehan, *A Bright Shining Lie: John Paul Vann and America in Vietnam* (New York, 1988), offers an epic account of shattered hopes and assumptions. Frances FitzGerald, *The Fire in the Lake: The Vietnamese and the Americans in Vietnam* (Boston, 1972), emphasizes cultural mis-understandings. Loren Baritz, *Backfire: A History of How American Culture Led Us into Vietnam and Made Us Fight the Way We Did* (New York, 1985), stresses the deep historical roots of American policy. General accounts include Guenter Lewy, *America in Vietnam* (New York, 1979); Stanley Karnow, *Vietnam: A History* (New York, 1991); and Marilyn Young, *The Vietnam Wars, 1945–1990* (New York, 1991). D. Michael Shafer, ed., *The Legacy: The Vietnam War in the American Imagination* (Boston, 1990), contains excellent essays on aspects of the war. For studies of the draft, see George Flynn, *Lewis B. Hershey, Jr.: Mr. Selective Service* (Chapel Hill, 1985); and Lawrence Baskir and William Strauss, *Second Chance: The Draft, the War, and the Vietnam Generation* (New York, 1978). Christian Appy, *Working-Class War: American Combat Soldiers in Vietnam* (Chapel Hill, 1993); and Wallace Terry, ed., *Bloods: An Oral History of the Vietnam War by Black Veterans* (New York, 1984), discuss recruitment and experiences in the field.

Backlash in the late 1960s: Aside from Rieder, *Canarsie*; Edsall, *Chain Reaction*; Matusow, *Unraveling*; and other books noted above, there are many sources that emphasize aspects of disillusion with liberalism, leading to political realignment after 1965. Among them are E. J. Dionne, Jr., *Why Americans Hate Politics* (New York, 1991); Kevin Phillips, *The Emerging Republican Majority* (New Rochelle, N.Y., 1969); David Farber, *Chicago '68* (Chicago, 1988); Lewis Chester et al., *An American Melodrama: The Presidential Election of 1968* (New York, 1969); William Chafe,

Never Stop Running: Allard Lowenstein and the Struggle to Save American Liberalism (New York, 1993), which has much to say about politics in 1967–68; Steven Gillon, *The Democrats' Dilemma: Walter F. Mondale and the Legacy of Liberalism* (New York, 1992); and Marshall Frady, *Wallace* (New York, 1968). Related books include Michael Novak, *The Rise of the Unmeltable Ethnics: Politics and Culture in the Seventies* (New York, 1972); Peter Steinfels, *The Neo-Conservatives: The Men Who Are Changing American Politics* (New York, 1979); and Jerome Himmelstein, *To the Right: The Transformation of American Conservatism* (Berkeley, 1990).

The early 1970s and Richard Nixon: A basic survey of the 1970s is Peter Carroll, *It Seemed Like Nothing Happened: The Tragedy and Promise of American Life in the 1970s* (New York, 1982). The most comprehensive source on the Nixon presidency is Stephen Ambrose, *Nixon: The Triumph of a Politician, 1962–1972* (New York, 1989), and *Nixon: Ruin and Recovery, 1973–1990* (New York, 1993). See also Joan Hoff, *Nixon Reconsidered* (New York, 1994), in some ways a revisionist account; and Stanley Kutler, *The Wars of Watergate: The Last Crisis of Richard Nixon* (New York, 1990), which is very critical of the President. Garry Wills, *Nixon Agonistes: The Crisis of the Self-Made Man* (Boston, 1970), is a provocative biographical study published early in Nixon's presidency. Herbert Parmet, *Richard Nixon and His America* (Boston, 1990), deals with some aspects of Nixon's career. Daniel Moynihan, *The Politics of a Guaranteed Income: The Nixon Administration and the Family Assistance Plan* (New York, 1973), is a detailed history of welfare reform efforts, by Nixon's domestic counselor. Aside from books on Vietnam noted above, the most deeply researched source for foreign policy in the Nixon years is Raymond Garthoff, *Détente and Confrontation: American-Soviet Relations from Nixon to Reagan* (Washington, 1985). Other books on foreign affairs include Robert Litwak, *Détente and the Nixon Doctrine: American Foreign Policy and the Pursuit of Stability, 1969–1976* (New York, 1984); and William Shawcross, *Sideshow: Kissinger, Nixon, and the Destruction of Cambodia* (New York, 1979), a sharp indictment of American policy-making. For economic troubles, see Calleo, *Imperious Economy*, and Harrison and Bluestone, *Great U-Turn*, noted above. Sources concerning American attitudes in the late 1960s and early 1970s include two books by Daniel Yankelovich: *The New Morality* (New York, 1974); and *New Rules: Searching for Self-Fulfillment in a World Turned Upside Down* (New York, 1981).

Index

Abernathy, Ralph, 583, 688
Abortion, 35, 360, 566n, 646, 711, 715, 781, 782
Acheson, Dean, 98n, 119, 194, 198; and Asia, 209, 241, 602; and Cuban missile crisis, 500, 501; and foreign policy, 103, 169; and hydrogen bomb, 174–75; and Korea, 209, 210, 227, 233, 234; and Nixon, 255, 434; *Present at the Creation*, 104; public service, 101-3; Republican hatred of, 201, 202, 284; and Truman Doctrine, 127–28; as Undersecretary of State, 112, 126; and Vietnam, 682, 683–84; White Paper (China), 171, 172
Adams, Sherman, 282, 415
Adorno, Theodor, 343, 344
"The Adventures of Ozzie and Harriet," 80, 351, 408, 454
Aerobics (Cooper), 713
"Affirmative Action," 642, 723, 728, 730, 732, 781, 782
The Affluent Society (Galbraith), 340, 408
African-Americans, 79, 331, 351, 452, 671, 675, 687; in armed forces, 16, 22; backlash against, 670–71; and busing/integration, 732–35; cultural life, 30, 31, 384, 655, 657; and education, 67, 69, 384, 386–95, 398, 411, 413–16; freedom rides (1947), 25; and housing, 26–27, 74–75, 336–37, 383–84, 474, 648, 664–65; and jobs, 19, 381, 478, 536, 723; in labor unions, 40, 382; lynching, 24–25, 381, 385; migration to North, 5, 19, 381, 385, 470, 654, 664–65; in politics/voting, 25, 29, 149, 151, 162, 413, 440, 579–88, 649–50, 653, 704, 705, 760, 764; population, 15, 151, 380; and poverty, 62–63, 311, 538, 673, 688; and racism, 21, 22, 24, 374, 384–85, 397, 468, 585–86; and rock 'n' roll, 371–72, 373; and segregation, 381, 384, 388–99, 400–406, 568; sit-ins, 430–33, 471; social/economic equality of, 638, 653; in professional sports, 382–83, 689; and urban riots, 448–49, 552, 588, 650, 653, 662–68; in Vietnam War, 614, 617; during World War II, 5, 23, 385. *See also* Civil rights movement
Agency for international Development (AID), 496
Agent Orange, 510, 595, 619
Agnew, Spiro T., 700–701, 702, 735–36, 737, 758, 776
Agriculture, 56, 63, 163, 166–67, 272, 274, 642–43
Aid to Families with Dependent Children (AFDC), 573, 575, 672–74, 720, 721n, 722n, 787n
AIDS, 788n
Alamagordo, 6, 108, 110
Alaska, 411
Alaska Native Claims Settlement Act, 723
Albany Movement, 472–73. *See also* Civil rights movement
Albert, Carl, 564
Alcatraz Island, 722

803

609; and France, 293–94; and Kennedy, 497–98, 510–11, 517; and SEATO, 296, strategic importance to U.S., 429; and Vietnam War, 605–6, 634, 756
Lardner, Ring, Jr., 190
Lasch, Christopher, 786
Last Tango in Paris (film), 711
The Late Great Planet Earth (Lindsey), 456
Latino-Americans, 375, 379, 380, 449, 452. *See also* Mexican-Americans
Lattimore, Owen, 199
Latvia, 85
Lawrence, David, 43
Lawrence, D. H., 359
"The Lawrence Welk Show," 454
Lawson, James, 431, 432
Leahy, William, 93
Leary, Timothy, 410, 445, 669
"Leave It to Beaver," 80, 351
Lebanon, 423
Lee, Rev. George, 395
Lee, Gypsy Rose, 238
Lee, Herbert, 472
Lehman, Herbert, 411
Lehrer, Tom, 408
LeMay, Curtis, 168, 287, 505, 605, 699, 702
Lemnitzer, Lyman, 495
Lerner, Max, 92, 94, 141, 159, 175
Lesbian rights. *See* Gay rights
Levison, Stanley, 475
Levitt, William, 72, 73, 75, 77, 184, 383
Levittown, 72–73, 74, 75, 76, 324, 340, 366, 369, 383
Lewis, John, 431, 469, 470, 483, 484, 580, 581, 655
Lewis, John L., 48–49, 50, 148, 162
Liddy, G. Gordon, 757, 772, 773, 774, 775, 779n
Life, 353
"The Life of Riley," 351n
"Life Is Worth Living," 237
Lilienthal, David, 94, 119, 174, 175, 254
Limelight (film), 239
The Limits of Growth (Club of Rome), 727, 728
Lincoln, Abraham, 522, 584
Lindsay, John, 700, 755
Lindsey, Hal, 456
Lippmann, Walter, 7–8, 625
Literacy, 27
Literature: during 1950s, 332, 341, 345, 358–60, 373–74, 409–10; during 1960s, 443–44, 448, 455; and Red Scare, 237, 262
Lithuania, 85
Little Richard, 371n

Liuzzo, Viola, 584
Lodge, Henry Cabot, Jr., 250, 251, 514, 593, 601, 683–84
Lolita (Nabokov), 359, 360
Lombardi, Vince, 454
The Lonely Crowd (Riesman & Glazer), 237, 338
Longworth, Alice Roosevelt, 158
The Love Bug (film), 454
Lovett, Robert, 99, 152, 154, 176, 177, 227, 419
Loving v. Virginia (1967), 640
Lowell, Robert, 690
Lowenstein, Allard, 552, 690, 691
Lowenthal, Max, 153
Loy, Myrna, 15
Loyalty oaths, 185–86, 263
Loyalty Review Board, 190–91
LSD, 445, 669, 710
Lubell, Samuel, 139, 247, 705–6
Lucas, Scott, 223
Luce, Clare Boothe, 360, 366
Luce, Henry, 7, 82, 91, 120, 171, 315, 785
Luck, Dr. J. Vernon, Sr., 453
Lucy, Autherine, 396
"The Lucy Show," 454
Lumumba, Patrice, 423
Lundberg, Ferdinand, 36
Luxembourg, 167n
Lynd, Robert, 22
Lynd, Staughton, 627

MacArthur, General Douglas, 217, 281; and Chinese offensive, 221–22; command during Korean War, 212, 213, 215, 232, 516; and Congress, 229–30; and Eisenhower, 245–47, 250; fired by Truman, 97, 198, 228–30; and Inchon, 216–17; and the press, 226–27; and Matthew Ridgway, 226; meeting with Truman, 220; and World War I, 246
Macdonald, Dwight, 244–45, 345, 347, 533
Maddox, Lester, 650
Magruder, Jeb Stuart, 772, 773, 775, 779n
Mailer, Norman, 408
Major League Baseball Players' Association, 452
The Making of the President 1960 (White, T.), 461
Malamud, Bernard, 345
Malaysia, 293
Malcolm X, 446, 483, 551–52, 686
Malenkov, Georgi, 175, 279
Maltz, Albert, 190
The Man with the Golden Arm (film), 347
The Man in the Gray Flanned Suit (Wilson, S), 338

Norway, 15, 167n
Novak, Robert, 529
Nuclear weapons, 236, 295; and arms race,
   418–20; and Cold War, 82, 83, 109–110,
   111, 115, 122, 135; and Cuban missile
   crisis, 498–502; development of, 8, 90,
   93, 110, 118–19; Eisenhower and, 248,
   287, 489; and Goldwater, 558; hydrogen
   bomb, 173–76; used against Japan, 5, 7,
   108–9, 110–11, 276; and Kennedy, 489–
   90; and Korean War, 207, 223–24, 228,
   233; non-proliferation treaty, 746; in popular culture, 408; and Quemoy and
   Matsu, 300, 301; and radiation, 276–77;
   SALT I, 747; Soviet development of,
   118, 169, 172, 176, 203, 205, 263; test
   ban, 424; testing of, 6, 107, 108, 110,
   168, 175, 176n, 276–77, 311, 424, 498,
   725; Truman and, 93, 99
Nuclear Weapons and Foreign Policy
   (Kissinger), 419
Nunn, Sam, 764n

O'Brien, Lawrence, 459, 544, 737, 772, 773
Ocean Hill–Brownsville, 662
Occupational Safety and Health Administra-
   tion (OSHA), 727
Odetta, 483
O'Donnell, Kenneth, 459
Office of Economic Opportunity (OEO),
   539–40, 541, 589, 590, 591, 641, 720
Office of Federal Contract Compliance, 642
Office of Price Administration, 144–45, 147
Office of Strategic Services (OSS), 134
Office of War Information, 77
Office of War Mobilization, 111
Oil, 784–85
O'Keeffe, Georgia, 37
Okinawa, 109
The Old Man and the Sea (Hemingway), 345
Omnibus Crime Control and Safe Streets
   Act (1968), 651
Omnibus Housing Act (1961), 466
On the Beach (film), 347
On the Road (Kerouac), 409–10
On the Waterfront (film), 326, 347
1,000,000 Delinquents (Fine), 369
O'Neill, William, 688
Operation Dixie, 52–53
Operation Mongoose, 498, 506, 507, 508,
   509, 519, 520
Oppenheimer, J. Robert, 6, 173–74, 264,
   268n
Organization of Afro-American Unity, 446.
   See also Malcolm X
Organization of American States (OAS),
   493, 502, 611

The Organization Man (Whyte), 338
Organization of Petroleum Exporting Coun-
   tries (OPEC), 784
The Origins of Totalitarianism (Arendt), 237
Orwell, George, 237
Oswald, Lee Harvey, 518, 519–20, 521
The Other America (Harrington), 444, 533–
   34
"Our Miss Brooks," 364–65

Pace College, 756
Page, Patti, 371
Pahlevi, Muhammad Reza Shah, 284–85
Pakistan, 296, 748
Palestine. See Israel
Palmer, A. Mitchell, 187
Parker, Charlie "Bird," 30
Parker, Mack Charles, 412
Parks, Rosa, 400–402, 404, 408
Paths of Glory (film), 347
Patterson, John, 547
Patton (film), 455, 718, 754
Patton, George C., 250, 287, 455
Paul, Alice, 37
Peace and Freedom party, 689, 704n
Peace Corps, 487, 517, 535
Peale, Rev. Norman Vincent, 332–33, 339,
   439, 456; A Guide for Confident Living,
   332; The Power of Positive Thinking,
   332
Pearson, Levi, 387, 399
Peck, Gregory, 347
Pells, Richard, 407
Pendergast, Thomas, 94, 96, 97,
   108
Pentagon Papers, 635, 756–57
People's Republic of China. See China
Pepper, Claude, 53, 120, 128, 156
Perkins, Frances, 36, 462n
Perkins, Milo, 542–43
Perkins, Tony, 365
"Perry Mason," 350
Peter, Paul and Mary, 445, 483, 583
Peterson, Esther, 462n
Petry, Ann, 27
"Petticoat Junction," 454
Peyton Place (Metalious), 359, 366
Philadelphia Plan, 723–25
Philippines, 212, 217–18, 296, 578
Phillips, Sam, 372
Pinky (film), 347
Playboy, 12, 358–59, 448
"Playhouse 90," 349
Pleiku, 610
Podhoretz, Norman, 651
Point Four, 167–68
Poitier, Sidney, 382

231, 243, 248, 466, 525, 719; and Red
Scare, 165–69, 184, 190, 240, 253; as
Senator, 94, 140; and Adlai Stevenson,
253–54; and Taft-Hartley Act, 51; vetoes,
142, 145, 240, 241; as Vice-President,
93, 123; and women, 35, 37, 366
Truman, Margaret, 128, 134, 159, 224
Truman Doctrine, 132, 133, 135; anti-
communism in, 129; and Freedom
Train, 184; introduced, 128; and Israel,
152; and Marshall Plan, 130; and Red
Scare, 190–91
Trumbo, Dalton, 190
Turkey: and Cuban missile crisis, 502, 504,
505, 507; and Great Britain, 169; and Is-
rael, 152; and NATO, 167n, 500; Soviet
influence in, 86, 87, 102, 118, 135; and
Stalin, 90, 111; U.S. aid to, 126–29
"Twenty-One," 351
Twining, Nathan, 294
Tydings, Millard, 199–200, 223

The Uncertain Trumpet (Taylor), 422
Unemployment, 4, 55, 61–62, 312, 451;
among African-Americans, 381; decline
of, 639; and Eisenhower administration,
272; increases in, 737–38, 783, 785; and
poverty, 536, 541; among women, 33,
40, 42
Union of Soviet Socialist Republics (USSR):
arms race with U.S., 419, 420, 421, 597,
769; and atomic bomb, 108–11, 118–19,
122, 169, 176, 178, 203, 746; and Berlin, 424; relations with China, 425, 487,
515; and Cuba, 427, 492, 600; and détente, 743, 748; economy of, 321; espionage, 263; and Hungary, 308–9; immigration to U.S. from, 15, 327; and
war with Japan, 108–11; Kennan and,
113–14; and Kissinger, 744; and Korea,
208, 209, 210, 221, 233–34; SALT I,
747; and Suez crisis, 307; Sputnik, 288n,
418, 419, 420, 421, 422; and thermonuclear weapons, 279–80; Truman and,
98, 105–6, 129; U.S. relations with, 4,
82–93, 100, 101, 109, 112, 124, 136,
284, 489, 498, 746, 748, 780, 784; and
Vietnam, 293, 597, 607, 609, 613; and
"wars of liberation," 488
United Automobile Workers (UAW), 39, 41–
43, 162, 382
United Farm Workers Organizing Committee, 642–43
United Fruit Company, 285
United Mine Workers, 48, 49, 147, 162
United Nations: and atomic energy, 119; and
China, 172; and Cuban Missile Crisis,

504; and European reconstruction, 130;
and Israel, 152, 784; and Korea, 211–16,
232, 236; and nuclear weapons, 279;
Roosevelt's support for, 92; Stevenson as
U.S. ambassador to, 25; and Taiwan,
747; and Truman Doctrine, 128
Universities. *See* Education
University of Alabama, 396, 481, 513n, 547
University of California, 185, 446, 624, 627,
629
University of Michigan, 562, 627
University of Mississippi, 477–78
University of New Mexico, 755
University of Southern California, 689
University of Washington, 185
University of Wisconsin, 627
Updike, John, 345
Urban League, 26, 185, 482
Urban renewal, 335
Urban riots, 5, 448–49, 552, 588, 650, 653,
662–68, 686, 690
U.S. v. Nixon (1974), 777
U-2 reconnaissance plane: and Cuban mis-
sile crisis, 499, 501, 502, 505, 507, 508;
shot down, 285, 425–27, 488; spying on
USSR, 420, 423

Van Doren, Charles, 351
Van Dusen, Henry Pitney, 343
Vandenberg, Arthur, 112, 127, 128, 129,
147
Vandenberg, Hoyt, 227
Venezia Giulia, 122
Venezuela, 509, 600
Veterans' Administration, 72, 73, 76,
315
Vietcong. *See* National Liberation Front
(NLF)
Vietnam: Buddhists in, 513–14; CIA and,
498; creation of, 293; destabilization of,
769; Eisenhower and, 266–99, 423, 428–
29; Goldwater and, 558; Johnson and,
300, 593–636; Kennedy and, 490, 509–
16; leadership of, 173, 207, 292, 597; reunification of, 296; U.S. forces in, 510,
511, 513, 595, 629–30. *See also* Indochina
Vietnam War, 80, 84, 95, 100, 241, 652,
769, 785, 787, 789; Accelerated Pacification Program, 684–85; African-Americans
serving in, 614, 654, 661; bombing, 595,
596, 599, 604–5, 610, 634, 684, 702,
750, 765; casualties, 4n, 595, 596, 599,
600n, 614, 615–16, 617, 630, 633, 654,
679–80, 766; China and, 746; cultural
misunderstanding, 618–19; domestic consequences of, 598–600, 637, 769–70;